E. Lippert · J. D. Macomber (Eds.)

Dynamics During Spectroscopic Transitions

Basic Concepts

With 214 Figures and 7 Tables

Springer Verlag
Berlin Heidelberg New York
London Paris Tokyo
Hong Kong Barcelona Budapest

Editors:

Prof. Dr. E. Lippert
Eichendorffstr. 4
D-87527 Sonthofen

J.D. Macomber†
3600 A Clayton Road
Concord, CA 94521, USA

ISBN-13:978-3-642-79409-4 e-ISBN-13:978-3-642-79407-0
DOI: 10.1007/978-3-642-79407-0

Library of Congress Cataloging-in-Publication Data
Dynamics during spectroscopic transitions : Basic concepts / E. Lippert, J.D. Macomber, eds.

ISBN-13:978-3-642-79409-4

1. Condensed matter – Optical properties. 2. Relaxation phenomena. 3. Spectrum analysis. 4. Theoretical Chemistry. 5. Quantum Theory. I. Lippert, Ernst. II. Macomber, James D., 1939–
QC173.458.O66D96 1995 530.4' 16–dc20

Softcover reprint of the hardcover 1st edition 1995

Typesetting: Macmillan India Ltd., Bangalore-25
SPIN: 10060515 51/3020 - 5 4 3 2 1 0 - Printed on acid-free paper

Preface*

"Even for the physicist [a] description in plain language will be a criterion of the degree of understanding that has been reached."

Werner Heisenberg,
Physics and Philosophy
(Harper & Bros., New York, 1958, p. 168).

"How does a photon get into an atom?" This question, or one similar to it is asked by every student of freshman chemistry or physics the first time he is shown a pair of horizontal lines connected by a vertical arrow, and is told that this picture represents a quantum transition. In particular, the student wants to know what mechanism the photon uses to cause the atom to accept the energy, how long the process takes, and what the atom "looks like" while it is going on. When the answer is not immediately forthcoming, the student is likely to assume that these matters will be explained later in a more advanced course.

If so, he is likely to be disappointed. Indeed, the subject may never be brought up at all. But after four years or more of exposure to energy level diagrams and arrows, and the introduction of terms like "interaction", "uncertainty principle," "quantum jumps", and "selection rules," the student comes to accept his ignorance. If he later becomes a teacher, his students will not hear a discussion of these matters from his lips, and the cycle will repeat.

Of course, there was a time when some of the leading scientific minds (e.g., Schrödinger and Heisenberg) were also asking how quantum transitions occurred. Not all of the questions proved to be answerable but, for those that were, answers were found. Later, experimental methods were devised to study systems during the absorption and emission processes, first in magnetic resonance and then in other kinds of spectroscopy in which masers and lasers were employed as light sources. Although the results are interesting and exciting, for some reason they are apparently not widely known. Hence, there is evidently a need for some way to bridge the gap between scientists and engineers with a conventional training in quantum mechanics to whom this material may be unfamiliar, and those who use these ideas daily while engaged in the study of coherent transient effects. This book is an attempt to fill this need.

---

*By permission taken from James D. Macomber: The Dynamics of Spectroscopic Transitions, John Wiley & Sons, Inc., New York 1976, ISBN 0-471-56300-5

The intended audience is advanced undergraduates, beginning graduate students, and practicing research scientists who use spectroscopy as a tool. Anyone who has had the first physics or chemistry course in quantum mechanics at the college junior or senior level should be able to read the book without difficulty. It is rather short, and heavy emphasis has been placed on very simple physical explanations, using the absolute minimum of mathematics. I hope that this book will improve the teaching of quantum mechanics, by providing a model for the transition process that can be more easily grasped by the student's physical intuition than the one based on energy level diagrams with arrows. Perhaps some readers will even develop research interests in coherent spectroscopy (nuclear magnetic resonance, electron paramagnetic resonance, lasers). If any of these things come to pass, I shall be very pleased....

JAMES D. MACOMBER

Baton Rouge, Louisiana
July 1975

Foreword

The text above is part of the preface by James David Macomber to his book on the dynamics of spectroscopic transitions. Professor Macomber and the publishing company John Wiley & Sons have been kind enough to provide us with an updated and revised edition.

Macomber's book has been reviewed as being clear in its presentation of coherent dynamical phenomena but seems to be lavish in its use of 5 chapters of introduction for 3 chapters of application.** I used it as my textbook 3 times in the years of 1979–1982 for introductory courses named *Physical Chemistry IV* for advanced students who intended to start with their own scientific work in different research groups dealing with magnetic resonance, lasers in chemistry, quantum statistics etc. at the Technische Universität Berlin, and I found even the 5 introductory chapters worth teaching. However in the meantime it appeared sensible to enlarge the part on applications from 3 to 8 chapters, even though the title of this book has been restricted to some selected topics of *dynamics during spectroscopic transitions*, especially of higher order processes.

Hopefully this has not changed the character of the book. Its aim remains the same: To introduce graduate students as well as scientific researchers to different fields even of quantum statistics, nonlinear optics magnetic resonance, etc. Anyone who is already working in one of these fields might be surprised how basically his field is related to some of the other fields. This textbook should be helpful for introductory courses into almost all the fields of its title.

Thanks are due to Dr. Rainer W. Stumpe at Springer-Verlag for his continuous help and suggestions for this new edition, Professor Christoph R. Bräuchle of the Ludwig-Maximilians-University Institute of Physical Chemistry for his encouragement, as well as to many other colleagues for their proposals concerning this new book and especially to all coauthors who helpfully followed my proposals to contribute to this rather interdisciplinary introduction. Now here it is.

Sonthofen, January 1995 ERNST LIPPERT

** Wunsch L, Neusser HJ (1976). Ber. Bunsenges Phys. Chem. 80: 1038.

List of Contributors

Professor Erkki Brändas
Department of Quantum Chemistry, University of Uppsala, P.O. Box 518, S-75120 Uppsala

Dr. Fred-Walter Deeg
Institut für Physikalische Chemie, Universität München, Sophienstraße 11, D-80333 München

Dr. Martin Goez
Institut für Physikalische und Theoretische Chemie der TU Braunschweig, Hans-Sommer-Straße 10, D-38106 Braunschweig

Professor G. Isoyama
Institute of Scientific and Industrial Research, Osaka University, 8-1 Mihogaoka, Ibaraki, Osaka 567, Japan

Dr. Albrecht Lau
Max-Born-Institut für Nichtlineare Optik und Kurzzeitspektroskopie, Geb. 19.8, Rudower Chaussee 6, D-12489 Berlin

Dr. Michael Pfeiffer
Max-Born-Institut für Nichtlineare, Optik und Kurzzeitspektroskopie, Geb. 19.8, Rudower Chaussee 6, D-12489 Berlin

Professor Makoto Watanabe
Research Institute for Scientific Measurements, Tohoku University, Sendai 980, Japan

Table of Contents

IV. Applications of LASER and SR Techniques

Key to Symbols

A	area of "pillbox" portion of sample
A_E	Einstein coefficient for spontaneous emission
A_1	first amplitude parameter in solution of Bloch equation for M_z or Δ
A_1', A_1''	real and imaginary parts of A_1
A_2	second amplitude parameter in solution of Bloch equation for M_z or Δ
$\|A\rangle$	a superposition ket
$[A]$	matrix of A
$[A]_T$	transpose of A matrix
$[A]^\dagger$	Hermitian adjoint of A matrix
$\mathbf{A}$	arbitrary vector in three-dimensional space
$\mathbf{A}\cdot\mathbf{B}$	scalar product of the vectors **A** and **B**
A_x, A_y, A_z	components of the vector **A**
$\mathbf{a}$	acceleration of a dipole in an inhomogeneous field
a_G	distance between $1/e$ point and center of Gaussian beam
a_j	jth expansion coefficient of the ket $\|a\rangle$
a_j^*	complex conjugate of a_j
$\|a_j\|^2$	complex square of scalar a_j
$a_j(t)$	jth superposition coefficient times jth time-dependent factor in eigenfunctions
$a_j^{(n)}$	$a_j(t)$ for the nth system in the ensemble
a_0	Bohr radius
$\|a\rangle$	ket a, representing a state of the system
$\langle a\|$	bra a, dual to $\|a\rangle$
$\|a\rangle\cdot\|b\rangle$	unsatisfactory scalar product in Hilbert space
$\langle a\|b\rangle$	satisfactory scalar product in Hilbert space
Δa	difference between successive expansion coefficients
α	eigenket of spin-$\frac{1}{2}$ particle with $m = \pm\frac{1}{2}$
(α, β)	scalar product of α and β
α_{jk}	phase of expansion coefficient c_{jk}
α_0	propagator in complex wave phase
α_1, α_2	orientation of the first and second pulses in the rotating frame

$B_x; B_y, B_z$	components of the vector **B**
B_1	first amplitude parameter in solution of Bloch equation for v or Ξ
B_1', B_1''	real and imaginary parts of B_1
B_2	second amplitude parameter in solution of Bloch equation for v or Ξ
$\|B\rangle$	a superposition ket, orthogonal to $\|A\rangle$
B	arbitrary vector in three-dimensional space magnetic induction vector
b_j	jth expansion coefficient of the ket $\|b\rangle$
b_{jk}	kth coefficient in the expansion of $\mathsf{D}\Phi_j$
$\|b\rangle$	ket b, representing a state of the system
$\langle b\|$	bra b, and dual to $\|b\rangle$
β	eigenket of spin-$\frac{1}{2}$ particle with $m = -\frac{1}{2}$
β_0	absorption coefficient
(β, α)	scalar product of β and α
C	in-phase component of polarization wave
c	speed of electromagnetic wave
$c_{A\pm}$	superposition coefficient for expansion of $\|A\rangle$ on measurement axis
c_j	jth superposition coefficient of the total wavefunction
c_{jk}	expansion coefficient of the jth perturbed wavefunction in terms of the kth unperturbed wavefunction
$\|c_{jk}\|$	magnitude of expansion coefficient, c_{jk}
c_{nj}	jth expansion coefficient in the superposition function of the nth system
c_0	speed of electromagnetic wave in vacuum
$c_{\pm j}$	expansion coefficient of $\| \pm \rangle$ in basis $\{\|j\rangle\}$
χ	magnetic susceptibility
χ', χ''	real and imaginary parts of magnetic susceptibility
$\chi_{xx}, \chi_{xy}, \chi_{yx}, \chi_{yy}$	elements of magnetic susceptibility tensor
χ_2	first nonlinear term in magnetic susceptibility
D	population difference between initial and final states
D_e	population difference between initial and final states at equilibrium
D_{jk}	jkth element of the D matrix
$D_{jk}^{(n)}$	jkth element of the D matrix of the nth system in the ensemble
D	electric displacement vector
D_n	D operator for the nth system in the ensemble
d	number density of photons
d_M	number density of photons absorbed by matter
d_R	number density of photons in the radiation field
Δ	ratio of actual population difference to that at equilibrium
δ	rate of change of phase with time, z constant

δ'	phase constant for interference beat of superposition wavefunction
δ_{jk}	Kronecker delta
δ_0	normalized off-resonance parameter
$\mathbb{S}$	sine of $U/2$, equal to sine of $\theta/2$
$E1, E2, \ldots$	electric monopole, dipole, etc.
E_1^0	maximum electric field amplitude
E_1	magnitude of electric field vector
$\lvert E\rvert$	real magnitude of complex field E
E_{atom}	magnitude of electric field felt by electron due to nucleus
$\mathbf{E}$	electric field vector
$\mathbf{E}_s$	Stark pulse
$\mathbf{E}_1$	electric field vector in oscillating wave
e	base of natural logarithms
$\mathscr{E}$	energy
$\{\lvert\mathscr{E}\rangle\}$	set of eigenvalues of the energy
$\mathscr{E}_{\text{atom}}$	total energy of the atom
$\lvert\mathscr{E}_j\rangle$	jth eigenket of the energy
$\mathscr{E}_{\text{photon}}$	energy per photon
$\mathscr{E}'_1, \mathscr{E}'_2$	perturbed energy eigenvalues
$\Delta\mathscr{E}$	uncertainty in energy difference between two states
$\Delta\mathscr{E}_{\text{atom}}$	change in energy of the atom produced by transition
ε	electric permittivity
ε_0	electric permittivity of field-free space
η	electric susceptibility
η', η''	real and imaginary parts of electric susceptibility
η_j	phase constant for jth eigenfunction plus jth superposition coefficient
η_2	first nonlinear term in electric susceptibility
F_x, F_y	complex amplitudes of applied field
$\lvert F_x\rvert, \lvert F_y\rvert$	magnitudes of field amplitudes
$\mathbf{F}$	applied field (electric or magnetic)
$\mathbf{F}_0$	force
$f(u)$	normalized line shape function
$f(z)$	an arbitrary state function
G	frequency-dependent gain
G_M	magnitude of magnetic field gradient
g	ratio of relaxation times, T_2/T_1
Γ	phase parameter in solution to Bloch equations
Γ', Γ''	real and imaginary parts of Γ
γ	magnetogyric (or electrogyric) ratio
H_{jk}	matrix element of the Hamiltonian operator
H_1^0	maximum magnetic amplitude of rf field
H_x, H_y, H_z	components of uniform induction field (laboratory coordinates)

H_0	magnitude of uniform induction field
$\bar{H}_0$	average magnitude of uniform induction field
H_1	magnitude of oscillating magnetic field, first three chapters; same for magnetic induction thereafter
$\mathbf{H}$	magnetic field vector, first three chapters; magnetic induction thereafter
$\mathbf{H}_A, \mathbf{H}_B$	inhomogenous magnetic fields in beam selectors
$\mathbf{H}_1$	magnetic field vector in oscillating wave, first three chapters; same for magnetic induction thereafter
H	Hamiltonian operator
H'	Hamiltonian term for a static perturbation
H_R	relaxation term in Hamiltonian operator
H_0	dominant static term in Hamiltonian operator
H_1	time-dependent perturbation term in Hamiltonian operator
h	Planck's constant
$\hbar$	Planck's constant divided by 2π
$\mathrm{Im}(a_j)$	imaginary part of a_j
i	square root of -1
ι	damping limit parameter
J	magnitude of irradiance vector
J_{rms}	root-mean-squared magnitude of irradiance vector
$J(x, y)$	beam profile
J_{12}	irradiance from two interfering sources
$\mathbf{J}$	*irradiance* vector
j	arbitrary integer
$\lvert j\rangle$	jth simultaneous eigenket of S^2 and S_x
k	arbitrary integer (in Chapters 3 and 8) the magnitude of $\mathbf{k}$
k_0	Boltzmann's constant
$\hat{k}$	unit vector in propagation direction
$\mathbf{k}$	propagation vector of electromagnetic wave
κ	absorption cross section
L	length of beam chamber
L_0	broad-band loss
L_x, L_y, L_z	components of angular momentum vector
$\mathbf{L}$	orbital angular momentum vector
L	angular momentum vector operator
$\mathsf{L}_x, \mathsf{L}_y, \mathsf{L}_z$	components of angular momentum vector operator
l	length of interaction region
l_C	length of beam path subject to C magnet
l_1	length of transition region
Δl	length of segment in interaction region
$\lvert\Lambda, m\rangle$	simultaneous eigenket of S^2 and S_z
Λ_j	jth eigenvalue of S^2, divided by $\hbar^2$
λ	wavelength

M	number of eigenkets in a complete set (number of eigenstates)
M_0	magnitude of equilibrium magnetization
M_x, M_y, M_z	components of the magnetization of the sample
$M1, M2, \ldots$	magnetic monopole, dipole, etc.
$\mathbf{M}$	magnetization (magnetic polarization) vector
$\mathbf{M}_0$	spontaneous magnetic polarization
m	general eigenvalue of $\mathbf{S}_z$, divided by $\hbar$
m_j	jth eigenvalue of $\mathbf{S}_z$, divided by $\hbar$
m_0	mass of particle
$\lvert - \rangle$	ket with lower eigenvalue
μ	magnetic permeability
μ_0	magnetic permeability of field-free space
μ_x, μ_y, μ_z	components of linear momentum vector
$\boldsymbol{\mu}$	linear momentum vector
$\boldsymbol{\mu}_j$	linear momentum vector of jth volume element
$\underline{\boldsymbol{\mu}}$	linear momentum vector operator
$\underline{\boldsymbol{\mu}}_x, \underline{\boldsymbol{\mu}}_y, \underline{\boldsymbol{\mu}}_z$	components of linear momentum vector operator
N	number of quantum systems in the ensemble (sample) number of measurements made on a system
N'	number of quantum systems per unit volume
N_s	number of segments in interaction region
n	arbitrary integer
n_R	refractive index
n_0	number of nodes in wavefunction (principal quantum number)
ν	normalized distance into sample
Ω	instantaneous phase of the precession of the magnetic dipole at time t
$\Omega_{jk}^{(n)}$	instantaneous phase of the off-diagonal element of the nth D matrix
$\Delta\Omega$	uncertainty in phase
ω	angular frequency of light wave
ω_0	natural frequency of a quantum system $\equiv$ beat frequency of superposition state
ω_1	precession frequency in field H_1 (twice the nutation frequency)
ω_0^0	center frequency of an inhomogeneously broadened line
ω_0^{S}	Stark-shifted absorption frequency
ω'	effective frequency of irradiation, due to fringing field
$\Delta\omega$	width of spectral line
$\Delta\omega_j^0$	difference between center frequency of jth isochromat and ω_0^0
o	polarization parameter in susceptibility tensor

o_E	polarization parameter in electric susceptibility tensor
o_M	polarization parameter in magnetic susceptibility tensor
P_0	equilibrium polarization of dipoles in field E_z
$\mathbf{P}$	electric polarization vector
$\mathbf{P}_0$	spontaneous polarization vector
p	magnitude of transition dipole moment (electric or magnetic)
p_E	certain component of electric dipole moment vector
P_M	certain component of magnetic dipole moment vector
	magnitude of dipole moment vector
$p_{u_1}, p_{u_2}, p_{u_3}$	components of dipole moment vector in particle coordinates
p_x, p_y, p_z	components of magnetic dipole moment vector in laboratory coordinates
$\mathbf{p}_E$	electric dipole moment vector
$\mathbf{p}_M$	magnetic dipole moment vector
p_E	electric dipole moment vector operator
p_M	magnetic dipole moment vector operator
¶	saturation/resonance denominator of susceptibility tensor
$¶_E$	saturation/resonance denominator of electric susceptibility tensor
$¶_M$	saturation/resonance denominator of magnetic susceptibility tensor
Φ	time-independent part of wavefunction
Φ'	time-independent part of perturbed wavefunction
ϕ	real phase constant of wave
ϕ'	azimulthal angle in laboratory space
ϕ_y, ϕ_z	phases of y and z components of wave
φ	laboratory angle between directions of propagation of pulses in photon echo experiments
Π	normalized amplitude of in-phase component of polarization wave
π	circumference-to-diameter ratio for circle
Ψ	total (superposition) wavefunction
Ψ_n	total (superposition) wavefunction for nth quantum system
ψ	wavefunction (eigenfunction)
ψ^*	complex conjugate of wavefunction
ψ_1, ψ_2	superposed eigenfunctions
ψ'_2, ψ'_2	perturbed eigenfunctions
$\|+\rangle$	ket with higher eigenvalue
Q	partition function
Q	operator that does not commute with R
q	independent variables other than time (e.g., x, y, and z)
q_j	jth eigenvalue of Q
q_0	charge on the proton
R	the physical property represented by R

$\bar{R}$	classical weighted average of R
R_{jk}	matrix element of R
R_1	twice the imaginary part of the off-diagonal elements of the density matrix
R_2	twice the real part of the off-diagonal elements of the density matrix
R_3	difference between the diagonal elements of the density matrix
R_3^e	difference between the diagonal elements of the density matrix at thermal equilibrium
$\langle R\rangle$	expectation value of R
$\langle\bar{R}\rangle$	ensemble average of the expectation value of R
$\langle R\rangle_n$	ensemble average of R for the nth system
R	operator representing some arbitrary physical property of the system
	operator of which a non-stationary state is an eigen-state
$\{R\}_t$	set of operators of which the superposition state is an eigenstate
$\mathrm{Re}(a_j)$	real part of a_j
r	radial distance
	general eigenvalue of R
$r(k)$	result of the kth measurement
$\{\lvert r\rangle\}$	set of eigenkets of R
r_j	jth eigenvalue of R
$\lvert r_j\rangle$	jth eigenket of R
$\mathbf{r}$	three-dimensional position vector
$\mathbf{r}_j$	distance from spin axis to jth volume element
r	three-dimensional position vector operator
$\underline{\rho}$	density operator
ρ_{jk}	element of the density matrix in the laboratory coordinate system
$\rho_{jk}^{\dagger}$	element of the density matrix in a rotating coordinate system
S	in-quadrature components of polarization wave
$\mathbf{S}$	total spin angular momentum vector
S	operator that commutes with R
	total spin angular momentum vector operator
$[\mathsf{S}, \mathsf{R}]$	commutator of S and R
$\mathsf{S}_x, \mathsf{S}_y, \mathsf{S}_z$	components of the spin angular momentum vector operator in the laboratory coordinate system
$\mathsf{S}_1, \mathsf{S}_2, \mathsf{S}_3$	components of the spin angular momentum vector operator in an arbitrary coordinate system
$\mathsf{S}_{\pm}$	raising and lowering operators for spin angular momentum
s	spin quantum number (maximum value of m)
$\{\lvert s\rangle\}$	set of eigenkets of S
s_j	jth eigenvalue of S

§	total phase constant for oscillating expectation values
Σ	sum of diagonal elements of density matrix
Σ^e	sum of diagonal elements of density matrix at thermal equilibrium
σ	electrical conductivity
σ_m	magnetic conductivity
T	Tesla
T	absolute temperature
$T(t)$	time-dependent part of the wavefunction
$T(0)$	time-dependent part of the wavefunction at $t = 0$
T_m	reciprocal half-width of a line that is both homogeneously and inhomogeneously broadened
T_s	spin temperature
T_{jk}	relaxation time of the jkth element of the density matrix
T_1	relaxation time of the diagonal elements of the density matrix of a two-level system
T_2	relaxation time of the off-diagonal elements of the density matrix for a two-level system
T'_2	reciprocal half-width of saturated line
T_2^*	reciprocal half-width of inhomogeneously broadened line
$\mathbf{T}$	torque vector
t	time
t'	interval between pulses in a pulse sequence
t_A	Beer's law absorption time
Δt	duration of radiated wave train
τ	normalized retarded time
Θ	normalized field intensity
Θ_0	normalized field intensity at entrance face
θ	angle between energy axis and axis of certainty parameter characterizing amplitudes of superposition coefficients
θ'	polar angle in laboratory space
$d\theta$	vector angular displacement
U	used in Chapter 8 instead of θ to avoid confusion with Θ
u	in-phase mode of bulk-resonance signal
$\hat{u}_1, \hat{u}_2, \hat{u}_3$	unit vectors in particle coordinates
ς	pulse duration
υ	velocity of propagation of self-induced transparency pulse
V	amplitude of perturbation potential difference
$V^\dagger$	maximum amplitude of perturbation
v	in-quadrature mode of bulk resonance signal speed of particle in beam
$\hat{v}_1, \hat{v}_2, \hat{v}_3$	unit vectors in measurement coordinate direction
$\Delta v_y, \Delta v_z$	uncertainty in velocity due to beam spreading
$\mathbf{v}$	velocity of particle in beam

W	total probability of absorption
W'	probability of absorption per absorber
$W(q, t)$	distribution function
W_j	statistical weight of jth result
w	work
X	operator representing the coordinate X
x	position coordinate along x axis
	direction of beam of particles
x'	x coordinate in rotating frame
x_j	jth eigenvalue of operator X
$[x_{jk}]$	matrix with elements x_{jk}
Δx	difference between successive eigenvalues of $\mathbf{X}$
$\hat{x}$	unit vector in the x direction of the laboratory coordinate system
Ξ	normalized amplitude of in-quadrature component of polarization wave
ξ_j	phase of the jth superposition coefficient
Y	operator representing the coordinate y
y	position coordinate along y axis
y'	y coordinate in rotating frame
y_0	argument of hyperbolic cosecant function
$\hat{y}$	unit vector in y direction
Z	impedance
Z_0	impedance of a vacuum
Z	operator representing the coordinate z
z	position coordinate along z axis
	direction of propagation of electromagnetic wave
Δz	thickness of "pillbox" section of sample
$\hat{z}$	unit vector in z direction
dz	vector linear displacement
ζ_j	phase constant for time-dependent part of jth wavefunction

I. Well-known Principles

1 Introduction[1]

J.D. Macomber

1.1 Electric Fields in Atoms and Waves

This book is about the interaction of light with matter. Nearly all of the phenomena that are produced by means of this interaction can be called *spectroscopic*, in the broadest sense of the term. Included are the ordinary processes of absorption and emission of radio waves, microwaves, and infrared, visible, and ultraviolet light, which leave the associated atoms, ions, or molecules chemically intact. Such phenomena are used by synthetic and analytic chemists in the identification and characterization of substances, and by physical chemists in elucidating molecular structure, determining bond strengths, and studying relaxation processes. When light in the optical region of the spectrum is employed, such effects provide the microscopic basis for theories of color and vision. The explanation of absorption and emission is the purpose of subsequent chapters.

First, what is meant by *light* and *matter*? Every student of elementary physics learns that a light beam is composed of electric and magnetic fields, $\mathbf{E}_l$ and $\mathbf{H}_l$, situated at right angles to one another, and oscillating sinusoidally in space and time. These oscillating fields may be decomposed mathematically into electromagnetic waves propagating in a direction $\mathbf{k}$ perpendicular to the $\mathbf{E}_l\mathbf{H}_l$ plane. At a given time for any given wave, the distance between successive peaks in the amplitude of either field (measured in the $\mathbf{k}$ direction) is called the *wavelength*, λ. Figure 1.1 is a schematic representation of an electromagnetic wave. The number of peaks that pass by a given point in space during a unit time interval is called the *frequency* of the wave, $\omega/2\pi$. (The symbol ω represents the angular frequency in radians per second.) The distance traveled by a given peak in unit time (also measured in the $\mathbf{k}$ direction) is called the *speed* of the wave, c. As in all wave phenomena, the wavelength, frequency, and speed are related by the formula

$$\lambda\left(\frac{\omega}{2\pi}\right) = c. \tag{1.1}$$

[1] With permission, substantially taken from James D. Macomber: The Dynamics of Spectroscopic Transitions, John Wiley & Sons, Inc., New York 1976, ISBN 0-471-56300-5

Electromagnetic waves differ from one another in both frequency and wavelength. When traveling in a vacuum, they all have the same speed, c_0:

$$c_0 = 2.99 \times 10^8 \text{ m s}^{-1} \text{ (meters per second).} \tag{1.2}$$

The scalar magnitudes, H_1 and E_1, of the *magnetic* and *electric fields* of the waves are also related:

$$E_1 = ZH_1. \tag{1.3}$$

Again, in a vacuum, the value of Z is the same for all waves, namely, Z_0:

$$Z_0 = 377 \text{ ohms.} \tag{1.4}$$

The constant Z_0 is called the *impedance* of free space. Although it is somewhat less familiar than c_0, its significance as a fundamental constant is comparable.

Because of the interrelationships described in Eqs. (1.1) and (1.3), two vector quantitites suffice to describe most of the interesting properties of each light wave. First, the *propagation vector*, **k**, gives the wavelength and direction of travel:

$$\mathbf{k} = \frac{2\pi\hat{k}}{\lambda} \tag{1.5}$$

($\hat{k}$ is a unit vector in the direction of **k**). Second, the electric field vector gives the amplitude and direction of polarization of the wave. The wave amplitude is important because it is directly related to the *irradiance*, **J**, associated with the wave:

$$\mathbf{J} = \frac{E_1^2\hat{k}}{Z}. \tag{1.6}$$

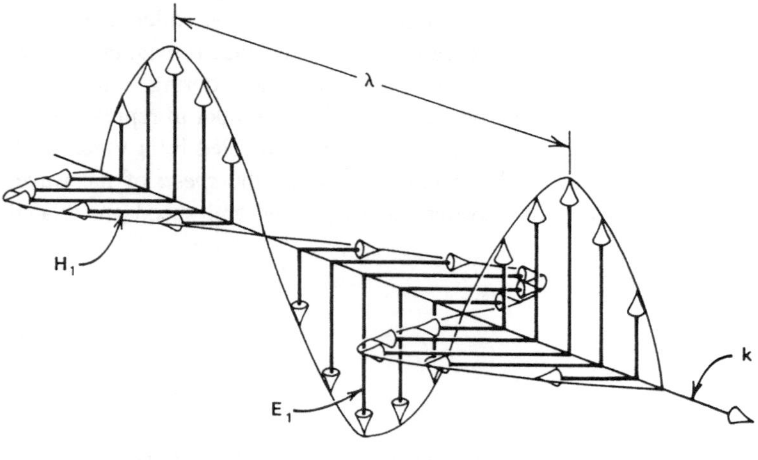

Fig. 1.1. Schematic representation of an electromagnetic wave

The dimensions of **J** are energy per unit time per unit area (in a plane perpendicular to $\hat{k}$). The quantity tabulated for most light sources is the root-mean-squared irradiance, J_{rms}:

$$J_{rms} \equiv \left[\frac{\omega}{2\pi}\int_0^{2\pi/\omega} J^2(t)dt\right]^{1/2}$$
$$= \frac{3(E_1^0)^2}{8Z}. \tag{1.7}$$

The peak value of E_1 is E_1^0.

Matter, on the other hand, consists of particles. Physicists have discovered a large menagerie of these, including such exotic beasts as neutrinos, mesons, and hyperons. Ordinary matter, however, consists of electrons, protons, and neutrons, and it will be sufficient to describe the interactions of electromagnetic radiation with these particles. In particular, it will be assumed that the protons and neutrons have been organized into one or more nuclei. The nuclei, in turn, have been surrounded by electrons to form atoms, molecules, or ions upon which the light waves impinge.

Since electrons and nuclei are charged, magnetic particles, it is not surprising that they interact with light. Indeed, the electrons and nuclei in atoms, molecules, and ions are held together by electric (and, to a lesser extent, by magnetic) fields. The static electric field intensity E_{atom} felt by an electron of charge $-q_0$ located a distance a_0 away from a nucleus of charge $+q_0$ is given by

$$E_{atom} = \frac{c_0 Z_0 q_0}{4\pi a_0^2}$$
$$\sim 5 \times 10^{11} \text{ V m}^{-1} \text{ (volts per meter)}. \tag{1.8}$$

It is instructive to compare E_{atom} with E_1^0, the electric field associated with a light wave. At a distance of 1 m from a 60 W (watt) light bulb (assumed for simplicity to produce waves of uniform frequency),

$$J_{rms} < \frac{60}{4\pi} \text{ W m}^{-2}. \tag{1.9}$$

The inequality is due to the fact that the efficiency of the bulb is less than 100%. Using Eqs. 1.7 and 1.9, one finds

$$E_1^0 = \sqrt{(2.67)(377)(60)/4\pi} = 70 \text{ V m}^{-1}. \tag{1.10}$$

(The value of Z for air is very close to that of Z_0.) By way of contrast, a modestly sized high-peak-power ruby laser may produce a flux as great as

$$J_{rms} \sim 10^{12} \text{ W m}^{-2}. \tag{1.11}$$

The corresponding peak electric field intensity is

$$
\begin{aligned}
E_1^0 &\sim \sqrt{(2.67)(377)(10^{12})} \\
&= 3 \times 10^7 \ \mathrm{V\ m^{-1}}.
\end{aligned}
\tag{1.12}
$$

Firing the laser beam through a converging lens with a focal length of ~0.2 m may reduce the beam diameter from 10 mm to 1 μm. This will bring the peak electric field intensity, E_1^0, into the same range as E_{atom}. Therefore atoms and molecules (e.g., air) at the focal point are ionized by the strong electric fields. The subsequent recombination of electrons with their ions produces a flash and a loud noise analogous to lightning and thunder. In some recent experiments, very powerful lasers have been fired at small solid targets containing lithium and deuterium. The electrons produced are so energetic that they heat and compress the target to the point where nuclear fusion takes place therein.

No such drastic processes will be considered in this book. Mathematically, this restriction can be expressed by the formula

$$E_1^0 \ll E_{\text{atom}}. \tag{1.13}$$

Moreover, there are some nondestructive processes that can take place at moderately high power,

$$E_1^0 \cong E_{\text{atom}}. \tag{1.14}$$

These are the interactions that produce changes in the frequency of the wave by nonlinear optical processes (see Chaps. 11, 12). They include frequency doubling (harmonic generation) and stimulated Raman and Brillouin scattering. The condition which remains is

$$E_1^0 \leqslant E_{\text{atom}}. \tag{1.15}$$

1.2 Photoionization and Nonresonant Scattering

Some photodestructive optical processes that can be produced at low power will also be eliminated from consideration. To see how these interactions occur, it is necessary to replace the wave picture of light with the photon picture, in which the light beam consists of a stream of photons with number density d moving with velocity $c\hat{k}$. (The dimensions of d are the number of particles per reciprocal volume.) Each photon carries an amount of electromagnetic energy $\mathscr{E}$, which can be calculated from the frequency of the associated wave:

$$\mathscr{E}_{\text{photon}} = \frac{h\omega}{2\pi}. \tag{1.16}$$

The constant h, called Planck's constant, is the same for all photons:

$$h = 6.62 \times 10^{-34} \text{ J Hz}^{-1} \text{ (joules per cycles per second).} \tag{1.17}$$

Therefore, from Eqs. 1.17 and 1.16 and the definition of d,

$$\mathbf{J}_{\text{rms}} = cdh\left(\frac{\omega}{2\pi}\right)\hat{k}. \tag{1.18}$$

The physical meaning of Eq. (1.18) is that even a rather feeble light beam (small J) can contain particles of high energy (large ω), albeit at low density (very small d).

Consider again Eq. (1.8) and calculate the binding energy of an electron with charge $-q_0$ to a proton with charge $+q_0$. At a separation distance of a_0, the energy will be $\mathscr{E}_{\text{atom}}$:

$$\begin{aligned}\mathscr{E}_{\text{atom}} &= \int_{r=a_0}^{\infty} (-q_0)\frac{c_0 Z_0 q_0}{4\pi r^2}dr = -\frac{1}{2}\frac{c_0 Z_0 q_0^2}{4\pi a_0} \\ &\sim -2 \times 10^{-18} \text{ J.}\end{aligned} \tag{1.19}$$

Whenever the energy of the photon exceeds that calculated in Eq. 1.19, it is possible for the photon to ionize the atom (ion or molecule) that contains the electron in question. The corresponding frequency is seen to be

$$\begin{aligned}\frac{\omega}{2\pi} &> \frac{-\mathscr{E}_{\text{atom}}}{h} \\ &\sim 3 \times 10^{15}\text{Hz.}\end{aligned} \tag{1.20}$$

In the case of molecules, light of about this frequency can produce dissociation of bonds or other kinds of photochemical reactions. These processes are also too drastic to be discussed in this book.

There are two ways for light to interact with matter that have not yet been considered. Atoms, ions, and molecules that are illuminated with electromagnetic radiation will necessarily receive energy from the beam. In the absence of the light, each electron (and, to a lesser extent, each nucleus) will circulate within the atom, ion, or molecule of which it is a part, because of the presence of the electric and magnetic fields of all of the other constituents. When the radiation arrives, the electrons will be pushed to and fro by the oscillating electric (and, to a lesser extent, the magnetic) fields in the waves. These pushes, together with the additional motion induced by them, constitute kinetic and potential energy for the atom, over and above that which it had in the absence of light. The quantum system may dispose of this excess energy either by reradiating it or by turning it into other forms.

Which method of dissipation is employed depends on whether or not the frequency of the light wave, ω, is close to one of the characteristic natural (quantum) frequencies of the atomic (molecular, ionic) system, ω_0. If the frequencies

ω and ω_0 do not match, the interaction may be termed "nonresonant." If the frequencies do match, the interaction is "resonant." The relationships between the characteristic energies of the atom and the energy of the photon for ionization, and for resonant and nonresonant interactions that do not result in ionization, are presented in Fig. 1.2. The amount of excess energy received by an atom, ion, or molecule when it interacts nonresonantly with an electromagnetic wave is ordinarily quite small, and it is usually disposed of with great efficiency by reradiation. The direction of reradiation by an individual atomic system need not be the same as the direction of the incident beam. For this reason, nonresonant interactions are frequently called *light scattering.* If a very large number of scatterers are present, there will in general be interference (constructive and destructive) between the waves reradiated by any pair of them. If there were such a thing as a scattering medium that was perfectly homogeneous within a volume $\sim \lambda^3$, the only scattered waves that would survive are those which propagated in the same direction as that of the incident beam. This scattering process (dominant in many real substances in spite of small inhomogeneities) is called *transmission.*

At the boundary between the two different homogeneous media, scattering can occur in directions other than the forward one. If the boundary is smooth

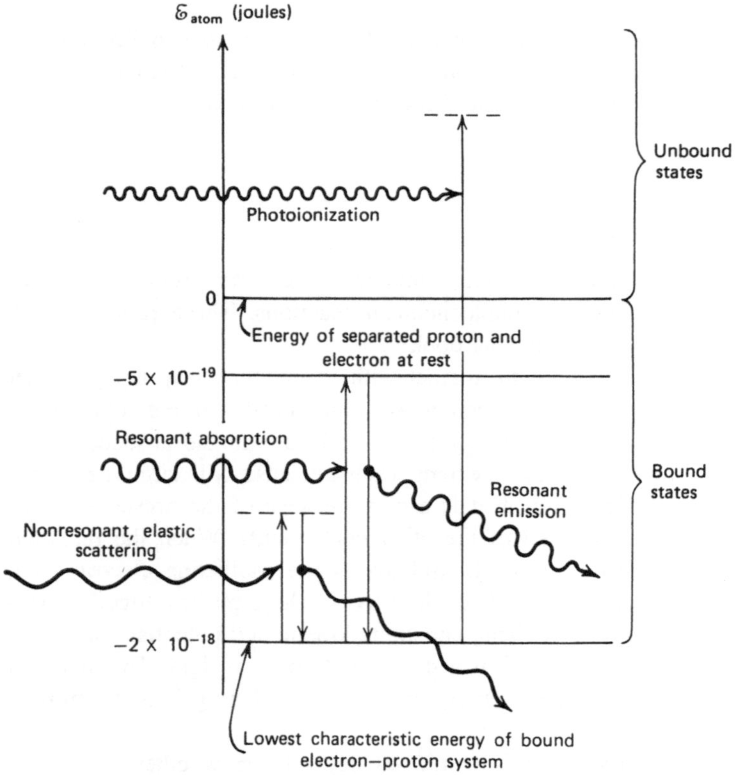

Fig. 1.2. Characteristic energies for a hydrogen atom, and the interactions between that atom and photons of various frequencies

(all irregularities $\ll \lambda$), constructive interference of the scattered light occurs in only two directions (reverse and forward) from the boundary; the resultant scattering processes are called *specular reflection* and *refraction*. If the boundary is rough on a wavelength scale, many different scattering directions are possible, and *diffuse transmission* and *reflection* result.

If the medium is inhomogeneous (particles of one kind embedded in a medium composed of a different substance), several different kinds of scattering processes may occur, depending on what the shapes of the embedded particles are, whether they are transparent or opaque, and whether they are larger or smaller than a wavelength. A continuum of scattering directions is possible. For example, spherical transparent particles with radii $> \lambda$ can produce rainbows (*Mie scattering*). If the inhomogeneities are periodically spaced at intervals $\sim \lambda$, the scattering occurs in discrete directions and the process is called *diffraction*. Diffraction also occurs around opaque objects and through holes in opaque objects, but the effects are not very dramatic unless the objects or the holes are of size λ. Randomly distributed density fluctuations (wherein the inhomogeneities are small) produce various kinds of weak scattering processes. If the density fluctuations do not move, *Rayleigh scattering* results. If weak density fluctuations propagate through the medium, the scattered light will be shifted in frequency because of the Doppler effect. This process is temed *Brillouin scattering*.

The degree of response of the electrons to the nonresonant electromagnetic wave is called *polarizability*. If the polarizability oscillates because of rotation and vibration of the molecules to which the electrons are bound, the scattered light will be amplitude modulated. The frequency of the scattered light will therefore be less or greater than that of the incident beam by the modulation frequency (sideband generation). This process is called *Raman scattering*. Although investigations of some of these phenomena, especially Brillouin and Raman scattering, can provide spectroscopic information (and are therefore sometimes called *Brillouin* and *Raman spectroscopy*), such studies are not spectroscopic in the sense intended here. (Garbuny's book [1] is a useful source of information about a wide range of optical phenomena.)

We will discuss, therefore, only basic principles of stimulated scattering methods in order to demonstrate some third and fourth order processes.

The reader might already be curious enough to learn what the term "nonlinear optics" is about. This term is used if the second or higher terms of the electric susceptibility are needed for the development of the electric polarization in a McLaurin series as a function of an electric field strength (Eq. 3.15).

1.3 Resonant Interactions and Spectroscopy

In this book the word *spectroscopic* is used only to describe resonant interactions between electromagnetic waves and matter. As stated previously, processes that lead directly to ionization and/or chemical reactions are excepted, as well as those that change the frequency of the wave. It is not intended, however, that processes which lead indirectly to chemical reactions (e.g., by subsequent decomposition of an excited molecule) be excluded. Similarly, resonant interactions will be termed "spectroscopic" even if some of the energy that was transferred from the electromagnetic wave to the atom, ion, or molecule eventually ends up as light of a different frequency (e.g., by phosphorescence or by Stokes-shifted fluorescence). It should be obvious that there cannot be a single hard and fast rule which will adequately classify every case that may arise. The intent of the definition, however, is to limit the discussion to interactions that are intrinsically reversible (elastic).

Resonant interactions are ordinarily much stronger than nonresonant ones. (A notable exception is the interaction of visible light with a metallic surface; nearly 100% nonresonant reflection can result.) Each characteristic frequency is actually a band of frequencies of width $\Delta\omega$ more or less symmetrically distributed about some center, ω_0. In many cases, the difference between two successive characteristic frequencies in a particular atomic system exceeds the band width of either of them. If one illuminates such a system by means of a light source of adjustable frequency, a gradual increase in ω will produce first one strong resonant interaction and then another. Between those two strong resonances, only weak nonresonant scattering processes will occur.

A graphical representation of such an experiment, with the magnitude of the response of the atomic system as the ordinate and the frequency of the illuminating light as the abscissa, is called a *spectrogram*. In the case just described, the spectrogram will consist of a series of well-spaced peaks, with the curve returning to the base line (parallel to the abscissa) in between. The peaks are called *spectral lines*. The process of obtaining (and interpreting) a spectrogram is known as *spectroscopy*, and the apparatus used to obtain the information presented on a spectrogram (and in some cases even to draw the curve) is a *spectrometer*. Typical components of a spectrometer are a source of light that produces many frequencies simultaneously, a dispersing element such as a prism or diffraction grating to separate the various frequencies produced by the source, a sample chamber, and a detector to measure the magnitude of the response of the sample to the source (see Fig. 1.3).

The resonant transfer of energy from a radiation field to matter (atoms, ions, or molecules) is called *spectroscopic absorption*; the resonant transfer of energy from matter to the radiation field is termed *spectroscopic emission*. If no nonradiative transfer of energy occurs in the time interval between absorption and subsequent emission, the overall process could properly be called *scattering*. Unlike the case of nonresonant scattering, however, a very large fraction of the incident electromagnetic energy participates. For this reason, other names are ordinarily used to describe the interaction (e.g., *resonance trapping*).

These two processes, absorption and emission, give their names to the two main types of spectroscopy. In absorption spectroscopy, the light source produces a well-defined pencil of parallel rays (the beam) which is incident upon a sample in thermal equilibrium at some preselected temperature. The atoms, ions, or

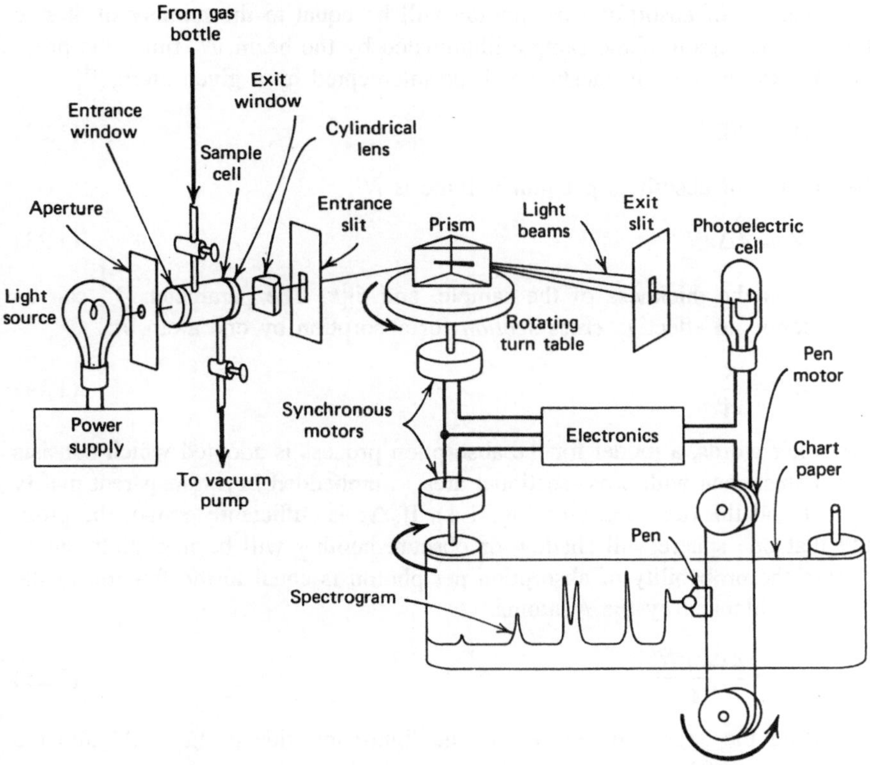

Fig. 1.3. Schematic diagram of a continuously recording absorption spectrophotometer

molecules in the sample absorb photons from the beam, and either reradiate them in some other direction by emission or turn the energy they represent irreversibly into heat. In either case, the intensity of the beam is much reduced by passage through the sample if its frequency satisfies the resonance condition with any of the atomic frequencies.

1.4 The Bouguer-Lambert-Beer Law

Consider a light beam of irradiance $J\hat{k}$ normally incident upon a circular region of the sample of area A. The number of photons removed from the beam in unit time, $-A\Delta J/h(\omega/2\pi)$, is equal to the number of photons which have struck the sample in that time, $AJ/h(\omega/2\pi)$, multiplied by the probability, W, that one photon striking A will be absorbed. Canceling $h(\omega/2\pi)$, one has

$$-A\Delta J = -WAJ. \tag{1.21}$$

The probability of absorption per photon will be equal to the number of atomic systems in the region of the sample illuminated by the beam, N, times the probability that the photon in question will be intercepted by a given atom, W':

$$W = NW'. \tag{1.22}$$

If the number of absorbers per unit volume is N',

$$N = A\Delta z N', \tag{1.23}$$

where Δz is the thickness of the sample, and $\hat{z}||\hat{k}$. The parameter W' can be used to define an effective *cross section* for absorption by one atom, κ:

$$W' \equiv \frac{\kappa}{A}. \tag{1.24}$$

In other words, a model for the absorption process is adopted which consists of N spheres, each with cross-sectional area κ, embedded in a transparent matrix of area A and thickness Δz (see Fig. 1.4). If Δz is sufficiently small, the probability that one sphere will shadow or obscure another will be negligibly small. Therefore the probability of absorption per photon is equal to the fraction of the total area A blocked by the N atoms:

$$W = \frac{A\Delta z\, \kappa N'}{A} \tag{1.25}$$

By inserting the expression for W on the right-hand side of Eq. 1.25 into the formula in Eq. 1.21 and canceling the A's, one obtains

$$-\Delta J = J\kappa\Delta z\, N'. \tag{1.26}$$

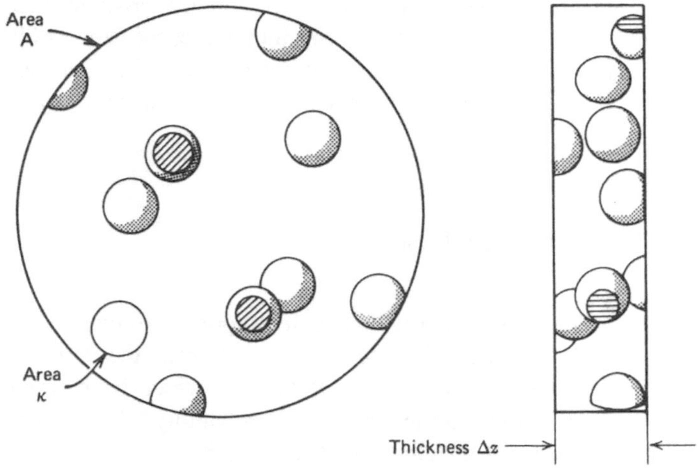

Fig. 1.4. Model for the absorption process used to derive the Bouguer-Lambert-Beer law

In the limit as Δz goes to zero, Eq. 1.26 can be written as a differential equation. This may be integrated to find the diminution in irradiance suffered by a light beam traveling through a thick sample composed of many such lamina stacked together:

$$J(z) = J(0)\exp(-\kappa N' z). \tag{1.27}$$

Equation 1.27 is commonly known as *Beer's law*. Historically, the discovery that a beam of light is attenuated exponentially with distance as it passes through an absorbing medium was made by Pierre Bouguer, a French pharmacist, in 1729, and independently by the English physicist J.H. Lambert in 1760. The discovery that the damping constant (the argument of the exponential, called the *absorption coefficient*) is proportional to the concentration of the absorbing species, N', was made in 1851 by the German, A. Beer.

In emission spectroscopy, the sample itself becomes the light source. The atoms, molecules, and ions of which the material is composed are heated in an oven or placed in an electric discharge. In this manner, the atomic systems acquire more energy than they would have if they were at thermal equilibrium with their surroundings. Some of the excess energy is radiated away in the form of light, which is separated into waves of different frequencies by the dispersing element and allowed to strike the detector. In principle, the absorption spectrogram for any substance should be the negative image of the emission spectrogram: dark absorption lines should appear on a bright background at exactly the same frequencies at which bright emission lines appear on a dark background. Whenever nonradiative processes for disposing of excess energy are rapid enough to compete with resonant emission, however, this rule of similarity between the two spectra becomes unreliable.

1.5 The Importance of Spectroscopic Transitions

To the scientist, the principal interest in the spectroscopic interaction between light and matter is that the characteristic resonant frequencies of atoms, molecules, and ions can be discovered thereby. Since each different atomic system has a characteristic pattern of spectral lines, the spectrogram becomes a "fingerprint" of that system: spectroscopy is therefore a vital part of modern analytical chemistry. The physical chemist and chemical physicist use spectroscopic information in the study of atomic and molecular processes that determine the characteristic frequencies and widths of the lines. These processes include transitions between various quantum states of the atoms and molecules representing different amounts of magnetic, rotational, and vibrational energy. Also studied are transitions between energy states corresponding to different spatial distributions of the electrons about the nuclear framework. Analogous information about the inner workings of the nuclei themselves can also be obtained spectroscopically.

Regardless of whether or not we share these scientific interests, spectroscopic transitions play a crucial role in our everyday lives. Our most valuable sense is sight, and almost everything we see is rendered visible by these interactions. There are exceptions, of course. The daytime sky is perceived as blue because of Rayleigh scattering; many white objects (e.g., clouds and milk) are seen because of Tyndall scattering; colorless transparent materials such as glass and water are visible because of the glints of light from their surfaces produced by nonresonant dielectric reflection; silvery surfaces show up because of metallic reflection; and some birds, fishes, and oil films display iridescent colors as a result of interference between the reflectances of closely spaced dielectric surfaces. But almost every other material in our environment is seen because some of the light striking it produces spectroscopic-type transitions therein. Examples are grass, clothing, wine, bricks, and human skin. Light absorbed in these processes is usually turned irreversibly into heat: the balance enters our eyes by reflection or

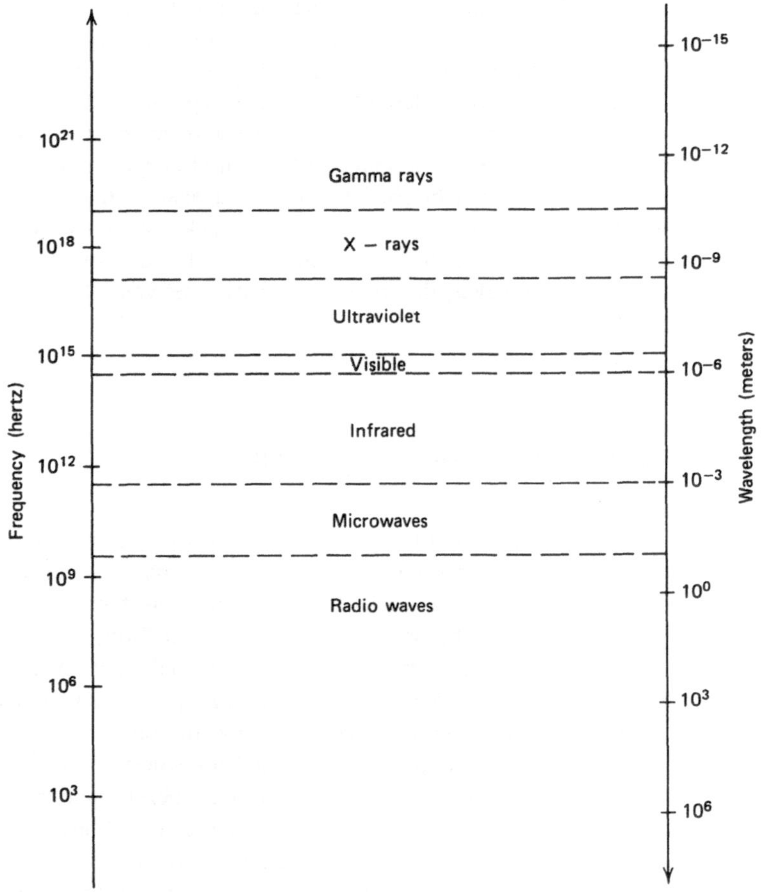

Fig. 1.5. The electromagnetic spectrum

transmission. Because absorption lines are frequently spread over only a relatively small range of frequencies, the transmitted and/or reflected light appears colored. Violet, indigo, blue, green, yellow, orange, and red are seen in turn as the frequency diminishes. If the most intense light striking the retina has a frequency of about 9×10^{14} Hz, violet is perceived. At 4×10^{14} Hz, red is seen. Many more spectroscopic transitions occur at higher frequencies and at lower frequencies than those of visible light. See Fig. 1.5.

1.6 The Question of Dynamics

The usual picture of spectroscopic transitions is a "thermodynamic" rather than a "kinetic" one. In other words, the emphasis is usually placed on the coincidence (resonance) between the energy of the photons being absorbed or emitted, $\mathscr{E}_{\text{photon}}$, and the difference in energy between the quantum states of the atom undergoing the transition, $\Delta\mathscr{E}_{\text{atom}}$:

$$\Delta\mathscr{E}_{\text{atom}} = \mathscr{E}_{\text{photon}}. \tag{1.28}$$

Seldom is one concerned about the dynamics of the absorption process. An exception, however, should be made: calculations of the overall absorption or emission rate (number of transitions per second, averaged over a large collection of identical quantum systems) produced by weak incoherent light are frequently performed. The parameters produced by means of such calculations (oscillator strengths, extinction coefficients) are used to predict the relative amplitudes and widths of absorption or emission lines that appear on spectrograms. But this is not the same thing as attempting to describe the mechanism by means of which photons are created or destroyed and their energy and momentum transferred to the quantum systems, nor do such calculations provide a description of the state of the quantum systems in the interval between the initiation and the termination of the transition process.

In one branch of spectroscopy, work has not been limited to the study of oscillator strengths and similar factors. Theorists exploring the behavior of magnetic particles (nuclei and electrons) interacting with radio waves and microwaves provided a detailed description of the dynamics of that process. This knowledge had little impact on the rest of spectroscopy, however, because it was generally considered that magnetic resonance was somehow a special case. The discovery that the formulas for nuclear magnetic resonance and electron paramagnetic resonance (nmr and epr) can be generalized to describe any quantum transition is relatively recent; a brief historical account is provided in the next two sections.

1.7 Early History of the Quantum Theory

In the seventeenth century Newton showed that two massive bodies, whenever attracted to one another by means of a force inversely proportional to the square of the distance between their centers, must necessarily move so that the path of one body is a conic section with the other body located at the principal focus of the conic. Newton further demonstrated that, if gravity is such a force, the observed motions of the planets around the sun correspond with the predictions of his theory with fantastic precision.

In 1785 Charles A. Coulomb showed that the electrostatic force between two charged particles is also an inverse-square force.[1] Soon after, it became known that matter is composed of atoms and, furthermore, that an atom is in turn composed of negatively charged electrons which can be removed by ionization and of a positively charged residue sufficient to render the atom as a whole electrically neutral.

As soon as Millikan was able to measure the charge on the electron in the early 1900s, the electronic mass could be calculated from the previously known charge-to-mass ratio. It then became clear that most of the mass of the atom is associated with the positively charged residue. In 1911 Rutherford showed that this residue is very small in comparison with the whole size of the atom, and named it the *nucleus*.

In 1913 Bohr reasoned that, since an atom is composed of several relatively light particles attracted to a heavy one by an inverse-square force, the atom must be very similar mechanically to the solar system. The electrons should orbit the nucleus in paths that are conic sections, just as the planets orbit the sun (see Fig. 1.6.)

Unfortunately, in one important respect, the Bohr model was inconsistent with the observed properties of atoms. In the 1860s James Clerk Maxwell had combined all the known laws of electricity and magnetism into a unified and consistent theory. An important consequence of this development was the prediction that a charged particle (e.g., an electron) undergoing accelerated motion (such as would be produced if the electron indeed moved in a Newtonian orbit about the nucleus) should radiate electromagnetic waves (light). This prediction is experimentally used e.g. in the production of synchrotron radiation by means of electron storage rings (see Chap. 13).

Bohr began his famous paper by showing that the energy of the electron in its orbit must eventually be completely radiated away by this mechanism. The atom should emit a continuous spectrum of electromagnetic waves at higher and higher frequencies as the electrons spiral into the nucleus, ceasing only at the point of complete atomic collapse. Experimentally, of course, atoms do not emit any light spontaneously at normal temperatures. When excited, they do radiate some of their excess energy electromagnetically in a continuum of frequencies. However,

---

[1] This fact had previously been discovered, but not published, by Henry Cavendish.

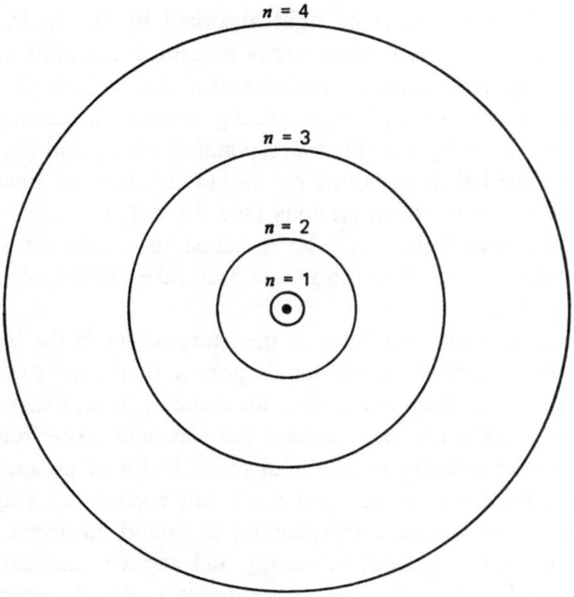

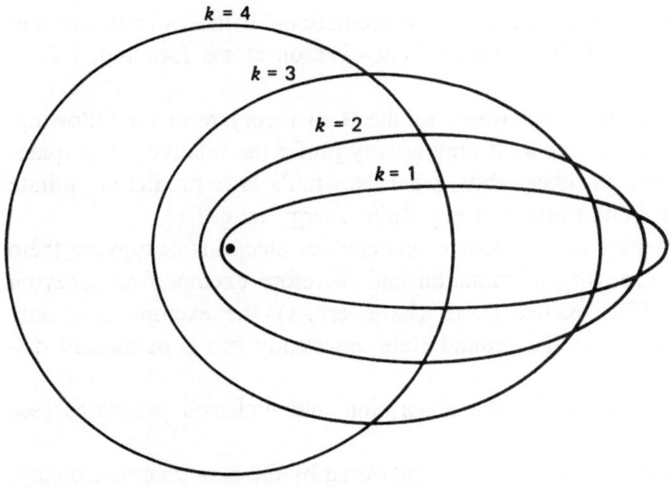

Fig. 1.6. Bohr orbits. At the top, orbits with angular momentum $k(\equiv l+1) = n$ are shown. At the bottom, orbits with $n = 4$ but differing in k are shown

as described in the portion of Sect. 1.1 devoted to emission spectroscopy, a very characteristic discrete spectrum of lines is produced as well. Under no conditions found on earth does a complete collapse of all of the electronic orbits occur.

Bohr was at a loss to explain this apparent contradiction between the laws of mechanics and the laws of electricity and magnetism. Nevertheless, he was able to calculate precisely the frequencies of light absorbed by the hydrogen atom by assuming that (1) the only Newtonian orbits permitted electrons were those with angular momenta equal to integral multiples of $\hbar$ (i.e., $h/2\pi$), (2) the electrons could move from one orbit to another by gaining or losing a quantity of energy equal to the difference in energies of the corresponding orbits, and (3) the necessary energy could be supplied or removed by the annihilation or creation of bundles of electromagnetic energy called photons (see Eq. 1.28).

In 1905 Einstein showed that light could be assumed to consist of such bundles. The relationship between this description and the wave picture of light was given previously in Eq. 1.16.

In Newtonian mechanics there are two types of planetary orbits. If the initial velocity of the planetary object relative to the sun is greater than the "velocity of escape" for the system (as sometimes is the case for comets), it will move in an open, or hyperbolic, path, make one pass around the sun, and never return. If, on the other hand, the initial velocity is less than the velocity of escape (as is the case for planets), the orbit will close upon itself and become an ellipse. In Bohr's theory, only the closed orbits, corresponding to bound electrons, are supposed to have discrete amounts (quanta) of energy and angular momentum, so that transitions between such orbits give rise to the discrete optical spectrum. A continuum of energies is allowed to the unbound electrons; transitions between their orbits, or between them and bound orbits, produce the continuous optical spectrum. The agreement between the predictions of the Bohr theory and experimental observations of the spectra of one-electron atoms (see Fig. 1.7) is extremely good.

Among the questions left unanswered by the Bohr theory were the following.

1. Why do electrons behave as if they greatly prefer the relatively few quantized orbits of Bohr over all other orbits, when Newton's laws predict an infinite number of permissible bound orbits in any finite energy range?

2. What special feature of the Bohr orbits enables electrons occupying them to be relieved of the necessity of radiation and therefore exempt from obeying Maxwell's equations? [For excited states (large orbits), the exemption is only temporary, but an electron in the ground state apparently has a permanent dispensation.]

3. What are the dynamics of the absorption and emission processes (see Sect. 1.6)?

All of these questions, and more, were answered by the *new quantum theory*, developed from 1925 to 1928 by Heisenberg and Schrödinger. The Schrödinger

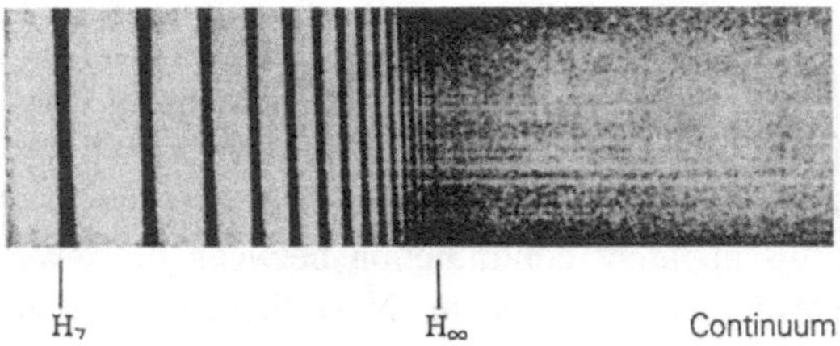

Fig. 1.7. The spectrum of hydrogen. Higher members of the Balmer series of the H atom in emission (final quantum state having $n = 2$), starting from the seventh line and showing the continuum [from Herzberg G (1927) Ann Phys (4) 84:565, Plate 14, Fig. 6]

formulation of this theory was based on an idea by De Broglie that matter, like light, has both a wave and a particle nature. Wave mechanics has proved to be an exceptionally useful tool in the development of relatively simple conceptual models for the structures of atoms and molecules. Consequently, it has been historically favored over the earlier Heisenberg formulation by atomic and solid-state physicists, as well as by physical chemists.

In wave mechanics, one describes the position of an electron by calculating a wavefunction $\psi(x, y, z, t)$, the complex square of which is equal to the probability of finding the electron in question at the position specified by the coordinates (x, y, z) at the time t. The Coulombic field set up by the positively charged nucleus creates a "trap" for the electron and localizes its probability wave in the immediate neighborhood. If the initial relative kinetic energy of the electron-nucleus system is sufficiently low, the electron will be caught in this trap. The

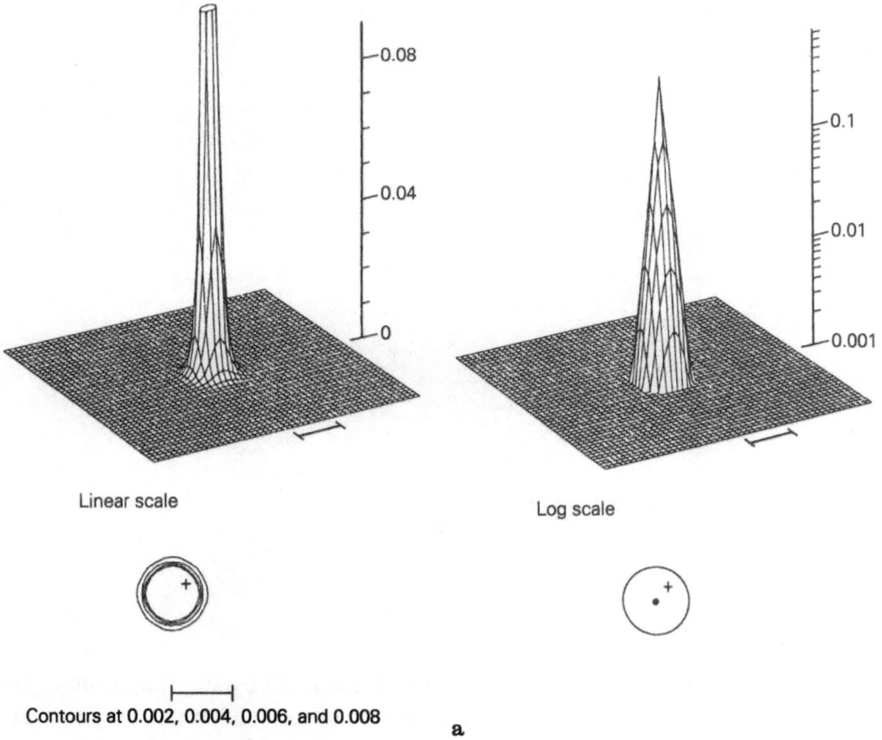

Fig. 1.8. Typical atomic orbitals from *Orbital and Electron Density Diagrams* (The Macmillan Co., New York, 1973), by A. Streitweiser, Jr. and P.H. Owens **a** 1*s* (Fig. 1.6 on P. 12 of the book cited). The plot at the upper left is a direct plot of ψ. A plane is clearly defined as a grid 64 atomic units (au) on a side. At the front of this grid a marked line is shown 10 au in length. The grid is the xy plane. For each point in this plane there is a corresponding value for ψ as given by the vertical scale at the right of the grid. For the 1*s* function, ψ has positive values throughout, so the figure plotted is entirely above the plane. The 1*s* orbital is spherically symmetric; Fig 1.8a represents the wavefunction in one plane and is not meant to be a "picture" of the entire orbital. **b** Same as 1.8a, but for $2p_x$ (Fig. 1.10 on p. 18 of the book cited)

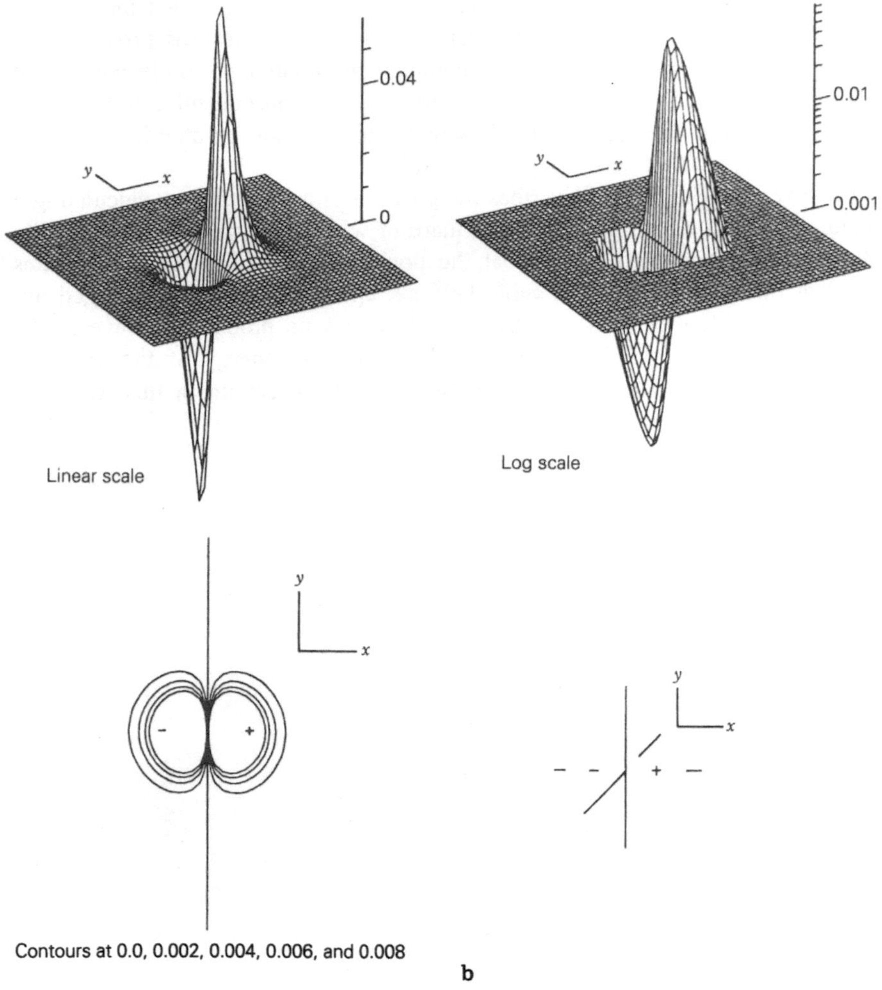

Fig. 1.8. (continued).

probability wave will be effectively tied down (localized) by the electrostatic potential in a fashion analogous to the tying down of a sound wave by an organ pipe. It is characteristic of bound oscillators that only discrete frequencies of oscillation (i.e., the fundamental mode and overtones or harmonics) can be sustained by them. The reason for the discrete frequencies is that an integral number of nodes for the wave must be located within the finite geometrical confines of the region in which it is bound.

In this way, wave mechanics answers the first question which the Bohr theory failed to answer: quantization is a natural consequence of the assumption that bound states for electrons in atoms are spatially localized, three-dimensional, standing probability waves with an integral number of nodes, n_0. In particular, the 1*s* state of hydrogen (*K* shell) is the fundamental mode of oscillation ($n_0 = 1$) for the probability wave of the electron. There is only one "node," the spherical one at $r = \infty$, in the 1*s* wave function; therefore there is only one way to construct such a wave (one orbital). For $n_0 = 2$ (*L* shell), one node is present in addition to the one at $r = \infty$. This mode of oscillation (first overtone) requires more energy to excite than does the fundamental (a fact that can readily be accepted by anyone who has ever played a wind instrument in an orchestra or band!). There are two shapes possible for the extra node, and these are consistent with the symmetry of the Coulombic potential field. First, it can be spherical, like the one at $r = \infty$. The resulting orbital is called 2*s*. (All waves with only spherical nodes are defined to be *s* orbitals). Next, it can be planar, passing through the nucleus. The resulting orbitals are called, collectively, the 2*p* subshell. (All waves with $n_0 - 1$ spherical nodes, plus one planar node, are called *p* orbitals.) There are three orbitals in the *p* subshell because there are three independent directions of orientation for a plane in a three-dimensional space. Similar explanations can be given for the second and higher overtones (*M*, *N*,... shells). Some typical orbitals are drawn in Fig. 1.8.

1.8 Dynamics During Spectroscopic Transitions[1]

As an introductory example for this field of quantum statistics let us start with a problem that, at first glance, seems to be easy to describe.

Consider a small hard-core molecule with an almost permanent electric dipole moment p_E fixed in it, e.g. CH_3Cl, CH_2Cl_2, or CH_3CN. In a low pressure gas phase it possesses rotational quantum states which are occupied in accordance with the Boltzmann distribution law and which allow for rotational lines by absorption in the far infrared (FIR) spectral region. The complete Hamiltonian H_c consists of the Hamiltonian H_0 for the unperturbed system and the Hamiltonian H_I for the interaction with a weak electric field

$$H_c = H_0 + H_I \tag{1.29}$$

$$H_I = -p_E E_0 \cos \omega t \tag{1.30}$$

The energies of the absorbed photons amount to

$$\hbar\omega_{if} = \varepsilon_f - \varepsilon_i \tag{1.31}$$

with ε_i and ε_f are the energies of the initial and the final state, respectivley, and $h = 2\pi\hbar$ Planck's constant.

---

[1] Contributed by the editor E. Lippert

In our example the size of the absorbing molecule (i.e. its mean diameter of about 10 Å $= 10^{-7}$ m) is much smaller than the wavelength of the absorbed photon, e.g. $\lambda = 0.3$ mm $= 3 \times 10^{-4}$ m. The time necessary for such an absorption process takes, therefore, at least

$$\tau_t = \lambda/c = 1/c\tilde{\nu} = 2\pi/\omega = 1 \text{ ps} = 10^{-12} \text{ s} \tag{1.32}$$

Let us now consider a dilute solution of the same polar molecule at the corresponding small concentration in an inert solvent like CS_2, CCl_4, or *n*-heptane at room temperature. Due to solute/solvent interactions there are no well-defined eigenvalues, and interaction with photons produces, therefore, no rotational lines but a continuous absorption band; instead of free rotation there occurs reorientational motion of all molecules of the interacting community of that solution. The mean energy of each degree of freedom amounts to

$$\Delta\mathcal{E} = k_B T/2 \tag{1.33}$$

where k_B is the Boltzmann constant, and T the absolute temperature.

The dissolved polar molecules are strongly coupled with the surrounding solvent molecules by a certain potential V which disturbs its rotational motion. Let τ_v be the characteristic time of the reorientational motions of p_E. Following Heisenberg's uncertainty relation let us assume

$$\tau_v = \hbar/\Delta\mathcal{E} = 2h/k_B T = 1 \text{ ps} \tag{1.34}$$

Indeed, this duration agrees in the order of magnitude with experimental values of the Debye relaxation time τ_D of such systems. Eventhough the amount p_E of the molecular dipole moment remains almost constant the vector p_E is not a constant in the picosecond time region but alters its direction.

For FIR absorption processes of the type we considered so far we notice that, in the order of magnitude,

$$\tau_t = \tau_v = \tau_D = 1 \text{ ps} \tag{1.35}$$

Under such conditions during the time of an absorption process the energy of the absorbed photon will be partly dissipated, i.e. irreversibly transformed into kinetic energy of neighbouring solvent molecules. In summing up we learn that FIR absorption irreducibly increases the heat, the temperature, and even the entropy of condensed matter. In terms of thermodynamics our system behaves as an open system.

Following van Vliet's [2] revisited Linear Response Theory (LRT) Eq. (1.29) must be generalized, i.e. the dissipation must be taken into account by adding a special term λV explicitely into the system Hamiltonian

$$H_{\text{open}} = H_0 + H_I + \lambda V \tag{1.36}$$

The factor λ controls the weight of coupling V, where $0 < \lambda \cong 0.1 < 1$, and V represents the many body operator as mentioned above (strictly speaking the fluctuations of H_0 during τ_t), and $H_I = -\Sigma m_j E(t)$, where m_j is the dipole moment operator of one molecular absorbing system that couples with the electric field $E(t)$.

The consequences of this generalization procedure are as follows. If the operator H_{open} is introduced into the time-dependent Schrödinger equation (Eq. 2.23) there remains [3]

$$\frac{\partial H_{open}}{\partial t} = -\dot{p}_E E - p_e \dot{E} \neq 0 \tag{1.37}$$

in contrast to isolated systems for which the so-called complete Hamiltonian of Eq. (1.29) should be used, yielding

$$\frac{\partial H_c}{\partial t} = \frac{\partial H_I}{\partial t} = 0 \tag{1.38}$$

since $p_E E = p_E E$. In the generalized approach [3] this condition is not fulfilled since even if the fluctuations $\partial \mathbf{p_E}$ are small the fluctuations of its time derivative $\partial \dot{\mathbf{p}}_\mathbf{E}$ could be rather large and only then the process becomes irreversible in

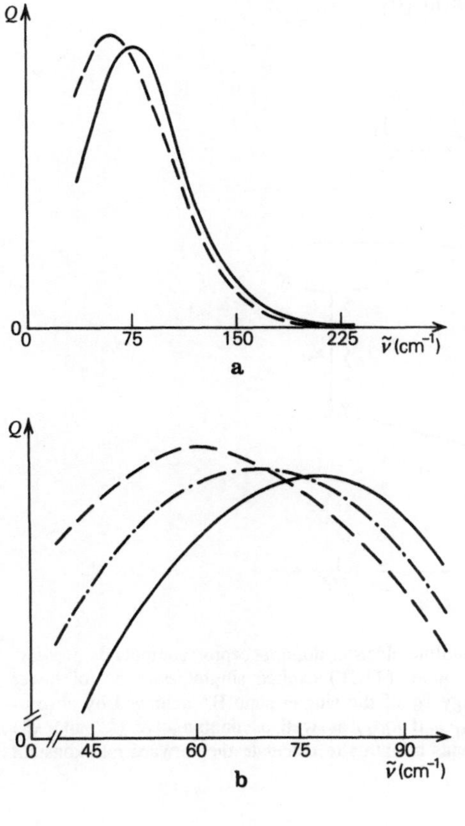

Fig. 1.9. a FIR absorption spectra of CH_3CN in CCl_4, concentration 0.19 mol/l at +60 °C ----, − 15 °C ---; **b** detail of Fig. 1.9a at +60 °C ----, +20 °C ----, and −15 °C --- [5]

the sense of the second law of thermodynamics and entropy production is a consequence of it.

The long-wavelength FIR absorption band of the solutions under consideration possess a rather strange behaviour since it shows a so-called anti-Boltzmann temperature dependence [4]; with increasing temperature the whole band is shifted to higher frequencies instead of being *red-shifted* (Fig. 1.9). This effect is possible according to Eq. (1.37): The absorption band seems to be due to the superposition of two quantum statistical contributions.

In order to try to give a vivid physical explanation we notice that with increasing temperature the number of collisions and the friction in the cages around the cores of the soluted molecules increase and, therefore, their effective mass, resulting in a slower reorientational motion. Hence the lower energy levels of the Hamiltonian of the absorbing system become more populated at the expense of the population of the higher energy levels. It has been shown that the non-diagonal part of the relevant density operator plays an important role on the FIR absorption process [5]. By applying the Complex Scaling Method (CSM) to the analysis of Prigogine's subdynamics it became clear that the concept of physical kinetics of an individually existing molecule in such condensed matter cannot be maintained since the molecules lose their individualities; in classical terms one should say that correlation patterns rather than molecules are meaningful physical concepts in the context under consideration [6].

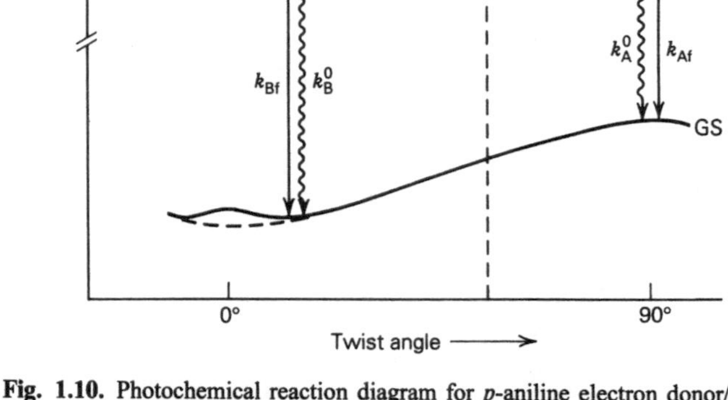

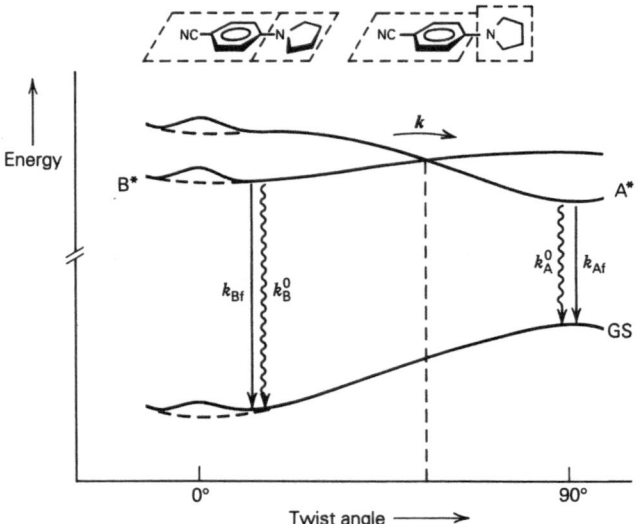

Fig. 1.10. Photochemical reaction diagram for *p*-aniline electron donor/acceptor compounds possessing an orthogonal twisted intramolecular charge-tranfer (TICT) excited singlet state A* of lower equilibrium energy ε_A as compared with the energy ε_B of the singlet state B* achieved by absorption. If both states experience both radiative (k_{Af} and k_{BAf}) as well as nonradiative (k_A^0 and k_B^0) deactivation then fluorescence intensity measurements allow us to calculate the forward rate konstant $\vec{k}$, activation energies, etc. [7]

Even though this problem cannot be dealt with in detail in this book some of the scientific foundations for its quantitative solution are presented in Chap. 5 where the density matrix will be introduced and in Chap. 7 where the entropy production in ensembles of irradiated systems will be discussed.

The above example illustrates the case $\tau_v \leqq \tau_t$ and is typical, therefore, for some areas of this book even though we will preferably deal with relaxation processes connected with the transitions themselves, i.e. the rearrangement of the electronic structure of atoms and molecules during absorption or emission in condensed matter.

Excluded from the contents are processes where $\tau_v > \tau_t$, occurring in the time between spectroscopic transitions such as photochemical *cis/trans*-isomerization of stilbene, or photophysics of internal twisting, e.g. molecular motion in biradicaloid aromatic compounds in liquid solutions between e.g. absorption and emission (Fig. 10). The study of the subnanosecond time dependence of the spectral quanta distribution of the emission eallows insight into the photophysics of these processes (for further introductory literature see [8]).

1.9 References

1. Garbuny M (1965) Optical Physics. Academic Press, New York
2. van Vliet KM (1978) J. Math. Physics 19:1345; (1979) J. Math. Physics 20:2573
3. Lippert E, Chatzidimitriou-Dreismann CA (1982) Intern. J. Quant. Chem., Quant. Chem. Symp. 16:183
4. Kroon SG, van der Elsken J (1967) Chem. Phys. Lett. 1:258
5. Lippert E, Chatzidimitriou-Dreismann CA and Naumann K-H (1984) Adv. Chem. Phys. 57:311
6. Chatzidimitriou-Dreismann CA, Brändas EJ (1988) Ber. Bunsenges. Phys. Chem. 92:549
7. Lippert E, Rettig W, Bonačić-Koutecký V, Heisel F and Miehé JA (1987) Adv. Chem. Phys. 68:1
8. Lippert E (1988) Ber. Bunsenges. Phys. Chem. 92:417

1.10 Problems

1.1. The vacuum wavelength, λ_0, of the radiation from a typical CO_2 laser is 10.6 μm. Calculate the frequency, $\omega/2\pi$, of these waves in hertz (cycles per second).

1.2. The root-mean-squared power in a CO_2 laser beam is 8.0 W, and the cross-sectional area of the beam is $1.0 \times 10^{-4} m^2$.
   a. What is the peak electric field amplitude, E_1^0, in volts per meter? Assume that the waves propagate through a vacuum.
   b. What is the peak magnetic field amplitude, H_1^0, in amperes per meter under the same conditions?

1.3. An ifrared absorption cell 0.10 m long is filled with SF_6 gas at a pressure of 0.20 torr (760 torr = 1 atm). A CO_2 laser beam enters the cell with an irradiance of $J_{rms} = 80$ W m^{-2} and leaves the cell with an irradiance of $J_{rms} = 32$ W m^{-2}.
   a. Calculate the absorption cross section of SF_6, κ, in square meters per molecule.
   b. Calculate the energy difference in joules per molecule between the two levels of the SF_6 molecules responsible for the absorption of 10.6 μm laser radiation.

2 Elementary Quantum Theory

J.D. Macomber

2.1 States and Their Properties: Operators and Kets

In order to understand the dynamics of spectroscopic transitions, it is convenient to review the elementary principles of quantum theory. The treatment that follows differs therefore somewhat in point of view from the one usually presented in a first course in quantum mechanics for chemists. A very straightforward, elegant, and detailed exposition of this material may be found in the book by Fano and Fano [1], as well as in other quantum mechanics textbooks.

A quantum system (e.g., an atom, ion, or molecule) behaves at any given time in a characteristic fashion determined by its state. (In a sense, the pattern of behavior *is* the state.) Ordinarily, a large number of different states are possible. The task of quantum mechanics is to enumerate them for each system of interest, and to calculate the physical and chemical properties of the system associated with each state. The first step is accomplished by assigning to each state a mathematical entity called a *ket*.

To each observable property of the system is assigned another kind of mathematical entity called an *operator*. The operators transform one ket into another. The transformation process is described algebraically:

$$\mathsf{R}|a\rangle = |b\rangle. \tag{2.1}$$

In Eq. 2.1, R is some operator; $|a\rangle$ and $|b\rangle$ are kets. Note that R is written to the left of $|a\rangle$. It is convenient to visualize kets as vectors in an abstract space, called *Hilbert space* representing an infinite number of dimensions. In this case, the operator R rotates and/or stretches $|a\rangle$ in order to turn it into $|b\rangle$ (See Fig. 2.1).

Sometimes the result of the operation of R upon a particular ket will be the same ket multiplied by a number, r:

$$\mathsf{R}|a\rangle = |b\rangle = r|a\rangle. \tag{2.2}$$

Kets that satisfy Eq. (2.2) are said to be *eigenkets* of R, and the constants r are termed the corresponding *eigenvalues*. It is possible to arrange all of the eigenkets of R into an ordered set, $\{|r\rangle\}$. (Two eigenkets that differ only by a constant multiplicative factor will not be counted as different.) The number of eigenkets in the set, M, may be different for different quantum systems – frequently M

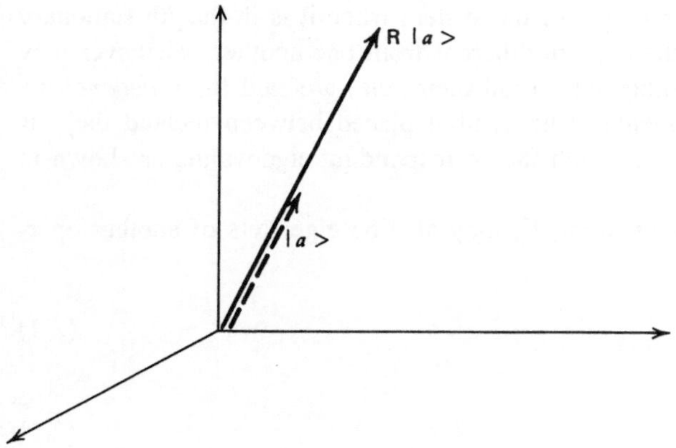

a

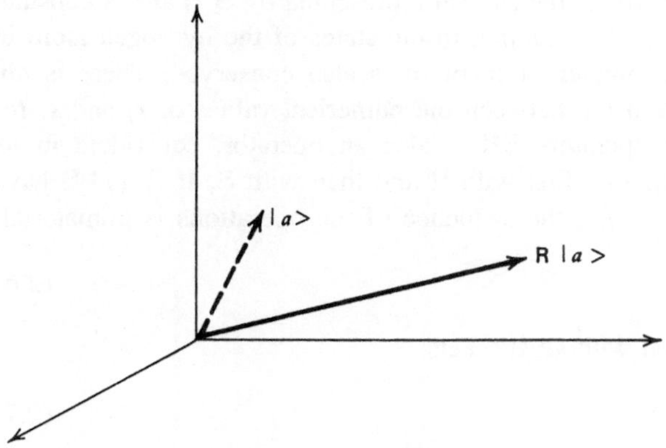

b

Fig. 2.1. Effect of operator R on ket $|a\rangle$. At the top, $|a\rangle$ is an eigenket of R; at the bottom, $|a\rangle$ is not an eigenket of R

is infinity. Eigenkets are of transcendent importance in quantum mechanics for several reasons.

1. If a system is in a state represented by an eigenket of R, the magnitude of the observable physical property represented by R is a constant equal to the corresponding eigenvalue. This is one of the reasons why states represented by eigenkets are sometimes called *stationary states.*[1] This term should *not* imply that the constituent parts of the quantum systems are not moving; it means merely that the property represented by R is a constant of the motion. In particular, if $|r_j\rangle$ is the jth member of the set $\{|r\rangle\}$ and R represents the electric dipole moment,

$$\mathsf{R}|r_j\rangle = r_j|r_j\rangle, \tag{2.3}$$

[1] Strictly speaking, the adjective "stationary" should be reserved for eigenstates of the Hamiltonian operator. See discussion in Sect. 2.3.

and r_j is the electric dipole moment of the system when it is in the jth stationary state (eigenstate). Usually, the r_j's are different from one another; whenever they are the same, the corresponding states (and their kets) are said to be *degenerate*. The label or name of an eigenket (the symbol placed between the $|$ and the $\rangle$) is frequently chosen to be identical with the corresponding eigenvalue, as shown in Eq. (2.3).

2. The eigenkets of one operator, R, may also be eigenkets of another operator, S:

$$\mathsf{S}|s_j\rangle = s_j|s_j\rangle, \tag{2.4}$$

and

$$\{|r\rangle\} = \{|s\rangle\}. \tag{2.5}$$

The operators R and S are said to have *simultaneous* eigenkets. This means that the observable physical property of the system represented by S is also a constant of the motion in these states. For example, in the states of the hydrogen atom in which energy is conserved, angular momentum is also conserved. There is not necessarily an obvious relationship between the numerical values of r_j and s_j for any j. The product of two operators SR is also an operator, equivalent in its effect on a ket to a transformation first with R and then with S. If R and S have simultaneous eigenkets (e.g., $|r_j\rangle$), the sequence of the operations is immaterial:

$$\mathsf{SR}|r_j\rangle = \mathsf{RS}|r_j\rangle. \tag{2.6}$$

Sometimes Eq. 2.6 is written without the kets:

$$\mathsf{SR} - \mathsf{RS} \equiv [\mathsf{S}, \mathsf{R}] = 0. \tag{2.7}$$

The difference $[\mathsf{S}, \mathsf{R}]$ is called the *commutator* of S and R. If the commutator is zero, the corresponding operators are said to commute.

3. The eigenkets of one operator R will not usually be satisfactory eigenkets for all of the other operators of interest. Let Q be an operator for which $\{|r\rangle\}$ is not the correct choice of eigenkets:

$$\mathsf{Q}|q_j\rangle = q_j|q_j\rangle, \quad 1 < j < M, \tag{2.8}$$

but

$$\{|r\rangle\} \neq \{|q\rangle\}. \tag{2.9}$$

This means that the observable physical property of the system represented by Q is *not* constant in states in which the property represented by R *is* constant, and vice versa. For example, there are no states of the hydrogen atom in which both energy and linear momentum are conserved. Nevertheless, $\{|q\rangle\}$ and $\{|r\rangle\}$ have the same number of members, M. If the operators R and Q do not have simultaneous eigenkets, they will fail to commute:

$$\mathsf{QR}|r_j\rangle \neq \mathsf{RQ}|r_j\rangle \tag{2.10}$$

or

$$[\mathsf{Q}, \mathsf{R}] \neq 0. \tag{2.11}$$

By far the most popular choice of a basis (complete set of eigenkets) to use in solving problems in quantum mechanics is the set of eigenkets of the Hamiltonian operator, H. (The physically observable property represented by the Hamiltonian operator is almost always the energy of the system.) It cannot be emphasized too strongly that this choice of basis is merely a custom, dictated by convenience in attacking the kinds of problems usually discussed in conventional courses in quantum mechanics. It would be an error to assume that quantum systems can exist in no states other than the ones in which the energy is a constant of the motion. If this were assumed, it would then be impossible to explain the rich variety of behavior displayed by quantum systems in nature.

2.2 Stationary States: Eigenkets and Eigenfunctions

The discussion given in Sect. (2.1) is the basis for a more precise statement of the two central tasks of quantum theory: (1) to find the ket, $|a\rangle$, that describes the system in the state of interest, and (2) to extract from this ket, by means of the corresponding operators, various physically observable properties of the system.

Suppose, for example, that the system consists of a single electron traveling through space, either freely or under the influence of some attractive or repulsive force. The position of this electron is specified by specifying its coordinates x, y, and z, measured along some convenient set of axes. To each one of these three coordinates corresponds an operator; let them be called X, Y, and Z. For simplicity, one can concentrate on X. The corresponding eigenvalues will be the possible results of a measurement of the x coordinate of the electron:

$$\mathsf{X}|x_j\rangle = x_j|x_j\rangle. \tag{2.12}$$

It is not likely that the system will be in an eigenstate of x; that is, the ket representing the actual state of the system will probably not satisfy Eq. (2.12). But whether or not the correct ket for the system *is* an eigenket of X, it will always be possible to represent that ket as a linear combination of such eigenkets:

$$|a\rangle = a_1|x_1\rangle + a_2|x_2\rangle + \cdots + a_M|x_M\rangle\,. \tag{2.13}$$

This is so because the eigenkets of any operator representing a physically observable property of the system form what is called a complete set. In Hilbert space, even though the vector $|a\rangle$ may not "point" exactly along any of the axes represented by $\{|x\rangle\}$, the "direction" along which it *does* point can always be specified by giving the "projections" of $|a\rangle$ along each of those axes; these projections are the a_j's (see Fig. 2.2).

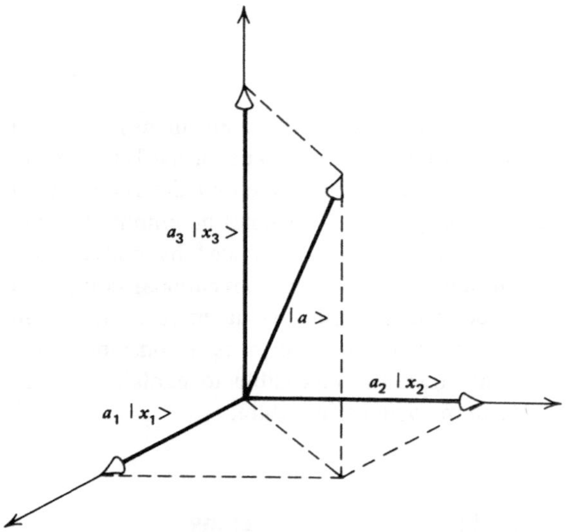

Fig. 2.2. Representation of ket $|a\rangle$ as a linear superposition of eigenkets $|x_1\rangle, |x_2\rangle$, and $|x_3\rangle$

It happens that there are an infinite number of possibilities of x_j; that is to say, $M = \infty$. Furthermore, the eigenvalues x_j form a continuum from $-L$ to $+L$, where $2L$ is the length of the portion of the x axis along which motion of the electron is possible. In other words,

$$x_{j+1} - x_j = \Delta x, \tag{2.14}$$

where Δx is a very small number.

It is also usually the case that the a_j's form a continuum:

$$a_{j+1} - a_j = \Delta a. \tag{2.15}$$

Whenever Eqs. (2.14) and (2.15) hold, it is convenient to define a continuous algebraic function, $\psi(x)$, from which all the a_j's can be calculated:

$$\psi_a(x_j) = a_j. \tag{2.16}$$

See Fig. 2.3.

The state of the system is completely specified by the a_j's, in accordance with Eq. (2.13). Furthermore, the effect of any operator on $|a\rangle$ will be to transform it into another ket, $|b\rangle$, in accordance with Eq. (2.1). Note that $|b\rangle$ will *also* be expressible in a linear combination of the members of the set of eigenkets of X:

$$|b\rangle = b_1|x_1\rangle + b_2|x_2\rangle + \cdots + b_M|x_M\rangle. \tag{2.17}$$

In general, the coefficients (projections) in the expansion of $|a\rangle$ will differ from those of $|b\rangle$,

$$a_j \neq b_j, \tag{2.18}$$

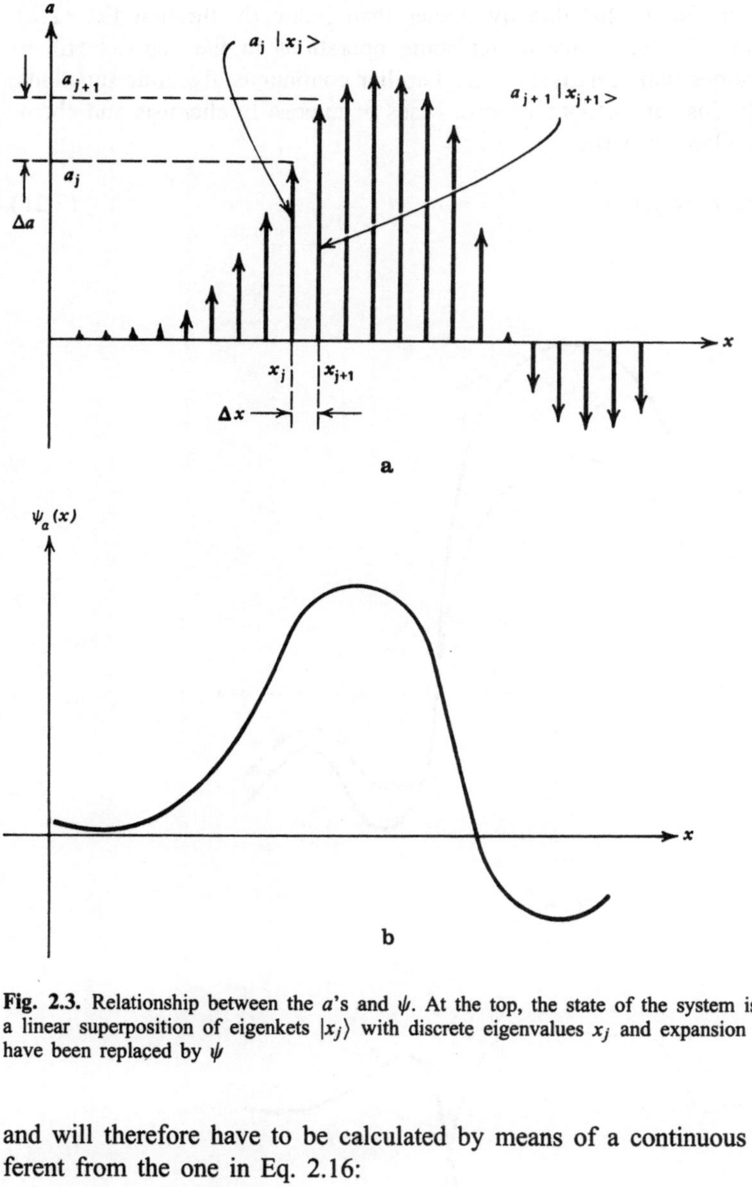

Fig. 2.3. Relationship between the a's and ψ. At the top, the state of the system is represented by a linear superposition of eigenkets $|x_j\rangle$ with discrete eigenvalues x_j and expansion coefficients a_j's have been replaced by ψ

and will therefore have to be calculated by means of a continuous function different from the one in Eq. 2.16:

$$\psi_b(x_j) = b_j. \tag{2.19}$$

Therefore the operation that transforms $|a\rangle$ into $|b\rangle$, Eq. 2.1, is completely specified by the change in the expansion coefficients:

$$\psi_a(x) \rightarrow \psi_b(x). \tag{2.20}$$

It is tempting, therefore, to seek a mathematical method of accomplishing the transformation in Eq. (2.20) directly, rather than indirectly through Eq. (2.1). This would have the advantage of replacing operations in the abstract Hilbert space by operations upon relatively more familiar continuous algebraic functions.

Fortunately this can be done in most cases of interest to chemists and chemical physicists. One can write

$$\mathsf{R}(x)\psi_a(x) = \psi_b(x), \tag{2.21}$$

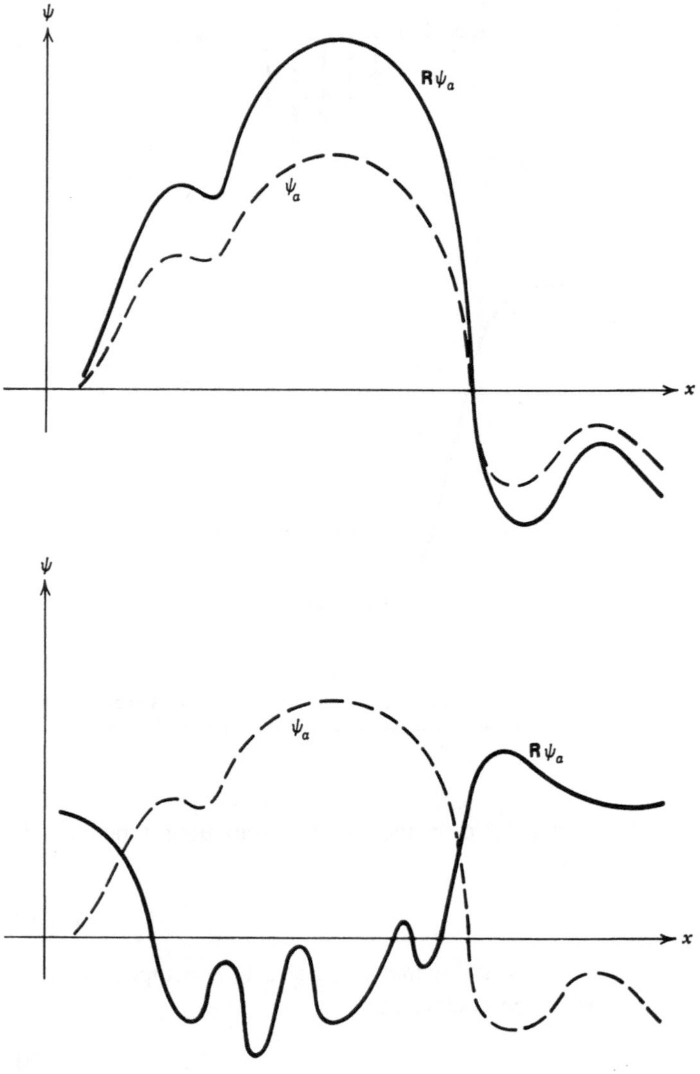

Fig. 2.4. Effect of operator R on wavefunction ψ_a. At the top, ψ_a is an eigenfunction of R; at the bottom, ψ_a is not an eigenfunction of R

and understand that the symbol R(x) in Eq. (2.21) accomplishes the same task in "function space" as the R of Eq. (2.1) does in Hilbert space. In other words, if ψ_a represents an eigenstate of the physically observable property R, then

$$\psi_b = r\psi_a, \tag{2.22}$$

where r is the same eigenvalue that appeared in Eq. (2.2). Equation (2.22) is also called an *eigenvalue equation*, and ψ_a is termed an *eigenfunction* of the operator R(x). These relationships are illustrated schematically in Fig. 2.4.

2.3 Schrödinger's Equation: Time Dependence of ψ

It has just been shown that to each ket (associated with motion of an electron) there corresponds a state function $\psi(x)$, which represents the coefficients of the expansion of that ket in terms of a linear combination of the eigenkets of X. More generally, in three dimensions, ψ is a continuous algebraic function of all three coordinate variables, x,y, and z, because the operators X, Y, and Z commute with one another and have simultaneous eigenkets. The calculation of ψ is equivalent to the determination of the corresponding ket and completely specifies the state being represented by that ket.

To accomplish the central tasks of quantum mechanics, as stated in the language of kets at the beginning of Sect. 2.2, it is necessary to calculate ψ and to find the operators in function space that will extract from it the properties of the system in the state represented by ψ. Recall from Chapter 1 de Broglie's suggestion that particles such as electrons have some of the properties of waves. Schrödinger, following up this suggestion, assumed that the ψ's had the mathematical form of amplitudes for these waves, oscillating in the three spatial dimensions as well as in time. He found the differential equation which must be satisfied by any ψ, in order that it properly represents the state of a quantum system in the way that was just described:

$$\mathsf{H}\psi = i\hbar\left(\frac{\partial\psi}{\partial t}\right)_q, \tag{2.23}$$

Where $i = \sqrt{-1}$. Henceforth the symbol q will stand for the entire collection of independent variables other than t (in this case the spatial coordinates, x, y, and z, of the electron).

The Hamiltonian operator, H, will in general depend on the variables q and t. In many cases of interest, however, the operator has no explicit time dependence. In such circumstances, it is possible to break Eq. (2.23) in two separate parts. To do this, it is assumed that the wavefunction is expressible as the product of two simpler functions. One function, T, depends only on the time. The other, Φ, depends on all of the other independent variables, q:

$$\psi(q,t) = \Phi(q)T(t). \tag{2.24}$$

The expression for ψ given in Eq. (2.24) may then be substituted into Schrödinger's Eq. (2.23). On the left-hand side, T is not affected by H because T and H depend on different variables. It may therefore be removed from the group of symbols to the right of H representing operands and placed to the left of H, becoming merely a scalar multiplier. On the right-hand side, Φ may be permuted with the operator $(\partial/\partial t)_q$ for the same reason. The result is expressed in Eq. (2.25):

$$T(t)\mathsf{H}(q)\Phi(q) = i\hbar\Phi(q)\left[\frac{\partial T(t)}{\partial t}\right]_q. \tag{2.25}$$

If both sides of Eq. (2.25) are divided by $\psi = \Phi T$, the left-hand side will depend only on q, while the right-hand side depends only on the time. If the spatial and temporal variables are truly independent of one another, there is only one way in which a function of q can be equal to a function of t: both must be equal to a scalar constant. By inspection of Eq. (2.25), it can be seen that this constant must have the units of energy. It will therefore be called $\mathscr{E}$. The Schrödinger wave equation has now indeed been broken into two parts as originally intended,

$$\frac{i\hbar}{T(t)}\frac{dT(t)}{dt} = \mathscr{E} \tag{2.26}$$

and

$$\frac{1}{\Phi(q)}\mathsf{H}(q)\Phi(q) = \mathscr{E}. \tag{2.27}$$

In mathematical language, $\mathscr{E}$ is called the *separation constant*. Equation (2.26) has the solution

$$T(t) = T(0)\exp\left(\frac{-i\mathscr{E}t}{\hbar}\right). \tag{2.28}$$

After multiplying both sides of Eq. 2.27 by ΦT, it may be seen that ψ is an eigenfunction of H and $\mathscr{E}$ is the corresponding eigenvalue. Physically then, $\mathscr{E}$ is the total energy of the system whenever it is in the state ψ, and is a *constant of the motion.*

In conclusion, if H is independent of time, the solutions to Schrödinger's Eq. (2.23) are the eigenfunctions of H. There are presumably several linearly independent solutions of Eq. (2.27), each of which may correspond to a different value of $\mathscr{E}$, and therefore there is more than one satisfactory ψ. In particular, according to the earlier notation, there are expected to be exactly M possible solutions, the members of the set $\{|\mathscr{E}\rangle\}$:

$$\mathsf{H}|\mathscr{E}_j\rangle = \mathscr{E}_j|\mathscr{E}_j\rangle, \quad 1 < j < M. \tag{2.29}$$

Futhermore, M must be ∞ in this case because (1) there are, by assumption, an infinite number of eigenstates of the position, and (2) the number of eigenstates does not change when one changes from one physically observable property (e.g., position) to another (energy). Since the time dependent portion of the corresponding eigenfunction (T_j) can be cancelled from both sides of Eq. (2.25), the principal emphasis in conventional courses in quantum mechanics is placed on finding the solution to Eq. (2.27). Sometimes sight is lost of the fact that the *complete* eigenfunction must include the time-dependent factor which appears in Eq. (2.26). Indeed, one edition of a respected standard college textbook of physical chemistry[1] does not even contain the full time-dependent Schrödinger Eq. (2.23). Only the formula for Φ, Eq. (2.27), is presented! This omission is fatal to an understanding of the process by means of which transitions occur between the eigenstates of H.

From Eqs. (2.24) and (2.28), one finds that

$$\psi_j(q,t) = \Phi_j(q)T_j(0)\exp\left(\frac{-i\mathscr{E}_j t}{\hbar}\right). \tag{2.30}$$

As stated in Chap 1, the distribution function (probability density) for finding the system at the coordinates q is given by the complex square of ψ. (It will be shown in the next section why this is so.) Note from Eq. (2.30) that, when this is done, the exponential terms in ψ and ψ^* (the asterisk denotes the complex conjugate) will cancel one another, leaving a function that depends on q alone. Thus the eigenstates of the Hamiltonian operator are stationary in two senses. Not only is the corresponding physically observable property of the system ($\mathscr{E}$) a constant of the motion (as would be true of the eigenstates of any operator), but also the probability density $\psi_j\psi_j^*$ is independent of time. This fact probably contributes to the unfortunate tendency to slight the importance of the $T_j(t)$ functions.

2.4 Kets and Bras: Orthogonality and Normalization

It is always possible to define the scalar product of two ordinary vectors. It is also possible to define the scalar product of two vectors in Hilbert space that represent states of quantum systems. For example, the dot product of two ordinary vectors,

$$\mathbf{A} = A_x\hat{x} + A_y\hat{y} + A_z\hat{z} \tag{2.31}$$

and

$$\mathbf{B} = B_x\hat{x} + B_y\hat{y} + B_z\hat{z}, \tag{2.32}$$

[1] Walter J. Moore (1962) *Physical Chemistry*, 3rd edn. Prentice-Hall, Englewood Cliffs, N.J., . Fortunately, the full equation does appear in the fourth since (1972) and subsequent editions.

is simply the scalar

$$\mathbf{A}\cdot\mathbf{B} = A_x B_x + A_y B_y + A_z B_z. \tag{2.33}$$

To obtain this result, it has been assumed that the vectors in the three different Cartesian directions, $\hat{x}$, $\hat{y}$, and $\hat{z}$, are orthogonal to one another:

$$\hat{x}\cdot\hat{y} = \hat{y}\cdot\hat{z} = \hat{x}\cdot\hat{z} = 0, \tag{2.34}$$

and normalized to unit length:

$$\hat{x}\cdot\hat{x} = \hat{y}\cdot\hat{y} = \hat{z}\cdot\hat{z} = 1. \tag{2.35}$$

It is necessary to generalize this procedure so that it will work in Hilbert space. Consider the two kets in Eqs. (2.13) and (2.17). The eigenkets of position, $|x_j\rangle$, are presumably analogous to the unit vectors, $\hat{x}$, $\hat{y}$, and $\hat{z}$. (Recall Fig. 2.2.) If they are orthogonal to one another and normalized to unit "length", the scalar product between $|a\rangle$ and $|b\rangle$ may be defined by

$$|a\rangle\cdot|b\rangle = a_1 b_1 + a_2 b_2 + \cdots + a_M b_M. \tag{2.36}$$

But the ket $|a\rangle$ is an eigenket of *some* operator, it matters not which. In that case, it will usually be normalized (e.g., to unity).[1] If Eq. (2.36) is used,

$$|a\rangle\cdot|a\rangle = a_1^2 + a_2^2 + \cdots + a_{M=1}^2. \tag{2.37}$$

It can be shown that Eq. (2.37) cannot be satisfied in general. In particular, suppose that the scalars a_j are complex numbers (note the i in Schrödinger's Eq. 2.23):

$$a_j = \mathrm{Re}(a_j) + i\,\mathrm{Im}(a_j). \tag{2.38a}$$

The scalars $\mathrm{Re}(a_j)$ and $\mathrm{Im}(a_j)$ are both real.

[It is also necessary to recall the definition of the complex conjugate of a_j,

$$a_j^* = \mathrm{Re}(a_j) - i\,\mathrm{Im}(a_j), \tag{2.38b}$$

which will be needed shortly.] The squares of these coefficients are therefore also complex:

$$a_j^2 = \mathrm{Re}^2(a_j) - \mathrm{Im}^2(a_j) + 2i\,\mathrm{Re}(a_j)\,\mathrm{Im}(a_j). \tag{2.39}$$

Yet it was decided at the outset (see Eq. 2.37) to define $|a\rangle$ in such a way as to make the sum of the squares be a real number, namely, 1. From Eq. (2.39) it seems that this goal cannot be accomplished by using the definition of the scalar product of two kets given in Eq. (2.36).

---

[1] In some cases, eigenkets cannot be normalized to unity. An important example is the set of eigenkets of momentum for a free particle. No such problems will be encountered, however, for the quantum systems described in this book.

This problem is solved by creating a new Hilbert space, called the *dual space*, also of M dimensions. To every ket $|a\rangle$ in the first space there corresponds another kind of vector, called a bra, $\langle a|$, in the dual space:

$$\langle a| = a_1^* \langle x_1| + a_2^* \langle x_2| + \cdots + a_M^* \langle x_M|. \tag{2.40}$$

Note that, in the expansion of $\langle a|$ in terms of the eigenbras of position $\langle x_j|$, the coefficients are the complex conjugates of those used in the expansion of the corresponding ket in Eq. (2.13). When scalar products are formed, a ket from one space is dotted into a bra from the other. This makes a bra(c)ket (the pun is intentional). In particular,

$$\langle a|a\rangle = a_1^* a_1 + a_2^* a_2 + \cdots + a_M^* a_M. \tag{2.41}$$

Note that each term in the sum of the right-hand side of Eq. (2.41) is automatically a real number,

$$a_j^* a_j = \mathrm{Re}^2(a_j) + \mathrm{Im}^2(a_j) \equiv |a_j|^2, \tag{2.42}$$

thereby guaranteeing that the scalar product of $\langle a|$ with $|a\rangle$ is also a real number.

If $|a\rangle$ and $|b\rangle$ are different eigenkets of some operator R with distinct eigenvalues ($r_a \neq r_b$), it can be shown that they are orthogonal to one another:

$$\langle a|b\rangle = a_1^* b_1 + a_2^* b_2 + \cdots + a_M^* b_M = 0. \tag{2.43}$$

Note that, when M is ∞ and both a and b form a continuum, the sums on the right-hand sides of Eqs. (2.41) and (2.43) become integrals:

$$\lim_{M\to\infty} \sum_{j=1}^{M} a_j^* a_j = \lim_{M\to\infty} \sum_{j=1}^{M} \psi_a^*(q_j)\psi_a(q_j) = \int_{-\infty}^{\infty} |\psi_a|^2 dq = 1 \tag{2.44}$$

and

$$\lim_{M\to\infty} \sum_{j=1}^{M} a_j^* a_j = \lim_{M\to\infty} \sum_{j=1}^{M} \psi_a^*(q_j)\psi_b(q_j) = \int_{-\infty}^{\infty} \psi_a^* \psi_b dq = 0. \tag{2.45}$$

Equation (2.44) has the following physical interpretation. The value of the integral is the probability of finding the system with coordinates q, summed over all q. This sum must then correspond to absolute certainty; the system will surely have its q *somewhere* in the range of $-\infty$ to $+\infty$. Since the orthogonality and normalization of state vectors in Hilbert space can be expressed by integrals of products of algebraic functions, it is convenient to say that the state functions themselves are orthogonal and normalized also.

It is also customary to normalize separately the factors Φ_j and T_j which appear in Eq. (2.30). Using Eq. (2.44), one obtains

$$\int_{-\infty}^{\infty} \Phi_j^*(q)\Phi_k(q)\, dq = \delta_{jk}, \tag{2.46}$$

where δ_{jk} is Kronecker's delta,

$$\delta_{jk} = \begin{cases} 1 & \text{if } j=k, \\ 0 & \text{if } j \neq k, \end{cases} \tag{2.47}$$

and

$$T_j^*(0)T_j(0) = 1. \tag{2.48}$$

The solution to Eq. (2.48) is

$$T_j(0) = \exp(i\zeta_j), \tag{2.49}$$

where ζ_j is a real constant.

2.5 Quantum Systems "in-between" Eigenstates

The eigenstates of the energy by no means exhaust the behavioral repertoire of a quantum system. If $\{\psi\}_R$ is the set of functions representing eigenstates of R, any linear combination of the members of that set $(\psi_1, \psi_2, \ldots)$ also represents possible states (see Eq. 2.13). Consider the state Ψ:

$$\Psi = c_1\psi_1 + c_2\psi_2 + \cdots + c_M\psi_M. \tag{2.50}$$

States such as Ψ may be called *nonstationary* because (1) when the system is in any one of them, the property represented by R will not be a constant of the motion; and (2) if R does not commute with H, the distribution function $\Psi^*\Psi$ will change with time.[1] The fact that functions of the type given in Eq. (2.50) are necessary and sufficient to represent all possible states of a quantum system is called the *principle of superposition*. Non-stationary states are therefore sometimes called *superposition states*. Equation (2.50) is trivial unless at least two of the c_j's are nonzero. On the other hand, to allow for more than two terms in the sum would greatly complicate the calculations that follow, without introducing any important additional physical insight into the problem. Therefore only the first two terms will henceforth be used.

Suppose that the superposition function Ψ is substituted into Schrödinger's Eq. (2.33). Is it true that

$$\mathsf{H}(c_1\psi_1 + c_2\psi_2) = i\hbar\left[\frac{\partial}{\partial t}(c_1\psi_1 + c_2\psi_2)\right]_q ? \tag{2.51}$$

[1] This statement is not true if all of the eigenfunctions included in Eq. (2.50) (i.e., those with nonzero coefficients) are degenerate.

The distributive law may be used on both sides of Eq. (2.51) to achieve

$$\mathsf{H}c_1\psi_1 + \mathsf{H}c_2\psi_2 \stackrel{?}{=} i\hbar\left[\frac{\partial(c_1\psi_1)}{\partial t}\right]_q + i\hbar\left[\frac{\partial(c_2\psi_2)}{\partial t}\right]_q. \tag{2.52}$$

Expressions for ψ_1 and ψ_2 may be found by combining Eqs. (2.30) and (2.49):

$$\psi_1 = \Phi_1 \exp\left[i\left(\zeta_1 - \frac{\mathscr{E}_1 t}{\hbar}\right)\right] \tag{2.53}$$

and

$$\psi_2 = \Phi_2 \exp\left[i\left(\zeta_2 - \frac{\mathscr{E}_2 t}{\hbar}\right)\right]. \tag{2.54}$$

Remember that the case in which H does not explicitly depend on time is being considered. Furthermore, c_1 and c_2 are assumed to be constants, independent of both q and t.

The first term on the left-hand side of Eq. (2.52) can then be rearranged:

$$\mathsf{H}c_1\psi_1 = c_1 \exp\left[i\left(\zeta_1 - \frac{\mathscr{E}_1 t}{\hbar}\right)\right]\mathsf{H}\Phi_1. \tag{2.55}$$

Next, Eq. (2.27) may be used to find

$$\mathsf{H}\Phi_1 = \mathscr{E}_1\Phi_1. \tag{2.56}$$

On the righ-hand side of Eq. (2.52), the first term becomes

$$i\hbar\left[\frac{\partial(c_1\psi_1)}{\partial t}\right]_q = i\hbar c_1\Phi_1\left(\frac{\partial\{\exp[i(\zeta_1 - \mathscr{E}_1 t/\hbar)]\}}{\partial t}\right)_q. \tag{2.57}$$

Next, the derivative on the right-hand side of Eq. (2.57) may be calcuated:

$$\left(\frac{\partial\{\exp[i(\zeta_1 - \mathscr{E}_1 t/\hbar)]\}}{\partial t}\right)_q = \frac{-i\mathscr{E}_1}{\hbar}\exp\left[i\left(\zeta_1 - \frac{\mathscr{E}_1 t}{\hbar}\right)\right]. \tag{2.58}$$

The corresponding expressions for the second terms on both sides of Eq. (2.52) are identical to Eqs. (2.56) and (2.58) except for the subscripts. These results may be substituted into Eq. (2.52):

$$c_1\exp\left[i\left(\zeta_1 - \frac{\mathscr{E}_1 t}{\hbar}\right)\right]\mathscr{E}_1\Phi_1 + c_2\exp\left[i\left(\zeta_2 - \frac{\mathscr{E}_2 t}{\hbar}\right)\right]\mathscr{E}_2\Phi_2 \stackrel{?}{=}$$

$$i\hbar c_1\Phi_1\left(\frac{-i\mathscr{E}_1}{\hbar}\right)\exp\left[i\left(\zeta_1 - \frac{\mathscr{E}_1 t}{\hbar}\right)\right] + i\hbar c_2\Phi_2\left(\frac{-i\mathscr{E}_2}{\hbar}\right)\exp\left[i\left(\zeta_2 - \frac{\mathscr{E}_2 t}{\hbar}\right)\right]. \tag{2.59}$$

It can be seen from Eq. (2.59) that the equality is satisfied. It has therefore been proved that a wavefunction formed by the linear superposition of two eigenfunctions of the Hamiltonian operator (Eq. 2.50) is as valid a solution to Schrödinger's Eq. (2.23) as either one of the two constituent eigen-functions separately. This proof can easily be extended to wavefunctions formed by the linear combination of any number of eigenstates. Therefore superposition states of quantum systems are as "real" (possible) in nature as are eigenstates.

Do wavefunctions representing superposition states (Eq. 2.50) also satisfy the time-independent portion of Schrödinger's Eq. (2.23),

$$\mathsf{H}(c_1\psi_1 + c_2\psi_2) \stackrel{?}{=} \mathscr{E}(c_1\psi_1 + c_2\psi_2). \tag{2.60}$$

The question of how the energy $\mathscr{E}$ should be evaluated in the above expression must be put aside for the moment. Suffice it to say that $\mathscr{E}$ is a nonzero scalar, as it must be in order to be a separation constant. The left-hand side of Eq. (2.60) is identical with the left-hand side of Eq. (2.51), so that it may be replaced by the left-hand side of Eq. (2.59). The right-hand side of Eq. (2.60) may be evaluated immediately from the distributive law. Equation (2.60) therefore becomes

$$c_1\exp\left[i\left(\zeta_1 - \frac{\mathscr{E}_1 t}{\hbar}\right)\right]\mathscr{E}_1\Phi_1 + c_2\exp\left[i\left(\zeta_2 - \frac{\mathscr{E}_2 t}{\hbar}\right)\right]\mathscr{E}_2\Phi_2 \stackrel{?}{=}$$
$$\mathscr{E}c_1\Phi_1\exp\left[i\left(\zeta_1 - \frac{\mathscr{E}_1 t}{\hbar}\right)\right] + \mathscr{E}c_2\Phi_2\exp\left[i\left(\zeta_2 - \frac{\mathscr{E}_2 t}{\hbar}\right)\right]. \tag{2.61}$$

It is obvious from inspection of Eq. (2.61) that the equality can be satisfied only if $\mathscr{E} = \mathscr{E}_1 = \mathscr{E}_2$. Therefore, if the superposition states formed by a linear combination of degenerate eigenstates of the Hamiltonian operator are excluded from the discussion, the equality expressed in Eq. 2.60 cannot hold. This means that the time-independent Schrödinger Eq. 2.56 has a very limited validity. In particular, if $M = 2$, and these are not energy degenerate, there are only two possible eigenstates of the system, represented by the wavefunctions that satisfy Eq. (2.29). By way of contrast, there are an infnite number of possible linear combinations of the form in Eq. (2.50). This is so because there are an infinite number of possible choices of c_1/c_2, each one of which represents a possible "real" and valid state of the quantum system. In general, then, Eq. (2.61) may be rewritten as follows:

$$\mathsf{H}\Psi \neq \mathscr{E}\Psi. \tag{2.62}$$

Some of the properties of these superposition states will now be explored: In the first place, the *distribution functions* for positions associated with these states are not stationary (unless the eigenstates of which they are composed are degenerate). To prove this, the distribution function can be calculated:

$$\Psi * \Psi = c_1^* c_1 \psi_1^* \psi_1 + c_2^* c_2 \psi_2^* \psi_2 + c_1^* c_2 \psi_1^* \psi_2 + c_2^* c_1 \psi_2^* \psi_1. \tag{2.63}$$

Recalling the rules about complex numbers and functions, expressed in Eqs. (2.38), (2.39), and (2.42), plus the fact that

$$a + a^* = 2\mathrm{Re}(a), \tag{2.64}$$

one may rewrite Eq. (2.63) to obtain

$$|\Psi|^2 = |c_1|^2|\psi_1|^2 + |c_2|^2|\psi_2|^2 + 2\mathrm{Re}(c_1^* c_2 \psi_1^* \psi_2). \tag{2.65}$$

Equations (2.53) and (2.54) supply the necessary expressions for the wave functions. Equation (2.65) becomes

$$|\Psi|^2 = |c_1|^2|\Phi_1|^2 + |c_2|^2|\Phi_2|^2 + 2\mathrm{Re}\{c_1^* c_2 \Phi_1^* \Phi_2 \exp[i(\delta' - \omega_0 t)]\} \tag{2.66}$$

In Eq. (2.66),

$$\delta' \equiv \zeta_2 - \zeta_1 \tag{2.67}$$

and

$$\omega_0 \equiv \frac{\mathscr{E}_2 - \mathscr{E}_1}{\hbar}. \tag{2.68}$$

The explicit time dependence of the third term on the right-hand side of Eq. (2.66) proves the assertion that the state represented by the wavefunction in Eq. (2.50) is not stationary. The probability distribution for a system in a stationary state is a "standing wave", $|\psi_j|^2$. Just as is the case in classical mechanics, a standing wave consists of two traveling waves having the same frequency but going in opposite directions (cf. the time dependences of ψ_j^* and ψ_j). Also just as in classical mechanics, whenever a system is excited in such a way that two modes of oscillation of different frequencies are present at the same time, one may expect to "hear" the beat frequency. In this case, $\mathscr{E}_1/\hbar$ and $\mathscr{E}_2/\hbar$ are the different modal frequencies, and ω_0 is the interference beat frequency. The distribution function in Eq. (2.66) then contains two constant terms (the first two on the right-hand side), one for each of the two eigenfunctions of which the total wavefunction is composed. It also contains the third term on the right-hand side; this describes the sinusoidally pulsating interference pattern between wavefunctions for the two eigenstates, which alternate between destructive and constructive interference at the frequency ω_0.

A relationship between the two constants, c_1 and c_2, which appear in Eqs. (2.50) and (2.66) can be discovered, requiring that Ψ be normalized to 1:

$$\int_{-\infty}^{+\infty} |\Psi|^2 dq = |c_1|^2 \int_{-\infty}^{+\infty} |\Phi_1|^2 dq + |c_2|^2 \int_{-\infty}^{+\infty} |\Phi_2|^2 dq$$

$$+\mathrm{Re}\left\{ c_1^* c_2 \exp[i(\delta' - \omega_0 t)] \int_{-\infty}^{+\infty} \Phi_1^* \Phi_2 dq \right\} = 1. \tag{2.69}$$

The integrals on the right-hand side of Eq. (2.68) may be evaluated from the orthogonal character of the Φ's—see Eq. (2.46). Equation (2.69) becomes

$$|c_1|^2 + |c_2|^2 = 1. \tag{2.70}$$

The most general solution to Eq. 2.70 is

$$c_1 = \left[\cos\left(\frac{\theta}{2}\right)\right]\exp(i\xi_1) \tag{2.71}$$

and

$$c_2 = \left[\sin\left(\frac{\theta}{2}\right)\right]\exp(i\xi_2). \tag{2.72}$$

The real scalars, θ, ξ_1, and ξ_2, replace the real and imaginary parts of the c's as the parameters that specify the superposition state. A physical interpretation of them will be given in Chaps 4 and 8. The total wavefunction in Eq. (2.50) can then be written as

$$\Psi(q,t) = \left[\cos\left(\frac{\theta}{2}\right)\right]\Phi_1(q)\exp\left[i\left(\eta_1 - \frac{\mathscr{E}_1 t}{\hbar}\right)\right]$$

$$+\left[\sin\left(\frac{\theta}{2}\right)\right]\Phi_2(q)\exp\left[i\left(\eta_2 - \frac{\mathscr{E}_2 t}{\hbar}\right)\right], \tag{2.73}$$

where η_j is defined by

$$\eta_j = \zeta_j + \xi_j, \quad j = 1, 2. \tag{2.74}$$

2.6 Eigenvalues and Expectation Values

Bras, kets, and wavefunctions for both stationary and nonstationary (superposition) states have been discussed. Next, a method for extracting information about the physical properties of a system from these entities must be found.

Suppose that the physical property of interest is represented by the operator R. Furthermore, suppose that the system is in a state represented by the ket $|a\rangle$. If the system is not in an eigenstate of R, it must be in a superposition of such states because the eigenstates of any operator constitute a complete set. Then the ket $|a\rangle$ can be written as a linear combination of the eigenkets of R, which shall be labelled by their corresponding eigenvalues, r_j:

$$|a\rangle = a_1|r_1\rangle + a_2|r_2\rangle + \cdots + a_M|r_M\rangle. \tag{2.75}$$

Equation (2.75) will be suitable for the description of $|a\rangle$ even if the system represented thereby is in an eigenstate (e.g., the 39th eigenstate) of R. In this case, it is simple to set all the a_j's equal to 0 except a_{39}, which can be made equal to 1.

Regardless of the number of nonzero terms in the sum, what will be the outcome of an experiment on the system in which the physical property R is measured? The answer to this question is decided by first operating upon $|a\rangle$ with the operator R and then forming the scalar product of the resultant ket with the bra $\langle a|$.

In the first step, the distributive law is used, plus the fact that operating with R commutes with multiplying by the scalar coefficients a_j:

$$\mathsf{R}|a\rangle = a_1\mathsf{R}|r_1\rangle + a_2\mathsf{R}|r_2\rangle + \cdots + a_M\mathsf{R}|r_M\rangle. \tag{2.76}$$

Since, on the right-hand side of Eq. (2.66), the operators R are operating only upon eigenkets, the results are known:

$$\mathsf{R}|a\rangle = a_1 r_1|r_1\rangle + a_2 r_2|r_2\rangle + \cdots + a_M r_M|r_M\rangle. \tag{2.77}$$

In the next step, the formation of the scalar product with $\langle a|$, the fact that the eigenkets of R are orthogonal to one another and normalized to unity is used:

$$\langle a|\mathsf{R}|a\rangle = |a_1|^2 r_1 + |a_2|^2 r_2 + \cdots + |a_M|^2 r_M. \tag{2.78}$$

If the system is in the 39th eigenstate, all of the terms on the right-hand side will be zero except the 39th, which will have the value r_{39}. Experimentally, the measurement process will produce a result equal to r_{39}. Furthermore, the measurement will be reproducible; for example, repeated measurements performed under identical conditions will produce the same result. On the other hand, if the system is in a superposition of eigenstates (two or more a_j's nonzero in Eq. 2.75), measurements of the property represented by R will not be reproducible. Each individual measurement will have, as its result, an eigenvalue corresponding to one of the eigen-functions included in the superposition wavefunction. There will be no way to predict, in advance of a particular measurement, which of the possible different eigenvalues will be obtainted therefrom. This fact is called the *uncertainty principle*. The probability that the result of the measurement will be that expected for the jth state will be $|a_j|^2$. Whenever the jth state shows up, the value of R obtained will be r_j. The expression on the right-hand side of Eq. (2.78) corresponds to the average result of a large number of such measurements. The average value of some physical property of a quantum system is called an *expectation value*. This special term is used to emphasize the distinction between the averaging process in classical mechanics, necessitated by experimental failure to adequately control all of the variables affecting the outcome of the measurement, and quantum-mechanical averaging, necessitated by fundamental laws of nature.

Equation (2.78) is identical with the formula for finding the weighted average of $\overline{R}$ in classical statistics:

$$\overline{R} = \frac{\sum_{j=1}^{M} W_j r_j}{\sum_{j=1}^{M} W_j}. \tag{2.79}$$

The M symbols W_j are the statistical weights to be assigned to each of the independent possibilities of R; it is easily seen that the $|a_j|^2$ terms in Eq. (2.78) play the same role as the W_j's in Eq. (2.79). The sum in the denominator on the right-hand side of Eq. (2.79) is usually 1 because the statistical weights are ordinarily normalized before use.

It sometimes happens in both classical and quantum statistics that there is a contnuum of possibilities. In such cases, the M statistical weights W_j are replaced by a continuous distribution function:

$$\overline{R(t)} = \frac{\int_{-\infty}^{+\infty} W(q,t)\,R(q)\,dq}{\int_{-\infty}^{+\infty} W(q,t)\,dq}. \tag{2.80}$$

In Eq. (2.80), R and W are both algebraic functions of the coordinate q. Again, the denominator on the right-hand side is ordinarily 1:

$$\lim_{n\to\infty} \sum_{j=1}^{M} |a_j|^2 r_j = \int_{-\infty}^{+\infty} |\psi_a(q,t)|^2 r(q)\,dq. \tag{2.81}$$

Therefore, from Eqs. (2.21), (2.45), and (2.78),

$$\langle R(t)\rangle \equiv \langle a|\mathsf{R}|a\rangle = \frac{\int_{-\infty}^{+\infty} \Psi^*(q,t)\mathsf{R}(q)\Psi(q,t)\,dq}{\int_{-\infty}^{+\infty} |\Psi(q,t)|^2 dq}. \tag{2.82}$$

Just as was the case in classical statistics, the denominator is usually 1. The operator R operates only on Ψ, the function that appears immediately to its right, in accordance with the usual custom. The symbol $\langle R\rangle$ is used to distinguish between the expectation value and the classical weighted average, $\overline{R}$.

It can be easily seen that, if Ψ is an eigenfunction of R, its time dependence will cancel that of Ψ^*. In such a case, $\langle R\rangle$ will depend on the time only if the operator R does as well. On the other hand, if Ψ is a linear combination of nondegenerate eigenfunctions of R (and therefore represents a superposition state for the system), $\langle R\rangle$ will be time dependent even if the operator R and the coefficients c_j are constants. For example, the wavefunction of Eq. (2.73) may be used in Eq. (2.82):

$$\langle R(t)\rangle = R_{11}\cos^2\left(\frac{\theta}{2}\right) + R_{22}\sin^2\left(\frac{\theta}{2}\right)$$
$$+2\cos\left(\frac{\theta}{2}\right)\sin\left(\frac{\theta}{2}\right)\mathrm{Re}\{R_{12}\exp[i(\S - \omega_0 t)]\}, \tag{2.83}$$

where

$$\S \equiv \eta_2 - \eta_1 \tag{2.84}$$

and

$$R_{jk} \equiv \int_{-\infty}^{+\infty} \Phi_j^*(q)\mathsf{R}(q)\Phi_k(q)\,dq. \tag{2.85}$$

The parameters θ, η, and ω_0 were defined in Eqs. (2.68), (2.71), (2.72), and (2.74). The symbol R_{jk} is sometimes termed the jkth matrix element of R.

Note that the results in Eq. (2.83) are completely general, in that they apply to any physically observable quantity represented by an operator which does not explicitly contain the time. Note also that, if two such operators commute, the expectation values of the physical properties that they represent must have the same kind of time dependence. The facts represented mathematically by Eq. (2.83) can be re-expressed in more physical terms. The averaged properties of a quantum system cannot depend on the time *unless* the system is in a superposition state.[1] The time dependence is due to the beating together of the two (or more) probability waves that have been superposed (compare Eqs. 2.66 and 2.83). The coefficients of the eigenfunctions, c_1 and c_2 (see Eqs. 2.71 and 2.72), must be *simultaneously* nonzero. In particular, suppose that $\theta = \pi/2$. Then $|c_1|^2=|c_2|^2 = \frac{1}{2}$. It is *not* correct (or, at least, it is very misleading) to describe the state represented by

$$\Psi = \frac{\psi_1 + \psi_2}{\sqrt{2}} \tag{2.86}$$

as one in which the system spends half of its time in the eigenstate represented by ψ_1 and the other half in the eigenstate represented by ψ_2. If the system represented by the function in Eq. (2.83) really did "hop back and forth" between the states represented by ψ_1 and ψ_2, there would never be an interval during which both wavefunctions would be present simultaneously, and therefore the beat term would always have zero amplitude. There would never be any time dependence

[1] Possible exceptions are provided by cases in which the operators themselves change with time.

in any physically observable property of the system, in the same way that one does not hear the beat between two notes played on a piano in succession. (The significance of the c's will be discussed in greater detail in Chap 4.)

2.7 Superposed Eigenfunctions and Perturbed Eigenfunctions

There is a superficial similarity between the defining equation for superposition states (Eq. 2.50) and equations that arise in the branch of quantum mechanics called "perturbation theory." However, *there is no similarity between superposition states and the "mixed states" encountered in perturbation theory.* To prove this, a brief review of the latter is now presented.

Let ψ_1 and ψ_2 be two eigenfunctions of a Hamiltonian, H_0, with eigenvalues $\mathscr{E}_1$ and $\mathscr{E}_2$. Suppose, however, that the problem of interest is one for which the Hamiltonian is

$$\mathsf{H} = \mathsf{H}_0 + \mathsf{H}'. \tag{2.87}$$

It may be difficult or inconvenient to start from scratch and solve the new problem, finding the new eigenfunctions, ψ_1' and ψ_2', and eigenvalues, $\mathscr{E}_1'$ and $\mathscr{E}_2'$, directly. There will be two differences between the set $\{\psi_j\}_{H_0}$ and the set $\{\psi_j'\}_{H_0+H'}$. The spatial distributions of the electron(s) will be different, and (because the corresponding eigenvalues are different) the time-dependent factors will also be different. In particular,

$$\psi_1(q, t) = \Phi_1(q)\exp\left(\frac{-i\mathscr{E}_1 t}{\hbar}\right), \tag{2.88}$$

$$\psi'^1(q, t) = \Phi_1'(q)\exp\left(\frac{-i\mathscr{E}_1' t}{\hbar}\right), \tag{2.89}$$

$$\Phi_1(q) \neq \Phi_1'(q), \tag{2.90}$$

and

$$\mathscr{E}_1 \neq \mathscr{E}_1'. \tag{2.91}$$

Similar inequalities will hold for ψ_2 and ψ_2'.

On the other hand, if $\Phi_1(q)$ and $\Phi_2(q)$ constitute a complete set for the first problem, the *location* of the electron in the second problem can be described by

$$\Phi_1'(q) = c_{11}\Phi_1(q) + c_{12}\Phi_2(q) \tag{2.92}$$

and

$$\Phi_2'(q) = c_{21}\Phi_1(q) + c_{22}\Phi_2(q). \tag{2.93}$$

Note especially that it is *not* true that

$$\psi_1'(q,t) = c_{11}\psi_1(q,t) + c_{12}\psi_2(q,t), \tag{2.94}$$

for example. This is so because the time dependence of ψ_1' is controlled by $\mathscr{E}_1'$, which in general will be different from the time dependence of ψ_1 and ψ_2, the latter being controlled by $\mathscr{E}_1$ and $\mathscr{E}_2$, respectively. In fact, Eqs. (2.92) and (2.93) have no particular quantum-mechanical significance because the problem for which Φ_1' and Φ_2' are appropriate is different from the one for which Φ_1 and Φ_2 are appropriate. The expansion of Φ_1' in terms of Φ_1 and Φ_2 was performed only for mathematical convenience; and other functions that are complete over x, y, z space would, in principle, do as well. In practice, Φ_1 and Φ_2 are convenient because they both happen to be eigenfunctions of H_0. By definition, ψ_1' and ψ_2' are eigenfunctions of H:

$$\mathsf{H}\psi_1' = \mathscr{E}_1'\psi_1' \tag{2.95}$$

and

$$\mathsf{H}\psi_2' = \mathscr{E}_2'\psi_2'. \tag{2.96}$$

Equations (2.89) and (2.92) may be substituted into Eq. (2.95):

$$(\mathsf{H}_0 + \mathsf{H}')(c_{11}\Phi_1 + c_{12}\Phi_2)\exp\left(\frac{-i\mathscr{E}_1't}{\hbar}\right) =$$

$$\mathscr{E}_1'(c_{11}\Phi_1 + c_{12}\Phi_2)\exp\left(\frac{-i\mathscr{E}_1't}{\hbar}\right) \tag{2.97}$$

If neither H_0 nor H' depends on the time, $\exp(-i\mathscr{E}_1't/\hbar)$ will commute with H on the left-hand side of Eq. (2.97) and subsequently be canceled with the same term on the right-hand side. Next, the distributive law and the fact the ψ_1 and ψ_2 are eigenfunctions of H_0 may be used to transform Eq. (2.97) into

$$c_{11}(\mathscr{E}_1\Phi_1 + \mathsf{H}'\Phi_1) + c_{12}(\mathscr{E}_2\Phi_2 + \mathsf{H}'\Phi_2) = c_{11}\mathscr{E}_1'\Phi_1 + c_{12}\mathscr{E}_1'\Phi_2. \tag{2.98}$$

Equation (2.98) can be transformed into two scalar equations by first multiplying from the left by Φ_1^* and integrating over coordinates, and then repeating the process using Φ_2^* instead. The results are as follows:

$$c_{11}(\mathscr{E}_1 + H_{11}') + c_{12}H_{12}' = c_{11}\mathscr{E}_1' \tag{2.99}$$

and

$$c_{11}H_{21}' + c_{12}(\mathscr{E}_2 + H_{22}') = c_{12}\mathscr{E}_1'. \tag{2.100}$$

The only integrations that must be performed to obtain Eqs. (2.99) and (2.100) are

$$H_{11}' \equiv \int \Phi_1^*(q)\mathsf{H}'(q)\Phi_1(q)dq, \tag{2.101}$$

$$H_{22}' \equiv \int \Phi_1^*(q)\mathsf{H}'(q)\Phi_2(q)dq, \tag{2.102}$$

and

$$H'_{12} \equiv \int \Phi_1^*(q) \mathsf{H}'(q) \Phi_2(q)\, dq, \tag{2.103}$$

This is so because Φ_1 and Φ_2 are orthogonal functions and because H' is Hermitian:

$$H'_{21} = H'^*_{12}. \tag{2.104}$$

(Equation 2.104 can be taken as the defining equation of a Hermitian operator.) It is also necessary that Φ'_1 and Φ'_2 be normalized. From this fact it follows that

$$|c_{11}|^2 + |c_{12}|^2 = 1 \tag{2.105}$$

and

$$|c_{21}|^2 + |c_{22}|^2 = 1. \tag{2.106}$$

Equations (2.99) and (2.100) can be summarized in matrix form:

$$\begin{pmatrix} 0 \\ 0 \end{pmatrix} = \begin{bmatrix} (\mathscr{E}_1 + H'_{11}) - \mathscr{E} & H'_{12} \\ H'_{21} & (\mathscr{E}_2 + H'_{22}) - \mathscr{E} \end{bmatrix} \begin{pmatrix} c_{11} \\ c_{12} \end{pmatrix} \tag{2.107}$$

The theory of linear equations may be used to show that the determinant of the 2×2 coefficient matrix on the left-hand side of Eq. (2.107) must be zero. This

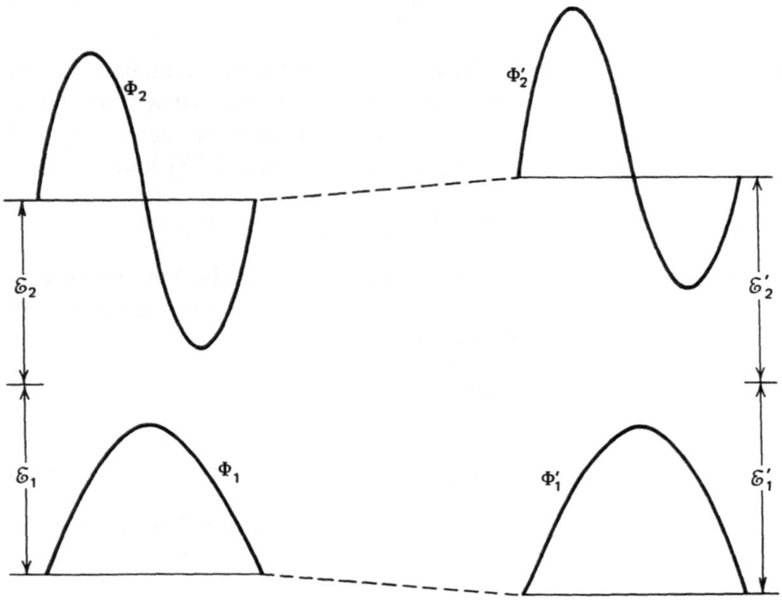

Fig. 2.5. Perturbation of the spatial part of the wavefunction by H′. On the left, the wavefunctions are unperturbed; on the right, perturbed

gives rise to what is called the *secular equation*:

$$[(\mathscr{E}_1 + H'_{11}) - \mathscr{E}][(\mathscr{E}_2 + H'_{22}) - \mathscr{E}] - H'_{12}H'_{21} = 0. \tag{2.108}$$

This equation is quadratic in $\mathscr{E}$, the only unkown, and will therefore have only two solutions. One of these will be the desired $\mathscr{E}'_1$, and the other will be $\mathscr{E}'_2$, the eigenvalue that is appropriate for the other wavefunction. Substitution of $\mathscr{E}'_1$ into Eqs. (2.99) and (2.100) will give two equations with four unkowns, the real and imaginary parts of c_{11} and c_{12}. If these coefficients are written as

$$c_{jk} = |c_{jk}|\exp(i\xi_{jk}), \tag{2.109}$$

Eqs. (2.105) and (2.106) may then be used to find $|c_{11}|$, $|c_{12}|$, and $\xi_{11} - \xi_{12}$.

Thus it can be seen that Eqs. (2.92) and (2.93) describe, *not* a superposition of two eigenstates to form a nonstationary state of a system with a Hamiltonian, H_0, but rather the spatially dependent parts of two new eigenfunctions, each appropriate for the new Hamiltonian, $\mathsf{H}_0 + \mathsf{H}'$ (see Fig. 2.5).

2.8 References

1. Fano U, Fao L. (1959) Basic physics of atoms and molecules. John Wiley, New York
2. Haken H, Wolf HC (1993) The physics of atoms and quanta, 3rd edn. Springer, Berlin Heidelberg New York

2.9 Problems

2.1. Prove that, if two operators R and S have simultaneous eigenkets, they commute.

2.2. Imagine a particle of mass m_0 constrained to move in one linear dimension, x. Examples are a bead on a taut string and a piston in a cylinder. Furthermore, imagine that the motion of the particle is confined to a small portion of the x axis, $0<x<L$ (e.g., a bead on a length of string that is knotted at both ends to prevent the bead from slipping off). The description of this system, called the one-dimensional infinite-square-well-potential problem, is discussed in many books on quantum mechanics. The spatial parts of the eigenfunctions of the Hamiltonian operator are

$$\Phi_n(x) = \begin{cases} \sqrt{2/L}\,\sin(n\pi x/L) & 0< x < L, \\ 0, & 0> x > L, \end{cases}$$

where n is a positive integer:

$$n = 1, 2, 3, \ldots, \infty.$$

The corresponding eigenvalues are

$$\mathscr{E}_n = \frac{n^2h^2}{8m_0L^2}.$$

a. Find an expression for the *total* eigenfunctions, $\psi_n(x,t)$, of the Hamiltonian operator for the bead on a string.

b. Find an expression for the total wavefunction, Ψ, representing the particle in a linear superposition of the nth and pth eigenstates of the energy. Use equal superposition coefficients,

$$c_n = c_p = \frac{1}{\sqrt{2}},$$

in order that the results may describe the "halfway" state.

2.3. Suppose that the bead in Problem 2.2 has an electric charge of $+q_0$, and the leftmost knot on the string, a charge of $-q_0$. Then the system will have an instantaneous electric dipole moment of $q_0 x\hat{x}$, where x is the instantaneous position of the bead and $\hat{x}$ is a unit vector in the x direction.
a. Calculate the expectation value of the dipole moment for the bead on the string when the system is in the $n = 1$ eigenstate of the energy. The operator for the x coordinate is x itself:

$$\mathsf{X} = x.$$

b. Calculate the expectation value of the dipole moment for the bead on the string when the system is in the "halfway" superposition of the $n = 1$ and $p = 2$ eigenstates of the energy.
c. Perform the same calculation as in Problem 2.3b, but use as the superposed states the $n = 1$ and $p = 3$ eigenstates of the energy.

2.4. Define $\mathcal{E}$ by the relation

$$2\mathcal{E} = \frac{(1^2 - 2^2)h^2}{8m_0L^2}.$$

Then, for the bead on the string described in Problem 2.2, set the zero of energy at the midpoint between the first and second eigenvalues:

$$\mathcal{E}_1 = -\mathcal{E}_2 = \mathcal{E}.$$

Suppose that a perturbation is applied to this system:

$$H'_{11} = H'_{22} = 0, \quad \text{and} \quad H'_{12} = H'_{21} = V\mathcal{E},$$

where V is a dimensionless parameter
a. Show that the perturbed energies are

$$\mathcal{E}'_1 = -\mathcal{E}'_2 = \mathcal{E}\sqrt{1 + V^2}.$$

b. Show that the spatial parts of the perturbed eigenfunctions of the new Hamiltonian operator are (aside from an arbitrary phase factor)

$$\Phi'_1 = \Phi_1 \cos\left(\frac{\theta}{2}\right) + \Phi_2 \sin\left(\frac{\theta}{2}\right)$$

and

$$\Phi'_2 = \Phi_1 \sin\left(\frac{\theta}{2}\right) + \Phi_2 \cos\left(\frac{\theta}{2}\right).$$

where

$$\frac{\theta}{2} = \tan^{-1}\left(\frac{V}{1 + \sqrt{1 + V^2}}\right).$$

c. Assume that $V = 0.5$. Then make energy level diagrams for both the perturbed and unperturbed cases, and sketch on each level the corresponding wavefunctions. Compare the results with those presented in Fig. 2.5.

3 Elementary Electromagnetic Theory

J. D. Macomber

3.1 Relationship Between Classical and Quantum-Mechanical Theories

Spectroscopy is the study of resonant, reversible exchanges of energy between chromophores (atoms, ions, and molecules) and an oscillating electromagnetic field. In Chap. 2, a theoretical description of the chromophores was presented. The purpose of this chapter is to provide a description of the electromagnetic field. The theoretical description of the chromophores, although elementary and incomplete, adhered strictly to the principles of the theory believed to be rigorously correct for such systems – namely, quantum mechanics. By way of contrast, it is proposed to describe the radiation field in accordance with classical electromagnetic theory, which is known to be only an approximation to the correct theory, with a limited range of validity. A thorough exposition of the classical theory may be found in many standard textbooks (e.g., the one by Stratton [1]).

Before beginning the classical treatment, it is appropriate to discuss briefly the rigorously correct theory of the electromagnetic field, quantum electrodynamics [2]. Many features of that theory of interest of spectroscopists are best discussed by contrasting them with corresponding features of the classical treatment. In classical electromagnetic theory, an arbitrary collection of charges and currents can serve as the source of electromagnetic waves. The dependence of the electric and magnetic fields produced by this source on space and time can be very complicated and must be analyzed by means of Maxwell's equations in their full generality. In spectroscopic experiments, however, the sample consists of absorbing quantum systems so situated in relation to the source that the fields of the latter take on a very simple form, namely, sinusoidal electromagnetic waves. Details of calculations based on this fact will be discussed in subsequent chapters. For the present purposes, it is sufficient to remember from Chap. 1 that each wave can be characterized by two parameters, its propagation vector, $\mathbf{k}$, and its electric field vector, $\mathbf{E}_1$. The wavelength may be obtained from the magnitude of $\mathbf{k}$, and the amplitude from the magnitude of $\mathbf{E}_1$. Finally, the velocity of propagation of these waves in a vacuum is c_0, a constant of nature. Ordinarily in spectroscopy, a sample is exposed simultaneously to several waves of differing $\mathbf{k}$ and $\mathbf{E}_1$. It is also ususally the case, however, that this complication can be ignored and the interaction of the sample with each wave treated separately.

In the quantum theory of radiation the wave picture is replaced by the photon picture, also introduced in Chap. 1. Each photon is a point particle of zero rest mass, containing electromagnetic energy hck with $k = 2\pi/\lambda = 2\pi\tilde{\nu} = 2\pi\nu/c = \omega/c$, and traveling in a direction parallel to **k** with a velocity c. In this chapter it will be convenient to allow the electric field amplitude to be a complex number. The number density of photons associated with a given wave must be proportional to $|E_1|^2$. Therefore

$$|E_1|^2 = \mathbf{E}_1^* \cdot \mathbf{E}_1, \tag{3.1}$$

rather than the E_1^2 used in Chap. 1, where E_1 was assumed to be real (Eqs. 1.6 and 1.18). For the rest of this chapter, E_1 will be written E.

3.2 Applicability of Classical and Quantum Theories

The statements made in Sect. 3.1 can be expressed in more formal language. Associated with each possible value of **k** is a mode of the radiation field. The energy in this mode is quantized in accordance with the formula

$$\mathscr{E}_{\mathbf{k},m} = \hbar ck(m + \tfrac{1}{2}), \quad m = 0, 1, 2, \ldots, \infty, \tag{3.2}$$

where m is the number of photons in the mode characterized by **k**. (Usually, the zero of energy is defined in such a way that the $\frac{1}{2}$ on the right-hand side of Eq. (3.2) does not appear.)

Note that, even in the total absence of photons ($m = 0$), there is still energy in the mode. A similar situation occurs in the quantum-mechanical treatment of matter, although it was not discussed in Chap. 2. An example is provided by the vibrational motions of a molecule. If such motions are adequately described by the harmonic oscillator model, the eigenvalues of the Hamiltonian operator are given by an expression identical in form to Eq. 3.2. In particular, there is an irreducible minimum vibrational energy, $\mathscr{E}_{\text{min}}$, where

$$\mathscr{E}_{\text{min}} = \frac{\hbar\omega}{2}. \tag{3.3}$$

This energy is present at equilibrium even at the absolute zero of temperature and is called the *zero-point energy* of the oscillator. Zero-point energy is required in order to satisfy the requirements of the uncertainty priciple.

Similarly, in Eq. (3.2) a zero-point energy is associated with the radiation field even in so-called *field-free* space. It must be included for the same reason that prevailed in the oscillator problem: it is required in order to satisfy the uncertainty principle. The corresponing zero-point electromagnetic fields perturb the motions of electrons in their courses around atomic nuclei. In particular, they thereby remove the degeneracy between the $2s$ and $2p$ states of the hydrogen atom, an effect called the *Lamb shift*. The effect of the zero-point fields of greatest general interest to spectroscopists is their power to force excited quantum systems to emit radiation. Such emission is called *spontaneous* emission. By way of contrast, emission forced by the radiation field in states where $m > 0$ is termed *stimulated* emission. No explanation of spontaneous emission on the basis of classical electromagnetic theory is as satisfying as this quantum-mechanical one.

A number of other electromagnetic phenomena are satisfactorily explained only by means of the quantum theory of radiation. For example, the distribution of energy among the different modes of an optical cavity at thermal equilibrium (Planck's law) implies a quantization of these modes. When this is used to explain spectroscopic transitions, such as those produced by absorption of energy from the **k**th mode of the radiation field, one imagines the field to be initially in an eigenstate of the energy characterized by the presence of m photons. After the transition has occured, the field contains $m - 1$ photons.

Unfortunately, the *dynamics* of these changes are not easy to picture from this viewpoint. During the transition process, the radiation field will not have a definite energy, and will therefore be in a superposition of states having definite numbers of photons [3].

3.3 When the Classical Theory May Be Used

A solution to the difficulty described in Sect. 3.2 is provided by the *correspondence principle*. Roughly speaking, this principle states that, whenever a quantum system is in a state associated with very large quantum numbers, that quantum system behaves in much the same way that it would if classical mechanics and classical electromagnetic theory applied instead. If the quantum system being considered happens to be the **k**th mode of a radiation field, it will be a good approximation to replace the photon stream by a classical electromagnetic wave if and only if m, the number of photons in that mode, is very large. (A more detailed and accurate statement of the necessary and sufficient conditions can be found in the work of Jaynes and Cummings [4]).

Such intense fields are easily obtained from conventional sources in radio and microwave spectroscopy (NMR and EPR), and in other regions of the spectrum wherever masers and lasers are employed. The classical theory can therefore be used without apology whenever such radiators are used. From Eq. 3.2 it can be seen that in such fields stimulated emission will be a more important mechanism for returning a chromophore to its ground state than spontaneous emission. The reason is that, in this case, $m \gg \frac{1}{2}$. Because the wave picture of light will be used throughout this book, spontaneous emission, whenever it has a measurable effect on the dynamics of the transition process, will be treated in an ad hoc fashion as just another *relaxation mechanism.*

Much of spectroscopy is performed using light sources very much weaker than those for which the correspondence principle permits the use of the semiclassical theory. Fortunately, the values of most of the measurable physical properties of quantum systems obtained by means of spectroscopic experiments performed in feeble light agree with the results obtained from experiments employing intense sources [4].

3.4 Waves and Particles and Their "Sizes"

The traditional experiment used to demonstrate the wave-particle duality of light is the diffraction experiment. Light from an incandescent source is filtered and collimated until all the waves that remain have very nearly the same **k**. These waves are used to illuminate a small aperture, and the transmitted light then falls upon a screen. If the smallest dimension of the aperture greatly exceeds a wavelength of the incident light, the edges of the aperture cast an apparently sharp shadow. (Even in this case, however, if one looks very closely at the edges of the shadow, they will not appear sharp). The pattern formed by the transmitted light on the screen is a well-defined image of the aperture and has very nearly the same dimensions. The illumination will be very nearly uniform over the image.

This experiment is analogous to firing a shotgun at a paper target through a knothole in a fence. If there are many pellets in the shell, and if the hole in the fence is smaller than the transverse scatter in pellet trajectories (but much larger than any one pellet), the pattern of pellets will form a resonably good image of the knothole (see Fig. 3.1). For this reason, the large-aperture experiment demonstrates the particle-like properties of light; photons, traveling on straight-line trajectories, form the same kinds of patterns that would be expected of shotgun pellets.

If the large aperture is replaced by one having at least one dimension comparable to λ, the pattern formed by the transmitted light becomes much larger than the aperture and exhibits bands or rings of alternating high and low intensity (see Fig. 3.2). This diffraction pattern is characteristic of waves and is due to the alternating constructive and destructive interference among them.

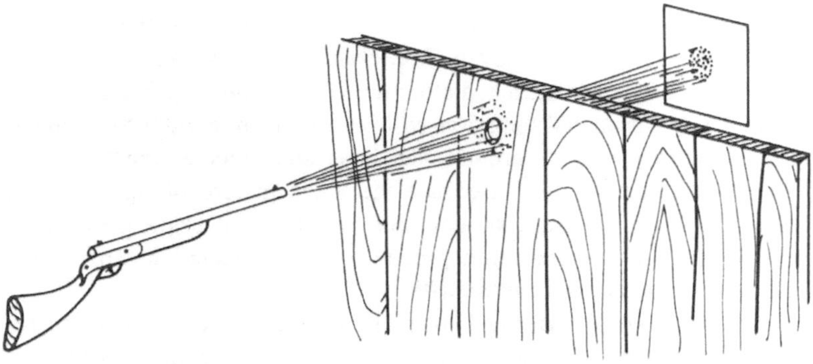

Fig. 3.1. Particles passing through an aperture make a fair image of the aperture

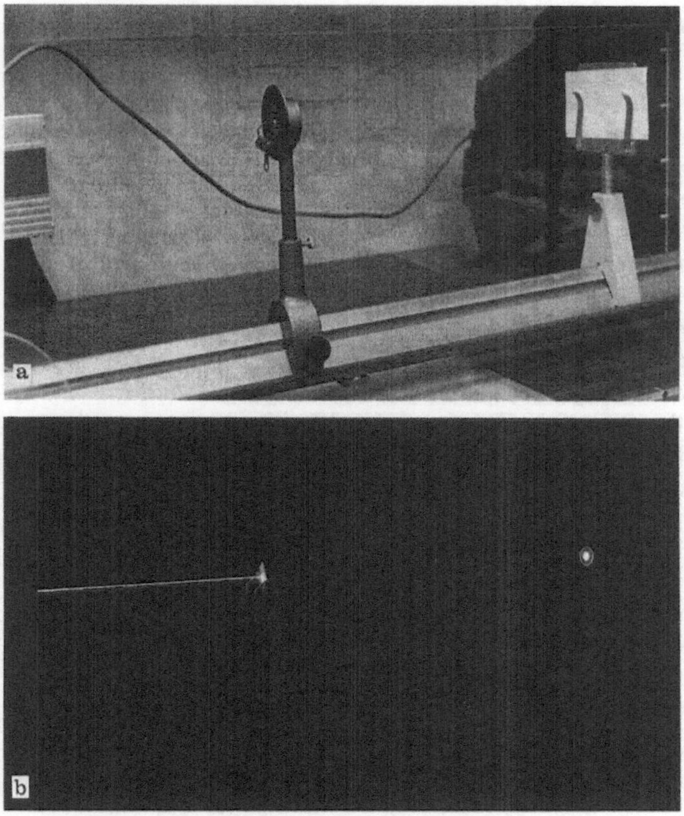

Fig. 3.2. Light passing through a small circular aperture. **a** A laser (He-Ne, $\lambda = 632.8$ nm) appears on left-hand edge of picture; a pinhole of 102-μm diameter appears near the center, and the spot illuminated by the beam appears on a white card at the right of the picture. All optical components are mounted on an optical bench placed in a Plexiglas trough. **b** With the room light out and the Plexiglas trough filled with fog (from dry ice in warm water) to make the beam visible by scattered light, the beam spreads into a Fraunhofer diffraction pattern on the card. This pattern illustrates the wave nature of light; the rings would be bigger but too faint to be seen if the pinhole were smaller. Photographs by Douglas M. Macomber

A less well-known but historically important equivalent set of experiments may be performed by substituting small disks, balls, and wires for circular and rectangular apertures. The images formed on the screen in these experiments will be identical with those formed in the ones described previously, except that light areas will appear where dark ones were seen before, and vice versa. For example, if a ball is used, the nature of the shadow formed on the screen depends on the ratio of the diameter of the ball to the wavelength of the light used to illuminate it. If the diameter is much greater than λ, the particle picture of light provides an accurate description of the shadow. If the diameter of the ball is less than (or comparable to) λ, the wave picture of light is appropriate instead.

3.5 Electromagnetic Waves and Quantum Systems: Size Ratios

The subject of this book is the interaction of light with quantum systems. It has already been decided to consider a light beam as a collection of electromagnetic waves rather than as a stream of photons. Is this choice consistent with the facts about light waves presented in Sect. 3.4? The quantum systems (e.g., atoms, ions, or molecules) are analogous to the balls in the experiment just described. In order to use the wave picture, then, it is necessary that the wavelength of light used to produce spectroscopic transitions exceed the radius of the quantum system being illuminated. Quantum systems found in nature vary greatly in diameter, and there are many kinds of spectroscopy, each of which employs electromagnetic radiation in a different region of the spectrum. It is not obvious that the necessary condition regarding the size ratio is always met, and for that reason the following discussion is presented.

First, note that very few "quantum systems" have diameters much larger than the diameter of a typical atom ($\cong 0.1$ nm). An obvious exception is provided by the spectroscopy of solids, where transitions between states in various energy bands may involve highly delocalized electrons. In these cases the corresponding wavefunctions may extend over a macroscopic portion of a crystal, and the following analysis will not apply. However, much of the spectroscopy of atoms, molecules, and ions in condensed phases can be adequately described by means of transitions between spatially localized wavefunctions. Here the dimensions of the electronic orbitals are of the same order of size as those that are appropriate for the same atoms, ions, and molecules isolated from one another as they would be in the gas phase. In these cases, therefore, the assumption fo a "ball" diameter of less than 1 nm is valid. It may be objected that some macromolecules (molecules containing many atoms) have radii of gyration much larger than 1 nm. However, the spectroscopic properties of such molecules are ordinarily dominated by small, relatively isolated clumps of atoms called *chromophoric groups.* Seldom does the diameter of an individual chromophore exceed 1 nm, so that this assumption holds even in these molecules.

Now it is appropriate to list the various common kinds of spectroscopy, in decreasing order of wavelength of light employed: nuclear magnetic and quadrupole resonance, electron spin resonance and rotational spectroscopy, vibrational spectroscopy, electronic spectroscopy, and nuclear spectroscopy. In the first and last named, the chromophore is the atomic nucleus, with a diameter of the order of 10 femtometers (10^{-15} m). Because of the size ordering in the above list, only nuclear spectroscopy need be considered. The gamma rays used in nuclear spectroscopy ordinarily have a much longer wavelength than 10 fm, so that the wave picture of light is appropriate for nuclear spectroscopy. For the three kinds of spectroscopy employing wavelengths between those of nuclear magnetic resonance and nuclear spectroscopy, the appropriate regions of the spectrum are the microwave, infrared, and visible-ultraviolet. The corresponding wavelengths are 10 mm, 10 μm, and 1 μm to 100 nm, all much larger than the "ball diameter" of

1 nm. According to these arguments, then, the wave picture of light is appropriate for all forms of spectroscopy with the possible exception of that of electronic transitions in the vacuum ultraviolet.

The latter possibility may be disposed of by the following argument. First, in Chap. 1, the scope of the book was limited to the study of transitions between *bound states* of the corresponding quantum systems. Any more drastic process (e.g., ionization) was called *destruction* rather than *spectroscopy*, at the risk of offending those who might object to such a restricted definition of the latter term. Next, an argument due to Heitler [2] can be used to calculate the shortest wavelength associated with the spectroscopic study of a bound electron in an atom. From Eqs. (1.1) and (1.6)

$$\mathscr{E}_{\text{photon}}(\max) = \frac{hc_0}{\lambda(\min)}. \tag{3.4}$$

From Eq. (1.28),

$$\mathscr{E}_{\text{photon}}(\max) = \Delta\mathscr{E}_{\text{atom}}(\max) \tag{3.5}$$

and

$$\Delta\mathscr{E}_{\text{atom}}(\max) = \mathscr{E}_{\text{atom}}(\max) - \mathscr{E}_{\text{atom}}(\min). \tag{3.6}$$

At the ionization limit for bound electrons,

$$\mathscr{E}_{\text{atom}}(\max) = 0. \tag{3.7}$$

Finally, $\mathscr{E}_{\text{atom}}(\min)$ was given in Eq. (1.19), where a_0 is now interpreted as the Bohr radius associated with the orbital in its normal (i.e., ground) state.

Equations (3.4) to (3.7) may be combined with Eq. (1.19) to yield

$$\frac{hc_0}{\lambda(\min)} = \frac{c_0 Z_0 q_0^2}{8\pi a_0}. \tag{3.8}$$

Therefore,

$$\frac{4\pi a_0}{\lambda(\min)} = \frac{Z_0 q_0^2}{2h} = \frac{1}{137}. \tag{3.9}$$

The particular dimensionless collection of fundamental constants that appears on the right-hand side of Eq. (3.9) is very famous. It is called α, the fine-structure constant, and is very nearly equal to 1/137.

It may be concluded, therefore, that in any experiment that can be described as spectroscopic the wavelength of electromagnetic radiation employed must exceed the diameter of the quantum system with which it interacts (the chromophore) by a factor of at least 137. Consequently, the description of the interaction of electromagnetic radiation and matter in spectroscopic experiments by means of the wave picture of light is consistent with both the correspondence principle (if the light is intense) and the results of experiments in physical optics.

3.6 Series Expansions for Electromagnetic Fields

Since the wave picture of light for spectroscopic experiments has been adopted, the next step is to recall that the field intensities in such waves vary in both space and time. A quantum system bathed in these fields will sense both of these variations. Because of the spatial modulation of the waves, the field intensity at any one instant of time felt by an atom, ion, or molecule will not be uniform. The nonuniformities, or gradients, in the field intensities from one side of the chromophore to the other will, however, be a small fraction of the rms values of these intensities. In fact, the fraction will be of the order of a_0/λ, which has been shown to be less than 1%. For this reason it is ordinarily a good approximation to neglect the spatial variation of field intensities in the electromagnetic waves that illuminate a quantum system in a spectroscopic experiment. Therefore, from the point of view of an atom, ion, or molecule, light consists of spatially uniform electric and magnetic fields, oscillating sinusoidally in time. In this regard, a quantum system in a light wave is like a cork bobbing up and down an ocean swell, but not being tipped or longitudinally displaced thereby.

This is very fortunate for anyone who wishes to understand the dynamics of spectroscopic transitions because there are only a limited number of ways in which a quantum system can interact with a spatially uniform field. Suppose that the propagation vector for the wave is parallel to the z axis, and that the atom, ion, or molecule in question is located at the origin. At the instant of time, the spatial dependence of the applied field in the vicinity of the origin may be expressed by a McLaurin series:

$$E(z,t) = E(0,t) + \left[\frac{\partial E(z,t)}{\partial z}\right]_{z=0} z + \frac{1}{2!}\left[\frac{\partial^2 E(z,t)}{\partial z^2}\right]_{z=0} z^2 + \cdots. \tag{3.10}$$

An equation identical in form to Eq. 3.10 describes the magnetic field, $H(z,t)$. These series converge in the case of interest ($z \cong a_0$), and do so uniformly, each term being about $1/137{\cdot}\pi$ of the previous one. The reason for the convergence can be seen by dividing the second term on the right-hand side of Eq. (3.10) by the first, replacing the differentials by finite differences. The result will be $(\Delta E/E){\cdot}(z/\Delta z)$. The maximum ΔE occurs when $\Delta z = \lambda/2$; in this case $|\Delta E/E| = 2$. With $z = a_0, (z/\Delta z) = 2/137$.

The atom, ion, or molecule, from the point of view of electromagnetic theory, consists in general of an arbitary collection of charges and currents which interacts with the externally imposed fields. It is possible to describe an arbitrary collection of charges and currents by a power series (of a different kind from the McLaurin expansion of the external field) called the *multipole expansion*. The details of this expansion are described in standard textbooks on electromagnetic theory.[1] It is sufficient here to give a qualitative description of the various terms that appear therein, and the implications of each to spectroscopy.

The charge distribution in the atom, ion, or molecule gives rise to the electric part of the multipole series. Each term in the electric part describes an asymmetry present in the charge distribution. The grossest asymmetry is described by the first term; successive terms represent increasingly subtle asymmetries, so that from a physical point of view this series also converges uniformly.

The first term, called the *electric monopole*, simply represents the net charge of the system. Therefore only ions have a nonzero electric monopole moment; for neutral atoms and molecules, the largest possible non-vanishing electric term is the electric dipole. Pure dipoles have equal numbers of positive and negative charges, but these are distributed in such a way that the "center of gravity" of the former is displaced from that of the latter. Polar molecules (e.g., HCl, N_2O, and H_2O) are those which, according to classical mechanics, would be expected to have permanent nonzero electric dipole moments. Molecular ions such as HCl^+ are expected to possess both electric monopole and dipole moments, and symmetric neutral molecules (e.g., H_2, CO_2, and CH_4), neither one. Just as a dipole may be formed from two equal and opposite monopoles displaced from one another, so the next term, the electric quadrupole, may be formed by two equal and oppositely directed dipoles with separated centers. Examples are CO_2 and PF_3Cl_2. These molecules always possess a positive or negative "waist" with head and feet of opposite polarity. The molecule-ion OCS^+ should possess all three of these electric moments. The octupole, hexadecapole, and the remaining 2^n-poles represent charge distributions so subtle[2] that they ordinarily have little

---

[1] The mathematical formalism associated with the multipole expansion is as follows. An arbitary microscopic volume element within the atom, ion, or molecule ($d\tau$) has a charge density ρ. The charge in this element, $\rho d\tau$, contributes a term to the electric potential at some point of observation. Let the point of observation be located at $\mathbf{r}$, the vector connecting it with $d\tau$; then the electric potential, V, is given by $(4\pi\varepsilon_0)^{-1}\int(\rho d\tau/r)$. The vector $\mathbf{h}$ connects $d\tau$ with the center of mass of the atom, ion, or molecule of which it is a part, and the vector connecting the center of mass with the point of observation is $\mathbf{R}$; $\mathbf{R} = \mathbf{h} + \mathbf{r}$. If the point of observation is very many atomic diameters away from the center of mass, r will differ only by a small amount from R, and the quantity $1/r$ can be expanded in a Taylor series in terms of $1/R$. The resultant expression for the electric potential is as follows:

$$V = (4\pi\varepsilon_0 R)^{-1}\sum_{n=0}^{\infty} R^{-n}\int \rho h^n P_n(\cos\theta)d\tau,$$

and the electric field $\mathbf{E}$ is the gradient of V. In this expression, θ is the angle between h and R, and the P_n are the Legendre polynomials. The successive terms in the expansion are contributions to the electric potential arising from projections of the successive electric multipole moment vectors upon the $\mathbf{R}$ axis. The electric field intensity is the gradient of this potential. A similar expansion for the *current* density (rather than the charge density) yields the magnetic multipoles.

[2] If the reader is puzzled by the use of the word "subtle", he should consider a bull and an earthworm. The earthworm is the sexual equivalent of a dipole (it functions one way on one end, and the other way on the other end). Surely the question of its sex is a more subtle one than that of the bull (monopole).

effect on the physical properties of the quantum systems that possess them. Some of the molecules displaying these moments are shown in Fig. 3.3.

As the charges in the atom, ion, or molecule circulate in their orbital paths, electric currents are set up that give rise to magnetic phenomena. The corresponding magnetic multipole series differs in two important respects from the electric one. In the first place, magnetic monopoles (e.g., an isolated "north" without a corresponding "south" nearby), unlike electric monopoles, may not exist in nature. Only one observation of a magnetic monopole has ever been reported [5], and the evidence presented in that case was rather indirect. In the second place, the effects that the possession of a magnetic multipole produce on the behavior of a quantum system are ususally less important than those produced by the corresponding electric monopole. For example, magnetic dipoles are usually comparable in importance to electric quadrupoles, magnetic quadrupoles to electric octupoles, and so on.

Finally, there is a rule that, for quantum systems in their lowest (ground) energy states, alternate multipoles are forbidden. Let the electric ones be designated by E, and the magnetic ones by M, followed by "1" for monopole, "2" for dipole, and so on. The allowed multipoles are $E1$, $M2$, $E4$, $M8$, $E16$, and so forth. Some readers may be surprised at the absence of $E2$, or the electric dipole moment, and ask, "What about the HCl molecule?" The answer is that for HCl in the gas phase "ground state" means "ground" in everything, including the rotational wavefunction. The instantaneous electric dipole moment of the molecule in such a state does exist, of course, but its orientation is completely uncertain. Hence there is a complete vector cancellation of this dipole when the expectation value is calculated. The situation is completely analogous to that of the hydrogen atom (the "protonium electride" molecule, it might be called), which also

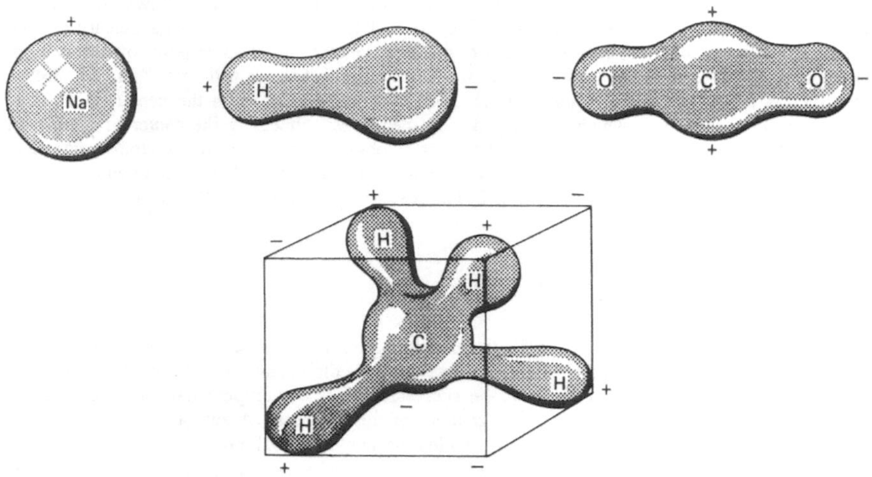

Fig. 3.3. Molecules with various electric multipole moments: Na^+, monopole; HCl, dipole; CO_2, quadrupole; CH_4, octopole

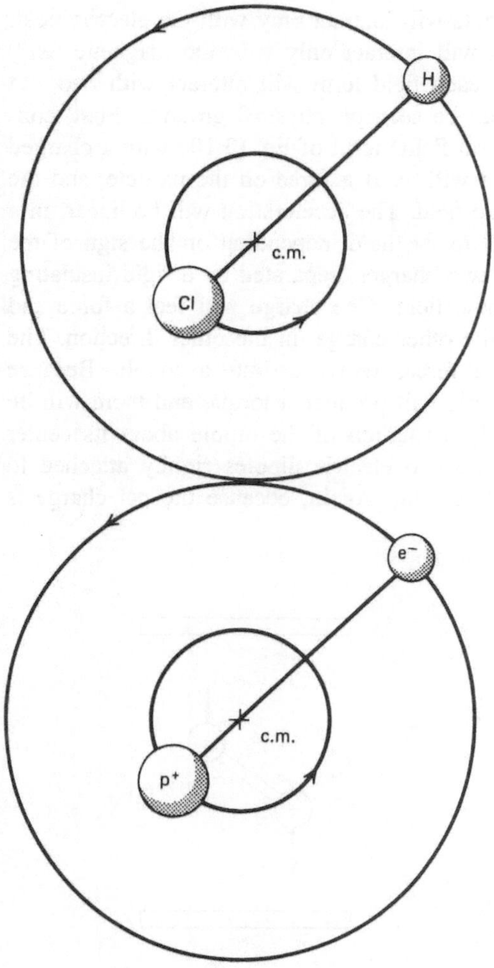

Fig. 3.4. Top: HCl in its ground state, with no electric dipole moment. Bottom: H atom in its ground state, with no electric dipole moment

lacks a permanent electric dipole moment in its 1*s* state, for the same reason (see Fig. 3.4).

3.7 Interactions Between Multipoles and Field Asymmetries

For the present discussion, the important point is that the behavior of an arbitrary collection of charges and currents in an arbitrary external field can be determined by considering separately the interaction of each one of the multipole moments of the former with each term in the McLaurin expansion of the latter (Eq. 3.10), and adding the results. If this is done, several important general priciples immediately come to light.

First, the electric multipole moments will interact only with the electric field, and the magnetic multipole moments will interact only with the magnetic field. Second, within any McLaurin series, each field term will interact with two and only two multipole moments. This can be seen on physical grounds. First, consider the interaction of the first (uniform field) term of Eq. (3.10) with a charged particle (electric monopole). The field will exert a force on the particle, and the latter will thereby be accelerated in the field. The acceleration will be linear, in a direction either parallel or antiparallel to the field, depending on the sign of the charge (see Fig. 3.5a). Now imagine two charges, separated by a rigid insulating rod (i.e. an electric dipole) in the same field. One charge will feel a force and be accelerated in one direction, and the other charge, in the other direction. The charges being equal and opposite, the forces will constitute a couple. Because the rod between them is rigid, the couple will produce a torque and there will be no net linear displacement, but merely a rotation of the dipole about its center of mass (see Fig. 3.5b). Next, imagine two electric dipoles rigidly attached to one another (a quadrupole) in the same field. Again, because the net charge is

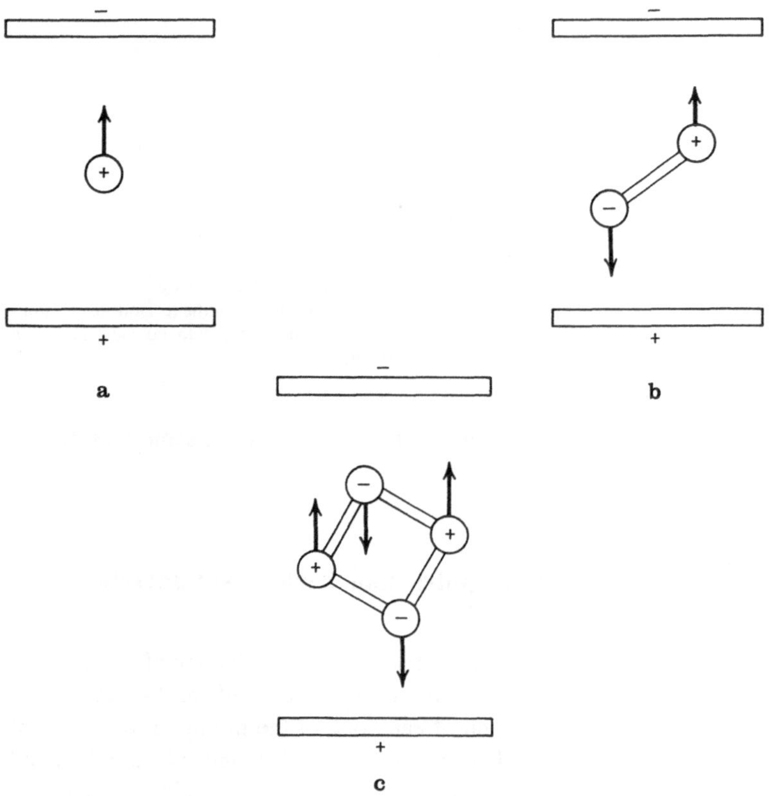

Fig. 3.5. Multipoles in a uniform field: **a** monopole, **b** dipole, **c** quadrupole

zero, the net linear force, acceleration, and displacement will all be zero. Both dipoles will feel equal torques, but, because their moments are oppositely directed, in opposite senses. Because of the rigid connection between them, no net angular acceleration or displacement will be possible (see Fig. 3.5c.) This shows that a quadrupole cannot be affected by a uniform field. The same analysis can be given for all higher electric multipole moments; because they represent even subtler asymmetries in the charge distribution, they will not be affected either. In summary, the first term in the McLaurin expansion of the electric field will exert a linear force on an electric monopole and an angular force (torque) on an electric dipole, and will have no influence on any higher moments. If the word "electric" were replaced by "magnetic" in the above sentence, it would still be true.

Now the second term in Eq. 3.10 – the uniform electric field gradient term, with contribution of zero to E at $z = 0$ – will be considered. If a point electric monopole (e.g., the center of charge of an ion) were located at $z = 0$, it would feel no force because the electric field is zero at that point. The equilibrium would be very unstable, of course, because even the slightest perturbation in the position of the monopole along the z axis would move it into a region of nonzero electric field. It would then be accelerated by the field into a region where the

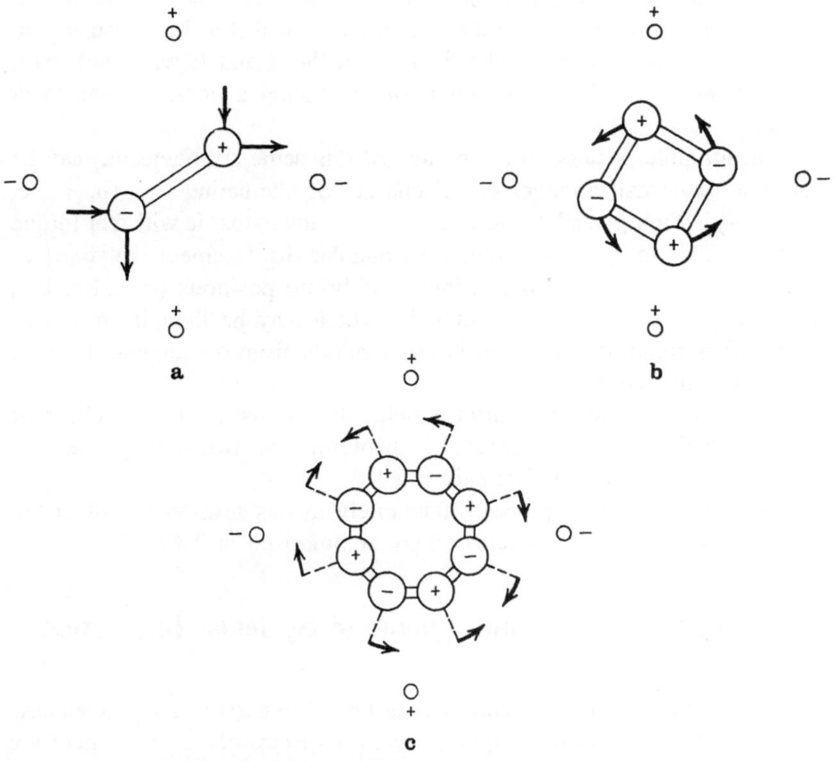

Fig. 3.6. Multipoles in a linear field gradient: **a** dipole, **b** quadrupole, **c** octopole

Table 3.1 Forces felt by a point multipole in an arbitrary field, F. L = linear acceleration, A = angular acceleration, N = no acceleration, U = unstable equilibrium

	$\partial^m F/\partial z^m$, $m =$				
2^n Pole, Order of $n =$	0	1	2	3	4 ...
0	L	U	U	U	U
1	A	L	U	U	U
2	N	A	L	U	U
3	N	N	A	L	U
4	N	N	N	A	L
5	N	N	N	N	A
6	N	N	N	N	N

force would be stronger yet, and so on until the monopole had been completely swept out of its original position.

Next consider a point dipole. It is convenient to imagine that the field gradient is produced by four wires perpendicular to the plane of the paper, and intersecting the paper at the corners of a square (see Fig. 3.6a). The wires are labeled as the points of the compass: N, E, S, and W. N and S are raised to a positive electric potential, and E and W are lowered to a negative potential. If the dipole is initially oriented to the northeast (e.g., positive "end" northeast and negative "end" southwest), it is clear that the dipole will feel a force, causing an acceleration towards the southeast. The fields from the S and E wires will pull, and those from the N and W wires will push, producing a linear displacement of the dipole.

A point quadrupole, located in the middle of this same arrangement, may be thought of as a very small arrangement of charges of alternating sign (n, + ; e, − ; s, + ; w, −). When placed at the center of this apparatus, it will feel torque about its center. This torque will produce an angular displacement (rotation) of the quadrupole toward one of two possible equilibrium positions (e.g., Ns, Ew, Sn, We) (see Fig. 3.6b). Finally, an octopole, which may be thought of as two quadrupoles with a common center displaced angularly from one another by 90°, will feel no net torque (Fig. 3.6c).

Just as was the case for the uniform field, the discussion of the effect of a uniform field gradient may be repeated, substituting the word "magnetic" for "electric," and the conclusion will remain true.

Evidently, it is possible to proceed indefinitely in this fashion for all of the terms in Eq. (3.10); the results anticipated are summarized in Table 3.1.

3.8 Electromagnetic Waves and Quantum Systems: Interactions

It has been shown that, within an accuracy of at least 136 parts in 137, the electric and magnetic fields produced by a light wave in a spectroscopic experiment are uniform; terms of the kind $(\partial^m/\partial z^m)_0$, with $m \neq 0$, can therefore be neglected. For this case, the results can be summarized in Table 3.2.

Table 3.2 Forces felt by a quantum system interacting spectroscopically with a light wave. L = linear acceleration, A = angular acceleration, N = no acceleration

	Kind of Multipole	
Order of 2^n pole, n =	Electric	Magnetic
0	L	L
1	A	A
2	N	N
3	N	N
4	N	N
5	N	N
6	N	N

Since there are apparently few magnetic monopoles in nature, the corresponding entry in Table 3.2 may be neglected. The electric monopole term applies only to ions; as a practical matter, the interaction of light from the sun with charged particles in the earth's ionosphere by means of this interaction helps shield us from harmful ultraviolet rays. Also, radio-wave messages may be bounced between remote points on the earth's surface from the underside of this same layer. Finally, the high reflectivity of metals is a well-known optical phenomenon produced by the interaction of conduction electrons with the uniform electric fields of light waves. Although any of these phenomena may occasionally be of practical importance to spectroscopists, none will be considered further in this book. Rather, attention will be concentrated on the spectroscopic interaction of electromagnetic radiation with neutral atoms and molecules, and with charged species (such as nuclei) in which the electric monopole interaction plays no more than a trivial role (e.g., NMR and EPR).

Finally, it should be noted that, in Maxwell's equations, the effects of matter on the fields are characterized by only three parameters. The electric conductivity, σ, has to do with the interaction of electric monopoles in the matter with the applied electric field. The electric permittivity, ε, is concerned with the interaction of electric dipoles in the matter with the applied electric field. The magnetic permeability, μ, is associated with the interaction of magnetic dipoles in the matter with the applied magnetic field. If magnetic monopoles abounded, one would have to introduce a new parameter, σ_m, magnetic conductivity, to characterize their interaction with the applied magnetic field; in this case, Maxwell's equations would assume an even more symmetric form.

In summary, *there are only three possible mechanisms by means of which electromagnetic radiation can interact with matter in a spectroscopic experiment*; that is to say, there are only three ways in which the matter can exchange energy with the radiation field. Suppose that the matter is to *receive* energy from the radiation field (absorption). The radiation field must do electromechanical *work* upon the matter (i.e., the quantum systems absorbing the energy). There are only two ways in which work can be done. First, a linear force, $\mathbf{F}_0$, may produce linear displacements, $d\mathbf{z}$. In that case, the work, w, is given by the elementary formula

$$w = \int \mathbf{F}_0 \cdot d\mathbf{z}. \tag{3.11}$$

Second, an angular force (torque), $\mathbf{T}$, may produce angular displacements, $d\boldsymbol{\theta}$; in that case, the work is given by

$$w = \int \mathbf{T} \cdot d\boldsymbol{\theta}. \tag{3.12}$$

The only way in which a uniform electric or magnetic field (e.g., the fields associated with a light wave in a spectroscopic experiment) can exert a linear force upon an arbitrary collection of charges and currents (e.g., a quantum chromophore) is through the corresponding monopole moment (σ). This interaction will play no more than a trivial role in the experiments which shall subsequently be described, as was stated previously.

The only way that a uniform electric or magnetic field can exert a torque upon an arbitrary collection of charges and currents is through the corresponding dipole moment (and therefore through the parameter ε or μ). For neutral atoms or molecules, then, *spectroscopic transitions are produced whenever the electric or magnetic fields of the light wave exert a torque upon the electric or magnetic dipole moments of the absorbing quantum systems.* These moments are the "handles" by means of which the radiation field "grasps" atoms or molecules with which it interacts. Whenever such a handle exists, the transition in question is called "electric (or magnetic) dipole allowed." Whenever such a handle is absent, the transition is termed "electric (or magnetic) dipole forbidden." To an accuracy of better than 136 parts in 137, *no other types of transitions are possible.*

This general a priori conclusion is not the only one that may be drawn about the interaction mechanism. It must now be remembered that, although the electric and magnetic fields which constitute a light wave may appear *spatially* unifrom to a quantum system in a spectroscopic experiment, they by no means appear *temporally* uniform. In fact, these fields completely reverse direction twice during each optical cycle, of duration $2\pi/\omega \equiv \lambda/c$. This fact implies that the dipole moment with which the field interacts must *also* oscillate in time, if the net energy transfer during one complete optical cycle is to be nonzero. The truth of this statement can be demonstrated by the following argument. The energy of a dipole $\mathbf{p}_E$ (or $\mathbf{p}_M$) in a field $\mathbf{E}$ (or $\mu_0\mathbf{H}=Z_0\mathbf{H}/c_0$) is given by

$$\mathscr{E} = -\mathbf{p}_E \cdot \mathbf{E}. \tag{3.13}$$

If $\mathbf{E}$ oscillates between $-|E|\hat{z}$ and $+|E|\hat{z}$ during each cycle, but $\mathbf{p}_E$ is constant, the time average of $\mathscr{E}$ in Eq. 3.13 will be zero.

In fact, it is easy to see that, for efficient energy transfer, the oscillation frequency of the dipole must be very nearly the same as the oscillation frequency of the field.

The problem of coupling two oscillators by matching their frequencies is well known to anyone who has ever pushed a small child on a swing. To increase the level of excitation of the child-swing oscillating system by drawing energy from the doting-parent oscillating system, two conditions must be satisfied. First, there must be a coupling mechanism (in this case, the contact force between the pusher's hand and the rear surface of the swing seat). Second, the pusher must synchronize his pushes with the natural period of oscillation of the pushee (determined in this case by the classical mechanics of pendulum motion) (see Fig. 3.7a). By "synchronization" is meant not only making the frequencies of the two oscillators equal, but also matching their phases. A 180° phase error in the push (pusher going forward while the swing system is moving backward) will result in energy being transferred from the child to the parent (the former ending up motionless, and the latter being knocked head over heels: see Fig. 3.7b), rather than vice versa.

If the pusher's hand is replaced by one of the fields of the light wave, the rear of the swing seat is replaced by the corresponding quantum-mechanical dipole moment, and the contact force is replaced by the electromagnetic force, the analogy between the two systems is evident. In the case of an atom or molecule undergoing a spectroscopic transition, the natural frequency is determined by quantum mechanics. The frequencies of the two oscillations must match one

a

Fig. 3.7. Energy transfer between oscillators: **a** parent doing work on child-in-swing **b** Child-in-swing doing work on parent

b

Fig. 3.7. (*Continued*)

another in quantum mechanics as well as in classical mechanics, and, depending on the relative phase of the oscillation of the field and of the dipole, either absorption or stimulated emission of radiation is obtained.

3.9 Complex Susceptibilities, Electric and Magnetic

The effects of the microscopic multipole interactions on the bulk electromagnetic properties of the system are discussed in this section. Imagine a substance consisting of charged paritcles in thermal and mechanical equilibrium. There will be no net flow of current in such a material in the absence of any external forces. If a uniform electric field **E** is applied to this substance, however, each of the charged particles will feel a force. Particles that are free to do so will move in response to this force and begin to flow through the material – the external field thus induces an electric current. The parameter σ is a measure of the mobility of charges; if it is zero (as will ordinarily be the case for substances discussed in this book), the electric monopoles in the material will not influence the electromagnetic behavior of the material.

Now imagine instead that the substance is composed of microscopic particles with permanent electric dipole (rather than monopole) moments. At thermal and mechanical equilibrium these dipoles will be oriented at random (provided that they do not interact strongly with one another). Therefore the vector sum of the electric dipole moments in any small (but macroscopic) volume element of the material will be zero, and no net electrical field will be associated with the sample.If an electric field is applied to this substance, each of the electric dipoles will feel a torque. Dipoles that are free to do so will move in response to this torque and will eventually line up with **E**. The vector sum of the dipole moments in small macroscopic volume elements of the material will no longer be zero, and a net electric field will be associated with the sample because of these dipoles. The external field thus induces a net dipole density (number of unit-sized microscopic dipole moments per unit volume) called the *electric polarization* of the sample and given the symbol **P**.

In addition to the orientation of permanent dipoles, other methods exist for producing polarization of the sample. Molecules having a nonuniform charge distribution can be stressed by the electric field, and the bonds of these molecules will stretch and bend in response to this stress. The results of the stretching and bending of the molecular framework will be to create molecular electric dipoles where none were present before, or to modify dipole moments already present. These created and/or modified moments will add up vectorially and contribute to **P**.

Finally, the entire positively charged nuclear framework can be displaced in one direction by the electric field, and the surrounding electrons displaced in the other. This also creates an atomic or molecular dipole moment in the direction of the field and adds to **P**.

All of the above three sources of polarization are important in determining the electromagnetic behavior of matter when the applied field **E** is static. If **E** oscillates in time, the dipoles (both permanent and induced) have some difficulty in keeping up because of the inertia of the particles to which they are attached.

The parameter ε is a measure of the ease with which the dipoles are induced by the field. From the argument given in the preceding paragraph, it may be concluded that there should be a relationship between ω (the frequency of the oscillation of **E**) and the magnitude of ε. In particular, ε should diminish in size as ω increases because of the diminishing ability of the atomic and molecular motions to respond. This is generally found to be the case, but there are other features of the ε versus ω curve that are not explained by the reasoning given above.

Superimposed on the relatively featureless background just described are a number of "spikes" – relatively large changes in ε occurring over a rather narrow frequency range. The center frequency of each of these spikes is located near that one of the resonant quantum transitions of the particle of which the sample is composed (ω_0). The molecular, atomic, or ionic dipole moments induced in these regions of the wave arise from the superposition of the quantum states associated with the transition. The oscillating applied field **E** first produces these dipoles and then interacts with them (in accordance with the mechanism described in Sect. 3.8) to transfer electromagnetic energy into or out of the sample material.

Regardless of the mechanism that gives rise to **P**, it is always possible to express this polarization mathematically in terms of the field **E**:

$$\mathbf{P} = \mathbf{P}(\mathbf{E}). \tag{3.14}$$

The function on the right-hand side of Eq. (3.14) can be expanded in a McLaurin series:

$$\mathbf{P} = \mathbf{P}_0 + \varepsilon_0\eta\mathbf{E} + \varepsilon_0\eta_2 E\mathbf{E} + \cdots, \tag{3.15}$$

where η, η_2, and so on are expansion coefficients to be discussed later. The first term on the right-hand side is nonzero only in special materials called *ferro-electrics*; and if conventional sources are used, the electric fields in light waves are much too small for the third and subsequent terms to have any practical importance. If very intense (e.g., laser) sources are used, it is not possible to assume a linear relationship between **P** and **E** (nonlinear optics), but for the moment this complication will be disregarded. The product $\varepsilon_0\eta$, which appears in the second term, may, for the present, be considered to be merely the conventional notation for $(dP/dE)_{E=0}$, and is a property of the polarizable material under consideration. The total field is the sum of the applied field and that which arises from the polarized matter. The quantities **P** and **E** cannot be added directly, of course, because they have different units; **P** has the units of dipole moments per unit volume, and **E** has the units of electric field strength. Conventionally the sum, **D**, is expressed in the same units as **P**:

$$\mathbf{D} = \varepsilon_0\mathbf{E} + \mathbf{P}. \tag{3.16}$$

The parameter ε_0, called the *electric permittivity of free space*, may be regarded at present as merely a conversion factor that transforms the field **E** into an equivalent electric dipole moment per unit volume. The quantity represented by **D** is termed the *electric displacement*; it is equal to the net effective polarization, which is the sum of the "genuine" polarization, **P**, and the equivalent polarization due to the driving field, **E**. The second term on the right-hand side of Eq. (3.15) may be substituted for **P** in Eq. 3.16 to obtain

$$\mathbf{D}_m = \varepsilon_0\mathbf{E} + \varepsilon_0\eta\mathbf{E}. \tag{3.17}$$

The terms proportional to **E** may be collected to obtain

$$\mathbf{D} = \varepsilon\mathbf{E}, \tag{3.18}$$

with

$$\varepsilon = \varepsilon_0(1 + \eta). \tag{3.19}$$

In Eqs. 3.18 and 3.19, ε is the total electric permittivity discussed previoulsy; it is the sum of two terms, the first representing the "permittivity of the vacuum," and the second, the effect of the polarizable medium. The quantity represented by the symbol η is called the *electric susceptibility* of the medium.

The equations corresponding to Eqs. (3.14) to (3.19) for magnetic phenomena are as follows:

$$\mathbf{M} = \mathbf{M}(\mathbf{H}), \tag{3.20}$$

$$\mathbf{M} = \mathbf{M}_0 + \chi\mathbf{H} + \chi_2 H\mathbf{H} + \cdots, \tag{3.21}$$

$$\mathbf{B}_m = \mu_0(\mathbf{H} + \mathbf{M}), \tag{3.22}$$

$$\mathbf{B}_m = \mu_0(1 + \chi)\mathbf{H}, \tag{3.23}$$

$$\mathbf{B}_m = \mu\mathbf{H}, \tag{3.24}$$

and

$$\mu = \mu_0(1 + \chi). \tag{3.25}$$

The units of **B** are lines of magnetic force per unit area (flux).

The names of the previously undefined symbols are as follows:

M, magnetic polarization (sometimes "magnetization");

μ_0, magnetic permeability of free space:

B, magnetic induction (in magnetic resonance, the symbol **H** is used for this);

and

χ, magnetic susceptibility.

The *spontaneous* magnetic polarization, $\mathbf{M}_0$, which is important for ferromagnetic materials, has obviously been discarded in Eq (3.23). The units of **H** and **M** are both magnetic dipole moments per unit volume. It seems unfortunate that the conventional definitions of **P** and **M** necessitate the omission of the factor ε_0 on the right hand side of Eq. 3.16 in front of the former, but retention of the factor μ_0 on the right hand side of Eq. (3.22) in front of the latter.

It is conventional to permit η and χ to be complex numbers:

$$\eta \equiv \eta' + i\eta'' \tag{3.26}$$

and

$$\chi \equiv \chi' + i\chi''. \tag{3.27}$$

3.10 Effect of Susceptibilities on Wave Propagation

Next, Maxwell's relations may be used to derive the wave equations,

$$\left(\frac{\partial^2 E}{\partial z^2}\right)_t - \varepsilon\mu\left(\frac{\partial^2 E}{\partial t^2}\right)_z = 0, \tag{3.28}$$

$$\left(\frac{\partial^2 H}{\partial z^2}\right)_t - \varepsilon\mu\left(\frac{\partial^2 H}{\partial t^2}\right)_z = 0. \tag{3.29}$$

In the interest of simplicity, attention will be concentrated on Eq. 3.28. A solution is assumed of the form

$$E = |E_0|\exp\{i[\phi + \omega(\alpha_0 z - t)]\}, \tag{3.30}$$

where $|E_0|, \phi, \omega$, and α_0 are all constants: all symbols except α_0 on the right-hand side of Eq. (3.30) represent real numbers. Equation 3.30 may be substituted into Eq. 3.28 to yield

$$(i\omega\alpha_0)^2 E + \varepsilon\mu(-i\omega)^2 E = 0. \tag{3.31}$$

Equation (3.31) is satisfied if and only if

$$\alpha_0^2 = \varepsilon\mu. \tag{3.32}$$

Equation 3.19 and 3.25–3.27 may be substituted into Eq. 3.32 to find an expression for α_0^2:

$$\frac{\alpha_0^2}{\varepsilon_0\mu_0} = [(1+\eta')(1+\chi') - \eta''(\chi'')] + i[(1+\eta')(\chi'') + \eta''(1+\chi')]. \tag{3.33}$$

To obtain α_0 itself, it is necessary to find the square root of a complex number. The answer is as follows. If

$$\sqrt{\alpha_0^2} \equiv u + iv \tag{3.34}$$

and

$$\alpha_0^2 \equiv U + iV, \tag{3.35}$$

then

$$u = \pm\frac{[(U^2+V^2)^{1/2} + U]^{1/2}}{\sqrt{2}} \tag{3.36}$$

and

$$v = \pm\frac{[(U^2+V^2)^{1/2} - U]^{1/2}}{\sqrt{2}}. \tag{3.37}$$

These formulas may be verified as follows. Substitute Eqs. 3.36 and 3.37 into Eq. (3.34) and square both sides. The result should be Eq. 3.35. Returning to Eq. (3.33), one finds

$$\begin{aligned}\frac{2[\mathrm{Re}(\alpha_0)]^2}{\varepsilon_0\mu_0} &= [(1+\eta')^2 + (\eta'')^2]^{1/2}[(1+\chi')^2 + (\chi'')^2]^{1/2} \\ &+ [(1+\eta')(1+\chi') - \eta''(\chi'')]\end{aligned} \tag{3.38}$$

and

$$\frac{2[\mathrm{Im}(\alpha_0)]^2}{\varepsilon_0\mu_0} = [(1+\eta')^2+(\eta'')^2]^{1/2}[(1+\chi')^2+(\chi'')^2]^{1/2}$$
$$-[(1+\eta')(1+\chi')-\eta''(\chi'')]. \tag{3.39}$$

Exactly the same results may be obtained from the wave equation for H, Eq. (3.29).

If the spectroscopic transition in question is electric dipole allowed, χ' and χ'' may usually be neglected. In this case, Eqs. (3.38) and (3.39) simplify to yield

$$\mathrm{Re}(\alpha_0) = \pm\frac{\sqrt{\varepsilon_0\mu_0}\sqrt{1+\eta'}}{\sqrt{2}}\left\{\left[1+\left(\frac{n''}{1+\eta'}\right)^2\right]^{1/2}+1\right\}^{1/2} \tag{3.40}$$

and

$$\mathrm{Im}(\alpha_0) = \pm\frac{\sqrt{\varepsilon_0\mu_0}\sqrt{1+\eta'}}{\sqrt{2}}\left\{\left[1+\left(\frac{n''}{1+\eta'}\right)^2\right]^{1/2}-1\right\}^{1/2}. \tag{3.41}$$

A further simplification is possible because ordinarily $[\eta''/(1+\eta')] \ll 1$, permitting the use of the series expansion

$$\sqrt{1+x} \cong 1+\frac{x}{2}, \quad x \ll 1: \tag{3.42}$$

$$\alpha_0 \cong \pm\sqrt{\varepsilon_0\mu_0}\sqrt{1+\eta'}\left\{\left[1+\frac{1}{8}\left(\frac{n''}{1+n'}\right)^2\right]+\frac{i\eta''}{2(1+\eta')}\right\}. \tag{3.43}$$

3.11 Absorption Coefficient and Refractive Index

Finally, the physical significance of the parameters that appear in Eq. 3.43 must be discovered. This may be done most easily by substituting Eq. 3.43 into the exponential expression for E in Eq. 3.30. First, let $\eta' = \eta'' = 0$. Since these terms are the only ones that represent the presence of matter, the resulting expression describes the properties of an electromagnetic wave in a vacuum:

$$E = |E|\exp\left[i(\phi \pm \omega\sqrt{\varepsilon_0\mu_0}\, z - \omega t)\right]. \tag{3.44}$$

Since ω is the angular frequency in radians per second,

$$\omega\alpha_0 \equiv k. \tag{3.45}$$

Therefore, Eqs. 1.1, 1.5, 3.3, and 3.45 may be used to show that

$$\sqrt{\varepsilon_0\mu_0} = \frac{1}{c_0}. \tag{3.46}$$

Equation 3.44 may be rewritten again to yield

$$E = |E|\exp\left\{\left[\phi \pm \frac{\omega}{c_0}(z \mp c_0 t)\right]\right\}. \tag{3.47}$$

The choice of algebraic signs in the exponent determines the direction of propagation. For example, set $\phi = 0$ and choose the upper sign. At location $z = 0$, the wave in Eq. 3.47 will have a "crest" ($E = |E|, H = |H|$) at $t = 0$. Anyone who wishes to "ride" upon this crest must "move" (change z) in such a way that the argument of the exponent remains zero for all time. The solution is $z = c_0 t$. It is just as easy to see that a wave traveling in the $-z$ direction is represented by the choice of the lower sign in Eq. 3.47.

If the matter is restored ($\eta', \eta'' \neq 0$) in Eq. 3.43, $\sqrt{\varepsilon_0 \mu_0}$ is multiplied by the factor $\sqrt{1+\eta'}$. From Eq. 3.46, it can be seen that this has the effect of reducing the velocity of propagation:

$$c_0 \rightarrow \frac{c_0}{n_R}, \tag{3.48}$$

where

$$n_R \equiv \sqrt{1+\eta'}. \tag{3.49}$$

The property represented by the symbol n_R is of course the *refractive index* of the medium.

Using Eqs. 3.46 and 3.49, one may rewrite Eq. (3.43) as

$$\alpha_0 = \frac{n_R[1 + (i\eta''/2n_R^2)]}{c_0}. \tag{3.50}$$

In Eq. (3.50), the upper algebraic sign has been chosen for concreteness, and it is assumed that $(\eta''/8n_R^2)^2 \ll 1$. Then Eq. (3.50) may be substituted into Eq. (3.30) to achieve the following expression for the electric field:

$$E = |E|\exp\left\{i\left[\phi + \omega\left(\frac{n_R}{c_0}\right)z - \omega t\right]\right\} \cdot \exp\left(\frac{-\omega\eta'' z}{2n_R c_0}\right). \tag{3.51}$$

Earlier, it was stated that the intensity of the light is proportional to the complex square of the electric field (Eq. 3.1). Therefore,

$$J \propto |E|^2 \exp\left(\frac{-\omega\eta'' z}{n_R c_0}\right). \tag{3.52}$$

This expression may be compared with Eq. (1.27) to show that Eq. (3.52) is, in fact, the Beer-Lambert law. The physical significance of η'' is now clear; it gives rise to the *absorption coefficient*, β_0 (see Eq. 1.27):

$$\beta_0 \equiv \frac{\omega\eta''}{n_R c_0} = N'\kappa. \tag{3.53}$$

In summary, a relationship has now been found between the real and imaginary parts of the electric susceptibility and the refractive index and absorption coefficient of the medium, respectively.

3.12 Phase Relationships: Absorption, Emission, and Dispersion

What is the appropriate relationship between the phase of the driving field, **E**, and the dipole moment being driven, $\mathbf{p}_E$? It may be remembered that this question motivated the above digression. The mathematics of this problem is most conveniently handled by the use of the complex number notation introduced in Chapter 2. (See Eqs. 2.38, 2.39, 2,42, and 3.1.)

Complex numbers sometimes seem very mysterious to those not accustomed to their use. For the moment, it is sufficient to consider a complex number, a, as merely an ordered pair of real numbers, Re(a) and Im(a). In particular, Eq. 3.30 represents

$$\mathrm{Re}(E) = |E| \cos[\phi + \omega(\alpha_0 z - t)] \tag{3.54}$$

and

$$\mathrm{Im}(E) = |E| \sin[\phi + \omega(\alpha_0 z - t)]. \tag{3.55}$$

The phase constant, ϕ, may be set equal to zero in Eqs. 3.54 and 3.55 without loss of generality in this case. Because E is a single quantity, in principle measurable in the laboratory, only one of the two expressions in Eqs. (3.54) and (3.55) can have physical significance. Conventionally, Re(E) is chosen: Im(E) would serve just as well, provided only that one were consistent throughout the calculation.

If Eqs. (3.26) and (3.30) is substituted into Eq. (3.15), keeping only the second term of the latter, the following expression for the oscillating polarization is obtained:

$$P = \varepsilon_0(\eta' + i\eta'')|E|\{\cos[\omega(\alpha_0 z - t)] + i\sin[\omega(\alpha_0 z - t)]\}. \tag{3.56}$$

To be consistent, Re (P) must be chosen as the physically observable polarization. Therefore, from Eq. (3.56),

$$\mathrm{Re}(P) = \varepsilon_0|E|\{\eta' \cos[\omega(\alpha_0 z - t)] - \eta'' \sin[\omega(\alpha_0 z - t)]\}. \tag{3.57}$$

From Eqs. (3.54) and (3.57) it can be seen that, whenever matter is polarized by an oscillating electric field, the resultant polarization wave is shifted in phase with respect to the driving field. The phase shift can be described by associating it with the real and imaginary parts of the electric susceptibility. The part of the polarization wave in phase with the driving field is proportional to η'; the part of the polarization wave in quadrature (90° out of phase) with the driving field is proportional to $\eta''[\cos(\theta + \pi/2) = -\sin\theta]$. The use of complex numbers is then merely a device to enable one to describe two properties of the polarizable medium, η' and η'', by a single symbol, η.

All of the above considerations are summarized in Fig. 3.8. First, let the oscillations of the dipole moments of the atoms or molecules in the spectroscopic sample be 90° out of phase (in quadrature) with the oscillations of the fields of the light wave. This corresponds to the choice $\eta' = 0$, $\eta'' > 0$, and will, as has

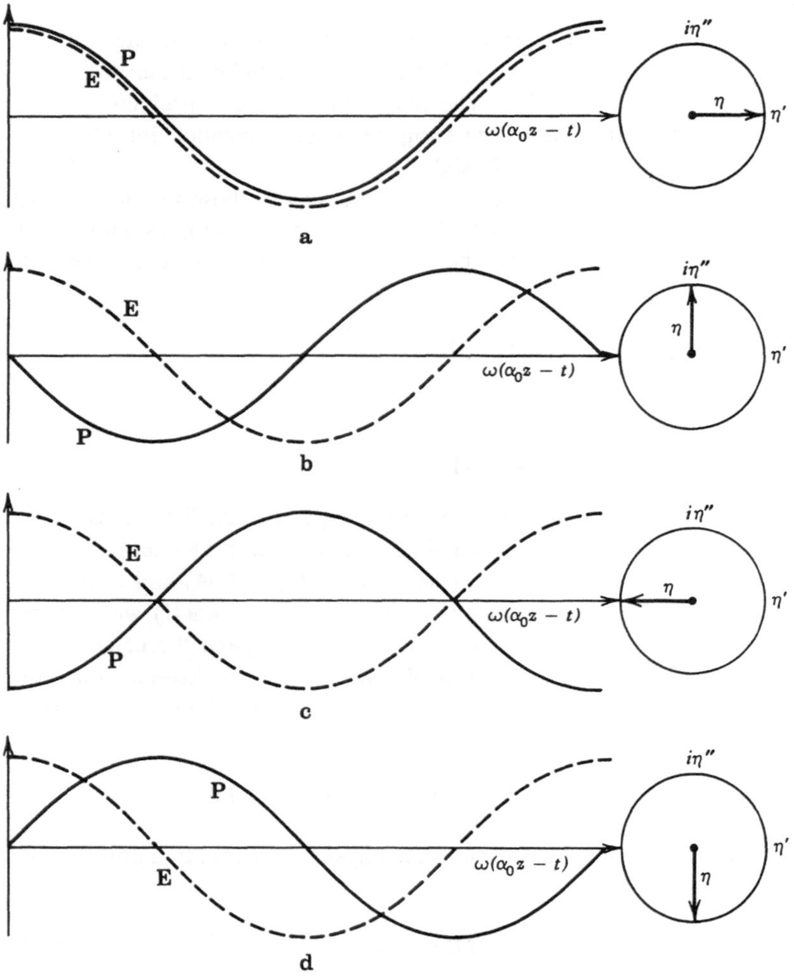

Fig. 3.8. Phase relationships between driving field and polarization waves for the following: **a** dispersion, **b** absorption, **c** anomalous dispersion, **d** stimulated emission

been shown, result in absorption of energy from the wave (see Eq. 3.52). The velocity of light through the medium will be the same as if no transitions were occurring in the medium (see Eqs. 3.48 and 3.49). The absorption coefficeint will be largest whenever the frequencies of the two oscillating systems are equal (Figure 3.8b).

Next, let the oscillations of the dipole moment be 0° out of phase (in phase) with the field oscillations. This corresponds to the choice $\eta' > 0$, $\eta'' = 0$. There will be no absorption (see Eq. 3.52), but there will be a marked reduction in the phase velocity of the light wave (see Eqs. 3.48 and 3.49). The index of refraction is frequency dependent because ε is frequency dependent, as stated in Sect. 3.10. When a light ray crosses the boundary between two substances having different refractive indices, it is bent (refracted) by an amount depending on the angle between the ray and the normal to the boundary, and also on the ratio of the n_R's. This is why a beam of white light (a mixture of waves with different frequencies) obliquely incident upon a prism is dispersed (separated) into rays of different colors (the constituent frequencies). For this reason, the dependence of the refractive index on frequency is sometimes called *dispersion* (Fig. 3.8a).

Next, let the dipoles oscillate 270° out of phase with the fields. This corresponds to $\eta' = 0$, $\eta'' < 0$. Inspection of Eq. 3.52 shows that, whenever this condition occurs, light will be amplified by the medium; the process is called *stimulated emission.* Just as was the case for absorption, the phase velocity of the wave is unaffected by the transition in question, and the magnitude of η'' depends on the frequency differences between the oscillators (Fig. 3.8d).

Finally, consider the case in which the dipoles are 180° out of phase with the driving fields. This corresponds to $\eta' < 0$, $\eta'' = 0$. Again, no energy will be transferred to or from the quantum system. The phase velocity of the wave will increase (see Eqs. 3.48 and 3.49) beyond the vacuum value, c_0. This does not violate the relativistic requirement concerning the velocity of light, which restricts only the group velocity (velocity of the center of the wave packet). The frequency dependence of the magnitude of η' reverses the order of the colors dispersed by a prism made of such a material. For this reason, this phenomenon is called *anomalous dispersion* (Fig. 3.8c).

It should be noted that "frequency difference," as used in the above paragraphs, refers to the difference between the *actual* frequency of the wave, ω, and the *natural* frequency of the dipole oscillators, ω_0. The acutal frequency of the dipole oscillators is the same as that of the driving field, ω (see Eq. 3.57). This will be true whenever the assumption of a linear relationship between E and P is valid (linear optics), save only for a small "pulling" effect due to the absorption (see the second term in square brackets on the right-hand side of Eq. 3.43).

It may occur to the reader that, insofar as the *total* magnitude of the dipole moment of an atom or molecule is fixed by quantum mechanics, the vector sum of the components of that dipole moment in phase and in quadrature should be constant. This implies that there ought to be a relationship between η' and η'' analogous to the Pythagorean theorem. Such a relationship, in fact, exists; it is called the *Kramers-Kronig relation.*

The reader may also ask, "Why does the most efficient energy transfer in spectroscopy occur when the oscillators are in quadrature, rather than in phase, the latter being the case for the swing and its pusher?" It will be seen later that this is so because of the torque developed is proportional to the vector (cross) product of the dipole moment and field vectors. It is well known that the cross product of two vectors is a maximum when the vectors are perpendicular.

The discussion of electromagnetic theory in this chapter is by no means exhaustive. It was intended merely to provide a review of the elementary ideas from that subject which will be required in subsequent chapters. A much fuller understanding can be gained by reading a good textbook on the subject [1].

3.13 References

1. Stratton JA (1941) Electromagnetic theory. McGraw-Hill, New York
2. Heitler W (1954) The quantum theory of radiation, 3rd edn. Clarendon Press, Oxford, p. 177
3. Glauber RJ (1963) Phys Rev 131: 2766
4. Jaynes ET, Cummings FW (1963) Proc. IEEE 51: 89
5. Price PB, Shirk EK, Osborne WZ, Pinsky LS (1975) Phys Rev Lett 35: 487

3.14 Problems

3.1. At a pressure of 1.0 torr. the unsaturated absorption coefficient, β_0, for SF_6 gas is 46 per meter for irradiation at a wavelength $\lambda = 10.6\ \mu$m. The index of refraction is $n_R = 1.00$, and the transition responsible for the absorption is electric dipole allowed.
a. Calculate the real part of the electric susceptibility of SF_6.
b. Calculate the imaginary part of the electric susceptibility of SF_6.

3.2. The electric field associated with a CO_2 laser beam ($\lambda = 10.6\mu$m) depends on distance and time in accordance with the expression

$$\mathbf{E} = \hat{x}|E|\mathrm{Re}\{\exp[i\omega(\alpha_0 z - t)]\}$$
$$= \hat{x}|E|\cos[\omega(\alpha_0 z - t)].$$

When this beam propagates through a sample characterized by an electric susceptibility

$$\eta = \eta' + \eta'',$$

it produces a polarization wave, $\mathbf{P}$:

$$\mathbf{P} = \hat{x}\varepsilon_0|E|\mathrm{Re}\{\eta\ \exp[i\omega(\alpha_0 z - t)]\}$$
$$= \hat{x}\varepsilon_0|E|\cos[\phi + \omega(\alpha_0 z - t)].$$

a. Write an expression for the phase shift of the polarization wave, ϕ, in terms of η' and η''.
b. Using the numerical values of η' and η'' calculated for SF_6 in Problem 3.1. calculate the numerical value of ϕ.

4 Interaction of Radiation and Matter

J.D. Macomber

4.1 Dipoles and Waves: the Semiclassical Theory

In Chap. 1 it was stated that atoms, ions, molecules and other quantum systems can absorb and emit electromagnetic radiation. Quantum-mechanical methods for calculating the properties of quantum systems were described in Chap. 2. In Chap. 3 it was explained that the most important method for the exchange of energy between radiation and matter is the exertion of torques by the uniform fields of the former upon the dipole moments of the latter. The purpose of this chapter is to link together the material presented in the first three chapters to produce a unified picture of spectroscopic transitions. Because the dipole moments will be calculated by means of quantum mechanics and the radiation fields by means of classical electromagnetic theory, the resultant description of the interaction process is called the *semiclassical* (or sometimes *neoclassical*) *theory of quantum transitions* [1].

It has been shown that the *sine qua non* of a spectroscopic transition is the existence of an electric or magnetic dipole moment, nonzero in magnitude, oscillating sinusoidally in time at a frequency very nearly equal to that of the electromagnetic wave. The question that arises is, "What dipole moment?"

Consider, for example, the hydrogen atom. In the Bohr theory, described in Chap. 1, the electron was supposed to move in an elliptical orbit around the nucleus. At each instant of time, the electron and the nucleus form an electric dipole of magnitude $q_0 r$, where $\mathbf{r}$ is the vector joining the two particles. As time passes, the dipole changes direction and magnitude periodically because the ellipse is a closed curve. Thus orbital motion of the electron generates an oscillating dipole moment. Furthermore, the frequencies of oscillation calculated from the Bohr theory are in the right range to explain the observed spectrum whenever $r \cong a_0$ (see Eq. 3.8). But it was explained in Chap. 1 how the acceleration of the charged electron as the dipole oscillates would necessarily produce the complete collapse of the hydrogen atom (and, by extension, all other atoms and molecules as well).

How is this problem resolved by the new quantum theory of Heisenberg and Schrödinger? In this theory, the energy and the electric dipole moment of the hydrogen atom are incompatible variables, in the sense described in Chap. 2; that is to say, the Hamiltonian operator H and the position operator r do not commute with one another in Hilbert space. At the beginning and the end of the transition process, the system is in an eigenstate of the energy (represented by H), and therefore is in a superposition state of the dipole moment (represented by $q_0\mathsf{r}$). If any attempt is made to measure the dipole moment of the atom in either the initial or the final state, the measurement will produce a definite result – the electron will be found *some*where. But *where* it will be found is unpredictable because of the uncertainty principle. Dipoles measured in this way are sometimes called "instantaneous" dipoles.

The best that can be done, then, is to calculate the average dipole moment obtained if one were to perform a large number of measurements upon the atom. Between each measurement of the dipole, it will be necessary to return the quantum system to the same initial state (i.e., an eigenstate of the energy). This average was named the "expectation value" in Chap. 2.

4.2 The Transition Dipole Moment of a Hydrogen Atom

Consider, for example, the first line in the Lyman series of the hydrogen spectrum (during which the atom undergoes a transition from the $1s$ to the $2p$ state). Is the expectation value of the electric dipole moment of a hydrogen atom in the $1s$ state nonzero? If so, will it oscillate at the correct frequency to explain the observed spectrum? The dipole moment operator is a vector quantity with three components (see Eq. 2.12):

$$\mathsf{p}_E = q_0\mathsf{r} = q_0(\mathsf{X}\hat{x} + \mathsf{Y}\hat{y} + \mathsf{Z}\hat{z}). \tag{4.1}$$

The expectation value of the x component may be calculated by means of Eq. 2.82:

$$\langle q_0 x\rangle_{11} \equiv q_0 \int d\mathbf{r}\psi^*_{1s}(\mathbf{r},t)\mathsf{X}\psi_{1s}(\mathbf{r},t), \tag{4.2}$$

where

$$\mathsf{X} = r\sin\theta'\cos\phi' \tag{4.3}$$

and

$$d\mathbf{r} = r^2 dr \sin\theta' d\theta' d\phi'. \tag{4.4}$$

(Primed symbols are used because the corresponding unprimed ones have been reserved for different quantities.) Any standard textbook on elementary quantum mechanics can supply the eigenfunctions of the Hamiltonian operator for hydrogen:

$$\psi_{1s}(\mathbf{r},t) = (\pi a_0^3)^{-1/2}\exp\left(\frac{-i\mathscr{E}_1 t}{\hbar}\right)\exp\left(\frac{-r}{a_0}\right). \tag{4.5}$$

where

$$a_0 = \frac{4\pi\varepsilon_0\hbar^2}{m_0 q_0^2}. \tag{4.6}$$

The eigenvalue of the energy, $\mathscr{E}_1$, was given in Eq. 1.19:

$$\mathscr{E}_1 \cong \mathscr{E}_{\text{atom}}. \tag{4.7}$$

When Eqs. (4.2)–(4.7) are combined, the resultant integration shows that

$$\langle q_0 x\rangle_{11} = 0. \tag{4.8}$$

The same result would have been obtained for $\langle q_0 y\rangle$ and $\langle q_0 z\rangle$. This explains why no energy is radiated by the hydrogen atom while it is in the $1s$ state, and therefore the electron does not spiral into the nucleus as described in Chap. 1. The collection of "instantaneous dipoles" associated with the $1s$ eigenfunction has a net vector sum of zero.

Similarly, it may be shown that the expectation value of the electric dipole moment of a hydrogen atom in its $2p_x$ state is zero. For example,

$$\langle q_0 x\rangle_{22} \equiv q_0 \int d\mathbf{r}\, \psi_{2p_x}^*(\mathbf{r},t)\mathbf{X}\,\psi_{2p_x}(\mathbf{r},t) = 0, \tag{4.9}$$

where

$$\psi_{2p_x}(\mathbf{r},t) = (32\pi a_0^5)^{-1/2}\exp\left(\frac{-i\mathscr{E}_2 t}{\hbar}\right)\exp\left(\frac{-r}{2a_0}\right) r\sin\theta'\cos\phi' \tag{4.10}$$

and

$$\mathscr{E}_2 = \frac{\mathscr{E}_1}{2^2}. \tag{4.11}$$

Equation (4.9) explains why the $2p_x$ state of hydrogen is also one of the stationary states: like the $1s$ state, it lacks a persistent electric dipole moment and therefore cannot generate or interact with electromagnetic waves. ("Persistent" is used here to mean only a nonzero expectation value; a persistent dipole is to be contrasted with an instantaneous one. The term "static" is reserved for a quantity that does not oscillate in time.)

The only thing that has not yet been calculated is the expectation value of the electric dipole moment of the hydrogen atom in a nonstationary state formed by the linear combination of $1s$ and $2p_x$ states (the $1s$ and $2p_x$ superposition state):

$$\Psi(\mathbf{r},t) = c_{1s}\psi_{1s}(\mathbf{r},t) + c_{2p_x}\psi_{2p_x}(\mathbf{r},t). \tag{4.12}$$

In accordance with Eq. (2.83),

$$\langle q_0 x\rangle_{\Psi\Psi} = \langle q_0 x\rangle_{11}\cos^2\left(\frac{\theta}{2}\right) + \langle q_0 x\rangle_{22}\sin^2\left(\frac{\theta}{2}\right)$$
$$+2\cos\left(\frac{\theta}{2}\right)\sin\left(\frac{\theta}{2}\right)\mathrm{Re}\{\langle q_0 x\rangle_{12}\exp[i(\S - \omega_0 t)]\}. \tag{4.13}$$

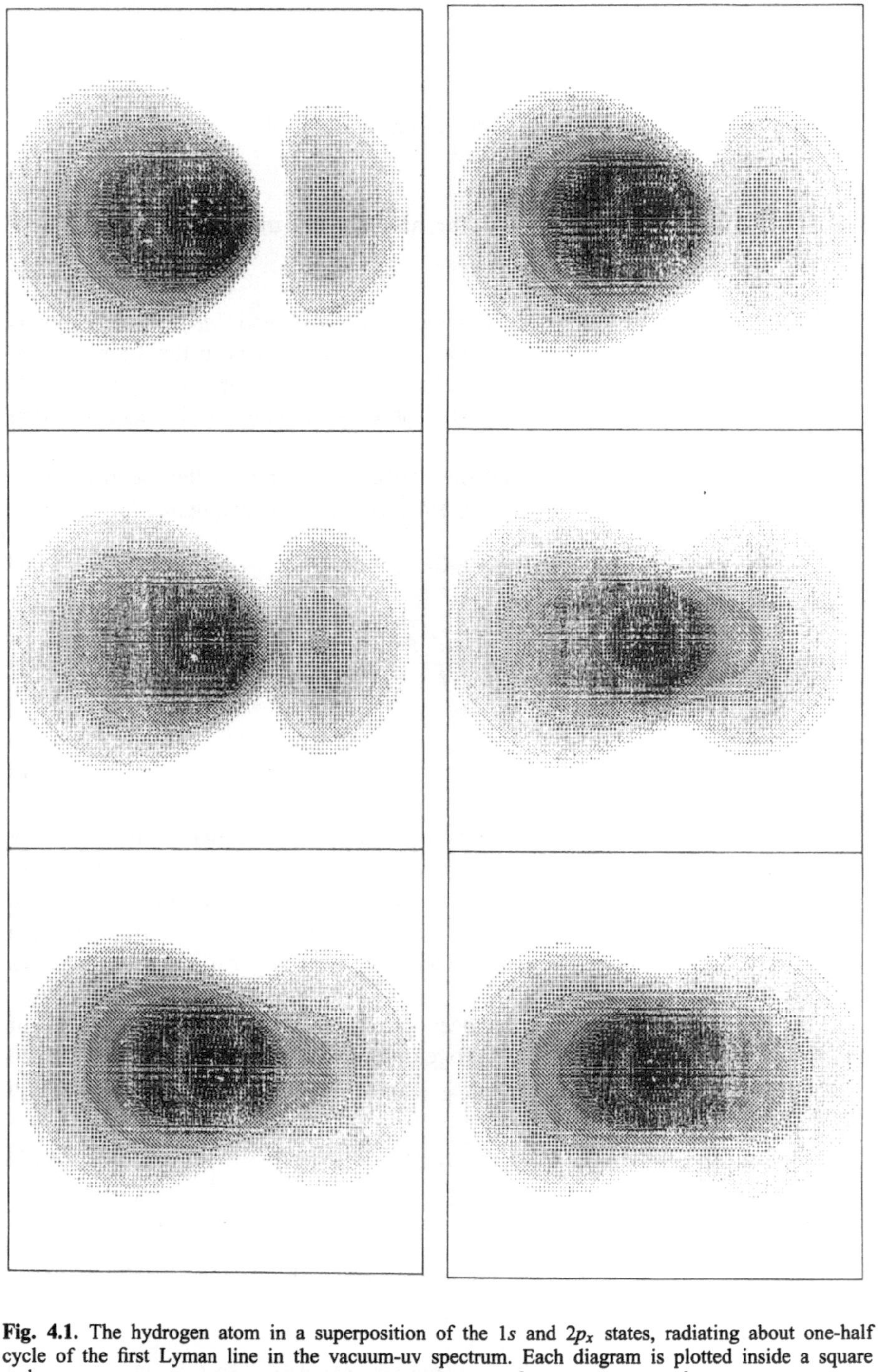

Fig. 4.1. The hydrogen atom in a superposition of the 1s and 2p_x states, radiating about one-half cycle of the first Lyman line in the vacuum-uv spectrum. Each diagram is plotted inside a square $13\frac{1}{3}a_0$ (Bohr radii) on a side. The probability density, $|\psi|^2$ (unit volume= a_0^3), in the xy plane is

From Eqs. 4.8 and 4.9, Eq. 4.13 reduces to

$$\langle q_0 x\rangle_{\Psi\Psi} = q_0 \sin\theta \cos(\S - \omega_0 t)\langle x\rangle_{12}, \tag{4.14}$$

where

$$\omega_0 = \frac{3m_0 q_0^4}{8(4\pi\varepsilon_0)\hbar^2}. \tag{4.15}$$

Also, from Eqs. 4.5 and 4.10,

$$\langle x\rangle_{12} = \frac{256}{243}\frac{a_0}{\sqrt{2}}. \tag{4.16}$$

It just so happens that the angular frequency of the electromagnetic wave radiated by hydrogen atoms in producing the first spectral line of the Lyman series in emission or absorption is correctly given by Eq. (4.15). Furthermore, the maximum magnitude of the dipole moment, which occurs when $\theta = 90°$ ($c_{1s} = c_{2p} = 1/\sqrt{2}$), implies a certain intensity for the resultant emission or absorption. The predicted and observed intensities also agree, within the limits of experimental uncertainties.

The nonzero expectation value of the electric moment of hydrogen in this superpostion state is therefore precisely the oscillating dipole that is called for in the semiclassical theory of quantum transitions. It may be termed the *transition dipole moment*.

Finally, it should also be noted that, as the emissive transition proceeds, c_{2p} decreases from 1 to 0. As soon as c_{2p} equals 0, the product $c_{1s}^* c_{2p} = \frac{1}{2}\sin\theta \exp[i(\xi_{2p} - \xi_{1s})]$ in Eq. 4.14 becomes 0 and the radiating dipole disappears. When two oscillators at different frequencies (such as adjacent piano strings) are excited simultaneously, one hears, in addition to the frequencies of the oscillators themselves, an interference "beat" note. The oscillation of the dipole moment of the superposition state is just such a beat. In an atomic emitter that has ceased

represented by dot density, d(unit area $= a_0^2$), according to

$$\begin{aligned}|\psi|^2 &= |\psi_{1s}(t) + \psi_{2p_x}(t)|^2\\ &= \frac{\exp[(15/4)^2 d]}{\pi\exp(13/2)}\\ &= 0.000479\exp(14.1d)\end{aligned}$$

The time intervals, Δt, between successive diagrams are chosen so that

$$\frac{[\mathscr{E}(2p_x) - \mathscr{E}(1s)]\Delta t}{\hbar} = \pi/6,$$

where

$$\Delta t = 3.38 \times 10^{-17}\ \text{s}$$

This figure was prepared with the assistance of Henry Streiffer using the facilities of the LSU computer center radiating, one of the two oscillators has died out, leaving only the other (the fundamental) – consequently, the beat note disappears. Throughout this process, the frequency of the beat note is constant, although its amplitude is continually changing, because the frequencies of the fundamental and harmonic that have interfered to produce it are constant. The electronic "motion", initially described by one standing wave pattern, has been converted into another standing wave pattern of a lower frequency. When standing wave oscillations of two different frequencies are simultaneously present in the same oscillator (as they are in the radiating atom), the interference beat note appears as a multiply reflected traveling wave. The traveling wave provides the mechanism by means of which the exess energy is transferred to the surroundings. A drawing of the hydrogen atom in the superposition state is presented in Fig. 4.1.

The Bohr frequency condition

$$\omega = \frac{(\mathscr{E}_2 - \mathscr{E}_1)_{\text{atom}}}{\hbar}, \tag{4.17}$$

was rationalized (in Chap. 1) in terms of thermodynamic necessity: the energy of the created (or annihilated) photon, in accordance with the Law of Conservation of Energy, must equal the amount of energy lost (or acquired) by the atom in the process. A kinetic interpretation of this, promised in Chap. 1, has just been presented. The frequency of electromagnetic radiation emitted by a dipole antenna must be precisely the frequency with which electric charges oscillate in that antenna. In absorption, the necessity of doing work by means of fields that exert a torque upon an oscillating dipole requires that the fields in question oscillate at the same frequency as the dipole being driven. It was pointed out in Chap. 3 that the driving field must "chase after" the driven dipole, so that the torque will always be exerted in the same sense. If there is a frequency difference between the two oscillators, the phase relationship that exists between them at any one time will reverse after a time interval of $(2\pi\Delta\omega)^{-1}$ seconds. Then the work done during one portion of the optical cycle will be undone in the subsequent portion (see Eq. 3.13).

The material of the preceding paragraphs constitutes a physical model for the dynamic interaction between radiation and matter in spectroscopic transitions; it may be summarized as follows. During the act of radiation or absorption of light by matter, electromagnetic waves constitute fields that are uniform in space but oscillating in time. These fields exert torques upon the oscillatory dipole moments (more precisely, upon the oscillatory expectation values of dipole moments) of quantum systems (e.g., atoms and molecules, or chromophoric portions of the latter). These dipole moments arise only if the systems are in states represented by a linear superposition of the nondegenerate eigenfunctions associated with the initial and final states. Net transfer of energy from one oscillating system to another (e.g., from the field to the matter) depends on the magnitudes of the corresponding fields and dipoles, the differences in their phases, and the departures of their frequencies from the natural (i.e., quantum) beat frequency of the superposed eigenstates.

4.3 Conceptual Problems with the Theory

The model of spectroscopic transitions just presented has some features that are easily grasped by physical intution, and it yields results in accord with experiments. The explanation does not rely on words such as *interaction, annihilation,* and *creation*. In subsequent chapters, it will be shown that this model also provides explanations in satisfactory agreement with experiment for even finer details of the absorption spectrum (e.g., line-widths), as well as for "exotic" phenomena (e.g., photon echoes and self-induced transparency).

It should be noted, however, that in the present status of the discussion several very important matters remain unexplained. For example, how does the transition process get started? Reconsider the problem of the Lyman spectrum of hydorgen. The initial state in emission will be $2p$; c_{1s} will be zero. From Eq. 4.13 it can be seen that there will be no oscillating dipole moment to generate the initial radiation field. The coefficient c_{2p} will be zero in the same formula in the case of absorption. There will therefore be no electromagnetic "handle" for the radiation fields to "grab" at the beginning of the absorption process. This difficulty could be resolved if every quantum system possessed a static dipole moment in the $\hat{z}$ direction while in an eigenstate. It would have to be static for the state to be stationary and to satisfy the requirement that oscillating portion of **P** lie solely in the xy plane. This $\hat{z}$ component could be an initial "handle" on the quantum system with which the **E** field could interact. In magnetic resonance spectroscopy, such a component of a dipole moment actually exists. It is, of course, magnetic rather than electric; for spin-$\frac{1}{2}$ particles it points in the z direction in one eigenstate (spin up) and in the $-z$ direction in the other (spin down).

In the case of electronic transitions, there is seldom a static $\hat{z}$ component of a dipole moment (cf. the case of hydrogen). In Chapter 10, it will be shown that a generalized transition dipole moment can be defined by analogy with the actual magnetic moment of an electron or a proton. The x' components of these moments give rise to the genuine "in-phase" oscillating component of **P** (dispersion); the y' components, to the genuine "in-quadrature" oscillating component of **P** (absorption); and the static z components, to a fictitous moment proportional to the difference between the squares of superposition coefficients, $|c_2|^2-|c_1|^2$. It will be shown in Chapter 10 that the sum of these three moments moves like a vector under the influence of the electromagnetic wave with which it interacts, allowing use of the entire formalism developed for magnetic resonance spectroscopy (where all three components are genuine dipoles).

One serious conceptual difficulty with this model can and should be addressed immediately – that of the relationship between the uncertainty principle and superposition states. As was noted in Chap. 2, people are sometimes confused about superposition states. Measurements performed on quantum systems yield only eigenvalues as results. This fact leads some persons to conclude that atoms, ions molecules, and so forth can exist only in the eigenstates which correspond to these eigenvalues. If this view is held, it is very difficult to explain the dynamics of spectroscopic transitions. Frequently it is asserted that systems change from one eigenstate of the energy to another by means of discontinuous processes called *quantum jumps*. Furthermore, the details of the jumping process are supposed to be fundamentally unknowable because of the uncertainty principle.

The point of view heretofore expressed in this book seems to be very different from the one that produced the idea of quantum jumps. Superposition states have been taken very seriously–indeed, they have been considered every bit as genuine eigenstates. The dynamics of the transition process have been explained by means of a model in which the electromagnetic waves operate upon the quantum system in a continuous fashion. The fields do work upon dipole moments that have nonzero expectation values only for superposition states. Indeed, there apparently have been no jumps at all in this picture.

How are these two points of view to be reconciled? What is the meaning of the superposition coefficients? How is the notion of a continuous process compatible with the discontinuties in the eigenvalues? *Are* there quantum jumps? How does the uncertainty principle affect the results of measurements performed on systems in superposition states? It is the purpose of subsequent sections of this chapter to answer as many of these questions as possible.

4.4 Quantum Jumps and the Uncertainty Principle

The first step in discussing these problems is to describe the transition of a quantum system from one eigenstate of the energy to another in a very careful way. The initial state will be numbered "1", and the final state, "2":

$$\mathsf{H}_0|1\rangle = \mathscr{E}_1|1\rangle, \tag{4.18}$$

$$\mathsf{H}_0|2\rangle = \mathscr{E}_2|2\rangle, \tag{4.19}$$

In accordance with the usual notation, $|1\rangle$ and $|2\rangle$ are orthogonal eigenkets of the Hamiltonian operator, H_0, with corresponding eigenvalues $\mathscr{E}_1$ and $\mathscr{E}_2$. The system will be prepared in such a way that it is definitely in state 1 at times $t < 0$ (something that can always be done).

At $t = 0$, a perturbation of constant "strength" V will be applied to the quantum system. The precise meaning of V will be discussed later. For the moment it will be sufficient to note that V has the dimensions of energy and is associated with the process of producing a transition of the system from state 1 to state 2 in a time Υ. In other words, the perturbation represented by V is switched off at $t = \Upsilon$, and for all $t > \Upsilon$ the system is definitely in state 2. Not all perturbations have this nice property of being able to complete their mission in a definite time interval, of course. (The ones described as "coherent," however, generally have this property.)

In the interval $0 \leqslant t \leqslant \Upsilon$, the state of the system is represented by a ket that will be labeled as $|A\rangle$, a linear superposition of the two relevant eigenkets of $\mathbf{H}_0$:

$$|A\rangle = c_{A1}|1\rangle + c_{A2}|2\rangle. \tag{4.20}$$

The coefficients c_{A1} and c_{A2} are continuous functions of the time. The coefficient c_{A1} goes from 1 to 0 in the interval $0 \leqslant t \leqslant \Upsilon$, and the coefficient c_{A2} goes from 0 to 1 in the same interval. At all times,

$$|c_{A1}|^2 + |c_{A2}|^2 = 1. \tag{4.21}$$

Equation 4.21 is required so that the superposition ket $|A\rangle$ will be properly normalized (see Eq. 2.70). As was shown in Chap. 2, Eq. 4.21 will be satisfied automatically by the choices

$$c_{A1} = \cos\left(\frac{\theta}{2}\right)\exp(i\xi_1) \tag{4.22}$$

and

$$c_{A2} = \sin\left(\frac{\theta}{2}\right)\exp(i\xi_2). \tag{4.23}$$

See Eqs. 2.71 and 2.72. It will be shown in Chapter 8 that

$$\theta = \frac{VT}{\hbar} = \gamma H_1 t \tag{4.24}$$

Perturbations which satisfy all of the above conditions are usually called 180° or π pulses (note what happens when $\theta = 180°$ in Eqs. 4.22 and 4.23). In addition to the previous assumption that relaxation can be neglected, it will also be assumed that the amount of the perturbation is "small":

$$|V| \ll |\mathscr{E}_2 - \mathscr{E}_1|. \tag{4.25}$$

It was noted earlier that a measurement of the energy of the system in the interval $0 < t < \Upsilon$ will give either the result $\mathscr{E}_1$ or $\mathscr{E}_2$. In particular, no value of the energy intermediate between $\mathscr{E}_1$ and $\mathscr{E}_2$ will ever be observed. Furthermore, the probability that $\mathscr{E}_1$ will be observed is $|c_{A1}|^2$, and the probability of finding $\mathscr{E}_2$ instead is $|c_{A2}|^2$. It was also mentioned that two conclusions are sometimes drawn from these facts. First, the quantum system proceeds from state 1 to state 2 by means of a quantum jump at some time in the interval $0 \leqslant t \leqslant \Upsilon$. Second, the uncertainty principle prevents an accurate prediction of the time at which the jump will occur. One may only state that $|c_{A1}(t)|^2$ is the probability that it will have already occured before the time in question. Quantum jumps of this kind will henceforth be called *Schrodinger jumps*, due to Schrödinger's criticism [2] of pictures of the transition process based upon them.

It is easy to see from this discussion why believers in Schrödinger jumps are likely to consider that the only genuine states of a quantum system are the eigenstates of the energy (in this case those represented by $|1\rangle$ and $|2\rangle$). In this view the superposition state has only a formal validity, in that it is a solution to Schrödinger's equation and identifies the initial and final states. Therefore c_1 and c_2 in Eq. 4.20 are viewed as "information variables." The actual state of the system is either 1 or 2, and there is no way of telling which; all that can be done is to measure the energy of the system and, in this way, remove the ambiguity.

The alternative interpretation of the facts about quantum transitions (i.e., the failure to observe any energy between $\mathscr{E}_1$ and $\mathscr{E}_2$ experimentally) begins with the assumption that the superposition state is a genuine state of the system. More precisely, in the process of undergoing the transition from state 1 and state 2, the system passes through a continuum of superposition states, designated collectively by $|A\rangle$. Each one of these is different from all the others and, in particular, different from the eigenstates represented by $|1\rangle$ and $|2\rangle$. Each of these A states is supposed to be a very bit as "real" as states 1 and 2. One of the properties shared by all the superposition states is that the eigenkets $|A\rangle$ which represent them are not eigenkets of the energy. This is the reason why no definite value between $\mathscr{E}_1$ and $\mathscr{E}_2$ is ever obtained when the property is measured in the interval $0 < t < \Upsilon$. Instead, *the act of measurement itself* precipitates a jump into either state 1 or state 2. This type of quantum jump was described by Heisenberg [3] in a reply to Schrödinger and will therefore henceforth be designated as a *Heisenberg jump*. The probability that a system in a superposition state will jump to stationary state 1 at time t is $|c_{A1}(t)|^2$; the probability that it will jump to state 2 instead is $|c_{A2}(t)|^2$.

Since both of these models, portrayed schematically in Fig. 4.2, predict results in complete accordance with the experiments described thus far, it might be thought that the distinction between Schrödinger and Heisenberg jumps is a meaningless one. It is possible, however, to imagine an experiment that would distinguish between these two kinds of quantum jumps. At the same time, the properties of the superposition states would be elucidated.

Suppose that $|A\rangle$ in Eq. 4.20 is an eigenket of an operator R representing some physical property of the system *other than the energy*, with corresponding eigenvalue r_A:

$$\mathrm{R}|A\rangle = r_A|A\rangle. \tag{4.26}$$

Furthermore, it will be supposed that there exists another eigenstate of R, represented by $|B\rangle$ with corresponding eigenvalue $r_B \neq r_A$:

$$\mathrm{R}|B\rangle = r_B|B\rangle. \tag{4.27}$$

Finally, it will be supposed that $|B\rangle$, like $|A\rangle$, can be represented as a linear combination of $|1\rangle$ and $|2\rangle$. In particular, suppose that

$$|B\rangle = -c_{A2}^*|1\rangle + c_{A1}^*|2\rangle, \tag{4.28}$$

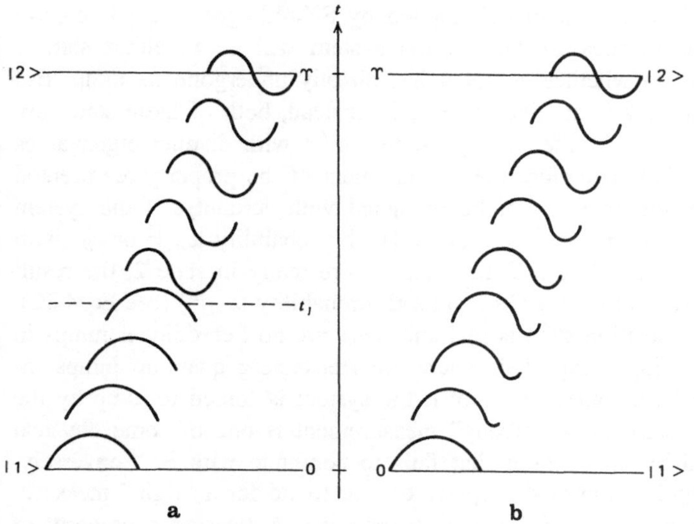

Fig. 4.2. Two different models for the change of a wavefunction with time during a quantum transition. **a** System hops abruptly from state 1 to state 2 at time t_j (Schrödinger jump); the time t_j cannot be predicted with certainty. **b** System evolves gradually from state 1 to state 2, as reflected by the continuous deformation of its wavefunction. No quantum jumps occur unless the transition process is interrupted by an energy measurement (Heisenberg jump)

where the coefficients c_{A1}^* and c_{A2}^* are the complex conjugates of those that appear in Eq. 4.20. It is easy to show that $|B\rangle$ is orthogonal to $|A\rangle$ and normalized to unity (see Eq. 4.21):

$$\langle B|A\rangle = 0 \tag{4.29}$$

and

$$\langle B|B\rangle = 1. \tag{4.30}$$

The coefficients for $|B\rangle$ in Eq. 4.28 have been chosen so that

$$|1\rangle = c_{A1}^*|A\rangle - c_{A2}|B\rangle \tag{4.31}$$

and

$$|2\rangle = c_{A2}^*|A\rangle + c_{A1}|B\rangle. \tag{4.32}$$

If a physical property represented by such an operator as R can be found (and if a device can be constructed in the laboratory to measure it), measurement of that property at time t will decide between Schrödinger and Heisenberg jumps. Suppose that the quantum model implied by Heisenberg jumps is the correct one. At time t, the system will be in the legitimate state A. Measurements of the property represented by R will therefore yield the result r_A, and never r_B.

Now suppose that the quantum model implied by Schrödinger jumps is correct instead. At the time of measurement, t, the system will be in either state 1 or state 2, according to whether or not it has already undergone its jump. But neither state 1 nor state 2 is an eigenstate of R. Instead, both of these states are linear combinations of two different eigenstates of R with distinct eigenvalues (see Eqs. 4.31 and 4.32). Therefore the measurement of the property represented by R will yield a result that cannot be predicted with certainty. If the system were really in state 1, the result would be r_A (with probability $|c_{A1}|^2$ or r_B (with probability $|c_{A2}|^2$) (see Eq. 4.31). If the system were really in state 2, the result would be r_A (with probability $|c_{A2}|^2$) or r_B (with probability $|c_{A1}|^2$) (see Eq. 4.32).

In fact, the superposition state is real and there are no Schrödinger jumps in nature, just as Schrödinger expected. There are Heisenberg quantum jumps instead, just as Heisenberg stated, but only if the system is forced to jump by the wrong kind of measurement. A "wrong" measurement is one of some physical property represented by an operator that fails to commute with R. Conversely, if the appropriate kind of physical property can be found for a "right" measurement (e.g., one represented by R), it will be possible to measure a property of a system "in between" eigenstates of the Hamiltonian without being subject to the limitations of the uncertainty principle. The wavefunction is changing continuously with time because the coefficients c_1 and c_2 are changing, as a result of the influence of the perturbation represented by V. Therefore the operator R will be "right" only at a particular time t. At any other t, a different operator will be the right one. In this way, an infinite number of operators $\{\mathsf{R}\}_t$ (not one of which commutes with any other or with the Hamiltonian) may be defined; these could be used to map out the progress of the system throughout the course of the quantum transition. In the final sections of this chapter it will be demonstrated by means of NMR pulse methods etc. that Schrödinger jumps do not occur in nature and that, therefore, spectroscopic transitions should be described by the superposition of the eigenkets of the initial and the final state during a well-defined transtition time, i.e. with time-dependent coefficients, and that the transition process can be interrupted by an energy measurement, i.e. by a Heisenberg jump.

4.5 The Spin-$\frac{1}{2}$ System

The quantum system selected for study in the *Gedanken* experiment is the electrically neutral, spin $s = \frac{1}{2}$ particle. This system has been mentioned before in this book and will be discussed again, in great detail, in Chap. 8. For the purposes of the present discussion, it will be sufficient to anticipate a few of the results of Chap. 8.

Spin-$\frac{1}{2}$ particles behave as if they have magnetic dipole moments of magnitude $\sqrt{3}\gamma\hbar/2$, where the constant γ is characteristic of the kind of system considered (e.g., silver atom, proton, or neutron). Since a magnetic dipole moment is a vector quantity, it has a direction as well as magnitude. The most convenient method for the experimental determination of both the magnitude and the direction of a vector is to measure the projection of the vector on each of three mutually perpendicular Cartesian axes. The orientation of the unit vectors in these directions, $\hat{x}$, $\hat{y}$, and $\hat{z}$, will be considered fixed (laboratory reference frame). In Chap. 3, the magnetic moment vector was labeled by the symbol $\mathbf{p}_M$. Therefore,

$$\mathbf{p}_M = p_x\hat{x} + p_y\hat{y} + p_z\hat{z}. \tag{4.33}$$

Suppose that there are a very large number of identical systems of this type in the laboratory, and that the magnetic moment vectors of these systems are oriented at random. What should be expected if the x component of magnetic moment of every system were measured? In classical mechanics, a continuous distribution of components ranging in magnitude from $+\sqrt{3}\gamma\hbar/2$ (corresponding to systems that happened to have their magnetic moments pointing "up" at the time of measurement) to $-\sqrt{3}\gamma\hbar/2$ (corresponding to systems that happened to have their magnetic moments pointing "down" at the time of measurement) would be expected. In fact, classical mechanics does not provide an appropriate description of these systems. The actual result, explicable only by means of quantum mechanics, is that the measured value of p_x for half of the systems is $+\gamma\hbar/2$, and for the other half, $-\gamma\hbar/2$. Precisely the same results are obtained by measurements of p_y and p_z.

The *length* of a vector p_M with these components is correctly given by the three-dimensionsal analog of the Pythagorean theorem:

$$\sqrt{p_x^2 + p_y^2 + p_z^2} = p_M. \tag{4.34}$$

In classical mechanics, the *direction* of the magnetic moment vector $\mathbf{p}_M$ is defined to be the direction along which the component of magnetization equals its total length:

$$\mathbf{p}_M = \sqrt{3}\frac{\gamma\hbar\hat{u}_1}{2}. \tag{4.35}$$

The unit vector $\hat{u}_1$, together with two other mutually orthogonal unit vectors $\hat{u}_2$ and $\hat{u}_3$, defines a coordinate system that is most convenient for describing the behavior of the particle (particle reference frame). The orientation of the axes of the $\hat{u}_1, \hat{u}_2, \hat{u}_3$ system with respect to the axes of the fixed $\hat{x}, \hat{y}, \hat{z}$ system will vary from particle to particle. In quantum mechanics, Eq. (4.34) is satisfied:

$$\sqrt{(\pm\gamma\hbar/2)^2 + (\pm\gamma\hbar/2)^2 + (\pm\gamma\hbar/2)^2} = \frac{\sqrt{3}\gamma\hbar}{2}, \tag{4.36}$$

but Eq. (4.35) is not quantum-mechanically valid in *any* coordinate system. (This fact will be proved in Chapter 8.)

A completely new way must be found to define the direction of the magnetic moment. Suppose that the particle were bathed in a uniform static field of induction, **H** (the symbol **B** is not used in magnetic resonance):

$$\mathbf{H} = H_x\hat{x} + H_y\hat{y} + H_z\hat{z}. \tag{4.37}$$

The energy $\mathscr{E}$ of a magnetic dipole in a uniform field is given by a scalar product like that of Eq. (3.13):

$$\mathscr{E} = -\mathbf{p}_M \cdot \mathbf{H}. \tag{4.38}$$

In the other words, the energy of the system (purely magnetic energy in this case) will now depend on the orientation. In particular, suppose that

$$\mathbf{H} = H_0\hat{u}_1. \tag{4.39}$$

Now one can write

$$\mathbf{p}_M = p_{u_1}\hat{u}_1 + p_{u_2}\hat{u}_2 + p_{u_3}\hat{u}_3, \tag{4.40}$$

and the energy becomes (from Eq. 4.38)

$$\mathscr{E} = -p_{u_1}H_0. \tag{4.41}$$

Because of the quantum-mechanical nature of $\mathbf{p}_M$, there can only be two possible eigenvalues of p_{u_1}:

$$p_{u_1} = \frac{\pm\gamma\hbar}{2}. \tag{4.42}$$

Hence, there are only two eigenvalues of the energy:

$$\mathscr{E} = \frac{\pm\gamma\hbar H_0}{2}. \tag{4.43}$$

Suppose that the system is put into the state having the lower eigenvalue. This could be done by placing the particle in thermal contact with a heat bath at a low temperature;

$$k_0T \ll \gamma\hbar H_0 \tag{4.44}$$

(k_0 is Boltzmann's constant, and T is the absolute temperature) and allowing a sufficiently long time to elapse:

$$t \gg T_1 \tag{4.45}$$

(T_1 is the relaxation time; see Chapter 5). In this state,

$$p_{u_1} = \frac{+\gamma\hbar}{2}. \tag{4.46}$$

In other words, the component of the magnetic moment parallel to the $\hat{u}_1$ axis will be nearly certain; repeated measurements of that component will yield reproducible results.

The same cannot be said of the u_2 and u_3 components. If the particle is put into the magnetic eigenstate represented by the eigenvalue in Eq. 4.46 by cooling it in a magnetic field in the u_1 direction, a measurement of p_{u_2} will yield the result $+\gamma\hbar/2$ half of the time and the result $-\gamma\hbar/2$ the other half. These two values will occur at random, and the result of any single measurement will be completely unpredictable. The same thing will happen if p_{u_3} is measured.

It is understood that the spin-$\frac{1}{2}$ particles must be returned to their initial states between successive measurements of any one component of $\mathbf{p}_M$. This might be accomplished by allowing each of them to "soak" in the field $\mathbf{H}$ for a time $t \gg T_1$ while again in contact with the heat bath. (This precaution may not be necessary after measurements of p_{u_1} because such measurements need not change the state of the particle.)

4.6 A Geometrical Model of the Transition Process

Imagine a pair of concentric spheres, the inner one of radius $\gamma\hbar/2$ and the outer of radius $\sqrt{3}\gamma\hbar/2$, as shown in Fig. 4.3. The magnetic moment of the particle can be represented by a vector $\mathbf{p}_M$, with its tail at the center of the spheres, and its tip somewhere on the larger sphere. The poles of both spheres are their intersection points with the z axis. Any attempt to measure $\mathbf{p}_M$ by measuring its projection along any direction in space will produce a component vector with its tail at the center and its tip on the smaller sphere.

It is therefore possible to single out a particular direction in space (call it $\hat{u}_1$) and to put the system in a particular eigenstate of the magnetization along that direction. One way of doing this, as was just mentioned, is to put a magnetic field $\mathbf{H}$ along that direction and to let the system sit for a long time in thermal contact with a low-temperature heat bath. (There are other ways of accomplishing the same task, but they need not be described here.) Now, if p_{u_1} is measured, the tip of the component vector will certainly be found at the point where the u_1 axis intersects the small sphere. The possible locations of $\mathbf{p}_M$ consistent with this fact will describe a cone of half angle

$$\cos^{-1}\left(\frac{1}{\sqrt{3}}\right) = 54°44'8.2'', \tag{4.47}$$

and the axis of this cone will be the u_1 axis. Therefore the locus of the head of $\mathbf{p}_M$ will describe a circle where the cone intersects the large sphere, and that circle in turn will lie in a plane tangent to the small sphere at the head of p_{u_1}. Finally, the locus of possible outcomes of measurements of p_{u_2} and p_{u_3} is a great circle on the small sphere in a plane perpendicular to $\hat{u}_1$.

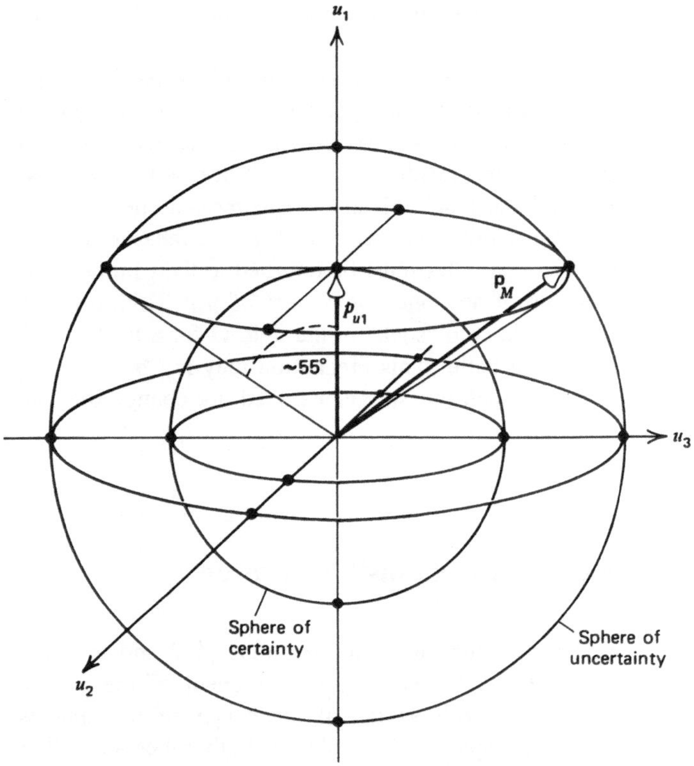

Fig. 4.3. Spheres of certainty and uncertainty for a spin-$\frac{1}{2}$ particle. The total magnetic moment vector $\mathbf{P}_M$ (of length $\sqrt{3}\gamma\hbar/2$) makes an angle of about 55° with its certain (measurable) component p_{u_1} (of length $\gamma\hbar/2$)

Since the only information availabe with certainty by direct experimental observation of this system can be represented by a component vector with its head on the small sphere, that sphere will be called the *sphere of certainty*. Since the exact location of the head of $\mathbf{p}_M$ on the larger sphere is uncertain, that sphere will be called the *sphere of uncertainty*. The difference in the radii of the two spheres is a consequence of the uncertainty principle, and is larger for spin-$\frac{1}{2}$ particles than for any other kind. [The ratio of the radii in the general case of a spin-s particle is $\sqrt{s(s+1)}/s$, and goes to 1 as s becomes very large. This fact is a consequence of the correspondence principle.] Also, p_{u_1} will be called the *certain component* of $\mathbf{p}_M$, since only that component can be measured many times in succession with reproducible results, and the u_1 axis will be termed the *axis of certainty*. (This axis is usually called the "axis of quantization," but since that term is sometimes also applied to the magnetic field axis, a new name has been invented for use in this discussion.)

After the static magnetic field has been used to line up p_{u_1}, it will be realigned with the z axis:

$$\mathbf{H} = H_0\hat{z}. \tag{4.48}$$

The result of this choice is that, whenever the system is in an eigenstate of the energy, the axis of certainty will be the $\hat{z}$ axis. Any measurement of p_z will yield a reproducible result. If the system is in the low-energy eigenstate, called 1, $p_z = \gamma\hbar/2$. If the system is in the high-energy eigenstate, called state 2, $p_z = -\gamma\hbar/2$. The results of the measurement of p_x and p_y (or the components along any other direction) will be indeterminate.

This information can now be used to visualize the difference between Heisenberg and Schrödinger jumps. Suppose that a spin-$\frac{1}{2}$ particle is put into state 1, and at time $t = 0$ the particle is suddenly embedded in a circularly polarized electromagnetic wave of frequency ω. In particular, suppose that the wave propagates along the $\hat{z}$ axis and has a spatially uniform magnetic amplitude H_1. (Since the dipole moment of the particle is purely magnetic, there is no need to be concerned with the electric amplitude.) The total magnetic field will therefore become

$$\mathbf{H} = H_0\hat{z} + H_1(\hat{x}\cos\omega t + \hat{y}\sin\omega t). \tag{4.49}$$

It will be shown in Chapter 8 that H_0 is associated with the H_0 term in the Hamiltonian operator, and H_1, with the pertubation represented by V – see Eqs. (4.18), (4.19), and (4.24). it will also be shown that the magnitude of the perturbation in this case will be

$$V = \gamma\hbar H_1. \tag{4.50}$$

In Eq. (4.50), it may be seen that V is the difference in the energies of the eigenstates of the same particle placed in a uniform static magnetic field of amplitude H_1. It will also be seen eventually that the general form of V for any spectroscopic transition is

$$V = 2|\mathbf{p}\cdot\mathbf{F}|. \tag{4.51}$$

The symbol $\mathbf{F}$ represents either the magnetic or the electric amplitude in the electromagnetic wave, depending on whether $\mathbf{p}$, the transition dipole moment, is electric or magnetic. By substituting Eq. (4.50) into Eq. (4.24) the time dependence of the parameter θ can be found:

$$\theta = \gamma H_1 t = \frac{\gamma t}{\gamma\hbar} V \tag{4.52}$$

If one remembers that the perturbation will be applied in the form of a pulse of duration Υ such that $c_{A1}(\Upsilon) = 0$ and $c_{A2}(\Upsilon) = 1$, it can be seen from Eqs. (4.22) and (4.23) that the necessary and sufficient condition is

$$\theta(\Upsilon) = n\pi. \tag{4.53}$$

where n is an odd number. With $n = 1$ (the simplest choice), Eqs. (4.52) and (4.53) can be combined to yield

$$\Upsilon = \frac{\pi}{\gamma H_1}. \tag{4.54}$$

A "typical" spectroscopic experiment on a spin-$\frac{1}{2}$ particle in a static magnetic field $H_0\hat{z}$ can now be described as follows.

1. The system is put into the low-energy eigenstate, so that the certain component of magnetic moment is parallel with the z axis (spin up). All other components of the magnetic moment (e.g., y and x) will be indeterminate.

2. An electromagnetic wave of the type described previously is turned on at time $t = 0$. The frequency of the wave is given by $\omega = \omega_0$ (see Eq. 2.68); from Eq. (4.43),

$$\omega_0 = \frac{(\gamma\hbar H_0/2) - (-\gamma\hbar H_0/2)}{\hbar} = \gamma H_0. \tag{4.55}$$

(These frequencies are ordinarily in the radio-wave region of the spectrum.)

3. The electromagnetic wave is turned off at time $t = \pi/\gamma H_1$. The system now will be in the high-energy eigenstate with the certain component of magnetic moment antiparallel to the z axis (spin down). All other components of the magnetic moment (e.g., y and x) will be indeterminate.

If the experiment described above is interrupted at some time in the interval $0 < t < \pi/\gamma H_1$, the H_1 wave turned off, and the z component of magnetic moment measured, that component will sometimes be found up and sometimes found down. The probability that "up" will result is $\cos^2(\gamma H_1 t)$, and the probability that "down" will be found in $\sin^2(\gamma H_1 t)$; this can be seen from Eqs. 4.22 and 4.23.

To distinguish between the two different kinds of quantum jumps, it is necessary next to find an infinite collection of operators, $\{\mathsf{R}\}_t$, associated with the infinite number of superposition states through which the particle passes from spin up to spin down. At a given time t, the superposition coefficients in Eqs. 4.22 and 4.23 uniquely define a state according to Eq. 4.20. This state should be an eigenstate of the property represented by the particular operator R associated with the time t.

The correct choice of that operator is the component of magneitc moment along the direction specified by θ' and Ω, hwere

$$\Omega = \S - \omega_0 t; \tag{4.56}$$

$\S$ is defined in Eq. (2.84), and ω_0 in Eq. 4.55. This component is the projection of the dipole moment on the axis of certainty. As will be shown in Chapter 8, the effect of the pertubation V is to tilt that axis away from the z axis through the angle θ'. It will also be shown that the "laboratory-space" angle θ' is always numerically equal to the "spin-space" angle θ. The projection of the axis of certainty upon the yx plane will make an angle Ω with the x axis. If the unit vector in the direction of the axis of certainty is called $\hat{u}_1$, the transition process causes the $\hat{u}_1, \hat{u}_2, \hat{u}_3$ coordinate system to rotate in the $\hat{x}, \hat{y}, \hat{z}$ frame until, at the time the pulse is terminated,

$$\hat{u}_1 = -\hat{z}. \tag{4.57}$$

The z axis, being the direction of the large static magnetic field, H_0, is fixed in the laboratory reference frame. The energy of the magnetic system is composed almost entirely of the energy associated with the component of magnetic dipole moment in the z direction because of the requirement

$$H_0 \gg H_1. \tag{4.58}$$

Equation (4.58) follows from the Eqs. (4.25), (4.50) and (4.55). Therefore the z axis remains the *energy* axis throughout the transition process. The relationships between the angles and vectors discussed in this section are shown in Fig. 4.4.

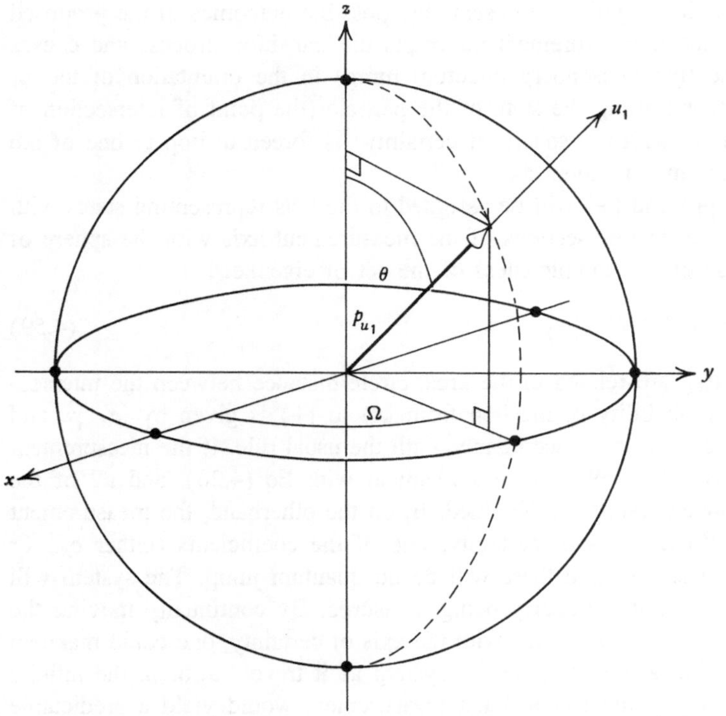

Fig. 4.4. Motion of the certain component of the dipole moment p_{u_1}, during the course of an uninterrupted spectroscopic trnasition. The u_1 axis is the axis of certainty, and the dotted line is a possible trajectory for the tip of p_{u_1} on the sphere of certainty

4.7 Quantum Jumps on the Sphere of Certainty

With this geometrical picture in mind, the consequences of believing in one kind of quantum jump or another can be explored. A belief in Schrödinger quantum jumps is equivalent to a belief that *no* experiment can have a certain outcome unless the u_1 axis is parallel or antiparallel to the z axis. If, on the other hand, one believes in Heisenberg jumps, one must introduce a third axis for complete visulaization of the experiment. This is the axis along which the projection of the magnetic moment operator is to be measured. The unit vector in the measurement direction will henceforth be designated as $\hat{v}_1$. In the absence of any disturbing measurement, the axis of certainty, $\hat{u}_1$, lines up parallel to $\hat{z}$, the magnetic field direction in the laboratory, at time $t < 0$; this fact was stated previously in Sect. 4.5. At $t = 0$, the axis of certainty rotates away from $\hat{z}$; its intersection point with the sphere of certainty is a point with a trajectory determind by Eqs. (4.52) and (4.56), starting at the "north pole" and ending at the "south pole". It follows from Heisenberg's view that each imaginable experiment to measure any Cartesian component of the magnetic dipole moment (in any coordinate system) can be specified by the orientation of the axis of measurement, v_1. The nature of this specification is such that the intersection of the v_1 axis with the sphere of certainty at two different points represents the possible outcomes of the proposed experiment. The act of measurement interrupts the transition process and causes an abrupt change (the Heisenberg quantum jump) in the orientation of the u_1 axis. The point representing the state of the particle (the point of intersection of the axis of certainty with the sphere of certainty) is forced to hop to one of the two "poles" of the measurement axis.

The symbol $|+\rangle$ and $|-\rangle$ will be assigned to the kets representing states with eigenvalues at opposite intersections of the measurement axis with the sphere of certainty. Because of the completeness of the set of eigenkets,

$$|A\rangle = c_{A+}|+\rangle + C_{A-}|-\rangle. \tag{4.59}$$

The coefficients $c_{A\pm}$ are related to the great-circle distance between the intersection points. The probability of the hop from $|A\rangle$ to $|+\rangle$ is given by $|c_{A+}|^2$, and from $|A\rangle$ to $|-\rangle$, by $|c_{A-}|^2$, in accordance with the usual rule. If the measurement axis is the z axis, Eq. (4.59) becomes identical with Eq. (4.20), and all of the previously mentioned results are obtained. If, on the otherhand, the measurement axis coincides with the axis of certainty, one of the coefficients (either c_{A+} or c_{A-}) in Eq. (4.59) is zero and there will be no quantum jump. The system will be in an eigenstate of the property being measured. By continually moving the measurement axis to keep it aligned with the axis of certainty, one could make an infinite number of measurements on the system as it travels through the infinite continuum of superposition states. Each measurement would yield a predictable outcome, so that the uncertainty principle would not limit the accuracy of the process. The system could be monitored during the entire transition process without disturbing its behavior in any way.

There is a one-to-one correspondence between each choice of V and the trajectory of the point (representing the superposition state associated with the ket $|A\rangle$) on the surface of the sphere of certainty. If one knew the physical situation well enough, the trajectory could be computed for any kind of quantum transtition. One could even discuss processes other than those produced by resonant interaction with electromagnetic radiation (spectroscopy), such as collisions. Although rather complicated spectroscopic trajectories have been described (see Fig. 4.5), no work known to this author up to 1976 had been done on any of the others.

A geometrical picture of the transition process has been developed in this section, and a description of how the behavior of the system expected for Schrödinger quantum jumps differs from that predicted for jumps of the Heisenberg variety has been given. In the next sections, a thought experiment based on this picture will be described. This experiment will decide the question of which kind of jump actually occurs in nature.

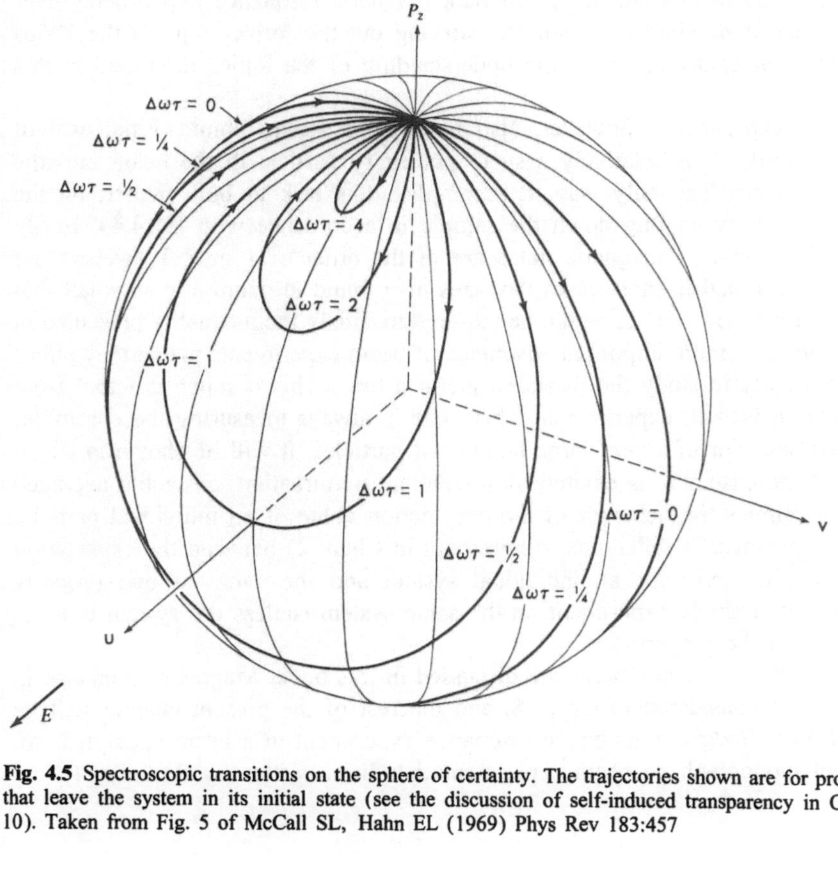

Fig. 4.5 Spectroscopic transitions on the sphere of certainty. The trajectories shown are for processes that leave the system in its initial state (see the discussion of self-induced transparency in Chapter 10). Taken from Fig. 5 of McCall SL, Hahn EL (1969) Phys Rev 183:457

4.8 Magnetic Resonance in Bulk and in Beams

The spectroscopic study of transitions between states representing different orientations of a magnetic dipole moment is called *magnetic resonance*. There are two important kinds of magnetic resonance experiments. The first takes place at low density in a molecular beam apparatus; the second, at high density in a stationary bulk sample.

Bulk experiments are superior in some ways to beam experiements for illustrating facts about transitions. In the crossed-coil bulk magnetic resonance spectrometer, the entire sample is enclosed within a radiofrequency receiver coil, so that the behavior of the quantum systems is monitored directly and continuously throughout the experiment. Usually it is not practical to monitor the behavior of a particle by means of direct observation during its flight through a beam apparatus, however. Instead the apparatus is designed in such a way that it deflects particles exhibiting undesired behavior, and transmits the others to a single detector located at the end of the flight path of the beam. The behavior of quantum systems is then *deduced* from the changes in beam current at the detector produced by changes in the state of the portion of the apparatus that interacts with the beam. The relative advantage of bulk magnetic resonance experiments over beams made it possible for scientists carrying out the former type in the 1950s to make great contributions to the understanding of the topics discussed in this book.

Beam experiments, however, also have some strong points. First, as will be seen shortly, it is relatively easy to put every particle in the beam into the same eigenstate. The only way to accomplish this task in bulk matter, on the other hand, is by cooling down the sample in accordance with Eq. 4.44. Easily obtainable laboratory magnetic fields are of the order of 1 or 2 T (webers per square meter), and in most cases the sizes of γ found in nature are so small that temperatures below 0.01 K would be required to satisfy the inequality presented in that equation. A more important advantage of beam experiments is that they afford the opportunity to study the particles one at a time. This is much different from the situation in bulk experiments, where one is always measuring the ensemble-averaged behavior of a very large number of particles. It will be shown in Chap. 5 that, if the ensemble is excited by a coherent perturbation, ensemble averaged behavior mimics the behavior of the expectation value of an individual particle. There is, however, a difference (pointed out in Chap. 2) between the expectation value of a property for an individual system and the value of that property measured in a single experiment on the same system (unless the system is in an eigenstate of the property).

Both kinds of experiments are discussed in this book. Magnetic resonance in bulk will be considered in Chap. 8, and the rest of the present chapter will be devoted to a *Gedanken* magnetic resonance experiment in a beam apparatus. An excellent monograph on beam experiments has been written by Ramsey [4].

4.9 The Stern-Gerlach Experiment

In a typical beam experiment (see Fig. 4.6) the atoms, ions, or molecules under study are vaporized in an oven. They are then allowed to escape from the oven through a hole (oriented to produce a stream of particles symmetrically distributed about the x axis of the laboratory coordinate system) into an evacuated chamber. Apertures are used to collimate the beam, eliminating systems with any appreciable component of velocity perpendicular to the beam axis. A one-to-one correspondence between the position of a particle in the apparatus and the time elapsed since it left the oven can be produced by giving all the particles in the beam very nearly the same speed. A pair of notched disks (or a spiral-grooved cylinder rotating about an axis parallel to $\hat{x}$) can be used to remove from the beam any quantum system having an undesired component of velocity along the beam direction. The particles that survive passage through the apertures and velocity selector have nearly identical velocities.

$$\mathbf{v} = v\hat{x}. \tag{4.60}$$

These particles enter the "interaction region" of the apparatus, where the desired experiment is performed upon them. All particles that are not removed from the beam by the experimental apparatus strike the detector.

As the beam particles pass into the interaction region, their angular momentum vectors are presumably oriented at random in the laboratory coordinate system. Each particle then passes between the pole faces of a horseshoe magnet (usually called *A magnet* by beam scientists). The pole faces of the A magnet are sculpted to produce a linear field gradient in the z direction:

$$\mathbf{H}_A = G_M z\hat{z}. \tag{4.61}$$

This gradient exerts a force upon an atom or molecule of mass m_0 to which the dipole is attached, and produces an acceleration $\mathbf{a}$ in accordance with the discussion given in Chap. 3:

$$\mathbf{a} = \frac{\hat{z}\, p_z G_M}{m_0}. \tag{4.62}$$

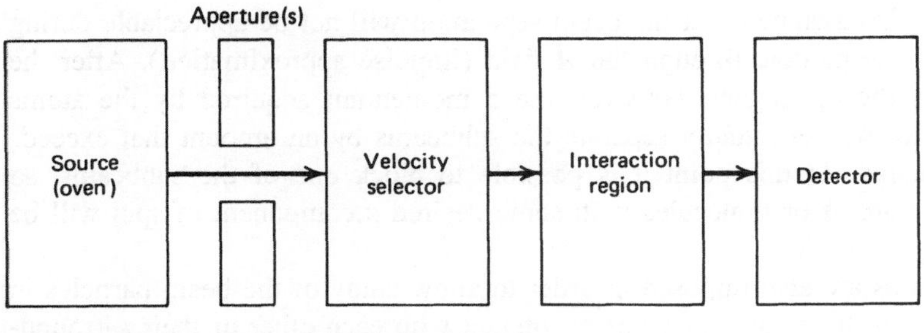

Fig. 4.6. Block diagram of a typical beam apparatus

If the length of the pole faces (measured in the x direction) is l_A, each particle will spend a time l_A/v in the inhomogeneous field. It will therefore acquire a component of momentum in the z direction, so that the total momentum,

$$\boldsymbol{\mu} = m_0 v\hat{x}, \tag{4.63}$$

becomes

$$\boldsymbol{\mu} = m_0 v \left[\hat{x} + \left(\frac{p_z G_M l_A}{m_0 v^2}\right)\hat{z}\right]. \tag{4.64}$$

The momentum increment produced by the A magnet will cause the beam to fan out in the xz plane. The atoms or molecules which had magnetic moments oriented in such a way that p_z was positive will be directed upwards, $(\|\hat{z})$, whereas those with negative z components will be directed downwards $(\| -\hat{z})$. The experiment just described was first performed by Stern and Gerlach [5], using silver atoms ($s = \frac{1}{2}$) as the beam particles. The detector was a cold glass plate oriented perpendicularly to the beam axis. Silver atoms that struck the plate condensed and eventually produced a mirror. The extent of the mirror in the z direction indicated the amount of deflection, and therefore the magnitude of p_z. Stern and Gerlach found that the distribution of beam trajectories in the "fan" was not continuous. Instead, as described in Sect. 4.5, there were only two beam components, corresponding to "spin up" and "spin down" – see Eq. (4.42). It would be hard to exaggerate the importance of this experiment in stimulating the development of quantum theory.

Several conditions must be imposed on the parameters of this experiment in order to make the mathematics tractable:

$$\Delta v_z, \Delta v_y \ll \frac{\gamma\hbar G_M l_A}{2m_0 v} \ll v \ll c. \tag{4.65}$$

The first condition in Eq. (4.65) ensures that the spreading of the beam (due to the residual momentum transverse to the beam direction, brought about by insufficient collimation) will not mask the separation of the incident beam into subbeams by the A magnet. The second condition ensures that the kinetic energy of the beam particle is not changed appreciably by the dipole-inhomogeneous field interaction, while the third obviates the necessity for consideration of relativistic effects. It is also assumed that the beam seperation will not be appreciable during the flight of the particle through the A field (impulse approximation). After the beam leaves the A magnet, however, the z momentum acquired by the atoms and molecules will eventually separate the subbeams by an amount that exceeds the beam radius. At this point it is possible to block one of the subbeams, so that only the atoms or molecules with some desired z component of spin will be transmitted.

Constraints are also imposed in order to allow study of the beam particles in isolation, before they have a chance to interact with each other or their surroundings. If such interactions were permitted, they would have the effect of permitting particles in state 2 to relax down into state 1 exponentially, with a characteristic time constant T_1, mentioned in Sect. 4.5. The total length, L, for the beam chamber is therefore chosen sufficiently short so that particles of speed v do not stay in the apparatus long enough to relax appreciably:

$$vT_1 \gg L. \tag{4.66}$$

4.10 State Selection in Beam Experiments

The experiments described in subsequent sections also take place in the apparatus diagramed in block form in Fig. 4.6, with one difference. The interaction region contains an A magnet similar to that described in the preceding sections, but is enlarged over that used by Stern and Gerlach to permit investigation of the properties of spin-oriented particles.

It is convenient to return the transmitted subbeams to the x axis for ease in subsequent experimentation or detection. This may be accomplished by causing the beam to pass through another inhomogeneous magnetic field, having its gradient oriented *anti*parallel to $\hat{z}$. The conventional designation for such a device is "B magnet":

$$\mathbf{H}_B = -G_M z\hat{z}. \tag{4.67}$$

(A combination of A and B magnets that will accomplish the desired refocusing of the beam is shown in Fig. 4.7.) In the region between the first and second

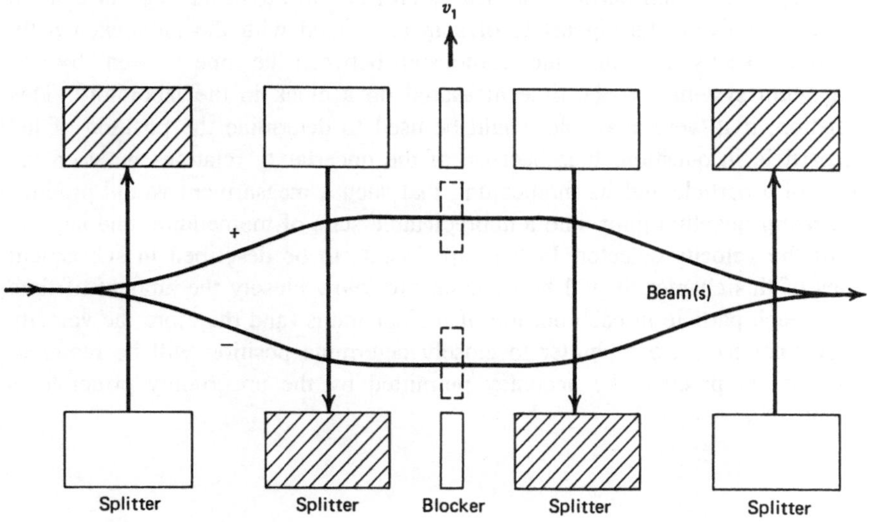

Fig. 4.7. Block diagram of a state selector; directions of field gradients are shown by *vertical arrows*

B magnets, the subbeams are well separated, so that it is possible to insert an obstruction into the path of either one without affecting the intensity of the other.

The direction of the field gradient in the A magnet at the entrance to this device is the axis of measurment, usually oriented parallel to the z axis:

$$\hat{v}_1 = \hat{z}_1. \tag{4.68}$$

If the subbeam deflected in the $+\hat{v}_1$ direction is blocked, all of the particles that survive passage through this device will have their certain components of spins oriented in the $-\hat{v}_1$ direction, and vice versa. This arrangement of magnetic field gradients and blockers will therefore be designated henceforth as a "Stern-Gerlach spin selector" ("SG" for short). From the point of view of believer in Heisenberg jumps, a particle propagating in the $\hat{x}$ direction that enters an SG with its certain component of spin pointing along an arbitrary direction $\hat{u}_1 \neq \hat{v}_1$ may have any one of six different experiences. Depending on how the two beam stops are placed, (1) it may pass through without change in spin direction (both beam stops withdrawn). The particle may encounter one beam stop placed to block the plus beam, precipitating a quantum jump into either the plus or minus state: in the former case, (2) it would be blocked, and in the latter, (3) transmitted. It may encounter the blocker in the path of the minus beam, which would also precipitate a quantum jump into eigenstates of the v_1 component of angular momentum. This time, a jump into the state represented by $|-\rangle$ would cause the particle to be removed from the beam (4), and a jump into the state represented by $|+\rangle$ would cause it to survive (5). Finally, if both blockers are inserted (6), it is certain that no particles will ever be transmitted.

The SG is therefore a device of the type that is required to put all of the quantum systems into the same inital state, namely, that represented by the wave function ψ_1. Each beam particle that leaves an SG will be, at that instant of time ($t = 0$), certainly in the eigenstate of spin associated with the unblocked path. One cannot exactly determine the relationship between the time t "seen" by any one quantum system and the time measured on a clock in the laboratory. This information, if it were available, could be used to determine the position of the beam particle in question. It is because of the uncertainty relations between the position of a particle and its momentum, that such a measurment would produce a Heisenberg quantum jump into a unpredictable state of momentum, undoing the work of the velocity selector. In the experiments to be described in subsequent sections of this chapter, it will be necessary to know closely the amount of time spent by each particle in each portion of the apparatus (and therefore the velocity must be known). Also a shutter to closely determine position will be required. Fortunately, in practice the accuracy permitted by the uncertainty principle is sufficient.

4.11 The Rabi Magnetic Resonance Experiment

The desired beam experiment must conform to the pattern described in Sect. 4.4: spin-$\frac{1}{2}$ particles must travel from the velocity selector into a transition region that has three different sections, as shown in Fig. 4.8. The initial and final state selectors can be SGs oriented with initial gradients ($\hat{v}_1$ axes) in the $+\hat{z}$ direction. Beam stops are inserted into the path of the minus beam in the inital SG and into the path of the minus beam in the final SG. Particles that exit from the intial detector are certainly in the plus or "spin-up" state, represented by $|1\rangle$. Any particle that changes its state in the transition region will surely be blocked by the final SG and will therefore not strike the detector.

The transition will not properly be called *spectroscopic* unless (1) the two states have different energies, and (2) the perturbation is an electromagnetic wave. To satisfy the first requirement, the entire interaction region will be inserted between the pole faces of a magnet that produces a uniform magnetic field in the z direction. This device is usually called a C magnet, and its magnitude is H_0, as described in Eq. 4.39. The C magnet therefore will give the two quantum states the energies expressed in Eq. 4.43. All of the particles that survive passage through the initial SG will be in the state having the lower of the two energies: recall Eqs. 4.41 and 4.46. (The magnetic energy added by the A and B magnets is zero because the beam propagates along the x axis, where $z = 0$. Therefore, form Eqs. 4.61 and 4.67,

$$\mathbf{H}_A = \mathbf{H}_B = 0. \tag{4.69}$$

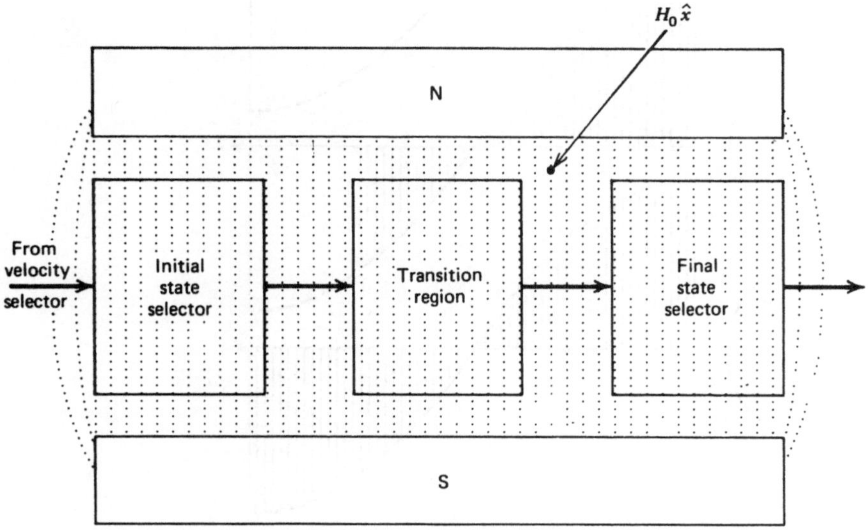

Fig. 4.8. Block diagram of interaction region

To satisfy the second requirement for a spectroscopic transition, an oscillating magnetic field is produced in the transition region at the frequency ω_0 (see Eq. 4.55). An rf oscillator operating at that frequency is connected to a transmitter coil elongated in the x direction, with its magnetic axis oriented along the y direction. The amplitude of the magnetic field produced by the coils is adjusted to $2H_1$ by controlling the power output of the oscillator. This oscillating field will

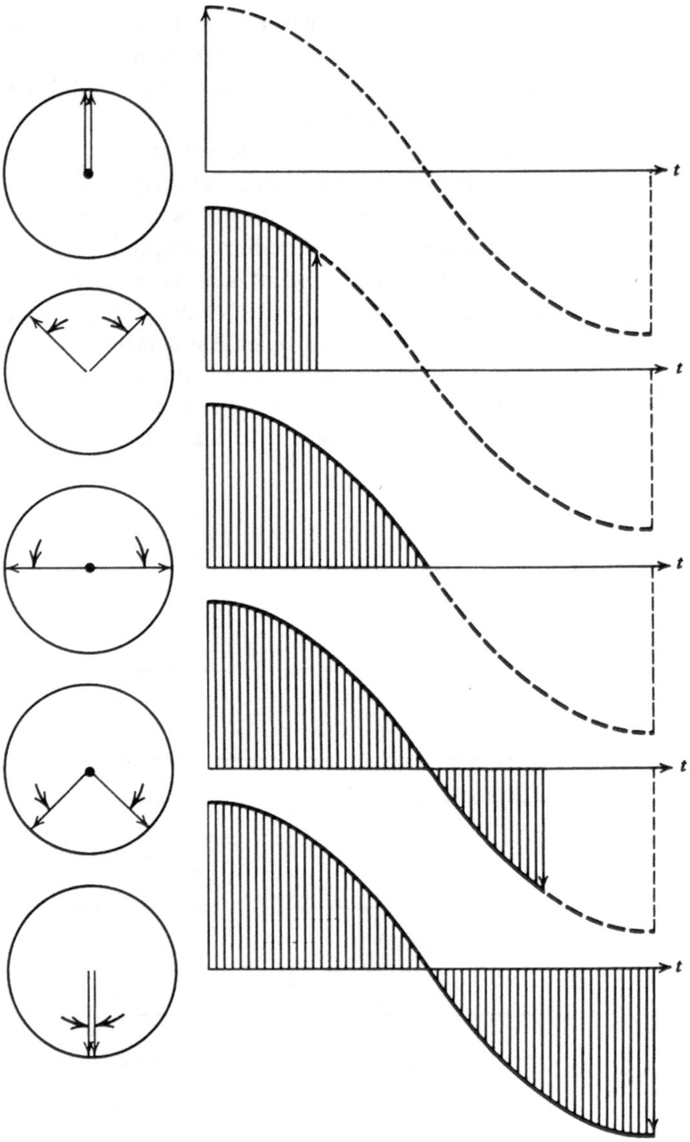

Fig. 4.9. A plane-polarized wave (amplitude shown on right) resolved into two circularly polarized waves (amplitude shown on left)

be equivalent to two rotating fields of constant amplitude H_1, one of which rotates in the same sense (clockwise or counterclockwise) as the axis of certainty and the other in the opposite sense (see Fig. 4.9). The influence of the latter rotating field will be negligible [6]; the former will produce all of the desired effects, as one can see from Eq. 4.49.

The time of flight of the particles through the transition region of length l will determine the irradiation period Υ:

$$\Upsilon = \frac{l}{v}. \tag{4.70}$$

Either the velocity v or the magnitude of H_1 should be adjusted until the irradiation period is sufficient to supply a π pulse to each particle, as indicated in Eq. 4.54.

When everything has been properly adjusted, each particle that leaves the initial SG will have rotated its axis of certainty by exactly 180° during its passage through the transition region and will be blocked with 100% probability by the final state selector. Equivalently, one may say that each quantum system has swallowed exactly one photon and has changed its state from that represented by $|+\rangle$ with energy $\mathscr{E}_1$, to that represented by $|-\rangle$ with energy $\mathscr{E}_2$. A believer in Schrödinger jumps may claim that the photon-swallowing act occurred at any arbitrarily chosen point along the x axis within the transition region ($0 \leqslant x \leqslant l$), and hence at any time in the interval $l/v(0 \leqslant v \leqslant \Upsilon)$. This apparatus therefore cannot distinguish between the two kind of jumps.

Beam experiments with C magnets and rf waves were developed by I. I. Rabi and his co-workers [7] to provide very accurate measurements of the magnetic moments of nuclei and molecules. The apparatus actually used was far simpler than that described above, however. Instead of eight inhomogeneous-field magnets and four beam stops to prepare and detect the initial and final quantum states, implied by the use of two SGs, Rabi employed only a single A magnet as the initial spin selector, and single B magnet as the final spin selector. By careful adjustment of the beam trajectories and detector position, such a device can even be made to work without using velocity selection. If SGs of the type described above are employed, however, velocity selection is required.

4.12 The Ramsey Separated Oscillating Fields Experiment

N.F. Ramsey [8] modified the Rabi magnetic apparatus in the following way. Instead of one transmitter coil for the $\mathbf{H}_1$ field (of length $l_1 = l_c$), he used two coils of length $l_1/2 < l/2$ – one located after the exit end of the initial SG and one just before the entrance end of the final SG.

Note that in both the Rabi experiment and in Ramsey's modifications thereof, the certain component of the magnetic dipole moment makes an angle $\theta' = 0$ with the z axis at the beginning of the $\mathbf{H}_1$ irradiation region and an angle $\theta' = \pi$

with the z axis at the end of that period. Recall from Section 4.6 that the angle θ' in laboratory coordinates happens to equal the "angle" θ that appears in the algebraic expressions for the superposition coefficients. This means that in both the Rabi and the Ramsey experiements, $\theta = \theta' = \pi/2$ at the half-irradiation point. The difference is that in the Rabi apparatus the beam particles are described by the wavefunction

$$\Psi = \frac{1}{\sqrt{2}}(\psi_1 + \psi_2) \tag{4.71}$$

at only one instant of time, $t = \Upsilon/2$, and at that instant the particles themselves are experimentally inaccessible. By way of contrast, in the Ramsey apparatus the axis of certainty for the spin of each particle is located in the xy plane over nonzero portions of the beam path, $l - l_1$, and over a correspondingly long interval of time, $(l - l_1)/v$. This space can be made long enough to accommodate additional experimental apparatus. The apparatus could be chosen to verify that the quantum systems are indeed in an eigenstate of "spin sideways" when they are in a superposition of eigenstates of "spin up" and "spin down".

Ramsey's purpose in designing the separated oscillating fields method was not to verify the existence of superposition states, however. He simply wanted to determine ω_0 to greater precision than was possible in the Rabi-type apparatus, and was able to show that merely separating the ω_1 fields would accomplish this purpose. The modification of the Ramsey apparatus required to disprove the Schrödinger jump picture of quantum transitions will be described in the next section.

4.13 A Thought Experiment

The desired experiment could, in principle, measure a property of a quantum system in a superposition state and not produce any quantum jumps. Such an experiment can be performed, using the Ramsey separated oscillating fields equipment, if some device that can measure the y component of magnetization is inserted into the space between the two oscillating fields.

The obvious choice for such an apparatus is an SG oriented along the y axis. If the atoms and molecules are still in either the spin-up or spin-down state (having undergone – or not undergone – their Schrödinger jumps), they would have a 50% probability of being deflected into the left-hand subbeam and a 50% change of being deflected into the right-hand one. On the other hand, if they have not undergone any jumps, they ought to be in an eigenstate of the y component of angular momentum. *Which* of these two eigenstates they would be in (i.e., whether they would be "spin left" or "spin right") depends on the algebraic sign of γ.

Unfortunately, the experiment just described would not work in the apparatus devised by Ramsey. The presence of the H_0 field in the space between the two irradiation regions causes components of the magnetic moment in the xy plane to precess about the z axis at the Larmor frequency. This means that the actual orientation of the axis of certainty in laboratory coordinates is given by

$$\hat{u}_1 = \cos\Omega \sin\theta' \hat{x} + \sin\Omega \sin\theta' \hat{y} + \cos\theta' \hat{z}, \tag{4.72}$$

where Ω was defined in Eq. (4.56). Since Ω changes very rapidly with time (ω_0 for protons exceeds 1 MHz even if H_0 is only 25 mT), the direction of the certain component of spin will reverse many, many times during the flight of the quantum system through the field of the A magnet. A large number of impulses $[\omega_0(l_C - l_1)/\pi v]$ will be delivered to each particle, half pushing it to the left and half to the right. The net y components of linear momentum acquired will therefore be zero, and the beam will not split into two subbeams.

To make the beam split into two subbeams, one with a trajectory to the left and one to the right in stationary laboratory coordinates, it is necessary to stop the Larmor precession. This can be done by replacing the C magnet of lenght l_C by two C magnets, each of lenght $l_C/2$. The first of these, in conjunction with the $\mathbf{H}_1$ field of length $l_1/2$, will produce the first $\pi/2$ pulse (i.e., $\theta' = \pi/2$ for particles leaving this portion of the apparatus). The angle θ' will remain at $\pi/2$ for the entire region of length $l_C - l_1$ in between the two C magnets. Furthermore, in this intermediate region, the angle Ω will remain constant because $\omega_0 = 0$. The amplitude of the field produced by the C magnet should be adjusted until the spins precess through $n + \frac{1}{2}$ Larmor half cycles:

$$\frac{2\omega_0}{l_C} = (n + \tfrac{1}{2})\pi. \tag{4.73}$$

Since the $\mathbf{H}_1$ field and the time of arrival of each particle may be synchronized by means of a shutter in the path of the beam, one can make $\S = \pi$. Therefore, in the intermediate region,

$$|\Omega| = (n - \tfrac{1}{2})\pi. \tag{4.74}$$

The result of these adjustments is to align the axis of certainty either parallel or antiparallel to the laboratory y axis:

$$\hat{u}_1 = \pm\hat{y}. \tag{4.75}$$

Blocking one of the two (left or right) subbeams should reduce the beam intensity to zero; blocking the other should produce no change in beam intensity. In the latter case, the effect of the sideways Stern-Gerlach apparatus in the intermediate region is to perform a measurement upon each beam particle without producing a Heisenberg quantum jump.

The spin-$\frac{1}{2}$ particles will then enter the field of the second C magnet and resume their precessional motion. The second $\mathbf{H}_1$ field will then produce the second $\pi/2$ pulse (i.e., $\theta = \pi$ for particles leaving this portion of the apparatus). It will probably be necessary to adjust the phase of the radio frequency used in the second oscillator to ensure that absorption of energy by the particles (rather than stimulated emission) will occur.

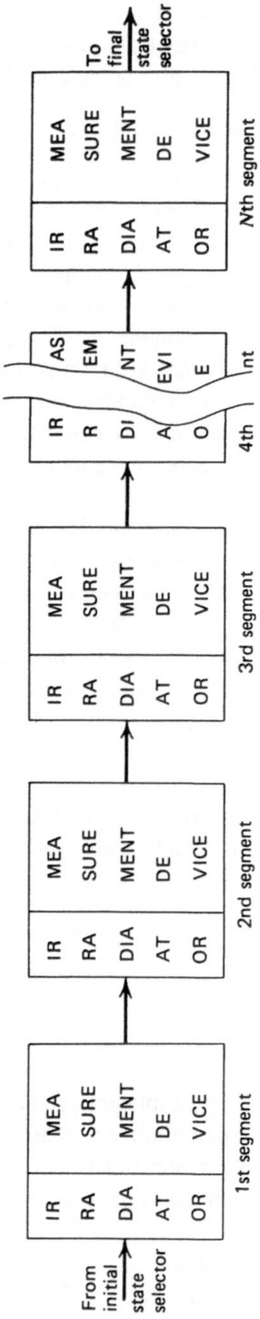

Fig. 4.10. Block diagram of transition region

Now imagine that the C magnet and the $\mathbf{H}_1$ field are divided into N_s segements of equal length. This would create $N_s - 1$ gaps between these sections, and a Stern-Gerlach apparatus could be inserted into each gap (see Fig. 4.10). The direction of the magnetic field gradient in successive SGs will make successively larger angles with the laboratory z axis, so that at each point it is aligned with the direction of the axis of certainty for the beam particles. If the direction of the field gradient in the jth SG is $\hat{u}_{1,j}$, then

$$\hat{u}_{1,j}\cdot\hat{u}_{1,j+1} = \cos\left(\frac{\pi}{N_s}\right) \tag{4.76}$$

and

$$\hat{u}_{1,j}\cdot\hat{z} = \cos\left(\frac{j\pi}{N_s}\right). \tag{4.77}$$

In each SG, the subbeam which intersects the negative branch of the $\hat{u}_1$ axis could be blocked without reducing the transmitted intensity. In this way, $N_s - 1$ measurements could be made upon each quantum system during its transition between the energy eigenstates represented by ψ_1 and ψ_2. Each measurement would have an eigenvalue as its result, so that no Heisenberg jumps would ever occur. A schematic drawing of a typical segement of the proposed apparatus appears in Fig. 4.11.

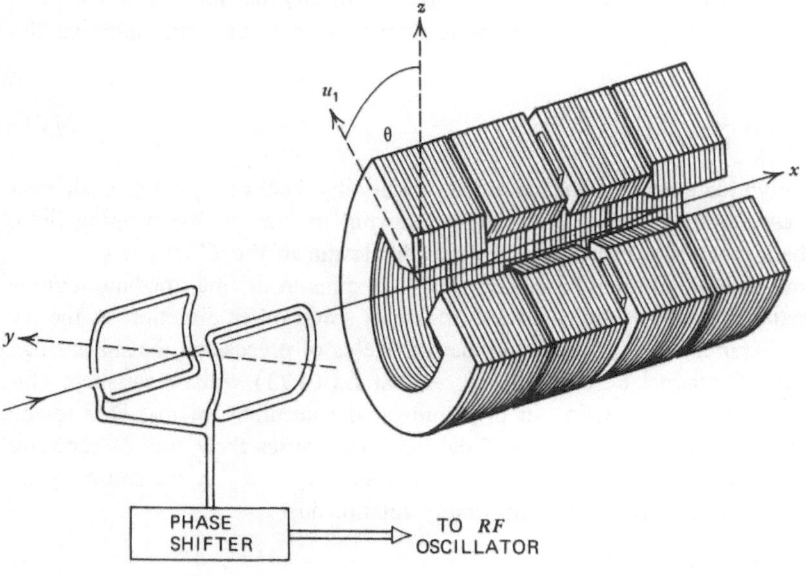

Fig. 4.11. Sketch of one segment of the transition region

4.14 Difficulties with the Proposed Experiment

There are two major practical problems with the *Gedanken* experiment proposed in the preceding section. The first of these can be solved by ensuring that the magnetic field intensity due to the C magnet increases (decreases) gradually from 0 to H_0 (H_0 to 0). An abruptly changing electric or magnetic field can be expressed as a sum of sinusoidally oscillating electromagnetic waves; the amplitude of these waves can be calculated by means of Fourier analysis. This analysis shows that the amplitudes are appreciable at frequencies as high as the reciprocal of the transit time of a quantum system through the "fringing field" region. If this region for the last C magnet (for example) begins at x_0 and ends at $x_0 + \Delta l$,

$$H_C = \begin{cases} \dfrac{H_0 x}{\Delta l}, & x_0 < x < x_0 + l, \\ H_0, & x > x_0 + l. \end{cases} \tag{4.78}$$

The oscillating fields then have frequencies

$$\omega' < \frac{v}{\Delta l} \tag{4.79}$$

and are capable of producing transitions (called *Majorana flops* [9]) whenever

$$\omega' \sim \omega_0. \tag{4.80}$$

Therefore the condition (called the *adiabatic condition*) that must be satisfied to keep the beam particles in whatever state they happen to be while entering the C magnet is

$$\Delta l > \frac{v}{\omega_0}. \tag{4.81}$$

An alternative statement of the adiabatic condition, therefore, is that each spin must precess through many Larmor cycles during its stay in the fringing field. The adiabatic condition can be met by careful design of the C magnets.

A more serious problem is ensuring that the spins in the intermediate regions point exactly along the y axis, instead of along some other direction in the xy plane. The requirement is that the number of cycles of precession during passage through each C magnet be controllable, so that Eq. (4.73) will be satisfied. Unfortunately there is a limitation in principle on the accuracy of ω_0. Due to the fact that the beam particles only "see" the field that causes their axes of certainty to precess for a time Υ/N_s, the energy difference $\mathscr{E}_2 - \mathscr{E}_1$ is uncertain by an amount $\Delta\mathscr{E}$, which satisfies the uncertainty relationship

$$\frac{\Upsilon \Delta\mathscr{E}}{N_s} > \frac{\hbar}{2}. \tag{4.82}$$

This leads to an uncertainty in the Larmor frequency;

$$\Delta\mathscr{E} = \hbar\Delta\omega_0. \tag{4.83}$$

These two equations can be combined to yield

$$\Delta\omega_0\Upsilon = 1. \tag{4.84}$$

Equation (4.84) can be obtained without the use of quantum mechanics; it is simply the Fourier relationship between the spread of frequencies $\Delta\omega_0$ in a wave packet which persists for a duration Υ. (Remember the previous discussion of the adiabatic theorem.) Since the quantum systems precess for a time Υ/N_s, the uncertainty in precession frequency leads to an uncertainty $\Delta\Omega$ in the accumulated phase:

$$\Delta\Omega = \frac{\Delta\omega_0\Upsilon}{2}. \tag{4.85}$$

The angle Ω in the regions between the C magnets therefore cannot be determined within limits narrower that about $\pm\frac{1}{2}$ radian. This means that the maximum intensity ratio of the left- and right-hand beams cannot be the desired 1.00:0.00. Instead, the best that can be expected will be of the order of

$$\cos^2\tfrac{1}{4} : \sin^2\tfrac{1}{4} : : 0.94 : 0.06. \tag{4.86}$$

The estimate given in Eq. 4.86 is decidedly optimistic.

These errors would occur at each SG in the transition region and accumulate in such a way that the transmitted intensity would be very low if N_s were large. A "perfect" experiment with a larger number of SG magnets is not necessary, however. If the results of a single measurement between C magnets (e.g., at the half-way point with a field gradient $\mathbf{H}_A = G_M y\hat{y}$) produced anything but a 50 : 50 beam, the hypothesis of Schrödinger jumps would be disproved.

4.15 The Bloom Transverse Stern-Gerlach Effect

The *Gedanken* experiment just described is supposed to prove that superposition states have properties that can be measured without producing quantum jumps. It is essential to the argument that the superposed states be eigenstates of the energy with different eigenvalues. Although the states in the proposed experiment did have different energies while they were in the fields produced by the C magnets, they had the same magnetic energy (namely, zero) while the crucial measurement was being performed. For this reason, it would be desirable to perform an experiment on spin-$\frac{1}{2}$ particles during their passage through the field of a C magnet and attempt to separate spin-left particles from spin-right particles in the rotating coordinate system. Myer Bloom and his co-workers [10] have accomplished this separation in a remarkable experiment demonstrating what they called the *transverse Stern-Gerlach effect.*

In the Bloom experiment, a beam of neutral potassium atoms was directed along the z axis of a solenoidal C magnet. The detector then recorded a beam profile having an intensity maximum on the axis (i.e., at the origin of the xy plane). Four wires were placed inside the solenoid parallel to the z axis in symmetrical positions (in the north, east, south, and west electric quadrupole arrangement described in Chap. 3) about the beam. An rf voltage applied alternately to these wires created an inhomogeneous magnetic field directed along the $\hat{x}'$ axis in a coordinate system rotating at the frequency ω. The magnetic field of the solenoid was increased until the Larmor frequency ω_0, equaled ω. At that point, the beam split into a number of subbeams corresponding to the number of eigenstates of the angular momentum of potassium atoms.

4.16 Quantum Jumps and Superposition States: Conclusion

If the experiment described in Sect. 4.13 is performed, the apparatus will have produced a transition of the system from state 1 to state 2, and will have performed N_s measurements upon the system during the process. If there were such things as Schrödinger quantum jumps in nature, the performance of these N_s experiments would have the effect of reducing the beam intensity nearly to zero.

The fact is that Schrödinger jumps do not occur in nature. The only quantum jumps that exist are those of the Heisenberg variety. Since the experiment was designed to avoid Heisenberg jumps, only those jumps necessitated by the considerations outlined in Sect. 4.14 will occur. As a consequence, the beam intensity will be minimally affected by the N_s measurements. If the experiment were actually performed, the failure of the beam intensity to diminish to zero could be taken as proof of the reality of superposition states.

In terms of the geometrical picture developed previously to describe the transition process, the rf irradiation of a quantum system in the first segment causes the point representing the state of the system to move away from the "north" pole on the sphere of certainty at the angular frequency $\omega_1 = \gamma H_1$. At the same time, the point circulates rapidly about the pole at the much larger rate ω_0 because of the field H_0. The resultant trajectory of the point is a tightly wound spiral on the sphere of certainty, originating at the "north" pole (see Fig. 4.12a). At the end of the irradiation period of the first segment, the particle passes out of the field H_0 and the precessional motion ceases. The representative point will come to rest on the sphere at the point $\theta = \pi/N_s$ and $\Omega = -\pi/2$ (See Eq. 4.72). The ray drawn from the sphere's center through to the representative point defines the axis of certainty, $\hat{u}_1$. The SG portion of the first segement is represented in this picture by a measurement axis, $\hat{v}_1$. Since $\hat{u}_1$ and $\hat{v}_1$ coincide, the measurement process does not alter the position of the representative point (see Fig. 4.12b). As the particle exists from the SG, the $\hat{v}_1$ axis disappears. The representative point resumes its spiral downward course as it becomes subject to the static and oscillatory magnetic fields associated with the second segment (see Fig. 4.12c).

Finally, the differences between the two kinds of jumps can be illustrated by analogy. Suppose that an extraterrestrial being who wishes to learn the history of this planet asks the question, "Who was Richard Evelyn Byrd?" A well-informed earthling can reply, "He was an explorer." "And what did he explore?" asks the being. "The polar regions of the planet," replies the earthling. The being has the power to look backward in time and therefore is able to see Byrd first at the North Pole and then at the South Pole. The being notes the change in Byrd's position and asks itself the question, "How did Byrd get from one pole to the other?" Since it knows that the poles are the intersection points of the earth's axis with the sphere representing the surface of the earth's crust (and since it has X-ray eyes), it carefully examines the interior of the earth along the axis over a period of time to see whether it can find Byrd in transit. All measurements performed along this axis have the result that Byrd is always either at the South Pole or the North Pole, but never in between. The being concludes on this basis that the mode of travel used by Byrd to progress from one pole to the other is a discontinuous process called a "quantum jump." Another extraterrestrial being might think to

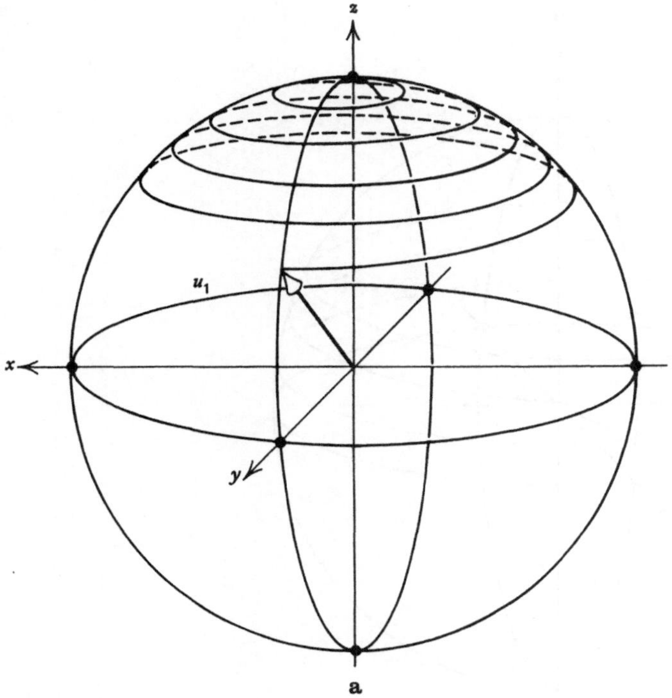

Fig. 4.12. **a** Spectroscopic trajectory of a precessing particle. **b** Measurement of the dipole moment of a nonprecessing particle **c** Resumption of precessional motion by the particle.

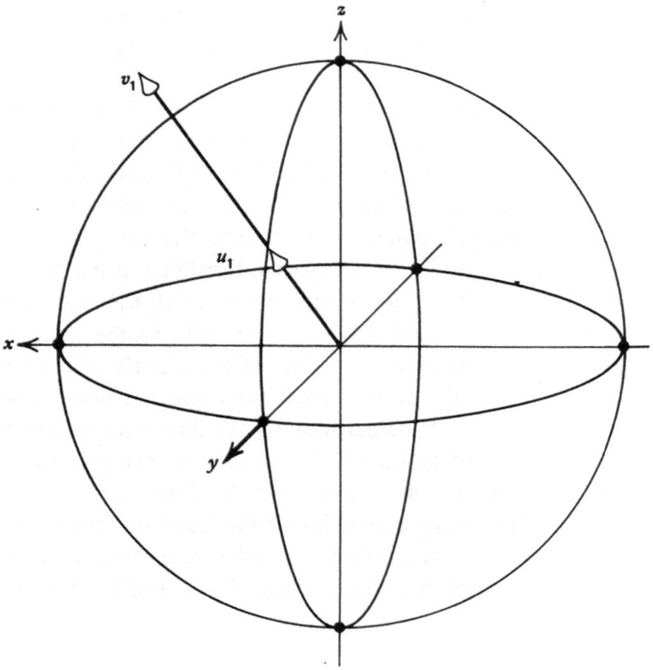

b

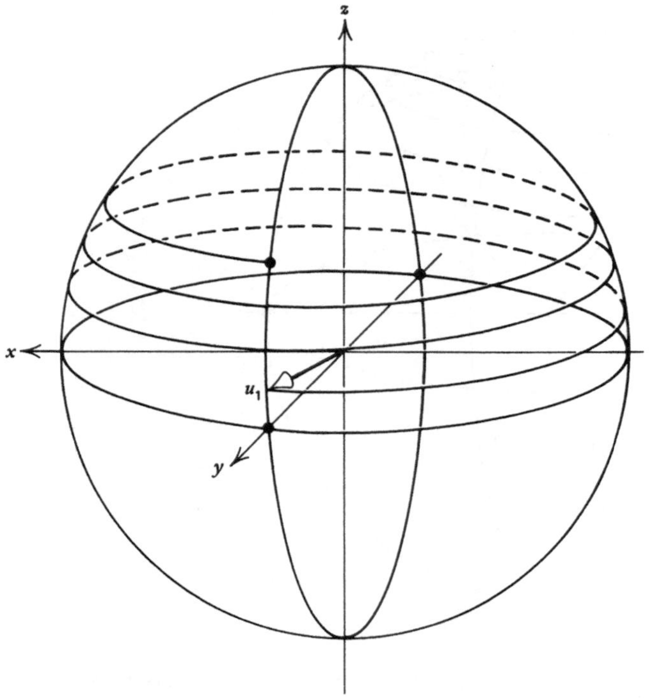

c

Fig. 4.12. (*Continued*)

look at other portions of the earth's surface during the transition process. If it did so, it might detect Admiral Byrd moving in a continuous trajectory upon that sphere from one pole to another, and therefore reach a different conclusion about the dynamics of polar exploration.

4.17 References

1. Jaynes ET, Cummings FW (1963) "Comparison of quantum and semiclassical radiation theories with application to the beam maser. Proc. IEEE 51: 89
2. Schrödinger E (1956) Are there quantum jumps?, What is life? (Doubleday Anchor Books, NY, p 132
3. Heisenberg W (1958) Physics and philosophy (Harper and Bros., New York)
4. Ramsey NF (1956) Molecular beams. Clarendon Press, Oxford
5. a. Stern O (1921) Zs Phys 7: 249
   b. Gerlach W, Stern O (1924) Ann Phys 74: 673
   c. Gerlach W (1925) Ann Phys 76: 163
6. a. Bloch F, Siegert A (1940) Phys Rev 57: 522
   b. Stevenson AF (1940) Phys Rev 58: 1061
7. a. Rabi II, Zacharias JR, Millman S, Kusch P (1938) Phys Rev 53: 318
   b. Rabi II, Millman S, Kusch P, Zacharias JR (1939) Phys Rev 55: 526
8. Ramsey NF (1950) Phys Rev 78: 695
9. Majorana E (1932) Nuovo Cimento 9: 43
10. a. Bloom M, Erdman K (1962) Can J Phys 40: 179
    b. Bloom M, Enga E, Lew H (1965) Can J Phys 45: 1481

4.18 Problems

4.1. At what temperature (kelvins) is $kT = \gamma\hbar H_0$ for protons ($\gamma/2\pi = 42$ MHz T^{-1}) in a magnetic field of 1.4 T?

4.2. Calculate $\langle x\rangle_{12}$, $\langle y\rangle_{12}$, and $\langle z\rangle_{12}$ for hydrogen in a "halfway" superposition of the $1s$ and $2s$ states ($c_1 \equiv c_{1s} = c_2 \equiv c_{2s} = 1/\sqrt{2}$).

$$\psi_{2s}(\mathbf{r}, t) = (32\pi a_0^3)^{-1/2} \exp\left(\frac{-i\mathscr{E}_2 t}{\hbar}\right)\left[2 - \left(\frac{r}{a_0}\right)\right] \cdot \exp\left(\frac{-r}{2a_0}\right).$$

4.3. Calculate the perturbation energy V (in joules) required to produce a π pulse of duration $\Upsilon = 1.0 \times 10^{-9}$ s.

4.4. What is the minimum angle between the total angular momentum and its certain component for $s = 1$? $s = 10$? $s = 100$?

4.5. Calculate the value of H_1 (in teslas) required to produce a π pulse for protons (γ in Problem 4.1) of duration $\Upsilon = 1.0 \times 10^{-6}$ s.

II. Quantum Statistics

5 Ensembles of Radiating Systems

J.D. Macomber

5.1 Reasons for the Use of Statistical Methods

The quantum systems that interact with light waves in spectroscopic experiments have been treated in preceding chapters as if these systems were to be studied one at a time. In practice, however, one ordinarily deals with macroscopic samples of matter, each of which consists of very large numbers ($\cong 10^{20}$) of microscopic chromophores. Statistical methods must be used in such cases to predict the behavior of the sample, and it is the purpose of this chapter to develop appropriate techniques.

It might be imagined that the statistical problem in quantum mechanics had already been solved when it was discovered how to calculate expectation values. If some physically observable property of the system is represented by the operator R, the expectation value of that property, $\langle R \rangle$, may be calculated by means of Eq. 2.78. The interpretation of $\langle R \rangle$ is that each of N repeated measurements of that property made upon a single system produces a result r_n, drawn from the set of M eigenvalues $\{r_n\}_M$. The average of these measurements approaches $\langle R \rangle$ as N becomes very large:

$$\lim_{N \to \infty} \frac{1}{N} \sum_{n=1}^{N} r_n = \langle R \rangle. \tag{5.1}$$

If a single measurement is made simultaneously upon each of N identical particles (the ensemble), the same average is to be expected. (The theoretical justification for the equality of these two averages is called the *ergodic hypothesis* in statistical mechanics.) If this were all there was to the problem, the ensemble average would in every case be the expectation value, and there would be no need for additional discussion of quantum statistical mechanics.

The N-particle ensemble average will not necessarily equal the N-measurement single-particle average, however, unless certain conditions are met. First, whenever repeated measurements are being made upon a single system, it is necessary to return that system exactly to its initial state between each pair of measurements. (Whenever this precaution is *not* taken, the measurements will be called successive rather than repeated.) The corresponding requirement in regard to the equivalent experiment in which simultaneous measurements are made upon a large number of identical systems is that each of these systems be in the same state.

These conditions (which make the collection of systems a *microcanonical ensemble*) are not always met in practice. Measurements are more commonly performed upon quantum systems, the states of which are determined by an equilibrium Boltzmann distribution among the several stationary energy levels associated with an absolute temperature $T > 0$. Collections of systems so distributed are called *canonical ensembles*. Whenever averages of physical properties are computed for such ensembles, one must perform both "quantum" averaging (i.e., calculations of expectation values), and "classical" or ensemble averaging over the various states that are present. The decision was made in Chap. 2 to indicate the former type of averaging by means of triangular brackets – for example, $\langle R \rangle$. The latter type of averaging will be represented by the symbol used in Chap. 2 for a strictly classical average, a superior bar (e.g., by $\bar{R}$), and the combination of the two types by both symbols (e.g., by $\langle \bar{R} \rangle$).

The main concern here is with systems that are in the process of changing their states. It may well be that the processes which are producing the changes of state do not affect every system in the same way. This means that, even if all of the systems start out in the same state, so that $\langle \bar{R} \rangle = \langle R \rangle$ at the start, they will very shortly be in different states. Therefore $\langle \bar{R} \rangle \neq \langle R \rangle$, and ensemble averaging will have to be used (in addition to the calculation of expectation values) in order to predict accurately the physical properties of the sample.

5.2 Coherent and Incoherent Perturbations

Perturbations that affect every system of the ensemble in the same way are called *coherent*. The most important examples for the purposes of this chapter are the following ones.

1. The oscillating electric and magnetic fields produced by inductors, capacitors, or antennas connected to tuned LC circuits. These produce coherent perturbations at radio frequencies and cause transitions between states representing different orientations of nuclear spins in beams (as described in Chap. 4) and bulk samples (nuclear magnetic and electric quadrupole resonance). They also produce transitions in beams between states representing different orientations of molecular rotation axes whenever a rotational magnetic moment is present (molecular magnetic resonance) or when a change in the orientation of an electric dipole or quadrupole moment is possible (molecular electric resonance).

2. The oscillating electric and magnetic fields produced in waveguides or cavities connected to klystrons or similar devices. These produce coherent perturbations at microwave frequencies and cause transitions between states representing different orientations of electron spins (electron paramagnetic, ferromagnetic, antiferromagnetic, and similar types of resonance experiments), and between states representing different molecular rotational frequencies (microwave rotational spectroscopy). They also produce transitions between states representing

different orbital patterns for charged particles about the lines of force in a magnetic field (cyclotron resonance).

3. The oscillating electric and magnetic fields produced in waveguides or microwave cavities by masers. These produce coherent perturbations at microwave frequencies and can therefore, in principle, be used to do any of the things that klystrons can do. In addition, they can produce useful power at some shorter wavelengths. so that the two sources together can also explore transitions between states representing different low-frequency vibrational states of molecules. These include inversion, ring puckering, wagging, and torsional oscillations.

4. The oscillating electric and magnetic fields produced in and radiated by the cavities of optical masers (lasers). These produce coherent perturbations in the near- and far-infrared, the visible, and the near-ultraviolet regions of the spectrum. They cause transitions between states representing different frequencies of stretching vibrations in molecules, and states representing different electronic states of all kinds in atoms, molecules, and ions.

It is likely that the list of sources and source wavelengths will be extended, and in the future many more areas of spectroscopy will have routine access to coherent sources e.g. synchrotron radiation, see Chap. 13. By way of contrast, Globars (used to produce infrared radiation) and heated filaments and electric discharges (used to produce incandescent and fluorescent light) give rise to incoherent radiation. The theoretical description of ensembles of systems irradiated by means of these sources must include ensemble averaging from the outset. Even the rf, microwave, maser, and laser sources are not *perfectly* coherent, and there are cases in which this fact has to be considered in computing the ensemble-averaged behavior of a system irradiated in such a way.

The most pervasive incoherent perturbations in spectroscopic experiments are relaxation processes. Their importance lies in the fact that whereas absorption or stimulated emission can always be brought about by means of a coherent perturbation (at least in principle), spontaneous emission and radiationless deexcitation nearly always occur as random processes. In other words, no two systems in a canonical ensemble necessarily "feel" each other or the heat bath in the same way at the same time, and therefore ensemble averaging is required to predict the effect of relaxation on the sample. Such processes may ordinarily be characterized by relaxation times T_1 and T_2, to be described later. A rigorous quantummechanical description of coherent states of the radiation field has been given by Glauber [1].

5.3 Strongly Coupled and Weakly Coupled Systems

It has been suggested that, in order to obtain good agreement between theory and experiment, calculations of expectation values must be supplemented by calculations of ensemble averages. The results of some ensemble-averaging processes can be anticipated. Consider a charged spinning particle, such as an electron or a proton. In Chap. 4 it was shown that such particles have a magnetic dipole moment, and there are only two eigenstates of the operator representing the projection of that dipole moment along any direction (e.g., the z axis of the laboratory coordinate system). The eigenvalues are $\pm\gamma\hbar/2$ (see Eq. 4.42). The magnetogyric ratio can be subscripted to indicate whether the spin of the electron (γ_e) or that of nucleus (γ_n) is being discussed.

Imagine a mole of metal atoms (e.g., thallium), each of which has a nucleus containing one unpaired proton. The total z component of nuclear magnetic moment of the *ensemble* should be the vector sum of all the individual nuclear moments, and therefore could range between $+N_0\gamma_n\hbar/2$ and $-N_0\gamma_n\hbar/2$ ($N_0 = 6\times 10^{23}$), depending on the distribution of systems between the two eigenstates. With a small magnetic field along the z direction to line up the spins, the value of $+N_0\gamma_n\hbar/2$ would correspond to the state of lowest energy, and therefore would be the limiting value of the z component of magnetic moment at low temperatures. Hence the ensemble-averaged nuclear magnetic moment would be $\gamma_n\hbar/2$. The ensemble-averaged electron magnetic moment under the same conditions would, however, *not* be $\gamma_e\hbar/2$. Instead, it would be very much smaller than that amount and perhaps would even have the opposite sign.

The nuclei, being separated from one another by about 10^4 nuclear radii, have very little spatial overlap among their wavefunctions. Wavefunctions of the unpaired electrons on each of the thallium atoms, however, are spread out over the entire metal crystal, so that there is a very large amount of spatial overlap among them. Whenever appreciable spatial overlap exists, the fundamental indistinguishability of the particles becomes very important, cooperative phenomena appear, and the ensemble-averaged properties cannot be computed by the simple statistical methods employed in classical mechanics. The two special types of quantum statistics developed to treat such cases are called *Fermi-Dirac* and *Bose-Einstein statistics.*

Most of the attention of this book will be focused on systems that do not have much spatial overlap of their wavefunctions. Indeed, it will be assumed that these systems are coupled together only in two ways. They all are being irradiated by the same radiation field, and they all are relaxing by transfer of energy of the same heat bath. Therefore it will not be necessary to use Bose-Einstein or Fermi-Dirac statistics, and ensemble averages can be computed by the straightforward techniques of classical statistics. In particular, the ensemble-quantum average of the property represented by the operator R can be calculated as follows:

$$\langle\bar{R}\rangle = \frac{1}{N}\sum_{n=1}^{N}\langle R\rangle_n. \tag{5.2a}$$

The conditions under which Eq. (5.2a) yields the correct ensemble-quantum average are as follows.

1. The wavefunctions for the constituent systems must have little spatial overlap.

2. The systems must interact with one another only weakly (this condition is not independent of the first).

Equation (5.2a) reduces to

$$\langle\bar{R}\rangle = \langle R\rangle \tag{5.2b}$$

if the following conditions are met.

3. Any perturbation that is applied to the system in order to produce absorption or stimulated emission must be coherent.

4. At the time that the perturbation is "switched on", all the systems of the ensemble must be in identical states (i.e., the internal degrees of freedom for each quantum system must be represented by the same wavefunction).

5. The experiment must not last longer than $\Upsilon \ll T_1, T_2$ (to be discussed later), so that relaxation effects will not be important.

A convenient technique for calculating average values when these five requirements are satisfied is presented in the next two sections.

5.4 Computing Expectation Values from Superposition Coefficients

Just as was the case in Chap. 4 (Eqs. 4.12 and 4.20), it is necessary to consider a quantum system that is in a superposition of eigenstates. Only two eigenstates will be considered, as before. Equations 2.24 and 2.50 may be combined to provide an expression for the wavefunction:

$$\Psi(q,t) = c_1(t)T_1(t)\phi_1(q) + c_2(t)T_2(t)\phi_2(q). \tag{5.3}$$

It is convenient to combine all of the time-dependent factors of each term on the right hand side of Eq. (5.3):

$$a_j(t) = c_j(t)T_j(t), \quad j = 1,2. \tag{5.4}$$

In Chaps. 2 and 4, superposition wavefunctions were used to calculate the expectation value of some physically observable property of the system, R (see Eqs. 2.83 and 4.13). The same thing must be done here. Equation 5.3 may be substituted into Eq. 2.82, using the notation introduced in Eq. 5.4, to produce

$$\langle R(t)\rangle_{\Psi\Psi} = |a_1(t)|^2 R_{11} + |a_2(t)|^2 R_{22} + 2\mathrm{Re}[a_1^*(t)a_2(t)R_{12}]. \tag{5.5}$$

Equation (5.5) expresses, in slightly modified form, exactly the same information as is contained in Eq. 2.83. The R_{jk}'s are defined in Eq. 2.85 where it may be seen that they all are independent of the time.

The products $a_j a_k^*$ can be arranged to form a 2×2 matrix, just as can the quantities R_{jk}. In the latter case, the matrix represents some operator (e.g., the dipole moment). It may be imagined that the matrix $a_j^* a_k$ also represents an operator, which may be called D:

$$D_{jk} \equiv a_j a_k^*. \tag{5.6}$$

Equation (5.5) can then be rewritten, replacing quantities that appear on the right-hand side by the elements of the corresponding matrices:

$$\langle R(t)\rangle_{\Psi\Psi} = D_{11}R_{11} + D_{22}R_{22} + D_{21}R_{12} + D_{12}R_{21}. \tag{5.7}$$

What are the states represented by $\psi_1 = T_1\Phi_1$ and $\psi_2 = T_2\Phi_2$ in Eq. 5.3? Suppose that they are eigenfunctions of an operator representing another physically observable property of the system (e.g., the energy). In that case, ψ_1 and ψ_2 are members of a complete set, $\{\psi\}_E$. (In the interest of simplicity, it will be assumed that they are the *only* members of the set $\{\psi\}_E$.) Each member of the set depends on the independent variables q and t and is factorable into two terms, $\Phi(q)$ and $T(t)$, as in Eq. 2.24. These facts imply that Φ_1 and Φ_2 together also constitute a complete set of functions that can be used in linear combination to represent any state of the system at any given point in time. The only limitation is that the values of the coefficients of Φ_1 and Φ_2 which are appropriate at one instant of time are, in general, not satisfactory at any other instant in time. This causes no problem, however, because the coefficients themselves can be considered to be continuous scalar functions of t.

What happens, then, if an operator operates upon Φ_1? The result will be another state function. This state function can be written as a linear combination of Φ_1 and Φ_2 because, as has just been stated, they constitute a complete set. In particular,

$$\mathsf{D}\Phi_1(q) = b_{11}(t)\Phi_1(q) + b_{12}(t)\Phi_2(q). \tag{5.8}$$

From the preceding discussion, it can be seen why b_{11} and b_{22} are, in general, time dependent. If both sides of Eq. 5.8 are multiplied by Φ_1^* and integrated over the entire range of q, Eq. 5.8 becomes

$$\int \Phi_1^*(q)\mathsf{D}\Phi_1(q)dq = \int b_{11}|\Phi_1(q)|^2 dq + \int b_{12}\Phi_1^*(q)\Phi_2(q)dq. \tag{5.9}$$

The coefficients b_{11} and b_{12} are independent of q and may be taken out of the integrals on the right-hand side. It has previously been assumed (Eq. 2.46) that the Φ's are orthogonal, and the integral on the left-hand side is, by definition, D_{11}. Therefore

$$D_{11} = b_{11}. \tag{5.10}$$

This entire process can be repeated using Φ_1^*, Φ_2, and Φ_2^* in place of Φ_1 on the left-hand side of Eq. 5.8, and Φ_1, Φ_2^*, and Φ_2 in place of Φ_1^* on the left-hand side of Eq. 5.9. The results can be expressed by a generalization of Eq. 5.10:

$$D_{jk} = b_{jk}, \quad j = 1, 2. \tag{5.11}$$

These results can be substituted back into Eq. (5.8):

$$\mathsf{D}\Phi_j = D_{j1}\Phi_1 + D_{j2}\Phi_2. \tag{5.12}$$

Now, the entire process described by Eqs. 5.8–5.12 is repeated, using the operator R in place of D. The result, analogous to Eq. 5.12, is

$$\mathsf{R}\Phi_k = R_{k1}\Phi_1 + R_{k2}\Phi_2. \tag{5.13}$$

Next, D operates upon both sides of Eq. 5.13:

$$\mathsf{DR}\Phi_k = \mathsf{D}(R_{k1}\Phi_1 + R_{k2}\Phi_2). \tag{5.14}$$

On the right-hand side of Eq. (5.14), the operator D commutes with the scalars R_{kj}, so that each of the two terms is of the form $R_{kj}\mathsf{D}\Phi_j$. Therefore the expression in Eq. (5.12) may be substituted into each term on the right-hand side of Eq. (5.14), to obtain

$$\mathsf{DR}\Phi_k = R_{k1}(D_{11}\Phi_1 + D_{12}\Phi_2) + R_{k2}(D_{21}\Phi_1 + D_{22}\Phi_2). \tag{5.15}$$

It was pointed out in Chap. 2 that the product of operators DR is itself an operator. Therefore an expression for $\mathsf{DR}\Phi_k$ might have been obtained using the procedure outlined in Eqs. 5.8–5.12. The result would have been as follows:

$$\mathsf{DR}\Phi_k = (\mathsf{DR})_{k1}\Phi_1 + (\mathsf{DR})_{k2}\Phi_2. \tag{5.16}$$

A comparison of the right-hand sides of Eq. 5.15 and 5.16, term by term, shows that

$$(\mathsf{DR})_{kj} = R_{k1}D_{1j} + R_{k2}D_{2j}. \tag{5.17}$$

Equation (5.17) is merely the general formula for computing an element of the product of two 2×2 matrices. If a system with more eigenstates were chosen ($M > 2$), Eq. 5.17 would have contained the general formula for computing an element of the product of two $M \times M$ matrices. All operators in quantum mechanics can be represented by matrices that obey these multiplication rules.

The reason for introducing matrices at this point, however, is rather special. The first and last terms on the right-hand side of Eq. 5.7 can be grouped together and compared with the right-hand side of Eq. 5.17. The second and third terms on the right-hand side of Eq. 5.7 can be similarly grouped and compared. It is seen that

$$\langle R(t)\rangle_{\Psi\Psi} = (\mathsf{DR})_{11} + (\mathsf{DR})_{22}. \tag{5.18}$$

Equation 5.18 is usually written as

$$\langle R(t)\rangle_{\Psi\Psi} = \mathrm{Tr}[\mathsf{DR}]. \tag{5.19}$$

The symbol Tr [x] (read "trace of x") means the sum of the diagonal elements of the square matrix [x] with elements x_{jk}.

The formula in Eq. 5.19 is remarkable for several reasons. In the first place, although it was suggested in Eq. 5.5 that the operator R represented the dipole moment of the system, no use was made of any particular properties of the dipole moment in order to derive the formula. Therefore Eq. 5.19 is valid for *any* operator R representing a physically observable property of the system (e.g., energy, angular momentum, or linear momentum). Next, in the particular picture of quantum mechanics that has been used, the operator R will not ordinarily explicitly contain the time. Since the basis set of functions used to form the matrix, namely, Φ_1 and Φ_2, have only time-independent members, the elements of the matrix of R found in this basis will also be time independent. All of the time dependence of the problem, both from the time-dependent parts of the complete wavefunctions ψ_1 and ψ_2 and from the c's, has been lumped together in the matrix of D. Finally, although it was suggested (in the discussion following Eq. 5.7) that Φ_1 and Φ_2 might be eigenfunctions of the energy operator, no use was made of any particular properties of those eigenfunctions to derive the formula. Therefore the validity of Eq. 5.19 is independent of the basis set used to form the matrices of D and R, save only that the *same* basis must be used for both.

A universal procedure for finding the time-dependent expectation value of any property has therefore been discovered. First, choose any complete orthonormal set of time-independent wavefunctions $\{\Phi_j\}$. In that basis, form the time-independent matrix of the operator R representing the physically observable property of interest. Also form the matrix of the time-dependent operator D in the same basis. Multiply the two $M \times M$ matrices together, and sum the M diagonal elements of the resultant $M \times M$ product matrix. The result will be the expectation value of the property of interest, with the correct time dependence. It was stated at the outset that this derivation would utilize a particular choice of M, namely two. A more general treatment, appropriate for any value of M, may be found in the book by Slichter [2].

5.5 Equations of Motion for the Operator D

It has been established that the operator D is a remarkably important one in time-dependent quantum mechanics. It is therefore of great interest to discover how to compute its matrix elements. This can be done by substituting from Eq. 5.3 into the time-dependent Schrödinger equation (Eq. 2.23), using the definition of the a_j's from Eq. 5.4 and the distributive law:

$$a_1 \mathsf{H}\Phi_1 + a_2 \mathsf{H}\Phi_2 = i\hbar\left[\Phi_1\left(\frac{\partial a_1}{\partial t}\right) + \Phi_2\left(\frac{\partial a_2}{\partial t}\right)\right]. \tag{5.20}$$

The fact that the coefficients a_j do not depend on spatial or spin coordinates (and are therefore not affected by the Hamiltonian operator, H) has been used on the left-hand side of Eq. 5.20.

Both sides of Eq. 5.20 may be multiplied by Φ_1^* and then integrated over q. The result is

$$i\hbar\left(\frac{\partial a_1}{\partial t}\right) = a_1 H_{11} + a_2 H_{12}. \tag{5.21}$$

If both sides had instead been multiplied by Φ_2^* and then integrated, the result would have been

$$i\hbar\left(\frac{\partial a_2}{\partial t}\right) = a_1 H_{21} + a_2 H_{22}. \tag{5.22}$$

To find a differential equation for a particular matrix element of D, the definition in Eq. (5.6) may be used:

$$\frac{\partial D_{11}}{\partial t} \equiv \frac{\partial(a_1^* a_1)}{\partial t}. \tag{5.23}$$

Therefore, from the chain rule,

$$\frac{\partial D_{11}}{\partial t} = \left[\frac{\partial a_1^*}{\partial t}\right] a_1 + a_1^* \left[\frac{\partial a_1}{\partial t}\right]. \tag{5.24}$$

The second term on the right-hand side of Eq. 5.24 may be obtained by multiplying Eq. 5.21 by $a_1^*/i\hbar$, and the first term, by multiplying the complex conjugate of Eq. 5.21 by $-a_1/i\hbar$. The results, when substituted into Eq. 5.24, yield

$$\frac{\partial D_{11}}{\partial t} = \frac{-a_1}{i\hbar}(a_1^* H_{11}^* + a_2^* H_{12}^*) + \frac{a_1^*}{i\hbar}(a_1 H_{11} + a_2 H_{12}). \tag{5.25}$$

On the right-hand side of Eq. (5.25), the definition of D_{jk} from Eq. (5.6) may be used again:

$$\frac{\partial D_{11}}{\partial t} = -\frac{1}{i\hbar}[(D_{11} H_{11}^* + D_{12} H_{12}^*) - (D_{11} H_{11} + D_{21} H_{12})]. \tag{5.26}$$

The Hamiltonian, H, like all operators in quantum mechanics, satisfies the Hermitian condition, Eq. 2.104. By using that fact and a slight rearrangement of terms, Eq. 5.26 may be rewritten as follows:

$$\frac{\partial D_{11}}{\partial t} = -\frac{1}{i\hbar}[(D_{11} H_{11} + D_{12} H_{21}) - (H_{11} D_{11} + H_{12} D_{21})]. \tag{5.27}$$

Each sum in parentheses on the right-hand side of Eq. 5.27 may be recognized (with the aid of Eq. 5.17) as the matrix element of the product of two matrices:

$$\frac{\partial D_{11}}{\partial t} = -\frac{1}{i\hbar}[(\mathsf{DH})_{11} - (\mathsf{HD})_{11}]. \tag{5.28}$$

Now, the term in square brackets on the right-hand side of Eq. 5.28 may be recognized (with the aid of Eq. 2.7) as a matrix element of the commutator of the operators D and H:

$$\frac{\partial D_{11}}{\partial t} = -\frac{1}{i\hbar}[\mathsf{D}, \mathsf{H}]_{11}. \tag{5.29}$$

More generally,

$$i\hbar \frac{\partial D_{jk}}{\partial t} = [\mathsf{H}, \mathsf{D}]_{jk}, \qquad 1 < j < M, \quad 1 < k < M. \tag{5.30}$$

It is customary to rewrite Eq. 5.30 as a relationship between the operators D and H themselves,

$$i\hbar\left[\frac{\partial D}{\partial t}\right] = [\mathsf{H},\mathsf{D}]. \tag{5.31}$$

It is to be understood that the matrix elements of Eq. 5.31 must be used in performing the indicated computations.

Equation 5.31, then, represents M^2 equations in the M^2 unknown matrix elements of D. All these equations are first-order differential equations, and therefore M^2 arbitrary constants will appear in their solutions. These M^2 constants must be fixed by boundary conditions, and this is ordinarily done by specifying the entire D matrix at time $t = 0$. Once the initial state is specified, one may obtain (in principle) exact expressions for the time dependence of the matrix elements of D. These formulas, in turn, may be used in calculating the product of the D matrix and that of any other operator. Once this product is obtained, the time-dependent behavior of the expectation value of any physically observable property may be calculated by means of Eq. 5.19.

It can be seen from this discussion that the material presented in this section is an outline of the general solution to any problem one might wish to undertake in time-dependent quantum mechanics (but only if Eqs. 5.2a and 5.2b are satisfied). The circumstances under which this condition is met are limited to single particles, or ensembles of particles satisfying all five of the requirements listed in Sect. 5.3. It is now necessary to generalize this treatment so that there will be adequate theoretical means for the description of ensembles which satisfy only Eq. 5.2a. Such ensembles fail to meet one or more of the requirements numbered 3, 4, and 5 in Sect. 5.3.

5.6 The Density Operator

In a canonical ensemble consisting of N identical quantum systems, the nth system has internal coordinates q_n and a wavefunction

$$\Psi_n(q_n,t) = c_{n1}\psi_1(q_n,t) + c_{n2}\psi_2(q_n,t). \tag{5.32}$$

Note that Eq. 5.32 is identical with Eq. 5.3 except for the additional subscript.

By following the procedure outlined in Sect. 5.5, an operator D_n with elements defined as in Eq. 5.6 may be obtained:

$$D_{jk}^{(n)} = a_j^{(n)} a_k^{(n)*}. \tag{5.33}$$

This operator can be used to compute expectation values of any physical quantity, for example, the quantity represented by R (see Eq. 5.19):

$$\langle R(t)\rangle_n = \mathrm{Tr}\,[\mathsf{D}_n\,\mathsf{R}]. \tag{5.34}$$

Note that the operator R is the same for every system in the ensemble and therefore need not bear the subscript n.

Because all ensembles with appreciable spatial overlap of wavefunctions representing different systems have been excluded from consideration, the computation of the ensemble average, $\langle\bar{R}\rangle$, is very simple:

$$\langle\bar{R}\rangle = \frac{1}{N}\sum_{n=1}^{N}\langle R\rangle_n. \tag{5.35}$$

Now Eq. 5.34 may be used on the right-hand side of Eq. 5.35:

$$\langle\bar{R}\rangle = \frac{1}{N}\sum_{n=1}^{N}\mathrm{Tr}[\mathsf{D}_n\,\mathsf{R}]. \tag{5.36}$$

Calculating the trace of a matrix is an operation that follows the distributive law (because the multiplication of matrices obeys that law):

$$\mathrm{Tr}(\mathsf{AB}) + \mathrm{Tr}(\mathsf{AC}) = \mathrm{Tr}[\mathsf{A}(\mathsf{B}+\mathsf{C})]. \tag{5.37}$$

Equation 5.37 may be used on the right-hand side of Eq. 5.36 to obtain

$$\langle\bar{R}\rangle = \mathrm{Tr}\left[\left(\frac{1}{N}\sum_{n=1}^{N}\mathsf{D}_n\right)\mathsf{R}\right]. \tag{5.38}$$

The result expressed algebraically in Eq. 5.38 can be stated in ordinary language as follows. The ensemble-quantum averaged value of R is the trace of the product of the matrices of R and $\underline{\rho}$, where the matrix of $\underline{\rho}$ is the linear average of the D_n matrices representing individual systems of the ensemble:

$$\underline{\rho} \equiv \frac{1}{N}\sum_{n=1}^{N}\mathsf{D}_n. \tag{5.39}$$

The operator $\underline{\rho}$ is called the *density operator*, and its matrix, the *density matrix*. It is easy to show that the equation of motion for $\underline{\rho}$ is the same as that for each D separately (see Eq. 5.31):

$$i\hbar\left(\frac{\partial\underline{\rho}}{\partial t}\right) = [\mathsf{H},\underline{\rho}]. \tag{5.40}$$

An equation like Eq. 5.40 occurs in classical statistical mechanics, where it is called the *Liouville equation.*

The average values of physical observables (e.g., the one represented by R) are calculated by means of a formula analogous to Eq. (5.34):

$$\langle\overline{R(t)}\rangle_N = \mathrm{Tr}[\underline{\rho}(t)\mathsf{R}]. \tag{5.41}$$

In summary, the procedure for calculating the result of any experiment by means of quantum mechanics is as follows. Choose an initial state for the ensemble by choosing the complex elements of the density matrix of the density operator, $\underline{\rho}$, at time $t = 0$. Using these M^2 constants [only $M(M-1)$ of them are independent because the diagonal elements must be real], solve Eq. 5.40 for the values of the matrix elements of $\underline{\rho}$ at the time of measurement, t. Calculate the time-independent matrix of the operator representing the desired physical property (e.g., R), using the same basis. Multiply the two matrices and sum the diagonal elements as indicated in Eq. 5.41; the result will be the experimental value of the desired property at time t, correctly averaged over all systems of the ensemble.

5.7 Properties of the Density Matrix

Before the calculations for various cases of interest are performed, some general remarks can be made. If the macrosystem (ensemble) is a closed one, the identification of the terms on the right-hand side of Eq. 5.39 with particular microsystems may not change with time. Since the trace of each of the $[\mathsf{D}_n]$ matrices is 1, the trace of $[\underline{\rho}]$ will also be 1, for all time. The invariance of the unit trace of $[\underline{\rho}]$ therefore reflects the law of conservation of matter in a closed macrosystem (ensemble).

The Hamiltonian matrix, which drives the density matrix in accordance with Eq. 5.40, ordinarily contains terms of three different kinds. One group of terms, called collectively H_0, is time independent and has relatively large matrix elements. It establishes the stationary states of the system (see Eqs. 4.18 and 4.19) and is therefore the object of greatest attention in conventional quantum mechanics. Another group of terms, called H_1 (or sometimes V– see Eqs. 4.24 and 4.25), is time independent and provides the mechanism for transitions to occur between the eigenstates of H_0. The third group of terms, called H_R, describes the processes by means of which the quantum systems can exchange with each other or with the heat bath. The H_R terms provide the mechanism by means of which the ensemble of systems will relax back to an equilibrium distribution among the states of H_0, whenever it is displaced from equilibrium by H_1.

The question of the choice of basis (complete set of eigenstates defined in Chap. 2) for the matrices $[\mathsf{D}_n]$, $[\mathsf{H}]$, and $[\mathsf{R}]$ has been left open. The only thing decided is that, for each system in the ensemble, the same basis must be used for all three matrices. The most convenient choice of basis is the set of eigenfunctions of the operator H_0. Since all of the $[\mathsf{D}_n]$'s are Hermitian, $[\underline{\rho}]$ will also be Hermitian. (This is the reason for the previously mentioned fact that each diagonal element is a real number.) The complex off-diagonal elements of $[\mathsf{D}_n]$ can be written (see Eq. 5.33) as

$$D_{jk}^{(n)} = |D_{jk}^{(n)}|\exp[i\Omega_{jk}^{(n)}]. \tag{5.42}$$

The corresponding off-diagonal element of $[\underline{\rho}]$ can therefore be written (see

Eq. 5.39):

$$\rho_{jk} = \frac{1}{N}\sum_{n=1}^{N} |D_{jk}^{(n)}|\{\cos[\Omega_{jk}^{(n)}] + i\sin[\Omega_{jk}^{(n)}]\}. \tag{5.43}$$

For a sufficiently large ensemble (N→∞), the sums in Eq. 5.43 can be replaced by integrals. For example, the first term can be written as

$$\frac{1}{N}\sum_{n=1}^{N} |D_{jk}^{(n)}|\cos[\Omega_{jk}^{(n)}] = \lim_{N\to\infty} \frac{1}{2}\int_{-\Omega}^{\Omega} |D(x)|\cos x\, dx. \tag{5.44}$$

One of the basic assumptions of statistical mechanics is the *principle of equal apriori probabilities.* This hypothesis states that all configurations of any system (or ensemble of systems) which have the same energy are equally likely at equilibrium. Since the energy of a state is not affected by the phase of the corresponding wavefunction, the principle of equal apriori probabilities will ensure that the phases Ω_{jk} of D_{jk} are distributed at random throughout the ensemble. The integral on the right-hand side of Eq. 5.44 represents the net effect of an infinite number of oscillations at an infinite number of frequencies, all with amplitudes bounded by $0 < |D(\Omega)| < 1$. (Remember the definition of D in Eq. 5.33, and the fact that $|a_j|$ must lie between $+1$ and -1.) In this jumble of oscillations, positive values of $\cos\Omega$ must be as numerous as negative ones, and the integral in Eq. 5.44 must therefore be zero. In other words, in an ensemble at equilibrium, all the off-diagonal elements of the density matrix must be zero, because of the destructive interferences among the corresponding elements of the constituent matrices. This result is called the *hypothesis of random phases.*

The elements of the density matrix that do not vanish when the ensemble represented by $\underline{\rho}$ is at thermodynamic equilibrium are the diagonal ones:

$$\rho_{jj} = \frac{1}{N}\sum_{n=1}^{N} D_{jj}^{(n)} = \frac{1}{N}\sum_{n=1}^{N} |a_j^{(n)}|^2. \tag{5.45}$$

It is obvious from inspection of Eq. 5.45 that the jth diagonal element at equilibrium merely represents the fraction of the systems of the ensemble which will give the answer E_j if their energy is measured. In other words, the diagonal elements are the energy-eigenstate-occupation probabilities, which, in an ensemble of semi-isolated systems, are given by the Boltzmann factors:

$$\overline{|a_j^{(n)}|^2} = \frac{\exp(-\mathscr{E}_j/k_0T)}{Q}, \tag{5.46}$$

where Q is the partition function:

$$Q \equiv \sum_{j=1}^{M} \exp\left(\frac{-\mathscr{E}_j}{k_0T}\right). \tag{5.47}$$

5.8 Effect of Relaxation on the Density Matrix

In this book, several different quantum-mechanical problems will be considered, and the corresponding H_0 and H_1 terms will be characterized explicitly. The same will not be done for H_R, however, because it is ordinarily much more complicated than the others. Instead, a phenomenological approach will be adopted as follows. First, the density matrix at equilibrium, $[\underline{\rho}^e]$, is given:

$$[\underline{\rho}^e] = \frac{1}{Q}\begin{bmatrix} \exp(-\mathscr{E}_1/k_0T) & 0 & \cdots & 0 \\ 0 & \exp(-\mathscr{E}_2/k_0T) & \cdots & 0 \\ 0 & 0 & & 0 \\ \vdots & \vdots & & \vdots \\ 0 & \cdots & \cdots & \exp(-\mathscr{E}_M/k_0T) \end{bmatrix}. \tag{5.48}$$

Next, it will be assumed that the perturbation represented by H_1 commences, drives the ensemble away from equilibrium, and then ceases to act. At this point, each element of the density matrix will begin to decay back to its equilibrium value:

$$\frac{d\rho_{jk}}{dt} = -\frac{(\rho_{jk} - \rho^e_{jk})}{T_{jk}}. \tag{5.49}$$

There may be terms in addition to the one shown on the right-hand side of Eq. 5.49, but the simplest assumption is to neglect them and imagine that all the decays are first order, with a rate constant T_{jk}^{-1}. Not all the T_{jk}'s are independent of one another; T_{jk} must equal T_{kj}, for example, because ρ_{jk} is merely the complex conjugate of ρ_{kj} and must therefore decay to equilibrium at the same rate (remember that $\underline{\rho}$ is Hermitian). There are also strictures upon the relaxation times for diagonal elements because, regardless of the choice of the set $\{T_{jk}\}_M$, one must always have

$$\sum_{j=1}^{M} \rho_{jj} = 1. \tag{5.50}$$

In the simplest nontrivial case, there are only two stationary states ($M = 2$) and therefore only two rate constants. The relaxation times for the diagonal elements must be exactly equal. Conventionally, this time is called the *longitudinal relaxation time* and has the symbol T_1. Since there are only two off-diagonal elements, $\rho_{12} = \rho^*_{21}$, there is only one possible additional relaxation time, conventionally called the *tranverse relaxation time*, T_2. It should also be remembered that, because of the hypothesis of random phases, $\rho^e_{jk} = 0$ for $j \neq k$.

In summary, the effect of H_R on the density matrix is equivalent to the addition of decay terms of the form shown in Eq. 5.49 to the rate equations. For a quantum system with only two eigenstates, the four differential equations for the elements of $[\rho]$ are as follows:

$$\frac{d\rho_{11}}{dt} = \frac{[(\mathsf{H}_0 + \mathsf{H}_1), \underline{\rho}]_{11}}{i\hbar} - \frac{\rho_{11} - \rho_{11}^e}{T_1}, \tag{5.51}$$

$$\frac{d\rho_{22}}{dt} = \frac{[(\mathsf{H}_0 + \mathsf{H}_1), \underline{\rho}]_{22}}{i\hbar} - \frac{\rho_{22} - \rho_{22}^e}{T_1}, \tag{5.52}$$

$$\frac{d\rho_{12}}{dt} = \frac{[(\mathsf{H}_0 + \mathsf{H}_1), \underline{\rho}]_{12}}{i\hbar} - \frac{\rho_{12}}{T_2}, \tag{5.53}$$

and

$$\frac{d\rho_{21}}{dt} = \frac{[(\mathsf{H}_0 + \mathsf{H}_1), \underline{\rho}]_{21}}{i\hbar} - \frac{\rho_{21}}{T_2}. \tag{5.54}$$

In obtaining Eqs. 5.51–5.54, use has been made of Eqs. 5.40, 5.49, and 5.50, the hypothesis of random phases, and the fact that the total Hamilitonian, H, is given by

$$\mathsf{H} = \mathsf{H}_0 + \mathsf{H}_1 + \mathsf{H}_R. \tag{5.55}$$

5.9 Equations of Motion for the Density Matrix

Equations 5.53 and 5.54 may be simplified further, using the facts that the basis of the density matrix is the set of eigenfunctions of H_0 and that the distributive law holds for commutators:

$$[(\mathsf{H}_0 + \mathsf{H}_1), \underline{\rho}] = [\mathsf{H}_0, \underline{\rho}] + [\mathsf{H}_1, \underline{\rho}], \tag{5.56}$$

$$[\mathsf{H}_0, \underline{\rho}] = \mathsf{H}_0\underline{\rho} - \underline{\rho}\mathsf{H}_0, \tag{5.57}$$

$$(\mathsf{H}_0\underline{\rho})_{pq} = \sum_{r=1}^{M} (\mathsf{H}_0)_{pr}\rho_{rq}, \tag{5.58}$$

and

$$(\mathsf{H}_0)_{pr} = (\mathsf{H}_0)_{pr}\delta_{pr}, \tag{5.59}$$

where δ_{pr} is the Kronecker delta defined in Eq. 2.47. Therefore

$$(\mathsf{H}_0\underline{\rho})_{pq} = (\mathsf{H}_0)_{pp}\rho_{pq} \tag{5.60}$$

and

$$[\mathsf{H}_0, \underline{\rho}]_{pq} = [(\mathsf{H}_0)_{pp} - (\mathsf{H}_0)_{qq}]\rho_{pq}. \tag{5.61}$$

It can now be seen that the choice of the set of eigenfunctions of H_0 as the bases for all of the matrices has had two useful consequences. First, it has made it possible to interpret the diagonal elements of $[\underline{\rho}]$ as the fractional occupancies of the various stationary states of H_0 by the quantum systems in the ensemble

(Eqs. 5.45 to 5.47). Second, the matrix of H_0 in this basis has elements of magnitude zero everywhere except on the principal diagonal (Eq. 5.59). The fact that the matrix of H_0 is diagonal permits a simplification of the $[\mathsf{H}_0, \underline{\rho}]$ term in the Liouville equation, as has been seen.

It is also possible to simplify the $[\mathsf{H}_1, \underline{\rho}]$ term slightly by insisting that this matrix of H_1 has nonzero elements only *off* the principal diagonal. If the operator H_1 as initially chosen has nonzero elements both on and off the diagonal, both H_0 and H_1 can be redefined in such a way that the new H_0 is the sum of the old H_0 plus the diagonal elements of the old H_1; the remaining off-diagonal elements constitute the new H_1:

$$[\mathsf{H}_1, \underline{\rho}] = \mathsf{H}_1 \underline{\rho} - \underline{\rho} \mathsf{H}_1, \tag{5.62}$$

$$(\mathsf{H}_1 \underline{\rho})_{pq} = \sum_{r=1}^{M} (\mathsf{H}_1)_{pr} \rho_{rq}. \tag{5.63}$$

For $M = 2$,

$$(\mathsf{H}_1 \underline{\rho})_{pq} = (\mathsf{H}_1)_{p1} \rho_{1q} + (\mathsf{H}_1)_{p2} \rho_{2q} \tag{5.64}$$

and

$$(\mathsf{H}_1)_{rs} = 0 \quad \text{unless } r \neq s. \tag{5.65}$$

Therefore

$$[\mathsf{H}_1, \underline{\rho}]_{11} = 2i \,\mathrm{Im}[(\mathsf{H}_1)_{12} \rho_{21}] = [\mathsf{H}_1, \underline{\rho}]_{22}, \tag{5.66}$$

and

$$[\mathsf{H}_1, \underline{\rho}]_{12} = (\mathsf{H}_1)_{12} (\rho_{22} - \rho_{11}) = -[\mathsf{H}_1, \underline{\rho}]^*_{21}. \tag{5.67}$$

Equations 5.56, 5.64, 5.66, and 5.67 may be used in Eqs. 5.51 to 5.54, together with

$$[(\mathsf{H}_0)_{pp} - (\mathsf{H}_0)_{qq}] \equiv \mathscr{E}_p - \mathscr{E}_q \tag{5.68}$$

and the definition of ω_0 in Eq. 2.68, to obtain

$$\frac{d\rho_{11}}{dt} = \frac{2\mathrm{Im}\,[(\mathsf{H}_1)_{12} \rho_{21}]}{\hbar} - \frac{\rho_{11} - \rho^e_{11}}{T_1}, \tag{5.69}$$

$$\frac{d\rho_{22}}{dt} = -\frac{2\mathrm{Im}\,[(\mathsf{H}_1)_{12} \rho_{21}]}{\hbar} - \frac{\rho_{22} - \rho^e_{22}}{T_1}, \tag{5.70}$$

$$\frac{d\rho_{12}}{dt} = i\omega_0 \rho_{12} + \frac{(\mathsf{H}_1)_{12}(\rho_{22} - \rho_{11})}{i\hbar} - \frac{\rho_{12}}{T_2}, \tag{5.71}$$

and

$$\frac{d\rho_{21}}{dt} = -i\omega_0 \rho_{21} - \frac{(\mathsf{H}_1)^*_{12}(\rho_{22} - \rho_{11})}{i\hbar} - \frac{\rho_{21}}{T_2}. \tag{5.72}$$

5.10 Coherence in Ensembles of Quantum Radiators

No further progress in the solution of Eqs. 5.69 to 5.72, the equations of motion for the elements of the density matrix, will be made in this chapter.

Instead, the dependence of the elements of both ($\underline{\rho}$) and [D_n] on the phases $\Omega_{jk}^{(n)}$, given in Eqs. 5.42 to 5.44 will be discussed. The definition of the elements of [D_n] (Eqs. 5.33 and 5.42) and the definition of the a's (Eq. 5.4) may be combined to obtain

$$D_{jk}^{(n)} = |c_j^{(n)}||c_k^{(n)}|\exp[i\Omega_{jk}^{(n)}]. \tag{5.73}$$

By analogy to Eq. (4.56),

$$\Omega_{jk}^{(n)} = \S_{jk}^{(n)} + \frac{(\mathscr{E}_j - \mathscr{E}_k)t}{\hbar}. \tag{5.74}$$

The behavior of the ensemble-averaged properties is very sensitively dependent on the values of the M phase constants, $\Omega_{jk}^{(n)}(0) = \S_{jk}^{(n)}$. It has already been stated that if they are distributed at random all of the off-diagonal elements of the density matrix will vanish. But suppose that they are all the same. This means that all of the matter waves are oscillating in step with one another, in a way which is analogous to the synchrony among coherent light waves. The importance of phase coherence among the wavefunctions of quantum systems can be seen by examining Eqs. 5.42–5.44.

It is a cliché of statistical mechanics that one cannot deduce all the properties of microsystems form the ensemble-averaged behavior of the macrosystem. The reason for this can be seen from Eq. (5.44). When the Ω's start to differ from one another, even very slightly, the ensemble-averaged properties of the sample begin to blur as the off-diagonal elements start to cancel out. The difference between ensemble-averaged properties and the properties of individual microsystems is always of the same nature: the microsystems have a greater repertoire and display a wider variety of behavior. The system can do everything that the ensemble can do, and more; but the ensemble cannot do anything that an individual system in the ensemble cannot do. This is due to the fact that the systems are only weakly interacting, so that the matrix [$\underline{\rho}$] is simply a sum of the matrices [D_n] (see Eq. 5.39).

Now it can be seen (from Eq. 5.43) that as the phases $\Omega_{jk}^{(n)}$ of the systems within the ensemble become more and more nearly the same, the density matrix approaches more and more closely the typical [D_n] matrix in all of its elements. With perfect coherence among the matter waves of the systems, perfect mimicry of the expectation value of behavior of each system by the entire ensemble will be achieved. In this way, microscopic (quantum) behavior of individual atoms, ions, and molecules will be "seen" in the macroscopic world. This is contrary to the claims of universal validity for the cliché which says that one cannot visualize quantum-mechanical behavior from macroscopic laboratory experiments. The cliché is wrong in principle, but usually correct in practice because it is ordinarily difficult in the laboratory to bring about the phase coherence necessary to produce counter-examples. At first sight, in fact, the reader might imagine that bringing about coherence between the phases of Avogadro's number of quantum systems would be completely impossible. Apparently, a quantum-mechanical version of Maxwell's demon would be required! However, Dicke [3] has described how the phases in an ensemble of excited quantum systems could spontaneously achieve coherence in a radiative process. His work gives a great deal of insight into the process by which laser action is initiated. Dicke introduced a new quantum number (the cooperation number) to describe the circumstances that give rise to these phenomena, and pointed several unique properties of the radiation that would be emitted by phase-correlated systems.

One of Dicke's results can be made plausible by an argument based on classical electromagnetic theory. The radiating atom has been viewed as a tiny dipole, oscillating at the beat frequency between two stationary-state wavefunctions. In order for emission to occur, it is necessary that the dipole first receive energy from some external source. Suppose that the driving field is represented in this case by a tiny transmitter connected to the dipole, as would be the case if the atom and its driver were really a tiny radio or television station. The resultant electromagnetic waves are broadcast in all directions (except along the dipole axis), just as one calculates in classical electromagnetic theory. Now, imagine a second antenna placed next to the first. This antenna is identical in every way to the first one and is driven by a transmitter at exactly the same frequency (there being no governmental regulatory agency in Hilbert space to prevent this). The waves radiated by these two dipoles will overlap one another spatially, and the corresponding electric and magnetic fields will constructively and destructively interfere. The resultant pattern of broadcast energy flux will no longer be as symmetric as it was when there was only one dipole; zones of "darkness" and "brightness" will appear. Precisely where these zones appear depends on the relative values of the phase constants of the two antennas.

More antennas can be added to the first two. If a pair of these is selected at random, and a point of observation is chosen, the electric fields $\mathbf{E}_1$ and $\mathbf{E}_2$ will add to form a resultant. The square of this resultant will be proportional to the irradiance, J_{12}, at the point of observation (see Eq. 1.6):

$$J_{12} \propto |\mathbf{E}_1 + \mathbf{E}_2| \cdot |\mathbf{E}_1 + \mathbf{E}_2| = |E_1|^2 + |E_2|^2 + 2|E_1||E_2|\cos\theta_{12}. \tag{5.75}$$

The angle θ_{12} in Eq. (5.75) depends on many different factors – the orientation of the two dipoles, the distance between them, and the location of the point of observation are examples. But it also depends on the difference between the phases of oscillation of the two dipoles, and this is the most important for the present discussion. While the antennas are being driven, this phase difference will be determined primarily by the difference in the phases of oscillation of the two transmitters. As will be seen in subsequent chapters, the phase of a transmitter affects the phase constant $\Omega_{jk}^{(n)}$ (see Eq. 5.74) of a particular off-diagonal element of the D_n matrix for the quantum system being driven thereby, and this in turn determines the phase of $\langle R \rangle_n$. In any event, when all relevant factors are taken into account, θ_{12} can be calculated. Suppose that $|E_1| = |E_2| = |E|$. Then, from Eq. 5.75,

$$J_{12} \propto 2|E|^2(1 + \cos\theta_{12}). \tag{5.76}$$

If $\theta_{12} = 180°$ (destructive interference), the resultant intensity will be 0. If $\theta_{12} = 90°$, the resultant intensity will be exactly twice that expected for a single dipole

$$J_1 = J_2 \propto |E|^2. \tag{5.77}$$

On the other hand, if $\theta_{12} = 0$ (constructive interference), J_{12} will be four times that expected from a single dipole.

These results can be generalized to N dipoles as follows:

$$J_N \propto \left| \sum_{p=1}^{N} \mathbf{E}_p \cdot \sum_{q=1}^{N} \mathbf{E}_q \right| = \sum_{n=1}^{N} |E_n|^2 + \sum_{p=1}^{N} \sum_{q=1}^{N} |E_p||E_q| \cos\theta_{pq}. \tag{5.78}$$

If it is assumed, as before, that all $|E_n|$ are equal, then

$$J_N \propto |E|^2 \left(N + \sum_{p=1}^{N} \sum_{q=1}^{N} \cos\theta_{pq} \right). \tag{5.79}$$

Complete destructive interference can be acheived in a number of ways. For example, $\theta_{pq} = 0°$ for $p + q$ even and $\theta_{pq} = 180°$ for $p + q$ odd will produce complete destructive interferencee if N is even, and nearly complete if N is odd. If, instead of complete destructive interference between the waves radiated by N dipoles, complete constructive interference is desired, it can be achieved by $\theta_{pq} = 0°$ for all p and q:

$$J_N \propto |E|^2 \, [N + N(N-1)] = N^2|E|^2. \tag{5.80}$$

It should be remembered that the values of θ_{pq} ordinarily depend on the point of observation, so that the radiation emitted from a particular array can vary from 0 to an amount proportional to $N^2|E|^2$, depending on the direction from which the array is viewed (see Fig. 5.1). Finally, if the values of θ_{pq} are distributed at random over the ensemble of N dipoles, the double sum in Eq. 5.78 is proportional to the average value of $\cos\theta$ over the interval $(0 \ldots 2\pi)$ and is therefore zero (see Eq. 5.44 et seq). Therefore, in this case,

$$J_N \propto N|E|^2. \tag{5.81}$$

Also, in this case, the directionality of the radiation is completely lost; the pattern shown in Fig. 5.1 would be replaced by a uniformly gray field.

The discussion concluded above must be summarized. The radiation emitted by N dipoles differs in character from the radiation emitted by one dipole. If N dipoles are driven by a common transmitter (or individual transmitters with

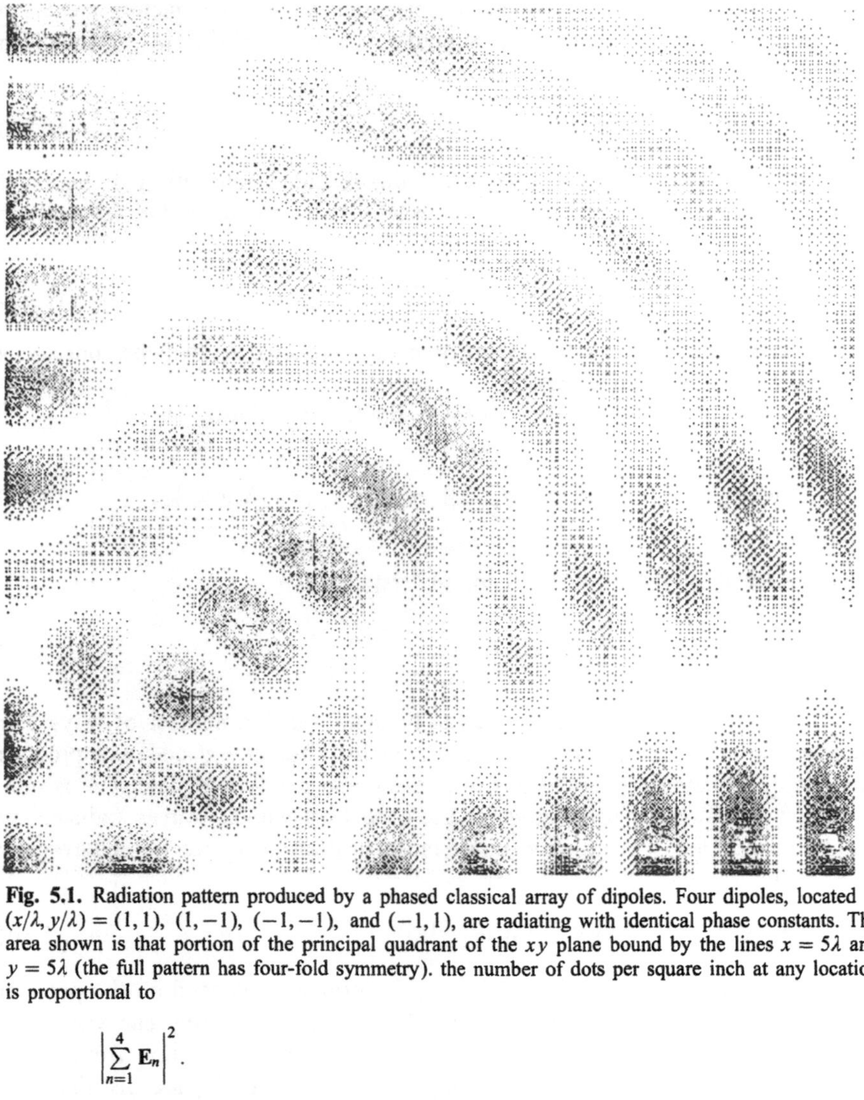

Fig. 5.1. Radiation pattern produced by a phased classical array of dipoles. Four dipoles, located at $(x/\lambda, y/\lambda) = (1,1)$, $(1,-1)$, $(-1,-1)$, and $(-1,1)$, are radiating with identical phase constants. The area shown is that portion of the principal quadrant of the xy plane bound by the lines $x = 5\lambda$ and $y = 5\lambda$ (the full pattern has four-fold symmetry). the number of dots per square inch at any location is proportional to

$$\left|\sum_{n=1}^{4} \mathbf{E}_n\right|^2 .$$

Note constructive interference along the axes and destructive interference elsewhere. The figure was prepared with the assistance of Henry Streiffer using the facilities of the LSU computer center

synchronized phases), the off-diagonal elements of the corresponding D_n matrices will also be synchronized and the off-diagonal elements of $\underline{\rho}$ will be nonzero. As a result, the dipole moments of the quantum systems in the ensemble will radiate electromagnetic waves having synchronized phases (coherent radiation). This radiation will be highly directional, because of the geometric factors that, in addition to the phase constants $\Omega_{jk}^{(n)}$, determine the angles θ_{pq}. It also will be very intense in the direction in which it is emitted; it will be N^2 times as intense as the radiation expected in that direction from a single dipole. An example is given in Fig. 5.2.

On the other hand, if the N dipoles radiate under the influence of transmitters having phases distributed at random, the phase constants $\Omega_{jk}^{(n)}$ of the off-diagonal elements of the associated $\mathbf{D}_n$ matrices will also be random. Therefore the corresponding elements of the density matrix $\underline{\rho}$ will vanish because of phase cancellation, and no net macroscopic (experimentally observable) oscillating polarization will be associated with the radiation process. The emitted electromagnetic waves will be incoherent, isotropic, and of modest intensity (N times that to be expected from a single quantum radiator).

The utilization of a phased array of dipoles to produce an intense unidirectional coherent beam of electromagnetic radiation is well known to electronics

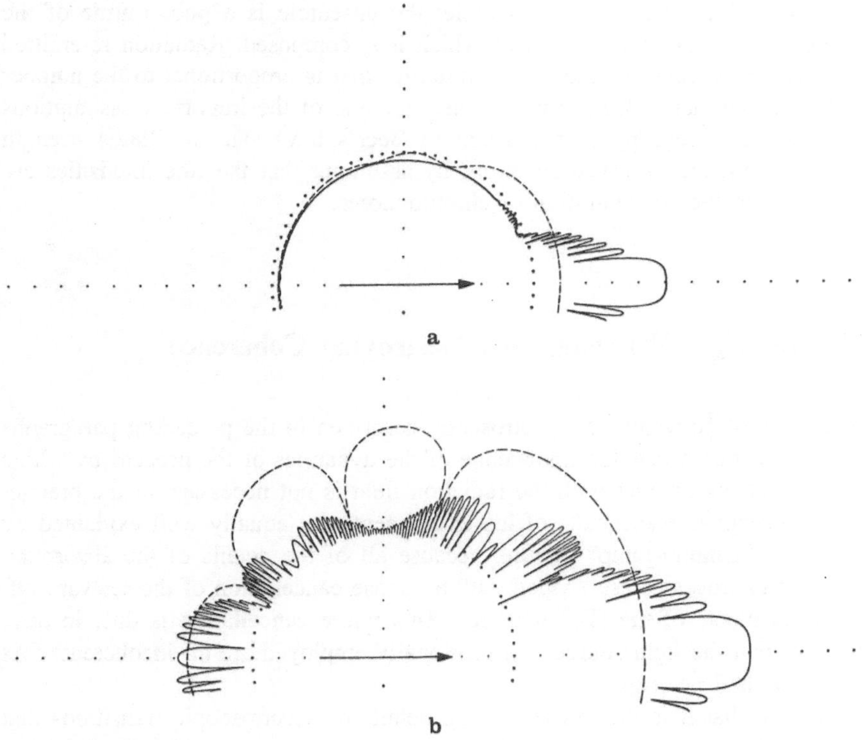

Fig. 5.2. a Radiation patterns from two right circular cylinders, one (solid line) of length $500\lambda/\pi$ and radius $5\lambda/\pi$, and the other (broken line) of length $5\lambda/\pi$ and radius $\lambda/2\pi$. Both have 61.2 excited chromophores per unit volume (λ^3). The dotted line is the incoherent radiation pattern from either cylinder when all of the atoms are excited. The radiation pattern is initiated by a weak coherent pulse propagating along the cylinder axis, shown by the arrow. **b** Same as (*a*), but with 61.2×10^4 chromophores per unit volume (λ^3). Taken from *Phys. Rev.* **A3**, 1735 (1971) by N.E. Rehler and J.H. Eberly

engineers in the macroscopic world; it is employed in radio astronomy and radio direction finding.

Most light sources in conventional emission spectroscopy produce an incoherent jumble of electromagnetic waves. After this light passes through a monochromater, it is analogous to a collection of N transmitters, each operating at the same frequency but with a different phase. The antennas in this case are microscopic quantum systems (atoms, ions, or molecules). The D_n matrix for each of these microsystems has nonzero off-diagonal elements, oscillating at frequencies corresponding to all of the absorption lines in the spectrum. Associated with the off-diagonal elements of $[\mathsf{D}_n]$ are oscillating microscopic dipole moments (e.g., $\langle p_E \rangle_n$) which interact with the incoming radiation fields. Since the driving electromagnetic waves are unsynchronized, the constants $\Omega_{jk}^{(n)}$ for the off-diagonal elements of $[\mathsf{D}_n]$ are randomly distributed throughout the ensemble. Therefore the density matrix associated with the ensemble $[\underline{\rho}]$ remains totally diagonal throughout the absorption process; the off-diagonal elements vanish because of phase cancellation. Consequently, there is no (measureable) macroscopic polarization of the spectroscopic sample; the ensemble is a poor mimic of the behavior of the quantum systems of which it is composed. Radiation re-emitted by the sample is isotropic and has an intensity that is proportional to the number of radiators. The latter feature is the basis for one of the important assumptions of emission spectroscopy (fundamental to Beer's law): The oscillator strength is derived from the observed spectrum by assuming that the line intensities are proportional to the concentration of chromophores.

5.11 Creating, Observing, and Destroying Coherence

The features of conventional spectroscopy described in the preceding paragraphs made it clear that a detailed knowledge of the dynamics of the process by which quantum systems interact with the radiation field is not necessary to the practicing spectroscopist. The results of his experiments are equally well explained by adopting a "quantum-jump" picture, because all of the details of the absorption and emission processes are "wiped out" by phase cancellation of the relevant off-diagonal elements of the $[\mathsf{D}_n]$ matrices. This phase cancellation is due, in turn, to the fact that the light sources conventionally employed are all incoherent – "As ye sow, so shall ye reap".

Suppose that it is desired to see the details of spectroscopic transitions that cannot be observed in conventional spectroscopic experiments. The first trick, as was suggested above is to induce phase correlations among the matter waves of the quantum systems in the ensemble. It was mentioned that such correlations can sometimes be produced spontaneously, but the usual method is to produce them by means of coherent radiation. Both of these methods will be discussed in more detail in subsequent chapters. For the purposes of the remaining portion of this chapter, it will be sufficient to suppose that an off-diagonal element of the density matrix (more properly, a pair of such elements, since ρ is Hermitian) has been made nonzero. Suppose that a coherent driving field (H_1) has been used to produce the nonzero element, and that it is now desired to study the behavior of the ensemble after this field has been turned off. There are several ways in which the action of the heat bath, represented by H_R, can return the ensemble to equilibrium (i.e., a Boltzmann distribution of quantum systems among the eigenstates of H_0). For simplicity, the quantum systems will be presumed to have only two energy levels. Equations (5.69–5.71) were developed for the purpose of describing the time development of the elements of the density matrix associated with an ensemble of such systems. Since H_1 is supposed to be absent, the only terms that remain in these equations are the relaxation terms, plus terms that are proportional to ω_0 and can be transformed away by mathematical methods to be described in subsequent chapters. The relaxation terms are characterized by the phenomenological constants T_1 and T_2, introduced in Eqs. (5.51–5.54), which will now be discussed in more detail.

The first discussion concerns T_1, the longitudinal relaxation time. It is possible to define a quantity T_s, with the dimensions of temperature, to characterize the diagonal elements of the density matrix:

$$T_s \equiv \frac{\mathscr{E}_2 - \mathscr{E}_1}{k_0 \ln(\rho_{22}/\rho_{11})}. \tag{5.82}$$

The quantity T_s defined in Eq. 5.82 has been dubbed the "spin temperature" of the ensemble and is to be contrasted with the temperature of the surrounding heat bath, as given by the Boltzmann formula in Eqs. (5.46) and (5.47):

$$T = \frac{\mathscr{E}_2 - \mathscr{E}_1}{k_0 \ln(\rho_{22}^e/\rho_{11}^e)}. \tag{5.83}$$

In Eqs. (5.82) and (5.83) as in Eqs. (5.46) and (5.47), k_0 is Boltzmann's constant. The difference between the spin temperature, T_s, and the actual temperature, T, is a measure of the energy received by the sample (because of excitation by means of H_1) in excess of the amount that one would expect if the ensemble were in thermal equilibrium with the heat bath. It can be seen, therefore, that T_1 characterizes the rate of flow of energy from the ensemble to the heat bath, and vice versa; it is, in effect, the "$1/e$ cooling (heating) time" required to bring T_s and T into coincidence. If the sample is excited by means of incoherent radiation (or any other random process), off-diagonal elements of the density matrix will never appear and T_1 will be the only measurable relaxation time in the system. One mechanism that the ensemble possesses for getting rid of excess energy is radiation (luminescence). It has already been noted that the intensity of the incoherent luminescence is proportional to N, the number of quantum systems in the ensemble. Now it can be seen that the excess of population in state 2 over that in state 1 will decay exponentially with a rate constant $1/T_1$. Therefore the net flux of electromagnetic radiation out of the system will also decay. For this reason, T_1 is called the "fluorescence lifetime" by conventional spectroscopists.

Now T_2, the transverse relaxation time, can be considered. At the end of the T_1 process, the sample will have returned to Boltzmann equilibrium; the net excess energy in the ensemble will be zero. At a sufficiently low temperature, $T \ll E_2/k_0$, it is accurate to say that every system in the ensemble eventually will be in one of the stationary states of the system–the ground state. Therefore each system in the ensemble, characterized in general by

$$\Psi_n = c_{1n}\psi_{1n} + c_{2n}\psi_{2n}. \tag{5.84}$$

will have $c_{2n} = 0$. Consequently, the off-diagonal elements of every $[\mathsf{D}_n]$ matrix, being proportional to the product $c_{1n}c_{2n}^*$, will also be zero. Since each element of the density matrix for the ensemble is proportional to the sum of the corresponding elements of the $[\mathsf{D}_n]$ matrices, $\rho_{jk}(j \neq k)$ will also be zero. Finally, it therefore follows that all T_1 processes are also T_2 processes, because T_1 relaxation is sufficient to produce all $\rho_{jk}(j \neq k)$ equal to zero.

It sometimes happens that the T_1 processes are the dominant ones, so that all T_2 relaxation occurs in this way. In such circumstances, $T_1 = T_2$. More generally, however, the universal relationship between the two times is

$$T_2 \leqslant T_1. \tag{5.85}$$

Equation 5.85 states that, although T_1 processes are *sufficient* to bring about T_2 relaxation, they are not necessary. It may be that other processes occurring within the ensemble will cause the decay of off-diagonal elements of the density matrix *faster* than the diagonal elements re-establish Boltzmann equilibrium.

One must return now to the picture of the phased array of dipoles. The cessation of H_1 is equivalent to disconnecting all the transmitters from the antennas abruptly. Each antenna will be "caught" with excess kinetic energy in the motions of its electrons, which it can discharge by continuing to radiate for a time T_1. Suppose that before the antennas complete this process, some other factor causes them to become out of phase with each other ($T_2 < T_1$). The dynamics of the radiation process subsequent to the disconnection of the transmitters will be as follows. Immediately after cessation of the driving forces, the antennas will continue to radiate an intense directional beam. As the synchronization forced on the antennas by the transmitters disappears, the radiators will gradually "forget" their timing because of T_2 processes. The phases of the systems will become "unglued" from one another, and the intensity and directionality of the radiation emitted by them will be lost. Eventually, the radiation from the array will be completely incoherent and isotropic, and will be produced at a greatly reduced rate until T_1 relaxation ends the entire process. It is as if all the radiators are soldiers, called out on a field at night and formed in line by a drill sergeant (the coherent driving field, or transmitter). The drill sergeant marches all of his soldiers (synchronizing their motions) by counting cadence. Disconnecting the transmitter (shooting the drill sergeant) will not produce an immediate breakup of the marching formation, because each soldier will have formed in his own mind a memory of the sergeant's cadence, All of these memories, however, are slightly imperfect and differ from one another. After a time T_2, the cumulative effects of the cadence errors will be quite noticeable; the marching file will have straggled out over the field in a most unmilitary fashion. All of this will ordinarily occur long before the soldiers cease marching, because it takes a long time (T_1) for any of them to drop from exhaustion or to desert the ranks. See Fig. 5.3.

The difference in intensity between the initial coherent radiation and the ultimate incoherent emission can be truly enormous for macroscopic samples of quantum systems – the expected factor-of-N enhancement of the former is of the order of 6×10^{23} (Avogadro's number). This fact is essential to the successful operation of crossed-coil NMR spectrometers, as the ordinary incoherent fluorescence from magnetized nuclei is far too weak to be detected experimentally. Ensembles with a high degree of phase correlation among the constituent systems are called "superradiant"; they are characterized by a high value of the "cooperation" quantum number. The decay of superradiant emission (transitions of the ensemble to states of lower cooperation number) in a time T_2 is called "free-induction decay" by NMR spectroscopists.

In the classical phased array described above, it has been assumed tacitly that the dominant T_1 mechanism was the ohmic electrical resistance of the antennas – friction between the electrons and the internal parts of their conductors. In spectroscopy, such relaxation processes, which convert excess ensemble energy directly into heat, are called "radiationless". In nuclear magnetic resonance it is indeed true that T_1 relaxation is completely dominated by radiationless processes;

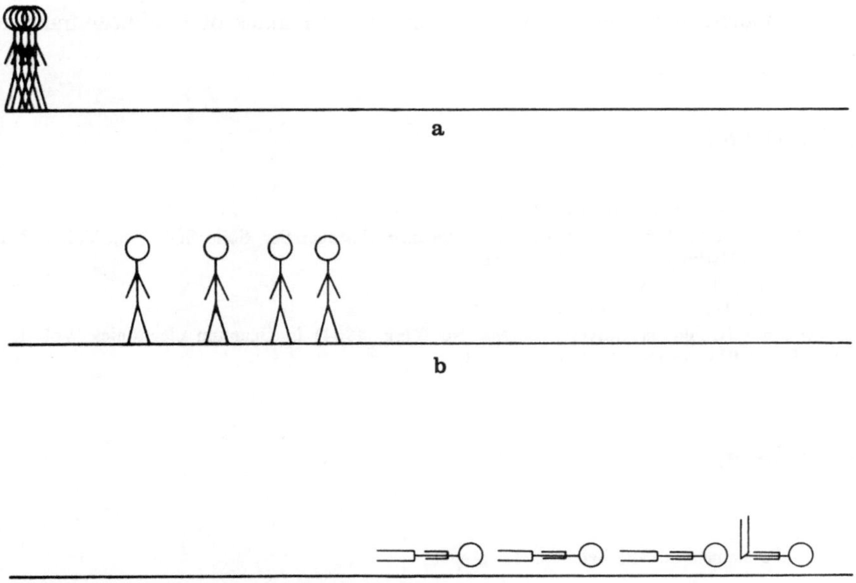

Fig. 5.3. a Coherently driven soldiers. **b** Soldiers marching freely, suffering T_2 relaxation. **c** soldiers after T_1 relaxation

spontaneous emission (incoherent fluoresence) is so slow that it would take literally forever ($\cong 10^{15}$ years) for Boltzmann equilibrium to be established by this means. In systems undergoing electronic transitions, emitting and absorbing light in the visible and ultraviolet regions of the spectrum, one commonly finds that incoherent spontaneous emission is rapid enough so that an appreciable fraction of the excitation energy of an ensemble can escape into the heat bath thereby, imagine how much more energy can be eliminated by coherent superradiant emission! Even in NMR spectroscopy, it sometimes happens that free-induction decay can compete with radiationless relaxation as a T_1 process. This effect is termed "radiation damping."

As will be seen in subsequent chapters, each increase in the relaxation rate $1/T_2$ causes the corresponding spectral line to increase in width. Line broadening due to radiation damping presents a problem in the design of commercial NMR spectrometers. It is solved by decreasing the Q of the receiver circuit; this, in effect, prevents a rapid escape of electromagnetic radiation from the sample. The magnitude of the broadening due to damping depends on the *square* of the concentration of nuclei in the samples, as is expected for a coherent process.

In this chapter, a powerful method for predicting the macroscopic behavior of microscopic quantum systems has been developed. The principle tool used in this method is the density matrix, which contains all of the information about the time-development of an ensemble in statistical quantum mechanics. The notion of phase-coherence among matter waves was introduced, as was the influence of perturbations and of relaxation processes. In the next chapter, the formalism of the density matrix will be applied in extensive fields of physical processes in condensed matter, and after that we shall discuss dynamics of nonlinear transitions.

5.12 References

1. Glauber RJ (1963), Phys Rev 131: 2766
2. Slichter CP (1990), Principles of magnetic resonance. Springer Ser. Solid State Sci., Vol. 1, 3rd ed., Springer, Berlin Heidelberg, New York
3. a. Dicke RH (1954), Phys Rev 93: 99
   b. Dicke RH (1957), J Opt Soc Am 47: 527
   c. Dicke RH (Columbia University Press, NewYork, 1964) in Quantum electronics, Vol. 1, Grivet P and Bloembergen N, Eds pp 35–53.

5.13 Problems

5.1

a. Find the elements of the matrix D for a spin

$s = \frac{1}{2}$ system in the state represented by

$$\Psi = \exp\left(\frac{i\omega_0 t}{2}\right)\cos\left(\frac{\theta}{2}\right)\alpha + \exp\left(\frac{-i\omega_0 t}{2}\right)\sin\left(\frac{\theta}{2}\right)\beta.$$

b. Show that, for $\theta = 62°28'34.3''$, the matrix becomes

$$[\mathsf{D}] = \frac{1}{1+e}\begin{bmatrix} e & \exp(\frac{1}{2} + i\omega_0 t) \\ \exp(\frac{1}{2} - i\omega_0 t) & 1 \end{bmatrix}.$$

c. What is the expectation value of the energy for this system?

5.2 Consider a particle that has only two eigenstates of the energy, $\mathscr{E} = kT$ and 0.

a. What is the density matrix for an ensemble of N such particles at equilibrium?

b. Calculate the total energy of the ensemble.

5.3

a. What is the density matrix for an ensemble of two level particles, all of which are in the gound state?

b. What is the density matrix for an ensemble of two level particles, all of which are in the excited state?

5.4 Solve Eqs. (5.69) to (5.72) for $\mathsf{H}_1 = 0$ and $\mathscr{E} = kT$. Use the boundary condition that, at time $t = 0$,

$$\rho(0) = \frac{1}{2}\begin{bmatrix} 1 & \exp(i\omega_0 t) \\ \exp(-i\omega_0 t) & 1 \end{bmatrix}.$$

6 Relaxation Processes and Coherent Dissipative Structures

E. Brändas

6.1 Dissipative Systems

A dissipative system is a system in which energy is dissipated. In the present Chapter our major concern will be focused around this concept. Although we will describe and emphasize concrete applications we will also occupy ourselves with a general view of the fundamental dilemma usually phrased as "why and how does there exist a unique privileged direction of time" and further the immediate consequences for scenarios related to the question of *reductionism versus holism*. In this context it is also appropriate to discuss the hierarchical character of subdynamics, timescales, entanglement in quantum mechanics, EPR correlations, nonlinear dynamics in non-equilibrium situations, reduced density matrices of complex systems etc.

It will be necessary to generalize microscopic time reversible dynamics towards an irreversible dynamical formulation, where the reversible Schrödinger- or Liouville equation is naturally embedded. This extension will be shown to follow naturally from complex scaling or the dilatation transformation. This simple generalization thus allows the incorporation of resonances or the so-called resonance picture of unstable states into a standard quantum mechanical framework without too many modifications, albeit on a much more general interpretative level.

Here the word dissipative system, or equivalently open system, will be used more or less interchangeably. The first concept, however, has aquired a rather special meaning due to the fundamental work of the Brussels School [1–4] while the latter has been recurrently used in more conventional contexts. Since our focus will be on the conceptual basis of a formulation starting at a microscopic level, i.e. involving nuclear, atomic and molecular state descriptions, we will make frequent references to the excellent monograph *Chemistry, Quantum Mechanics and Reductionism* by Primas [5], where these aspects are thoroughly discussed.

Let us begin by attempting a definition of a *dissipative system.*

Definition: *A dissipative system is a system in which there exists a flow of entropy due to exchanges of energy or matter with the environment.*

For more details, see Prigogine et al. [1–4]. One can also define the non-isolated system or *open system* from the view-point of its time evolution properties [5].

Definition: *If the state of the system at time $t > t_0$ is not a function of the state at t_0 alone then one speaks of an open system.*

It is clear that these definitions, as mentioned in passing, do not refer to simple systems in isolation. Rather one considers systems with at least some degree of complexity. Primas [5] gives the following illustration to the concept. For instance the (normally) very simple (iron) weight may be a very complicated and frustrating piece of matter for a shot putter, while the collector of scrap-iron has no special relation to it. Hence the environment of the system must enter a *relevant* description, see more below, and the nature of the complexity thus arises from the number of ways in which we are able to interact with the system.

Thus without going into details, we have at our disposal a physical theory of motion for the density matrix ρ given by the Liouville equation

$$i\frac{\partial \rho}{\partial t} = \hat{L}\rho \tag{6.1}$$

with the initial condition

$$\rho(t) = \rho(0); \quad t = 0. \tag{6.2}$$

where the Liouville operator is the commutator $[H, \ ]$ with the hamiltonian H as developed in previous Chapters. For simplicity we have also put Planck's constant divided by 2π equal to one, i.e. $\hbar = 1$.

In principle Eq. (6.1) refers to reversible microscopic processes, but as we will see, there are ways of converting the reversible group evolution to a contractive description with a preferred time direction.

Before we make room for some more mathematics it is worth while spending some more time to get a more "realistic flavor" of a dissipative system. Obviously every phenomenon – and associated level of organization – by necessity requires its own kind of state description, as an open dynamical nonlinear dissipative system, with an appropriate language for every level. Hence a certain hierarchical order of arrangements, principles, rules etc. are prevalent. Primas [5], see also the work of the Brussels Schools [1–4], speaks of hierarchical systems in the following way:

A *complex system* can be decomposed into an ascending family of successive (more encompassing) subsystems such that every lower level is subordinated by an *authority relation* to the next level, i.e. into levels of organization.

As an example of this hierarchy one can mention [5]: molecules – cells – tissues – organisms – breeding populations – species – social systems – ecosystems. Further examples are:

Zoology: A molecular description for the understanding of the behavior of honeybees?

Chemistry: Characterization of molecules – substances and the meaning of temperature.

Water: The possibility of explaining the properties of steam, ice and liquid water from the concept of a water molecule.

Fig. 6.1. Quantum zoology

Superconductivity: ODLRO (off-diagonal long-range order) and the coupling of fermions to bosons.

In the examples above it is evident that every category of complexity has its own phenomenological level of description, and that a higher level in a hierarchy always has a (much) longer reaction time than a level classified as lower. As an example consider timescales in biology:

a) biochemical scale (fraction of seconds)
b) metabolic scale (order or minutes)
c) epigenetic scale (several hours)
d) development scale (days or years)
e) evolutionary scale (10^3–10^6 years)

The possibility of separating out different timescales is an underlying, absolutely necessary condition for the theoretical understanding and explanation of any natural phenomena. As a simple example we consider the correlation time τ_{corr} given by the uncertainity principle as

$$\tau_{\text{corr}} \simeq \frac{\hbar}{k_B T} \tag{6.3}$$

where k_B is Boltzmann's constant and T the temperature. At room temperature τ_{corr} is about 10^{-14} s. This means that a microscopic system in contact with a heat bath is subject to about 10^{14} "butts or kicks" per second and this is certainly of relevance for the dynamics on the next level of description. Take for instance the molecular reorientation time or the proton relaxation time characteristic of fundamental picosecond processes or other dynamically induced fluctuations on the microsecond scale. Here the different magnitudes of τ_{corr} and the higher level relaxation time τ_{rel} represents the *reduction ascending* going from one level to the next. When timescales referring to separated levels are about to coalesce, one would expect level interference and selforganization and/or ordering of new structures and phases. These structures would then, in the therminology of the Brussels School, be called *dissipative structures.*

6.2 Entanglement and Interference Effects

In contrast to classical theories, quantum mechanics predicts an entanglement of a system with its surroundings. One of the most well-known examples of this entanglement is the Pauli antisymmetry principle for fermions – or the symmetry principle for bosons.. Hence if one consider a system S consisting of two microscopic subsystems S_A and S_B with state spaces $\mathscr{H}_A$ and $\mathscr{H}_B$, then the state vector $\Psi \in \mathscr{H}$ where $\mathscr{H} = \mathscr{H}_A \otimes \mathscr{H}_B$ is *never* represented by the product

$$\Psi = \Psi_A \Psi_B; \quad \Psi_I \in \mathscr{H}_I; \quad I = A, B \tag{6.4}$$

Instead one needs to represent the state vector by the sum

$$\Psi = \sum_n \sum_m \omega_{nm} \Psi_{A,n} \Psi_{B,m} \tag{6.5}$$

It is customary to refer to a diagonal (natural) representation of Eq. (6.5), i.e.

$$\Psi = \sum_n \omega_n \Psi_{A,n} \Psi_{B,n} \tag{6.6}$$

as the natural expansion, but this implies that the matrix ω must be representable as a diagonal matrix. Although we have not explicitly introduced new notations, the quantities in Eq. (6.6) are in fact the transformed ones.

The famous mirror theorem [6, 7] for transformations T_{AB} and T_{BA} between $\mathscr{H}_A$ and $\mathscr{H}_B$ and vice versa, see also general discussions by Löwdin [8] and Coleman [9], implies identical non-vanishing canonical forms for T_{AB} and T_{BA} and hence a direct mirror symmetry. This is also in accord with the general theorem stating that any matrix ω can always be transformed to complex symmetric form, i.e. $\omega_{kl} = \omega_{lk}$, via a similarity transformation. For a direct elegant proof of this theorem see Gantmacher [10], and for additional constructions see Reid and Brändas [11].

Note that group theoretical decompositions of the product representation (6.5) may imply many dimensional irreducible representations. Also the choice of $\mathscr{H}_A$ and $\mathscr{H}_B$ as complex conjugate Hilbert spaces may, in its dilated form, lead to nondiagonalisability of the matrix ω.

We can now define exactly (or mathematically) the precise meaning of an *entangled system.*

Definition: *If the natural expansion (6.6) contains more than one non-vanishing eigenvalue, (and/or for symmetry reasons the matrix ω can not be diagonalized), then we speak of an entangled system.*

It is important to realize that weakly interacting (or even non-interacting) systems *with a common ancestor* are in general correlated or entangled. A famous example of this is given by EPR paradox [12], see e.g. d'Espagnat [13] for a complete discussion and appraisal of this important result.

We will discuss the general nondiagonal case further below with particular reference to a general time irreversible subdynamics framework. Here we will briefly proceed from the natural expansion (6.6), since some fundamental conclusions can still be made of systems subject to conservative time evolutions. Thus all statistical predictions for the outcome of experiments restricted to S_A will be expressed by the reduced density operator

$$\mathscr{D}_A = \mathrm{Tr}_{\mathscr{H}_B}\{|\Psi\rangle\langle\Psi|\} = \sum_n |\omega_n|^2 |\Psi_{A,n}\rangle\langle\Psi_{A,n}| \tag{6.7}$$

and such measurements can at most determine $|\omega_n|$ (and $\Psi_{A,n}$). Hence maximum information accessible from measurements on both subsystems can be represented as

$$\mathscr{D}' = \sum_n |\omega_n|^2 |\Psi_{A,n}\Psi_{B,n}\rangle\langle\Psi_{A,n}\Psi_{B,n}| \tag{6.8}$$

while complete information is given by

$$\mathscr{D} = |\Psi\rangle\langle\Psi| = \sum_n \sum_m |\Psi_{A,n}\Psi_{B,n}\rangle\omega_n\omega_m^*\langle\Psi_{A,m}\Psi_{B,m}| \tag{6.9}$$

Note that $\mathscr{D}$ and $\mathscr{D}'$ differ by the interference terms (given by $n \neq m$). It is precisely these interference terms that are responsible for the non-local behavior of quantum mechanics as discussed by Einstein, Podolsky and Rosen [12]. If these effects are smeared out by the experimenter he or she cannot find any EPR-correlations, since the latter as already mentioned are true interferences. An illustrative example, where phase averaging in a macroscopic phenomenological situation would be totally destructive, is the coherence effect in a superconductor. Hence quantum systems may display holistic character even if their subparts are separated by macroscopic dimensions. We will discuss this issue separately in a later section.

6.3 The Ensemble for an Ideal System

In quantum statistics any measurement, corresponding to an observable $\mathscr{A}$ obtains from

$$\langle\mathscr{A}\rangle = \mathrm{Tr}\{\mathscr{A}\Gamma\} \tag{6.10}$$

where Γ is the relevant *System Operator* for the studied process. The fundamentals of this approach goes back to the classical treatise of J. von Neumann [14].

Before we give the main characteristics of this formulation we will briefly review the situation for a class of systems called *ideal systems*, defined by a linear superposition of hamiltonians, where each subhamiltonian describes a finite (usually small) set of degrees of freedom and with no interaction between the corresponding subsystems. An an example we may think of a system of N independent identical particles, with the energy E in a given configuration given by

$$E = \sum_{j=1}^{N} \varepsilon^{(j)} \tag{6.11}$$

where $\varepsilon^{(j)}$ is the energy level of particle j.

The correct partition function (referring of N indistinguishable particles with n_k characterizing the occupation number of a given distribution of particles among the various single particle states m_k) is then ($\beta = \frac{1}{k_B T}$)

$$\mathscr{Z} = \sum_{m^{(1)}} \dots \sum_{m^{(N)}} \frac{n_k! n_l! \dots}{N!} \exp\left(-\beta \sum_j \varepsilon^{(j)}\right) \tag{6.12}$$

where $m^{(j)}$ denotes a complete set of quantum numbers of the level of the single particle or molecule j. In the classical regime most of the single particle states are empty and a few contain one molecule, i.e. $\langle n_k \rangle \sim 1$ for some configurations while the average value of n_k is much smaller than 1 (the latter means that the constraints of quantum statistics can be forgotten here). One thus obtains ($\mathscr{V}$ is the volume, see below)

$$\mathscr{Z} = (N!)^{-1} \sum_{m^{(1)}} \dots \sum_{m^{(N)}} \exp\left(-\beta \sum_j \varepsilon^{(j)}\right) \tag{6.13}$$

Hence the partition function for our *ideal system* in the *Boltzmann approximation* (note that the present picture obtains as a common limit of the *Bose-Einstein* and *Fermi-Dirac statistics*) is the product of N factors:

$$\mathscr{Z} = (N!)^{-1} \prod_j^N \left[\sum_{m^{(j)}} \exp(-\beta \varepsilon^{(j)})\right] \tag{6.14}$$

or since all factors are equal

$$\mathscr{Z} = (N!)^{-1} \mathscr{Z}_1^N \tag{6.15}$$

with $\mathscr{Z}_1 = \Sigma_m \exp(-\beta\varepsilon)$ and the summation carried out over all levels of a single particle. Using the Stirling formula $\ln(N!) \sim N\ln(\frac{N}{e})$ the free energy obtains as

$$\mathscr{F}(T, \mathscr{V}, N) = -N k_B T \ln[e N^{-1} \mathscr{Z}_1(T, \mathscr{V})] \tag{6.16}$$

We will take this evaluation one step further and consider the total energy for each particle written as

$$\varepsilon = \varepsilon_{\text{tr}} + \varepsilon_i \tag{6.17}$$

where ε_{tr} is the energy associated with the center-of-mass motion of the molecule, and ε_i corresponds to the internal degrees of freedom (rotation, vibration, electronic excitation, etc.). Again we can factor the partition function

$$\mathscr{Z}_1 = \mathscr{Z}_{1,\text{tr}} \mathscr{Z}_i. \tag{6.18}$$

and consider the translational motion of the molecule with total mass m and with kinetic energy given by

$$\varepsilon_{tr} = \frac{p_x^2 + p_y^2 + p_z^2}{2m} \equiv \frac{p^2}{2m}. \tag{6.19}$$

As usual in this context, we follow a proposition of Planck, i.e. converting the continuous classical phase space into a "cellular" quantum mechanical phase space, and enclose the system in a box of volume $\mathscr{V} = L^3$. From our knowledge of elementary quantum mechanics we obtain, and make use of, the eigenvalues of the momentum operator as

$$p_{in} = \hbar \left(\frac{2\pi}{L}\right) n_i; \quad n_i = 0, \pm 1, \pm 2 \ldots \quad i = x, y, z \tag{6.20}$$

to obtain the translational partition function as

$$\mathscr{Z}_{1,tr} = \sum_{n_x} \sum_{n_y} \sum_{n_z} \exp\left\{-\beta \frac{h^2}{2mL^3}(n_x^2 + n_y^2 + n_z^2)\right\} \tag{6.21}$$

If the spacings between successive momentum levels are small (compared to $(2mk_BT)^{\frac{1}{2}}$ or

$$\frac{h}{L(2mk_BT)^{\frac{1}{2}}} \ll 1 \tag{6.22}$$

then the summation (6.21) may be replaced by an integration, which yields

$$\mathscr{Z}_{1,tr} = \frac{L^3}{h^3} \int \int \int_{-\infty}^{+\infty} dp_x dp_y dp_z \exp\left\{-\frac{(p_x^2 + p_y^2 + p_z^2)}{2mk_BT}\right\} = \mathscr{V}(2\pi m k_B T h^{-2})^{\frac{3}{2}}$$

or introducing the *thermal de Broglie wavelength* $\Lambda = (\frac{h^2}{2\pi m k_B T})^{\frac{1}{2}}$

$$\mathscr{Z}_{1,tr} = \mathscr{V}\Lambda^{-3}. \tag{6.23}$$

In general the equation of state for a system in equilibrium is given by

$$(T, N \text{ const.})\ \mathscr{P} = -\frac{\partial \mathscr{F}}{\partial \mathscr{V}} = \frac{k_BT}{\mathscr{Z}} \frac{\partial \mathscr{Z}}{\partial \mathscr{V}}; \ \mathscr{F} = -k_BT \ln \mathscr{Z} \tag{6.24}$$

where e.g. the pressure $\mathscr{P}$ is related to the appropriate partition function $\mathscr{Z}$. Other thermodynamic properties can be obtained similarly.

It is customary to expand the equation of state in the dimensionless parameter δ (or the density n)

$$\delta = g^{-1}\Lambda^3 n$$

where g is the degeneracy parameter. In the Boltzmann approximation $\delta \ll 1$ and the form of the expansion above – usually applied in the case of imperfect gases – is called *a virial expansion*. For quantum situations $\delta \gg 1$ and consequently the difference between the fermion and the boson gas is exhibited through typical quantum effects. We will come back to the question of the correct quantum statistics further below. Nevertheless it is clear that the size of the *thermal de Broglie wavelength* in comparison with the interparticle distances is indicative whether or not quantum effects should be of crucial importance. For general systems (c.f. liquids) there is the well-known *criterion* separating the classical and the quantum cases:

Criterion: *The necessary condition for the classical regime to hold is that the mean separation l of particles (or molecules) should be much larger than its thermal de Broglie wavelength, i.e. $l \gg \Lambda$.*

It is easy to see that this criterion is equivalent to the previously mentioned condition, namely that the mean occupation number for the single particle states should be much smaller than 1.

Although we may here be far away from considering any ideal system, we observe that the parameter δ (or Λ in Eq. (6.23)) "survives" the extension to the imperfect gas in the high temperature approximation – or for constant T the low density approximation. A large value of δ may also necessitate the specification of appropriate Bose-Einstein or Fermi-Dirac statistics. In many cases (e.g. liquids) it will furthermore be necessary to deal with non-equilibrium situations where the criterion above *cannot* be met. Hence one finds that Λ may serve as an indicator of a "relevant size" for the inclusion of quantum delocalization effects. We will subsequently modify the description, when necessary, to a set of problems, where quantum statistics must explicitly enter the picture because the *thermal de Broglie wavelength* is too large for a classical description to be valid, and where far-from-equilibrium conditions may prevail.

Let us now go back to our ideal system defined by Eqs. (6.11–23) and try to obtain a more explicit form of the free energy and consequently also the entropy, i.e.

$$\mathscr{F}(T,\mathscr{V},N) = -N k_B T \ln \left[e \frac{\mathscr{V}}{N} \left(\frac{2\pi m k_B T}{h^2} \right)^{\frac{3}{2}} \mathscr{Z}_i \right] \tag{6.25}$$

in which the internal partition function is given by (g_i is the degeneracy parameter)

$$\mathscr{Z}_i = \sum_{\varepsilon_i} g_i \exp\left(-\frac{\varepsilon_i}{k_B T} \right)$$

Finally we obtain an expression where the volume and temperature dependence are explicitly separated out

$$\mathscr{F}(T,\mathscr{V},N) = -N\left(k_B T \ln \frac{\mathscr{V}}{N} + k_B T + \frac{3}{2} k_B T \ln(2\pi m k_B T h^{-2}) - a_i(T) \right)$$

with $a_i(T) = -k_B T \ln \mathscr{Z}_i(T)$. Note that the entropy obtains as

$$S = -\left(\frac{\partial \mathscr{F}}{\partial T}\right)_{\mathscr{V},N} \tag{6.26}$$

As an example one might consider an ideal system of non-interacting diatomic molecules in which translational, rotational and vibrational degrees of freedom are independent. From the rigid rotator model and the associated quantum mechanical eigenvalues one can determine the rotational free energy, the internal energy, the specific heat and the contribution to the entropy from the rotational degrees of freedom, see problem 1 in Sect. 6.11.

6.4 The Density Matrix

We will here briefly review some fundamental properties of density matrices and system operators. For more details we refer to Löwdin [15] and references therein, see also earlier Chapters. The (full) Nth order density matrix $\Gamma^{(N)}$ for a *pure state* derived from the N – particle Schrödinger equation

$$H\Psi = E\Psi \text{ or } H\Psi = i\hbar\frac{\partial\Psi}{\partial t}$$

is by definition

$$\Gamma^{(N)} = |\Psi\rangle\langle\Psi| = \Psi(x_1,x_2,\ldots,x_N)\Psi^*(x'_1,x'_2,\ldots,x'_N). \tag{6.27}$$

In the time dependent case above there may also be an explicit time dependence, i.e. $\Gamma^{(N)} = \Gamma^{(N)}(t,t')$. For reasons that will be obvious further below we will consider the time independent case to start with. We will also consider a quantum mechanical observable $\mathscr{A}$ satisfying the eigenvalue relation

$$\mathscr{A}\Psi = \lambda\Psi; \quad \lambda = \lambda^* \tag{6.28}$$

Since we will consider a system consisting of N fermions subject to the *Pauli principle* (yielding an immediate representative of $\mathscr{A}$)

$$\mathscr{P}\Psi = (-1)^{p(\mathscr{P})}\Psi, \tag{6.29}$$

where $p(\mathscr{P})$ is odd or even depending on the parity of $\mathscr{P}$. Equation (6.29) implies that Ψ is antisymmetric in the combined space-spin coordinate $(x = \vec{r},\varsigma)$.

It will be assumed that the reader is familiar with the six basic postulates of quantum mechanics, which are usually adopted according to the so-called Copenhagen interpretation. In our axiomatic review we will summarize the properties of $\Gamma^{(N)}$ and the immediate consequences that follow and can be mathematically proved [15]. Note also that we will adhere to the von Neumann interpretation, but since the difference is quite subtle we will not say more about this here except than give the four relevant axioms, see also Primas [5] for comparative discussions. The density matrix Γ as defined in Eq. (6.27) satisfies [15]:

1 $\Gamma = \Gamma^{\dagger} \Leftrightarrow \langle\mathscr{A}\rangle = \mathrm{Tr}\{\mathscr{A}\Gamma\} = \text{real}$

2 $\Gamma \geqslant 0 \Leftrightarrow (\Delta\mathscr{A})^2 = \langle\mathscr{A}^2\rangle - \langle\mathscr{A}\rangle^2 \geqslant 0$

3 $\mathrm{Tr}\{\Gamma\} = 1 \Rightarrow \mathrm{Tr}\{\mathscr{A}\Gamma\}\text{exists}$

4 $\Gamma^2 = \Gamma \Leftrightarrow \text{pure state}$

If (1–3) is fulfilled, but not necessarily (4) one says that Γ is a *System Operator*. As quantum mechanics is based on the concept of a linear space and characterized by a superposition principle so is the operator space in general and the system operators in particular. One way to phrase this is to say that Γ belongs to a convex set $\mathscr{C}$, i.e. $\Gamma \in \mathscr{C}$ if

$$\Gamma_1 \in \mathscr{C}; \quad \Gamma_2 \in \mathscr{C} \Rightarrow \Gamma \in \mathscr{C}$$

$$\Gamma = \Gamma_1 c_1 + \Gamma_2 c_2; \quad c_1 \geqslant 0; \quad c_2 \geqslant 0; \quad c_1 + c_2 = 1.$$

If condition (4) is fulfilled, i.e. $\Gamma_1 = \Gamma_2 = \Gamma$ then Γ is a *homogeneous ensemble* or a *limit point* of $\mathscr{C}$, and otherwise one speaks of an *inhomogeneous one* corresponding to an *interior point* of $\mathscr{C}$.

Although the phenomena we will be looking into involves a large (even infinite) number, s, of degrees of freedom it may nevertheless be possible to reduce the description by *reduction ascending* to a finite (even small) number p. To do this we will introduce a series of density matrices of various orders [8]:

$$\Gamma^{(p)} = \binom{N}{p} \int \Psi^*(x_1,\ldots,x_p,\ x_{p+1},\ldots,x_N)$$
$$\times \Psi(x'_1,\ldots,x'_p,x_{p+1},\ldots,x_N)dx_{p+1}\ldots dx_N$$
$$\mathrm{Tr}\{\Gamma^{(p)}\} = \binom{N}{p} \tag{6.30}$$

A particularly useful quantity is obtained from $p = 2$. This follows from the simple observation that the calculation of any average value of a sum of two-particle operators (like the standard Coulomb many-body hamiltonian) in reality only requires the knowledge of $\Gamma^{(2)}$, i.e. for $p = 2$ one obtains

$$E = \frac{\mathrm{Tr}\{H_{\text{red}}\Gamma^{(2)}\}}{\mathrm{Tr}\{\Gamma^{(2)}\}} = \mathrm{Tr}\{H_1\Gamma^{(1)}\} + \mathrm{Tr}\{H_{12}\Gamma^{(2)}\} \tag{6.31}$$

where the *reduced hamiltonian* corresponding to the total many-body hamiltonian

$$H = H_0 + \sum_{i=0}^{N} H_i + \sum_{i<j}^{N} H_{ij} \tag{6.32}$$

is given by

$$H_{\text{red}} = H_0 + NH_1 + \binom{N}{2} H_{12} . \tag{6.33}$$

In the present example above $\Gamma^{(2)}$ derives from the homogeneous ensemble (6.27). In this case $\Gamma^{(p)}$ is *N-representable*. Of course any Γ can be reduced by taking the appropriate partial trace. If Γ is inhomogeneous one says the $\Gamma^{(p)}$ is *ensemble representable*. In general Γ may be neither.

In the theory of *subdynamics* [1–4, 16] one realization of the projective decomposition of *the system* and its *environment* is via partial tracing. Other choices concern of the diagonal and off-diagonal parts of the density matrix in *the appropriate physical representation*. The equilibrium situation is then characterized by a diagonal system operator whose off-diagonal correlations have *died out*. If this does not "happen" we are in some sense *far-from equilibrium*. In general it is impossible to define a non-equilibrium partition function. However, the concept of a *reduced distribution function* (derived from $\Gamma^{(p)}$) *is valid in equilibrium as well as outside*, and hence it gives a *deeper meaning* to the usage of *reduced system operators*.

6.5 Variational Principles and the Negentropy

We will here for completeness briefly discuss the quantum analog of the various types and properties of statistical ensembles, the concept of entropy, free energy, etc. For more details we refer to Löwdin's work on trace algebra [15]. In the previous section we introduced the notion of a convex set $\mathscr{C}$ of density operators Γ. We remind the reader that a system operator is defined by the first three conditions of the last paragraph. We also recall that an ensemble is *inhomogeneous* if the matrix ω defined in

$$\Gamma = \sum_{k,l} |\Psi_k\rangle \omega_{kl} \langle \Psi_l| = |\Psi\rangle \omega \langle \Psi| \tag{6.34}$$

has more than one nonzero eigenvalue (or in an extended framework, see below, cannot be diagonalized). Since Γ is a system operator one knows that $\Gamma = \Gamma^\dagger$; $\text{Tr}\{\Gamma\} = 1$; and $\Gamma \geqslant 0$. The fulfilment of the fourth condition, i.e. $\Gamma^2 = \Gamma$ then implies further that the ensemble is homogeneous.

In order to obtain quantitative statements from our system operator description we now introduce a convex functional F on (the convex set) $\mathscr{C} = \{\Gamma\}$, i.e.

$$F\{\Gamma\} = \text{complex number} = \text{Tr}\{f(\Gamma)\}, \tag{6.35}$$

where $f(x)$ is a convex function on the interval $\mathscr{I} = [0,1]$ conveniently chosen so that $f(0) = 0$. As an example of a convex function on $\mathscr{I} = [0,1]$ one might mention $f = x^2$, or $f = x \log x$. (The definition implies that $f'' \geqslant 0;\ x \in \mathscr{I}$). Since by definition

$$f(\Gamma) = \sum_k f(\lambda_k)\mathscr{O}_k \tag{6.36}$$

where $\mathscr{O}_k$ are the (one-dimensional) eigenprojectors associated with the eigenvalues λ_k (degeneracies can be delt with, see further below)

$$\Gamma = \sum_k \lambda_k \mathscr{O}_k = \sum_k \Gamma_k \tag{6.37}$$

one finds that

$$F\{\Gamma\} = \sum_k f(\lambda_k)\mathrm{Tr}\{\mathscr{O}_k\} = \sum_k f(\lambda_k) \tag{6.38}$$

From the convexity of the functional follows (for any system operator) that

$$F\{\sum_k c_k\Gamma_k\} \leqslant \sum_k c_k F\{\Gamma_k\}; \quad c_k \geqslant 0, \sum c_k = 1 \tag{6.39}$$

with the components $\Gamma_k = \lambda_k \mathscr{O}_k$. Hence $F\{\Gamma\}$ is maximum on the "border" of $\mathscr{C}$, i.e. for the limit points $\Gamma^2 = \Gamma$, one obtains

$$F\{\Gamma_k\} = \mathrm{Tr}\{f(\Gamma_k)\} = f(\lambda_k) \leqslant f(1)$$

This leads to a simple bound for F in (6.39) and the conclusion that there is a variational principle associated with the functional F. Furthermore it is easy to show [15], that the *only* functional which is *additive* so that

$$\Gamma = \Gamma_A \Gamma_B$$

and

$$F\{\Gamma\} = F_A\{\Gamma_A\} + F_B\{\Gamma_B\}$$

derives from $f(x) = cx \log x$ where the constant c can be identified with the *Boltzmann constant* k_B, see (6.42) below. Hence we can define the *negentropy* S through the functional F or

$$F\{\Gamma\} = c\mathrm{Tr}\{\Gamma \log \Gamma\} = -\mathscr{S} \tag{6.40}$$

Note that there are other definitions of the entropy using a different choice of convex (Lyapunov) functions, see the work of the Brussels School [1–4] for more details. We will, however, not dwell onto the more advanced analysis of subdynamics, which in addition to the complications arising from non-equilibrium situations and broken time symmetries, also should combine classical and quantum features in a rigorous and unified way. For some new trends in this direction using modern dilation transformations, see Obcemea and Brändas [16]. We will return to these questions in a later section.

The variational property of the functional F thus means that we can try to minimize F or maximize $\mathscr{S}$, i.e.

$$\mathscr{S} = -k_B \mathrm{Tr}\{\Gamma \log \Gamma\} = -k_B \langle \log \Gamma \rangle \tag{6.41}$$

subject to the constraints

$$\mathrm{Tr}\{\Gamma\},\ \mathrm{Tr}\{H\Gamma\} = \langle H \rangle = \text{constant}$$

and of course

$$\Gamma^\dagger = \Gamma, \text{ and } \Gamma \geqslant 0\,.$$

The solution to this optimization problem is the well-known *canonical ensemble*, or rather the system operator characterizing the canonical ensemble, given by

$$\Gamma = \frac{\exp(-\beta H)}{\mathscr{Z}} \tag{6.42}$$

$$\mathscr{Z} = \mathrm{Tr}\{\exp(-\beta H)\}; \quad \beta = \frac{1}{k_B T}\,.$$

from which the *free energy* $\mathscr{F}$, *the internal energy* $\langle H \rangle$ and $\mathscr{S}$ can be determined through the relation

$$\mathscr{F} = \langle H \rangle - \mathscr{S} T = -k_B T \log[\mathrm{Tr}\{\exp - \beta H\}]\,. \tag{6.43}$$

The *equation of state* for a system in equilibrium follows finally from

$$\mathscr{P} = -\frac{\partial \mathscr{F}}{\partial \mathscr{V}} = \frac{k_B T}{\mathscr{Z}} \frac{\partial \mathscr{Z}}{\partial \mathscr{V}}\,. \tag{6.44}$$

c.f. Sect. 6.3. and particularly Eq. (6.24).

6.6 The Extreme Case

It is intuitively obvious, from the previous section, that the actual (and possible) size of the eigenvalues of the density matrix should be of fundamental importance for the physical properties of the system. For a long time this problem was overlooked simply because one believed that, in analogy with the case of the first order reduced density matrix, see Eq. (6.30) with $p = 1$, that all nonzero eigenvalues λ_k should fulfil

$$0 < \lambda_k \leqslant 1 \tag{6.45}$$

This condition is in a deeper sense a direct consequence of the Pauli antisymmetry principle for fermions. The configuration interaction procedure of a manybody fermionic system of particles may thus be analysed in terms of its first order reduced density matrix and subsequent eigenvalues and eigenfunctions (natural spinorbitals). In the special case of all nonezero $\lambda_k = 1$ one obtains the well-known Fock-Dirac density matrix $\rho = \Gamma^{(1)}$ characterized by

$$\rho^2 = \rho; \quad \mathrm{Tr}\rho = N \tag{6.46}$$

where N is the number of fermions. We refer to the classic paper by Löwdin [8] for the complete treatment. Note also that the normalization for $\Gamma^{(2)}$ from (6.30) requests [8]

$$\mathrm{Tr}\{\Gamma^{(2)}\} = \binom{N}{2} \tag{6.47}$$

which differs from the notation of Yang [17] and Coleman [9]. The unit trace normalization is here reserved for the system operator $\Gamma = \Gamma^{(N)}$, while the reduced density matrices are normalized to $\binom{N}{p}$, i.e. for $p = 2$ the total number of possible pairings and so on. Although it is important to know the normalization it is of course necessary to state that the results presented here do not in any way depend on this choice.

The properties of $\Gamma^{(2)}$, in comparison with $\Gamma^{(1)}$, however, is now much more complex, since N fermions may under specific conditions *condense* into $N/2$ bosons. If this is energetically favorable then $\Gamma^{(2)}$ should exhibit a macroscopically large eigenvalue, of size $N/2$, i.e. corresponding to half the number of fermions in the system. This result was first communicated by Yang [17] in his celebrated paper on off-diagonal long-range order (ODLRO) of reduced density matrices. Although he did not give the full mathematical machinery of the proof, the bound for an even number of fermions was fully demonstrated. More importantly he showed that the result may lead to a new order or thermodynamic phase, i.e. he established for the first time the connection between ODLRO and the nonclassical phenomena of superfluidity and superconductivity.

Independently, a detailed mathematical proof, including bounds for both even and odd values of N, was given by Sasaki [18]. The story behind the growing awareness of the precise features of these theorems in connection with the general N-representability problem, i.e. the determination of the mathematical conditions for a density matrix to correspond to a N-particle wavefunction, was amply described by Coleman [9] – who had also noted the importance and possibility of a condensation of fermions into bosons. This anticipation led him to formulate, and introduce, so-called wavefunctions of *extreme type* and the subsequent structure of the associated second order reduced fermion density matrix. In the following this will be referred to as *the extreme case*.

Before we continue to discuss this important theorem, we will consider a simple example of a density matrix that have played a fundamental role in the development of superfluid theories and in some sense can be thought of as a forerunner to ODLRO. In 1956, Penrose and Onsager [19] attempted a microscopic description of the phenomenon of superfluidity by suggesting that the boson first order reduced density matrix ρ should be of the form

$$\rho(\vec{r},\vec{r}\,') \equiv \Psi^*(\vec{r})\Psi(\vec{r}\,') + \text{incoherent terms} \tag{6.48}$$

The incoherent terms should vanish for large $|\vec{r} - \vec{r}\,'|$. The first term which remains finite even for $|\vec{r} - \vec{r}\,'| \to \infty$; corresponds to the coherent contribution and what Yang [17] calls *off-diagonal long-range order* (ODLRO).

As the theory of superconductivity requires the second order reduced fermion density matrix for a fermionic system, which under specific conditions might condensate into a system of (quasi) bosons, one might here equivalently refer to either a fermionic two-matrix or speak of a bosonic one-matrix. One should also note that the appropriate quantum statistics, due to the requested N-representability of the density matrix should contain the correct symmetries and correlations of the full N-particle wavefunction as to be reflected in the (effective,"macroscopic") wave function displayed in Eq. (6.48)

To set up a general coherence model, which will show the explicit connection of the Penrose-Onsager ansatz with the extreme case and ODLRO, we will assume that our system consists of N particles of fermionic character and that they under some specific conditions may "condensate" into $N/2$ pairs or quasibosons. At the moment we will leave any specific physical mechanism aside and return to this question later. Note also that our focus on the second order reduced density matrix *does not imply* that we are in any way restricting our analysis to non-interacting particle pairs. On the contrary, complicated correlations between pairs, paired pairs etc. will be included.

In order to describe these paired degrees of freedom we introduce a geminal (two-particle) basis of order s denoted by $|(i)\rangle, i = 1,2,\ldots,s$. In general $s \geqslant N/2$ can be infinitely large. However, if $s = N/2$ we realize that we can accommodate a perfect pairing between s different (linearly independent) spatial orbitals of α- and β-spin, respectively. This would correspond to the well-known Hartree-Fock independent particle model. In general we will assume that $2s > N$. In the theoretical analysis one usually takes the (thermodynamic) limit and /or $s \to \infty$. For the moment we will keep s finite and deal with the size of s and the appropriate limiting procedure separately. We will furthermore assume that each vector $|(i)\rangle$ is localized somewhere in our system and that the $N/2$ pairs, transitions, excitations, fluctuations etc. (depending on the physical context) are *equally distributed* over all the units $|(i)\rangle, i = 1,2,\ldots,s$, which are in accordance with the basic ideas of equilibrium (and quasi-equilibrium) statistical mechanics.

We will now set up the appropriate density matrix for the physical situation above assuming the form (6.48). We will for simplicity be a bit intuitive here, see e.g. Brändas and Chatzidimitriou-Dreismann [20] for more details. Since our chosen normalization is different from that of Penrose and Onsager [19] we rewrite (6.48) as

$$\Gamma^{(2)} = \Gamma_L^{(2)} + \text{incoherent terms} \tag{6.49}$$

$$\Gamma_L^{(2)} = |g\rangle\lambda_L\langle g|; \quad |g\rangle = \sum_i^s \frac{1}{\sqrt{s}}|(i)\rangle \tag{6.50}$$

where Eq. (6.50) defines the coherent delocalized geminal g, c.f. the Penrose-Onsager ansatx. As we will see the trace condition (6.47) will essentially determine λ_L, see below.

The second "incoherent term" (the nomenclature is historic rather than physical) consists of two parts. One, the small component, which we denote by $\Gamma_S^{(2)}$ representing the correlated (but not coherent) localized units $|(i)\rangle$, $s = 1, 2, \ldots, s$ of rank $s-1$ and one less interesting part, the tail component, denoted by $\Gamma_T^{(2)}$ which is a diagonal projector γ_T of rank

$$\binom{2s}{2} - s = 2s(s-1)$$

with a $2s(s-1)$ degenerate eigenvalue λ_T and with trace

$$\binom{N}{2} - \frac{N}{2} = \frac{N(N-2)}{2}$$

consisting of all the "unpaired" possibilities. Note that there are $\binom{N}{2}$ possible pairings but only $N/2$ pairs. We thus write for the second part of (6.49), or the incoherent terms consisting of $\Gamma_S^{(2)}$ and $\Gamma_L^{(2)}$, i.e.

$$\Gamma_S^{(2)} + \Gamma_T^{(2)} = \lambda_S \sum_{i,j}^s \left| (i)\rangle \left(\delta_{ij} - \frac{1}{s} \right) \langle (j) \right| + \lambda_T \gamma_T \tag{6.51}$$

We note that the part $\Gamma_S^{(2)}$ is not diagonal, in contrast to $\Gamma_T^{(2)}$. This simply follows from the fact that the localized units in the representation of $\Gamma_S^{(2)}$ must be made orthogonal to the geminal (or superfluid effective wave function). This is the reason why we must subtract $1/s$ from δ_{ij} in Eq. (6.51).

The determination of λ_S and λ_T is now very simple in the present (extreme) model. First we compute λ_S by distributing the number of possible pair-pair correlations, $\binom{\frac{N}{2}}{2}$, equally over the number, $s(s-1)/2$, of possible configurational pair-pair interactions, i.e.

$$\lambda_S = \binom{s}{2}^{-1} \binom{\frac{N}{2}}{2} = \frac{N(N-2)}{4s(s-1)} \tag{6.52}$$

From the condition that the trace of Γ_T equals $N(N-2)/2$ and that λ_T is $2s(s-1)$ degenerate follows immediately that $\lambda_S = \lambda_T$. To obtain λ_L we further need the relation for the trace of $\Gamma^{(2)}$ over the pair subspace. This is easily seen from the observation that this trace is independent of s. In the independent particle approximation, where $s = N/2$, $\lambda_L = \lambda_S = \lambda_T$ one finds

$$\mathrm{Tr}\{\Gamma_L^{(2)} + \Gamma_S^{(2)}\} = \lambda_L + (s-1)\,\lambda_S = \frac{N}{2} \tag{6.53}$$

from which follows that (valid for all $s > N/2$)

$$\lambda_L = \frac{N}{2} - \frac{N(N-2)}{4s} \tag{6.54}$$

We also conclude from above that in the limit of large s the weight factors follow the law $\lambda_L \to \frac{N}{2}$ and $\lambda_S \to 0$.

In summarizing the present development, we make the following observations:

first, we have started with the phenomenological ansatz of Penrose and Onsager
second, we have demonstrated that this defines uniquely (in a natural pairing model) the large component $\Gamma_L^{(2)}$, the small component $\Gamma_S^{(2)}$ and the tail componenent $\Gamma_T^{(2)}$ with the respective eigenvalues uniquely determined from the trace (normalization) conditions
third, the present *extreme ensemble* thus allows a possible description for a system that makes a transition from a state of independent uncorrelated units to a state of coherent units via various degrees of correlated intermediate transients, where full coherence is given by $\Gamma_L^{(2)}$, competing correlated motions are represented by $\Gamma_S^{(2)}$ and the remaining background "noise" is residing in $\Gamma_T^{(2)}$
fourth, the formulation is highly nonlinear and
fifth, it applies to far-from equilibrium situations.

To justify some of our claims above we will finally attempt to show the relation between the presently obtained *extreme ensemble* and ODLRO, i.e. that

$$\Gamma^{(2)} = \Gamma_L^{(2)} + \Gamma_S^{(2)} + \Gamma_T^{(2)} \tag{6.55}$$

where the large, small and tail components are given by

$$\Gamma_L^{(2)} = |g\rangle\lambda_L\langle g|; \quad \lambda_L = \frac{N}{2} - \frac{N(N-2)}{4s} \tag{6.56}$$

$$\Gamma_S^{(2)} = \lambda_S \sum_{i,j}^{s} \left| (i)\rangle \left(\delta_{ij} - \frac{1}{s} \right) \langle (j) \right|; \quad \lambda_S = \frac{N(N-2)}{4s(s-1)} \tag{6.57}$$

$$\Gamma_T^{(2)} = \lambda_T \gamma_T; \quad \lambda_T = \frac{N(N-2)}{4s(s-1)} \tag{6.58}$$

actually, via Coleman's extreme case, give long-range order.

This subsection will be somewhat special and may therefore be skipped. However, for those with particular interest in coherent states etc. we recommend a closer look at these issues. Since the present demonstration will be quite short, we refer the reader to the original literature [9, 17, 18], see also the review by Brändas and Chatzidimitriou-Dreismann [21].

Since the N-particle projection of the celebrated BCS-function [22] of classical superconductors is an *anti-symmetrized geminal power* (AGP), we will briefly review the fundamentals of the AGP state. Note that this is the only reason for considering AGP's (for the definition see below). We are in no way assuming our system to be in any definite quantum state described by Schrödinger wavefunction. Our interest in this matter stems from the very simple and physically imaginative structure of the extreme case and as such it motivates a formulation, where dissipativity may enter the picture. We will return to this aspect in connection with the dynamical extension of quantum mechanics via the complex dilation technique in the next section.

Proceeding we note that antisymmetrized geminal powers lead to a very simple and sparse matrix representation of the two particle density operator $\Gamma^{(2)}(x_1,x_2|x_1',x_2')$ known as the "box and tail" [9]. Focussing on this simplifying feature of the AGP N-particle wavefunction, we first consider, c.f. Eq. (6.50), the geminal g of rank s expanded in a basis of $r = 2s$ orthogonal spin orbitals, $(2s \geqslant N)$

$$g(1,2) = \sum_{i=1}^{s} g_i|\phi_i, \phi_{i+s}\rangle \tag{6.59}$$

where $|\phi_i, \phi_{i+s}\rangle$ is a (normalized) Slater determinant of the spin orbitals ϕ_i and ϕ_{i+s}. From g, one may construct and define the AGP as (with $N = 2m$ for simplicity) the following antisymmetrized product

$$|g^{N/2}\rangle = [S_{N/2}]^{-1/2} \sum_{\substack{j_r<j_{r+1}\\ 1\leqslant r\leqslant N/2}} g_{j_1}g_{j_2}\cdots g_{j_{N/2}}|\phi_{j_1}\phi_{j_1+s}\cdots\phi_{j_{N/2}}\phi_{j_{N/2}+s}\rangle \tag{6.60}$$

where the normalization integral $S_{N/2}$ is given by the symmetric sum

$$S_{N/2} = \sum_{\substack{j_r<j_{r+1}\\ 1\leqslant r\leqslant N/2}} n_{j_1}n_{j_2}\ldots n_{j_{N/2}}; \quad n_j = |g_j|^2 \tag{6.61}$$

The one-particle reduced density operator $\Gamma^{(1)}(g^{N/2})$ associated with the AGP is then given by

$$\Gamma^{(1)}(g^{N/2}) = \sum_{j=1}^{s} \lambda_j^{(1)}\{|\phi_j\rangle\langle\phi_j| + |\phi_{j+s}\rangle\langle\phi_{j+s}|\} \tag{6.62}$$

with

$$\mathrm{Tr}\,\Gamma^{(1)}(g^{N/2}) = N. \tag{6.63}$$

The eigenvalues $\lambda_i^{(1)}$ of $\Gamma^{(1)}(g^{N/2})$ given by

$$\lambda_i^{(1)} = [S_{N/2}]^{-1} n_i \frac{\partial S_{N/2}}{\partial n_i}. \tag{6.64}$$

are doubly degenerate, corresponding to the natural spin orbitals (NSO) ϕ_i and ϕ_{i+s}. In passing one should realize that $\Gamma^{(p)}(g), p = 1, 2$ are different from $\Gamma^{(p)}(g^{N/2})$, albeit related. For instance

$$\Gamma^{(2)}(g) = |g\rangle\langle g| \tag{6.65}$$

$$\mathrm{Tr}\,\Gamma^{(2)}(g) = 1 = \sum_{i=1}^{s} n_i \tag{6.66}$$

and

$$\Gamma^{(1)}(g) = \sum_{j=1}^{s} n_j\{|\phi_j\rangle\langle\phi_j| + |\phi_{j+s}\rangle\langle\phi_{j+s}|\} \tag{6.67}$$

$$\mathrm{Tr}\,\Gamma^{(1)}(g) = 2\,. \tag{6.68}$$

Hence, while $\Gamma^{(1)}(g)$ and $\Gamma^{(1)}(g^{(N/2)})$ have the same NSO's their doubly degenerate occupation numbers are different as shown in Eqs. (6.64) and (6.61). Although $|g\rangle$ is a natural geminal of $\Gamma^{(2)}(g)$ it is not in general an eigengeminal to $\Gamma^{(2)}(g^{N/2})$, except in a very important case namely for the extreme geminal, see below. To see this we consider the representation of $\Gamma^{(2)}(g^{N/2})$ in the basis

$$\{\{|\phi_i, \phi_{i+s}\rangle,\ 1 \leqslant i \leqslant s\},\ \ \{|\phi_i, \phi_j\rangle,\ 1 \leqslant i \leqslant j \leqslant 2s;\ i + s \neq j\}\}$$

which leads up to a very simple "box and tail" form

$$\begin{pmatrix} b_{11} & b_{12} & \dots & b_{1s} & 0 & 0 & \dots & 0 & 0 \\ b_{21} & b_{22} & \dots & b_{2s} & 0 & 0 & \dots & 0 & 0 \\ \vdots & \vdots & \ddots & \vdots & \vdots & \vdots & & \vdots & \vdots \\ b_{s1} & b_{s2} & \dots & b_{ss} & 0 & 0 & \dots & 0 & 0 \\ 0 & 0 & \dots & 0 & t_{12} & 0 & \dots & 0 & 0 \\ 0 & 0 & \dots & 0 & 0 & t_{13} & \dots & 0 & 0 \\ \vdots & \vdots & & \vdots & \vdots & \vdots & \ddots & \vdots & \vdots \\ 0 & 0 & \dots & 0 & 0 & 0 & \dots & t_{(2s-2)2s} & 0 \\ 0 & 0 & \dots & 0 & 0 & 0 & \dots & 0 & t_{(2s-1)2s} \end{pmatrix} \tag{6.69}$$

The "box", which has dimensions s, has the elements

$$b_{j,j+s;k,k+s} \equiv b_{jk} = [S_{N/2}]^{-1} g_j g_k^* \frac{\partial^2 S_{N/2+1}}{\partial n_j\, \partial n_k};\quad 1 \leqslant j \neq k \leqslant s; \tag{6.70}$$

and

$$b_{j,j+s;j,j+s} \equiv b_{jj} = [S_{N/2}]^{-1} n_j \frac{\partial S_{N/2}}{\partial n_j} = \lambda_j^{(1)}; \quad 1 \leqslant j \leqslant s \tag{6.71}$$

where in Eq. (6.70) $S_{N/2+1}$ is the symmetric function of order $N/2+1$. The "tail", which has the dimension $2s(s-1)$, has the elements

$$\begin{aligned} t_{ij} &= [S_{N/2}]^{-1} n_i n_j \frac{\partial^2 S_{N/2}}{\partial n_i \partial n_j} \\ &= \lambda_i^{(1)} \lambda_j^{(1)} + \frac{1}{2} \left\{ n_j \frac{\partial \lambda_i^{(1)}}{\partial n_j} + n_i \frac{\partial \lambda_j^{(1)}}{\partial n_i} \right\} \\ & \quad 1 \leqslant i < j \leqslant 2s; \quad i + s \neq j \end{aligned} \tag{6.72}$$

Instead of constructing the general BCS-function [22] and taking the thermodynamic limit, i.e. letting N (and the volume) go to infinity we will take a closer look at the finite N case. The exact form of $\Gamma^{(2)}(g^{N/2})$ was, as previously mentioned, thoroughly studied by Coleman [9]. The upper bound on $\lambda^{(2)}$ was obtained by Yang [17] (for N even) and by Sasaki [18] (for N odd). The upper limit to $\lambda^{(2)}$ was thus found to be

$$\lambda_i^{(2)} = \begin{cases} \leqslant (N-1)/2 & \text{if } N \text{ odd (Sasaki);} \\ < N/2 & \text{if } N \text{ even (Yang).} \end{cases} \tag{6.73}$$

Coleman then introduced the following concept:

Definition: $g^{N/2}$ *(N even) is said to be extreme if all occupation numbers* n_j *of* $\Gamma^{(1)}(g)$ *are equal.*

Using Eqs. (6.64), (6.70–72), the "box and tail" becomes in the extreme case:

$$b_{jk} = b = \lambda^{(1)}(1 - \frac{N-2}{2(s-1)} = \frac{N(2s-N)}{4s(s-1)}, \quad 1 \leqslant j \neq k \leqslant s \tag{6.74}$$

$$b_{jj} = \lambda^{(1)} = \frac{N}{2s}, \quad 1 \leqslant j \leqslant s \tag{6.75}$$

$$t_{ij} = \frac{N(N-2)}{4s(s-1)} = \lambda^{(1)} - b, \quad 1 \leqslant i < j \leqslant 2s; \quad i + s \neq j. \tag{6.76}$$

The characteristic equation for the "non-diagonal" part becomes

$$(\lambda^{(1)} + (s-1)b - \lambda^{(2)})(\lambda^{(1)} - b - \lambda^{(2)})^{(s-1)} = 0$$

with a "large eigenvalue" given by

$$\lambda_1^{(2)} = \lambda_L^{(2)} = \lambda^{(1)} + (s-1)b = \frac{N}{2} - \frac{N(N-2)}{4s} \tag{6.77}$$

and an $(s-1)$ degenerate eigenvalue

$$\lambda_j^{(2)} = \lambda_D^{(2)} = \lambda^{(1)} - b = \frac{N(N-2)}{4s(s-1)} \quad j = 2\ldots s \tag{6.78}$$

Hence the "box and tail" becomes in the extreme case precisely the representation previously discussed in Eqs. (6.55–58). This can be seen by writing out $\Gamma^{(2)}(g^{N/2})$ explicitly. i.e.

$$\Gamma^{(2)}(g^{N/2}) = \frac{N}{2s}\sum_{i=1}^{s} |\phi_i,\phi_{i+s}\rangle\langle\phi_i,\phi_{i+s}| + \frac{N(2s-N)}{4s(s-1)}\sum_{i\neq j} |\phi_i,\phi_{i+s}\rangle\langle\phi_j,\phi_{j+s}| +$$

$$+\frac{N(N-2)}{4s(s-1)}\sum_{\substack{i<j\\ i+s\neq j}} |\phi_i,\phi_j\rangle\langle\phi_i,\phi_j| = \Gamma_L^{(2)} + \Gamma_D^{(2)} \tag{6.79}$$

where

$$\Gamma_L^{(2)} = |\Phi_L\rangle\lambda_L^{(2)}\rangle\Phi_L|; \quad \Phi_L = \sum_{i=1}^{s}\frac{1}{\sqrt{s}}|\phi_i,\phi_{i+s}\rangle = g \tag{6.80}$$

$$\lambda_L^{(2)} = \frac{N}{2} - \frac{N(N-2)}{4s}$$

and

$$\Gamma_D^{(2)} = \lambda_D^{(2)}\left\{\sum_{i,j} |\phi_i,\phi_{i+s}\rangle\left(\delta_{ij} - \frac{1}{s}\right)\langle\phi_j,\phi_{j+s}| + \sum_{\substack{i<j\\ i+s\neq j}} |\phi_i,\phi_j\rangle\langle\phi_i,\phi_j|\right\}$$

$$\lambda_D^{(2)} = \frac{N(N-2)}{4s(s-1)}. \tag{6.81}$$

From Eq. (6.80) it is easily seen that even if the basis $|\phi_i,\phi_{i+s}\rangle$ is localized in some sense, one finds that the *extreme* geminal g is delocalized. Hence in the limit of large s this delocalization remains which also is a reflection of the properties of the off-diagonal matrix elements in the "box". Note also that this limit obtains through the existence of a large eigenvalue, proportional to the number of particles, which also means that the off-diagonal parts of $\Gamma^{(2)}$ (in the coordinate representation) remains finite in the thermodynamic limit, i.e. $N \to \infty$, $V_N \to \infty$ with $\frac{N}{V_N}$ is finite. In these limits the large eigenvalue approaches the upper bound in Eq. (6.73). This completes the connection between the extreme case and ODLRO.

The results quoted here can be strictly formulated into a theorem due to Coleman [9]:

Theorem: *"The geminal g is an eigenfunction of $\Gamma^{(2)}(g^{N/2})$ with a non-vanishing eigenvalue if and only if g is of extreme type, i.e. the eigenvalues of $\Gamma^{(1)}(g)$ are all equal."*

Since Yang's assumption of ODLRO leads to a large eigenvalue of the 2-matrix, flux quantization and the Meissner effect (i.e. that the magnetic field is exponentially damped inside the superconductor), it is intuitively clear that, any N-particle projection of the coherent Fock representation of let us say the BCS function, or the N-representable component of the Penrose-Onsager ansatz, should display extreme behavior or more strongly expressed, that the extreme condition for g is a necessary condition for the development of ODLRO. However, it is also immediately clear that the reverse is not true, i.e. the extreme case it not sufficient for ODLRO in the sense that $\Gamma^{(2)}$ must exhibit a macroscopically large eigenvalue. Of course this sentence is only meaningful if we consider *not* to take the thermodynamic limit and thus allowing ourselves to situations outside statistical equilibrium. Since this is often a cause for misunderstanding we cannot overemphasize the importance of this distinction in connection with the interpretation of our microscopic approach to dissipative system and structures.

To continue from here we use (6.55–58) or (6.79–81) to simply obtain the expression for the correlation energy, neglecting $\Gamma_T^{(2)}$ and assuming that the zero- and one-body parts of (6.32) corresponds to the independent particle model, as

$$\begin{aligned}\left\langle \sum_{i<j}^{N} H_{ij} \right\rangle &= E_{\text{corr}} = \text{Tr}\{H_{12}\Gamma^{(2)}\} \\ &= \lambda_L \langle g|H_{12}|g\rangle + \lambda_S \left[\sum_{i,j}^{s} \langle (i)|H_{12} \left(\delta_{ij} - \frac{1}{s} \right) |(j)\rangle \right] \end{aligned} \tag{6.82}$$

We have thus explicitly demonstrated that a particular realization of the Penrose-Onsager ansatz yields a direct connection with Yang's fundamental concept of ODLRO and allows identification with the extreme case of Coleman. However, as stressed above, by not considering the thermodynamic limit, we will show that our extreme ensemble (6.55–58) may contain additional information about the interplay between coherences and correlations or phrased differently, between correlations and specific physical mechanisms. This will lead to the concept of a coherent-dissipative structure.

6.7 Resonance Picture of Unstable States

In order to treat the dynamics of a quantum mechanical system we need to consider the time dependent Schrödinger (or Liouville) equation, see Eq. (6.1) or Eq. (6.27). The usual ansatz is a factorization of the corresponding wavefunction into a part that depends on the spatial coordinates (and/or spin) and a time dependent part leading up to the time independent Schrödinger equation etc. This leads to the concept of stationary states and unitary time evolution as we have seen in previous Chapters.

One of the crucial facts of life, however, is that we are all subject to an unavoidable arrow of time. A more mathematical expression of this fundamental property of the universe is that unitary group evolution, as expressed in the primary laws of physics – based on some time independent hamiltonian; it is not necessary to go into the intricacies of fundamental symmetries in particle physics here – somewhere during the transition from the microscopic to the macroscopic level exhibits a dissipative, semigroup evolution, which is tantamount to a broken time symmetry. A rather simple trick to deal with this situation in quantum mechanics was advocated by George Gamow [23] in 1928. His theory, based on complex resonance waves, known today as Gamow waves, was successfully applied to the phenomenon of radioactivity. The complex resonance theory, although very successful in its later multidisciplinary applications in physics and chemistry, was always looked upon with scepticism and awe by more traditional scientists. Perhaps rightly so, since these Gamow resonances were not, strictly speaking, embodied by the axioms of quantum mechanics and hence restricted to the "unphysical sheet", see more below.

Nevertheless the simple reason that resonant Gamow vectors, although having many attributes of a bound state like localization inside a potential barrier etc., do play the role of a second class citizen, is that they are not subject to ordinary (square integrable) boundary conditions. Instead they have the "nasty" property of *diverging with increasing values of the radial coordinate.* This has naturally led to problems, both computationally as well as theoretically. Not only are the Gamow waves terrible to handle numerically but they are, as already mentioned above, outside the axiomatic framework of quantum mechanics. Despite their successful stage appearance, they were both difficult to calculate and interprete.

This situation changed dramatically around 1970, when the celebrated theorem due to Balslev and Combes [24] was published. A parallel development was

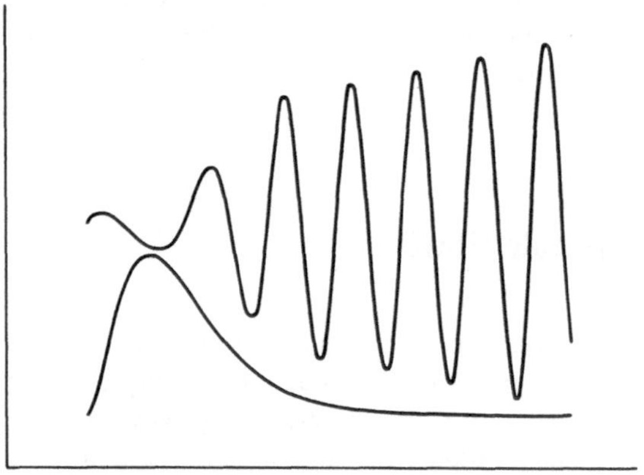

Fig. 6.2. Real part of a Gamow wave for the potential $V = V_0 r^2 e^{-r}$. V_0 is a positive constant, r is the radial coordinate and the diverging Gamow wave is displayed for the resonance closest to the real energy axis

independently given by van Winter [25]. For an elegant and insightsful explanation and appraisal for non-mathematicians we refer to Simon [26].

Since we will not have time and space to go into detail here, we will refer to a recent symposium where most of these issues were discussed in detail [27]. Briefly the mathematical theorem [24] concerns the spectrum of a dilated operator. As we will see this will allow us to rigorously employ (analyse and compute!) Gamow waves and related dynamical quantities. The "trick" is carried out via an (unbounded!) transformation usually referred to as the Complex Scaling Method (CSM) and the gist of the argument thus revolves around proper analytic extensions to the complex energy plane.

Since CSM plays such a fundamental role in our present study we will give a short review of some of its more puzzling aspects. It is not improper to consider the deformation of the coordinates into complex valued degrees of freedom as an attempt to analytically continue quantum mechanics beyond its conventional domain of a time reversible microscopic formulation into a valid theory of a more general nature, where irreversibility is naturally imbedded in the dynamics. The meaning of this somewhat cryptic statement will hopefully be clarified in more detail below.

Let us introduce some general mathematical concepts needed for the actual extension. Although our technique emanates from the pioneering work of Balslev and Combes [24] we will here follow closely the development given in the proceedings from the Lertorpet symposium [27]. It is customary to introduce the N-body hamiltonian as $H = T + V$, where T is the kinetic energy operator and V the interaction potential. Since H is an unbounded operator, albeit bounded from below, we need to restrict the domain (the space of allowed vectors on which H applies) somewhat. Let $\mathscr{H}$ denote the Hilbert space, complete with respect to the standard norm of square integrability, i.e. $\Phi \in \mathscr{H}$ with $\|\Phi\|_{L^2} < \infty$. Then the domain of H is given by

$$\mathscr{D}(H) = \{\Phi \in \mathscr{H}, \quad H\Phi \in \mathscr{H}\} \tag{6.83}$$

We will basically consider interactions such that the unboundedness of H arises entirely from the kinetic energy operator T. The potential V will be further specified below, however at the present stage it will be sufficient to treat the case with $\mathscr{D}(H) = \mathscr{D}(T)$.

A key quantity in the CSM formulation is the scaling operator

$$U(\theta) = \exp(iA\theta) \tag{6.84}$$

where

$$A = \frac{1}{2}\sum_{k=1}^{N}[\vec{p}_k\vec{x}_k + \vec{x}_k\vec{p}_k] \tag{6.85}$$

is the generator of the scaling transformation, θ is a parameter, which may be real or complex and $\vec{x}_k$ and $\vec{p}_k$ are the coordinate and momentum of particle k. If

$\theta \in R$, then U is a unitary operator, defining the dilation group, with $\mathscr{D}(U) = H$, which effectuates the scaling

$$U(\theta)\Phi(x_1,\ldots,x_N) = \exp\left(\frac{3N}{2}\theta\right)\Phi(e^\theta x_1,\ldots,e^\theta x_N) \tag{6.86}$$

By considering complex scalings, i.e $U(\eta)$ with $\eta = |\eta|e^{i\theta}$ (note that our parameter in U now refers to the whole analytic parameter η rather than the dilation group parameter θ) Balslev and Combes [23] proved that for certain classes of hamiltonians, i.e. so-called dilatation analytic hamiltonians, the continuous spectrum changed under the complex deformation and became rotated down -2θ in the complex energy plane, thereby opening up sectors on the "unphysical" Riemann sheet, where possible finite dimensional "resonance eigenvalues" would appear. They also demonstrated the non-existence of singularly continuous spectra in these cases, as well as invariance of the exposed spectrum (including bound states) to variations of θ.

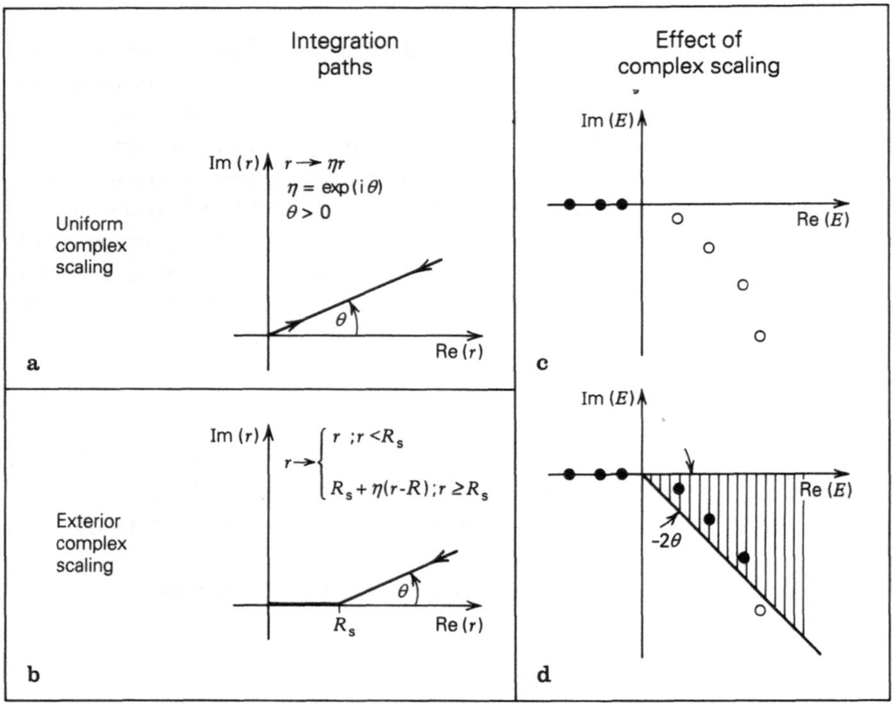

Fig. 6.3. Effects from distortion of real integration path out into the complex plane. **a** Uniform scaling. **b** Exterior complex scaling. **c** Without complex scaling only bound states are accessible (filled circles). **d** Scaling (**a** or **b**) rotates continuum by the angle $-2x$ (argument of scale factor) uncovering previously unaccessible resonances, i.e. making those resonances in the shaded sector equivalent with the bound states. Others, denoted by unfilled circles remain "out of reach". Inspite of the magic appearance of resonances in the simplified example above, the discrete spectrum is not dependent on the values of the scale factor once it is uncovered

Without going into too many mathematical details, we will thus consider $\eta = |\eta|e^{i\theta}$, with $0 \leqslant \theta < \theta_0$. Generally θ_0 depends on the potential, for instance in the Coulomb case no limit is invoked, while in some other cases, as will be seen below it is natural to have $\theta_0 \leqslant \frac{\pi}{4} - \delta$. We also introduce Ω as

$$\Omega = \{\eta,\ |\arg(\eta)| < \theta_0\} \tag{6.87}$$

Furthermore we make the decomposition $\Omega = \Omega^+ \cup \Omega^- \cup R$, where the real axis $R = R^+ \cup R^- \cup \{0\}$ partitions Ω into its upper and lower parts in the complex plane. In what follows we will assume that the interaction V is a sum of two body interactions V_{ij} such that V_{ij} is Δ_{ij} -compact in $L^2(R^3)$, and furthermore that $V_{ij}(\eta) = U_{ij}(\eta)V_{ij}U_{ij}^{-1}(\eta)$, $\eta \in R^+$, has a compact analytic extension to Ω^+. The definition of the dilatation analytic family of operators $H(\eta)$ is then given by

$$H(\eta) = U(\eta)HU^{-1}(\eta),\ \ \eta \in R^+ \tag{6.88}$$

and

$$H(\eta) = \eta^{-2}T + V(\eta),\ \ \eta \in \Omega^+ \tag{6.89}$$

In the expression above it is notable that $H(\eta)$ is obtained in two steps. First via a unitary transformation to a scaled representation and thereafter followed by an analytic continuation to Ω^+. Although this is a mathematically rigorous procedure it is preferable, as we will see below, to consider the unbounded similarity transformation $U(\eta)$, $\eta \in \Omega^+$, directly. We will then introduce the domain $\mathcal{N}(\Omega)$ as the subset

$$\mathcal{N}(\Omega) = \{\Phi,\ \Phi \in \mathcal{H};\ U(\eta)\Phi \in \mathcal{H};\ \eta \in \Omega\} \tag{6.90}$$

We furthermore complete the space with respect to the norm

$$\int_{-\theta_0}^{+\theta_0} \|U(\eta)\Phi\|^2_{L^2} d\theta = \|\Phi\|^2_{N_{\theta_0}} \tag{6.91}$$

Since we also want the first and second partial derivatives to satisfy Eq. (6.91) it is natural to define the spaces $\mathcal{N}^{(i)}_{\theta_0}$, $i = 0, 1, 2$, analogously. We are now in the position to make an alternative definition of the analytic family $H(\eta)$, i.e.

$$H(\eta) = U(\eta)HU^{-1}(\eta);\ \ \mathcal{D}(UHU^{-1}) = \mathcal{N}^{(2)}_{\theta_0} \tag{6.92}$$

and

$$H(\eta) = \eta^{-2}T + V(\eta);\ \ \mathcal{D}(UHU^{-1}) \to \mathcal{D}(T) \tag{6.93}$$

Note that $\mathcal{N}^{(i)}_{\theta_0}, i = 0, 1, 2$ are Hilbert spaces, which allow $H(\eta)$ to be interpreted as a similarity transformation of the self-adjoint unscaled operator H. We also note that $U(\eta)$ exhibits the "star-unitary" property

$$U^{\dagger}(\eta^*) = U^{-1}(\eta) \tag{6.94}$$

which can be easily proven from

$$\langle\Phi|\Psi\rangle = \langle\Phi(\eta^*)|\Psi(\eta)\rangle = \langle\Phi|U^{\dagger}(\eta^*)U(\eta)|\Psi\rangle \tag{6.95}$$

which holds for all $\Phi, \Psi \in N_{\theta_0}$.

In the definition presented above, leading of course to the same analytic family $H(\eta)$, we can see that the two steps involved in Eqs. (6.92–93) are different in comparison to Eqs. (6.88–89). Here the first step consists of restricting $\mathscr{H}$ to a smaller domain $N_{\theta_0}^{(2)}$ for which the scaling $U(\eta)$ is defined for all η such that $\arg(\eta) < \theta_0$. After making the complex rotation, i.e. changing the parameter θ in e^{θ} from real to complex, one completes the subset (6.90) (dense in $\mathscr{H}$) to $\mathscr{D}(T)$), which means that completion is made with respect to the "standard" L^2-norm for the functions and its first and second partial derivatives.

One can understand the abstract transformations above in a very simple way. Consider e.g. the integral

$$I(b) = \int_0^{+\infty} e^{-br} dr = \frac{1}{b} \tag{6.96}$$

whose simple analytic form trivially allows analytic continuation to negative values of b. In the latter case the integral is not convergent even if $I(b) = \frac{1}{b}$ for all $b \neq 0$. If $b \neq 0$ is complex (or negative) so that the integrand in (6.96) does not vanish for $r \to \infty$, then one may consider the trick of defining a complex integrated path with θ sufficiently large so that

$$I = \lim_{R\to\infty} \int_0^{Re^{i\theta}} e^{-br} dr = \frac{1}{b} \tag{6.97}$$

The complex path can hence be used to explicitly compute a numerical value to the integral I even when its analytical form is unknown.

However, when evaluating matrix elements, using a complex deformation as in Eq. (6.97) a word of warning should be issued. Again we approach a subsection that can be skipped by the general reader, but since the interest in the complex energy plane may arise with some delayed priority we leave the discussion here for digestion when necessary.

For instance if one does not realize that the spaces $\mathscr{N}_{\theta_0}^{(i)}, i = 0, 1, 2$, have a finer topology than $\mathscr{H}$ then paradoxical situations may occur. Consider e.g. the resolvent $(H - \lambda I)^{-1}$, which for $\lambda^* \neq \lambda$ is a bounded operator in $\mathscr{H}$. *This is not true in* $\mathscr{N}_{\theta_0}$, We will exemplify this as follows (I am greatly indebted to B. Nagel for this particular example). In momentum space we obtain with $H_0 = k^2; \Phi = \pi^{-\frac{1}{4}} e^{-\frac{k^2}{2}}, \eta = |\eta| e^{i\theta}, \theta_0 \leqslant \frac{\pi}{4} - \delta$ with $\delta > 0$ the following result

$$(U(\eta)\Phi) = \pi^{-\frac{1}{4}} \frac{1}{\sqrt{\eta}} e^{-\frac{k^2}{2\eta^2}} \tag{6.98}$$

and

$$\|U(\eta)\Phi\|_{L^2}^2 = \frac{1}{\sqrt{\cos(2\theta)}} \tag{6.99a}$$

$$\|U(\eta)\Phi\|_{N_{\theta_0}}^2 = \int_0^{\frac{\pi}{2}} \frac{d\theta}{\sqrt{\cos\theta}} < \infty \tag{6.99b}$$

Since $0 < \theta_0 \leqslant \frac{\pi}{4} - \delta$ then $k^n e^{-\frac{k^2}{2}} \in \mathcal{N}_{\theta_0}$ for all n, and hence $\Phi \in \mathcal{N}_{\theta_0}^{(\infty)}$. Now it is easy to show that if $-2\theta_0 < \arg\lambda < 0$, then $\chi = (H_0 - \lambda I)^{-1} e^{-\frac{k^2}{2}} \notin \mathcal{N}_{\theta_0}$. One obtains

$$(U(\eta)\chi)(k) = \frac{1}{\sqrt{\eta}} (\eta^{-2}k^2 - \lambda)^{-1} e^{-\frac{k^2}{2\eta^2}} \tag{6.100}$$

and

$$\begin{aligned}\|\chi\|_{L^2}^2 &= 2(\cos 2\theta)^{\frac{3}{2}} \int_0^\infty \frac{e^{-k^2} dk}{(k^2 - \lambda^* \cos 2\theta e^{-2i\theta})(k^2 - \lambda \cos 2\theta e^{2i\theta})} \\ &= [\lambda \cos 2\theta e^{2i\theta} = k_0^2 e^{2i\varepsilon}; \quad k_0 = |\lambda|^{\frac{1}{2}} (\cos 2\theta)^{\frac{1}{2}}] \\ &= 2(\cos 2\theta)^{\frac{3}{2}} \int_0^\infty \frac{e^{-k^2} dk}{|k + k_0 e^{i\varepsilon}|^2 (k - k_0 e^{i\varepsilon})(k - k_0 e^{-i\varepsilon})}\end{aligned} \tag{6.101}$$

It is obvious that the divergence of the integral above as $\varepsilon \to 0$ determines whether $\|\chi\|_{N_{\theta_0}}^2$ is finite or not. For small ε one finds that

$$\begin{aligned}\|\chi\|_{L^2}^2 &\sim \frac{e^{-k^2}}{4k_0^2} \int_{-\delta}^{\delta} \frac{dx}{(x - i\varepsilon k_0)(x + i\varepsilon k_0)} \\ &\sim \pi \frac{e^{-k_0^2}}{4k_0^3 |\varepsilon|}\end{aligned} \tag{6.102}$$

which since $\int_{-\delta}^{\delta} \frac{d\varepsilon}{|\varepsilon|} = \infty$ implies that $\chi \notin \mathcal{N}_{\theta_0}$.

This simple example clarifies the following somewhat surprising situation. In accordance with Eq. (6.95) we write the following matrix element with respect to the operator H, utilizing Eqs. (6.92–94) with $\Phi, \Psi \in \mathcal{N}_{\theta_0}^{(2)}$, such that

$$\langle \Psi | H | \Psi \rangle = \langle \Psi(\eta^*) | H(\eta) | \Psi(\eta) \rangle \tag{6.103}$$

holds. It is important to realize that $H(\eta)\Psi(\eta) = U(\eta)HU^{-1}(\eta)U(\eta)\Psi = U(\eta)(H\Psi)$ are all valid operations since $\Psi \in \mathcal{N}_{\theta_0}^{(2)}$ implies $H\Psi \in \mathcal{N}_{\theta_0}$ and hence Eq. (6.103) reduces to Eq. (6.95).

The invariance (6.103) shows that the matrix element is not altered by the complex deformation, if Ψ, Φ are in the right domain, see above. However, comparing e.g. (6.96) and (6.97), the scaled integral to the right in (6.103) may have a finite value for a suitable η even when the unscaled integral to the left may not exist. Hence the right hand side of (6.103) exists for a more general class of functions than those that are L^2 in its unscaled form or more precisely those that belong to $\mathcal{N}^{(2)}_{\theta_0}$. Thus we see that complex scaling allows us to extend quantum mechanics beyond its conventional domain in that it assigns well defined meaning to vectors and matrix elements with respect to operators (resolvents), requiring in analogy with Eq. (6.103) the existence of the scaled representation for some nonreal η.

Before we continue, we want to emphasize that we have implicitly assumed that $V\Psi \in \mathcal{N}_{\theta_0}$ for $\Psi \in \mathcal{N}^{(2)}_{\theta_0}$. Even if this may be a restriction on V in general one can circumvent this problem by considering more general deformations like e.g. exterior scaling etc. These are, however, technical points that will not alter our general considerations and conclusions as well as physical interpretation.

With the preceeding development in mind, it is now easy to avoid the following paradoxical situation. If we employ Eq. (6.103) with H replaced by the resolvent operator one may ask whether the following relation with $\eta \in \Omega^+, \lambda^* \neq \lambda$

$$\langle \Phi | (H - \lambda I)^{-1} | \Psi \rangle = \langle \Phi(\eta^*) | (H(\eta) - \lambda I)^{-1} | \Psi(\eta) \rangle \tag{6.104}$$

holds for all $\Phi, \Psi \in \mathcal{N}_{\theta_0}$. The puzzle consists in that the left hand side of Eq. (6.104) is always finite for λ complex, while the right hand side may become infinite if λ is a complex eigenvalue of the complex rotated operator $H(\eta)$, which is a possibility according to the Balslev-Combes theorem. Now we can return to our analysis carried out in Eq. (6.102). For instance with the choice $H = H_0 = k^2$ and $\Psi = \Phi$ with $-2\theta_0 < \arg \lambda < 0$ it was demonstrated that $\chi = (H - \lambda I)^{-1}\Phi \notin \mathcal{N}_{\theta_0}$ and hence Eq. (6.95) (with χ replacing Ψ) is *not* applicable and the equality (6.104) is *not* valid. Nevertheless the scaled matrix element in Eq. (6.104) exists for e.g. $2\theta_0 > \arg \lambda > 0$, and is equal to the unscaled integral and more importantly has a meromorphic continuation into the sector given by $-2\theta_0 < \arg \lambda < 0$.

For simplicity we have displayed a simple situation with one threshold at zero, but the generalization to the general manybody situation should be obvious.

It is also clear that we have considered the Hilbert spaces $\mathcal{N}^{(i)}_{\theta_0}, i = 0, 1, 2$ in order to find convenient domains for the unbounded (in $\mathcal{H}$) similitude $U(\eta), \eta \in \Omega$. After appropriate deformations completion with respect to the standard L^2 norm is made. In this manner we arrive at the formal eigenvalue relation $(\eta = |\eta| e^{i\theta})$

$$H(\eta)\Psi(\eta) = \varepsilon(\eta)\Psi(\eta) \quad \eta \in \Omega^+ \tag{6.105}$$

with $\arg \eta$ sufficiently large to uncover the resonance ε. The conjugate of Eq. (6.105) becomes

$$\overline{H(\eta)}\ \overline{\Psi(\eta)} = \overline{\varepsilon(\eta)}\ \overline{\Psi(\eta)} \quad \eta \in \Omega^- \tag{6.106}$$

with the involution $\overline{A(\eta)} = A^*(\eta^*)$ being introduced to include also (complex) optical potentials. Obviously Eqs. (6.105) and (6.106) in conjunction with (6.103) motivates the construction

$$\langle \overline{\Psi(\eta^*)} | H(\eta) | \Psi(\eta) \rangle = \varepsilon(\eta) \langle \overline{\Psi(\eta^*)} | \Psi(\eta) \rangle \tag{6.107}$$

from which stationary variational principles can be derived in almost the same fashion as in ordinary quantum mechanics, with the important distinction that the extremum property of the principle has been lost. Note also that if H satisfies $H^* = H$ then there is no restriction *from the variational point of view* to assume that $\Psi^* = \Psi$ and then the construction (6.107) defining a trial $\varepsilon(\eta)$ based on a trial $\Psi(\eta)$ and corresponding $\overline{\Psi(\eta^*)} = \Psi(\eta)^*$ becomes complex symmetric, c.f. instance the Gantmacher theorem [10,11]. However, in many cases it may be useful to analyse the wavefunction in terms of nonreal spherical harmonics $Y_{l,m}(\vartheta, \varphi)$, or other convenient nonreal representations. The complex symmetry is thus a convenience which may or may not be used, albeit it can always be imposed without restricting the formulation. We will impose it below for simplicity.

Since η sometimes is used primarily as a numerical convergence factor we may replace the explicit η-dependence for the quantity A by replacing $A(\eta)$ by A^c. Hence we write

$$H^c \Psi^c = \varepsilon \Psi^c \tag{6.108}$$

and for the complex conjugate equation

$$H^{c*} \Psi^{c*} = \varepsilon^* \Psi^{c*} \tag{6.109}$$

and

$$\langle \Phi^{c*} | H^c | \Psi^c \rangle = \varepsilon \langle \Psi^{c*} | \Psi^c \rangle \tag{6.110}$$

Even if this simple notation is very appealing in that almost any standard quantum mechanical technique can be taken over provided it is appropriately modified, one should note that it results in a formulation that goes beyond conventional quantum mechanics "on the real axis". The most direct consequence is, as mentioned, the appearance of complex resonance eigenvalues and associated Gamow vectors. A closer study of the full generalized spectral properties of the complex deformed problem shows that these eigenstates essentially deflate the (generalized) spectral density, giving in an asymptotic sense a decomposition of the continuum into resonances and background. This has the important consequence that each Gamow vector represents a well-defined section of the continous spectrum associated with the unscaled self-adjoint problem. In other words one can say that

the Gamow representation condensates an infinite dimensional Hilbert space associated with a particular spectral part of the continuum into a finite dimensional linear space of suitable Gamow vectors.

This condensation will play a fundamental role for our theory. There are, however, two important problems that need immediate attention in order to proceed, namely
first, since the general case may lead to matrices that cannot be diagonalized, we must be able incorporate so-called Jordan blocks into our theory (this will turn out to be a blessing in disguise.)
second, we also need to introduce these Gamow-like resonances into a noncontradictory formulation of the appropriate Liouville equation (this turns out to be nontrivial).

The consequences of the CSM extension to the dissipative subdynamics will be quite surprizing. In the next paragraph we will see how the two aforementioned problems lead to rather specific conditions which are applicable to concrete physical situations. There are also another class of problems that can be studied, i.e. the general question of integrability in theoretical physics, but this will not be further mentioned here.

Before we end this section, we will briefly comment on how it may happen that the matrix cannot be diagonalized and what a Jordan block means in this connection. Consider the very simple (complex) symmetric matrix $\mathbf{Q}$ given by

$$\mathbf{Q} = \frac{1}{2}\begin{pmatrix} 1 & -i \\ -i & -1 \end{pmatrix} \tag{6.111}$$

By inspection we see that the square of Eq. (6.111) is the zero matrix, but since the rank (the number of linearly independent vectors that constitute the matrix) is one then $\mathbf{Q} \neq 0$. Hence $\mathbf{Q}$ is similar to

$$\mathbf{C} = \begin{pmatrix} 0 & 1 \\ 0 & 0 \end{pmatrix}$$

i.e that there exists an invertible matrix $\mathbf{B}$, so that the triangular matrix $\mathbf{C}$ is related to the symmetric $\mathbf{Q}$ via

$$\mathbf{Q} = \mathbf{B}^{-1}\mathbf{C}\mathbf{B} \tag{6.112}$$

Rather than giving the explicit form here, we will see below that this problem has a very simple general solution. We will also see that the vectors that constitute $\mathbf{B}$ or $\mathbf{B}^{-1}$ will have some specific physical properties which may be directly related to the extreme case or Yang's ODLRO.

In order to generalize the discussion above, we will focus on the general classical canonical form of any finite dimensional matrix. We assume that the reader is familiar with the simple fact that a matrix with distinct eigenvalues (corresponding to one dimensional linearly independent eigenvectors) can always be transformed, via a similarity transformation, of type (6.112), to diagonal form. The problem thus arises when degeneracies appear. If the matrix represents a normal operator, i.e. an operator which commutes with its own adjoint, then one can also prove diagonalizability of the matrix. However, for general matrices this is no longer true and there may appear blocks, of various dimensions, in the matrix corresponding to a particular degenerate eigenvalue. The large dimension of such a Jordan block defines the Segrè characteristic for this particular degenerate eigenvalue, and this will be a key quantity in what follows.

Since one can easily prove that any matrix can be transformed into a triangular form, it is obvious that any general matrix can also be transformed into complex symmetric form provided that there exists a similarity transformation of type (6.112) between the triangular- and the symmetric block. It suffices to focus on a particular degenerate eigenvalue and without restriction we can further put this eigenvalue equal to zero in the analysis. Assuming that the corresponding Segrè characteristic is s, the dimension of the largest Jordan block of the (zero) degenerate eigenvalue, it is clear that the rank of this nondiagonalizable block must be $s-1$. This follows easily, since the zero eigenvalue implies that there must be a linear relation between the columns that constitute the matrix or block. The fact that the matrix, taken to the power $s-1$, is different from zero (but equals zero for higher powers) means that the nondiagonalizable blocks always can be transformed to a classical canonical (Jordan) form with ones in the entries above, or below, the diagonal.

In a previous study Reid and Brändas [11] found a very simple complex symmetric form of this Jordan block, i.e.

$$q_{kl} = \left(\delta_{kl} - \frac{1}{s}\right)\exp\left(i\pi\frac{k+l-2}{s}\right), \tag{6.113}$$

where s is the dimensionality of the Jordan block and

$$1 \leqslant k, l \leqslant s\,. \tag{6.114}$$

with the property that the matrix in (6.113) is *similar* to

$$\mathbf{C} = \begin{pmatrix} 0 & 1 & 0 & \dots & 0 \\ \vdots & \ddots & \ddots & \dots & \vdots \\ & & \ddots & \ddots & 0 \\ \vdots & & & \ddots & 1 \\ 0 & \dots & & \dots & 0 \end{pmatrix} \tag{6.115}$$

One can further prove that the matrix defined in (6.113) ($s = n$), see Exercise Problem 4 in Sect. 6.11, is similar to (6.115) by considering the matrix $\mathbf{B}$ given by ($\omega = \exp\left(\frac{i\pi}{n}\right)$)

$$\mathbf{B} = \frac{1}{\sqrt{n}}\begin{pmatrix} 1 & \omega & \omega^2 & \dots & \omega^{n-1} \\ 1 & \omega^3 & \omega^6 & \dots & \omega^{3(n-1)} \\ \dots & \dots & \dots & \dots & \dots \\ 1 & \omega^{2n-1} & \omega^{2(2n-1)} & \dots & \omega^{(n-1)(2n-1)} \end{pmatrix} \tag{6.116}$$

or equivalently given by $b_{kl} = \omega^{(2k-1)(l-1)}$. One realizes that $\mathbf{B}$ is unitary and then it is simple to compute and demonstrate that $\mathbf{Q} = \mathbf{B}^{-1}\mathbf{C}\mathbf{B}$.

The transformation above is quite interesting in its own right, since it shows that a real triangular matrix and a complex symmetric matrix are connected through a unitary transformation. This will be of fundamental importance in what follows below.

6.8 Resonances and Dissipative Dynamics

Returning to the introductory remarks as well as to the development of previous Chapters, it was emphasized that the dynamics during a spectroscopic transition is derived from the equation of motion for the density matrix ρ given by the Liouville-von Neumann equation

$$i\hbar \frac{\partial \rho}{\partial t} = \hat{L}\rho \tag{6.117}$$

which together with some given initial conditions defines a unitary time-reversible evolution. Note that we have inserted $\hbar$ for completeness and further indicated the superoperator $\hat{L}$ with a hat to distinguish it from the Schrödinger differential operator as in the hamiltonian formulation.

Our direct interest concerns the density matrix ρ, or its reduced system operator projections to be defined later, and we will first show how the complex scaling method (CSM) can be incorporated. This will necessitate a little bit of algebra, which, as we will see later, will turn out to be very convenient for us.

To allow for different choices of representation we define the superoperator $\hat{P}$ by

$$\hat{P} = A\rangle\langle + \rangle\langle B \tag{6.118}$$

where the choice of A and B will be made below and $\rangle\langle$ is short for the "bra" and "ket" components of the density matrix ρ. Thus Eq. (6.118) is equivalent to

$$\hat{P}\rho = A\rho + \rho B$$

and with this notation it is straight-forward to show that the definition (6.118) leads to

$$e^{\hat{P}} = e^{A}\rangle\langle e^{B} \tag{6.119}$$

or

$$e^{\hat{P}} = e^{A}\rho e^{B}$$

and from Eq. (6.117) that

$$\partial \rho = dA\rho + \rho dB \tag{6.120}$$

There are basically two choices of A and B and $\hat{P}$ that correspond to physically meaningful realizations. Note that we are now focusing on dilation analytic extension or a CSM framework for the Gamow-type solutions of Eq. (6.108), see previous paragraph.

1. The first choice of operators concerns time evolution of the Liouville equation [16]. One obtains $\hat{P} = -i\hat{L}t, A = -iHt = B^\dagger$ which leads to

$$e^{-i\hat{L}t}\rangle\langle = |e^{-iHt}\rangle\langle e^{-iHt}| \tag{6.121}$$

and

$$i\frac{\partial \varrho}{\partial t} = \hat{L}\varrho \tag{6.122}$$

For a particular component, using the CSM construction [16], one gets with

$$\varrho_{kl}^c = |\Psi_k^c\rangle\langle\Psi_l^c| \tag{6.123}$$

and

$$H^c\Psi_k^c = \varepsilon_k\Psi_k^c \tag{6.124}$$

with

$$\varepsilon_k = E_k - i\varepsilon_k \tag{6.125}$$

the following eigenvalue relation

$$e^{-i\hat{L}^c t}\varrho_{kl}^c = e^{\{-i(E_k-E_l)-(\varepsilon_k+\varepsilon_l)\}t}\varrho_{kl}^c \tag{6.126}$$

This choice is characterized by the fact that Eq. (6.126) contains energy differences, while the widths are added. Note also that ϱ in Eq. (6.123) for $k = l$ is *not* a projector, i.e. $\varrho^2 \neq \varrho$, which makes the interpretation fundamentally different i.e. in connection with direct probability interpretations etc.

2. The second choice concerns the Boltzmann factor containing $\beta = \frac{1}{k_B T}$, where k_B is the Boltzmann constant and T the absolute temperature. Here the representation obtains from $\hat{P} = -\beta\hat{L}$ and $A = -\beta H = B$ satisfying

$$-\frac{\partial \rho}{\partial \beta} = \hat{L}_B\rho \tag{6.127}$$

with the eigenvalue relation (note the complex conjugate sign in the bra-position)

$$e^{-\beta\hat{L}_B^c}|\Psi_k^c\rangle\langle\Psi_l^{c*}| = e^{-\beta\{(E_k+E_l)-i(\varepsilon_k+\varepsilon_l)\}}|\Psi_k^c\rangle\langle\Psi_l^{c*}| \tag{6.128}$$

Note that $\rho_{kk}^c = |\Psi_k^c\rangle\langle\Psi_k^{c*}|$ here *is a (nonselfadjoint) projector* i.e. $\rho^{c2} = \rho^c$, while

$\rho^{c\dagger} \neq \rho^c$. In addition to this feature we also see that *both widths and energies are added* in Eq. (6.128). In what follows we will for convenience let $\beta \rightarrow \frac{\beta}{2}$ so that our formulation agrees with the standard definition of the Boltzmann factor, i.e.

$$e^{-\frac{\beta}{2}\hat{L}_B^c}\rho_{kk}^c = e^{-\beta\varepsilon_k}\rho_{kk}^c \tag{6.129}$$

We observe that $\hat{L}$ in Eq. (6.122) is defined by

$$\hat{L} = H\rangle\langle - \rangle\langle H \tag{6.130}$$

while $\hat{L}_B$ in Eq. (6.127) is given by

$$\hat{L}_B = H\rangle\langle + \rangle\langle H \tag{6.131}$$

and the fundamental difference between ϱ and ρ in the two formulations above. In the undilated formalism they are of course identical if restricted to the same dynamical framework.

As we have stated at the end of the previous section, complex scaling leads to a certain asymptotic decoupling of the localized resonance state from the background. This is also very easy to visualize from the physical point of view. By scaling the coordinates so that the appropriate outgoing Gamow waves become square integrable, precisely means that the environment will be shielded from the system in a manner precisely given by the CSM deformation. The incorporation of complex scaling into the present superoperator picture therefore gives, both in a physical as well as a mathematical sense, a valid subdynamical picture where the correlations from the environment can be preserved as much as needed for the dissipative state to "survive". The consequences of this "philosophy" will be explicitly explored below.

Although complex scaling exhibits the above mentioned localization properties, it is important to remember that irreducible non-diagonal structures, so-called Jordan blocks, may appear in contrast to the conventional formulation "on the real axis". Here this will be a desired "complication" since it will in fact be associated with the selforganization of new coherence patterns of a nonlocal, nonlinear quantum statistical origin.

In the treatment above time- and temperature dependences are formulated through *different realizations* and hence a mixed representation is not strictly valid in a more rigorous framework as we will see in the following section. Nevertheless it is sometimes convenient to consider time and temperature combined into a complex variable. Although this is popular and often used as a practical device a warning for extending this view too far here is issued. In line with this restriction we will give a simple account of the relations between the resolvent and the evolution operator. Despite our emphasis on density matrices and the Liouville equation we will given some of our formulas in the hamiltonian form.

Temporarily we will put $\hbar = 1$, which means that the hamitonian (or equivalently the Liouvillian) expresses energies (or energy differences) in frequency

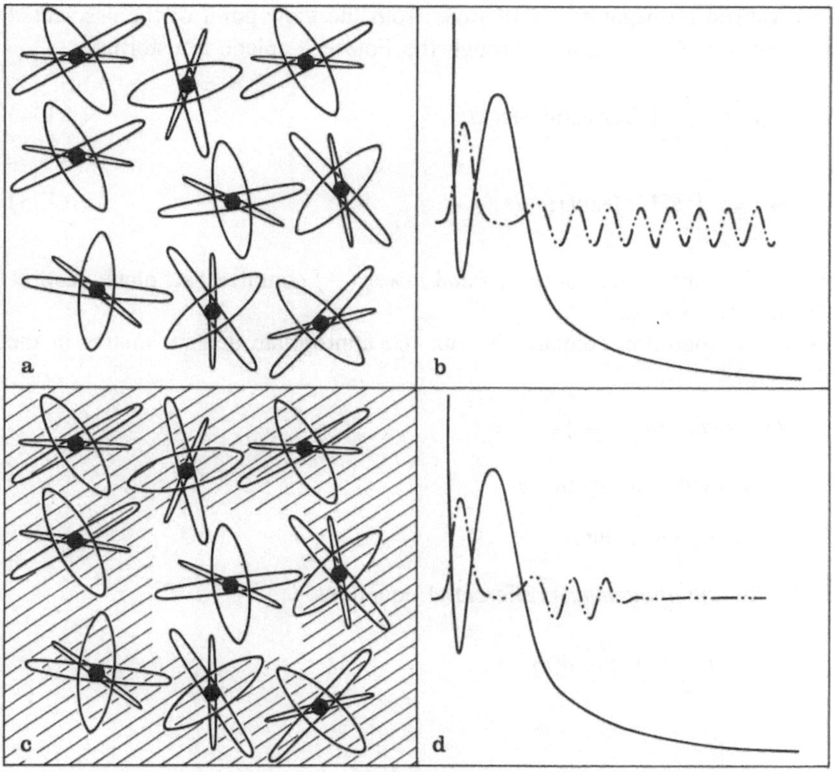

Fig. 6.4. Example of the physical effects of (exterior) complex scaling.
a Ensemble of atoms or molecules before scaling. **b** Display of asymptotic form of quasibound orbitals represented by an outgoing Gamow wave (real part). **c** By exterior scaling around the center of our atomic system we focus on the internal structure of the quasibound system. **d** The orbital wavefunction has now turned into a square integrable one, with the environmental atoms or molecules "shielded away" by the complex scaling. Adapted from Ref. [27]

units. Although we will write down the general propagator equations for a hamiltonian operator, it is not difficult to realize the same abstract equations for a general operator, provided there is a well-defined mathematical structure behind the analytic continuation into the complex energy plane. Thus we could immediately exchange the operator H below for $\hat{L}$ as given by the CSM modification due to Obcemea and Brändas [16]. To explicitly keep the time directions (as well as associated analyticity requirements for the resolvent) we will define the *retarded-advanced* ($\pm$) evolution operator and the associated resolvents as

$$\mathscr{G}^{\pm}(t) = \mp i\vartheta(\pm t)\exp(-iHt) \tag{6.132}$$

$$\mathscr{G}(z) = (zI - H)^{-1} \tag{6.133}$$

where $\vartheta(x) = 1$ for $x \geqslant 0$, and zero for $x < 0$. The advantage of using the retarded-advanced propagators are obvious from the view-point of the associated resolvents, since they are related through the Fourier-Laplace transforms

$$\mathscr{G}^{\pm}(t) = \frac{1}{2\pi} \int_{C\pm} \mathscr{G}(z) \exp(-izt) dz \tag{6.134}$$

$$\mathscr{G}(z) = \int_{-\infty}^{+\infty} \mathscr{G}^{\pm}(t) \exp(izt) dt \,. \tag{6.135}$$

The contour $C^{\pm}$ runs in the upper (+) and lower (−) complex half plane, respectively, from $-\infty$ to $+\infty$.
With the initial condition (replace Ψ with the appropriate density matrix in the Liouville case)

$$\Psi(t) = \Psi_0; \text{ for } t = 0$$

and the rule of evolution given by

$$\Psi^{\pm}(t) = \pm i \mathscr{G}^{\pm}(t) \Psi_0$$

one obtains the *inhomogeneous* differential equations

$$\left(i \frac{\partial}{\partial t} - H \right) \mathscr{G}^{\pm}(t) = \delta(t)$$

and

$$(z - H)\Psi(z) = \begin{cases} +i\Psi_0 & z \text{ in upper half plane, } t > 0; \\ -i\Psi_0 & z \text{ in lower half plane, } t < 0; \end{cases}$$

The interpretation of a complex resonance, c.f. discussions around so-called Gamow waves [23] usually evolves as follows. For $t > 0$ one obtains e.g. the probability amplitude

$$a(t) = \langle \Psi_0 | \Psi^{+}(t) \rangle = +i \langle \Psi_0 | \mathscr{G}^{+}(t) | \Psi_0 \rangle$$

$$a(t) = \langle \Psi_0 | \frac{i}{2\pi} \int_{C+} (zI - H)^{-1} \exp(-izt) dz | \Psi_0 \rangle \tag{6.136}$$

from which one deduces via a quick detour into the "unphysical Riemann sheet" that

$$a(t) = \sum_k |\langle \Psi_0 | \Psi_k \rangle|^2 \exp(-iE_k t) + \text{background contributions.} \tag{6.137}$$

Since a self-adjoint operator only exhibits a real spectrum the true interpretation of Eq. (6.137) should be that of a complex amplitude resting inside the the circle with radius $\max|a(t)|$ in the complex plane. However, if one introduces the concept of a *complex resonance eigenvalue* (the motivation will come later)

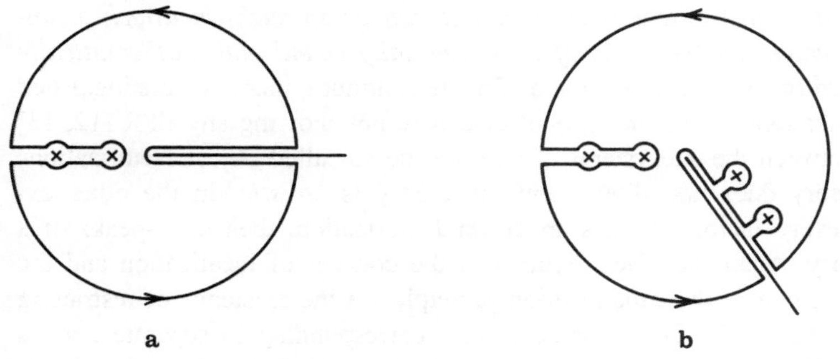

Fig. 6.5. Integration contour for the evaluation of the integral 6.136. **a** before complex scaling, **b** after complex scaling

and assumes that there is, in some sense, an expansion formula of the conventional "spectral resolution type" one obtains (superficially), with $\varepsilon_k = E_k - i\frac{\Gamma_k}{2}$ (and $R_k = \langle\Psi_0|\Psi_k\rangle$) using the residue theorem on Eq. (6.136)

$$|a(t)|^2 = |R_k|^2 \exp(-\Gamma_k t) + \ldots \tag{6.138}$$

The precise meaning of Eq. (6.137) is usually made rather vaguely. First of all so-called interference terms are neglected, which may or may not be important but there is also the problem of small times and long time tails, which do not subscribe to the formula above. Nevertheless it is possible to identify the evolution according to Eq. (6.137) for some relevant time intervals of some given prepared system (as in radioactive decay).

Despite the hand-wavy nature of the present description and the nonrigorous mathematical setting, we can already here introduce the concept of a *level width*

$$-2\Im\varepsilon_k = \Gamma_k \tag{6.139}$$

through our knowledge of the complex eigenvalue given by CSM. The contour integration can basically be carried out rigorously, however, the interpretation of the corresponding (sub)dynamics is still an open field with several views under discussion. We will return to this issue later in connection with irreversible processes of steady state situations and coherence effects and relaxation phenomena of open systems, e.g. population inversion in the operation of lasers and masers etc.

6.9 The Coherent-Dissipative Ensemble

It is clear from the present discussion that the nonseparable nature of the universe leads to a shocking shift of paradigm in comparison to that stimulated by the traditions of classical mechanics. As expressed by Primas [5] *The nonseparability implied by quantum mechanics is anything but an unsatisfactory feature:*

it is a triumph of theoretical science that it can grasp such an utterly counterintuitive phenomenon in terms of a conceptually sound and mathematically well-formulated theory. Nevertheless, as Primas continues, there is a distinguished class of factorizations of the universe of discourse not showing any EPR [12, 13] correlations between the subsystems. These are the so-called object factorizations and if the theory does not allow them the theory is *holistic*. In the other extreme, i.e. if every factorization is an object factorization, then one speaks of a separable theory. Again our focus returns to the concept of localization and the possible break down of the superposition principle via the existence of restricting superselection rules [28]. For instance objects corresponding to separate masses or charges, and/or at different times or temperatures belong to distinct superselection sectors. We can add a superselection rule also for CSM, i.e. for objects at different dilation angles.

Wick, Wightman and Wigner [28] gives the following definition of a superselection rule:

Definition: *A superselection rule operates between subspaces if there are no spontaneous transitions between their state vectors (i.e. a selection rule operates between them) and if, in addition to this, there are no measurable quantities with finite matrix elements between their state vectors.*

We have here adopted the customary interpretation of a selection rule as operating between subspaces of the total Hilbert space if the state vectors of each subspace remain orthogonal to all state vectors of the other subspaces as long as the system is isolated.

To connect with this discussion and to set the stage for the present discussion we begin by referring back to the extreme case discussed in Sect. 6.6., Eqs. (6.56–57). Neglecting the "tail" component, the reduced density matrix in the pair basis $|(i)\rangle$ $i = 1, \ldots s$, to be denoted by $|\mathbf{h}\rangle$, is given by Eqs. (6.56–57). Since $\Gamma^{(2)}$ is not diagonal in this subspace, we can, in principle, represent the "box" component with any orthogonal set of vectors subject to the appropriate space of the actual eigenvalue. Hence we may introduce for convenience the transformations

$$|\mathbf{h}\rangle\mathbf{B} = |\mathbf{g}\rangle = |g_1, g_2, \ldots, g_s\rangle \tag{6.140}$$

and

$$|\mathbf{h}\rangle\mathbf{B}^{-1} = |\mathbf{f}\rangle = |f_1, f_2, \ldots, fs\rangle \tag{6.141}$$

where we will use the basis $|\mathbf{f}\rangle$ later. We also see that g_1 is the extreme geminal previously called g, while the others g_i, $i = 2 \ldots s$ are given by the corresponding column vectors of (6.116). Thus we have, neglecting $\Gamma_T^{(2)}$,

$$\Gamma^{(2)} = \Gamma_L^{(2)} + \Gamma_S^{(2)} = \lambda_L|g_1\rangle\langle g_1| + \lambda_S \sum_{k=1}^{s}\sum_{l=1}^{s} |h_k\rangle\left(\delta_{kl} - \frac{1}{s}\right)\langle h_l|$$

$$= \lambda_L|g_1\rangle\langle g_1| + \lambda_S \sum_{k=2}^{s} |g_k\rangle\langle g_k| \tag{6.142}$$

It follows from Eq. (6.142) that the extreme case implies a separation of phases, such that the large macroscopic eigenvalue displays full coherence via the totally symmetric representation $|g_1\rangle$, while the other part, corresponding to the small eigenvalue λ_S, yields phases associated with the odd powers of $\omega = \exp(\frac{i\pi}{s})$. This will be significant below. It is also interesting to note that the phases, defined above lead to a very important property, i.e. that for an extreme interaction $\mathscr{V}_{12}$, such that all matrix elements in the $|\mathbf{h}\rangle$ basis are the same irrespective of sign, one obtains

$$\langle g_i|\mathscr{V}_{12}|g_j\rangle = 0, \quad \text{all } i, j \text{ except } i = j = 1. \tag{6.143}$$

which in contrast, i.e.

$$\langle g_1|\mathscr{V}_{12}|g_1\rangle \neq 0 \tag{6.144}$$

may yield substantial contributions to an energy expression (with $H_{12} = \mathscr{V}_{12}$) from the large component in $\Gamma^{(2)}$ through (neglecting $\Gamma_T^{(2)}$)

$$\mathrm{Tr}\{H_{12}\Gamma^{(2)}\} = \lambda_L\langle g_1|H_{12}|g_1\rangle + \lambda_S \sum_{k=2}^{s} \langle g_k|H_{12}|g_k\rangle \tag{6.145}$$

For stability reasons the interaction in Eq. (6.145) must be negative in order to contribute to a stable state below the energy threshold at zero energy. Further, by letting g_1 be replaced by some other g_i, $i \geqslant 2$ in Eq. (6.145), which is tantamount to defining a new interaction $\mathscr{V}_{12}^p$ with constant absolute value for all matrix elements (in the $\mathbf{h}$ basis as before) but containing specific phases, we can rewrite Eq. (6.145) by simply replacing H_{12} with $\mathscr{V}_{12}^p$. The component $\Gamma_L^{(2)}$ will now give a zero contribution to the energy and the nonzero contributions in the small component will still be small since λ_S is very small. Energy stabilization must then come from the one-body part in Eq. (6.31), in order for ODLRO to develop. This could very well be a possibility in high temperature superconductivity (HTSC) where nonadiabatic features (i.e. the breakdown of the separation of electronic and nuclear motion) may dominate. We will return to this in the next Chapter.

Thus we may suspect that the extreme case, which under favourable circumstances develops into ODLRO, also will be destroyed by fluctuations promoting the importance of $\Gamma_S^{(2)}$ in relation to $\Gamma_L^{(2)}$. Since any measurement related to ODLRO is unable to give well-defined matrix elements between the "macroscopic wave function" obtained from $|g_1\rangle$ and the "correlation space" related to $\Gamma_S^{(2)}$, we may conclude that there is here a superselection rule at work and hence the linear superposition principle is violated.

For this reason we will return to the concept of the canonical ensemble, see (6.42). Since we will implement CSM together with the reduced density matrix formalism, it is important to point out that condensed (dissipative) quantum systems cannot be properly treated as an ensemble of isolated small systems. It should also be remembered that the canonical density operator (6.42) is strictly confined to physical systems which are in thermal equilibrium with their environment.

For a general ensemble representable density operator ρ we have in general

$$\rho = \sum_{k,l} |\Psi_k\rangle c_{k\,l} \langle \Psi_l| \tag{6.146}$$

where $\{|\Psi_k\rangle\}$ is a set of appropriate (many-particle) states and $c_{kl} = c^*_{lk}$ are complex numbers commensurate with the given conditions, see section 6.5. In the special case where these states are eigenfunctions of the system hamiltonian H, i.e. $H\Psi_k = E_k\Psi_k$, (6.42) the *canonical* (diagonal) density operator is well defined through the well known formula

$$\rho^{\mathrm{d}}_{\mathrm{can}} = \Gamma = \frac{1}{Z}\sum_k |\Psi_k\rangle e^{-\beta E_k} \langle \Psi_k| \tag{6.147}$$

where as before $\beta = \frac{1}{k_B T}$ (k_B is the Boltzmann constant) and Z is the appropriate normalization factor or partition function.

In the usual representation (before applying CSM) H is of course a selfadjoint operator with real eigenvalues E_k. It is then trivial to see that the characteristic Boltzmann factor $e^{-\beta E_k}$ is obtained by the action of the superoperator with $\frac{1}{Z}e^{(-\beta \hat{H})}$ with $\hat{H} = \frac{1}{2}\hat{L}_B$, see Eq. (6.129), on the component $\rho_{k\,k} = |\Psi_k\rangle\langle\Psi_k|$, i.e.

$$e^{-\beta \hat{H}} \rho_{k\,k} = e^{-\beta E_k} \rho_{k\,k} \tag{6.148}$$

In this equation, $\hat{H}$ is the energy superoperator, see previous paragraph or e.g. Prigogine [4]. The physical meaning of the superoperator in Eq. (6.148) is naturally that it produces the well known Boltzmann factors. However, if applied to a general ket-bra operator of type $\rho_{kl} = |\Psi_k\rangle\langle\Psi_l|$, $k \neq l$, we obtain in a consistent manner purely quantum mechanical off-diagonal contributions to the density operator. i.e. correlations of a pure quantum mechanical character. We thus obtain

$$e^{-\beta \hat{H}} \rho_{k\,l} = e^{-\frac{\beta}{2}H} |\Psi_k\rangle\langle\Psi_l| e^{-\frac{\beta}{2}H} = e^{-\frac{\beta}{2}(E_k+E_l)} \rho_{kl} \tag{6.149}$$

from which one obtains the off-diagonal terms corresponding to Eq. (6.147).

The construction gives (as it should) the following natural properties:

(1) the density operator is self-adjoint, i.e. $\rho^\dagger = \rho$

(2) the correct diagonal elements are proportional to $e^{-\beta E_k}$.

Let us summarize the situation in the following way and let

$$\rho_{\mathrm{m}} \equiv \sum_{k,l} |\Psi_k\rangle c_{k\,l} \langle \Psi_l| \tag{6.150}$$

with (for simplicity)

$$c_{k\,k} \equiv \text{const., all } k, \tag{6.151}$$

and

$$c_{kl} = c^*_{lk}, \quad (l \neq k) \tag{6.152}$$

be a "non-thermalized" density operator. (Clearly, this operator corresponds to a microcanonical-type ensemble.) The canonical density operator ρ_{can} is then obtained through

$$\rho_{\text{can}} = \frac{1}{Z} e^{-\beta \hat{H}} \rho_{\text{m}} \tag{6.153}$$

It follows, by disregarding the off-diagonal elements of ρ_{can}, that one can obtain the very well known "diagonal" part as given in Eq. (6.147), which – per definition – contains no information about EPR-correlations between different states.

Thus far no complex scaling (CSM) has been considered. Using the results of previous paragraphs, however, this extension is easily implemented, because of the simple analytic form of the "Boltzmann" energy superoperator. We need in principle only to know how to scale the hamiltonian H. From Sect. (6.7) we then obtain:
(1) The complex scaled hamiltonian H^{c} is given by a similarity transformation

$$H^{\text{c}} \equiv UHU^{-1} \tag{6.154}$$

where the operator U, is the *unbounded* dilation operator fulfilling the condition

$$(U^\dagger)^{-1} = U^* \tag{6.155}$$

in the appropriate domain.
(2) The hamiltonian H^{c} is complex symmetric in the same domain i.e.

$$(H^{\text{c}})^\dagger = (H^{\text{c}})^* \tag{6.156}$$

This simple property is of particular importance in the context under consideration, because it allows us to construct the complex scaled canonical operator $\rho^{\text{c}}_{\text{can}}$ in a very simple way. Due to the simple analytic form of the exponential operator one finds immediately

$$[e^{-\beta H}]^{\text{c}} = e^{-\beta H^{\text{c}}} \tag{6.157}$$

Using Eq. (6.128), remembering the factor 1/2, where E_k, $\varepsilon_k =$ real and $\varepsilon_k \geqslant 0$ one finds

$$e^{-\frac{\beta}{2} H^{\text{c}}} |\Psi^{\text{c}}_k\rangle\langle\Psi^{c*}_l| e^{-\frac{\beta}{2} H^{\text{c}}} \equiv \omega_{kl} |\Psi^{\text{c}}_k\rangle\langle\Psi^{c*}_l| \tag{6.158}$$

with ω_{kl} given, see Eq. (6.128), as

$$\omega_{kl} = e^{-\frac{\beta}{2}\{(E_k+E_l)-i(\varepsilon_k+\varepsilon_l)\}} \tag{6.159}$$

This shows immediately that the density operator

$$\rho^{c} \equiv \sum_{k,l} |\Psi_k^c\rangle \omega_{k\,l} \langle \Psi_l^{c*}| \tag{6.160}$$

is trivially complex symmetric, i.e.

$$\rho^{c\dagger} = \rho^{c*} \tag{6.161}$$

and this is in agreement with (6.156), since ρ^c is per definition an operator function of H^c. We also see trivially that all the results concerning the complex scaled quantities coalesce to "unscaled" results in the "limit" $\varepsilon_k \rightarrow 0$.

Finally we will formulate the analogue of Eq. (6.147) in the CSM representation. The density operator ρ_{m}^{c} corresponding to Eq. (6.150) may simply be represented as

$$\rho_{\mathrm{m}}^{c} = \sum_{k,l} |\Psi_k^c\rangle c_{k\,l} \langle \Psi_l^{c*}| \tag{6.162}$$

with $c_{k\,l} = c_{l\,k}$, which indeed is a complex symmetric operator. The corresponding complex scaled canonical density operator ρ_{can}^{c} is then obtained through the relation

$$\rho_{\mathrm{can}}^{c} = \frac{1}{Z} e^{-\frac{\beta}{2}H^{c}} \rho_{\mathrm{m}}^{c} e^{-\frac{\beta}{2}H^{c}} \tag{6.163}$$

From the consideration above one finds that this operator is also complex symmetric. From Eqs. (6.158), (6.159) and (6.162) we obtain

$$\rho_{\mathrm{can}}^{c} = \frac{1}{Z} \sum_{k,l} c_{k\,l} \omega_{k\,l} |\Psi_k^c\rangle \langle \Psi_l^{c*}| \tag{6.164}$$

where $\omega_{k\,l}$ is given by Eq. (6.159) above, which gives the explicit form of the canonical operator needed in the derivations below.

Proceeding now to present the formal conditions that accompany the spontaneous creation of microscopic coherent-dissipative structures, we will focus on a particular subspace related to a highly degenerate energy, which without restrictions can be put equal to zero. Further, since we are far from equilibrium, we will also consider the hierarchy of reduced density matrices encountered in Sect. 6.4. The possible development of ODLRO implies that $\Gamma_L^{(2)}$ and $\Gamma_S^{(2)}$ belong to different superselection sectors. Overlapping these sectors are commensurate with the destruction of the coherent state represented by $\Gamma_L^{(2)}$.

We thus concentrate on the thermalization of $\Gamma_S^{(2)}$. The three main formulas that are needed in the following are, in addition to Eqs. (6.164), (6.113) and (6.142), i.e.

$$q_{k\,l} = \left(\delta_{k\,l} - \frac{1}{s}\right) \exp\left(i\pi \frac{k+l-2}{s}\right) \tag{6.165}$$

and

$$\Gamma_S^{(2)} = \lambda_S \sum_{i,j}^{s} |h_i\rangle \left(\delta_{ij} - \frac{1}{s} \right) \langle h_j| \tag{6.166}$$

see also Eqs. (6.56–58).

Applying CSM to Eq. (6.166) utilizing the replacements $|h_k\rangle \to |h_k^c\rangle$, is nothing but a special case of the complex scaled "microcanonical" operator ρ_m^c with the coefficients

$$c_{k\,l} \equiv \lambda_S \left(\delta_{k\,l} - \frac{1}{s} \right) \tag{6.167}$$

In what follows, the absolute values of the numbers $c_{k\,l}$, which represent quantum correlations, do not obey any restriction, and thus – if desired – they may also be considered to be arbitrarily small. We can immediately see that $\Gamma_S^{(2)c}$ is also complex symmetric, as one would expect. We are now ready to ask what happens to $\Gamma_S^{(2)}$ if the systems of interest are subject to the aforementioned ensemble formalism. Making Eq. (6.166) subject to the transformation

$$\gamma \equiv \frac{1}{Z} e^{-\frac{\beta}{2} H_p^c} \Gamma_S^{(2)^c} e^{-\frac{\beta}{2} H_p^c} \tag{6.168}$$

Here, H_p^c strictly speaking should represent the complex scaled reduced hamiltonian H_p derived from Eq. (6.32–33), i.e.

$$H_p = \frac{1}{(N-1)}(h_1 + h_2) + h_{12} \tag{6.169}$$

Using Eq. (6.164) and the additional assumption that the (real) energies of all the pairs are equal, i.e.

$$E_k = E = 0, \quad (k = 1, 2, \ldots, s) \tag{6.170}$$

we immediately obtain

$$\gamma \equiv \sum_{k,l} \gamma_{k\,l} |h_k^c\rangle \langle h_l^{c*}| \tag{6.171}$$

with

$$\gamma_{k\,l} = \frac{\lambda_S}{Z} \left(\delta_{k\,l} - \frac{1}{s} \right) e^{i\frac{\beta}{2}(\varepsilon_k + \varepsilon_l)} \tag{6.172}$$

We thus make the following crucial observation. The matrix elements $\gamma_{k\,l}$ are formally similar to the matrix elements of the matrix **Q**, see Eq. (6.113). Precisely, if we require the validity of the thermal *"quantum" conditions*

$$\pi \frac{k-1}{s} = \frac{\beta}{2} \varepsilon_k, \quad (k = 1, 2, \ldots s) \tag{6.173}$$

we obtain the important equality

$$\gamma_{k\,l} = \frac{1}{Z}\lambda_D^{(2)} q_{k\,l} \equiv \text{const}\cdot q_{k\,l} \tag{6.174}$$

In other words if the energy widths ε_k and the energies E_k of the complex scaled pairs $|h_k^c\rangle$ fulfill the conditions (6.170) and (6.173), then the matrix elements of the density operator γ constitute a Jordan block $\mathbf{C_S(0)}$ as given by Reid and Brändas [11]. The density operator γ represents an irreducible (non-diagonalizable) part of $\Gamma^{(2)}$ which again derives from a canonical density operator appropriately reduced. Note also that $\Gamma^{(2)}$ before reduction, thermalization and development of ODLRO is (wavefunction) N-representable and hence allows the CSM construction. In what follows, we will refer to the irreducible unit defined by (6.172–174) as a *coherent-dissipative structure*, which will account for a variety of cooperative or synergetic phenomena of interest. This remark concludes the desired formal derivations.

6.10 References

1. Glansdorff P, Prigogine I (1971) Thermodynamic theory of structure, stability, and fluctuations. Wiley, New York
2. Nicolis G, Prigogine I (1977) Self-organization in non-equilibrium systems. Wiley, New York
3. Prigogine I (1980) From being to becoming. Freeman WH, San Francisco
4. Nicolis G, Prigogine I (1989) Exploring complexity. Freeman WH, New York
5. Primas H (1983) Chemistry, quantum mechanics and reductionism. Springer, Berlin Heidelberg New York
6. Schmidt E (1907) Math Ann 63: 433
7. Carlson BC, Keller JM (1961) Phys Rev 121: 659
8. Löwdin P-O (1955) Phys Rev 97: 1474
9. Coleman AJ (1963) Rev Mod Phys 35: 668
10. Gantmacher FR (1959) The theory of matrices. Vol II, Chelsea Publishing Company, New York
11. Reid CE, Brändas EJ (1989) "On a theorem for complex symmetric matrices and its relevance in the study of decay phenomena," Brändas E, Elander N Eds, Lecture notes in physics 325: p. 475–483
12. Einstein A, Podolsky B, Rosen N (1935) Phys Rev 47: 777–780
13. d'Espagnat B (1976) Conceptual foundations of quantum mechanics. Benjamin, London
14. von Neumann J (1932) Mathematische grundlagen der quantenmechanik. Springer, Berlin Heidelberg New York
15. Löwdin P-O (1977) Int J Quantum Chem 12: Suppl 1, 197
16. Obcemea CH, Brändas EJ (1983) Ann Phys 151: 383
17. Yang CN (1962) Rev Mod Phys 34: 694
18. Sasaki F (1965) Phys Rev 138: B 1338
19. Penrose O, Onsager L (1956), Phys Rev 104: 576
20. Brändas EJ, Chatzidimitriou-Dreismann CA (1991) Int J Quant Chem 40: 649
21. Brändas, Chatzidimitriou-Dreismann CA (1989) "Creation of long range order in amorphous condensed systems," Brändas E, Elander N Eds, Lecture notes in physics 325: p. 486–533
22. Bardeen J, LN Cooper Schrieffer JN (1957) Phys Rev 108: 1175
23. Gamow G (1928) Z Phys 51: 204
24. Balslev E, Combes JM ((1971) Commun Math Phys 22: 280
25. van Winter C, (1974) Math Anal 47: 633
26. Simon B (1973) Ann Math 97: 247

27. Brändas E, Elander N (Eds) (1989) Resonances – The unifying route towards the formulation of dynamical processes – foundations and applications in nuclear, atomic and molecular physics, lecture notes in physics, Vol 325: Springer, Berlin Heidelberg
28. Wick GC (1952) Wightman AS, Wigner EP, Phys Rev 88: 101

6.11 Problems

1. The quantum mechanical eigenvalues of the rigid model, corresponding to a gas of polar molecules is given by $\varepsilon_{\text{rot}} = (\frac{\hbar^2}{2\mathscr{I}})j(j+1), j = 0, 1, 2\ldots$, where $\mathscr{I}$ the moment of inertial. Determine the rotational partition function, the (rotational) free energy, and the (rotational) specific heat. Assume that the translational, rotational and vibrational degrees of freedom are independent and replace the summation over j with an integral in the high temperature limit.
   a. What happens with the energy distribution when T increases?
   b. In the far-infrared (FIR) absorption band of dilute solutions of acetonitrile (CH_3CN) in n-heptane a red shift in the temperature dependence of the absorption cross section (the band shifts to lower frequencies with increasing temperature) has been observed. What should be the conclusion(s)?
2. Derive the formula
$$E = \text{Tr}\{H_1\Gamma^{(1)}\} + \text{Tr}\{H_{12}\Gamma^{(2)}\}$$
   i.e. Eq. (6.31) from Eqs. (6.30) and (6.32–33).
3. The following argument is sometimes employed as a proof of the nonexistence of nonreal resonance eigenvalues in connection with complex scaling of the Hamiltonian $H(1)$:
   Assume an eigenvalue in the spectrum of $H(\eta)$, i.e. $H(\eta)\Psi(\eta) = E\Psi(\eta)$, with $\eta = \exp(i\theta)$. Apply the (inverse) coordinate transformation $\eta^* = \exp(-i\theta)$. From the new eigenvalue relation one concludes that $H(1)\Psi(1) = E\Psi(1)$, i.e. an eigenvalue relation corresponding to the unscaled Hamiltonian. Since the selfadjoint operator $H = H(1)$ has only real eigenvalues which stay invariant under scaling complex resonance eigenvalues are ruled out.
   Find the mistake in this "derivation".
4. Compute $\mathbf{Q} = \mathbf{B}^{-1}\mathbf{CB}$, where the matrices $\mathbf{Q}$, $\mathbf{B}$ and $\mathbf{C}$ are given by Eqs. (6.113), (6.115–116).
5. Carry through the proof of the (algebraic) factorization property Eq. (6.119), i.e.
$$e^{\mathscr{P}} = e^A \times e^B, \quad \mathscr{P} = A \times + \times B$$
   where $\times$ may be realized as in Sect. 6.8.

7 Applications of CSM Theory

E. Brändas

7.1 Occurrence of Coherent-Dissipative Structures

We will now proceed to discuss some recent applications of the present description of quantum correlation effects of disordered condensed matter using the theoretical development in the previous Chapter, i.e. in terms of *coherent dissipative structures*. The present view-point has led to the study of resonances in quantum chemistry, e.g. in atomic, molecular and solid state theory but recent emphasis on collective non-linear effects has produced many new surprising and unexpected applications to physical chemistry and the physics of disordered condensed matter. Predictions and theoretical interpretations have been made, see below, and to recapitulate the situation we will start by stressing the following fundamental points:

1. by refering to a density matrix, which subscribe to the general decomposition, see the previous chapter on the second order reduced density matrix and the *extreme case*, as (neglecting the "tail")

   $$\Gamma^{(2)} = \Gamma_L^{(2)} + \Gamma_S^{(2)} + (\Gamma_T^{(2)})$$

   where the first part is the "large component" associated with coherence and the possible development of ODLRO and the second "small part" relates to the correlation sector,
2. by extending the quantum mechanical formulation through the theory of complex scaling (CSM), so that irreversibility is naturally embedded in the dynamics from the beginning, and simultaneously, through the reduction above, to far from equilibrium situations,
3. by considering the thermal quantum correlations obtained from the thermalization of the reduced density matrix $\Gamma^{(2)}$,
4. by showing that these thermal quantum correlations can *not* refer to a wave function, like those at $T = 0$ K,
5. and by *not* considering any specific physical mechanism, except the general perturbational influence given by universal, environmental quantum correlations as exhibited through the *extreme case* previously described.

These characteristics will essentially be of importance in connection with applications to open systems in far-from-equilibrium situations and we will define a system where this is relevant as a *coherent dissipative structure* [1–3].

In bypassing we also mention that the following examples and experimental contexts have been confronted:

1. proton transfer processes in water and aqueous solutions [4],
2. anomalous H^+ conductance of H_2O/D_2O mixtures [4, 5],
3. ionic conductance of molten alkali chlorides [6],
4. quantum correlations in high-T_c Cu-O-superconductors [7, 8]
5. quantum correlations as shown by the spin-waves of *Gd* far above T_c [9],
6. the fractional quantum Hall effect [10],
7. conductance background effects in high-quality tunnel junctions [11],
8. proton dynamics in DNA, see relevant work in [12],
9. spontaneous and stimulated emission of radiation in masers [13].

In this review we will briefly look at (1), (2), (4), (6) and (9). The main features of the cases (1–9) above, are that of suggesting examples of a *coherent-dissipative structure*, i.e. as a time-irreversible organized form, created by thermal correlations at $T \neq 0$, exchanging energy with its environment and with a critical size characterized by $s = s_{\min}$. These structures provide precursors for the coherent state which may or may not develop into ODLRO for example in superconductivity or superfluidity. Note that the precursor here should not be identified with a traditional (stable) structure, since it is connected with a fundamentally different concept, see above. For instance, the traditional view of superconductivity as a basic pairing phenomenon, where the stability of the coherent state essentially refers to a pair breaking process, may not always be accurate. Since the temperature domain, over which the phase transition occurs, is much larger for high temperature superconductors (HTSCs) than for low temperature superconductors (LTSCs), the precursor notion may therefore be of direct importance in HTSC, where nonconventional energy gaps, saturation effects and other non-LTSC behaviour are prominent.

The concrete applications (1–9) further illustrate the general theoretical result that the spatial extension of a coherent dissipative structure depends (to a large degree) linearly on the minimal dimension in the physical state space, and the thermal de Broglie wavelength of the elementary quantum system. It was first suggested heuristically [4, 21] that the geometric linear size of such a "structure" should be given as

$$d_{\min} = F \cdot s_{\min} \cdot \Lambda \tag{7.1}$$

where Λ is the relevant quantum mechanical size e.g. the de Broglie wavelength and F is a functional that depends on the Hamiltonian and possibly on some other parameters such as the temperature T etc. In fact by replacing H_{12} in (6.145) with $|\vec{x}_1 - \vec{x}_2|$, where $\vec{x}_i$ denotes the (fermionic) coordinate i, it is possible to make a more a rigorous derivation of this result provided the identification $\Lambda = \langle|\vec{x}_1 - \vec{x}_2|\rangle$ is made. From this an expression for the functional F can in principle be explicitly written out [10].

Finally we note the possibility of interpreting the coherences/correlations above from two different angles, either by referring to light carriers moving in a nuclear skeleton, like in superconductivity, or as nuclei moving in the field of defect or excess electrons. Simplifying a bit we might say that ODLRO in the former corresponds to condensation in coordinate space and in the latter in momentum space.

7.2 Proton Transfer Processes in Water and Aqueous Solutions

One of the oldest and most important problems in the physical chemistry of water is the evaluation of the rate constants k_1 and k_2 characterizing the chemical equilibrium between H_3O^+ and OH^- with H_2O, see e.g. the path breaking work of Eigen and De Maeyer [14]. A few years after the aforementioned fundamental self-dissociation studies, Meiboom [15] showed that these reaction rates, k_i and activation energies E_i, with $i = 1,2$ could be determined by NMR spectroscopic methods.

These chemical reactions play a fundamental role in many biological processes. The study of proton transfer reactions in water is also of importance to the understanding of the excess or anomalous conductivities of the hydronium and hydroxyl ions in water and aqueous solutions. Fast proton transfer in water is often considered to be involved as part of the actual reaction scheme [14]. Recently Hertz [16] presented a detailed analysis of a series of different experimental methods (NMR, X-ray and neutron scattering, etc.) used to detect the H^+ or the H_3O^+ ion in aqueous solution directly. This analysis revealed that, thus far, none of the considered experimental investigations had been able to detect directly the so-called H^+ or the H_3O^+ particle [16].

The traditional view of this problem is that the entity H^+ is postulated to exist – at least for short times – and to correspond to some fast "moving" proton. In this context the well-known Grotthus mechanism, see [17] for a detailed discussion, is believed to explain classically the high excess conductivity of H^+ and OH^- in aqueous solution. However, the allowance of some short cut for the distance that the proton has to move in order to transport current, provided by the hydrogen bonds of water (see the basic volumes of Schuster, Zundel and Sandorfy [18] on the hydrogen bond in this context), was given an exhaustive analysis by Hertz et al. [17]. The surprising conclusion was made that the physical object usually defined to be H^+ or H_3O^+ cannot be considered – thus far – to represent a particle in the conventional sense. This finding led Hertz to the conclusion that the object we call H^+ ion in aqueous solutions really is a dynamical property of the solution [16]. Note that this analysis was carried out entirely within classical mechanics. Since no delocalization effects, typical of quantum mechanical processes, were part of this framework the aforementioned conclusion is clearly remarkable.

From the quantum mechanical point of view it appears that the H-constituents forming the H^+ ions are indistinguishable from those belonging to the water molecules and being in the vicinity of the ions. This consideration is motivated by the fact that the thermal de Broglie wavelength, Λ_{H+}^{dB}, of the "quasi-free" proton is about 1 Å at room temperature. This is large enough to find, in most cases, "water protons" at a distance of about Λ_{H+}^{dB} around each H^+. This fact may lead to typical delocalization and/or interference effects characteristic of quantum theory. From the hypothesis that the fermionic entity H^+ due to quantum correlations may form *coherent dissipative structures* in aqueous solutions follows two important predictions which appears to be in contradiction to all known conventional theories.

The first one connects the proton transfer rates, k_i, $i = 1, 2$ with the excess ionic conductivities of H^+ and OH^-, λ_{H+}^e and λ_{OH-}^e, in water. The second one concerns the decrease of the conductance λ_{H+} in equimolar H_2O/D_2O mixtures. We will first look at the excess ionic conductivities defined conventionally as

$$\lambda_{H+}^e = \lambda_{H+} - \lambda_{X+} \tag{7.2}$$

with $X^+ = K^+$ or Na^+, and

$$\lambda_{OH-}^e = \lambda_{OH-} - \lambda_{Cl-} \tag{7.3}$$

where λ_X represents the experimentally measured ionic conductance of the ion X in water [19].

The conventional treatment of the connection under consideration is based on the well established equations of Nernst

$$\lambda^e = \frac{qD}{k_B T} \tag{7.4}$$

and Einstein

$$D = \frac{\langle x^2 \rangle}{6\tau_{rel}} \tag{7.5}$$

where q is the elementary charge, D the diffusion coefficient describing charge transport due to proton transfers, τ_{rel} the average lifetime of a H_3O^+ or OH^- ion and $\langle x^2 \rangle$ the average of the charge displacement accompanying a proton transfer. In a simple model, one may identify $\langle x \rangle$ with the mean distance between two oxygen atoms of water molecules [15].

From these equations and the standard relation [15] for the relaxation times associated with the reaction rate k_1 and the concentration $[H_2O]$

$$\frac{1}{\tau_{rel\ H+}} = k_1 \cdot [H_2O] \tag{7.6}$$

one obtains

$$\lambda_{H+}^e \cdot T = C \cdot k_1 \tag{7.7}$$

where C is a temperature independent constant. A corresponding equation holds for OH^-, and finally one gets

$$\frac{\lambda_{H+}^e}{\lambda_{OH-}^e} = \frac{k_1}{k_2} \tag{7.8}$$

which represents the desired connection. As usual this relation can be put in the standard Arrhenius form

$$k_i = C_i \cdot e^{-E_i/RT} \; (i = 1,2) \tag{7.9}$$

in terms of the molar energy and the gas constant R or expressed as

$$\log\left(\frac{\lambda^e_{\mathrm{H}+}}{\lambda^e_{\mathrm{OH}-}}\right) = C - \frac{(E_1 - E_2)}{RT} \tag{7.10}$$

The classical derivations leading up to Eqs. (7.8–7.10) are definitely in disagreement with the corresponding predictions of our CSM theory of quantum correlations and in disagreement with recent experiments utilizing the ^{1}H-NMR spin-echo technique [20].

The aforementioned data of [19] for the ionic conductances in water yield the classically predicted value

$$\frac{k_1}{k_2} \approx 2.35 \text{ at } T = 25\,^\circ\mathrm{C} \tag{7.11}$$

obtained from Eq. (7.8), and

$$E_1 - E_2 \approx -2.0 \text{ kJ/mol for } T = 15\,^\circ, \ldots, 55\,^\circ\mathrm{C} \tag{7.12}$$

which follows from Eq. (7.10), for the considered difference of the activation energies.

Using the formula (6.173) of the previous section, i.e. the quantization conditions for the largest size $s = s_{\mathrm{min}}$ of the *coherent dissipative structure* (Jordan block), obtained for $k = 2$, gives

$$\frac{2\pi k_B T}{\varepsilon_2} = s_{\mathrm{min}}$$

which with the usual definition of the relaxation time $\tau_{\mathrm{rel}} = \hbar/\varepsilon_2$ becomes

$$s_{\mathrm{min}} = \frac{4\pi k_B T}{\hbar}\tau_{\mathrm{rel}} \tag{7.13}$$

Eq. (7.13) in combination with Eq. (7.1) yields

$$\frac{d_{\mathrm{min\,H}+}}{d_{\mathrm{min\,OH}-}} = \frac{F_{\mathrm{H}+}}{F_{\mathrm{OH}-}} \cdot \frac{\Lambda^{\mathrm{dB}}_{\mathrm{H}+}}{\Lambda^{\mathrm{dB}}_{\mathrm{OH}-}} \cdot \frac{\tau_{\mathrm{rel\,H}+}}{\tau_{\mathrm{rel\,OH}-}} \tag{7.14}$$

It was earlier suggested [21] that the conductance should appear to be proportional to the spatial dimension d_{min}. This assumption yields immediately

$$\frac{\lambda^e_{\mathrm{H}+}}{\lambda^e_{\mathrm{OH}-}} = \frac{F_{\mathrm{H}+}}{F_{\mathrm{OH}-}} \cdot \sqrt{\frac{m_{\mathrm{OH}-}}{m_{\mathrm{H}+}}} \cdot \frac{k_2}{k_1} \tag{7.15}$$

using Eqs. (7.6), (7.13) and with the thermal de Broglie wave length given by

$$\Lambda^{\mathrm{dB}}_{\mathrm{X}} = \hbar\sqrt{\frac{2\pi}{m_{\mathrm{X}} k_B T}} \tag{7.16}$$

It turns out, however, that we can give this result a more concrete derivation, since one can determine the conductance directly from (6.145) with $H_{12} = |\vec{v}_1 - \vec{v}_2|$, where $\vec{v}_i$ is the velocity of the ith particle. Thus one obtains, remembering that the one particle occupation number $\lambda_X^{(1)} = N_X/2s$ for $s = s_{\min X}$, is given by Eqs. (6.45) and (6.75), cf. also Eq. (7.1),

$$\lambda_X^e \propto \sum_{i<j} \langle |\vec{v}_i - \vec{v}_j| \rangle \approx \mathrm{Tr}\{|\vec{v}_1 - \vec{v}_2| \Gamma_L^{(2)}\} = \lambda_X^{(1)}\Big(1 + \tfrac{1}{s} - \lambda_X^{(1)}\Big) \cdot s \langle |\vec{v}_1 - \vec{v}_2| \rangle \tag{7.17}$$

Neglecting $s^{-1} = (s_{\min X})^{-1}$ in Eq. (7.17), one finds Eq. (7.15) by introducing a typical relative velocity, i.e. the root-mean-square,

$$v_{\mathrm{rms}} = \langle |\vec{v}_1 - \vec{v}_2| \rangle = \sqrt{\frac{3k_BT}{m_X}} \tag{7.18}$$

By further assuming that the degree of correlation is similar, i.e. $\lambda_{H+}^{(1)} \approx \lambda_{OH-}^{(1)}$ or (note that an equivalent argument also holds for the molten salts [6])

$$F_{H+} = F_{OH-} \tag{7.19}$$

we obtain

$$\frac{\lambda_{H+}^e}{\lambda_{OH-}^e} = \sqrt{\frac{m_{OH-}}{m_{H+}}} \cdot \frac{k_2}{k_1} \tag{7.20}$$

It should be pointed out that Eq. (7.19) represents a physical assumption based on reasonable considerations concerning the extension and the dynamics of quantum correlations around each classically described relaxing center. Nevertheless the result was derived and predictions made [21] already before experimental verification [20].

As in the classical derivation, see Eqs. (7.6–7.10), one can convert Eq. (7.20) to the form

$$\log\left(\frac{\lambda_{H+}^e}{\lambda_{OH-}^e}\right) = C' + \frac{E_1 - E_2}{RT} \tag{7.21}$$

With the aid of the precise data of Ref. [19] for the ionic conductivities, we predicted the values [4, 21]

$$\frac{k_1}{k_2} \approx 1.75 \text{ at } T = 25\,^\circ\mathrm{C}$$

and

$$E_1 - E_2 = \begin{cases} +2.1 \text{ kJ/mol} & \text{for } X^+ = K^+ \\ +1.9 \text{ kJ/mol} & \text{for } X^+ = Na^+ \end{cases} \tag{7.22}$$

in the aforementioned temperature range $T = 15$–$55\,^\circ$C

It is notable that the results obtained from the present theory for a *coherent dissipative system*, Eqs. (7.20–7.21) is fundamentally different from those of the conventional theory (7.8–7.9). As a consequence of the reciprocal reaction rate quotient, as compared with the "classical" case, the predicted numerical value of $E_1 - E_2$, Eq. (7.10), even differs by sign from the predicted results (7.22). As the existing data for this difference, at the time, exhibited a considerable scattering, new high precision experiments utilizing the ^{1}H-NMR spin-echo techniques, was carried out in the laboratory H. G. Hertz [20]. The experimental data [20] are graphically presented in Fig. 7.1, together with the classical and CSM-theoretical predictions based on the same high precision conductivity data [19]. Note that we have omitted the experimental values of k_1 and k_2 at $T = 5.3\,°C$, since – to our knowledge – no experimental value of λ_{OH^-} around $5\,°C$ exists in the literature, and due to the well-known anomaly that water exhibits at $4\,°C$. For the temperature range $T = 10.7 - 58.8\,°C$ we obtained [4] the experimental value

$$E_1 - E_2 = +(1.9 \pm 0.5)\ \text{kJ/mol} \tag{7.23}$$

Including also the data point at $T = 5.3\,°C$ one obtains $E_1 - E_2 = +1.3$ kJ/mol,

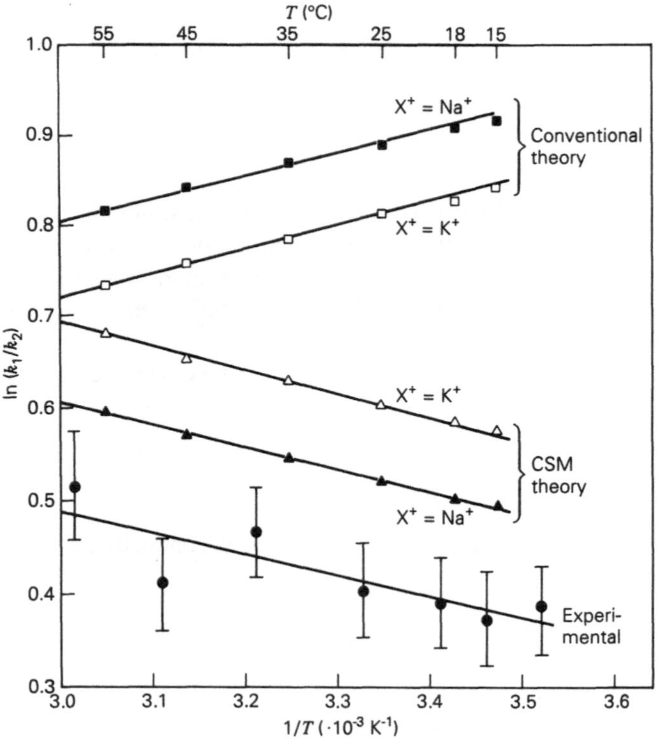

Fig. 7.1. Graphical representation of the quantity $\log(k_1/k_2)$ as a function of the inverse temperature. Shown are (i) the predictions of conventional theory and CSM-theory, and (ii) the experimentally (NMR) determined values of Ref. [20] for the temperature range $T = 10.7 \ldots 58.8\,°C$. The data points were calculated with the aid of the high-precision conductivity data of Ref. [19] and Eqs. (7.8) and (7.20). The reference cation (K^+ and Na^+) used by the calculation of the H^+-excess conductivity, Eq. (7.2), is shown on the graphs. (Reproduced from Ref. [4])

see Ref. [20]. In any case, however, the above experimental results confirm the positive sign of $E_1 - E_2$, thus being in clear disagreement with the predictions of the classical theory.

As we have mentioned the theory of coherent dissipative structures applies directly to fermionic systems. Since the H^+ and OH^- ions are fermions it would be interesting to study the corresponding reactions and activation energies characterizing D^+ transfer in D_2O. The *bosonic* character of D^+ might lead to deviations from the preceding results. There exists unfortunately very few results for the D_2O case that can be interpreted by the present theory. The reason is that these D-NMR experiments are more difficult to do than the corresponding H-NMR experiments due to the smallness of the appropriate spin-spin coupling constants in the former case. Even if some experiments [22] seem to be in line with the aforementioned qualitative expectation [23], there seems to be at present little chance of obtaining reliable reproducible data for D^+ transfer reactions in D_2O.

As we have already stressed many times, one cannot attribute the usual probability interpretation to a coherent dissipative structure. There is also the question whether the description refers to electronic or nuclear motion [23] or both, see comments above. Generally speaking the dynamical features of the nuclei are given by the potentials associated with the corresponding electronic distributions. Hence, there is necessarily a correlation between the nuclear and electronic degrees of freedom, which is playing a vital role in the conductance measurements affecting both the charge and the mass. These considerations lead us to the second prediction mentioned above, namely the anomalous decrease of H^+ mobility in H_2O/D_2O mixtures, cf. Zundel [18], particularly Sect. 15.8.1 on conductivity changes with salt concentrations.

The expected deviation from the proton exchange data, by introducing the (bosonic) D^+ ion, i.e. measuring the conductance properties of H_2O/D_2O mixtures, was quantified and predicted in a simple "volume" model [24], and verified experimentally by Wiengärtner and Dreismann [3, 5]. The argument runs as follows. If the well-known high H^+ conductance, λ_{H^+}, in liquid water is caused by the assumed specific quantum interference effects, then there must be an anomalous decrease of λ_{H^+} in H_2O/D_2O mixtures due to the so-called mass and spin superselection rules, see [5] in the previous Chapter. In these mixtures possible quantum interferences between appropriate protonic states become disrupted by deuterons belonging to D_2O, HDO or D^+ ions and being near or between the considered protons.

The simple volume model [3, 24] gives a crude estimate of the decrease of λ_{H^+} in an 1:1 mixture of H_2O/D_2O. By letting V_O and V_H be the volume parts being occupied by oxygen and hydrogen, respectively then the total volume associated with pure water (consisting of classical bodies) is given by $V = V_O + V_H$. In the equimolar mixture we obtain $V = V_O + V_H^* + V_D^*$ with essentially $V_H^* = V_D^*$. The coherent-dissipative structure around an H^+ ion may be imagined as a sphere of radius r, in the case of pure H_2O, and r^*, in the mixture. Since, as we have repeatedly pointed out here, the actual mobilities are in fact linearly related with the spatial size r or r^* of the coherent-dissipative structures, one gets from ($V_O \approx V_H$)

$$\frac{r}{r^*} = \left(\frac{V_O + V_H^*}{V_O + V_H}\right)^{1/3} \equiv \left(\frac{V - V_D^*}{V}\right)^{1/3} \tag{7.24}$$

that $r/r^* \approx 0.9$ and thus an anomalous decrease of the H^+ mobility (or conductance) of about –10%.

This prediction was tested experimentally [5], for a discussion see also [4], where molar conductances, Λ, of different HCl/DCl and KCl solutions in H_2O/D_2O mixtures were measured [5]. The experimental results are summarized in Fig. 7.2.

Firstly let us consider the conductivity of KCl solutions. The conductances of KCl solutions in H_2O/D_2O mixtures are found to depend almost linearly on the D-atom fraction, X_D, of the solvent. This linearity appears to be independent of the concentration of the measured solutions (C = 0.01–0.1 mol/l); see [5] for details. This result is as expected from classical electrochemical theory [19], because it is experimentally well established that the fluidity of the considered

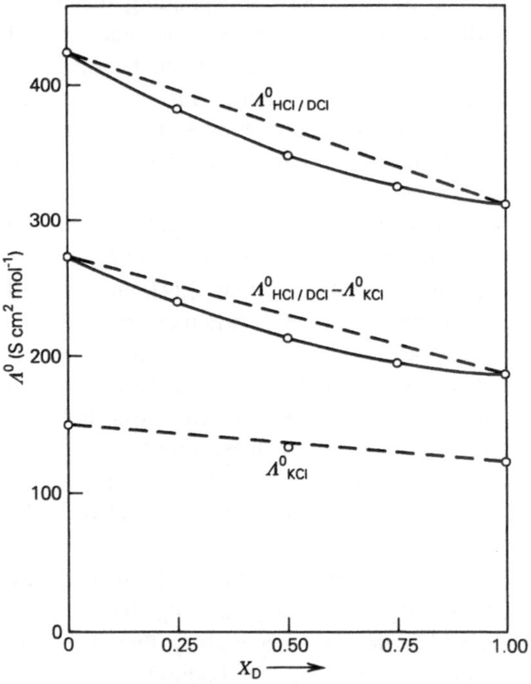

Fig. 7.2. Molar conductances of HCl/DCl, and of KCl, in H_2O/D_2O mixtures at infinite dilution and $T = 25°C$ as a function of the mole fraction X_D of deuterium. Also shown is the excess conductance, determined from the difference between the data for HCl/DCl and KCl. *Solid* and *broken lines* are guides to the eye. Error bars are smaller than the size of each data point. (Reproduced from Ref. [5])

mixtures depend almost linearly on X_D and ionic conductances are, to a very good approximation, directly proportional to the fluidity of the solvent (Walden's rule).

Secondly, let us consider the conductivity data for HCl/DCl in the studied H_2O/D_2O mixtures. Figure 7.2 shows the resulting conductancies at infinite dilution, $\Lambda°$, plotted against the mole fraction X_D. It is seen that at intermediate solvent compositions, the curve lies distinctly below the straight line connecting the limiting values in pure H_2O, where $X_D = 0$, and D_2O, where $X_D = 1$.

Defining the deviation of the measured conductance of a mixture, of concentration C, $\Lambda(X_D, C)$ from that of the linear interpolation between the values of the two pure solutions ($X_D = 0$ and 1), $\Lambda_{lin}(X_D, C)$, as

$$\Delta\Lambda(X_D, C) = \Lambda(X_D, C) - \Lambda_{lin}(X_D, C) \tag{7.25}$$

where $\Lambda_{lin}(X_D, C) = (1 - X_D) \cdot \Lambda(0, C) + X_D \Lambda(1, C)$ and the corresponding relative deviation by

$$\Delta^{rel} \equiv \Delta\Lambda(X_D, C)/\Lambda_{lin}(X_D, C) \tag{7.26}$$

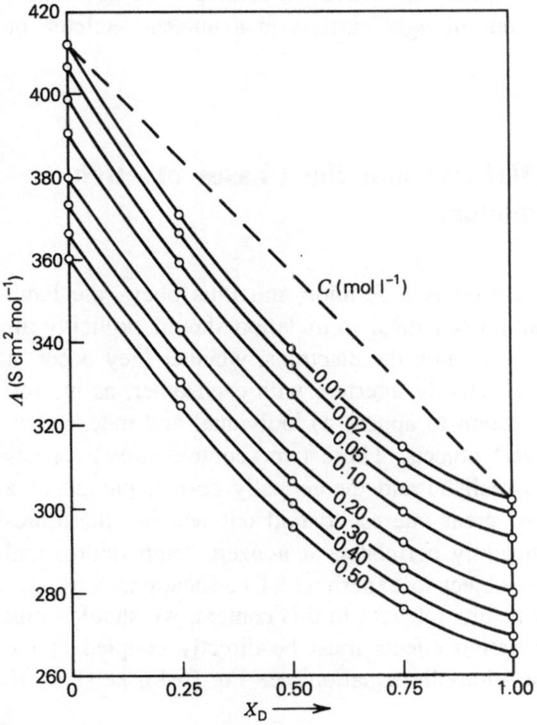

Fig. 7.3. Graphical representation of the experimentally determined molar conductances of HCl/DCl in H_2O/D_2O mixtures at 25 °C as a function of solvent composition (expressed by the atom fraction X_D of deuterium) and of acid concentration C. (Reproduced from Refs. [4, 5])

we are able to formulate the experimental results in more quantitative terms. Thus the relative deviation at the equimolar solvent composition was found [5] to be $\Delta^{\rm rel}(0.5, C \rightarrow 0) \approx -5.1\%$. Transforming the experimental data to excess conductances, which is defined by the difference of the data obtained for HCl/DCl and KCl, the corresponding anomalous decrease of the excess conductance at $X_{\rm D}$ becomes -7.7%, cf Fig. 7.2.

The analysis [5] shows that the magnitude of the anomalous decrease under consideration is independent of the acid concentration $C = 0.01$–0.5 mol/l, cf. Fig. 7.3. The experimental data also seem to confirm the expectation that the maximum of the predicted anomalous decrease of the H^+/D^+ appears at $X_{\rm D} = 0.5$. Note also that the conductances of KCl solutions in H_2O/D_2O mixtures follow the linear law obtained from standard electrochemical theory. At the same time it is the almost linear dependence of $\Lambda^0_{\rm KCl}$ on the mole fraction $X_{\rm D}$ (in connection with the linear dependence of the fluidity of H_2O/D_2O on $X_{\rm D}$) that establishes the observed "anomalous" decrease of $\Delta\Lambda$ of excess molar conductance of HCl/DCl as a specific property of the H^+ and D^+ ions and the H_2O/D_2O solvent.

The aforementioned theoretical and experimental investigations seem to indicate that coherent-dissipative structures may play an important role in the dynamics of H^+-transport and H-bond formation in physical chemical and biological systems. We will now proceed to applications where ODLRO is more developed, i.e. in connection with the behaviour of light carriers in a nuclear skeleton or background.

7.3 The Development of ODLRO and the Phases of High Temperature Superconductors

Since the light carrier, i.e. the electron is a fermion and thus obeys the Pauli principle, it is obvious that a quantum statistical formulation should explicitly incorporate Fermi statistics. At the same time the electrons, whether they occur in a beam at relativistic energies or effectively interfere with each other, as in condensed matter systems, sometimes seem to appear as individual and independent (albeit correlated on a higher level) objects. These two (contradictory) aspects are nevertheless fundamental to a realistic and theoretically correct picture of a dissipative structure. It is hence of great interest to find out whether the aforementioned fermions are also sufficiently correlated at nonzero temperatures and thus their very specific properties subject to experimental verification.

Before we quote some recent results relevant in this context, we should point out that any organization or correlation effects must be directly coupled to the phase space dynamics through the Liouville equation, see Eq. (6.1), i.e.

$$i\frac{\partial \rho}{\partial t} = \hat{L}\rho$$

As already stated $\hat{L}$ is the commutator with the Hamiltonian ($\hbar$ should be inserted unless we deal directly with energies in terms of frequencies) in the quantum case, and $i\hat{L}$ is the Poisson bracket in the classical formulation. In the previous section, we started to develop a micro-dynamical picture of correlated fermions. Further, we discussed, see Sect. 6.6, the well-known fact that these correlations manifest themselves as the appearance of occupation numbers $\lambda^{(1)}$ fulfilling $0 \leqslant \lambda^{(1)} \leqslant 1$, (due to the Pauli principle), see also Eq. (6.45). For weakly correlated particles these occupation numbers are close to one, but in strongly correlated situations, e.g. in high-T_c superconductivity, (HTSC), or in the fractional quantum Hall effect, (FQHE), these occupancies may deviate significantly from unity.

We also demonstrated, from a simple application of fermionic statistics, that the system might, under optimal conditions, lead to the so-called extreme case of maximum coherence and correlation, (see previous Chapter, Coleman [9], which in addition may or may not develop into off-diagonal long-range order (ODLRO) (Yang [17], Chap. 6). These correlations and coherences might extend over macroscopic dimensions and thus become a true manifestation of quantum effects at the most conspicuous and concrete level.

In the correlated (extreme) case the occupation number was shown to be given by

$$\lambda^{(1)} = \frac{N}{2r}$$

where N is the number of fermions and $2r$ is the number of available spinorbitals. Equivalently r is the rank of the fermion (geminal) pair subspace. It is now a simple task to set up a "correlation matrix", cf. the discussion in Sect. 6.6, which describes quantum mechanically the dynamics of an electron in one level as correlated with or stimulated by other electronic excitations. The off-diagonal (density matrix) element is simply given by the product of the probability of the correlated electron being in the appropriate spinorbital, i.e. $\frac{N}{2r}$ with the probability of the connecting spinorbital being empty, i.e. $\frac{N}{2r} \cdot \frac{(2r-N)}{2(r-1)}$.

The detailed statistical analysis was carried out in [1], so we will not repeat it here. It is important to observe, however, that we are indeed considering fermions as well as the fermion pair subspace together with the associated second order reduced density matrix. Since fermion pairs behave as (quasi) bosons we find, from the extreme case, a mixed statistics, where the main matrix block allows the nonlinear intermixing between actual coherences and associated correlations. The full treatment displays additional dissipation and possibly self-organization.

Returning to the "correlation matrix" above it is a simple matter to diagonalize it. This leads, see problem in Sect. 7.8.1, to a large eigenvalue λ_{L} and a small one λ_{S} given by

$$\lambda_{\mathrm{S}} = \frac{N(N-2)}{4r(r-1)}; \quad \lambda_{\mathrm{L}} = \frac{N}{2} - (r-1)\lambda_{\mathrm{S}}. \tag{7.27}$$

For the identification with Coleman's extreme case or with Yang's concept of ODLRO, see Chap. 6.

In Sect. 6.9 we started by introducing a localized basis of fermion pair functions of rank r given by $\mathbf{h} = |h_1, h_2, \ldots, h_r\rangle$. We may here choose it as a real localized set of spin paired reference determinants but in general it is important to realize that a dynamic formulation necessitates transformations to complex Gamow like representation, see Sect. 6.7. It was also introduced a delocalized coherence basis $\mathbf{g}$ and a complex correlation basis called $\mathbf{f}$ via

$$|\mathbf{h}\rangle\mathbf{B} = |\mathbf{g}\rangle = |g_1, g_2, \ldots, g_r\rangle \tag{7.28}$$

and

$$|\mathbf{h}\rangle\mathbf{B}^{-1} = |\mathbf{f}\rangle = |f_1, f_2, \ldots, f_r\rangle \tag{7.29}$$

with $\mathbf{B}$ given by

$$\sqrt{(r)} \cdot \mathbf{B} = \begin{pmatrix} 1 & \omega & \omega^2 & \ldots & \omega^{r-1} \\ 1 & \omega^3 & \omega^6 & \ldots & \omega^{3(r-1)} \\ \ldots & \ldots & \ldots & \ldots & \ldots \\ 1 & \omega^{2r-1} & \omega^{2(2r-1)} & \ldots & \omega^{(r-1)(2r-1)} \end{pmatrix} \tag{7.30}$$

and $\omega = \exp(i\frac{\pi}{r})$.

The (relevant part of) the reduced second order density matrix for the correlated electron problem reads, see previous Chapter,

$$\Gamma^{(2)} = \Gamma^{(2)}_{\mathrm{L}} + \Gamma^{(2)}_{\mathrm{S}} = \lambda_{\mathrm{L}}|g_1\rangle\langle g_1| + \lambda_{\mathrm{S}}\sum_{k=1}^{r}\sum_{l=1}^{r}|h_k\rangle\left(\delta_{kl} - \tfrac{1}{r}\right)\langle h_l| \tag{7.31}$$

with $Tr\{\Gamma^{(2)}\} = \frac{N}{2}$ i.e. the correlation matrix is normalized to the number of fermionic pairs. In passing we mention that we have neglected the "unpaired" contribution $\Gamma^{(2)}_{T}$, to the full $\Gamma^{(2)}$ above.

It is interesting to note that, for a given value of r, i.e. $2r$ much greater than N, it follows that λ_{L} first "grows" linearly with N until the quadratic term, see Eq. (7.27), starts contributing. The parabola "ends" for $2r = N$, where coherence and delocalization (in the sense given above) no longer persists. We have put the words "grows" and "ends" within quotation marks to issue a warning that changes in N and/or r are not to be interpreted as in linear theories, i.e. to see what happens with the system when certain parameters vary. Here the formulation is *manifestly nonlinear* and therefore N and/or r are not at "our disposal" since they simply reflect that all parts of the system are intrinsically coupled as a dissipative structure.

We have also demonstrated how Eq. (7.31), via complex dilations and thermalization, assumes a Jordan block structure which for the transformed Γ_{S} (we will not introduce a new notation here, although the transformation, as stressed earlier, is nonunitary) takes the simple form, i.e.

$$\Gamma_{\mathrm{S}} = \lambda_{\mathrm{S}}\sum_{k=1}^{r-1}|f_k\rangle\langle f_{k+1}| = \lambda_{\mathrm{S}}J \tag{7.32}$$

with the important property

$$J^r = 0, \quad J^{r-1} \neq 0. \tag{7.33}$$

As mentioned above we may take our reference space real and let all dynamic complexities be invoked through effective reduced Hamiltonians (and/or Liouvillians) and associated partitioning techniques. We must remember, however, that in this formulation the Liouvillian is no longer self-adjoint and hence, as pointed out in the previous Chapter, there is the possibility of a spontaneous breaking of time reversibility. We will also see what consequences the present extension has for the Liouville phase space density or rather the associated space of reduced density matrices.

We now turn to the problem of the creation of ODLRO. First it should be pointed out that coherent-dissipative structures always have a *finite lifetime.* In contrast, the superconducting state, corresponding to ODLRO, or the coherent superselection sector, has practically an infinite lifetime. Therefore coherent-dissipative structures and superconducting states represent *qualitatively different* organized forms of matter. Yet one organized form may destroy the other depending on the environmental conditions. We will now adopt this view-point in connection with some remarks on the recently discovered high temperature superconductors (HTSCs). While thermal correlations can be constructive as building up precursors for ODLRO in certain experimental situations, they also *mutatis mutandis* becomes destructive in the context of ODLRO, cf. high-T_c superconductivity.

High temperature superconductivity (HTSC) was discovered by Bednorz and Müller [25] in 1986, when they succeeded to vary the composition and the thermal treatment of Ba-La-Cu oxide thereby shifting the onset of the resistivity drop to a temperature above 30 K. After showing the presence of the Meissner-Ochsenfeld effect the discovery was quickly accepted and less than two years later awarded with the Nobel prize in Physics for 1987. In the meantime the critical temperature was raised to 92 K substituting the "larger" Lanthanum with the "smaller" Yttrium [26].

There are many differences between the traditional low-temperature superconductors (LTSC) and the HTSCs. For a rather recent appraisal, see the excellent review by Burns [27] and Table 1. We will briefly mention a few of them *viz.*:

1. exceedingly high T_c values
2. strong anisotropies
3. small coherence length
4. anomalous energy gap
5. indication of universal linear relations between T_c and carrier concentration (per effective mass)
6. indication of saturation or universal branching off, in the aforementioned relation.

There has indeed been a considerable discussion – not to mention controversies – regarding the relevant physical mechanisms involved in the HTSCs. First

Tabel 1. List of high-temperature superconductors as compiled by Burns [27]. T_c values, number of CuO planes in the unit cell as well as some notations used in the literature are given. For more details see [27]

Formula	T_c(K)	n	Notations	
$(La_{2-x}Sr_x)CuO_4$	38	1	La(n=1)	214
$(La_{2-x}Sr_x)CaCu_2O_6$	60	2	La(n=2)	–
$Tl_2Ba_2CuO_6$	0–80	1	2-Tl(n=1)	T12201
$Tl_2Ba_2CaCu_2O_8$	108	2	2-Tl(n=2)	T12212
$Tl_2Ba_2Ca_2Cu_3O_{10}$	125	3	2-Tl(n=3)	T12223
$Bi_2Sr_2CuO_6$	0–20	1	2-Bi(n=1)	Bi2201
$Bi_2Sr_2CaCu_2O_8$	85	2	2-Bi(n=2)	Bi2212
$Bi_2Sr_2Ca_2Cu_3O_{10}$	110	3	2-Bi(n=3)	Bi2223
$(Nd_{2-x}Ce_x)CuO_4$	30	1	Nd(n=1)	T′
$YBa_2Cu_3O_7$	92	2	Y123	YBCO
$YBa_2Cu_4O_8$	80	2	Y124	–
$Y_2Ba_4Cu_7O_{14}$	40	2	Y247	–
$TlBa_2CuO_5$	0–50	1	1-Tl(n=1)	T11201
$TlBa_2CaCu_2O_7$	80	2	1-Tl(n=2)	T11212
$TlBa_2Ca_2Cu_3O_9$	110	3	1-Tl(n=3)	T11223
$TlBa_2Ca_3Cu_4O_{11}$	122	4	1-Tl(n=4)	T11234
$CaCuO_2$	–	1	n=∞	–
$(Nd,Ce,Sr)CuO_4$	30	1	–	T*
$(Ba_{0.6}K_{0.4})BiO_3$	30	–	–	BKBO

of all there is the semantic problem whether it is appropriate to use the term "BCS superconductor" in this context. In a strict sense, this should only be used in reference to those cases that fits the criteria developed in the classical BCS paper of 1957 ([22], Chap. 6). This nomenclature would further lead to the "weak-coupled BCS", where the electron-phonon interaction is weak (but attractive), with the fairly straight forward extension to (isotropic) strong-coupled BCS being quite obvious. However, as the mechanism behind the phenomenon of high-temperature superconductivity is sofar unknown, it is clear that a satisfactory treatment of the facts listed above should, in some general way, be answered, explained or at least indicated as a realistic possibility. Although many (but not all!) scientists in the field tend to view HTSC as building upon some not yet understood pair interaction, there are indications, see again differences listed above, that the pairing could not be (essentially) phonon mediated. Since magnetism and superconductivity are strongly interrelated, the plethora of mechanisms involve antiferromagnetism, semiconduction, magnons, polarons, spinons, holons, anyons etc. For an entertaining and insightsful description on "consensus and no consensus" on HTSC in the mid 1991, see the discussions between the leading proponents of condensed matter physics P.W. Anderson and R. Schrieffer [28]. It thus becomes clear that the traditional BCS model is not applicable to HTSC, and that some general pair correlation scheme should be needed for the development of ODLRO. In this extended sense a more generally developed BCS theory may merit the use of the term, yet the semantic problem exists.

Before we leave this rather controversial issue we will also mention some recent very remarkable developments. Since the experimental data, referring to HTSC, is accumulating at a rate that requires almost quixotical efforts to keep up with the flood, there are still many different opinions and conflicting interpretations even within the general BCS view-point. For instance, it is believed that hole carriers are responsible for superconductivity in the high-T_c cuprates and further that the pairing should be s-wave, Anderson [28], see also in [29], or speculated to be d-wave see Pines in [29]. A very beautiful experiment [30], i.e. phase coherence measurements in bimetallic dc SQUIDs (Superconducting Quantum Interference Device) and tunnel junctions, made from single crystals of $YBa_2Cu_3O_{7-\delta}$, (YBCO) and thin films of the conventional s-wave superconductor Pb, gave rather strong evidence for the existence of a $d_{x^2-y^2}$ pairing state.

On the other hand, a recent analysis [31] of iodine intercalation in $(BiO)_2$ bilayers of the bismuth cuprates indicates that the effects on critical temperatures in these systems are neither caused by an increase of hole concentration in the CuO_2 layers nor a weakened inter-cell coupling between these layers. Here an indirect exchange pairing mechanism (s-waves) suggest that the high-T_c cuprates and also alkali-doped C_{60}'s "belong to the same family of superconductors [31, 32]". Moreover, the mechanism refers exclusively to indirect couplings of (delocalized) conduction electrons via closed shell cations, in analogy with the old superexchange idea used for describing magnetic insulators.

The above mentioned features are indeed very interesting and require further analysis and development. Here we will only comment on some very general observations that seem to indicate that general thermal quantum correlated fluctuations yield a good first order description of HTSCs, which may give general qualitative information of the HTSC phenomena, which together with specific mechanisms may give quantitative information as well.

In our opinion, it is therefore not enough to find particular mechanisms, be it derived from crude coulombic forces, magnetic interactions, Cooper pairing or based on Hubbard hamiltonians etc. Rather we need to know the general environmental perturbations of the actual manybody fermionic system in order to understand e.g. the formation or destruction of ODLRO as well as the associated anomalous behaviour listed above. In this initial phase we are not primarily interested in the precise pair mechanism, but rather how the physical interaction is perturbed by the universal thermally activated quantum correlations. We have discussed this in some detail elsewhere [1, 7, 8]. We will here briefly mention the points listed as [4–6] above as an illustration to what we mean.

The energy gap, point [4], associated with the stability of the (high-T_c cuprates) superconducting state has been a controversial issue, with several values for the gap, referring to the various directions, in the ab-plane (CuO_2 planes) or the c-axis (perpendicular) given in the range of [3–9] $k_B T_c$. Also the Uemura universal linear relation [33], see Fig. 7.4, has been a much debated issue. We are here referring to the well-established μSR-experimental results of Uemura et al. [33], which reveals a remarkable universal relation between the transition temperature, T_c, and the carrier density over effective mass, for a large

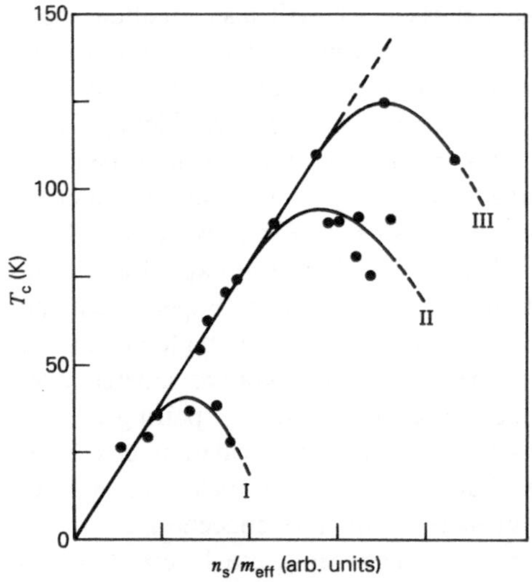

Fig. 7.4. Schematic representation of the effect of the superconducting transition temperature T_c plotted versus carrier concentration n_s divided by the effective mass m^*. For details see Uemura et al. in Ref. [33]

number of different classes of HTSCs and other unconventional superconductors. To these belong all cuprate, bismutate, certain organic, different Chevrel-phase, heavy fermion systems as well as the new buckminsterfullerenes. These experiments also display a second surprising observation, i.e. above certain critical values of the carrier density there seem to be some saturation with the above compounds all appearing to branch off from the linear relation in very much the same way. These two effects seem to be independent of the actual interaction mechanisms referred to above.

To give a simple first order qualitative interpretation of these experimental facts we turn directly to our density matrix theory. In the previous sections we derived the decomposition Eq. (7.31), (neglecting $\Gamma_T^{(2)}$), see also Eqs. (6.55–6.58). Here we will particularly focus on the "large component" given by

$$\Gamma_L^{(2)} = |g\rangle\lambda_L\langle g|; \quad \lambda_L = \frac{N}{2} - \frac{N(N-2)}{4r} \tag{7.34}$$

As already emphasized, we will not explicitly expose the pair forming mechanism except denoting it by $\mathscr{V}_{12}$. The energy gap 2Δ of the superconductor being in a physical state represented by the density matrix $\Gamma_L^{(2)}$ is then expressed by the formula

$$2\Delta = |\mathrm{Tr}\{\mathscr{V}_{12}\Gamma_L^{(2)}\}| = \lambda_L|\mathrm{Tr}\{\mathscr{V}_{12}|g\rangle\langle g|\}| \tag{7.35}$$

In this formula only the coherent part of the density operator $\Gamma^{(2)}$ appears, since the supercurrent is known to be representable by a "macroscopic pure state".

Applying first the CSM theory and then thermalization, where the latter non-unitary transformation introduces appropriate temperature dependencies, see e.g.

Eqs. (6.168–6.174), to the neglected correlation sector, represented by

$$\Gamma_S^{(2)} = \lambda_S \sum_{i,j} |(i)\rangle(\delta_{ij} - \tfrac{1}{r})\langle(j)|; \quad \lambda_S = \tfrac{N(N-2)}{4r(r-1)} \tag{7.36}$$

specific Jordan blocks may occur corresponding to transient structures, precursors, with phase correlation times τ_k given by Eq. (6.173) and with a minimum size given by Eq. (7.13). Therefore to break a certain cluster, corresponding to a Jordan block of order $r = s$ we use Eq. (7.13), which for $s = 2$ gives with the characteristic energy $\hbar/\tau_{\text{char}}$,

$$\hbar/\tau_{\text{char}} = \frac{4\pi k_B T_0}{s} = 2\pi k_B T_0 \tag{7.37}$$

where T_0 is the corresponding characteristic temperature. We now introduce a *critical temperature* T_c of the superconductor, i.e. the temperature at which relevant clusters – we assume Bose condensation of correlated units beyond the initial pairing – of size 2, 4, 8, 16,...,2^n become destroyed. From the formula (7.37), for the characteristic energy, the energy gap 2Δ simply may be written as

$$2\Delta = \hbar/\tau_{\text{rel}} = 2\pi S k_B T_c; \quad S = \sum_{k_{\min}}^{k_{\max}} \frac{1}{2^k} \tag{7.38}$$

for some $k_{\min}$ and $k_{\max}$. Together with Eq. (7.35) one obtains

$$2\pi S k_B T_c = |\text{Tr}\{\mathscr{V}_{12}\Gamma_L^{(2)}\}| \tag{7.39}$$

and furthermore (with r being a "carrier reservoir" index as well as the dimension of the geminal-pair subspace)

$$T_c = \frac{1}{2\pi S k_B} |\langle g|\mathscr{V}_{12}|g\rangle|\lambda_L = \text{const}\left\{\frac{N}{2} - \frac{N(N-2)}{4r}\right\} \tag{7.40}$$

These simple formulae represent the main results in the present context and they suggest immediate three physical predictions and interpretations:
(1) Formula (7.40) predicts energy gaps in the range (breaking clusters of order $s = 2$ or $s = 2, 4$ etc.)

$$\frac{2\Delta}{k_B T_c} = (2-3)\pi \sim (6.3 - 9.4) \tag{7.41}$$

but other possibilities are also imagined, depending on the experimental situation.
(2) Formulae (7.40) also predicts a *general linear dependence* of T_c on the carrier number $N/2$, if the latter number is sufficiently small (i.e., for $N^2 \ll r$).
(3) Additionally, for larger values of $N/2$ one obtains a *general negative quadratic deviation* from the above linear dependence, which is due to the last term appearing in λ_L in Eqs. (7.40) and (7.34).

Note the difference between the BCS theory and the present description in terms of a precursor state. For instance if we consider an explicit pair breaking mechanism, we must instead explicitly consider the relation between s in

Eq. (7.37) and the number of fermions $N = 2n$. The number of possible pair energies and associated pair space dimension (equal to the bosonic degrees of freedom), see also the discussion in the next section, then becomes

$$s = \binom{N}{2}$$

The energy gap, see Eq. (7.38), then is obtained from

$$S = \sum_{k_{\min}}^{k_{\max}} \binom{2n}{2}^{-1} \tag{7.38'}$$

with $k_{\min} = 2$. Note that $k_{\max} \to \infty$ gives $S = 2\log 2 - 1 \approx 0.3864$. A realistic estimate of the corresponding energy gap from, let us say $N = 8-10$ or summing the first three or four terms in (7.38′) yields $S \approx 0.269 - 0.291$ and

$$\frac{2\Delta}{k_B T_c} = 3.4 - 3.6 \ (S = 0.386 \text{ corresponds to } 4.8)$$

in almost perfect correspondence with weak- and strong-coupled BCS superconductors [27].

These surprising quantitative results of our general CSM theory are also found to be in remarkable agreement with very recent experimental results on HTSCs:

(4) The energy gap in the superconducting ab-plane of high-T_c Cu-O superconductors appears to be

$$\frac{2\Delta}{k_B T_c} \approx 7 - 8$$

and these results clearly disagrees with the standard BCS value of 3.5. There are some disagreements, however, on the precise non-BCS behavior depending on which experiment one is analysing, e.g. infrared reflectivity or transmission, uv-photoemission etc. We will not expound further on this controversy here except pointing out that our correlation theory favors such a complex behavior, and that our derivation stems from the fact that we have assumed the existence of a precursor in contradistinction to the BCS model [1, 7, 8].

(5) Our predictions also agree with the experimentally established *universal linear relation* of Uemura et al. [33].

(6) The predicted negative quadratic deviation is further in qualitative agreement with the extensive experimental data just referred to [33]. Indeed, this deviation or branching off seems to be a universal feature of the presented data, rather than just a "disturbing accompanying effect" being due to e.g. higher carrier concentration, phase separation etc., and thus there is no clear "mechanistic" explanation for the suppression or saturation of T_c in this region.

In connection with the aforementioned universal correlation data, we have assumed that each of the different specimens of the 214 series, denoted by I,

123 and 2212 by II, and the 2223 systems denoted by III, see Table 1 and Fig. 7.5, can be controlled by a given $r = s_{\min}$ which will act as a reservoir of available states. In cuprate superconductors the physical interpretation is that the charge carriers are holes (p-type) and in this case each class is predicted to be controlled by its hole reservoir content. The Uemura results relate T_c with the carrier concentration per effective mass.

From Eq. (7.40), we further obtain, with V and m^* being the volume and the effective mass respectively, $n_c^{(*)}$ the carrier concentration (per effective mass), and $n_r^{(*)}$ the concentration (per effective mass; large N and r) of available fermion states (virtual particle concentration),

$$T_c = \text{const}\ V\left(n_c - \frac{n_c^2}{n_r}\right) \tag{7.42}$$

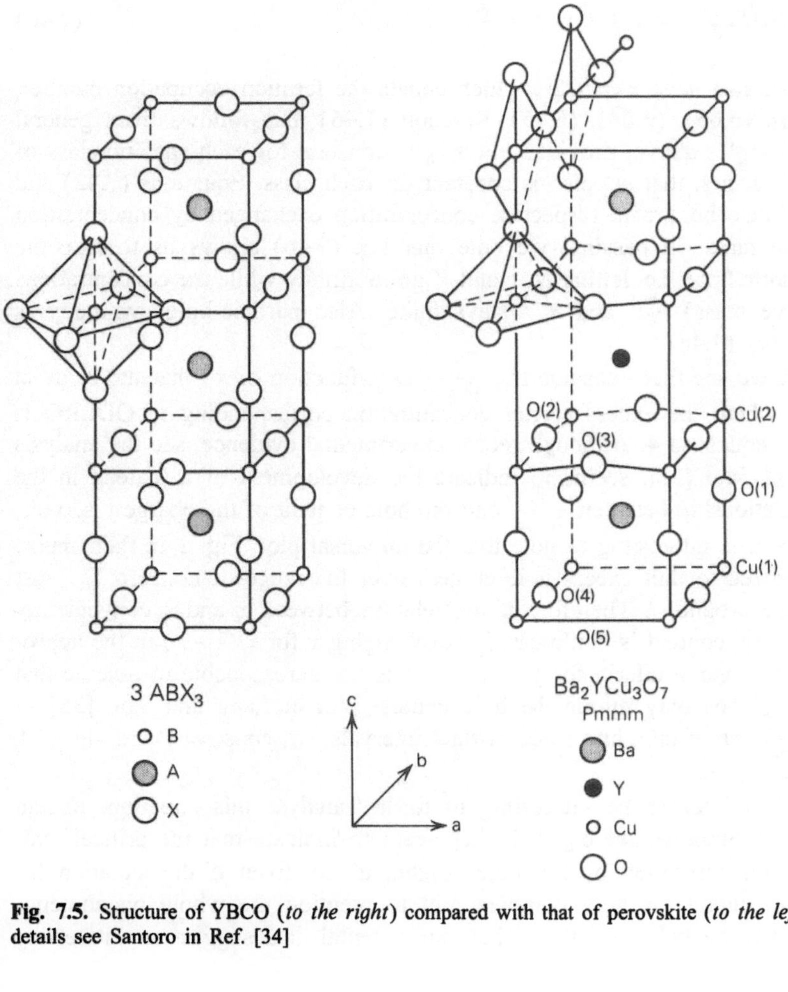

Fig. 7.5. Structure of YBCO (*to the right*) compared with that of perovskite (*to the left*). For more details see Santoro in Ref. [34]

or

$$T_c = \text{const } Vm^*\left(n_c^* - \frac{n_c^{*2}}{n_r^*}\right) \tag{7.43}$$

where $n_c = N/V$, $n_c^* = n_c/m^*$ and $n_r = 2r/V$, $n_r^* = n_r/m^*$. It simply follows that the maximum critical temperature $T_{c,\max}$ (note that $n_c^{(*)}$ varies, while $n_r^{(*)}$ and m^* is assumed to be constant for each class of superconductor) is given by

$$T_{c,\max} = \text{const } V\frac{n_r}{4} \tag{7.44}$$

for

$$n_c = \frac{1}{2}n_r \tag{7.45}$$

The universal plot of Uemura et al. [33] now writes ($0 \leqslant n_c \leqslant n_r$)

$$T_c = T_{c,\max} \cdot 4(x - x^2);\; x = \frac{n_c}{n_r} \tag{7.46}$$

Note that we also have $x = N/2r$, which equals the fermion occupation number, see e.g. Eqs. (6.65), (6.64), (6.75). Relation (7.46) thus follows from general physical principles above, provided $r = s_{\min}$ is constant for each specific class of HTSCs and further, that m^* also is constant for each class. Equations (7.42) and (7.44–7.46) also hold with respective concentration exchanged by concentration per effective mass. In passing, we note that Eq. (7.46) allows us to take the thermodynamic limit, i.e. letting N, r and V go to infinity while the concentrations (per effective mass) $n_c^{(*)}$ and $n_r^{(*)}$ stays finite. Also particle-hole symmetry is restored in Eq. (7.46).

Finally, we see that – considering $\frac{T_c}{T_{c,\max}}$ as a function of x – that the slope at $x = 0$, i.e. where the excess carrier concentration corresponding to ODLRO is building up, equals to 4. Although recent experimental evidence, see the analysis by Zang and Sato [35], seems to indicate the development of a plateau in the universal relationship between $\frac{T_c}{T_{c,\max}}$ and the hole content of the charge reservoir, see Fig. 7.6, it is interesting to note that the universal plot, Fig. 1 in their report [35], normalized to unit excess hole content over the range of nonzero T_cs, just gives a slope around 4. Therefore if the relation between x and a conveniently normalized hole content is nonlinear, i.e. exchanging x for $x/(1-x)$ in the appropriate interval (and similarly for $y = 1-x$) it is not unreasonable to assume that the relation (7.46) may mimic the hole content plot of Zang and Sato [35] so that the $\frac{T_c}{T_{c,\max}}$ curve falls into three distinct intervals, $4x$, const $= 1$ and $4(1-x)$, respectively.

It would of course be interesting to further analyse this situation. Recent results and discussions, see e.g. [31, 32], seem to indicate that the critical temperature is not controlled by the hole content alone. Even if the situation for each high-T_c cuprate is very complex, not to mention the whole phenomenological picture, we believe that reliable experimental data suggest an interesting

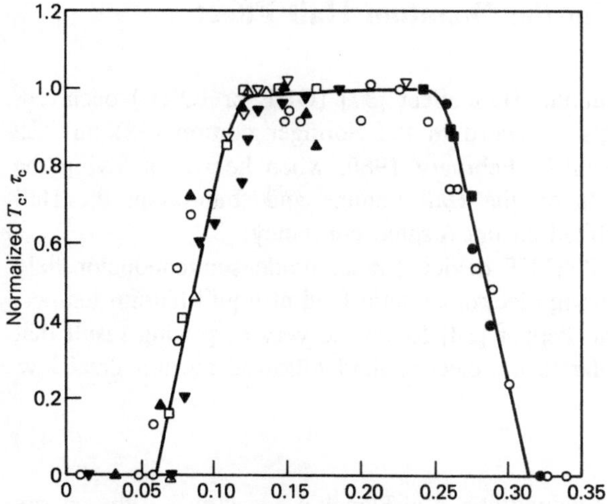

Fig. 7.6. Schematic representation of universal relation between normalized T_C and hole content for cuprate superconductors. For more details see Ref. [35]

universal behaviour and possible hole content relations. It is therefore an encouraging feature of our approach that we are able to capture a reasonable first order qualitative picture which could be further developed in combination with specific mechanisms.

Since the variation in $T_{c,\max}$ in our formulation depends on r, which in turn depends on the effective reduced mass m^* (larger $r = s_{\min}$ yields smaller m^*) [8], one may conclude that "lighter carriers" also should yield a higher $T_{c,\max}$.

Before leaving this subject, it should be stressed, see [34] for details, i.e. that integrating the current density expression over any closed path leads via application of Stokes' theorem to the famous result of flux quantization. The fluxoid quantization demonstrates beautifully how much charge is responsible for the flux through the surface surrounded by the above mentioned path. This picture suggests a phenomenological description in terms of vortices and corresponding dynamics. These vortices are subject to the Lorentz force leading to flux flow. Vortex interactions, flux pinnings, creeps etc. have been very successful in the phenomenological understanding of general aspects of superconductivity. A good discussion and historical account of the various phases of the phenomenological development of superconductivity, particularly of the golden years between the world wars has recently been given by Dahl [36].

This concludes the HTSC review. In the next section we will see how the present picture offers a unified description also on the fractional quantum Hall effect (FQHE).

7.4 Fractional Statistics in the Quantum Hall Effect

The birth of the (integer) quantum Hall effect [37] (QHE or IQHE) occurred, according to K. von Klitzing's foreword in the Springer edition [38] on *The Quantum Hall Effect*, one night in February 1980, when he was studying the influence of localization effects on the Hall voltage and found that the Hall resistance on the plateax exhibited an unexpected constancy.

By observing a silicon MOSFET device (metal oxide semiconductor field effect transistor) placed in a strong electromagnetic field at liquid helium temperatures, von Klitzing, Dorda and Pepper [38] found the very surprising result that the current density perpendicular to the electric field followed the universal law

$$\sigma_{xy} = -\frac{ie^2}{h} \tag{7.47}$$

where σ_{xy} is the off-diagonal conductivity, h is Planck's constant, $-e$ the charge of the electron and i a small integer. In passing we note an important geometric feature, i.e. that in two dimensions size factors cancel out so that "ivities" and "ances" coincide. The law, (7.47), with (7.48) below, states that the Hall resistance R_{H} exactly equals h/e^2i. Since the fine structure constant α, the coupling between matter and the electromagnetic field, is *exactly* expressed in terms of e^2/h, the latter is determined with the same certainty.

While σ_{xy} above quantizes in units of e^2/h the corresponding diagonal conductivity was found to vanish, i.e.

$$\sigma_{xx} = 0. \tag{7.48}$$

Since Eqs. (7.47–7.48) imply nearly complete freedom from dissipation it is clear that there is a close connection between QHE and superconductivity.

As the background for these comments we point out the following scenario. In the presence of a magnetic field **B**, chosen to be in the z-direction, the electrons are subject to the well-known Lorentz force. Together with the electric field this yields the current density from which the conductivity tensor, σ, is given by

$$\sigma = \rho^{-1} = \begin{pmatrix} \rho_{xx} & \rho_{xy} \\ \rho_{yx} & \rho_{yy} \end{pmatrix}^{-1} = \begin{pmatrix} \rho_0 & B/nec \\ -B/nec & \rho_0 \end{pmatrix}^{-1} = \begin{pmatrix} \sigma_{xx} & \sigma_{xy} \\ \sigma_{yx} & \sigma_{yy} \end{pmatrix} \tag{7.49}$$

where σ is the matrix inverse of the resistivity tensor ρ, $-e$ the electronic charge, c the velocity of light and n here the number of electrons. Also $\sigma_{xx} = \sigma_{yy}$ and $\sigma_{xy} = -\sigma_{yx}$.

From the standard semiclassical picture one expects the Hall resistance to be a linear function of the field $B = |\mathbf{B}|$. In other words, since the off-diagonal term of ρ defines the Hall resistivity (depending also on the sign of the carriers!), one gets trivially the relation between the normalized inverse Hall resistance h/e^2R_H and the (normalized) density nhc/eB to be a line with unit slope. As already mentioned, there appeared, instead of the expected behaviour, a series of steps very accurately given by integers and later by odd fractions, i.e. the integer or fractional quantum Hall effect (IQHE, FQHE), see also Fig. 7.7.

In the derivation of the quantum Hall resistance $R_H(i) = V_H/I = h/e^2 i$ where V_H is the appropriate (transverse) voltage difference between contacts, one usually makes the assumption of a perfectly homogeneous medium with no imperfections or localization effects and no scattering processes of any kind as well as "naturally" taking the temperature to be at absolute zero. Further by studying the problem of two-dimensional spinless electrons in a perpendicular magnetic field in the Landau gauge, see e.g. Prange and Girvin [38] for details, one finds the so-called Landau (harmonic oscillator) levels for the motion along the y-direction (where the vector potential has been put equal to zero). For the other "nonlocalized" (x-)direction one may impose periodic boundary conditions. This implies that the number of states per unit area in a (non-relativistic) Landau level becomes precisely eB/hc. Note also that the states, as mentioned, are extended in the x-direction and confined in the y-direction.

7.4.1 The Fractional QHE

While these studies were intensely carried forward another surprize entered the stage. Two years after the QHE discovery, Tsui, Störmer and Gossard [39] found a dip in the resistivity tensor corresponding to fractional values $f = \frac{1}{3}$ of i (in units of e^2/h) Fig. 7.7. They appeared in some nearly ideal GaAs-$Al_xGa_{1-x}As$ high quality heterojunctions at very low temperature, and this was the start of the anomalous or fractional quantum Hall effect, FQHE. The observation of these additional Hall plateax at f equals $\frac{1}{3}$, $\frac{2}{3}$, $\frac{2}{5}$ etc. with even denominators $\frac{1}{2}$, $\frac{1}{4}$ etc. notably missing, has triggered new developments and stimulated a healthy cooperation between different areas of physics [38].

It was first believed that one had found a new state of matter, i.e. the long-sought-after Wigner-electron-solid [40]. For instance, a new electronic state commensurate with the fraction $\frac{1}{3}$ should then be consistent with triangular crystal symmetry, favored when the unit cell area of the lattice is a multiple of the area of a magnetic flux quantum. Thinking along those lines as well as extending the "Landau picture" above, using fundamental gauge invariance principles, etc. Laughlin [41] proposed an explanation of the FQHE based on fractionally filled Landau levels.

Although recent work, in terms of two-dimensional topological units, called *anyons* [42, 43] and subject to fractional statistics (neither bosons or fermions), allows deep analogies to be made between the new FQHE and the Landau level interpretation, there are some basic unresolved problems which concern basic symmetries, collective effects and long-range interactions, see Girvin in [38].

We approach here a rather strange situation. We started out, see the previous section, by considering an open or complex system, which during the transition from the microscopic level to a mesoscopic- or macroscopic one, by necessity

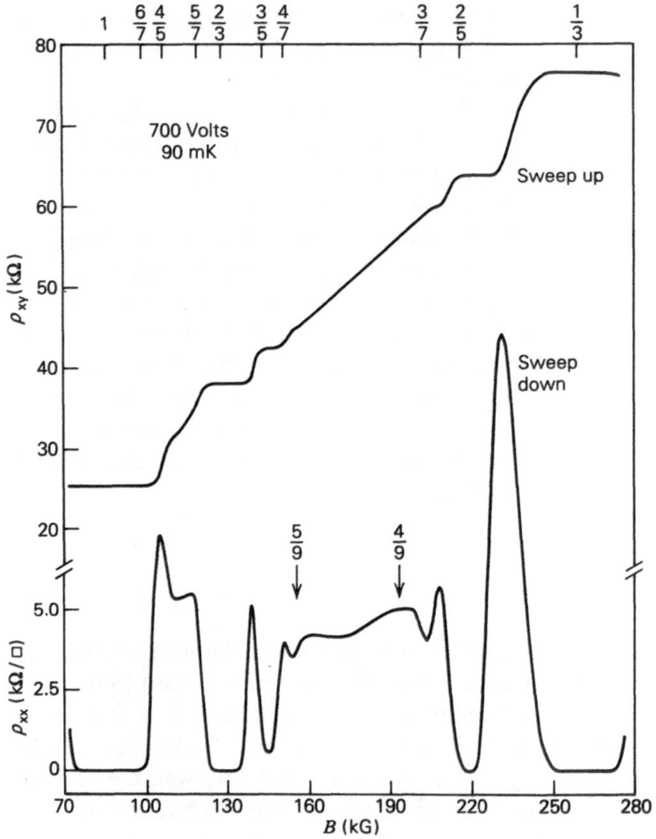

Fig. 7.7. Diagonal and nondiagonal resistivity at 90 mK, for a sample which shows the FQHE at fractions 1/3, 2/3, 2/5, 3/5 3/7, 4/7/ 4/9, and 5/9 (Chang in Ref. [38])

has lost some of its fundamental symmetries. Further by considering here the super-selection sector corresponding to ODLRO, see also specific density matrix representations related to the extreme case, we have to return to the fundamental question of how to deal with conservation and/or spontaneous breaking of symmetries, like time reversibility, gauge invariance, parity, chirality etc. We will, however, not consider this question. Here we will instead directly analyse and describe the FQHE from our presently developed machinery. Since we have reported these results previously [10] we will only make a brief account of the situation. The results will then be reformulated and reinterpreted so as to make it more adapted to the Wigner crystallization hypothesis.

7.4.2 Quantum States

It has been observed, see Thouless in [38], that there seem to exist two different sorts of quantum numbers in condensed matter: *those related to symmetry and those that are determined by topology. ... if the quantization of Hall conductance has to be fitted into one of these two classes of quantum numbers, it is obviously more attractive to fit it into the class of topological quantum numbers.*

This topological argument brings us back to the relation between QHE and (high or low temperature) superconductivity (HTSC or LTSC). In SC one finds, below the critical temperature T_c, that the magnetic flux is quantized in units of $\Phi_B = hc/2e$. The analogy between QHE and SC is in fact even stronger, since the development of ODLRO will play the same fundamental role in both cases.

From the theoretical point of view, electron Coulomb interactions, and corresponding correlations, should enter any realistic description of QHE, or at least provide some initial motivation. Depending on insurmountable difficulties, however, there does not presently seem to exist a complete or satisfactory alternative to the simplified incompressible quantum fluid picture, which allows a systematic account of fundamental principles and topological interpretations. We will attempt to give one here and it will be demonstrated that the observed fractional Hall plateaux, are manifestations of strongly correlated objects within the off-diagonal long-range order (ODLRO) superselection sector. Further this interpretation yields $(n-1)/(2n-1)$, $n = 2, 3, \ldots$ the most stable fractions. The recently observed fractional Hall states, with even numerators, in bilayer two-dimensional electron systems, can also be considered here, but we will leave this issue aside [10].

At first sight the theoretical problems seems formidable. A general correlation approach motivates a many-body Hamiltonian starting point. Further, the correct Hamiltonian should also contain appropriate vector potentials commensurate with the experimental situation. Gauge symmetry considerations, earlier mentioned, are vital as well as other characteristic features of dynamical nonstationary systems. Instead of considering a topological framework, including the specific vector potential, referring to a problem of determining the populations of various Landau levels, we will start in the other end, i.e. with a standard many-body formulation based on a general crude Coulomb Hamiltonian. We stress that the dynamic features of the problem can be extracted by using the complex scaling method, CSM. Note also that complex scalings violate gauge invariance. Hence the gauge is fixed from the beginning.

Parallel to this formulation we remind the reader of the reduced formulation of the many-body problem via the hierarchy of reduced fermion density matrices with the extreme case as the platform for ODLRO, see previous section. Our strategy is then to show that strongly correlated fermions, subject to Coulomb forces, each fermion entangled with its environment and described by the appropriate component of the second order reduced density matrix, may appear as certain odd fractions in the Hall experiment. Our view is therefore that these fractions manifest themselves as strongly correlated objects which become observable through the external magnetic field in the Hall experiment, i.e. when matched with the appropriate Landau level fillings.

7.4.3 The Many-Body System

Let us first discuss the many-body (fermionic) characteristics of a system described by the Hamiltonian

$$H = \sum_{i}^{N} h_i + \sum_{i<j}^{N} h_{ij} = H_0 + V, \tag{7.50}$$

where h_i and h_{ij} are one particle and two particle operators respectively. It is convenient to let the first term, H_0, correspond to independent particles moving in some appropriate nuclear background, and the second term, V, to the so-called correlation contributions to the energy. In the "crude case" mentioned above this would involve Coulombic interparticle interactions. In general we will be interested in N-particle problems with $N = 2, 4, 6, \ldots$ etc., but for the time being we will concentrate on a particular projection onto $N = 2n$ particles. Note that the total energy for a many-body (fermionic) system can essentially be written

$$E = E_0 + E_{\text{corr}}$$

where $E_0 = \sum_i \varepsilon_i^0 = \sum_i \langle i|h_1|i\rangle$ is the zero order energy for the noninteracting zero order Hamiltonian H_0 in Eq. (7.50) above with $|i\rangle$ being the ith spin orbital making up the zero order $N = 2n$ determinant of occupied orbitals, and

$$E_{\text{corr}} = \sum_{i<j} \varepsilon_{ij}$$

with

$$\varepsilon_{ij} = \sum_{\mu<\nu} \langle ij|h_{12}(1 - P_{12}|\mu\nu\rangle C_{ij}^{\mu\nu}. \tag{7.51}$$

The correlation energy arises from the two-body terms in V above with $|\mu\rangle$ denoting the so-called excited or unoccupied spin orbitals and $C_{ij}^{\mu\nu}$ being the correct expansion coefficients of the doubly excited part of the full many-body wavefunction and P_{12} is the permutation operator, for details see [10] and references therein.

We can also write the energy as a sum of $s = \binom{N}{2}$ pair energies

$$E = \sum_{i<j} (\varepsilon_{ij}^0 + \varepsilon_{ij}); \; \varepsilon_{ij}^0 = \frac{1}{(N-1)}(\varepsilon_i^0 + \varepsilon_j^0). \tag{7.52}$$

To each pair energy there is also a corresponding geminal (two-particle function) given by

$$|(ij)\rangle = \frac{1}{\sqrt{2}}(1 - P_{12})\{|ij\rangle + \sum_{\mu<\nu} |\mu\nu\rangle C_{ij}^{\mu\nu}\}. \tag{7.53}$$

Formulas (7.52–7.53) are completely general and apply therefore to LT- and HTSC. Alternatively we can also look at the corresponding energy obtained from the N-particle projection of the BCS wavefunction with $N = 2n$. From Eq. (6.31) in Chap. 6, we obtain the simple relation

$$\langle H \rangle_{\text{Av}} = \text{Tr}\left\{ \left[\frac{1}{(N-1)}(h_1 + h_2) + h_{12} \right] \Gamma^{(2)} \right\} \tag{7.54}$$

with $\Gamma^{(2)}$ being the second order reduced density matrix definea by ($p = 2$) Eq. (6.30).

In order to emphasize the present view of QHE as related to LTSC or HTSC, we refer once again to the so-called *extreme case*, see Chap. 6. In this context we learned that the eigenvalues of $\Gamma^{(1)}(g^{\frac{N}{2}})$ are degenerate and given by

$$\lambda^{(1)} = \frac{N}{2r}, \tag{7.55}$$

where $2r$ is the rank of $\Gamma^{(1)}(g^{\frac{N}{2}})$, and that the largest eigenvalue of $\Gamma^{(2)}(g^{\frac{N}{2}})$ becomes

$$\lambda_{\text{L}}^{(2)} = \frac{N}{2}\left\{1 - \frac{(N-2)}{2r}\right\} \tag{7.56}$$

Note that r is also the rank of the pair space over which the maximization of the pair occupation number yields Eq. (7.56).

We have emphasized that the eigenvalues of $\Gamma^{(1)}(g^{\frac{N}{2}})$, satisfy $0 < \lambda^{(1)} \leqslant 1$ due to the Pauli principle, and that they are all equal to one in the independent particle picture. Usually the largest occupation numbers of the first order reduced density matrix, corresponding to a "realistic configuration interaction expansion" of an atomic or molecular system, are close to unity, however, in a strongly correlated situation like in QHE or SC, these occupation numbers may be a fraction that deviates significantly from unity.

In the strongly coherent situation, with almost all the fermionic particles paired up into quasibosonic degrees of freedom, one finds that the condensate is represented by the largest pair occupation number, i.e. the eigenvalue $\lambda_{\text{L}}^{(2)}$ of the second order density matrix, becomes very close (for large r) to $N/2 = n$, cf. development of off-diagonal longe-range order, ODLRO. In other words the extreme situation corresponds to the case where the factorizable part of $\Gamma^{(2)}$ is maximized. Deviations from $N/2$ signifies various levels of correlation and less coherent motion.

In the independent particle limit all eigenvalues of $\Gamma^{(2)}(g^{\frac{N}{2}})$ become degenerate and equal to one. Based on these limits it is natural to derive from Eq.

(7.56) that the corresponding pair correlation or pair correlation contribution g_{12} should be given as

$$g_{12} = \lambda_{\mathrm{L}}^{(2)}/n - 1/r. \tag{7.57}$$

As a check we see that $g_{12} \to 1$ as $r \to \infty$ and $g_{12} = 0$ for $r = N/2 = n$. As a further test one obtains $g_{12} = 1 - \lambda^{(1)}$.

7.4.4 High Correlation

Let us now consider a system of highly correlated fermions, like the one in a QHE or SC experiment. Since ODLRO may be possible if the temperature is low enough for arbitrarily large values of r, it will be energetically favourable for the system to approach the extreme configuration. Equating the energy formulas Eqs. (7.50–7.53) with the expression Eqs. (7.54–7.57), one may conclude that an extreme configuration, corresponding to stabilization and energy lowering, assumes for $r = s = \binom{N}{2} = n(2n-1)$.

Note that r is also the rank of the pair- or geminal space as discussed in connection with Eq. (7.56). From this follows the important result that the fractional occupation number $\lambda^{(1)}$ becomes with $N = 2n$

$$r = n(2n-1); \; \lambda^{(1)} = \frac{N}{2r} = \frac{1}{(2n-1)}, \quad n = 1, 2, \ldots \tag{7.58}$$

yielding the fractions 1, $\frac{1}{3}$, $\frac{1}{5}$ and so on. As discussed before, in connection with HTSC, we find that particle-hole symmetry is lost.

This result is really encouraging since the odd fractions appear from a very simple analysis of the extreme situation. There is, however, no topological insight offered by this energy stabilization analysis, and in order to understand how these fractions manifest themselves, we need to find the relation with the corresponding filling factors of the Landau theory. This will be given further below but first we will improve the discussion leading up to Eq. (7.58).

Since we have used an energy argument for the derivation of Eq. (7.58), one may try to derive additional fractions by noting that any expectation value of the two-body operator $W = \sum_{i<j} v_{ij}$ can be written (in the extreme situation) as

$$\langle W \rangle = \mathrm{Tr}\{v_{12}\Gamma^{(2)}\} \sim \lambda_{\mathrm{L}}^{(2)} \langle v \rangle \tag{7.59}$$

where $\langle v \rangle$ is the expectation value of v_{12} taken over the (extreme) geminal $1/\sqrt{r}\sum|(ij)\rangle$ with i and $j = i + r$ suitably paired. Further, since the trace in Eq. (7.59) obtains as the "large eigenvalue" $\lambda_{\mathrm{L}}^{(2)}$ times the aforementioned quantum mechanical average, one might consider varying N for each fixed given values of r so as to optimize $\lambda_{\mathrm{L}}^{(2)}$ which, for $v_{12} = h_{12}$, corresponds to finding the optimal energy. This yields

$$r = N - 1 = 2n - 1; \ \lambda^{(1)} = \frac{N}{2r} = \frac{n}{2n-1}. \tag{7.60}$$

As mentioned earlier [10] the above presented objects obey fractional statistics. Identification with models based on topological arguments also suggests the standard treatment of FQHE as a *two-dimensional electron system* (2DES). The optimization leading to Eq. (7.60) therefore refers to a single layer. Other choices or r should be possible in a multilayer system [10]. Note also that Eq. (7.60) in the asymptotic (large N) limit displays "supersymmetry", i.e. the number of bosonic degrees of freedom r approximately equals the number of fermions N. Traces of this symmetry can e.g. be seen in the Uemura plots [33].

Finally, the (normalized) pair occupation g_{12} is obtained from Eqs. (7.56–7.57), i.e.

$$g_{12} = 1 - \lambda^{(1)} = \frac{n-1}{2n-1}. \tag{7.61}$$

Our interpretation is thus that $\lambda^{(1)}$ gives the fraction of the fermion(s) that do not participate in pair formations, while g_{12} gives the remaining fraction contributing to the coherent superfluid wavefunction.

The following interpretation is suggested. For a fully coherent situation with $r \to \infty$ one finds that $\lambda^{(1)} \to 0$ and $g_{12} \to 1$. In analogy with the analysis of the previous section, we conclude that the two-dimensional projection of our strongly correlated system, via a suitable electromagnetic field, results in a manifestation of the extreme case through suitable Landau fillings. Hence, for finite values of r all fractions mentioned above appear.

It should therefore be natural to associate the fractions in Eq. (7.61) with the primary sequence of filling fractions (with increasing N) $\frac{1}{3}$, $\frac{2}{5}$, $\frac{3}{7}$ etc. One would hence expect all odd fractions as mentioned above to appear and a "spectroscopy" as proposed by Laughlin, in [38], to be worked out in detail based on energy and/or localization sum rules.

We may also note that some of the large fractions observed [38] can also be associated with the large eigenvalue $\lambda_L^{(2)}$ of $\Gamma^{(2)}$, see Eq. (7.56). This yields fractions of type $\frac{4}{3}$, $\frac{5}{3}$, $\frac{7}{3}$, $\frac{8}{5}$ etc. as well as large even fractions $\frac{5}{2}$, $\frac{7}{2}$, $\frac{9}{4}$ etc. for various choices of N and r.

We also mentioned above that large even fractions (from $\lambda_L^{(2)}$) occur by relaxing (7.60), i.e. let us say by considering a double-layer 2DES. The fraction $\frac{5}{2}$ in $\lambda_L^{(2)}$ occurs for $N = r = 8$. In this case one may even get $g_{12} = \lambda^{(1)} = \frac{1}{2}$, which corresponds precisely with the new *double-layer* 2DES FQHE state recently reported by Suen et al. [44] and Eisenstein et al. [45].

7.4.5 Wigner Solids

We will now reinterpret the present result in terms of a topological or geometric picture, thereby tying this discussion to the elusive Wigner crystal hypothesis.

First we may directly realize that the choice $r = s = \binom{N}{2}$ for the number of correlations between N fundamental fermionic objects bears an analogy with the construction of braids, c.f. how the bobbins are moved in lacemaking by means of half- and whole stitches in terms of crosses and twists. However, a more direct pictorial realization of the extreme case may add a different flavour to the subsequent "order" as well as indicating a possible intuition as regards the projection of fractionality. First a brief introduction.

We will associate an N-fermion system with N vertices in an $N-1$-dimensional space, and we will need the following definitions:

Polytope: A generalization of the terms point, segment, polygon and polyhedron in higher space. A polytope can be subdivided into a number of congruent 'characteristic' simplexes. A simplex is a generalization of an interval (1-simplex), triangle (2-simplex), tetrahedron (3-simplex), etc. A p-simplex Fig. 7.8, is defined through its $p+1$ vertices and is a convex set of $\mathbf{R}^m$, $m \geqslant p$. One may also introduce the notation α_n for a regular n-simplex, e.g. an equilateral triangle for $n = 2$, etc. Further by extending a tetrahedron by joining a fifth point (outside its 3-space) we can for instance construct a pentatope, and similarly, hexatopes, heptatopes and so on. For more details on regular polytopes and their history we refer to Coxeter [46].

In order to investigate whether it is possible to associate an ordered configuration of fermions in a higher dimensional space with a regular polytope, we will focus on the extreme case and then speculate whether this "organization" may have something to do with the elusive Wigner solid. Note that the present interpretation, although stimulated by quantum mechanics, has a specific "classical" flavour.

As a starting point we choose $N = 2n$, $n = 2, 3, \ldots$ fermions, subject to the Pauli principle and a further commensurate set of reduced density matrices $\Gamma^{(p)}$ with $p = 1, 2, \ldots, N$, see Eqs (6.27), (6.30) and (6.69–6.72). Since our discussion refers to situations where ODLRO may develop, it is a reasonable assumption to restrict our formulation to the extreme case Eqs. (6.56–6.58). A generalized ordering in a higher dimensional space would then be developed in terms of the number of vertices defined by N (fermionic) particles, defining at most an $N-1$ dimensional space (N – 1-simplex) with $s = \binom{N}{2}$ edges, $\binom{N}{3}$ triangles, $\binom{N}{4}$

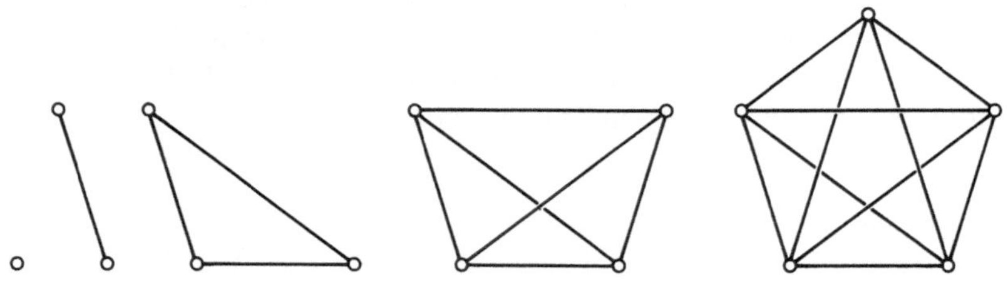

Fig. 7.8. Perspective view of the first regular p-simplexes. Note that the equilateral triangle has been distorted so as to emphasize its occurrence as a face of the next one, and so on

tetrahedrons, etc. Note that the extreme case corresponds to a highly correlated situation (involving all N particles), so it is not possible to reduce the $N-1$-simplex (spanned by its N vertices) to lower dimensions. It also follows that the occupation number is $\lambda^{(1)} = N/2s$, see e.g. (6.75), which with $s = \binom{N}{2}$ equals $1/(2n-1)$, $n = 1,2,\ldots$. We earlier analysed FQHE along this approach [10] and found excellent agreement with experiment. However, as this analysis was more algebraic than topological, we will rederive some of these results within the present topological framework and see how a possible "ordered structure" of the fermions, might be deduced from a simple generalization of polyhedrons to a higher dimensional (geometric) space. We will suggest this to be related to a (generalized) Wigner solid.

Thus for $N = 4$ we associate the "structure" with 4 points, which defines a tetrahedron in a 3-dimensional space with 6 edges and 4 equilateral triangles. Since the Hall effect amounts to a 2-dimensional projection (in geometrical space), corresponding to the appropriate Landau levels, we will be interested in how these 4 particles (originally fermions), ordered through the extreme situation, will distribute and manifest themselves in the experiment. The extreme situation with the inherent statistics, suggests that the 4 particles, represented by the vertices of the tetrahedron, distribute equally on the 4 triangles. Each triangle will precisely correspond to one particle and with three vertices the particle fractionizes as found by Tsui et al. [39].

To continue by analogy, we thus need to look at the equal distribution of $N = 2n$ particles, associated with a regular $N-1$-simplex with one particle in each regular $N-2$-simplex (defined by $N-1$ vertices). A two dimensional projection may at most give an $N-1$-polygon over which the particle fractionizes. Hence for $N = 6$, i.e. for an extreme ordering corresponding to 6 points, we obtain a hexatope with 15 edges, 20 triangles, 15 tetrahedrons and 6 pentatopes. The six particles distribute equally over the six pentatopes. Every pentatope has 5 vertices (over which the fermion fractionizes) and consists of 10 edges, 10 triangles and 5 tetrahedrons. However, since 6 independent pentatopes correspond to 60 edges but in the hexatope only produce 15 different edges, every edge in the hexatope must be met by 4 surfaces or double as many as in the tetraeder. A 2-dimensional projection of the pentatope (containing each a particle), see also the analogy above, may at most be a pentagon. This suggests 1/5 fractions and since every edge in the hexatope is met by twice as many surfaces as in the tetraeder case, the two-dimensional projection of a hexatope, with each particle confined to one pentatope, may thus produce 2/5 fractions, but other multiples of 1/5 are of course also possible.

It is now clear how the extreme ordering (or generalized Wigner solid) appears. For $N = 8$, i.e. 8 points defining the octatope with 28 edges, 56 triangles, 70 tetrahedrons, 56 pentatopes, 28 hexatopes and 8 heptatopes, we will have precisely one particle distributed over each heptatope (with 7 vertices and 21 edges). We conclude that a 2-dimensional projection may at most produce heptagons and 1/7 fractions. In analogy with the above, we find that 8 independent heptatopes correspond to 168 edges, but with only 28 edges available in the octatope each edge meets 6 surfaces or three times as many as in the tetraeder. Hence a 3/7 fractions would be enhanced but also other fractions derived from 1/7 may occur.

Although the correct statistics was given at the outset, the extreme situation, the fractionizations and projections yields topological units with fractional statistics of anyon type. We will not go into these details here except suggesting that the present visualisation of coherent-dissipative structures seem to favour a Wigner-solid-type ordering as originally proposed by Tsui et al. [39].

7.5 Spontaneous and Stimulated Emission of Radiation in Masers

As we have seen from the previous Chapter a new component in modern natural sciences have entered the stage, namely the field of complexity. Complex phenomena are seen every day, yet the simplifications introduced to define a physical system lead us most of the time to a limited "linear thinking". The novelty and importance of a more basic view-point has been particularly emphasized and developed by the Brussels School, see Chap. 6, [1–4]. The new paradigm, following Kuhn [47], forces us to rethink the way we look at reality and to focus on the understanding of complex systems and cooperative behavior [48].

We have also seen how many surprizing applications can be obtained from the present theory. We have emphasized that our subdynamics view follow from a microscopic approach, based on the complex scaling method, (CSM), a fundamental hierarchy of reduced density matrices in the so-called extreme configuration, an algebraic structure involving Jordan forms and associated strongly singular distributions. Since the theoretical development here concerns complexity rather than a quantal or classical formulation we will rewrite the equations so that Planck's constant will not appear, except that it may be easily included when quantum conditions have to be explicitly considered. In doing so we will treat the case of a relativistic beam of electrons subject to a constant magnetic field and an applied auxiliary rf field. This involves not only gyrotron design in the Cyclotron Maser Concept, (CMC), but also a complex system lending itself to simultaneous classical and quantum descriptions as far as particle-field interactions are concerned.

7.5.1 Einstein Relation

One of the key issues in the development of a satisfactory theory for the interaction between light and matter is the notion of spontaneous and stimulated emission. There are many treatments and discussions on this problem which started already with Einstein's famous rederivation of Planck's radiation law and the introduction of the famous A and B coefficients, see e.g. Bohm [49] for more details.

In a "classical" description of black-body radiation at thermal equilibrium, the intensity $I(\nu)$ of radiation per unit solid angle and any given polarisation is given by

$$I(\nu) = \frac{h\nu^3}{c^2}(e^{\frac{h\nu}{k_B T}} - 1)^{-1} \tag{7.62}$$

where ν is the appropriate frequence for a transition, h is Planck's constant and c the velocity of light, and $k_B T$ as before.

To calculate the probability of emission one usually assumes a Maxwellian distribution for the probability p_n of finding an atom in a given stationary state n and then one proceeds to obtain the sought-after probability by adding two parts, i.e. an induced or stimulated part $AI(\nu)$, proportional to $I(\nu)$, and the spontaneous contribution, B, independent of the intensity. The equilibrium condition for transitions between stationary states n and m then reads

$$p_n AI(\nu) = p_m(B + AI(\nu)); \;\; p_m/p_n = \exp - (h\nu/k_B T) = p \tag{7.63}$$

from which one gets $B + AI(\nu) = AI(\nu)\exp(\frac{h\nu}{k_B T})$ or

$$\frac{B}{A} = \frac{h\nu^3}{c^2} \tag{7.64}$$

We have put "classical" within quotation marks, since the famous Einstein relation between spontaneous and stimulated emission involving the intensity $I(\nu)$ and the famous A and B coefficients, i.e.

$$\frac{AI(\nu)}{B} = \frac{1}{\exp(h\nu/k_B T) - 1} = \delta \tag{7.65}$$

is equivalent to the (quantum) Planck radiation law for energy quantized harmonic oscillators.

7.5.2 Cyclotron Maser Concept

With this short introduction in mind we will quickly review the present situation. Since lasers have been discussed in other Chapters, we will briefly concentrate on the maser (microwave amplification by the stimulated emission of radiation) mechanism.

In a maser, i.e. in the application of electron beams for the generation and amplification of electromagnetic radiation it is desirable to have the (relativistic) particle beam emit energy to the resonator field, while in an accelerated beam one would like to have essentially the converse situation, which hopefully should lead to an improved energy spread in electron energy (or velocity).

The advent of intense pulsed relativistic electron beams has renewed interest in the cyclotron maser and corresponding instability questions. At the same time the invention of electron cooling has suggested new ways of improving the properties of circulating particle beams. The combination of the two ideas, namely to force radiation cooling due to enhanced emission of cyclotron radiation may lead to new maser cooling principles which, if applicable to circulating particles in a storage ring, should lead to dramatically refined and organized beam qualities.

A historic account should start with the classical linear mechanism for a possible electron cyclotron maser, first proposed by Twiss [50] in 1958, followed, one year later, by Schneider's [51] quantum mechanical (linear) formulation based on the relativistic Landau oscillator. The general nonlinear theory, including the importance of relativistic effects associated with gyrating electrons about an external magnetic field, was finally reviewed and thoroughly treated by Sprangle and Drobot [52].

The observed maser action, see [53] for the earliest definite confirmation of the electron maser mechanism, is mainly due to relativistic corrections, which remove the degeneracy in the level spacings for free electrons in a magnetic field. If higher magnetic moments are in sufficient population excess, the system can support enhanced stimulated emission of electric dipole radiation at the electron cyclotron frequency.

The work of Schneider [51] indicated that a net stimulated emission should be possible in certain regimes provided the linear mechanism for the maser instability is valid. Starting from Schneider's result, Ikegami [54] developed the "cyclotron maser cooling" (CMC) principle by assuming that the loss of transverse energy through cyclotron radiation should equal the radiated net transfer of power in the quantum mechanical linear theory.

A simple schematical setup for the present discussion is given by Fig. 7.9. Since the maser action is associated with relativistic effects we have to distinguish between the relativistic gyrofrequency, which depends on the increase of particle mass m^* due to the transverse kinetic energy of the beam particles, and the nonrelativistic frequency $\omega_c = eB/m_0$, compare also discussion of the QHE, where $-e$ is the electron charge, B the magnitude of the magnetic field and m_0 the electron rest mass.

The nonlinearity introduced by the relativistic dependence leads directly to the famous gainfunction, see Fig. 7.10. The present format will not allow a detailed derivation of this result, however, a few comments can be made. First a classical dynamical formulation follows directly from the relativistic Hamiltonian

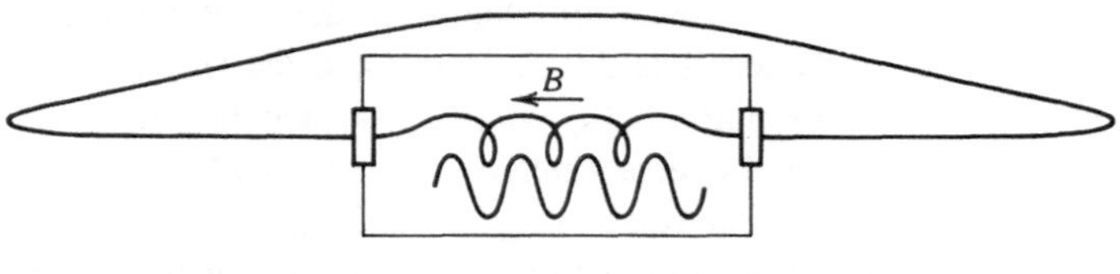

Fig. 7.9. The Cyclotron Maser Concept (CMC) applied to a circulating beam. The beam is deflected before and after the "CMC" section. The particles gyrate in the static and homogeneous field along the main direction of the originial beam propagation. An oscillating electric field oriented perpendicular to the constant field interacts with the gyrating beam particles

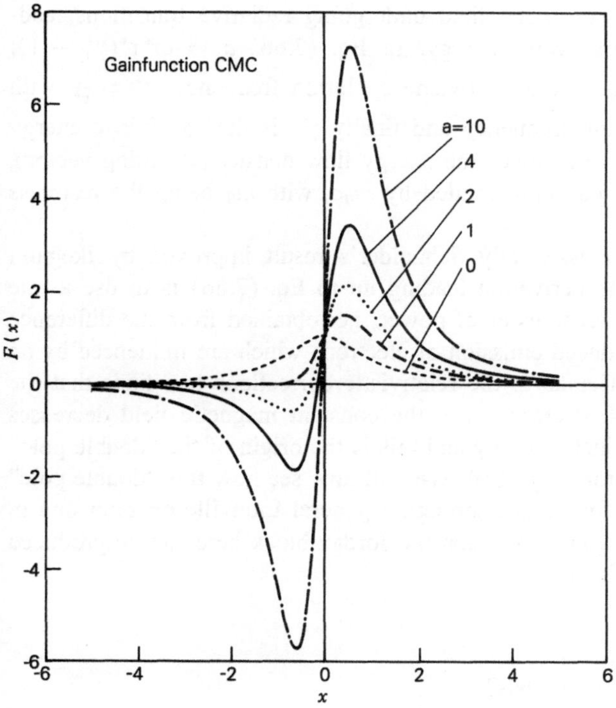

Fig. 7.10. A normalized gainfunction displayed for different values of the parameter a. x is the difference between the gyro-frequency and the frequency of the auxilliary rf field, multiplied by the lifetime, see text

described by the potentials $(\vec{A}, \phi)$ i.e.

$$H = \sqrt{m^2c^4 + \left(\vec{P} - \frac{e}{c}\vec{A}\right)^2 c^2} + e\phi$$

where the vector potential $\vec{A} = A_y\hat{y}$ contains the appropriate components, i.e. the constant magnetic field in let us say z-direction and the field distribution for a TE_{10p} wave with the indices $(1, 0, p)$ denoting the particular external cavity mode.

The first predictions related to a possible "negative absorption" were, as stressed already, given by the classical analysis due to Twiss [50] and by Schneider's [51] celebrated quantum mechanical treatment of the relativistic Landau oscillator perturbed by an oscillating electromagnetic field. The latter result is so fundamental that the paper should be studied carefully by anybody interested in CMC. Suffice it to say here that one can derive the rate formula

$$\frac{d\gamma_\perp^*}{dt^*} \propto \tau \left\{ \frac{1}{1+x^2} + \frac{2ax}{(1+x^2)^2} \right\} \tag{7.66}$$

for particles gyrating in a magnetic field undergoing radiative transitions leading to a change of the transverse energy. In Eq. (7.66) $a = \omega_{\rm L}^* \tau^* (\gamma_\perp^* - 1)$; $x = \tau^*(\omega_{\rm L}^* - \omega)$, $\omega_{\rm L}^* = \frac{\omega_c^*}{\gamma_\perp^*}$ is the relativistic cyclotron frequency, $\tau^* = \frac{1}{\nu^*}$ with ν^* being the phase bunching frequency and finally $\gamma_\perp^*$ is the relativistic energy factor. The "constant" above involves the energy flow density (Poynting vector), the classical radius of the particles divided by $m_0 c$, with m_0 being the restmass and c the speed of light.

The formula above is essentially Schneider's result improved by Ikegami [54]. The main trick in the derivation leading up to Eq. (7.66) is to use a rate formula derived from the net transfer of power, i.e. obtained from the difference between absorption and induced emission of electrons which are influenced by an alternating electric field. Thanks to the relativistic corrections one finds that the energy level spacing for free electrons in the constant magnetic field decreases for increasing transverse kinetic energy and this is the origin of the "double pole" behaviour of the gainfunction, Fig. 7.10. We will now see how this "double pole" behaviour can be directly formulated through a general Liouville operator and in doing this it is quite surprizing to see that the Jordan block here can be produced by a relativistivic mechanism!

7.5.3 Liouville Formulation of CMC

Since the CMC problem is formulated in the Liouville picture we will transfer the previous correlation matrix formulation, Eqs. (7.27–7.33), see also the introductory discussion in Sect. 7.3, to a super-operator framework which exhibits the same analogous structure as above.

Since our aim is to obtain a Liouville equation valid in the classical as well in the quantum domain, we will interchangingly proceed to classical limits when appropriate. We will also couple the formulation to a suitably defined impedance or gainfunction.

We will denote the density matrix corresponding to a transition (system absorbs) between the Landau levels $i+1$ and i by $\|h_{k(i)}\rangle\rangle = |\{i+1\}(k)\rangle\langle i(k)|$, where k denotes a particular transition involving Landau levels i and $i+1$ characterized by the angular frequency $\omega_{i+1,i} \approx \omega_{\rm L}$ given by the eigenvalue (of the "zero order" Liouvillian) $\hbar\omega_{i+1,i} = E_{i+1} - E_i$.

In the same way as before we obtain a super-operator transformation between the transition matrices $\|\mathbf{h}\rangle\rangle, \|\mathbf{f}\rangle\rangle$ and $\|\mathbf{g}\rangle\rangle$ via the $r \times r$ transformation matrices $\mathbf{B}^{-1}$ and $\mathbf{B}$, respectively. In the basis $\|\mathbf{f}\rangle\rangle$ the transitions are correlated and in $\|\mathbf{g}\rangle\rangle$ coherent according to the properties of respective transformations. For example, $\|g_1\rangle\rangle$ describes simultaneous coherent transitions by the frequence $\omega_{i+1,i}$, while $\|f_1\rangle\rangle$ represents transitions that are phase correlated through the factors exp $(\frac{i\pi}{r}(k-1)), k = 1,2,\ldots r$. Note also that $\langle\langle\mathbf{g}\|$ and $\langle\langle\mathbf{f}\|$ describe coherent and correlated emissions with the eigenvalues $\hbar\omega_{i,i+1,} = E_i - E_{i+1}$ of the corresponding (zero order) Liouvillian.

Absorption and emission are therefore intrinsically connected through the bra-ket correlated transformations based on $\|\mathbf{h}\rangle\rangle\mathbf{B}$ and $\|\mathbf{h}\rangle\rangle\mathbf{B}^{-1}$. We also remark that the usual transformation property for timereversal, i.e. $t \to -t$; $\hat{L} \to -\hat{L}$, will not work here, since we will have an intrinsic coupling with the environment through the new terms, see below, obtained from the dissipative dynamics.

In the present formulation the scalar product hence becomes

$$\langle\langle h_{k(i)}\|h_{l(j)}\rangle\rangle = Tr\{h^{\dagger}_{k(i)}h_{l(j)}\} \tag{7.67}$$

One should observe here that in general one must use the generalized scalar product for $\|\mathbf{h}\rangle\rangle$ complex

$$\langle\langle h^{*}_{k(i)}\|h_{l(j)}\rangle\rangle = Tr\{h^{*}_{k(i)}{}^{\dagger}h_{l(j)}\} \tag{7.67$'$}$$

Since our reference basis is chosen real we may not dwell on this complication here. Suffice it to say that meaningful scalar products should in general be taken between components of the two conjugate Hilbert spaces and all vectors in the bra position should have a complex conjugate sign to show that we take a component from the dual space, see Chap. 6, [11, 27] for details. Note that we are here treating non-selfadjoint Liouvillians with real reference spaces, and that unitary transformations in this context therefore could lead to nonsymmetric Jordan blocks. Since the density matrix in the extreme case (the analogy with spontaneous and stimulated transitions will be made below) develops a large component and a small one, it is clear that the corresponding super selection sectors become entangled during thermalization. The occurrence of Jordan blocks thus implies important consequences for the dissipative dynamics.

We will now proceed to construct the relevant Liouvillian $\hat{L}$ commensurate with the above mentioned Jordan block structure. Note that we will obtain a time independent effective Liouvillian even if the original Liouvillian may contain time dependent parts. We will return to this question below. In analogy with the previous section we write (note that Γ here means the resonance width related to the characteristic (collision or phase debunching) lifetime of the state via $\frac{\hbar}{\Gamma} = \tau$ not to be confused with the density matrix which have indices S or L)

$$\hat{L} = (z_{\mathrm{L}} - i\Gamma)\hat{I} + \hbar\omega_{\mathrm{S}}\hat{J} \tag{7.68a}$$

$$\hat{I} = \sum_{k=1}^{r}\|g_k\rangle\rangle\langle\langle g_k\| = \sum_{k=1}^{r}\|f_k\rangle\rangle\langle\langle f_k\|, \; \hat{J} = \sum_{k=1}^{r-1}\|f_k\rangle\rangle\langle\langle f_{k+1}\| \tag{7.68b}$$

In Eq. (7.68a) $z_{\mathrm{L}} = \hbar\omega_{\mathrm{L}}$ with ω_{L} being the characteristic angular frequency asssociated with a specific transition in question, and finally ω_{S} to be determined. One possibility would be to define a relevant time scale through $\hbar\omega_{\mathrm{S}} = \Gamma$. This choice follows from the uncertainty relation and it has some experimental justification [11].

Another choice would be to deduce a relation (from inspection with Eq. (7.31), i.e. that the large component (in the identity operator) scales with the

Jordan block component as

$$\omega_S = \omega_L \cdot \frac{\lambda_S}{\lambda_L} \tag{7.69}$$

which means that the quotient between ω_S and ω_L can be expressed in terms of N, the number of particles in the system and r, the dimension of the pairspace, see Eq. (7.27). This might be relevant in connection with spontaneous and stimulated emission. Both choices may be mutually compatible but in general we must check that

$$\omega_S = 1/\tau \tag{7.70}$$

where $\tau = \hbar/\Gamma$ is the microscopic time scale commensurate with the relevant relaxation time.

Our density matrix is now properly specified. Although this construction appears quite ad hoc – our intention is clearly not to derive the details of the full Liouvillian – but rather to see how the second term in (7.68a) arises from first principles and further to demonstrate the subsequent consequences with respect to its Liouvillian properties.

In the discussion of the exact underlying principles for this construction we must realize that the Liouvillian in e.g. the CMC case will be time dependent through a perturbing electromagnetic rf field. Of course one might consider to derive appropriate equations from general partitioning techniques with the attempt to continue analytically the nonlinear equations via well-defined regularizations or dilations. This complication can be dealt with by considering general correlation functions and associated spectra, but here we will only briefly express the abstract resolvent equations for a general Liouvillian referring to a particular timescale, where explicit time dependencies on a shorter scale have been averaged out.

The full nonlinearity of the problem can now in principle be explicitly specified giving the precise condition for the characteristic angular frequency as well as the one determining $\Gamma = \frac{\hbar}{\tau}$. This follows from the unique analytic continuation of the Liouville problem, projected onto $\hat{P}$

$$\{\hat{P}\hat{\mathscr{L}}\hat{P} + \hat{\Psi}(z) - z\hat{\mathscr{I}}\}\|f_l\rangle\!\rangle = 0; \; l = 1, 2, \ldots, r. \tag{7.71a}$$

in which $\hat{P} = \sum_{k=1}^{r} \|g_k\rangle\!\rangle\langle\!\langle g_k\| = \sum_{k=1}^{r} \|f_k\rangle\!\rangle\langle\!\langle f_k\|$ with $\hat{P} + \hat{Q} = \hat{\mathscr{I}}$ and

$$\hat{\Psi}(z) = -\hat{P}\hat{\mathscr{L}}\hat{Q}(\hat{Q}\hat{\mathscr{L}}\hat{Q} - z\hat{\mathscr{I}})^{-1}\hat{Q}\hat{\mathscr{L}}\hat{P} \tag{7.71b}$$

In the equations above the Liouvillian $\hat{\mathscr{L}}$ describes (in principle) the complete physical mechanisms, including correlations, relativistic effects etc. Further $\hat{\mathscr{I}}$ is the complete unit operator and Eqs. (7.71a–b) give the condition for ω_L and Γ including in principle all the non-linearities of the problem. The construction in Eq. (7.68) is obtained from Eq. (7.71) by the choice $\hat{P} = \hat{I}$ and $\hat{L} = \hat{P}\hat{\mathscr{L}}\hat{P} + \hat{\Psi}(z_L - i\Gamma)$. See also Chap. 6, [16] for more details.

Before we continue to demonstrate how the associated dynamics of the present model evolves, we will go back to the particular situation experienced in CMC. At first we note the enormous degeneracies of the Landau oscillator problem, which, however, are coupled to decreasing level spacings as energy increases. Secondly, this implies that stimulated absorption and emission must be correlated so that excited oscillators in some specified energy levels should be phase correlated with the corresponding oscillators, already situated at higher energies, with appropriate de-excitation properties for stimulated emission to occur. Obviously this collective effect could lead to decrease of energy spread and thus organization. The intriguing possibility that appears here is the following. The "classical" Maxwellian probability is given by p in Eq. (7.63). Further Eq. (7.65) couples Planck's radiation law with Einstein formula which indeed transcends to the quantum regime. The interesting analogy with the previous derivations follows now by considering the correlation matrix, cf. discussions in the beginning of sect. 7.3, with diagonal elements p and offdiagonal elements $p(1-p)$, where the latter corresponds to the phase correlated coupling between a particular oscillator being excited and another one being de-excited.

It is a remarkable coincidence that this simple correlation matrix has the same algebraic structure as the one appearing in condensed matter, see sect. 7.3, albeit for a weakly correlated situation. Thus we obtain

$$\lambda_{\mathrm{L}} = sp - (s-1)p^2, \ \lambda_{\mathrm{S}} = p^2 \tag{7.72a}$$

where in the weakly correlated regime $s \approx r \approx N/2$. From relation Eq. (7.69) one obtains

$$\omega_{\mathrm{S}}/\omega_{\mathrm{L}} = \delta/s \approx 2 \cdot \delta/N \tag{7.72b}$$

Eqs. (7.68–7.72) will be shown to lead to a general impedance or gainfunction, cf. the classical formulation of CMC [50–54], but at the same time coupling the formulation, through the present dissipative subdynamics model, to a (self-) organization property.

7.5.4 Self-Organization

We will now see how the Liouvillian, obtained in the previous section directly connects a suitable gainfunction with organization through the mechanism of spontaneous and stimulated transitions. We will start with the general relation between the propagator and resolvent ($t > 0$ and C^+ contour in the upper half plane and $\hbar = 1$, see Chap. 6,

$$e^{-i\hat{L}t} = \frac{i}{2\pi} \int_{C+} (z - \hat{L})^{-1} e^{-izt} dz \tag{7.73}$$

Equation (7.73) is convenient when working directly with angular frequencies and inverse lifetimes. For the remainder of this section we will, however, insert $\hbar$ again in order to connect with the Liouvillian of the previous section. Regarding the aforementioned problem of timescales, it is easy to see that one can replace the lefthand side of Eq. (7.73) with a general expression corresponding to a more complicated correlation function. Through appropriate dispersion relations a relevant timeindependent operator can be constructed so that Eq. (7.73) and in principle Eq. (7.71), can be expressed without contradictions, albeit within a generalised analytical context.

Thus leaving this technicality aside as solved in principle, the resolvent becomes

$$(z-\hat{L})^{-1} = \frac{1}{z-(\hbar\omega_{\rm L}-i\Gamma)}\hat{I} + \frac{\hbar\omega_{\rm S}}{(z-(\hbar\omega_{\rm L}-i\Gamma))^2}\hat{J} + \cdots \frac{(\hbar\omega_{\rm S})^{r-1}}{(z-(\hbar\omega_{\rm L}-i\Gamma))^r}\hat{J}^{r-1} \tag{7.74}$$

and the propagator

$$\exp\left(-\frac{i\hat{L}t}{\hbar}\right) = \exp(-i\omega_{\rm L}t)\exp\left(\frac{-\Gamma t}{\hbar}\right)\cdot\exp(-i\omega_{\rm S}\hat{J}t)\hat{I} \tag{7.75}$$

with

$$\exp(-i\omega_{\rm S}\hat{J}t) = \left(1 - i\omega_{\rm S}t\hat{J}\cdots + \frac{(-i\omega_{\rm S}t)^{r-1}}{(r-1)!}\hat{J}^{r-1}\right)$$

The formula can be obtained either from simple matrix algebra or from the general resolvent propagator expression above.

Let us briefly look at the transitions described by $\|f_1\rangle\!\rangle$ correlated with the transitions $\|f_k\rangle\!\rangle$ characterized by the operator $\hat{J}^{k-1}$ with $k = 2,3,\ldots,r$. Note that the characteristic (collision) lifetime τ is still given by $\frac{\hbar}{\Gamma}$. The time rule for, let us say, transitions involving $\|f_1\rangle\!\rangle$ and $\|f_k\rangle\!\rangle$ are given by (note that the present resonance model is only realistic or relevant "in between" small and large times, i.e. does not describe the evolution for very short or very large times compared to multiples of $\frac{\hbar}{\Gamma}$)

$$N(t) = \left|\left\langle\!\!\left\langle f_1 \left\| \exp\left(-\frac{i\hat{L}t}{\hbar}\right)\right\| f_k\right\rangle\!\!\right\rangle\right| \propto t^{k-1}\exp\left(-\frac{\Gamma t}{\hbar}\right) \tag{7.76a}$$

which yields

$$dN \propto t^{k-2}\left(k-1-\frac{\Gamma t}{\hbar}\right)\exp\left(-\frac{\Gamma t}{\hbar}\right)\cdot dt. \tag{7.76b}$$

Equations (7.76a–b) show that $dN > 0$ for $t < (k-1)\cdot\frac{\hbar}{\Gamma}$, $k = 2,3,\ldots,r$. Thus the occurrence of Jordan blocks in the Liouville picture leads to increase of $N(t)$ for the times specified above, i.e. an increase of number of particles in the correlated state given by $\|f_1\rangle\!\rangle$ by correlated transitions from all other states $\|f_k\rangle\!\rangle, k = 2,\ldots,r$. In general there may be a certain balance between increase and decrease in coherence, correlation and dissipation. Since there is an overall decay out of $\hat{I}$ given by

$$N(t) = \left| \left\langle\!\left\langle f_1 \right\| \exp\left(-\frac{i\hat{L}t}{\hbar}\right) \left\| f_1 \right\rangle\!\right\rangle \right| \propto \exp\left(-\frac{\Gamma t}{\hbar}\right) \tag{7.77a}$$

yielding the standard exponential decay rule

$$dN \propto \left(-\frac{\Gamma}{\hbar}\right) \exp\left(-\frac{\Gamma t}{\hbar}\right) \cdot dt. \tag{7.77b}$$

the summing up of all contributions or transitions to (or from) the correlated state $\|f_1\rangle\!\rangle$ may lead to dissipative structures which during certain timespans of order $r\tau$ have increasing order etc., and in this sense we may speak of self-organisation on a microscopic level "created" by the Jordan block term in $\hat{L}$. In the following subsection we will derive the relation between these Jordan blocks and general impedance or gainfunctions.

7.5.5 Connection with the Gainfunction

A typical gainfunction in the cyclotron maser problem was given by Schneider [51], see Fig. 10. Based on Schneider's work, Ikegami [54] expanded the expression for the radiated net transfer of power in the quantum mechanical linear theory as well as attempted a derivation based on classical expressions involving specifically obtained friction terms.

We may also compare the Schneider expression where $a = Q\frac{W}{m_0c^2}$, $Q = \omega_0\tau$ and $x = (\omega_{i+1,i} - \omega)\tau$, with τ being the characteristic time, W the kinetic energy of the electron and ω_0 the cyclotron frequency.

From Eq. (7.71) we obtain the nonlinear equations determining $\hbar\omega_{\mathrm{L}} - i\Gamma$. The analytic continuation – necessary to obtain the complex solutions corresponding to the full subdynamics – rests on the simple dispersion relation

$$\Psi(\hbar\omega+i0) = \mathscr{P}(\hat{P}\hat{\mathscr{L}}\hat{Q}(\hbar\omega-\hat{Q}\hat{\mathscr{L}}\hat{Q})^{-1}\hat{Q}\hat{\mathscr{L}}\hat{P}) - i\pi\delta(\hat{P}\hat{\mathscr{L}}\hat{Q}(\hbar\omega-\hat{Q}\hat{\mathscr{L}}\hat{Q})\hat{Q}\hat{\mathscr{L}}\hat{P})$$

with $\mathscr{P}(\Psi)$ and $\delta(\Psi)$ denoting the principal part and the delta function contribution of Ψ respectively.

Using the Obcemea-Brändas construction, Chap. 6, [16], the $\delta(\Psi)$ contribution can be retrieved from the corresponding Hamiltonian dynamics via a dispersion relation for $\bar{H}(E + i0) = H + HG(E + i0)H$, where $G(E + i0)$ is a suitably defined reduced resolvent and $H = H_0 + V$ with V a suitable optical potential. For a sufficiently monochromatic beam of wavepackets φ with kinetic energies given by H_0, one can identify $\bar{H}(E + i0)$ with the t-matrix, i.e. $\bar{H}(E + i0) = H_0 + V + VG(E + i0)V = H_0 + t(E + i0)$ with G now being the full resolvent $G = (E + i0 - H)^{-1}, t = 1/2\pi(\hat{S} - \hat{\mathscr{I}})$ and $\hat{S}$ being the scattering matrix.

Eigenvalues of $\bar{H}(E)$ are then given by $E_i(E) - i\varepsilon(E)$ with

$$E_i(E) = \langle\varphi_i|H_0 + V + \mathscr{P}(V(E - H)^{-1}V)|\varphi_i\rangle$$

and

$$\varepsilon_i(E) = \pi(\varphi_i|V\delta(E - H)V|\varphi_i).$$

Further the construction amounts to finding an analytic continuation based on $\hat{\mathscr{L}}$ with eigenvalues

$$E_i(E) - E_j(E) - i(\varepsilon_i(E) + \varepsilon_j(E)).$$

We hence note that $\varepsilon_i(E) = \pi\{-\Im t_{\varphi_i\varphi_i}\} \propto \sum_\gamma \sum_{i\to\gamma}$, where $\sigma_{i\to\gamma}$ is the cross section for the transition $i \to \gamma$ and the sum is the total cross section. The construction above further demonstrates that the analytical structure based on the Liouvillian and originating from dilatation analyticity on the Hamiltonian level not only implies a nonfactorizable entanglement between the system and its environment but also includes a new eigenvalue spectrum. It is easy to see that the Obcemea-Brändas construction suggests the existence of Liouvillian eigenvalues which cannot be traced back to e.g. differences between (dilated) Hamiltonian spectra or corresponding integrable quantities in the Poisson formulation. The fact that correlations and relativistic effects may lead to Jordan blocks imply that higher order singularities, or related products of δ-functions and its derivatives in the corresonding dispersion relations, indeed signify a breakdown of traditional descriptions and the emergence of complex behavior and possible self-organization.

Therefore the fundamental nonlinearities of the CMC problem require extra work in the degeneracy space $\hat{I}$. Under "normal" conditions, i.e. with no Jordan blocks present, the dispersion relation for $(z - \hat{L})^{-1}$ should give the projector associated with $\hat{P}$, see Eq. (7.71), and hence trivially only "heating" of the particle beam is possible. We therefore first conclude that the absorption curve for the electron beam in the CMC experimental setup, see e.g. Fig. 7.10, should follow a Lorentzian curve with $a = 0$.

Next we need to study the resolvent $(z - \hat{L})^{-1}$, including the Jordan block term, in order to find the appropriate evolution commensurate with the degenerate root manifold spanned by $\|\mathbf{f}\rangle\rangle$. Thus we obtain from Eq. (7.74)

$$(z - \hat{L})^{-1} = \frac{\tau}{\hbar}\left\{\frac{1}{(y+i)}\hat{I} + \frac{\omega_S\tau}{(y+i)^2}\hat{J} + \cdots\right\} \tag{7.78}$$

with $y = \tau(z/\hbar - \omega_L)$. Intuitively we might conclude that we only have to put $y = -x$, with x defined in connection with the previous identification with Schneider's gaincurve or Fig. 7.10. Even if the contributions to the rate, see below, will look similar in nature, there will be some important differences. First of all it is important to realize that ω_L, obtained from Eq. (7.71) *is not equivalent to* Ikegami's ω_L^*, or the gyrofrequency. Our ω_L is the energy parameter divided by $\hbar$ and thus contains relativistic corrections, with the leading terms obtained from the relativistic Schrödinger (or Klein-Gordon) equation, small collective effects, and possibly contains a weak interaction with the longitudinal degrees of freedom. Furthermore, there are higher order terms in the resolvent expansion Eqs. (7.74–7.75) which might completely blur out the simple and clear cut information produced by a "well-behaved" gainfunction. It is nevertheless an interesting coincidence that such a gainlike function indeed can be obtained here. To see that consistent gains can be obtained, which has the desired domain of negative absorption, we briefly return to our scattering theoretical review above.

Since the cross section (or associated spectral density) refers to taking the imaginary parts of the factors in front of $\hat{I}$, $\hat{J}, \ldots, \hat{J}^r$ (the correct spectral density would of course be derived from a correlation function of the type discussed earlier, but from the mathematical point of view the present development is essentially correct) we can find the contributions directly, i.e. from the first term

$$\frac{d\gamma_\perp^*}{dt^*} \propto \langle\langle f_1 \| \Im\{-(z-\hat{L})^{-1}\} \| f_1 \rangle\rangle = \frac{\tau}{\hbar} \frac{1}{(y^2+1)} \tag{7.79}$$

and from the second term (and similarly for higher orders)

$$\frac{d\gamma_\perp^*}{dt^*} \propto \langle\langle f_1 \| \Im\{-(z-\hat{L})^{-1}\} \| f_2 \rangle\rangle = \frac{\tau}{\hbar} \left\{ \left(\frac{2\omega_S\tau \cdot y}{(y^2+1)^2} \right) \right\} \tag{7.80}$$

We note that for $\omega_L > z/\hbar$, or $y < 0$, we obtain a negative absorption which is characteristic of a gainfunction. We can also compare with Schneider's and Ikegami's expressions and it might be tempting to estimate the value of $a = \omega_S\tau$, determined from Eq. (7.72b), i.e.

$$a = \omega_S\tau = \omega_L\tau \cdot 2\delta/N$$

which is related to Schneider's relativistic factor $\omega_0\tau \cdot W/m_0c^2$($W$ is the kinetic energy of the electrons) or Ikegami's parametrization.

The gainfunction has been a key quantity in accelerator physics, since it suggests a possibility to improve the quality of the accelerated beam. It is important, however, to realize that this property alone may not coincide with cooling, since it does not guarantee organization or increase of phase space density.

Nevertheless, the absolute size of $a = \omega_S\tau$ may in the present context, note that (7.70) can be overcome by $\tau \to r\tau$ from (7.76), still be of fundamental importance for any cooling effects to appear. This conclusion is only indirectly drawn from Eq. (7.80), since $|a|^2$ appears in the proportionality factor of Eq. (7.76), see also (7.75).

7.5.6 Conclusion

The present approach was also, see earlier sections, applied to condensed matter situations, where quantum effects were prominent, see earlier sections. The point, we want to make in this final subsection on CMC, concerns the possibility of making a simultaneous classical and quantum formulation. We have emphasized the cyclotron maser concept as a convenient example where both aspects seem compatible and we have derived a Liouville like operator whose time evolution was explicitly carried out and interpreted in terms of an organizational property.

To combine these aspects further we will give the equations in the "classical" form promised in the introduction of Sect. 7.5. This is simply done by introducing dimensionless quantities in the standard way, i.e.

$$\hat{\mathscr{P}} = \frac{\tau}{\hbar}\hat{L} \tag{7.81}$$

and with $z = \hbar\omega, \omega_L = \omega_0$

$$y = \tau\left(\frac{z}{\hbar} - \omega_0\right) \tag{7.82}$$

It follows directly from Eqs. (7.81–7.82) and the properties of $\hat{\mathscr{J}}$ that the propagator can be written as $(n = r)$

$$\exp(-i\hat{\mathscr{P}}t/\tau) = \exp(-i\omega_0 t)\exp(-t/\tau)\sum_{k=0}^{n-1}(-it/\tau)^k\frac{1}{k!}\hat{\mathscr{J}}^k \tag{7.83}$$

and

$$(\omega\tau\hat{\mathscr{J}} - \hat{\mathscr{P}})^{-1} = \sum_{k=1}^{n}(y+i)^{-k}\hat{\mathscr{J}}^{k-1} \tag{7.84}$$

For simplicity, we define ω_0 as the resonance frequency, and ω_0^{-1} the short time scale, τ the phenomenological time scale and $\omega_S = (\tau)^{-1}$ the "mismatch frequency" that may lead to chaotic behavior or organization, depending on the situation. Planck's constant can be included without problems when quantum conditions have to be explicitly considered.

Thus the operator $\hat{\mathscr{J}}$ "condenses" n singular points into one higher order singularity. In principle, this set of points could have an infinite number of elements condensed to an essential singularity and as such it may be speculated whether one could interpret these considerations as the sought-after connection between chaos in the classical and quantum domains.

7.6 Final Conclusions

Fundamental physics has usually been developed according to the very large, as in astro- or space physics, or to the very small, as in particle, nuclear, high energy and quantum physics. However, as we have stressed here, a new overlapping component emerges in modern natural sciences namely, the very complex, involving e.g. chaos, self-organization, nonlinear dynamics and nonequilibrium phenomena. The field of Complexity is rapidly becoming a target for contemporary research and its relevance for many fundamental questions is currently on the rise.

The fundamental nature of complex phenomena is something we see every day, yet the simplifications, introduced to define a physical system, lead us most of the time into a limited "linear thinking". The novelty and importance of a more basic view-point has been particularly emphasized and developed by the Brussels School under the leadership of I. Prigogine and by the new field of synergetics pioneered by H. Haken. Further attention has been given to the field through the European network *nonlinear phenomena and complex systems* organized by G. Nicolis.

We have considered here a general formulation of a complex system within the subdynamics framework. The derivations were suggested from recent work using the Complex Scaling Method, CSM, of quantum mechanical systems with dilatation analytic perturbations in Liouville space. Our approach is mathematically similar to the Prigogine subdynamics decomposition of the Liouville operator for systems having absolutely continuous spectra. However, the question of irreversibility and the approach to equilibrium are somewhat different. A specific point is the occurrence of Jordan blocks in the dilated equations, and the suggested connection with microscopic self-organization. Applications with respect to both classical and quantum situations have been considered here in some detail.

Obviously the present direction penetrates fundamental areas like material science, big science, biophysics, information technology, and meteorology and the bridge between generic and strategic research is getting stronger every day.

We have described the theoretical development leading up to a simple Liouville like equation for a dissipative (open) system. Since the formulation has emphasized complexity rather than a quantal or classical angle, Planck's constant was removed in the general equations, and we showed that a particular (super) operator

$$\hat{\mathscr{P}} = (\omega_0\tau - i)\hat{\mathscr{I}} + \hat{\mathscr{J}} \tag{7.85}$$

and its propagator

$$\exp(-i\hat{\mathscr{P}}t/\tau) \tag{7.86}$$

could be obtained from general (first) principles. In Eqs. (7.85–7.86) ω_0 is the resonance frequency and t is the time, scaled in relation to τ, the corresponding phenomenological relaxation lifetime. Further notation involves $\hat{\mathscr{I}}$ as the unit operator, and $\hat{\mathscr{J}}$ representing the interaction with the environment. The latter was shown to have a very curious property with no counterpart in standard linear equilibrium dynamics. One could also think of the product $\omega_0\tau$ as a "Q-value" of a "resonating cavity", $\tau^{-1} = \omega_S$ as a mismatching frequency, and ω_0^{-1} as the fast time variable. Obviously this "Q-value" could be very large compared to unity and hence one might question the importance of the perturbation $\hat{\mathscr{J}}$. Nevertheless the operator $\hat{\mathscr{J}}$ induced new dynamical features and suggested a connection between microscopic subdynamics and self-organization. Note also the following difference. In condensed matter $\hat{\mathscr{J}}$ derives from correlations, while in CMC $\hat{\mathscr{J}}$ mainly concerns relativistic effects. Compare also the formulation of the resonance paradox, problem 6.11.3. and the CMC paradox, problem 7.8.4.

We have indicated how Eqs. (7.85–7.86) could be obtained from first principles. We have referred to CSM, to the occurrence of Jordan blocks, the relevant algebraic structure, and the analytic structure suggested by dilatation analyticity. In doing so we have developed a theorem for a non-Abelian representation. The electron cyclotron maser was used as a specific example of a system, where subsequent formulations of classical and quantum formulations coexist, but we have also treated applications to condensed matter, e.g. HTSC and FQHE as well as proton mobility in water, as examples of complex systems where collective effects lead to interesting predictions and consequences.

It is obvious that the new complexity paradigm should lead to new interpretations and the way we look at our world. Obviously this would involve a new consensus among scientists how to approach probabilistic formulations, the paradox of irreversibility versus dynamics, the measurement problem and Bell's subsequent anathema, *Against 'measurement'* [55].

The success of quantum mechanics as the most correct formulation, so far, in accomodating atomic and subatomic systems and most dramatically tested by Bell's celebrated theorem, cannot hide the fact that the issue of quantum mechanics as a complete theory is still open. These questions have been debated at various intensities ever since Bohr and Einstein and for a recent appraisal of these questions from Einstein-Podolsky-Rosen through Bell's inequality to the Kochen-Specker Paradox and the (quantum) logical consequences, we refer to M. Redhead's award winning essay on *Incompleteness, Nonlocality, and Realism* [56]. It is not a far-fetched speculation that new ideas which involve a generalized probabilistic interpretation incorporating analytic, algebraic and topological structures in a complexity framework can lead us to a more satisfactory description of the world which we inhabitate, or in the words of Kuhn [47], to a new paradigm.

7.7 References

1. Brändas EJ, Chatzidimitriou-Dreismann CA (1991) Int J Quant Chem 40: 649
2. Brändas EJ, Chatzidimitriou-Dreismann CA, Brändas E, Elander N (1989) Eds., Lecture Notes in Physics 325, p. 486–533
3. Chatzidimitriou-Dreismann CA (1991) Adv Chem Phys 80: 201
4. Chatzidimitriou-Dreismann CA, Brändas EJ (1991) Ber Bunsenges Phys Chem 95: 263
5. Weingärtner H, Chatzidimitriou-Dreismann CA (1990) Nature 346: 548
6. Chatzidimitriou-Dreismann CA, Brändas EJ (1989) Ber Bunsenges Phys Chem 93: 1065
7. Brändas EJ, Chatzidimitriou-Dreismann CA (1991) Ber Bunsenges Phys Chem 95: 462
8. Karlsson E, Brändas EJ, Chatzidimitriou-Dreismann CA (1991) Physica Scripta 44: 77
9. Chatzidimitriou-Dreismann CA, Brändas EJ, Karlsson E (1990) Phys Rev Rapid Communications B 42: 2704
10. Brändas EJ (1993) Ber Bunsenges Phys Chem 97: 55
11. Chatzidimitrious-Dreismann CA, Brändas EJ (1992) Physica C 201: 340
12. Brändas EJ (1993) Ed. Proceedings of the Nobel Satellite Symposium on Resonances and Microscopic Irreversibility, Uppsala, December 1991, Int J Quant Chem 46: Nr. 3

13. Brändas EJ (1993) Proceedings from the BEAM-COOL Workshop, Montreaux, October 4–8
14. Eigen M, De Maeyer L (1958) Proc R Soc (London) 247: 505
15. Meiboom S (1961) J Chem Phys 34: 375
16. Hertz HG (1987) Chemica Scripta 27: 479
17. Hertz HG, Brown BM, Müller KJ, Maurer R (1987) J Chem Ed 64: 777
18. Schuster P, Zundel G, Sandorfy C (1976) The hydrogen bond, vols I–III North Holland, Amsterdam
19. Robinson RA, Stokes RH (1970) Electrolytic solutions, Butterworths, London
20. Pfeifer R, Hertz HG (1990) Ber Bunsenges Phys Chem 94: 1349
21. Chatzidimitriou-Dreismann CA, Brändas EJ (1990) Int J Quant Chem 37: 155
22. Halle B, Karlström G (1983) J Chem Soc Faraday Trans 2, 79: 1031
23. Brändas EJ, Chatzidimitriou-Dreismann CA (1989) Int J Quant Chem Symp 23: 147
24. Chatzidimitriou-Dreismann CA (1989) Int J Quant Chem Symp 23: 153
25. Bednorz JG, Müller KA (1986) Z Phys B 64: 189
26. Chu CW (1987) Proc Natl Acad Sci USA, 84: 4681
27. Burns G (1992) High-temperature superconductivity. An Introduction, Academic Press, Inc., New York
28. Anderson PW, Schrieffer R (1991) Physics Today June
29. Bedell KS, Coffey D, Meltzer DE, Pines D, Schrieffer JR (1990) (Eds.), High Temperature Superconductivity: Proceedings, Addison-Wesley, Redwood City, Calif
30. Wollman DA, Van Harlingen DJ, Lee WC, Ginsberg DM, Leggett AJ (1993) Phys Rev Letters 71: 2134
31. Jansen L, Block R (1993) Physica A 198: 551
32. Jansen L, Chandran L, Block R (1993) Chem Phys 176: 1
33. Uemura YJ et al. (1989) Phys Rev Lett 62: 2317
34. Lynn JW (ed) (1990) High temperature superconductivity. Springer, Berlin Heidelberg New York
35. Zhang H, Sato H (1993) Phys Rev Lett 70: 1697
36. Dahl PF (1992) Superconductivity. Its historical roots and development from mercury to the ceramic oxides, American Institute of Physics, New York
37. von Klitzing K, Dorda G, Pepper M (1980) Phys Rev Lett 45: 494
38. Prange RE, Girvin SM (1990) The quantum hall effect second edition, Springer, Berlin Heidelberg New York
39. Tsui DC, Störmer HL, Gossard AC (1982) Phys Rev Lett 48: 1559
40. Wigner E (1934) Phys Rev 46: 1002
41. Laughlin RB (1983) Phys Rev Lett 50: 1395
42. Leinaas JM, Myrheim J (1977) Nouvo Cimento 37 B 1
43. Wilczek F (1982) Phys Rev Lett 49: 1
44. Suen YW, Engel LW, Santos MB, Shayegan M, Tsui DC (1992) Phys Rev Lett 68, 1379
45. Eisenstein JP, Boebinger GS, Pfeiffer LN, West KW, Song He (1992) Phys Rev Lett 68: 1383
46. Coxeter HSM (1973) Regular Polytopes, Dover Publications, Inc., third edition
47. Kuhn TS (1962) The structure of scientific revolutions, The university of Chigago Press; second edition, enlarged (1970)
48. Haken H, Mikhailov A (1993) Interdisciplinary approaches to nonlinear complex systems, Springer Series in Synergetics, Vol. 62
49. Bohm D (1951) Quantum Theory, Prentice-Hall, Inc., New York
50. Twiss RQ (1958) Aust J Phys 11: 564
51. Schneider J (1959) Phys Rev Lett 2: 504
52. Sprangle P, Drobot AT (1977) IEEE Trans Microwave Theory Tech 25: 528
53. Hirshfeld JL, Wachtel JM (1964) Phys Rev Lett 12: 533
54. Ikegami H (1990) Phys Rev Lett 64, 1737 (1990); ibid 2593
55. Bell J (1990), Against 'measurement', Physics World, Volume 3, No 8, 33, August
56. Redhead M (1987) Incompleteness, Nonlocality, and Realism, Clarendon Press, Oxford

7.8 Problems

1. Find the eigenvalues and eigenvectors corresponding to the $r \times r$ matrix where
   a) the diagonal elements are all equal to p and the off-diagonal elements are all $p(p-1)$, see Sect. 7.5.
   b) the diagonal elements are all equal to $\frac{N}{2r}$ and the off-diagonal elements are all $\frac{N}{2r} \cdot \frac{(2r-N)}{2(r-1)}$, see Sect. 7.3.
2. Find the components of the conductivity tensor, defined as the inverse of the resistivity tensor, Eq. (7.49). Complete the scenario presented in the introduction of Sect. 7.4.
3. Carry through the topological description offered in connection with the Wigner solid, see Sect. 7.4, for $N = 10$
4. In Sect. 7.6. a gainfunction for the relativistic treatment of an electron beam in a constant magnetic field and subject to a resonator rf field was derived. This scenario was referred to as the cyclotron maser concept, CMC. Now two simple statements can be made:
   i) a full dynamical description, the details are not important for the argument, but Schneider's treatment [51] may be consulted, will not lead to increase of phase space density. This is undeniably true since the dynamics derive from a timedependent Hamiltonian for which Liouville's theorem is satisfied. Hence there can be no cooling associated with CMC.
   ii) we have shown in these two chapters that a consistent subdynamics formulation can be obtained provided the Hamiltonian description is generalized to a more general framework, incorporating the resonance picture of unstable states, cf. problem 6.3. In the above mentioned section we have shown explicitly that the energy gainfunction, associated with CMC and mainly a result of relativistic effects, can lead to self-organization. Organization means loss of entropy and CMC should therefore in the present context lead to cooling.

   Find the solution to the CMC-paradox.

III. Gyrating Dipole Moments

8 Basic Principles of Magnetic Resonance

J.D. Macomber

8.1 Operators Representing Orbital Angular Momentum

The spinning charged particle having an angular momentum quantum number of $\frac{1}{2}$ was introduced in Chap. 4 to illustrate the quantum-mechanical nature of spectroscopic transitions. This system is particularly suitable for such a purpose because the component of the transition dipole moment in any direction in space can be made an eigenvalue property by choosing an appropriate state. In Chap. 1 it was noted that the ideas presented in this book were historically first discovered and applied in the study of such systems. For these reasons, and because of the remarkable simplicity of the quantum-mechanical calculations in this case, the first detailed application of material developed in the first five chapters will be made to quantum transitions of spin-$\frac{1}{2}$ particles in a magnetic field (magnetic resonance spectroscopy). Chapter 4 was devoted to magnetic resonance in molecular beams, so that the quantum systems could be studied one at a time. By way of contrast, in this chapter the emphasis will be placed on spin-$\frac{1}{2}$ particles in bulk samples, and the statistical methods introduced in Chap. 5 must be employed.

The discussion will be prefaced by a review of various facts about angular momentum. In classical mechanics, the angular momentum $\mathbf{L}$ of a particle about a point is given by the cross product of the vector joining the point to the particle, $\mathbf{r}$, and the linear momentum vector of the particle, $\boldsymbol{\mu}$ (do not confuse $\boldsymbol{\mu}$ with the magnetic permeability, μ_0):

$$\mathbf{L} = \mathbf{r} \times \boldsymbol{\mu}. \tag{8.1}$$

(see Fig. 8.1). Each of these three vectors may be written in terms of its Cartesian components:

$$\mathbf{L} = L_x\hat{x} + L_y\hat{y} + L_z\hat{z}, \tag{8.2}$$

$$\mathbf{r} = x\hat{x} + y\hat{y} + z\hat{z}, \tag{8.3}$$

and

$$\boldsymbol{\mu} = \mu_x\hat{x} + \mu_y\hat{y} + \mu_z\hat{z}. \tag{8.4}$$

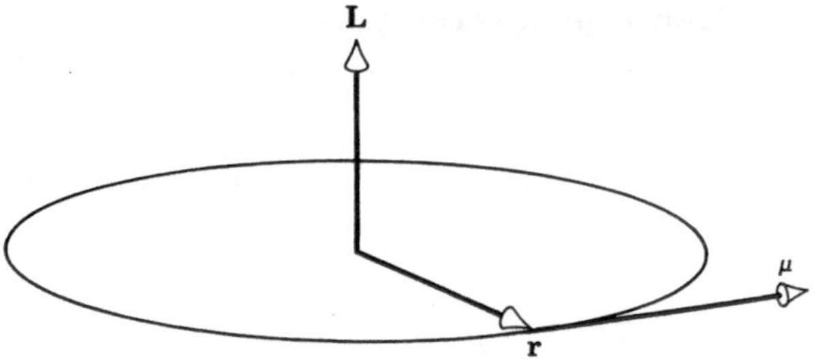

Fig. 8.1. Vector relationship between orbital angular momentum, **L**, linear momentum $\boldsymbol{\mu}$, and position **r**

According to the determinantal rule,

$$\mathbf{r} \times \boldsymbol{\mu} \equiv \begin{vmatrix} x & y & z \\ \mu_x & \mu_y & \mu_z \\ \hat{x} & \hat{y} & \hat{z} \end{vmatrix}. \tag{8.5}$$

By comparing the results of expanding the right-hand side of Eq. 8.5 term by term with the right-hand side of Eq. 8.2, the relationships

$$L_x = y\mu_z - z\mu_y, \tag{8.6}$$

$$L_y = z\mu_x - x\mu_z, \tag{8.7}$$

and

$$L_z = x\mu_y - y\mu_x. \tag{8.8}$$

may be found. Equations 8.6–8.8 can be used to find the operators corresponding to the three components of angular momentum (associated, e.g., with the motion of an electron around a nucleus) by the following simple substitutions:

$$\underline{\boldsymbol{\mu}}_x \equiv -i\hbar\left(\frac{\partial}{\partial x}\right)_{y,z,t}, \tag{8.9}$$

$$\underline{\boldsymbol{\mu}}_y \equiv -i\hbar\left(\frac{\partial}{\partial y}\right)_{x,z,t}, \tag{8.10}$$

$$\underline{\boldsymbol{\mu}}_z \equiv -i\hbar\left(\frac{\partial}{\partial z}\right)_{x,y,t}, \tag{8.11}$$

$$\mathsf{X} = x, \tag{8.12}$$

$$\mathsf{Y} = y, \tag{8.13}$$

$$\mathsf{Z} = z, \tag{8.14}$$

$$\mathsf{L}_x = L_x, \tag{8.15}$$

$$\mathsf{L}_y = L_y, \tag{8.16}$$

and

$$\mathsf{L}_z = L_z. \tag{8.17}$$

After these substitutions have been made, Eq. (8.2) may be interpreted as the definition of a vector operator, with its three components being ordinary scalar operators defined by Eqs. 8.6–8.17.

It is interesting to find the commutators of the three orbital angular momentum operators. For example, since the commutator rule obeys the distributive law,

$$\begin{aligned}[\mathsf{L}_x, \mathsf{L}_y] &= [(\mathsf{Y}\underline{\mu}_z - \mathsf{Z}\underline{\mu}_y), (\mathsf{Z}\underline{\mu}_x - \mathsf{X}\underline{\mu}_z)] \\ &= [\mathsf{Y}\underline{\mu}_z, \mathsf{Z}\underline{\mu}_x] - [\mathsf{Y}\underline{\mu}_z, \mathsf{X}\underline{\mu}_z] - [\mathsf{Z}\underline{\mu}_y, \mathsf{Z}\underline{\mu}_x] + [\mathsf{Z}\underline{\mu}_y, \mathsf{X}\underline{\mu}_z].\end{aligned} \tag{8.18}$$

It is easy to see that

$$[\mathsf{Y}\underline{\mu}_z, \mathsf{Z}\underline{\mu}_x] = \mathsf{Y}\underline{\mu}_x[\underline{\mu}_z, \mathsf{Z}], \tag{8.19}$$

$$[\mathsf{Y}\underline{\mu}_z, \mathsf{X}\underline{\mu}_z] = 0, \tag{8.20}$$

$$[\mathsf{Z}\underline{\mu}_y, \mathsf{Z}\underline{\mu}_x] = 0, \tag{8.21}$$

and

$$[\mathsf{Z}\underline{\mu}_y, \mathsf{X}\underline{\mu}_z] = \mathsf{X}\underline{\mu}_y[\mathsf{Z}, \underline{\mu}_z]. \tag{8.22}$$

The results in Eqs. 8.19–8.22 may be substituted into Eq. (8.18) to produce

$$[\mathsf{L}_x, \mathsf{L}_y] = (\mathsf{X}\underline{\mu}_y - \mathsf{Y}\underline{\mu}_x)[\mathsf{Z}, \underline{\mu}_z]. \tag{8.23}$$

The first factor in parentheses on the right-hand side of Eq. (8.23) is simply L_z; see Eq. (8.8). The commutator in square brackets may be evaluated by letting it operate upon some state function $f(z)$:

$$\begin{aligned}[\mathsf{Z}, \underline{\mu}_z]f &= -i\hbar\left[z\frac{\partial f}{\partial z} - \frac{\partial}{\partial z}(zf)\right] \\ &= -i\hbar\left[z\frac{\partial f}{\partial z} - \left(f + z\frac{\partial f}{\partial z}\right)\right] \\ &= i\hbar f.\end{aligned} \tag{8.24}$$

Now that $f(z)$ has done its job, it may be removed from Eq. (8.24) and the resultant equality substituted into Eq. (8.23):

$$[\mathsf{L}_x, \mathsf{L}_y] = i\hbar \mathsf{L}_z. \tag{8.25}$$

Similarly, one can show that

$$[\mathsf{L}_y, \mathsf{L}_z] = i\hbar \mathsf{L}_x \tag{8.26}$$

and

$$[\mathsf{L}_z, \mathsf{L}_x] = i\hbar \mathsf{L}_y. \tag{8.27}$$

Equation (8.25) can be remembered easily because the subscripts of the operators in it are in alphabetical order; Eqs. (8.26) and (8.27) can be obtained from Eq. (8.25) by a cyclic permutation of these subscripts.

8.2 Operators Representing Spin Angular Momentum

In the quantum theory of the electron developed by Pauli,[1] a triplet of operators, S_1, S_2, and S_3, were discovered which obeyed the same commutative relations that L_x, L_y, and L_z obeyed. In other words,

$$[\mathsf{S}_1, \mathsf{S}_2] = i\hbar \mathsf{S}_3. \tag{8.28}$$

Also, the two other equations obtained from Eq. (8.28) by cyclic permutation of the subscripts are obeyed. Therefore it seemed reasonable to define a new vector operator:

$$\mathsf{S} = \mathsf{S}_x\hat{x} + \mathsf{S}_y\hat{y} + \mathsf{S}_z\hat{z}, \tag{8.29}$$

where x, y, and z have replaced 1, 2, and 3 as subscripts. The physical property represented by the operator S is called the *spin angular momentum* of the electron.

It was recognized that there must be something rather unusual about the operator S. In particular, the connection between the orbital angular momentum operator L and the vector operators r and $\underline{\boldsymbol{\mu}}$ is clearly defined. It is, in fact, the same relationship found in classical mechanics, as shown in Eqs. 8.1–8.8. If an electron were a particle that had a nonzero volume (e.g., a small sphere), one could divide up this volume into N infinitesimal volume elements, the jth having a linear momentum $\boldsymbol{\mu}_j$ and located a distance $\mathbf{r}_j$ from the center of mass (see Fig. 8.2). Then

$$\mathbf{S} = \sum_{j=1}^{N} \mathbf{r}_j \times \boldsymbol{\mu}_j. \tag{8.30}$$

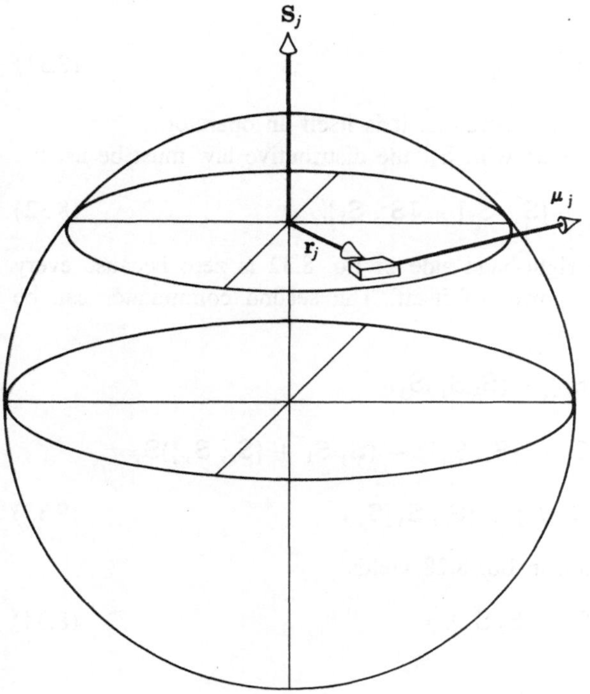

Fig. 8.2. Vector relationship between spin angular momentum S, and the positions $\mathbf{r}_j$ and linear momenta $\boldsymbol{\mu}_j$ of differential mass elements for a classical sphere

The spin angular momentum would then be exactly like the momentum associated with the rotation of the earth on its axis, for which an equation like Eq. (8.30) holds. This would complete the analogy between the motion of the earth around the sun and the motion of the electron around a nucleus. Just as **L** determines the length of the "year" of both earth and electron, **S** would determine the "day" of both. Although this may still be a useful analogy, one cannot write an equation of the form of Eq. 8.30 for an electron. The reason is that the electron acts as if it were a mathematical point, and hence the properties $\mathbf{r}_j$ and $\boldsymbol{\mu}_j$ of the infinitesimal volume are undefined.

Nevertheless, it will be shown that all of the most important measurable physical consequences of L lie in the commutative relations expressed in Eqs. 8.25–8.27. It has therefore been decided to adopt the commutative relations as the fundamental *definition* of angular momentum, rather than Eqs. (8.1) and (8.30) (which were utilized for that purpose in classical mechanics). This is in accord with a time-honored tradition in physics – old definitions are generalized so that new discoveries can be fitted into theoretical frameworks which have proved their utility in the past.

If S is a vector in the laboratory coordinate system, it ought to be possible to define its length. Actually, it is found to be more convenient to deal with the square of the length of S,

$$\mathsf{S}^2 \equiv \mathsf{S}\cdot\mathsf{S} = \mathsf{S}_x^2 + \mathsf{S}_y^2 + \mathsf{S}_z^2\,. \tag{8.31}$$

Since S^2 is a sum of products of operators, it is itself an operator.

To find the commutator of S^2 with S_x, the distributive law must be used:

$$[\mathsf{S}^2, \mathsf{S}_x] = [\mathsf{S}_x^2, \mathsf{S}_x] + [\mathsf{S}_y^2, \mathsf{S}_x] + [\mathsf{S}_z^2, \mathsf{S}_x]. \tag{8.32}$$

The first commutator on the right-hand side of Eq. 8.32 is zero because every operator commutes with any power of itself. The second commutator can be written, using the associative law, as

$$\begin{aligned} [\mathsf{S}_y^2, \mathsf{S}_x] &= \mathsf{S}_y(\mathsf{S}_y\,\mathsf{S}_x) - (\mathsf{S}_x\,\mathsf{S}_y)\mathsf{S}_y \\ &= \mathsf{S}_y(\mathsf{S}_x\,\mathsf{S}_y - [\mathsf{S}_x, \mathsf{S}_y]) - (\mathsf{S}_y\,\mathsf{S}_x + [\mathsf{S}_x, \mathsf{S}_y])\mathsf{S}_y \\ &= -(\mathsf{S}_y[\mathsf{S}_x, \mathsf{S}_y] + [\mathsf{S}_x, \mathsf{S}_y]\mathsf{S}_y). \end{aligned} \tag{8.33}$$

Using the commutative relation in Eq. 8.28 yields

$$[\mathsf{S}_y^2, \mathsf{S}_x] = -i\hbar(\mathsf{S}_y\,\mathsf{S}_z + \mathsf{S}_z\,\mathsf{S}_y). \tag{8.34}$$

Next,

$$\begin{aligned} [\mathsf{S}_z^2, \mathsf{S}_x] &= -(\mathsf{S}_z[\mathsf{S}_x, \mathsf{S}_z] + [\mathsf{S}_x, \mathsf{S}_z]\mathsf{S}_z) \\ &= -i\hbar(\mathsf{S}_z\,\mathsf{S}_y + \mathsf{S}_y\,\mathsf{S}_z). \end{aligned} \tag{8.35}$$

Equations 8.34 and 8.35 may be substituted into Eq. 8.32 to give

$$[\mathsf{S}^2, \mathsf{S}_x] = 0. \tag{8.36}$$

Similarly, it can be shown that

$$[\mathsf{S}^2, \mathsf{S}_y] = 0 \tag{8.37}$$

and

$$[\mathsf{S}^2, \mathsf{S}_z] = 0\,. \tag{8.38}$$

8.3 Eigenkets of Spin; Raising and Lowering Operators

In Chap. 2 it was shown that a single set of eigenkets will suffice for two operators that commute with one another. The operator S^2 will commute with any of the three components, S_x, S_y, or S_z. The latter, however, do not commute with each other. Physically, this means that the system may move in such a way that the length of the total angular momentum, plus the length of any one Cartesian component, will be conserved in a stationary state. Whichever component is picked to be a constant of the motion, the other two components will not be constant. Therefore there are three choices of a complete set of linearly independent eigenkets of spin: one is the set of eigenkets of $\mathbf{S}_x$; the second, of $\mathbf{S}_y$; and the third, of $\mathbf{S}_z$. Whichever is picked, it will automatically be an eigenket of $\mathbf{S}^2$. In what follows, the set of eigenkets of $\mathbf{S}_z$ is selected arbitrarily as the representation.

Let us choose a particular member of this set (call it $|j\rangle$) and operate upon it with $\mathbf{S}_z$. Since $\mathbf{S}_z$ represents angular momentum, its eigenvalues will have to have the *units* of angular momentum. This will be taken care of automatically if each eigenvalue is written as a product of a dimensionless constant with $\hbar$;

$$\mathbf{S}_z|j\rangle = \hbar m_j|j\rangle . \tag{8.39}$$

Similarly, the eigenvalues of $\mathbf{S}^2$ will have to have units of angular momentum squared:

$$\mathbf{S}^2|j\rangle = \hbar^2\Lambda_j|k\rangle . \tag{8.40}$$

The allowed values of m and Λ for all members of the set $\{|S_z\rangle\}$ plus the number of states M must be determined.

What happens when $\mathbf{S}_y$ and $\mathbf{S}_x$ operate upon $|j\rangle$ must also be discovered. As things turn out, it is more convenient to attack this question first. Rather than investigate the problem directly, it is simpler to define two new operators,

$$\mathbf{S}_+ \equiv \mathbf{S}_x + i\,\mathbf{S}_y \tag{8.41}$$

and

$$\mathbf{S}_- \equiv \mathbf{S}_x - i\,\mathbf{S}_y, \tag{8.42}$$

and see what *they* do to $|j\rangle$. It is easy to see that if what $\mathbf{S}_+$ and $\mathbf{S}_-$ do to $|j\rangle$ is known, what $\mathbf{S}_y$ and $\mathbf{S}_x$ do can be discovered by inverting Eqs. 8.41 and 8.42. All that is known at this stage is that the result of operating upon a ket with any operator is to turn that ket into another ket:

$$\mathbf{S}_+|j\rangle = |+\rangle \tag{8.43}$$

and

$$\mathbf{S}_-|j\rangle = |-\rangle. \tag{8.44}$$

The results in Eqs. (8.43) and (8.44) are disconcertingly vague because there are literally an infinite number of possibilities for $|+\rangle$ and $|-\rangle$. The only thing $|\pm\rangle$ *cannot* be is simply a constant multiplied by $|j\rangle$ itself. This is known because $\mathbf{S}_+$ and $\mathbf{S}_-$ are linear combinations of $\mathbf{S}_y$ and $\mathbf{S}_x$, and neither of the latter commute with S_z. Since $|j\rangle$ is an eigenket of S_z, it will not be simultaneously an eigenket of S_+ or S_-. In general, then,

$$|\pm\rangle = c_{\pm 1}|1\rangle + c_{\pm 2}|2\rangle + \cdots + c_{\pm N}|N\rangle, \tag{8.45}$$

where at least one $c_{\pm k}$ $(k \neq j)$ is nonzero.

The *simplest* possibility for $|\pm\rangle$ would be for all but one of the $c_{\pm k}$'s in Eq. (8.45) to be zero. In that case, Eqs. (8.43) and (8.44) could be written as follows:

$$\mathsf{S}_+|j\rangle = c_{+k}|k\rangle, \qquad k \neq j, \tag{8.46}$$

and

$$\mathsf{S}_-|j\rangle = c_{-l}|l\rangle, \qquad l \neq j. \tag{8.47}$$

Perhaps the nicest thing about the hypotheses in Eqs. 8.46 and 8.47 is that they can be tested. If $|+\rangle$ and $|-\rangle$ were just other members of the same set of eigenkets from which $|j\rangle$ was drawn, they would have to satisfy the same equations that gave $|j\rangle$ its identity in the first place. In particular, then, the tests are

$$\mathsf{S}_z|\pm\rangle \stackrel{?}{=} \text{const.} \times |\pm\rangle \tag{8.48}$$

and

$$\mathsf{S}^2|\pm\rangle \stackrel{?}{=} \text{const.} \times |\pm\rangle. \tag{8.49}$$

It is also necessary that

$$|\pm\rangle \neq 0, \tag{8.50}$$

because even though $|\pm\rangle = 0$ satisfies Eqs. (8.48) and (8.49), it corresponds to no particle at all being present and is therefore of no interest.

Applying the test in Eq. 8.49 first yields

$$\begin{aligned}\mathsf{S}^2|\pm\rangle &= \mathsf{S}^2\mathsf{S}_\pm|j\rangle \\ &= \mathsf{S}^2(\mathsf{S}_x \pm i\mathsf{S}_y)|j\rangle \\ &= (\mathsf{S}^2\mathsf{S}_x \pm i\mathsf{S}^2\mathsf{S}_y)|j\rangle \\ &= (\mathsf{S}_x\mathsf{S}^2 \pm i\mathsf{S}_y\mathsf{S}^2)|j\rangle \\ &= (\mathsf{S}_x \pm i\mathsf{S}_y)\mathsf{S}^2|j\rangle.\end{aligned} \tag{8.51}$$

In the above, the distributive law has been used, plus the fact that S^2 commutes with S_y and S_x (see Eqs. 8.36 and 8.37).

The fact that multiplication by a scalar commutes with all other operators may be applied to combine Eqs. (8.40) and (8.51):

$$\begin{aligned}\mathbf{S}^2|\pm\rangle &= \mathbf{S}_\pm\hbar^2\Lambda_j|j\rangle \\ &= \hbar^2\Lambda_j\mathbf{S}_\pm|j\rangle \\ &= \hbar^2\Lambda_j|\pm\rangle. \end{aligned} \tag{8.52}$$

Also used were the definitions of $\mathbf{S}_\pm$ and $|\pm\rangle$, as expressed in Eqs. (8.41–8.44). It can be seen from Eq. (8.52) that the wavefunctions $|+\rangle$ and $|-\rangle$ indeed pass the test expressed in Eq. (8.49).

The test in Eq. (8.48) may now be applied:

$$\begin{aligned}\mathbf{S}_z|\pm\rangle &= \mathbf{S}_z\mathbf{S}_\pm|j\rangle \\ &= \mathbf{S}_z(\mathbf{S}_x \pm i\mathbf{S}_y)|j\rangle \\ &= (\mathbf{S}_z\mathbf{S}_x \pm i\mathbf{S}_z\mathbf{S}_y)|j\rangle. \end{aligned} \tag{8.53}$$

The operator products in parentheses to the right of the equality sign in Eq. (8.53) may be rewritten with the aid of the commutative relations; Eq. (8.53) then becomes

$$\begin{aligned}\mathbf{S}_z|\pm\rangle &= (\{\mathbf{S}_x\mathbf{S}_z + [\mathbf{S}_z, \mathbf{S}_x]\} \pm i\{\mathbf{S}_y\mathbf{S}_z - [\mathbf{S}_y, \mathbf{S}_z]\})|j\rangle \\ &= \{(\mathbf{S}_x \pm i\mathbf{S}_y)\mathbf{S}_z + ([\mathbf{S}_z, \mathbf{S}_x] \mp i[\mathbf{S}_y, \mathbf{S}_z])\}|j\rangle \\ &= [\mathbf{S}_\pm\mathbf{S}_z + (i\hbar\mathbf{S}_y \mp i^2\hbar\mathbf{S}_x)]|j\rangle \\ &= [\mathbf{S}_\pm\mathbf{S}_z \pm \hbar(\mathbf{S}_x \pm i\mathbf{S}_y)]|j\rangle \\ &= \mathbf{S}_\pm(\mathbf{S}_z \pm \hbar)|j\rangle \\ &= \mathbf{S}_\pm(\mathbf{S}_z|j\rangle \pm \hbar|j\rangle). \end{aligned} \tag{8.54}$$

Equation (8.39) may now be used to rewrite Eq. (8.54) as follows:

$$\begin{aligned}\mathbf{S}_z|\pm\rangle &= \mathbf{S}_\pm(\hbar m_j|j\rangle \pm \hbar|j\rangle) \\ &= \mathbf{S}_\pm[\hbar(m_j \pm 1)]|j\rangle \\ &= \hbar(m_j \pm 1)\mathbf{S}_\pm|j\rangle \\ &= \hbar(m_j \pm 1)|\pm\rangle. \end{aligned} \tag{8.55}$$

It can be seen from Eq. (8.55) that the kets $|+\rangle$ and $|-\rangle$ indeed pass the test expressed in Eq. (8.48).

In summary, not only has it been discovered that the result of operating upon $|j\rangle$ with $S_+(S_-)$ is another member of the set of eigenkets of S_z, but also it has been learned *which* member of the set – the member with the same value of Λ as $|j\rangle$, but with a value of m higher (lower) by one unit. For this reason, S_+ and S_- are called *raising* and *lowering operators*. A notation frequently used for this ket is

$$|j\rangle \equiv |\Lambda, m\rangle. \tag{8.56}$$

Equations (8.46) and (8.47) therefore may now be rewritten as follows:

$$\mathsf{S}_\pm|\Lambda, m\rangle = c_{\pm,\ m\pm 1}|\Lambda, m \pm 1\rangle. \tag{8.57}$$

8.4 Number of States, Normalization, and Eigenvalues

The facts expressed in Eq. (8.57) make it possible to start with any one of the eigenkets $|\Lambda, m\rangle$ representing a particular stationary state of the system, and to generate from it the eigenkets representing all other stationary states having angular momentum vectors of the same length (Λ), but different projections of these vectors on the z axis (m). For example, by repeated operation using S_+,

$$\begin{aligned}\underbrace{\mathsf{S}_+\mathsf{S}_+\cdots\mathsf{S}_+}_{u\text{ times}}|\Lambda, m\rangle &\equiv \mathsf{S}_+^u|\Lambda, m\rangle \\ &= c_{+,m+1}c_{+,m+2}\cdots c_{+,m+u}|\Lambda, m+u\rangle\end{aligned} \tag{8.58}$$

Common sense indicates that there must be some limit to the process expressed in Eq. (8.58); surely the component of the angular momentum along the z axis cannot exceed the length of the total vector. In other words, there must be some limit upon $m + u$. Similarly,

$$\mathsf{S}_-^v|\Lambda, m\rangle = c_{-,m-1}c_{-,m-2}\cdots c_{-,m-v}|\Lambda, m-v\rangle; \tag{8.59}$$

there must be a limit on $m - v$ also.

The easiest way to obtain automatic upper and lower bounds on m would be to have $c_{+,m}$ be zero if m becomes too high and $c_{-,m}$ be zero if m becomes too low. It will be necessary to compute these quantities anyway, in order to normalize the wavefunctions generated by means of Eq. (8.57). First, $\mathsf{S}_-\mathsf{S}_+$ is computed using the definitions in Eqs. (8.41) and (8.42), the commutator in Eq. (8.28), the definition of S in Eq. (8.31), and the distributive law:

$$\begin{aligned}\mathsf{S}_-\mathsf{S}_+ &= (\mathsf{S}_x - i\mathsf{S}_y)(\mathsf{S}_x + i\mathsf{S}_y) \\ &= (\mathsf{S}_x^2 + \mathsf{S}_y^2) + i(\mathsf{S}_x\mathsf{S}_y - \mathsf{S}_y\mathsf{S}_x) \\ &= (\mathsf{S}^2 - \mathsf{S}_z^2) - \mathsf{S}_z\hbar.\end{aligned} \tag{8.60}$$

Then $|+\rangle$ is normalized, using Eqs. (8.43) and (8.57):

$$(\mathbf{S}_+|\Lambda, m\rangle)^\dagger \mathbf{S}_+|\Lambda, m\rangle = |c_{+,m+1}|^2\langle\Lambda, m+1|\Lambda, m+1\rangle. \tag{8.61}$$

The superscript $\dagger$ means that the ket vector in parentheses should be replaced by the corresponding bra. To discover the identity of this bra, one must make use of what is called the *Hermitian property of Hilbert space* (defined previously in Eq. 2.104). It can be shown that, because of this property, corresponding bras and kets are Hermitian adjoints of one another. When operators, bras, and kets are represented by matrices (as was done, e.g., in Chap. 5), their Hermitian adjoints can be defined very simply. The transpose $[A]_T$ of any row, column, or square matrix $[A]$ is formed by interchanging rows and columns. In other words, the elements of the nth row of $[A]$ become the elements of the nth column of $[A]_T$. The Hermitian adjoint, $[A]^\dagger$, of any matrix $[A]$ is formed by replacing each element of $[A]_T$ by its complex conjugate. Using this definition and Eqs. (8.41) and (8.42), one can show that the Hermitian adjoint of $\mathbf{S}_+$ is $\mathbf{S}_-$. Assume that $|j\rangle$, the starting eigenket, is already normalized. Then, from Eq. (8.61),

$$\begin{aligned} |c_{+,m+1}|^2 &= (\mathbf{S}_+|\Lambda, m\rangle)^\dagger \mathbf{S}_+|\Lambda, m\rangle \\ &= ((\Lambda, m|\mathbf{S}_-)\mathbf{S}_+|\Lambda, m\rangle. \end{aligned} \tag{8.62}$$

Next, the associative and distributive laws are used, and Eq. (8.60) is substituted into Eq. (8.62):

$$\begin{aligned} |c_{+,m+1}|^2 &= \langle\Lambda, m|\mathbf{S} - \mathbf{S} + |\Lambda, m\rangle \\ &= \langle\Lambda, m|\mathbf{S}^2|\Lambda, m\rangle - \langle\Lambda, m|\mathbf{S}_z^2|\Lambda, m\rangle - \hbar\langle\Lambda, m|\mathbf{S}_z|\Lambda, m\rangle. \end{aligned} \tag{8.63}$$

Equations (8.39) and (8.40) are then used to find

$$\begin{aligned} |c_{+,m+1}|^2 &= \hbar^2\Lambda - \hbar^2 m^2 - \hbar^2 m \\ &= \hbar^2[\Lambda - m(m+1)]. \end{aligned} \tag{8.64}$$

Similarly, starting with the operator $\mathbf{S}_+\mathbf{S}_-$, one has

$$|c_{-,m-1}|^2 = \hbar^2[\Lambda - m(m-1)]. \tag{8.65}$$

Whenever the expressions in square brackets on the right-hand sides of Eq. (8.64) and (8.65) are *not* zero, Eq. (8.50) is satisfied. This means that $|+\rangle$ and $|-\rangle$ have completely satisfied the tests laid down for them, and they indeed qualify as members of the set $\{|S_z\rangle\}$, as hoped. Whenever they *are* zero, they provide boundaries for the allowed values of m in Eq. (8.57). From Eq. (8.64),

$$\Lambda - m_{\text{MAX}}(m_{\text{MAX}} + 1) = 0. \tag{8.66}$$

From Eq. (8.65),

$$\Lambda - m_{\mathrm{MIN}}(m_{\mathrm{MIN}} - 1) = 0. \tag{8.67}$$

From Eqs. (8.66) and (8.67), a relationship is obtained between m_{MAX} and m_{MIN}:

$$m_{\mathrm{MAX}}(m_{\mathrm{MAX}} + 1) = m_{\mathrm{MIN}}(m_{\mathrm{MIN}} - 1). \tag{8.68}$$

The numerical values of m_{MAX} and m_{MIN} are still not known. It is known from Eq. 8.58, however, that it must be possible to start at some arbitrarily chosen m and, by repeated applications of the $\mathbf{S}_+$ operator, reach m_{MAX} in an integral number of steps:

$$m_{\mathrm{MAX}} = m + u. \tag{8.69}$$

Similarly, according to Eq. (8.59), it must be possible to start at the same point, step down an integral number of times by applying $\mathbf{S}_-^v$, and arrive at m_{MIN}:

$$m_{\mathrm{MIN}} = m - v. \tag{8.70}$$

Equation (8.70) may be subtracted from Eq. (8.69) to find

$$m_{\mathrm{MAX}} - m_{\mathrm{MIN}} = u + v \equiv q. \tag{8.71}$$

Since u and v are integers, q must also be an integer. Next, Eq. (8.71) can be substituted into Eq. 8.68 to obtain

$$(m_{\mathrm{MAX}})^2 + m_{\mathrm{MAX}} = (m_{\mathrm{MAX}} - q)^2 - (m_{\mathrm{MAX}} - q). \tag{8.72}$$

Equation (8.72) may be easily solved by

$$m_{\mathrm{MAX}} = \frac{q}{2}. \tag{8.73}$$

Therefore m_{MAX} and all other values of m are either integers or half integers, depending on whether q is even or odd.

The fact that m_{MAX} must be either integral or half integral, as indicated by Eq. (8.73), severely restricts the possible solutions to Eq. (8.68). In fact, there are only two corresponding possibilities for m_{MIN}:

$$m_{\mathrm{MIN}} = m_{\mathrm{MAX}} + 1 \tag{8.74}$$

and

$$m_{\mathrm{MIN}} = -m_{\mathrm{MAX}}. \tag{8.75}$$

The "solution" in Eq. (8.74) may be disregarded because it contradicts the very concepts of minimum and maximum. If m_{MAX} is denoted by the symbol s, in accordance with the usual custom, it is found from Eq. (8.66) that

$$\Lambda = s(s + 1). \tag{8.76}$$

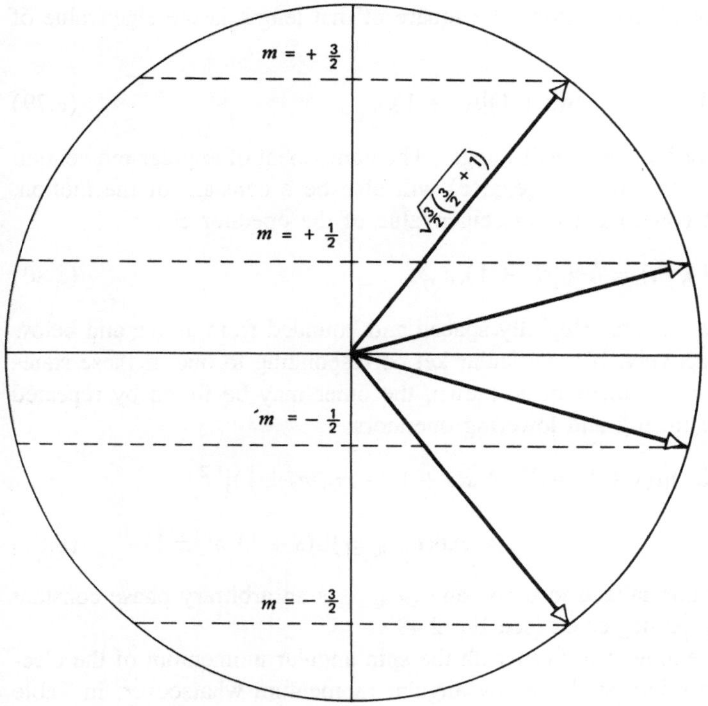

Fig. 8.3. Quantum states of a spinning particle; relationships between m and s for $s = \frac{3}{2}$

The same result may be obtained from Eqs. (8.67) and (8.75). A schematic presentation of the relationship between m and s is given in Fig. 8.3.

8.5 Spinning Particles in Nature

In summary, the following has been learned. Whenever one finds in quantum mechanics a system described within a Hilbert space containing three operators, S_x, S_y, and S_z, obeying the rule

$$[\mathsf{S}_x, \mathsf{S}_y] = \mathsf{S}_z \tag{8.77}$$

and cyclic permutations thereof, one has found a system that can exist in states represented by

$$|j\rangle \equiv |s(s+1), m_s\rangle. \tag{8.78}$$

When the system is in any of these states, the length of the angular momentum vector is a constant of the motion. The square of that length is the eigenvalue of the operator $\mathbf{S}^2$,

$$\mathbf{S}^2|s(s+1), m_s\rangle = \hbar^2 s(s+1)|s(s+1), m_s\rangle, \tag{8.79}$$

where s is a positive integer or half integer. The component of angular momentum along one of the Cartesian axes (e.g., z) will also be a constant of the motion. The length of that component is the eigenvalue of the operator $\mathbf{S}_z$,

$$\mathbf{S}_z|s(s+1), m_s\rangle = \hbar m_s|s(s+1), m_s\rangle, \tag{8.80}$$

where the values of m_s are integrally spaced and bounded from above and below by s and $-s$, respectively. If a particular ket corresponding to one of these states of stationary angular momentum is known, the other may be found by repeated applications of the raising and lowering operators:

$$\begin{aligned}(\mathbf{S}_x \pm i\mathbf{S}_y)|s(s+1), m_s\rangle =& \hbar[s(s+1) - m_s(m_s \pm 1)]^{1/2} \\ & \times \exp(i\zeta_{s,m_s\pm 1})|s(s+1), m_s \pm 1\rangle,\end{aligned} \tag{8.81}$$

where all kets are normalized to unity, and $\zeta_{s,m_s\pm 1}$ is an arbitrary phase constant that can ordinarily be neglected (see Eq. 2.49).

These results, although derived with the spin angular momentum of the electron specifically in mind, apply to any angular momentum whatsoever. In Table 8.1 the symbols used in different systems for the quantum numbers are employed to display the otherwise identical formulas.

It has already been mentioned that a clear classical analog of spin angular momentum is possible only if the electron is imagined to be a ball of nonzero volume. Then the ball could be subdivided into infinitesimal volume elements, and $\mathbf{S}$ could be calculated by means of Eq. (8.30). Of course, each of these infinitesimal volume elements would have associated with it not only a tiny chunk of mass, but a a tiny bit of negative electrical charge as well. The rotation of

Table 8.1. Similarities among quantum systems possessing angular momentum

Type of Angular Momentum	Length of Total Vector	Length of z Component	Restrictions on Total Length	Restrictions on Length of z Component
Electron spin	$\hbar[s(s+1)]^{1/2}$	$\hbar m_s$	$s = \frac{1}{2}$	$m_s = \pm\frac{1}{2}$
Proton spin	$\hbar[I(I+1)]^{1/2}$	$\hbar m_I$	$I = \frac{1}{2}$	$m_I = \pm\frac{1}{2}$
Electron orbit	$\hbar[l(l+1)]^{1/2}$	$\hbar m_l$	$l = 0, 1, \ldots, \infty$	$m_l = -l, -l+1, \ldots, l$
Total electron	$\hbar[j(j+1)]^{1/2}$	$\hbar m_j$	$j = l + s$	$m_j = -j, -j+1, \ldots, j$
Total H atom	$\hbar[F(F+1)]^{1/2}$	$\hbar m_F$	$F = j + I$	$m_F = -F, -F+1, \ldots, F$
Molecular rotation	$\hbar[J(J+1)]^{1/2}$	$\hbar M$ (laboratory axis)	$J = 0, 1, \ldots, \infty$	$M = -J, -J+1, \ldots, J$
		$\hbar K$ (molecular axis)		$K = -J, -J+1, \ldots, J$

the ball would therefore carry these elements of charge around the axis in circular loops, and associated with each loop would be a magnetic field. Therefore a rotating charged sphere should have a magnetic dipole moment. If the infinitesimal charges are rigidly fixed to the infinitesimal masses that constitute the sphere, the dipole moment vector and the angular momentum vector will be parallel (or antiparallel, depending on the algebraic sign of the charge) (see Fig. 8.4).

As was previously stated, electrons are points, not balls. This fact does not prevent them from having spin angular momentum, however, nor does it prevent them from having magnetic dipole moments. Other particles of importance in this chapter (in addition to electrons) are atomic nuclei. All nuclei with nonzero spin angular momenta also have magnetic moments due to their charge. Even the neutrons, of charge zero, have magnetic moments. To explain this, it is imagined that neutrons are composed of both positive and negative charges in equal

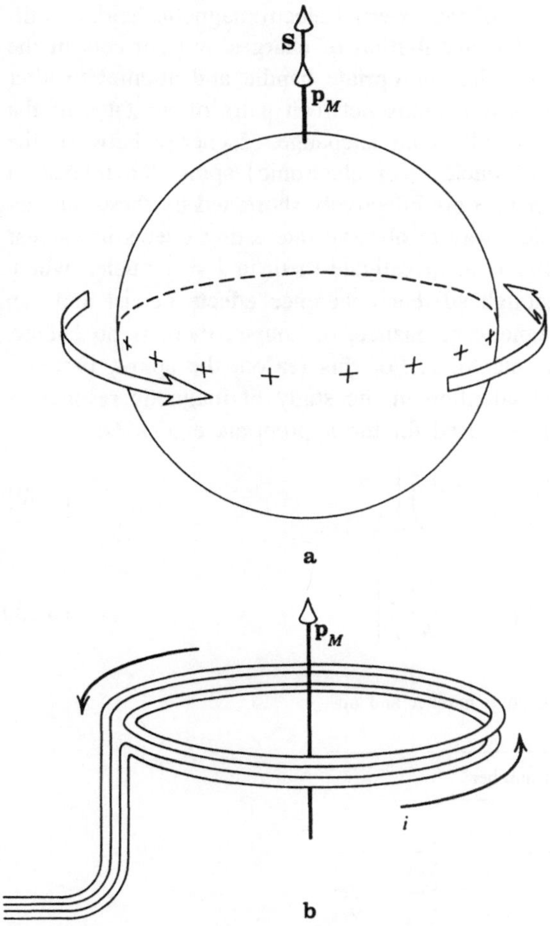

Fig. 8.4. a Magnetic moment of a classical spinning charged sphere. **b** Magnetic moment of a coil of wire carrying a current

amounts, with the negative charges located nearer the surface of the sphere than the positive charges. In this way, the negative charges have greater tangential velocities than the positive charges and, therefore, give rise to greater equivalent currents. The area enclosed by the equivalent "loop of wire" is greater for charges on the outside of the neutron. Since the magnetic field associated with the rotation of a charge is proportional to the product of the current and the loop area, the magnetic field associated with the negative charges of the neutron is bigger than that associated with the positive charges. The former cancels the latter and gives rise to a net magnetic moment for the neutron comparable in magnitude to nuclear moments, but opposite in sign to the proton. The relationship between the electromagnetic multipole moments possessed by a nucleus and the magnitude of the associated angular momentum is shown in Table 8.2.

Each nonzero multipole moment higher than the monopole, or charge, constitutes a "handle" on the nucleus which may be "grasped" by the appropriate term in the Taylor series expansion of the external electromagnetic field, as outlined in Chap. 3. Fluctuations of the distribution of charges and currents in the environment of each nucleus grasp the appropriate handle and attempt to alter the spin orientation. This produces transitions between pairs of m states of the quantum systems. Since the net result is an exchange of energy between the surroundings and the ensemble of nuclear (or electronic) spins, the relaxation times, T_1 and T_2, discussed in Chap. 3 are effectively shortened by these interactions. As a consequence, the easiest way to observe interesting effects in nuclear magnetic resonance experimentally is to investigate them in $I = \frac{1}{2}$ nuclei, where the relaxation times are long enough so that coherence effects persist and can be observed. In electron paramagnetic resonance, of course, there is no choice, $s = \frac{1}{2}$ is a fixed property of the electrons. For this reason, the spin-$\frac{1}{2}$ particles have been the subject of special attention in the study of magnetic resonance, and the special symbols α and β are used for the appropriate eigenkets:

$$|1\rangle = |\tfrac{3}{4}, \tfrac{1}{2}\rangle = \alpha \exp\left[-i\left(\zeta_1 + \frac{\mathscr{E}_1 t}{\hbar}\right)\right] \tag{8.82}$$

and

$$|2\rangle = |\tfrac{3}{4}, -\tfrac{1}{2}\rangle = \beta \exp\left[-i\left(\zeta_2 + \frac{\mathscr{E}_2 t}{\hbar}\right)\right]. \tag{8.83}$$

Table 8.2. Relationship between spin quantum number and nuclear multipole moment

	Spin quantum number				
Moment	0	$\frac{1}{2}$	1	$\frac{3}{2}$	2
Electric monopole	Yes	Yes	Yes	Yes	Yes
Magnetic dipole	No	Yes	Yes	Yes	Yes
Electric quadrupole	No	No	Yes	Yes	Yes
Magnetic octopole	No	No	No	Yes	Yes
Electric hexadecapole	No	No	No	No	Yes

8.6 The Effect of a Static Magnetic Field

To proceed with this analysis, it is necessary to obtain the Hamiltonian operator for the spin-$\frac{1}{2}$ particle. The operator corresponding to the magnetic dipole, p_M, should be proportional to the angular momentum operator:

$$\mathsf{p}_M = \gamma \mathsf{S}. \tag{8.84}$$

The magnetogyric ratio, γ, was introduced in Chap. 4, where it was defined by Eq. (4.35). Classically, the energy of a magnetic dipole in a uniform induction field, **H** (**B** in all but magnetic resonance), is given by

$$\mathscr{E} = -\mathbf{p}_M \cdot \mathbf{H}. \tag{4.38}$$

By the usual procedure in quantum mechanics, the Hamiltonian is constructed from the classical expression for the energy by replacing all quantities appearing therein by the corresponding operator equivalents:

$$\mathsf{H} = -\gamma \mathsf{S} \cdot \mathbf{H}. \tag{8.85}$$

In accordance with Eq. (5.55), the Hamiltonian is broken up into three terms. The first of these, H_0, should be associated with the establishment of the stationary states of the system:

$$\mathsf{H}_0 = -\gamma \mathsf{S} \cdot \mathbf{H}_0. \tag{8.86}$$

In conformance with the previously adopted convention, the z component of angular momentum is the one that is invariant in the stationary states of the system. This implies that the coordinate system should be oriented so that $\mathbf{H}_0$ points along the z axis (see Eq. 4.48):

$$\mathbf{H}_0 = H_0 \hat{z}. \tag{8.87}$$

Therefore, using Eqs. (8.29) and (8.87), one may rewrite Eq. (8.86) to obtain

$$\mathsf{H}_0 = -\omega_0 \mathsf{S}_z. \tag{8.88}$$

The symbol ω_0 used in Eq. (8.88) was defined previously in Eq. 4.55. The first step in solving the problem is to determine the energy eigenvalues, $\mathscr{E}_1$ and $\mathscr{E}_2$. From Eq. (8.82),

$$\begin{aligned} \mathsf{H}_0|1\rangle &= \mathscr{E}_1|1\rangle \\ &= -\gamma H_0 \mathsf{S}_z|\tfrac{3}{4}, \tfrac{1}{2}\rangle \\ &= -\gamma H_0 \hbar(\tfrac{1}{2})|1\rangle . \end{aligned} \tag{8.89}$$

Therefore

$$\mathscr{E}_1 = \frac{-\hbar\omega_0}{2}. \tag{8.90}$$

Similarly, from Eq. (8.83),

$$\mathscr{E}_2 = \frac{+\hbar\omega_0}{2}. \tag{8.91}$$

These results were presented without proof in Eq. (4.43).

The orientation of the spin magnetic moment when the particle is in a superposition state will now be calculated. The ket for the particle in this state can be written, as suggested by Eqs. (2.73), (8.82), (8.83), (8.90), and (8.91):

$$|A\rangle = \cos\left(\frac{\theta}{2}\right)\left\{\alpha\exp\left[i\left(\eta_1 + \frac{\omega_0 t}{2}\right)\right]\right\} + \sin\left(\frac{\theta}{2}\right)\left\{\beta\exp\left[i\left(\eta_2 - \frac{\omega_0 t}{2}\right)\right]\right\}. \tag{8.92}$$

To calculate the orientation of the magnetic moment of the system when in the state described by Eq. (8.92), the expectation value of each of the three components, p_x, p_y, and p_z, of the magnetic moment vector, $\mathbf{p}_M$, must be calculated using Eq. (8.84). In Chap. 4 the sense in which the results of such a calculation can indeed be called the orientation of *the* magnetic moment was discussed. It was found that not the orientation of the true (total, instantaneous) dipole moment, but only what was called the *certain component* thereof, would be obtained. This component is of interest because of the fact that, in the course of its interaction with the radiation field, the spin behaves as if this "component" receives the torque and is therefore responsible for the transition.

In performing these calculations, the formula given in Eq. (2.83) may be used. For example,

$$\langle p_x\rangle = \gamma\left[(\mathsf{S}_x)_{11}\cos^2\left(\frac{\theta}{2}\right) + (\mathsf{S}_x)_{22}\sin^2\left(\frac{\theta}{2}\right) + 2\cos\left(\frac{\theta}{2}\right)\sin\left(\frac{\theta}{2}\right)\mathrm{Re}\{(\mathsf{S}_x)_{12}\exp[i(\S - \omega_0 t)]\}\right]. \tag{8.93}$$

Equations similar to Eq. (8.93) hold for $\langle p_y\rangle$ and $\langle p_z\rangle$.

Evidently, it will be necessary to compute the elements of the S_x, S_y, and S_z matrices in the basis $\{\alpha, \beta\}$. First, Eqs. (8.41) and (8.42) must be solved for the desired S_y and S_x operators in terms of the raising and lowering operators:

$$\mathsf{S}_x = \frac{\mathsf{S}_+ + \mathsf{S}_-}{2} \tag{8.94}$$

and

$$\mathbf{S}_y = \frac{\mathbf{S}_+ - \mathbf{S}_-}{2i}. \tag{8.95}$$

From Eq. (8.81), ignoring phase constants, it can be computed that, for an $s = \frac{1}{2}$ particle,

$$\mathbf{S}_+\alpha = \mathbf{S}_-\beta = 0, \tag{8.96}$$

$$\mathbf{S}_-\alpha = \hbar\beta, \tag{8.97}$$

and

$$\mathbf{S}_+\beta = \hbar\alpha. \tag{8.98}$$

Note that α and β are orthonormal kets:

$$(\alpha, \alpha) \equiv \langle \tfrac{3}{4}, \tfrac{1}{2} | \tfrac{3}{4}, \tfrac{1}{2} \rangle = 1, \tag{8.99}$$

$$(\beta, \beta) \equiv \langle \tfrac{3}{4}, -\tfrac{1}{2} | \tfrac{3}{4}, -\tfrac{1}{2} \rangle = 1, \tag{8.100}$$

and

$$(\beta, \alpha) = (\alpha, \beta) \equiv \langle \tfrac{3}{4}, \pm\tfrac{1}{2} | \tfrac{3}{4}, \pm\tfrac{1}{2} \rangle = 0. \tag{8.101}$$

This fact, plus Eqs. (8.94) to (8.98), may be used to calculate

$$(\mathbf{S}_x)_{11} = (\mathbf{S}_x)_{22} = (\mathbf{S}_y)_{11} = (\mathbf{S}_y)_{22} = 0, \tag{8.102}$$

and

$$(\mathbf{S}_x)_{12} = i(\mathbf{S}_y)_{12} = \frac{\hbar}{2}. \tag{8.103}$$

Also, from Eq. (8.80), for an $s = \frac{1}{2}$ particle,

$$\mathbf{S}_z\alpha = \frac{\hbar\,\alpha}{2} \tag{8.104}$$

and

$$\mathbf{S}_z\beta = \frac{\hbar\,\beta}{2}. \tag{8.105}$$

Therefore,

$$(\mathbf{S}_z)_{11} = -(\mathbf{S}_z)_{22} = \frac{\hbar}{2} \tag{8.106}$$

and

$$(\mathbf{S}_z)_{21} = (\mathbf{S}_z)_{12} = 0. \tag{8.107}$$

Next, Eqs. (8.102), (8.103), (8.106), and (8.107) are substituted into Eq. (8.93) and its analogs. The results are

$$\langle p_y \rangle = \gamma\hbar \cos\left(\frac{\theta}{2}\right)\sin\left(\frac{\theta}{2}\right)\cos(\S - \omega_0 t), \tag{8.108}$$

$$\langle p_x \rangle = \gamma\hbar \cos\left(\frac{\theta}{2}\right)\sin\left(\frac{\theta}{2}\right)\sin(\S - \omega_0 t), \tag{8.109}$$

and

$$\langle p_z \rangle = \frac{\gamma\hbar[\cos^2(\theta/2) - \sin^2(\theta/2)]}{2}. \tag{8.110}$$

Finally, the substitutions

$$p_M = \frac{\gamma\hbar}{2} \tag{8.111}$$

and

$$\Omega = \S - \omega_0 t \tag{8.112}$$

(see Eq. 4.56) are made into Eqs. 8.108–8.110 to produce

$$\langle \mathbf{p}_M \rangle = p_M \cos\theta \hat{z} + p_M \sin\theta \cos\Omega \hat{x} + p_M \sin\theta \sin\Omega \hat{y}. \tag{8.113}$$

It can be seen from Eq. (8.113) that $\langle \mathbf{p}_M \rangle$ is a vector of length $p_M = \gamma\hbar/2$, making an angle θ with respect to the z axis, and having its projection on the xy plane at an angle $\Omega = \S - \omega_0 t$, measured counterclockwise from the x axis. In

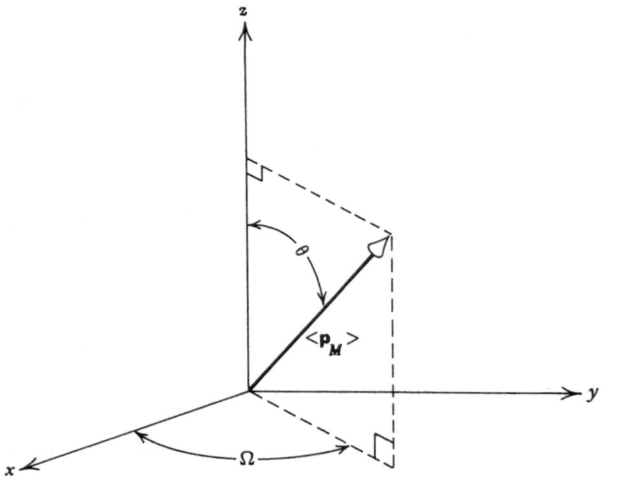

Fig. 8.5. Expectation value of the spin-angular-momentum vector of an $s = \frac{1}{2}$ particle.

other words, $\langle \mathbf{p}_M \rangle$ precesses around the z axis at a constant frequency, ω_0 (called the *Larmor frequency*), in such a way that its length is constant. The cone angle is determined by the relative amounts of α and β in the superposition. The vector $\langle \mathbf{p}_M \rangle$ is shown in Fig. 8.5 (note the similarity to Fig. 4.4).

In accordance with what was stated previously in Chap. 4, $\langle \mathbf{p}_M \rangle$ carries around with it another vector of length $\sqrt{3}\,\gamma\hbar/2$. This vector is the total angular momentum. Its location cannot be determined precisely, because of the uncertainty principle, but one can be sure that it is somewhere on a cone whose apical angle is 54°44′8″ and whose axis is $\langle \mathbf{p}_M \rangle$. The location of $\langle \mathbf{p}_M \rangle$, on the other hand, can be determined as precisely as one desires; no limit is imposed on the accuracy of the measurement of Ω or θ by the uncertainty principle.

8.7 The Effect of an Oscillating Magnetic Field

The second term in the Hamiltonian operator, Eq. 5.55, will be discussed next. This term, H_1, describes the interaction between the spinning particle and the electromagnetic radiation. It was stated in Chap. 2 that this interaction must proceed by means of the fields associated with the light waves pushing monopoles and/or twisting dipoles. Every particle under consideration in this chapter has a magnetic dipole moment and will therefore feel a torque in the spatially uniform magnetic field **H**. Every one but the neutron also has an electric monopole moment and will be pushed by the spatially uniform electric field, **E**.

These fields will have a large effect on quantum systems containing spin-$\frac{1}{2}$ particles only if the motions they excite oscillate at frequencies near those of the corresponding quantum-mechanical transitions (resonance). The torques exerted by **H** (in typical large laboratory magnetic fields of $\sim$1 T) excite precessional motions in nuclei and electrons that will be at resonance with the oscillations of **H** at radio-wave or microwave frequencies. The pushes of **E** upon nuclei will tend to excite displacement motions of the nuclear framework (e.g., molecular rotation and vibration). The quantum-mechanical transition frequencies for these motions are much too high for resonance with rf waves to occur. (Microwaves can produce rotational transitions in gas-phase molecules, but only liquid and solid samples are of concern here.) The pushes of **E** upon electrons will tend to excite spatial redistributions of their probability densities (e.g., electronic transitions) at frequencies even higher than those of molecular rotation and vibration. Because only rf and microwave-frequency "light" will be considered in this chapter, only **H** is important; the influence of the **E** field may be neglected.

It can be seen from Eqs. 8.108 and 8.109 that the oscillations of the magnetic dipole moment vector occur in the xy plane, perpendicular to the direction of the field $H_0\hat{z}$. Therefore, in order to drive these dipoles, the oscillations of the magnetic vector associated with the light wave must occur in this same plane. Since electromagnetic waves are transverse (like water waves) rather than longitudinal (like sound waves in air), the oscillating field vectors are perpendicular to the direction of propagation (see Fig. 1.1). Therefore the light waves must propagate along the z axis of the coordinate system (i.e., parallel or antiparallel to $\mathbf{H}_0$).

In practice, of course, the wavelength of the radiation employed in nuclear magnetic resonance is too long for anything remotely resembling propagation to occur. The rf oscillations are delivered to the sample by means of a transmitter coil, wound in such a way as to produce a magnetic field that is spatially uniform in the neighbourhood of the sample. The axis of this coil is usually parallel to the x axis of the laboratory coordinate system. The resulting field oscillations are equivalent to those that would be produced by a linear superposition of travelling waves (phased to produce standing wave patterns) propagating in all directions in the yz plane. In electron spin resonance, the wave length of the radiation employed is small in comparison to the dimensions of the apparatus, but frequently still large in comparison to the sample. The radiation is piped about in waveguides rather than by means of wires because the electrical resistance of the latter is inconveniently high at microwave frequencies. In nuclear magnetic resonance the sample is usually supplied with radiation by a pair of Helmholz coils; in electron paramagnetic resonance the sample is placed in a microwave cavity at the end of a waveguide.

Regardless of the considerations outlined in the preceding paragraphs, it is accurate to describe the total magnetic field felt by the sample by means of the equation

$$\mathbf{H}(t) = H_0\hat{z} + H_1(t)\hat{x} \tag{8.114}$$

in any magnetic resonance experiment. The expression in Eq. (8.114) may be substituted into Eq. 4.38 to obtain a classical expression for the energy:

$$\mathscr{E} = -[p_z H_0 + p_x H_1(t)]. \tag{8.115}$$

By analogy to Eq. (8.85), the corresponding Hamiltonian operator is obtained from Eq. (8.115):

$$\mathsf{H} = -\gamma[H_0\,\mathsf{S}_z + H_1(t)\,\mathsf{S}_x]. \tag{8.116}$$

A comparison of Eq. (8.116) with Eqs. (5.55) and (8.86) permits identification of the term in the former that corresponds to the perturbation H_1:

$$\mathsf{H}_1 = -\gamma H_1(t)\,\mathsf{S}_x. \tag{8.117}$$

Can Eq. (8.117) be questioned by offering an alternative explanation for Eq. (8.116)? The field vector $\mathbf{H}$ presented in Eq. (8.114) is the vector sum of components in the $\hat{x}$ and $\hat{z}$ directions. It therefore describes a vector in the xz plane (see Fig. 8.6). If H_1 oscillates sinusoidally with time, then $\mathbf{H}$ oscillates back and forth about the z axis. What is to prevent the magnetic moments from simply following these oscillations, waving back and forth as reeds in a breeze, without any

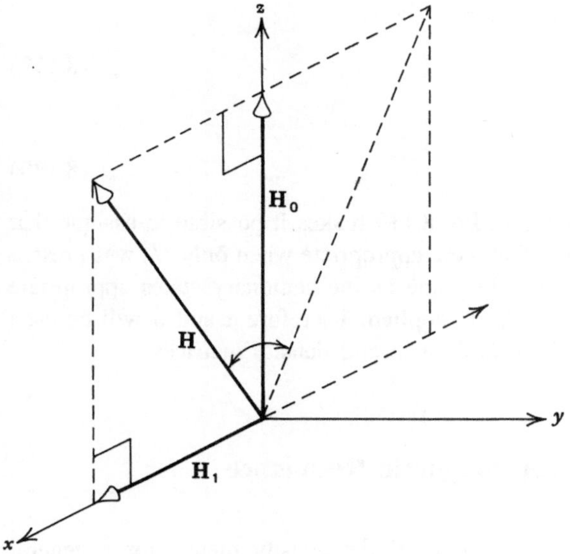

Fig. 8.6. Static and oscillating magnetic fields in a spin-resonance experiment

transitions occurring? Imagine, for example, that the big magnet used to produce $\mathbf{H}_0$ is mounted on a disk-shaped platform in the laboratory xz plane, and that the disk undergoes rapid, small-amplitude oscillations. This arrangement would produce a magnetic field of the same form as that expressed in Eq. (8.114) if $\hat{x}$, $\hat{y}$, and $\hat{z}$ are understood to be fixed laboratory coordinates.. Would these oscillations be capable of producing transitions just as an electromagnetic field is supposed to do?

These objections are legitimate – there are circumstances in which the field $H_1(t)$ does *not* produce transitions, and there are also circumstances in which a mere change in the relative coordinates of the sample and the magnet that produces $\mathbf{H}_0$ *does* produce transitions. Any oscillations (in magnitude or direction) of the total magnetic field $\mathbf{H}$, regardless of how they are produced, may or may not cause transitions to take place; what actually happens in any particular case is determined by the *adiabatic theorem*. This theorem can be stated precisely and proved rigorously, but for the moment it will be sufficient to state it loosely. If H_1 is small in comparison to H_0, *and* if H_1 oscillates slowly in comparison with the Larmor precession frequency, transitions will not occur: the dipoles will "follow" the vector sum of the fields, $\mathbf{H}$. If H_1 is large *or* oscillates at a frequency near $\omega_0 = \gamma H_0$, *or* both, transitions will occur. The discussion that follows will be restricted to the latter case.

If the electromagnetic oscillations H_1 are sinusoidal (i.e., if the "light" is monochromatic),

$$H_1(t) = H_1^0 \cos \omega t. \tag{8.118}$$

As has just been stated, the assumptions

$$H_1^0 \ll H_0 \tag{8.119}$$

and

$$\omega \cong \gamma H_0 \tag{8.120}$$

will be made. The condition stated in Eq. 8.119 makes it possible to assume that the stationary states of the system that were appropriate when only H_0 was present (namely, α and β) are very nearly the same as the stationary states appropriate for the system when both H_0 and H_1 are applied. Therefore α and β will be used as the representation for both the Hamiltonian and density matrices.

8.8 The Density Matrix for Magnetic Resonance

The equations of motion for the elements of the density matrix for a general two-level system were presented in Chap. 5 (Eqs. 5.69–5.72). The analysis was interrupted because an explicit expression for the elements of the matrix representing H_1 was lacking at that time. Now that the correct form for the H_1 operator (Eq. 8.117) and the matrix elements of S that appear therein (Eq. 8.102 and 8.103) have been presented, it can be shown that

$$(\mathsf{H}_1)_{12} = (\mathsf{H}_1)_{21} = -\frac{\gamma\hbar H_1(t)}{2} \tag{8.121}$$

and

$$(\mathsf{H}_1)_{11} = (\mathsf{H}_1)_{22} = 0. \tag{8.122}$$

Therefore, from Eqs. (5.69–5.72) and the above,

$$\frac{d\rho_{11}}{dt} = -\gamma H_1(t)\,\mathrm{Im}\,(\rho_{21}) - \left(\frac{\rho_{11} - \rho_{11}^e}{T_1}\right), \tag{8.123}$$

$$\frac{d\rho_{22}}{dt} = \gamma H_1(t)\,\mathrm{Im}\,(\rho_{21}) - \left(\frac{\rho_{22} - \rho_{22}^e}{T_1}\right), \tag{8.124}$$

$$\frac{d\rho_{12}}{dt} = i\omega_0\rho_{12} + \frac{i\gamma H_1(t)(\rho_{22} - \rho_{11})}{2} - \frac{\rho_{12}}{T_2}, \tag{8.125}$$

$$\frac{d\rho_{21}}{dt} = -i\omega_0\rho_{21} - \frac{i\gamma H_1(t)(\rho_{22} - \rho_{11})}{2} - \frac{\rho_{21}}{T_2}. \tag{8.126}$$

Equations 8.123–8.126 can be simplified by doing a little thinking. It is already known that all magnetic moments in the sample are precessing in cones about the z axis at the Larmor frequency, ω_0. This fast motion in the laboratory coordinate system interferes with visualization of the problem and may be eliminated by transforming everything into a rotating coordinate system:

$$\rho_{12}^{\dagger} = \rho_{12}\exp(-i\omega_0 t) \tag{8.127}$$

and

$$\rho_{21}^{\dagger} = \rho_{21}\exp(+i\omega_0 t). \tag{8.128}$$

From this it can be calculated that

$$\frac{d\rho_{12}^{\dagger}}{dt} = \frac{d\rho_{12}}{dt}\exp(-i\omega_0 t) - i\omega_0\rho_{12}\exp(-i\omega_0 t) \tag{8.129}$$

and

$$\frac{d\rho_{21}^{\dagger}}{dt} = \frac{d\rho_{21}}{dt}\exp(i\omega_0 t) + i\omega_0\rho_{12}\exp(i\omega_0 t). \tag{8.130}$$

Also, it is useful to define

$$\omega_1 = \gamma H_1^0 \tag{8.131}$$

and use the expression for $H_1(t)$ given by Eq. 8.118 to write

$$\frac{d\rho_{11}}{dt} = -\omega_1 \cos\omega t\, \mathrm{Im}[\rho_{21}^{\dagger}\exp(-i\omega_0 t)] - \left(\frac{\rho_{11} - \rho_{11}^{e}}{T_1}\right), \tag{8.132}$$

$$\frac{d\rho_{22}}{dt} = \omega_1 \cos\omega t\, \mathrm{Im}[\rho_{21}^{\dagger}\exp(-i\omega_0 t)] - \left(\frac{\rho_{22} - \rho_{22}^{e}}{T_1}\right), \tag{8.133}$$

$$\frac{d\rho_{12}^{\dagger}}{dt} = \frac{i\omega_1 \cos\omega t(\rho_{22} - \rho_{11})}{2}\exp(-i\omega_0 t) - \frac{\rho_{12}^{\dagger}}{T_2}, \tag{8.134}$$

and

$$\frac{d\rho_{21}^{\dagger}}{dt} = -\frac{i\omega_1 \cos\omega t(\rho_{22} - \rho_{11})}{2}\exp(i\omega_0 t) - \frac{\rho_{21}^{\dagger}}{T_2} \tag{8.135}$$

Next, Eqs. (8.132) and (8.133) may be added and subtracted from one another:

$$\frac{d(\rho_{22} + \rho_{11})}{dt} = -\frac{[(\rho_{22} + \rho_{11}) - (\rho_{22}^{e} + \rho_{11}^{e})]}{T_1} \tag{8.136}$$

and

$$\frac{d(\rho_{22} - \rho_{11})}{dt} = +\, 2\omega_1 \cos\omega t\, \mathrm{Im}[\rho_{21}^{\dagger}\exp(-i\omega_0 t)] - \frac{[(\rho_{22} - \rho_{11}) - (\rho_{22}^{e} - \rho_{11}^{e})]}{T_1}. \tag{8.137}$$

Also, Eqs. (8.134) and (8.135) may be added and subtracted:

$$\frac{d(\rho^{\dagger}_{12}+\rho^{\dagger}_{21})}{dt} = \left[\frac{i\omega_1 \cos\omega t(\rho_{22}-\rho_{11})}{2}\right][\exp(-i\omega_0 t) - \exp(i\omega_0 t)] - \frac{(\rho^{\dagger}_{12}+\rho^{\dagger}_{21})}{T_2} \tag{8.138}$$

and

$$\frac{d(\rho^{\dagger}_{12}-\rho^{\dagger}_{21})}{dt} = \left[\frac{i\omega_1 \cos\omega t(\rho_{22}-\rho_{11})}{2}\right][\exp(-i\omega_0 t) + \exp(i\omega_0 t)] - \frac{(\rho^{\dagger}_{12}-\rho^{\dagger}_{21})}{T_2}. \tag{8.139}$$

Next, four real quantities are defined, using notation suggested by Dicke [2]:

$$R_1 \equiv \frac{\rho^{\dagger}_{12}-\rho^{\dagger}_{21}}{i} = 2\mathrm{Im}(\rho^{\dagger}_{12}), \tag{8.140}$$

$$R_2 \equiv \rho^{\dagger}_{12}+\rho^{\dagger}_{21} = 2\mathrm{Re}(\rho^{\dagger}_{12}), \tag{8.141}$$

$$R_3 \equiv \rho_{11}-\rho_{22}, \tag{8.142}$$

and

$$\Sigma \equiv \rho_{11}+\rho_{22}. \tag{8.143}$$

Equations 8.140–8.143 may be substituted into Eqs. 8.138–8.139. The results are as follows:

$$\frac{d\Sigma}{dt} = -\frac{(\Sigma-\Sigma^e)}{T_1}, \tag{8.144}$$

$$\frac{dR_3}{dt} = +\omega_1 \cos\omega t(R_2 \sin\omega_0 t + R_1 \cos\omega_0 t) - \left(\frac{R_3 - R_3^e}{T_1}\right), \tag{8.145}$$

$$\frac{dR_2}{dt} = -\omega_1 \cos\omega t R_3 \sin\omega_0 t - \frac{R_2}{T_2}, \tag{8.146}$$

and

$$\frac{dR_1}{dt} = -\omega_1 \cos\omega t R_3 \cos\omega_0 t - \frac{R_1}{T_2}. \tag{8.147}$$

Note that Σ is proportional to the total number of systems in the ensemble (the trace of the density matrix). Since the trace is invariant (matter being conserved), the right-hand side of Eq. (8.144) is zero. Therefore the solution to the equation is

$$\Sigma = \text{const.} = 1. \tag{8.148}$$

Note that R_3 is proportional to the difference in population between the ground and excited states. At ordinary (positive) spin temperatures, R_3 is positive. For inverted systems (pumped to negative spin temperatures), R_3 will be negative. A physical interpretation of R_1 and R_2 will be obtained shortly.

Next, the simplification of Eqs. 8.145–8.147 will be continued. Trigonometric identities may be used to rewrite the time-dependent terms:

$$\begin{aligned}\frac{dR_3}{dt} &= \frac{\omega_1}{2}R_2\{\sin[(\omega+\omega_0)t] - \sin[(\omega-\omega_0)t]\}\\ &+\frac{\omega_1}{2}R_1\{\cos[(\omega+\omega_0)t] + \cos[(\omega-\omega_0)t]\}\\ &-\left(\frac{R_3-R_3^e}{T_1}\right),\end{aligned}\tag{8.149}$$

$$\frac{dR_2}{dt} = -\frac{\omega_1 R_3\{\sin[(\omega+\omega_0)t] - \sin[(\omega-\omega_0)t]\}}{2} - \frac{R_2}{T_2},\tag{8.150}$$

and

$$\frac{dR_1}{dt} = -\frac{\omega_1 R_3\{\cos[(\omega+\omega_0)t] + \cos[(\omega-\omega_0)t]\}}{2} - \frac{R_1}{T_2}.\tag{8.151}$$

Remember that it already has been assumed that ω is very close to ω_0. This means that terms which oscillate at the difference frequency ($\cong 0$) are very much slower than those which oscillate at the sum frequency ($\cong 2\omega_0$). What will be the effect of the $2\omega_0$ terms? They push the elements of the density matrix first in one direction and then in the other, reversing sign tens (even hundreds) of millions of times a second under usual experimental conditions. The effect of these terms must be very small (rotating wave approximation). On the other hand, the slowly varying terms can have some long-run effect on the magnitude of R_1, R_2, and R_3. Therefore

$$\frac{dR_1}{dt} \cong \frac{-\omega_1 R_3 \cos[(\omega-\omega_0)t]}{2} - \frac{R_1}{T_2},\tag{8.152}$$

$$\frac{dR_2}{dt} \cong \frac{+\omega_1 R_3 \sin[(\omega-\omega_0)t]}{2} - \frac{R_2}{T_2},\tag{8.153}$$

and

$$\begin{aligned}\frac{dR_3}{dt} &\cong \frac{\omega_1\{R_1\cos[(\omega-\omega_0)t] - R_2\sin[(\omega-\omega_0)t]\}}{2}\\ &-\frac{R_3-R_3^e}{T_1}.\end{aligned}\tag{8.154}$$

One final rotation of the coordinate system may be made:

$$R_1 = S\cos\delta t + C\sin\delta t\tag{8.155}$$

and

$$R_2 = S\sin\delta t + C\cos\delta t \tag{8.156}$$

where

$$\delta \equiv \omega - \omega_0. \tag{8.157}$$

From Eq. (8.155),

$$\frac{dR_1}{dt} = \left(\frac{dS}{dt} + \delta C\right)\cos\delta t - \left(-\frac{dC}{dt} + \delta S\right)\sin\delta t. \tag{8.158}$$

From Eq. (8.152),

$$\frac{dR_1}{dt} = -\frac{\omega_1 R_3\cos\delta t}{2} + \frac{S\cos\delta t}{T_2} - \frac{C\sin\delta t}{T_2}. \tag{8.159}$$

Since $\cos\delta$ and $\sin\delta$ are linearly independent functions. the only way that Eqs. 8.158 and 8.159 can both be true is for the coefficients of these functions to be equal in both:

$$\frac{dS}{dt} + \delta C = -\frac{\omega_1 R_3}{2} + \frac{S}{T_2} \tag{8.160}$$

and

$$-\left(-\frac{dC}{dt} + \delta S\right) = -\frac{C}{T_2}. \tag{8.161}$$

Equations (8.160) and (8.161) could also have been obtained from Eqs. (8.153) and (8.156).

Equations (8.155) and (8.156) may be substituted into Eq. (8.154). Then

$$\frac{dR_3}{dt} = \frac{\omega_1 S}{2} - \left(\frac{R_3 - R_3^e}{T_1}\right). \tag{8.162}$$

8.9 The Ensemble-Averaged Magnetization

Equations 8.160–8.162 together with all the changes of variables and transformations of coordinate systems that have been performed, contain the answers to all the questions that could possibly be asked about an ensemble of spin-$\frac{1}{2}$ particles. For example, the x, y, and z components of magnetization of the sample will be calculated. If N' is the number of spins per unit volume, the magnetization is given by

$$\mathbf{M} = N'(\langle\overline{p_x}\rangle\hat{x} + \langle\overline{p_y}\rangle\hat{y} + \langle\overline{p_z}\rangle\hat{z}). \tag{8.163}$$

Equation 5.41 may be used to calculate $\langle\overline{p_y}\rangle$:

$$\langle\overline{p_y}\rangle = \mathrm{Tr}\left\{\begin{bmatrix}\rho_{11} & \rho_{12}\\ \rho_{21} & \rho_{22}\end{bmatrix}\begin{bmatrix}(\mathbf{p}_y)_{11} & (\mathbf{p}_y)_{12}\\ (\mathbf{p}_y)_{21} & (\mathbf{p}_y)_{22}\end{bmatrix}\right\}$$
$$= \rho_{11}(\mathbf{p}_y)_{11} + \rho_{12}(\mathbf{p}_y)_{21} + \rho_{21}(\mathbf{p}_y)_{12} + \rho_{22}(\mathbf{p}_y)_{22}. \tag{8.164}$$

Since

$$\mathbf{p}_y = \gamma\mathbf{s}_y, \tag{8.165}$$

and the matrix elements of S_y are available in Eqs. (8.102) and (8.103),

$$\langle\overline{p_y}\rangle = \frac{i\gamma\hbar(\rho_{12} - \rho_{21})}{2}. \tag{8.166}$$

From Eqs. (8.127) and (8.128),

$$\langle\overline{p_y}\rangle = \frac{i\gamma\hbar[\rho_{12}^{\dagger}\exp(i\omega_0 t) - \rho_{21}^{\dagger}\exp(-i\omega_0 t)]}{2}. \tag{8.167}$$

From Eqs. (8.140) and (8.141),

$$\langle\overline{p_y}\rangle = \frac{-\gamma\hbar(R_1\cos\omega_0 t + R_2\sin\omega_0 t)}{2}. \tag{8.168}$$

From Eqs. (8.155) and (8.156),

$$\langle\overline{p_y}\rangle = -\frac{\gamma\hbar\,[S(\cos\delta\cos\omega_0 t - \sin\delta\sin\omega_0 t) + C(\sin\delta\cos\omega_0 t + \cos\delta\sin\omega_0 t)]}{2}. \tag{8.169}$$

Using trigonometirc identities and Eq. 8.157 yields

$$\langle\overline{p_y}\rangle = \frac{-\gamma\hbar(S\cos\omega t + C\sin\omega t)}{2}. \tag{8.170}$$

Similarly,

$$\langle\overline{p_x}\rangle = \frac{\gamma\hbar(-S\sin\omega t + C\cos\omega t)}{2}. \tag{8.171}$$

and

$$\langle\overline{p_z}\rangle = \frac{\gamma\hbar R_3}{2}. \tag{8.172}$$

The components of the expectation value of the magnetic dipole moment presented in Eqs. (8.170–8.172) can be compared with the components of the magnetic field that have produced these expectation values (see Eqs. 8.114, 8.117, and 8.118) in order to make a physical interpretation of the quantities S and C. The field $H_1(t)$ corresponds, as has been stated previously, to an electromagnetic wave propagating in the z direction and polarized so that the magnetic field oscillations occur exclusively in the xy plane. As was stated in Chap. 4, a plane-polarized wave may be thought of as consisting of two counterrotating circularly polarized waves of equal amplitude $H_1^0/2$, as shown in Fig. 4.9. Because the magnetogyric ratio of the spin-$\frac{1}{2}$ particle has a definite algebraic sign (assumed to be positive in this example), the Larmor precession of the magnetic moments occurs in only one sense (i.e., counterclockwise) about the z axis. Therefore only one circularly polarized component of H_1 can effectively "chase" the spins around. In the rotating coordinate system introduced by transformations given in Eqs. (8.127) and (8.128) the magnetization vector (Eq. 8.163) and the appropriate component of H_1 move very slowly. The former exerts a torque upon the latter, doing rotational work, and transfers energy from the electromagnetic field to the ensemble of spins. The other rotating component of H_1 oscillates in this frame at very nearly twice the Larmor frequency, producing only a small jiggling motion (sometimes called *zitterbewegung*) in the tip of the precessing magnetization vector, and is neglected. C is proportional to the component of magnetization in phase with the driving field; S is proportional to the component in quadrature with the same; R_3 is a measure of the remaining z component of magnetization, which changes only slowly in time in both the laboratory and rotating coordinate systems. Equations 6.160 to 6.162, when multiplied on both sides by $N'\gamma\hbar/2$ to give the quantities that appear there in the units of magnetization, were first derived by Felix Bloch [3] and are called the *Bloch equations* in his honor.

According to the conventional notation,

$$u \equiv \frac{N'\gamma\hbar C}{2}, \tag{8.173}$$

$$v \equiv -\frac{N'\gamma\hbar S}{2}, \tag{8.174}$$

$$M_z \equiv \frac{N'\gamma\hbar R_3}{2}, \tag{8.175}$$

and

$$M_0 \equiv \frac{N'\gamma\hbar R_3^e}{2}. \tag{8.176}$$

Then Eqs. 8.160–8.162 become

$$\frac{du}{dt} = -\delta v - \frac{u}{T_2}, \tag{8.177}$$

$$\frac{dv}{dt} = \delta u + \frac{\omega_1 M_z}{2} - \frac{v}{T_2}, \tag{8.178}$$

and

$$\frac{dM_z}{dt} = -\frac{\omega_1 v}{2} - \left(\frac{M_z - M_0}{T_1}\right). \tag{8.179}$$

Also, form Eqs. 8.163 and 8.170–8.175,

$$\mathbf{M}(t) = M_z\hat{z} + (v\sin\omega t + u\cos\omega t)\hat{x} + (v\cos\omega t - u\sin\omega t)\hat{y}. \tag{8.180}$$

The effect of the second coordinate rotation in Eqs. 8.155 and 8.156 was to transform the problem into a reference frame rotating at the frequency ω rather than ω_0 (interaction representation). The quantities M_z, u and v are the components of the magnetization vector which lie along the x', y', and z axes fixed in that rotating frame. In the same frame, the "correct" circularly polarized component of the electromagnetic wave is represented by a vector of constant length $H_1/2$ oriented along the x' axis. This physical interpretation will facilitate the solution to the Bloch equations 8.177–8.179 in several cases of interest.

8.10 Solutions to the Bloch Equations

Suppose that the ensemble sits for a time much longer than the relaxation time $T_1 > T_2$, in the absence of an rf field. The absence of an rf field means that $\omega_1 = 0$, and δ (not being defined) may be taken as zero without any loss of generality. The only terms that survive on the right-hand side of Eqs. 8.177–8.179 are the damping terms due to relaxation. The solutions are as follows:

$$u(t) = u(t')\exp\left[\frac{-(t-t')}{T_2}\right], \tag{8.181}$$

$$v(t) = v(t')\exp\left[\frac{-(t-t')}{T_2}\right], \tag{8.182}$$

and

$$M_z(t) - M_0 = [M_z(t') - M_0]\exp\left[\frac{-(t-t')}{T_1}\right]. \tag{8.183}$$

The values of the constants $u(t'), v(t')$, and $M_z(t')$ are not known, but that is not important because the system has sat for so long that $t - t' \to \infty$. The exponential terms therefore damp out the right-hand sides of Eqs. 8.181–8.183, giving the results

$$u = 0, \tag{8.184}$$

$$v = 0, \tag{8.185}$$

and

$$M_z = M_0. \tag{8.186}$$

Equations 8.184–8.186 are the conditions for equilibrium for an ensemble of magnetic two-level systems.

Next, suppose that an rf field $H_1(t)$ is turned on suddenly at time $t = 0$, and the behavior of the ensemble is examined at times very much shorter than T_1 and T_2, so that the relaxation will not yet have affected the magnetization. Furthermore, let the field oscillate at exact resonance with the Larmor frequency of the sample,

$$\omega = \omega_0. \tag{8.187}$$

The Bloch equations become

$$\frac{du}{dt} = 0, \tag{8.188}$$

$$\frac{dv}{dt} = \frac{\omega_1 M_z}{2}, \tag{8.189}$$

and

$$\frac{dM_z}{dt} = -\frac{\omega_1 v}{2}. \tag{8.190}$$

Equation 8.188 can be solved immediately:

$$u(t) = u(0) = 0. \tag{8.191}$$

The second equality in Eq. (8.191) comes from Eq. (8.184). To solve Eqs. (8.188) and (8.190), it is necessary to differentiate one with respect to time and substitute the other:

$$\frac{d^2v}{dt^2} = +\frac{\omega_1}{2}\frac{dM_z}{dt} = +\frac{\omega_1}{2}\left(-\frac{\omega_1}{2}\right)v \tag{8.192}$$

and

$$\frac{d^2M_z}{dt^2} = -\frac{\omega_1}{2}\frac{dv}{dt} = -\frac{\omega_1}{2}\left(+\frac{\omega_1}{2}\right)M_z. \tag{8.193}$$

This uncouples the two equations. The solutions to Eqs. 8.192 and 8.193 are

$$v = A_1 \sin\left(\frac{\omega_1 t}{2}\right) + A_2 \cos\left(\frac{\omega_1 t}{2}\right) \tag{8.194}$$

and

$$M_z = B_1 \sin\left(\frac{\omega_1 t}{2}\right) + B_2 \cos\left(\frac{\omega_1 t}{2}\right). \tag{8.195}$$

The relationships among the coefficients A_1, A_2, B_1, and B_2 can be discovered by substituting Eqs. (8.194) and (8.195) into the original Eqs. (8.189) and (8.190):

$$\omega_1 A_1 \cos\left(\frac{\omega_1 t}{2}\right) - \omega_1 A_2 \sin\left(\frac{\omega_1 t}{2}\right) = \omega_1 B_1 \sin\left(\frac{\omega_1 t}{2}\right) + \omega_1 B_2 \cos\left(\frac{\omega_1 t}{2}\right). \tag{8.196}$$

Since the cosine and sine functions are independent of one another, Eq. (8.196) can be satisfied only if

$$A_1 = B_2 \tag{8.197}$$

and

$$A_2 = -B_1. \tag{8.198}$$

Now, Eqs. (8.197) and (8.198) may be substituted into Eqs. (8.194) and (8.195), and boundary conditions used in Eqs. 8.184–8.186 to find

$$v(0) = A_1 \sin(0) + A_2 \cos(0) = 0 \tag{8.199}$$

and

$$M_z(0) = -A_2 \sin(0) + A_1 \cos(0) = M_0. \tag{8.200}$$

Therefore

$$A_2 = 0 \tag{8.201}$$

and

$$A_1 = M_0. \tag{8.202}$$

Finally, combining Eqs. 8.194, 8.195,. 8.197, 8.201, and 8.202 yields

$$v = M_0 \sin\left(\frac{\omega_1 t}{2}\right) \tag{8.203}$$

and

$$M_z = M_0 \cos\left(\frac{\omega_1 t}{2}\right). \tag{8.204}$$

Here is the physical interpretation for Eqs. 8.203 and 8.204. In the frame rotating at $\omega = \omega_0$ at equilibrium, the spins do not precess about z. But a magnetic field is defined in terms of the torque it exerts upon the magnetic dipole moment, and an object having angular momentum that experiences a torque *must* process. It follows that in the rotating frame there *is* no field $H_0\hat{z}$; it has been transformed away. When H_1 is turned on, half of it appears as a vector of fixed orientation along the x' axis. Since this is the only magnetic field present in the rotating frame at exact resonance, the magnetization **M** of magnitude M_0, initially aligned with the z axis by relaxation processes, begins to precess around it. The precession of **M** around the field $H_1^0\hat{x}'/2$ occurs in a clockwise direction about the x' axis in the $y'z'$ plane. The frequency with which **M** precesses about $\mathbf{H}_1$ in the rotating frame is $\gamma H_1^0/2$. In the laboratory frame, of course, v precesses about the z axis at the frequency $\gamma H_0 \gg \gamma H_1^0/2$ (see Fig. 8.7). (Note the similarity to Fig. 4.4.) Since an oscillating or rotating dipole radiates electromagnetic waves, the sample will begin to "broadcast" a radio frequency signal in the laboratory. The amplitude of the radiated waves will be proportional to the square of v, the magnitude of the oscillating component of the magnetization. What has just been stated in words can be expressed mathematically by substituting Eqs. 8.191, 8.203, and 8.204 into Eq. 8.180:

$$M(t) = M_0\left[\cos\left(\frac{\omega_1 t}{2}\right)\hat{z} + \sin\left(\frac{\omega_1 t}{2}\right)\sin\omega t\,\hat{x} + \sin\left(\frac{\omega_1 t}{2}\right)\cos\omega t\,\hat{y}\right]. \tag{8.205}$$

Note that the torque exerted by H_1 produces a slow motion of the magnetization in a direction that is perpendicular to the fast motion produced by the torque due to H_0. In the classical theory of gyroscopes, a slow wobble of the

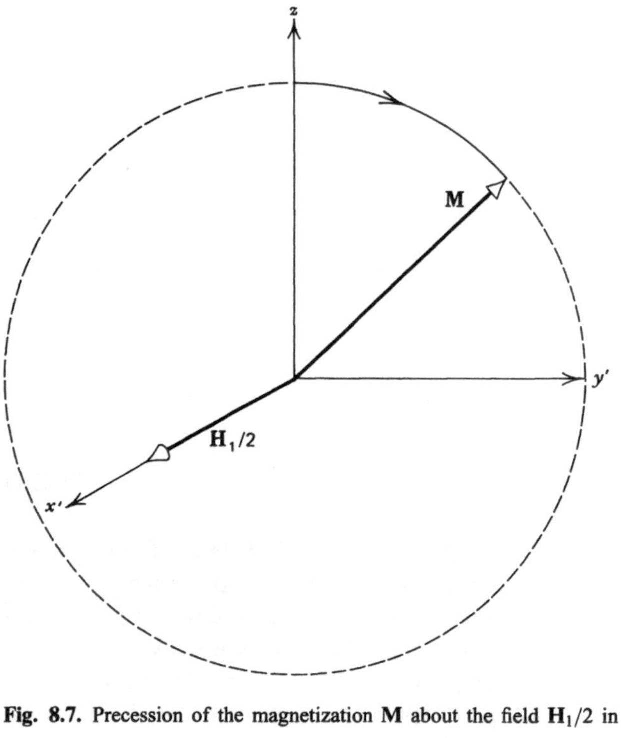

Fig. 8.7. Precession of the magnetization **M** about the field $\mathbf{H}_1/2$ in the rotating coordinate system

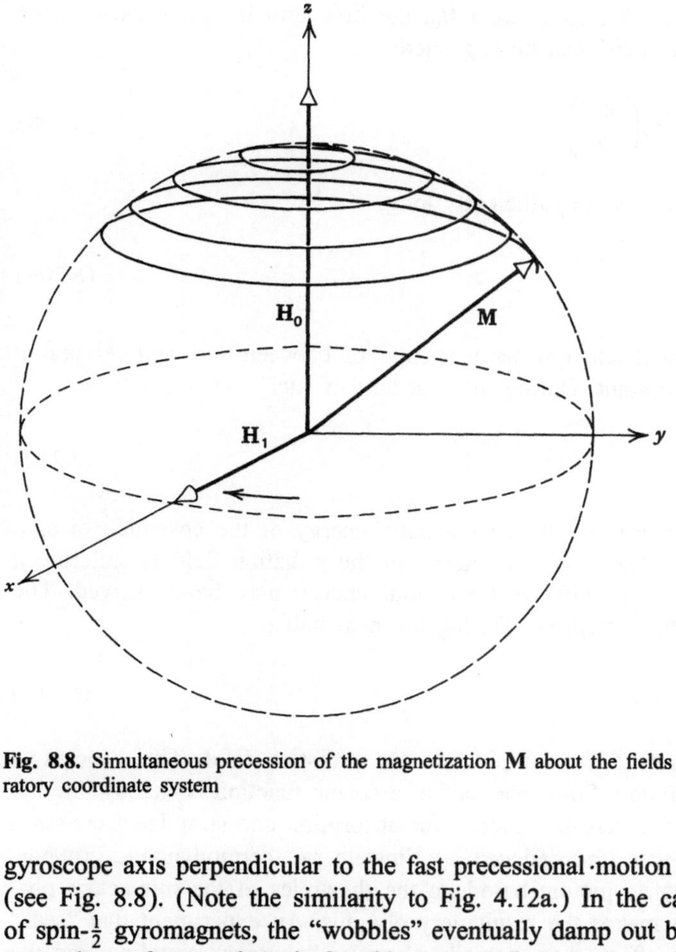

Fig. 8.8. Simultaneous precession of the magnetization **M** about the fields $\mathbf{H}_0$ and $\mathbf{H}_1$ in the laboratory coordinate system

gyroscope axis perpendicular to the fast precessional motion is called a *nutation* (see Fig. 8.8). (Note the similarity to Fig. 4.12a.) In the case of the ensemble of spin-$\frac{1}{2}$ gyromagnets, the "wobbles" eventually damp out because of relaxation terms heretofore neglected; for this reason, the response of the sample to the abrupt switching on of H_1 is called the *transient nutation effect*. It was first observed by H.C. Torrey [4] for nuclear spins.

8.11 Absorption and Stimulated Emission: Free-Induction Decay

The magnetic energy density of the ensemble can be computed by means of

$$\mathscr{E} = -\mathbf{M}\cdot\mathbf{H}. \tag{8.206}$$

Equation 8.206 is analogous to Eq. 8.115. Using Eq. 8.205 for **M** and Eq. 8.114 for **H**, one finds

$$\mathscr{E} = -M_0\left[\sin\left(\frac{\omega_1 t}{2}\right)\sin\omega t H_1^0\cos\omega t + \cos\left(\frac{\omega_1 t}{2}\right)H_0\right]. \tag{8.207}$$

Because H_1^0 is very much smaller than H_0, the first term in parentheses on the right-hand side of Eq. 8.207 can be neglected:

$$\mathscr{E} \cong -M_0 H_0 \cos\left(\frac{\omega_1 t}{2}\right). \tag{8.208}$$

If the relaxation times are sufficiently long,

$$T_1, T_2 \gg \frac{1}{\omega_1}, \tag{8.209}$$

the $\cos(\omega_1 t/2)$ term will undergo many excursions between $+\,1$ and -1 before damping becomes important. During the first half of such a cycle,

$$0 < \frac{\omega_1 t}{2} < \pi. \tag{8.210}$$

During the interval in Eq. 8.210, the magnetic energy of the ensemble of spin systems is increasing. Therefore the energy in the radiation field is suffering a corresponding decrease in order that the total energy may be conserved. The process corresponds to *absorption*. During the next half cycle,

$$\pi < \frac{\omega_1 t}{2} < 2\pi, \tag{8.211}$$

energy flows back out of the sample into the radiation field. This process is called *stimulated emission*. Since one half of a cosine function is exactly like the other, it is seen that the "cross sections" for absorption and stimulated emission are equal. This fact was first deduced by Einstein on thermodynamic grounds, but the simple geometric picture based on the dynamics of the interaction process presented above makes the result very obvious. An experiment that "sees" the transient nutation effect, then, actually observes "photons" crawling into the quantum systems and out again and over and over, until relaxation sets in. The amount of time required for either process can be calculated from

$$\frac{\omega_1 t}{2} = \pi. \tag{8.212}$$

The relationship between $\mathbf{H}_1$ and $\mathbf{M}$ during transient nutations is shown schematically in Fig. 8.9. The question of how such an experiment would be performed will be deferred for the moment. Suppose that, at some time Υ during the course of a Torrey-type experiment,

$$0 < \Upsilon \ll T_1, T_2, \tag{8.213}$$

$H_1(t)$ was turned off as suddenly as it was turned on. This would mean that the system had been irradiated with a brief pulse of electromagnetic radiation. The

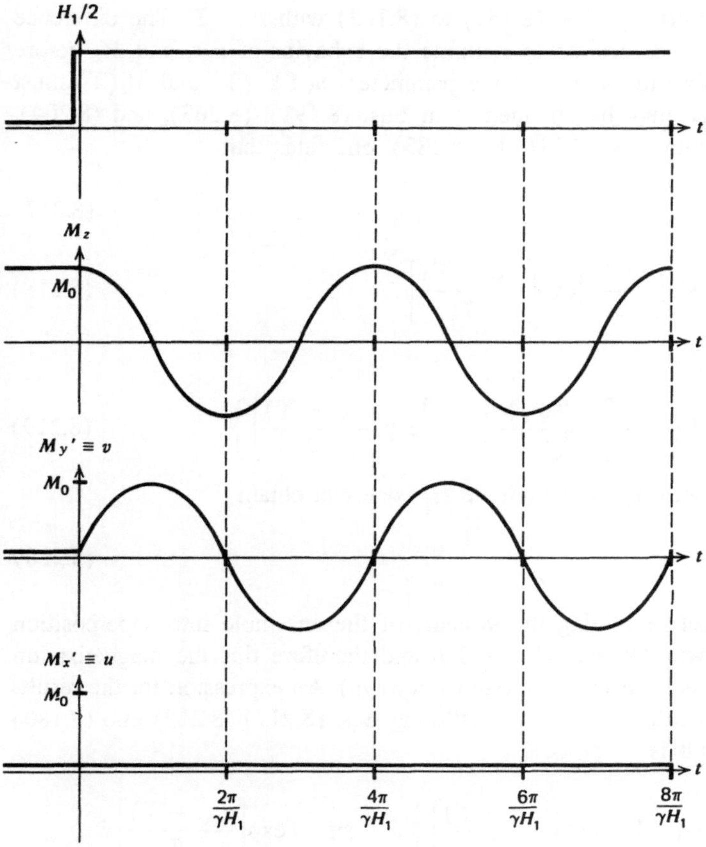

Fig. 8.9. Relationship between H_1 and M_x', and M_y', and M_z in the transient nutation effect.

Bloch equations (Eqs. 8.177 to 8.179) which apply during the subsequent time interval are

$$\frac{du}{dt} = -\frac{u}{T_2}, \tag{8.214}$$

$$\frac{dv}{dt} = -\frac{v}{T_2}, \tag{8.215}$$

and

$$\frac{dM_z}{dt} = -\frac{M_z - M_0}{T_1}, \tag{8.216}$$

Equations (8.214) to (8.216) are identical with those used at the start of the discussion of the values of u, v and M_z obtained at thermodynamic equilibrium;

the solutions are given by Eqs. (8.181) to (8.183) with $t' = \Upsilon$. The difference in this case is that it is desired to examine the behavior of u, v, and M_z before $t \to \infty$, and therefore the values of the parameters $u(\Upsilon), v(\Upsilon)$ and $M_z(\Upsilon)$ must be computed. These may be obtained from Eqs. (8.191), (8.203), and (8.204). Substituting them into Eqs. (8.181) to (8.183), one finds that

$$u(t) = 0, \tag{8.217}$$

$$v(t) = M_0 \sin\left(\frac{\omega_1 \Upsilon}{2}\right) \exp\left[-\frac{(t-\Upsilon)}{T_2}\right]. \tag{8.218}$$

and

$$M_z(t) = M_0\left\{1 + \left[\cos\left(\frac{\omega_1 \Upsilon}{2}\right) - 1\right] \exp\left[-\frac{(t-\Upsilon)}{T_1}\right]\right\}. \tag{8.219}$$

By properly choosing the amplitude H_1^0, one can obtain

$$\frac{\omega_1 \Upsilon}{2} = \frac{\pi}{2}. \tag{8.220}$$

which has the effect of putting the systems of the ensemble into superposition states exactly halfway "between" α and β and therefore tips the magnetization vector completely into the xy plane (spin sideways). An expression for the resulting magnetization can be found by substituting Eqs. (8.217) (8.219) into (8.180) as before. The result is

$$\mathbf{M}(t) = M_0\left(\left\{1 - \exp\left[-\frac{(t-\Upsilon)}{T_1}\right]\right\}\hat{z} + \sin\omega t \exp\left[-\frac{(t-\Upsilon)}{T_2}\right]\hat{x} + \cos\omega t \exp\left[-\frac{(t-\Upsilon)}{T_2}\right]\hat{y}\right). \tag{8.221}$$

Note that the x and y components of magnetization die away exponentially in a characteristic time T_2; the z component of magnetization, however, grows exponentially back to its equilibrium value in a time T_1. Because the xy magnetization induced originally by H_1 is no longer being driven by that field, the behavior of a sample after irradiation by a brief intense electromagnetic pulse is called *free-induction decay*. It was first observed E. Hahn [5], in an ensemble of nuclear spins. The relationship between $\mathbf{H}_1$ and $\mathbf{M}$ during free-induction decay is shown schematically in Fig. 8.10a; experimental results are displayed in Fig. 8.10b.

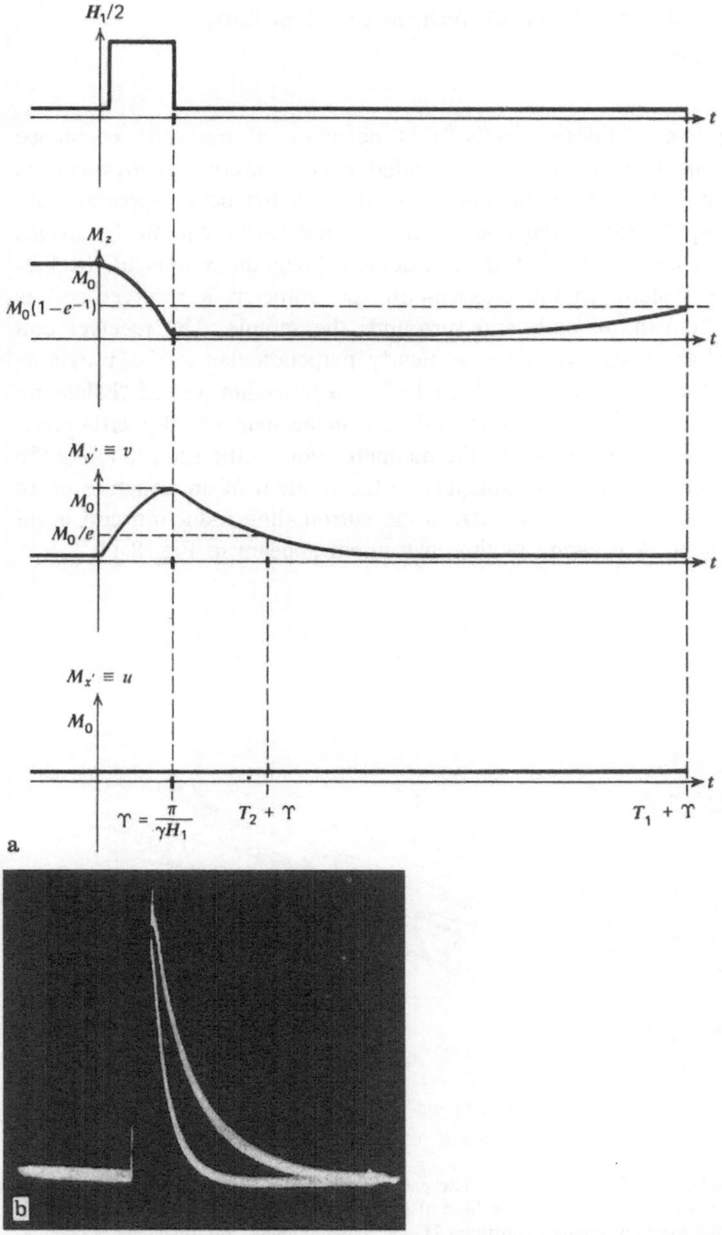

Fig. 8.10. **a** Relationship between H_1 and M_x', M_y' and M_z in free-induction decay **b** The first experimental observation of free-induction decay by E.L. Hahn [(1950) Phys Rev 77:297, Fig. 1]. The time constant for the decay, T_m, is a combination of the true relaxation time, T_2, and the inverse width of the distribution of Larmor frequencies, T_2^*;$1/T_m = 1/T_2 + 1/T_2^*$. The quantum systems are protons in water excited at exact resonance ($\delta = 0$) by means of a 90° pulse (not shown). The oscillogram is a double exposure; the longer decay corresponds to a T_2 of 1.8 ms, the shorter to 0.4 ms, and $T_2^* = 0.46$ ms for both. The relaxation time was adjusted by adding varying amounts of unpaired electron spins [Fe^{3+} ions from $Fe(NO_3)_3$]

8.12 The Crossed-Coil Nuclear Magnetic Resonance Spectrometer

It is now appropriate to discuss methods of detection of magnetic resonance signals. For nuclear spins, an instrument called a *crossed-coil spectrometer* is commonly employed. In this instrument, as well as in the beam apparatus described in Chap. 4, the radio frequency is transmitted to the sample by means of Helmholtz coils, aligned so that H_1 is oriented along the x axis of the laboratory coordinate system. In the crossed-coil spectrometer, a receiver coil is placed inside the transmitter coils and surrounds the sample. The receiver coil is mechanically aligned with its plane as nearly perpendicular to the y axis as possible. By adjusting field deflection "paddles" in a procedure called "balancing the probe," the transmitter and receiver coils are made magnetically orthogonal to one part per million. Oscillations of the magnetization of the sample along the y axis induce a voltage across the coil, just as the rotation of an armature of an electric generator produces a voltage across the surrounding induction coil in an electrical power plant. A drawing of this instrument appears in Fig. 8.11.

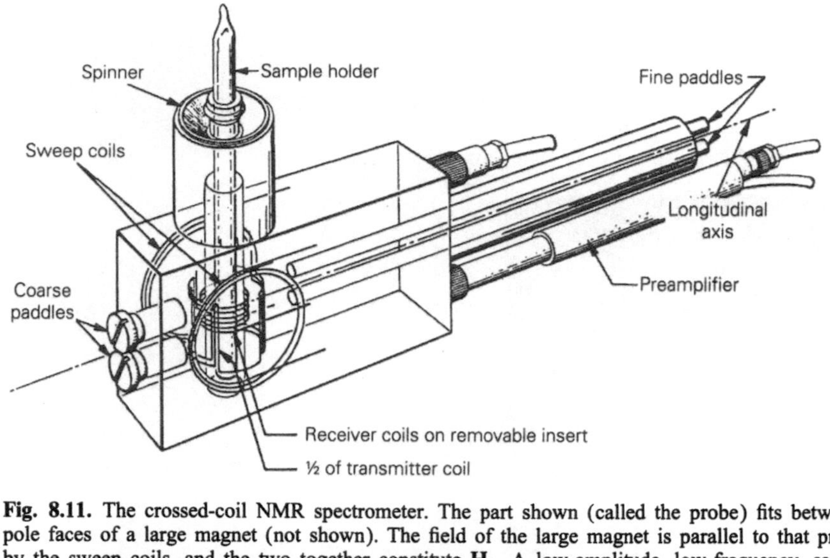

Fig. 8.11. The crossed-coil NMR spectrometer. The part shown (called the probe) fits between the pole faces of a large magnet (not shown). The field of the large magnet is parallel to that produced by the sweep coils, and the two together constitute $\mathbf{H}_0$. A low-amplitude, low-frequency, sawtooth-shaped current signal is applied through the lowest cable to the sweep coils. This causes the output of the receiver coil to periodically produce the slow-passage signal (u or v). This signal passes through the preamplifier, then to an rf detector, and then to the vertical deflection plates of an oscilloscope. (The horizontal plates are driven by the sweep generator.) An rf transmitter supplies a signal at the frequency ω through the uppermost cable to the transmitter coil, which then produces $\mathbf{H}_1$. The spinner averages out the spatial inhomogeneities in $\mathbf{H}_0$. Photograph courtesy of Norman S. Bhacca. This drawing is a positive print of Fig. 3 on p. 58 of *NMR and EPR Spectroscopy* (Pergamon Press, London, 1960), by Varian Associates

The amplitude of the oscillations in the voltage signal is directly proportional to the magnitude of M_y and has the same frequency. Generally, the electronics introduce a shift in the phase of the detected voltage relative to H_1:

$$V \propto v\cos(\phi + \omega t) - u\sin(\phi + \omega t). \tag{8.222}$$

This voltage is "mixed" with a reference signal derived from the same crystal-controlled oscillator that supplied H_1; the output of the mixer is porportional to the product of the inputs:

$$\begin{aligned} V_{\text{out}} &\propto V \cdot V_{\text{ref}} \\ &\propto H_1^0 \cos\omega t[v\cos(\phi + \omega t) - u\sin\phi + \omega t)] \\ &= \frac{H_1^0}{2}\{v[\cos(\phi + 2\omega t) + \cos\phi] - u[\sin(\phi + 2\omega t) + \sin\phi]\}. \end{aligned} \tag{8.223}$$

The output signal is then "detected" by passing it through an electronic element with a response time too slow to respond to oscillations as fast as ω:

$$V_{\text{det}} \propto H_1^0(v\cos\phi - u\sin\phi). \tag{8.224}$$

Because the magnitude of ϕ is controllable by the experimenter, the detected signal may be adjusted to be proportional to either u or v alone (e.g., by choosing $\phi = 3\pi/2$ or $\phi = 0$, respectively) or to any desired combination thereof according to Eq. (8.224). An oscilloscope or a strip-chart recorder may be used to display V_{det}, as desired.

This electronic gadgetry is called collectively a *phase-sensitive heterodyne detector*. It would also be possible to utilize a *homodyne detector*; this name means that the signal is its own reference in the mixer:

$$\begin{aligned} V_{\text{out}} &\propto V \cdot V_{\text{ref}} \\ &\propto v^2\cos^2(\phi + \omega t) + u^2\sin^2(\phi + \omega t)c - 2uv\cos(\phi + \omega t)\sin(\phi + \omega t) \\ &= \frac{v^2}{2}\{1 + \cos[2(\phi + \omega t)]\} + \frac{u^2}{2}\{1 - \cos[2(\phi + \omega t)]\} \\ &\quad -uv\sin[2(\phi + \omega t)]. \end{aligned} \tag{8.225}$$

In this case, after the high-frequency components are averaged to zero in the detector portion of the apparatus, the resultant signal is

$$V_{\text{det}} \propto u^2 + v^2. \tag{8.226}$$

Note that phase information is not obtainable from homodyne detection: Eq. 8.226 is independent of ϕ. The signal is always proportional to the square of the magnetization, and hence to the total power radiated by the sample; it is the *photon density*, one might say. It can be shown that photomultipliers and other detectors used in conventional spectroscopy effectively produce a homodyne signal as they are ordinarily employed.

More can be said about the photon picture of the detection process. Because the receiver coil is oriented to "see" magnetic oscillations along the y axis, it will pick up photons propagating along any direction in the xz plane. The transmitter coil, on the other hand, broadcasts in the yz plane. The intersection of these two planes is the z axis. The probability that the propagation vector of a transmitted photon lies exactly along that axis is small because the radiation is distributed isotropically in the yz plane. This is the reason why, in a properly balanced probe from which the sample has been removed, the transmitter can be operated at full power and yet no detectable amount of rf radiation will enter the receiver coil. This is exactly the opposite of what one would expect in an absorption sepctrometer, in which the signal detected is of maximum intensity in the absence of the sample. In such an instrument, the sample is placed between the source and the detector on the optic axis. The interaction of radiation and matter in the sample has the effect of removing photons from the incident beam and *reducing* the intensity of the radiation received by the detection system. Therefore the crossed-coil nmr spectrometer is not an absorption spectrometer and must be an *emission* spectrometer.

There are two kinds of emission, spontaneous and stimulated. One feature of stimulated emission noted by Einstein is that a stimulated photon always propagates in the same direction as the photon that does the stimulating. In otherwords, the beam of electromagnetic radiation produced by stimulated emission from matter is always collinear with the beam that is incident upon the sample. But, because of the geometry of a crossed-coil NMR spectrometer, the receiver coil detects only radiation propagating at right angles to the incident beam. Therefore the receiver coil cannot see either the incident photons or the photons stimulated by the incident beam. This means that the emission detected in a crossed-coil NMR spectrometer is spontaneous emission.

The geometrical arguments given in the preceding paragraph can be supplemented by temporal ones. If the light source is turned on abruptly in an absorption spectrometer, the signal appears with maximum intensity immediately. (There is actually a small delay due to the fact that perhaps 100 ps is required for the incident light to propagate from the source to the detector, but this can be neglected completely in NMR experiments.) Stimulated emission also appears immediately (although not necessarily at its maximum intensity). The signal in an absorption spectrometer ceases the instant the incident beam is turned off. Stimulated emission also disappears at the same instant that the exciting beam is extinguished. By way of contrast, in an NMR experiment, when the incident H_1 field is turned on the signal starts from zero and gradually builds to its maximum value as $\sin(\omega_1 t/2)$–see Eq. (8.205) and remember that the signal is proportional to M_y. After the exciting electromagnetic radiation is turned off, the signal does not disappear instantaneously; instead, M_y decays exponentially with a time constant T_2, which may be as long as several seconds. This can be seen from Eq. (8.221).

It has therefore been proved, by both geometric and temporal arguments, that the signal produced in a crossed-coil NMR spectrometer is due to spontaneous emission. But this spontaneous emission is very peculiar. A branch of conventional spectroscopy is devoted to the measurement of fluorescent lifetimes, an area of study that is superficially very similar to the study of free-induction decay in nuclear magnetic resonance. In both cases a more or less exponential decay of luminescent intensity following the pulse is observed. If one calculates the intensity of spontaneous fluorescence to be expected from a nuclear spin system (which may be done by calculating the Einstein A coefficient), one finds that the signal should be so feeble as to be unobservable. Furthermore, the fluorescent lifetime (the time constant for the exponential decay) is simply the time required for the population difference between initial and final states to return to the value expected for a Boltzmann distribution at the temperature of the sample. This is the time T_1. In fact, the intensity of the signal is *not* undetectably low, and the decay constant is T_2, which may be several orders of magnitude less than T_1–see Eq. (8.221) and again remember that the *signal* is M_y. The reason for this paradoxical behavior of the free induction is that the spontaneous emission detected in a crossed-coil NMR spectrometer is coherence brightened, as discussed in Chap. 4. (Some of the ideas presented in this section were published previously by the author [6].)

8.13 Steady-State Magnetization: Curie's Law

Now steady-state solutions to the Bloch equations will be discussed. These are appropriate for calculating the appearance of the ordinary NMR spectrum familiar to analytical chemists. In this case the field H_1 is so weak that energy is removed from the ensemble by relaxations as fast as it can be supplied by the absorption of electromagnetic radiation. Therefore

$$\frac{du}{dt} = \frac{dv}{dt} = \frac{dM_z}{dt} = 0. \tag{8.227}$$

Equations (8.177) to (8.179) become merely linear algebraic equations rather than differential equations and can be written in matrix form:

$$\begin{bmatrix} 0 \\ 0 \\ (M_0/T_1) \end{bmatrix} = \begin{bmatrix} +(1/T_2) & +\delta & 0 \\ -\delta & +(1/T_2) & -(\omega_1/2) \\ 0 & +(\omega_1/2) & +(1/T_2) \end{bmatrix} \begin{bmatrix} u \\ v \\ M_z \end{bmatrix}. \tag{8.228}$$

The solution to the system of equations represented by Eq. 8.228 can be found by standard methods (e.g., using determinants):

$$u = \frac{\delta T_2[-(\omega_1/2)]M_0}{\P} \tag{8.229}$$

$$v = -\frac{(\omega_1/2)M_0}{\P}, \tag{8.230}$$

$$M_z = \frac{[1+(\delta T_2)^2]M_0}{\P}, \tag{8.231}$$

where

$$\P \equiv 1 + \left(-\frac{\omega_1}{2}T_1\right)\left(-\frac{\omega_1}{2}T_2\right) + (\delta T_2)^2. \tag{8.232}$$

The calculation of M_0 from Eqs. (8.176), (8.142), (5.47), (5.48), (8.90), and (8.91) is straightforward:

$$\begin{aligned} M_0 &= \frac{N'\gamma\hbar R_3^e}{2} \\ &= \frac{N'\gamma\hbar(\rho_{11}^e - \rho_{22}^e)}{2} \\ &= \frac{N'\gamma\hbar}{2}\left[\frac{\exp(-\mathscr{E}_1/k_0T) - \exp(-\mathscr{E}_2/k_0T)}{\exp(-\mathscr{E}_1/k_0T) + \exp(-\mathscr{E}_2/k_0T)}\right] \\ &= \frac{N'\gamma\hbar}{2}\tanh\left(\frac{\hbar\omega_0}{2k_0T}\right). \end{aligned} \tag{8.233}$$

Finally, by substituting Eqs. 8.229–8.233 into Eq. 8.180, one may obtain

$$\begin{aligned} \mathbf{M}(t) = \frac{N'\gamma\hbar\tanh(\hbar\omega_0/2k_0T)}{2[1+(\omega_1^2/4)T_1T_2+\delta^2T_2^2]}\Big\{ & -\frac{\omega_1T_2}{2}[(\sin\omega t + \delta T_2\cos\omega t)\hat{x} \\ & +(\cos\omega t - \delta T_2\sin\omega t)\hat{y}] + (1+\delta^2T_2)\hat{z}\Big\}. \end{aligned} \tag{8.234}$$

Several features of this equation deserve special mention. First, note that the amplitude of the magnetization is proportional to the number of spin-$\frac{1}{2}$ particles in the ensemble. Since the radiated power is porportional to the square of the amplitude, it is also proportional to the square of the number of radiators. This quadratic dependence of the radiant intensity on the number of quantum systems is a feature of coherent spontaneous emission, as has been previously stated. Second, it may be seen that the magnetization is proportional to the magnitude of the magnetic dipole moment of a representative spin, $\gamma\hbar/2$. Third, it may be noted that the magnetization is proportional to the fractional excess of population in the ground state at equilibrium, $\tanh(\hbar\omega_0/2k_0T)$. The usual situation in magnetic resonance spectroscopy is that $\hbar\omega_0/k_0T \ll 1$ for ordinary temperatures. Therefore

$$\tanh\left(\frac{\hbar\omega_0}{2k_0T}\right) \cong \frac{\hbar\omega_0}{2k_0T}. \tag{8.235}$$

Remember that ω_0 is proportional to H_0. In the absence of an oscillating field, H_1, Eq. (8.235) implies that the equilibrium magnetization of a sample is inversely proportional to the temperature – this is called *Curie's law of paramagnetism* [7] (see Fig. 8.12). Nuclear paramagnetism is so weak that the detection of NMR signals requires very sensitive electronics. The fourth feature of note in Eq. (8.234) is that the sizes of both the x and the y components of magnetization are proportional to ω_1, which in turn, it should be remembered, is proprotional to the strength of the rf field H_0^1 for values of $\omega_1 T_2 \ll 1$. If the field is made too intense, however, the $\omega_1^2 T_1 T_2$ term in the denominator causes all three components of magnetization to diminish. This phenomenon, called *saturation*, is due to the fact that, at high power levels, photons are absorbed at such a rate that relaxation cannot keep up. The population difference between the two energy levels, small to begin with, is reduced still further until absorption and stimulated emission become equally likely and no net energy transfer between sample and radiation field is possible. Under these conditions, the Beer-Bouguer-Lambert law discussed in Chap. 1 cannot describe the absorptivity. This is so because an implicit assumption in Beer's law is that each chromophore has returned to its ground state before the next photon arrives, making the absorption coefficient independent of the illuminating intensity.

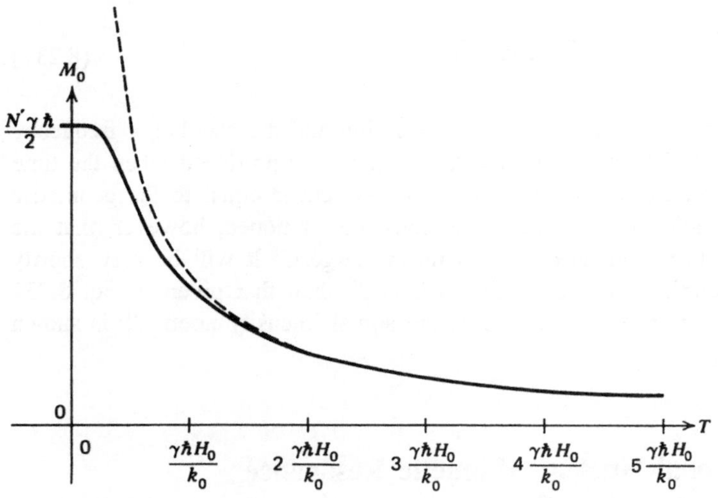

Fig. 8.12. The temperature dependence of the equilibrium magnetization, M_0. The *solid curve* is the exact solution (from Eq. 8.233 of the text). The *dotted curve* is the Curie's law approximation from Eq. 8.235. Note that the two are practically indistinguishable for $T > 2\gamma\hbar H_0/k_0$. Under typical conditions in nuclear magnetic resonance (protons in water, using magnetic field $H_0 = 1$ T at 300 K), $2\gamma\hbar H_0/k_0 \cong 4 \times 10^{-3}$ K

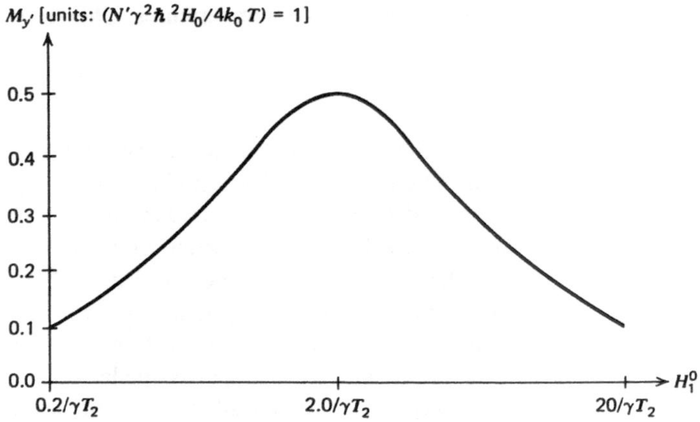

Fig. 8.13. The intensity dependence of the signal magnetization $M_y' = v$. It has been assumed that the spin systems are at exact resonance ($\gamma H_0 = \omega$) and that the relaxation times are equal ($T_1 = T_2$). Note that the location of the maximum is given correctly by Eq. 8.237 of the text. If mks units are used for $M, (N'\gamma^2\hbar^2 H_0/4k_0 T) \cong 0.3$ A m^{-1} in a typical case for nuclear magnetic resonance (protons in water, using a magnetic field $H_0 = 1$ T at 300 K). Also for protons, $1/\gamma T_2 \cong 4 \times 10^{-6}$ T ($\sim$ 40 mG) if $T_2 = 10^{-3}$ s

The value of H_1^0 that is optimum for detecting magnetic resonance is determined from

$$\frac{d}{dH_1^0}\left[\frac{H_1^0}{1 + \delta^2 T_2 + \gamma^2 (H_1^0)^2 T_1 T_2/4}\right] = 0. \tag{8.236}$$

The solution is

$$H_1^0 = \frac{2}{\gamma}\sqrt{(1/T_1 T_2) + \delta^2 (T_2/T_1)}. \tag{8.237}$$

At exact resonance between the Larmor precession and the oscillating frequency of the field, Eq. (8.237) states that the largest signal is produced when the time required for a photon to crawl into a quantum system is equal to the geometric mean of the relaxation times. The reader must be cautioned, however, that the word "optimum" has been used here to mean "largest." It will be seen shortly that for high-resolution work a much smaller H_1^0 than that given in Eq. 8.237 should be employed. The dependence of the signal intensity upon H_1^0 is shown graphically in Fig. 8.13.

8.14 Conventional Nuclear Magnetic Resonance Spectroscopy: Slow Passage

The fifth feature of Eq. 8.234 that should be examined is the dependence of M_x, M_y, and M_z on the frequency difference, δ. First, notice that for very large values of $|\delta|$ both M_x and M_y go to zero and M_z approaches M_0. Physically, this means that, if one is trying to drive an oscillator which has a natural frequency of ω_0, one ought to choose a driver that oscillates at very nearly the same frequency, $\omega \cong \omega_0$. There are two principal methods of observing the resonance. One is to slowly change ω, the frequency of the rf oscillator, keeping the magnetic field H_0 constant. The other is to keep ω constant and slowly vary the Larmor frequency by changing the magnetic field H_0. Since the response of the system depends only on the difference $\delta \equiv \omega - \omega_0$, the two methods produce results that are essentially identical.

By adjusting the phase of the detection system, one may obtain either the u-mode signal,

$$u_{\text{det}} \propto \frac{-\delta T_2}{1 + (\omega_1^2/4)T_1T_2 + \delta^2 T_2^2}, \tag{8.238}$$

or the v-mode signal.

$$v_{\text{det}} \propto \frac{1}{1 + (\omega_1^2/4)T_1T_2 + \delta^2 T_2^2}. \tag{8.239}$$

It is convenient to factor out the saturation term from the resonance denominator on the right-hand sides of Eqs. 8.238 and 8.239;

$$1 + \frac{\omega_1^2}{4}T_1T_2 + \delta^2 T_2^2 = \left(1 + \frac{\omega_1^2}{4}T_1T_2\right)[1 + \delta^2 (T_2')^2], \tag{8.240}$$

where

$$T_2' \equiv T_2 \left(1 + \frac{\omega_1^2}{4}T_1T_2\right)^{-1/2}. \tag{8.241}$$

If the u-mode signal is fed to the vertical deflection plates of an oscilloscope and the signal that drives δ to the horizontal plates, the resultant trace on the screen will be shown by the dashed curve in Fig. 8.14. If the v mode is selected instead, the oscilloscope trace will look like the solid curve in Fig. 8.14. For $\omega_1^2 T_1 T_2 \ll 1$, the amplitude of the v-mode signal will be 1 and the full width between half-amplitude points [i.e., $2\Delta\omega \equiv \delta(v_{\text{det}} = \frac{1}{2}) - \delta(v_{\text{det}} = -\frac{1}{2})$] will be $2/T_2$.

The effect of saturation on the line shape is twofold. The first effect, as has been mentioned, is that the amplitude is diminished by the factor $[1 + (\omega_1^2/4)T_1T_2]$. The second effect of saturation is a broadening of the lines by a factor of $[1 + (\omega_1^2/4)T_1T_2]^{1/2}$. The broadening (sometimes called power-broadening) may be thought of as being due to the uncertainty relation,

$$\Delta\mathscr{E}\Delta t \cong \hbar. \tag{8.242}$$

By substituting $\Delta\mathscr{E}\hbar\Delta\omega$ and $\Delta t = T_2'$, one achieves the relationship just described for a bell-shaped curve of the v type. Such a curve is called a *Lorentzian line shape function*. If T_2 may be thought of as the "lifetime" of a freely oscillating

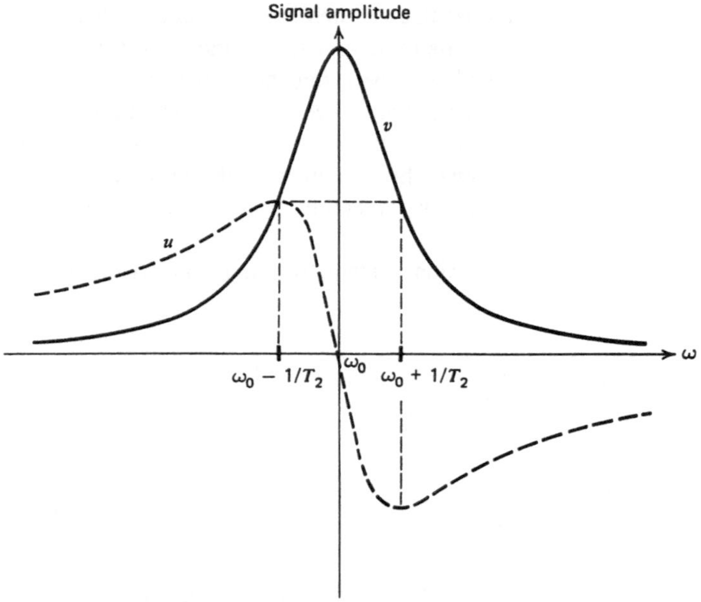

Fig. 8.14. The frequency dependence of the signal magnetization $M_x' \equiv u$ and $M_y' \equiv v$. The signal amplitude is in arbitrary units, and the frequency ω is in rad s $^{-1}$. Notice that the half width at half height is $1/T_2$

quantum system, T_2' is to be regarded as the lifetime of a driven oscillator. The radiation field produces stimulated emission that increases the decay rate from the excited states of the quantum systems. It should be remembered that the relaxation time T_2 is, properly speaking, the lifetime of the phase memory of the ensemble, which may be less than the actual state lifetime. Phase interruption processes will be discussed in more detail in Chap. 10.

It is important to remember the relationship between the magentization and the driving fields, which was discussed Chap. 3. Note that Eq. (8.118) is analogous to Eq. (3.54):

$$H_1(t) = \mathrm{Re}[H_1^0 \exp(-i\omega t)]. \tag{8.243}$$

It has been assumed already that H_1^0 is real (see Eqs. 8.123 and 8.124) and therefore analogous to $|E_0| \exp[i(\phi + \omega\alpha_0 z)]$ with $\phi + \omega\alpha_0 z = 0$. From Eq. (3.57), discarding all but the terms of interest, one obtains

$$\mu_0 \mathrm{Re}(M) = \mathrm{Re}[\,\chi H_1^0 \exp(-i\omega t)] \tag{8.244}$$

if the z component of magnetization is neglected. For the x and y components. Eq. 8.244 may be expressed in tensor (matrix) form:

$$\mu_0 \mathrm{Re}\begin{pmatrix} M_x \\ M_y \end{pmatrix} = \mathrm{Re}\left[\begin{pmatrix} \chi_{xx} & \chi_{xy} \\ \chi_{yx} & \chi_{yy} \end{pmatrix}\begin{pmatrix} H_1^0 \exp(-i\omega t) \\ 0 \end{pmatrix}\right]. \tag{8.245}$$

Using Eq. (3.27) in 8.245, one obtains two equations analogous to Eq. (2.115):

$$\mu_0 \mathrm{Re}(M_x) = \mathrm{Re}[(\chi'_{xx} + i\chi''_{xx})H_1^0 \exp(-i\omega t)]$$
$$= H_1^0(\chi'_{xx}\cos\omega t + \chi''_{xx}\sin\omega t) \tag{8.246}$$

and

$$\mu_0 \mathrm{Re}(M_y) = \mathrm{Re}[(\chi'_{yx} + i\chi''_{yx})H_1^0 \exp(-i\omega t)]$$
$$= H_1^0(\chi'_{yx}\cos\omega t + \chi''_{yx}\sin\omega t). \tag{8.247}$$

Equations (8.246) and (8.247) may be compared with Eq. (8.234) term by term to find (remember that H_1^0 is the magnetic induction here, not the field)

$$\chi_{xx} = -\frac{\mu_0 N'\gamma^2\hbar T_2 \tanh(\hbar\omega_0/2k_0T)}{4[1+\gamma^2(H_1^0/2)^2T_1T_2+\delta^2T_2^2]}(\delta T_2 + i) \tag{8.248}$$

and

$$\chi_{yx} = -\frac{\mu_0 N'\gamma^2\hbar T_2 \tanh(\hbar\omega_0/2k_0T)}{4[1+\gamma^2(H_1^0/2)^2T_1T_2+\delta^2T_2^2]}(1 - i\delta T_2). \tag{8.249}$$

It has already been explained why the susceptibility is anisotropic; H_1 can be thought of as two circularly polarized waves of equal amplitude $H_1^0/2$, only one of which rotates in the proper sense to have a long-term effect on the magnetic moments. As soon as the static magnetization M_0 is torqued away from the z axis by the H_1 field, it is carried around at the frequency ω_0 by the Larmor precession of the spins. This naturally results in a magnetization that is also circularly polarized. Hence both x and y components are present, even though they are lacking for the H_1 field itself. It will be shown in Chap. 10 that, in laser experiments, left-hand circularly polarized light and right-hand circularly polarized light are ordinarily equally effective in interacting with the sample. In such cases, waves of both senses of circular polarizaion are induced in the material, and their y components cancel vectorially, resulting in an overall isotropic susceptibility of $2\chi_{xx}$.

In Chap. 3 it was proved that the real part of the susceptibility gives rise to a change in the phase velocity of light waves. The amount of change varies with frequency. It may be seen from Eq. (8.24) that this phenomenon is associated with the u mode (the x' component of magnetization in the rotating coordinate system). The v mode is produced by the imaginary part of the susceptibility; it lies along the y' axis in the rotating frame and, as will be remembered from Chap. 3, is associated with the absorption of energy from the beam by the sample. The vector sum of u and v in the rotating frame shifts as one sweeps through resonance with the quantum transitions. At large values of δ, u is much larger than v so that the light wave passing through the medium is slowed (or, for $\delta > 0$, accelerated) without being attenuated. The phase relationship between a driving field and the response it induces in matter was shown schematically in Fig. 3.8.

A graphical display of the magnitudes of u and v as a function of δ is shown in Fig. 8.14. This is analogous to the case of transparent materials like water and glass in the visible region of the spectrum. In the neighborhood of $\delta = \pm 1/T_2'$, the magnitudes of u and v are comparable; their vector sum has swung 45° away from $H_1^0/2$ (the x' axis). At exact resonance, u disappears completely; the absorption is a maximum, and the phase velocity of light is the same as it would be if the transition were not taking place at all.

It can also be seen (from Eqs. 8.248 and 8.249) that the susceptibility depends on the magnitude of H_1^0, and therefore the relationship between **M** and **H** is nonlinear. This prevents the Kramers-Kronig relations from being valid except in the limit $H_1^0 \ll 1/\gamma\sqrt{T_1 T_2}$.

Historically, the first experimental detection of electron [8] and nuclear [9] resonances used the slow-passage technique.

8.15 Equivalence of Transient and Steady-State Methods

In this book, the primary emphasis will be placed on the description of the transient response of an ensemble of quantum systems of a brief, intense pulse of coherent light, rather than on the steady-state solutions to the Bloch equations usually emphasized in conventional NMR spectroscopy. It should be noted that for each type of steady-state experiment one can find an equivalent transient one. For example, the unsaturated slow-passage spectrum in Eq. 8.234 is simply the Fourier transform of the free-induction decay signal given by Eqs. 8.218 and 8.220 [10]. Also, the slow-passage spectrum of a sample that is strongly saturated by additional irradiation at the center frequency of the line is the Fourier transform of the transient nutation signal [11].

Relaxation times may be measured by either type of experiment. For example, T_2 is the inverse of the half width of the v-mode spectrum between half-amplitude points; it is also the $1/e$ decay time of the free-induction tail following a $\pi/2$ pulse. If one saturates the slow-passage spectrum with an intense field at fixed frequency to the point at which the signal vanishes, one can be sure that the z component of magnetization is also zero. After the saturating field is turned off, the z component will recover with a time constant T_1. Subsequent repeated rapid scans of the spectrum with a weak frequency-swept field will reveal a gradual return of the signal to its initial strength, because the M_x and M_y components are proportional to M_z. In this way, one can determine T_1 [12].

As for transient methods, it already has been mentioned that T_2 can be determined directly from the free-induction decay signal following a $\pi/2$ pulse. One can also determine T_1 by applying a π pulse. Before the π pulse, the sample is at equilibrium and therefore has a magnetization $M_0\hat{x}$. After the π pulse, the magnetization will be $-M_0\hat{x}$. (All pulses are presumed to be very brief in com-

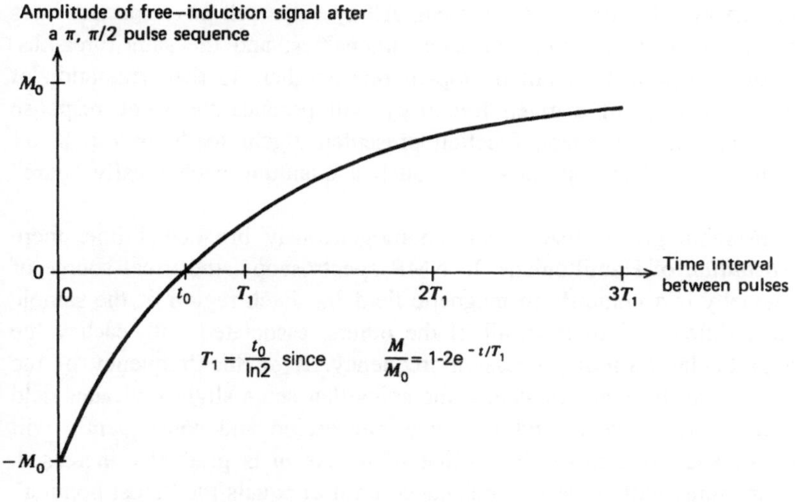

Fig. 8.15. The determination of T_1 by series of $\pi, \pi/2$ pulse sequences

parison with either T_1 or T_2.) As the sample sits in the heat bath, it will relax; the magnetization vector will shrink along the negative branch of the axis, pass through zero, and then regain its equilibrium value exponentially in a time T_1. How far this relaxation has proceeded after a given time interval t' can be determined by measuring the length of the magnetization vector at that time. This measurement may be performed by applying a $\pi/2$ pulse to the sample; whatever M_z happens to be, it will be tipped into the xy plane and will generate a signal, which will then decay in a Time T_2. The initial amplitude of this signal will be equal to that of M_z just before the $\pi/2$ pulse was applied. If one then be equal to that of M_z just before the $\pi/2$ pulse was applied. If one then waits for a time much longer than T_1, the sample will completely recover to equilibrium and the experiment can be repeated, this time using a different value of t'. By plotting the initial signal amplitudes versus pulse separation t', the entire decay process can be mapped out and T_1 determined. The signal produced following the $\pi/2$ pulse is displayed versus t' in Fig. 8.15.

8.16 Spin Echoes

The most dramatic transient coherence effect in all magnetic resonance spectroscopy is the phenomenon of spin echoes, also discovered by Hahn [13]. To explain this elegant experiment, it is necessary to discuss the difference between a homogeneous and an inhomogeneous broadening of spectral lines. "Homegeneous" contains the prefix *homo*, meaning "same." An ensemble of quantum systems all of which have exactly the same Hamiltonian will have a spectrum which looks just like that of a single system. All systems will have spectral lines with the same center frequencies, the same intensities, and the same breadths. These individual lines will all fall on top of one another, so that irradiation of the ensemble with light of a given frequency will produce the same response for each system. The line shape function presented algebraically in Eq. 8.234 and drawn in Fig. 8.14 is appropriate for such a quantum-mechanically "pure" ensemble.

In an ensemble giving rise to an inhomogeneously broadened line, there exists a *distribution* of Hamiltonians. In NMR spectroscopy, the usual source of the inhomogeneity is a nonuniform magnetic field H_0. Each region of the sample sees a slightly different field from all of the others; associated with each value of H_0 is a particular Larmor precession frequency. If ω, the frequency of the radio wave, is swept through resonance, the spins that see a slightly weaker field than the average will have a slightly slower precession and consequently will come into resonance at a rather low value of ω. As ω is gradually increased, more and more spins will come into resonance until ω equals the "most popular" (modal) precession frequency in the sample. Ordinarily, the mean value of ω_0 is also the modal frequency because the inhomogeneity in H_0 is usually distributed symmetrically about the mean, $\bar{H}_0$. As ω is increased beyond that point, fewer and fewer spins are in exact resonance, until finally ω exceeds the Larmor frequency for even the most strongly magnetized quantum systems.

The resultant spectral line may be regarded as the sum of the homogeneously broadened lines due to all regions of the sample, and the intensity of each such homogeneously broadened components is proportional to the number of spins in that region. Since the magnetic field distribution is continuous, the homogeneously broadened components will overlap and blur together, forming one broad line. The overall line shape will be given mathematically by the convolution integral of the homogeneous line shape with the distribution function for H_0 within the sample. The homogeneously broadened components of an inhomogeneously broadened line are shown schematically in Fig. 8.16a.

One way to discover whether or not a given line is inhomogeneously broadened is by means of a steady-state experiment. If one irradiates the sample with a very intense monochromatic beam of light somewhere in the neighborhood of the line's center, one will of course saturate the sample. If the line is homogeneously broadened, each system in the ensemble will be saturated to the same degree (with an efficiency that decreases with increasing $|\delta|$), and consequently the entire spectrum will broaden and diminish in intensity (a fact that one can discover by rapidly scanning through the line with a weak field after the saturating field has been extinguished). If, however, the line is inhomogeneously broadened, there will be exact resonance with only a few of the systems of the ensemble, that is, those with Larmor frequencies very close to ω. All of the other spins will be saturated to a much lesser extent because they are so far off resonance; see Eq. 8.237. A subsequent weak field scan will reveal a dip in the profile of the spectral line in the neighborhood of ω where the saturation is strongest; the rest of the line will also be reduced in intensity, but to a much lesser extent. The

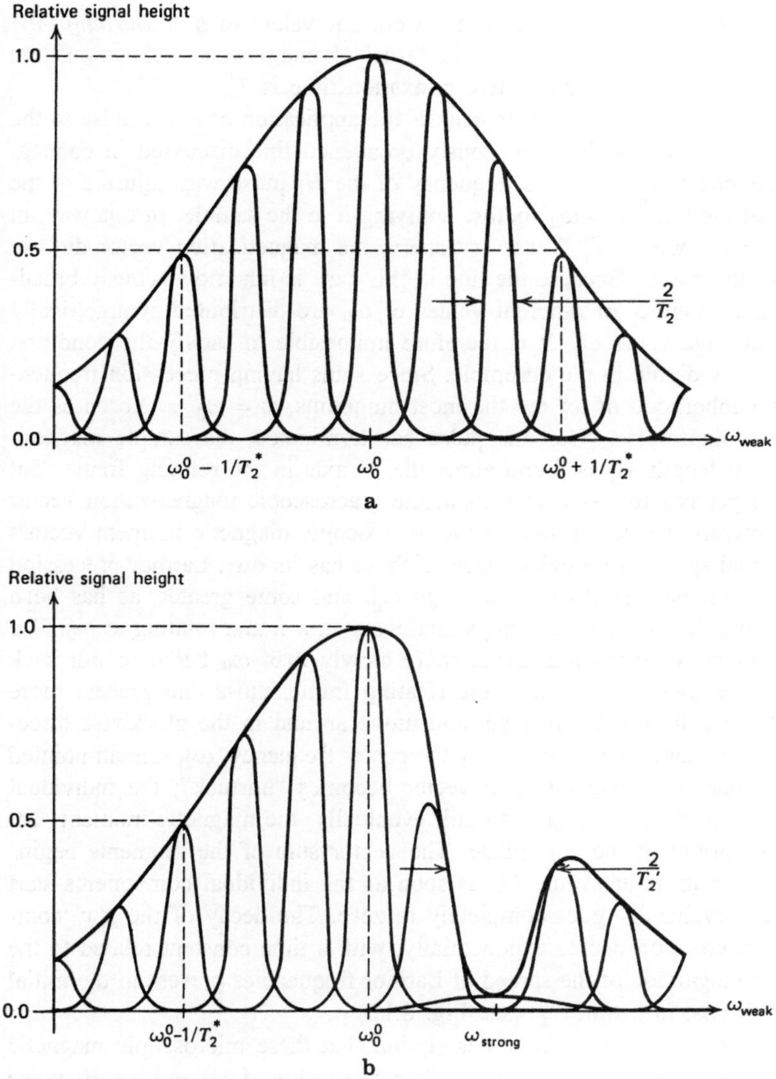

Fig. 8.16. An inhomogeneously broadened line and its homogeneously broadened components: **a** unsaturated and scannned with a weak field, and **b** saturated with a strong field at ω_{strong} near $\omega_0^0 + 1/T_2^*$, then scanned with a weak field

width of the hole "eaten" or "burned" into the line will be of the order of $2/T_2'$, and, if H_1 is removed, it will "heal" in a time $\cong T_1$ [12]. See Fig. 8.16b. The overall width of an inhomogeneously broadened line will be roughly the sum of the widths of the two distribution functions that were "convolved" in the making of it, and is sometimes expressed by the notation $2/T_m$, $1/T_m = 1/T_2 + 1/T_2^*$.

The Hahn spin echo experiment is the transient equivalent of *hole burning* [15] because it also enables one to measure the "true" (homogeneous) relaxation time T_2 in spite of the fact that the effective relaxation time is T_m.

The first step in producing a spin echo is the application of a $\pi/2$ pulse to the sample. In the case of the homogeneously broadened line discussed in connection with free-induction decay, the frequency of the H_1 pulse was adjusted to the exact center of the line ($\omega = \omega_0$) before applying it to the sample. In this way, in the rotating frame, where $H_1^0/2$ was stationary, the magnetization vector did not precess about the z axis. Because the line in this case is inhomogeneously broadened, an infinite number of different values of ω_0 are distributed symmetrically about some average value ω_0^0. It is therefore impossible to satisfy the condition $\omega = \omega_0^0$ for every dipole in the ensemble. Since spins having precession frequencies in the neighborhood of ω_0^0 are the most numerous, $\omega = \omega_0^0$ is chosen as the best possible compromise. After the pulse has terminated, the sample magnetization vector of length M_0 is found along the y' axis in the rotating frame. But this condition persists for only an instant; the macroscopic magnetization vector is, after all, merely the vector sum of the microscopic magnetic moment vectors of the individual spinning particles. Each of these has its own Larmor precession frequency, ω_0 – some less than the average ω_0^0, and some greater, as has been stated. Therefore they cannot all remain stationary in a frame rotating at ω_0. The magnetic moments of spins precessing more slowly than ω_0 begin to fall back in the counterclockwise direction in the rotating frame; those that precess more rapidly begin to gain, on the average, and move around in the clockwise direction. Only the few that precess at exactly the center frequency, ω_0, remain pointed along the y' axis. The magnetization vector becomes "unstuck"; the individual spins leak from it in both directions until, eventually, the magnetic moments are distributed uniformly in the $x'y'$ plane. The vector sum of the moments begins diminishing from its initial value M_0 as soon as the individual components start to fan out and eventually goes completely to zero. The decay of the $x'y'$ component of magnetization occurs exponentially, with a time constant related to the inverse of the magnitude of the spread of Larmor frequencies present in the initial distribution, that is, in a time T_2^* (see Fig. 8.17).

It may be seen from the above description that these microscopic magnetic moments are the "soldiers" described in Chap 5 (see Fig. 5.5), and the H_1 pulse is the "drill sergeant." In the laboratory frame, the drill sergeant makes all of the soldiers precess in step about H_0. As soon as he ceases counting cadence, however, each soldier precesses to the beat of his own internal "drummer." As a natural consequence of this individualism, all semblance of a proper military formation soon disappears. The only modification of the analogy suggested in Chap 5 is that now the soldiers must parade on a circular track.

The next step in the process is to call the sergeant out of retirement and have him issue the command, "To the rear! March!" to the soldiers. This is accomplished by applying a π pulse to the sample after a time t':

$$T_2 \gg t' \gg T_2^* . \tag{8.250}$$

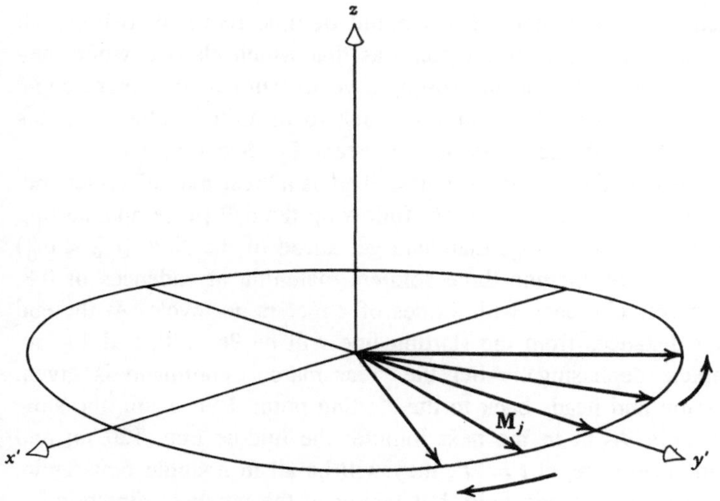

Fig. 8.17. Free-induction decay due to the dephasing of isochromats ($\mathbf{M}_j$) of an inhomogeneously broadened line, in a coordinate system rotating at frequency of its center (ω_0^0)

The appropriate circularly polarized component of the wave associated with the second pulse will occur along the x'' axis of the rotating coordinate system. Both pulses can be created by a single oscillator that operates continuously. The radio frequency is normally prevented from reaching the transmitter coils by means of an electronic shutter called a *gate*. When it is desired to irradiate the sample, the gate is opened for the appropriate interval. If a gated oscillator is used, all pulses will occur along the same axis in the rotating frame ($x'' = x'$).

The effect of the π pulse will be to invert the y' coordinate of each individual magnetic moment, e.g.,

$$\langle p_{y'}(t > t')\rangle_j = -\langle p_{y'}(t < t')\rangle_j, \quad \text{for all } j. \tag{8.251}$$

If one imagines a sphere of radius $\gamma\hbar/2$ aligned with the x' axis, the tip of the certain component of each magnetic moment vector will be located somewhere on a great circle (e.g., the 0° and 180° lines of longitude) at the start of the pulse. During the pulse, each microscopic vector will travel to the opposite meridian along a parallel of latitude. In particular, the spins that precess about the z axis at the frequency ω_0^0 initially parallel to the y' axis will travel along the equator and end up pointing in the negative y' direction. The drill sergeant again retires (the H_1 pulse has been completed), and each soldier resumes marching to his internal cadence (each spin precesses at its characteristic Larmor frequency, ω_0). The spins that precess slower than the average continue to slip in the counterclockwise direction in the rotating frame, and those that precess faster than ω_0^0 continue to move in a clockwise direction. Before the π pulse, these motions caused the spins to get further and further out of "formation." Now, however, the same motions bring the spins closer and closer together toward the $-y'$ axis, where the spins precessing at exactly ω_0^0 await them. The amount of time required to bring all of the spins back together is exactly the same as that which elapsed while they were coming apart, namely, t'. The macroscopic vector sum of the microscopic moments then grows exponentially from zero back to its initial value, M_0, this time along the $-y'$ axis, with the same time constant T_2^*. See Fig. 8.18.

This is the reason why the process was described as a "rear march" command. Picture the soldiers starting in a single rank, following the $\pi/2$ pulse and starting to straggle out as the fast ($\omega_0 > \omega_0^0$) marchers get ahead of the slow ($\omega_0 < \omega_0^0$) ones. For simplicity, let us imagine three soldiers marching at cadences of 0.8, 1.0, and 1.2 Hz, respectively, each with strides of 2 meters per cycle. At the end of $t' = 1$ min, their distances from the starting line will be 96, 120, and 144 m, respectively (complete dephasing). After the "rear march" command is given, each reverses direction and heads back to the starting point. But, again, the slow walker will march only 96 m in the next minute; the middle one, 120 m; and the fast one, 144 m. Therefore, at $t = 2t'$, they will be all in a single rank again, at the same position on the parade field, but facing in the opposite direction.

After the spins have recollected themselves along the $-y'$ axis, they will immediately begin to come apart again. The macroscopic magnetization will again decay exponentially to zero with the same time constant, T_2^*. This spontaneous recovery and subsequent decay of the $x'y'$ components of magnetization is called a "spin echo." The peak magnitude of the echo will not be exactly the same as the initial magnetization following the $\pi/2$ pulse, however. The echo amplitude will, in fact, be somewhat less than M_0 because of the fact that the parade field is not perfectly level; some of the soldiers will have stumbled slightly or bumped into one another in the time interval $2t'$. Such phase interruption processes are the military analogy to "true" T_2 processes, which produce uncontrollable and irreversible dephasing of the magnetic moment vector in the $x'y'$ plane. Soldiers that have stumbled or bumped, just like spins that have undergone transverse relaxation, will no longer be on exactly the same schedule that they would have

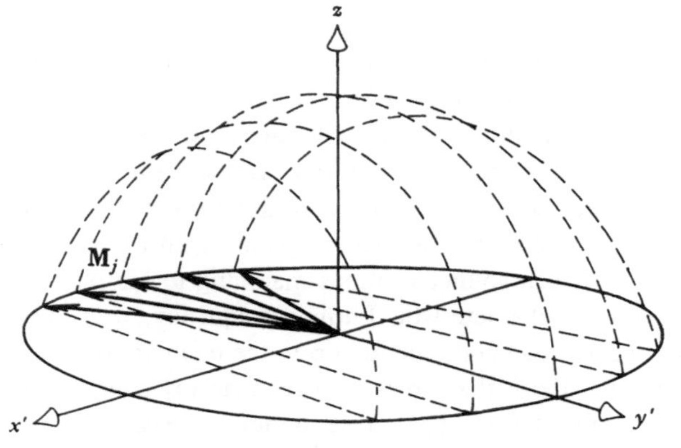

Fig. 8.18. Formation of a spin echo due to refocusing of isochromats ($\mathbf{M}_j$) along the $-y'$ axis by means of a 180° pulse applied along the $+x'$ axis

followed on an ideal parade ground, and will show up either too early or too late to contribute their portion of the echo. For $T_2 \gg T_2^*$, the echo will have an amplitude only very slightly less than M_0 if the condition in Eq. 8.250 holds. This experiment can be repeated for various values of t'; a plot of the echo amplitude versus t' will enable one to determine the time constant for irreversible (homogeneous) dephasing – the true T_2.

The relationship between $\mathbf{H}_1$ and $\mathbf{M}$ during a spin echo is shown schematically in Fig. 8.19; experimental results are presented in Figs. 6.20a and 6.20b. The $\pi/2, \pi$ pulse sequence for the determination of T_2 is a natural companion to the $\pi, \pi/2$ pulse sequence for the determination of T_1 described in Sect. 8.15, but the two differ in an important way.

In the determination of T_1, one *must* let a time $t \gg T_1$ elapse between pulse pairs. In the determination of T_2, an alternative system is possible; one may apply the pulse sequence $\pi/2(0), \pi(t'), \pi(3t'), \pi(5t')\ldots$. The symbol in parentheses after each pulse designation is the time when the pulse in question must be applied – "rear march" may be given a number of times to the same sample without waiting for a return to equilibrium. The successive echoes that are produced by this (at times $2t', 4t', 6t'$,etc.) become weaker and weaker, their amplitudes being $M_0 \exp(-2t'/T_2), M_0 \exp(-4t'/T_2), \ldots$. The pulse sequence described above was invented by Carr and Purcell [14], who used it to study self-diffusion in liquids (diffusion in an inhomogeneous H_0 field produces an irreversible dephasing and hence adds to the $1/T_2$ term in the exponential decay of the echo amplitudes).

The subjects presented in this chapter by no means exhaust the list of coherent transient effects that have been observed by ingenious experimentalists working in the fields of EPR and NMR spectroscopy, but should give some idea of the power of these methods. It should be mentioned that the standard

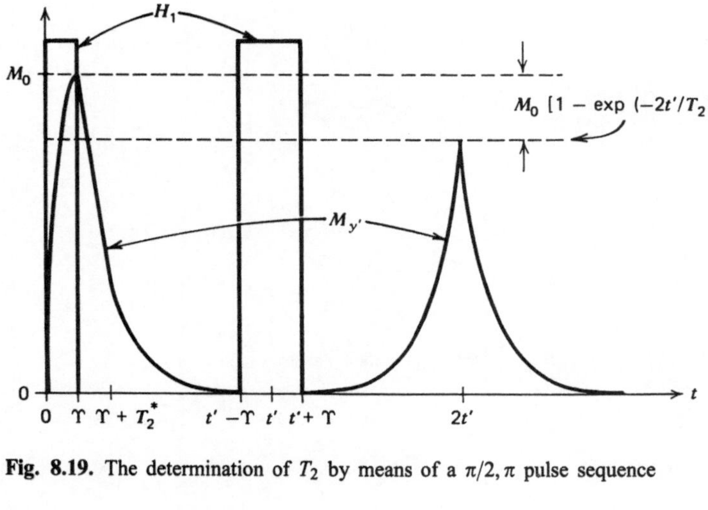

Fig. 8.19. The determination of T_2 by means of a $\pi/2, \pi$ pulse sequence

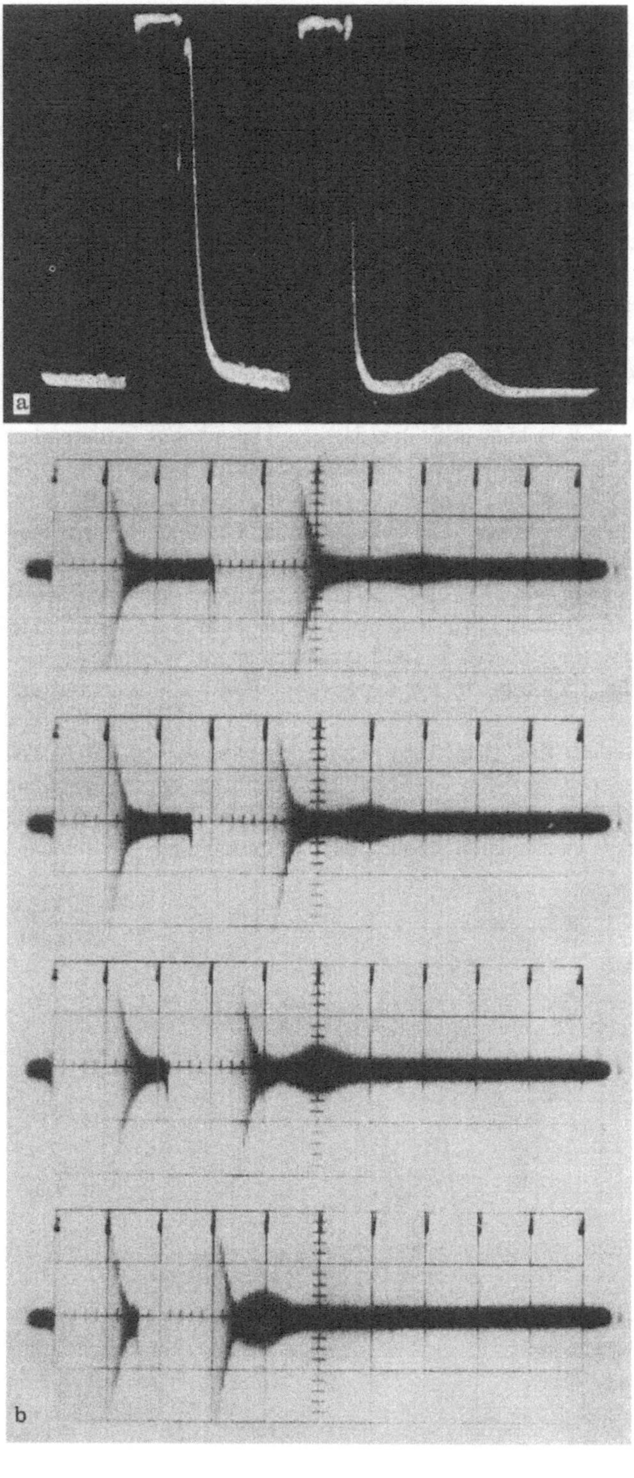

reference work on nuclear magnetic resonance is the book by Abragam [15]; a comprehensive work on electron spin resonance by Poole is also available [16]. For information on dynamics during transitions connected with two-dimensional NMR spectroscopy see Ref. 17, 18, 19.

8.17 References

1. Pauli W (1927) Z Phys 43: 601
2. Dicke RH (1954) Phys Rev 93: 99
3. Bloch F (1946) Phys Rev 70: 460
4. Torrey HC (1949) Phys Rev 76: 1059
5. Hahn EL (1950) Phys Rev 77: 297
6. Macomber JD (1968) Spectrosc Lett 1: 131; see also Spectrosc Lett 1: 265
7. Pierre Curie (1895) Propriétés Magnétiques des Corps a Diverses Températures (Gauthier-Villars, Paris), Chap. III, pp. 47–66
8. a. Zavoisky EK (1945) J Phys USSR 9: 211
   b. Zavoisky EK (1945) J Phys USSR 9: 245
   c. Zavoisky EK (1946) J Phys USSR 10: 197
9. a. Purcell EM, Torrey HC, Pound RV (1946) Phys Rev 69: 37
   b. Bloch F, Hansen WW, Packard M (1946) Phys Rev 69: 127
10. Lowe IJ, Norberg RE (1957) Phys Rev 107: 46
11. Macomber JD (1968) Appl Phys Lett 13: 5
12. Bloembergen N (1961) Nuclear Magnetic Relaxation (Benjamin WA, New York)
13. Hahn EL (1950) Phys Rev 80: 580
14. Carr HY, Purcell EM (1954) Phys Rev 94: 630
15. Abragam A, The Principles of Nuclear Magnetism (Oxford University Press, London, 1961).
16. Poole CP Jr, Electron Spin Resonance (Interscience Publishers, New York, 1967).
17. Ernst RR, Bodenhausen G, Wokaun A (1988) Principles of NMR in one and two dimensions (Clarendon Press, Oxford)
18. Friebolin H (1991) Basic One- and Two-Dimensional NMR Spectroscopy (Verlag Chmie, Weinheim)
19. Sanders JKM, Hunter BK (1993) Modern NMR Spectroscopy, a Guide for chemists (2nd ed., Oxford Univ. Press)

Fig. 8.20. a The first observation of spin echoes by E.L. Hahn [*Phys. Rev.* **80**, 580 (1950), Fig. 11.] The chromophores are protons in paraffin; the echo lasts for 14 μs. The rf pulses, about 25 μs wide, cause some blocking of the rf amplifier used in the detection system. **b** The first observation of electron spin echoes. The sample consisted of a solution of Na in NH_3. The transition was at exact resonance with the radiation field ($\delta = 0$), $\omega_0/2\pi = 17.4$ MHz ($H_0 = 6.2 \times 10^{-5}$ T). Each major division on the horizontal axis represents 1 μs. The 90° rf burst starts with the left edge of the oscillogram reticle and actually lasts 0.3 μs, although in this photo it appears to be prolonged to about 1.5 μs by residual ringing of the tuned pickup coil. As the 180° burst is moved to the right, the echo moves twice as far to the right. The signal is displayed without rf detection, with a sweep so fast that the actual oscillations at 17.4 MHz may be discerned. By way of contrast, only the envelope is displayed in (*a*), above. This photograph is a negative print of Fig. 5 in the paper by R.J. Blume [(1958) *Phys. Rev.* **109**: 1867

8.18 Problems

8.1 a. Prove that $[\mathbf{S}^2, \mathbf{S}_y] = 0$.
b. Calculate $[\mathbf{S}_+\mathbf{S}_-]$.

8.2 A matrix is Hermitian if it is equal to its Hermitian adjoint. Show that the matrices $[S_x]$, $[S_y]$, $[S_z]$ and $[S^2]$ are Hermitian for a spin $s = \frac{1}{2}$ particle, but $[S_+]$ and $[S_-]$ are not.

8.3 For a spin $s = 1$ particle the eigenkets are $|2,1\rangle$, $|2,0\rangle$, and $|2,-1\rangle$. Calculate the results of operating with $\mathbf{S}_+$ and $\mathbf{S}_-$ upon these kets, using Eq. 8.81.

8.4 The operator that represents the component of angular momentum along the axis oriented at (θ, Ω) is

$$\mathbf{S}_{u_1}(\theta,\Omega) = \mathbf{S}_x \sin\theta\cos\Omega + \mathbf{S}_y\sin\theta\sin\Omega + \mathbf{S}_z\cos\theta$$

Show that the ket

$$|(\theta,\Omega)\rangle = e^{ib}\cos\left(\frac{\theta}{2}\right)\alpha + e^{i(b+\Omega)}\sin = \left(\frac{\theta}{2}\right)\beta$$

is an eigenket of $\mathbf{S}_{u_1}(\theta,\Omega)$.

8.5 Plot M_x, M_y, and M_z as functions of t for the nutation effect (use Eq. 8.205). Use $M_0 = 1\,\mathrm{Am}^{-1}$, $\gamma/2\pi = 42\ \mathrm{MHzT}^{-1}$, $H_0 = 14$ mT, $H_1 = 1.4$ mT, and $\omega = \omega_0$. Let t run from 0 to 20 μs. Label the time axes to show where absorption and stimulated emission occur.

8.6 Plot M_x, M_y, and M_z as functions of t for free-induction decay (use Eq. 8.221). Use $M_0 = 1\,\mathrm{Am}^{-1}$, $\gamma/2\pi = 42\,\mathrm{MHzT}^{-1}$, $H_0 = 14$ mT, $T_2 = 5$μs, and $\omega = \omega_0$. Let t run from 0 to 20 μs.

8.7 Calculate the Curie's law paramagnetism (Eqs. 8.233 and 8.235) due to protons in water at 300 K. Use $\gamma/2\pi = 42\ \mathrm{MHzT}^{-1}$, $d = 1.0\times10^3\ \mathrm{Kg\,m}^{-3}$, M (gram-molecular weight) $= 0.018\ \mathrm{Kg\,mol}^{-1}$, and $H_0 = 1.0$ T.

8.8 a. Calculate the value of H_1^0 (teslas) that produces the maximum signal from a sample of protons ($\gamma/2\pi = 42\ \mathrm{MHzT}^{-1}$) in which the relaxation times are $T_1 = T_2 = 1$ s. Assume that the $\mathbf{H}_1$ field is in exact resonance with the Larmor precession of the spins (Eq. 8.237).
b. Calculate the effective relaxation time T_2' for such a system from Eq. 8.241.
c. Calculate $\mathrm{Im}(\chi_{xx})$ from Eq. 8.429 at 300 K, assuming exact resonance. What are the units?
d. Calculate the Beer's law absorption coefficient corresponding to the χ in Problem 8.8c, using relationships found in Chap. 3.
e. Evaluate the integral

$$I(\omega) = \frac{M_0}{2\pi}\int_0^\infty \exp\left(\frac{-t}{T_2^*}\right)\exp(i\omega t)dt.$$

9 Spin Dynamics and Radical Reactions

M. Goez

Key to Symbols

a	hyperfine coupling constant
β	Bohr magneton of electron
ε	Parameter ($\varepsilon = \pm 1$) denoting cage or escape product
g	g-factor of electron
H_{H}	Hamiltonian of hyperfine interaction
$\mathsf{H}_{\mathsf{H,\,a}}$	antisymmetric secular part of Hamiltonian of hyperfine interaction
$\mathsf{H}_{\mathsf{H,\,nonsec}}$	nonsecular part of Hamiltonian of hyperfine interaction
$\mathsf{H}_{\mathsf{H,\,s}}$	symmetric secular part of Hamiltonian of hyperfine interaction
$\mathsf{H}_{\mathsf{H,\,sec}}$	secular part of Hamiltonian of hyperfine interaction
$\mathsf{H}_{\mathsf{Z,\,a}}$	antisymmetric part of the Hamiltonian of the nuclear Zeeman interaction
$\mathsf{H}_{\mathsf{Z,\,e}}$	Hamiltonian of electron Zeeman interaction
$\mathsf{H}_{\mathsf{Z,\,n}}$	Hamiltonian of nuclear Zeeman interaction
$\mathsf{H}_{\mathsf{Z,\,s}}$	Symmetric part of the Hamiltonian of the nuclear Zeeman interaction
I	operator of nuclear spin
J	exchange integral
J	nuclear spin-spin coupling constant
μ	Parameter ($\mu = \pm 1$) denoting precursor spin multiplicity in a radical pair reaction
μ_e	magnetic moment of electron
μ_n	magnetic moment of nucleus (proton)
Q	matrix element of singlet-triplet mixing
$\mathscr{R}$	density matrix written as a vector
$\lvert s\rangle$	nuclear singlet state
$\lvert S\rangle$	electronic singlet state
S	operator of electron spin
$\mathsf{S}_{\mathsf{tot}}$	operator of total electron spin of a system
$\mathsf{S}_{\mathsf{x}}, \mathsf{S}_{\mathsf{y}}, \mathsf{S}_{\mathsf{z}}$	components of electron spin operator
$\lvert t\rangle$	nuclear triplet state

$\|T\rangle$	electronic triplet state
$\|t_{+1}\rangle, \|t_0\rangle, \|t_{-1}\rangle$	sublevels of nuclear triplet state
$\|T_{+1}\rangle, \|T_0\rangle, \|T_{-1}\rangle$	sublevels of electronic triplet state

9.1 Molecules in Doublet States

An electron possesses a spin of magnitude $(\sqrt{3}/2)\hbar$, just like a proton. Associated with this is a magnetic moment μ_e that is conventionally written as

$$\mu_e = -g\beta\mathsf{S}, \tag{9.1}$$

where S is the operator for the electron spin. As the minus sign shows, μ_e points in the opposite direction as the spin, owing to the negative charge of the electron. The strength of the magnetic moment, which is about 658 times that of a proton, is expressed in terms of the Bohr magneton β of the electron, $\beta = 9.27 \times 10^{-26}$ J/Gauss, and a number g. This so-called g-factor amounts to 2.0023 for a free electron. If the electron is confined in an organic molecule containing no transition elements, g is only slightly higher or lower, usually by much less than one percent. Its value is characteristic for its surroundings. (Actually, g is a tensor, but in liquid solution rapid molecular motions average out all anisotropy effects.)

We recall that, in contrast to Eq. (9.1), the relation between magnetic moment μ_n and spin I of a nucleus is usually expressed with its gyromagnetic ratio γ,

$$\mu_n = \gamma\mathsf{I}. \tag{9.2}$$

This difference in the notation of basically the same relationship, which is simply due to historical reasons, is unfortunate. Yet, for compatability with other texts in this field we shall stick to this convention.

As you will remember from your lessons in elementary chemistry, the majority of chemical species possess an even number of electrons. These tend to occupy the available energy levels in pairs, so their spins have to be antiparallel owing to the Pauli principle. As a result, these compounds are diamagnetic. However, there are also molecules that contain an odd number of electrons, and are therefore paramagnetic. The simplest case is that of just one unpaired electron. Such a species is called a radical. A few radicals are quite stable, but mostly they are highly reactive intermediates that can only persist in frozen matrices of inert solvents, where they have little chance of meeting a reaction partner by diffusion. If one wants to study radicals in liquid solution, one typically generates them as transient species by cleavage of a chemical bond.

When we put a radical into a static magnetic field H_0, which as usual we take along the z-axis, we force the spin of the unpaired electron to be quantized in the direction of the field. If the molecule does not contain any nuclei with magnetic moments, the spin Hamiltonian of this system is given by the Zeeman term $\mathsf{H}_{\mathsf{Z,e}}$,

$$\begin{aligned}\mathsf{H}_{\mathsf{Z,e}} &= -\mu_e \cdot H_0 \\ &= +g\beta H_0 \mathsf{S}_z \ .\end{aligned} \tag{9.3}$$

Only two energy levels exist, with eigenfunctions $|\alpha\rangle$ and $|\beta\rangle$ and corresponding eigenvalues s_z of $+\frac{1}{2}$ and $-\frac{1}{2}$. Consequently, we denote this species as a doublet. In contrast to a proton, the most stable state of the electron in the field H_0 is $|\beta\rangle$, because μ_e is aligned opposite to S. An oscillating magnetic field of frequency $g\beta H_0/h$ perpendicular to H_0 will induce transitions between the two eigenstates, and thus give rise to the EPR (electron paramagnetic resonance) spectrum of this system, which consists of a single line.

Let us now suppose that our radical contains one magnetic nucleus of spin $\frac{1}{2}$, say a proton. We then also have to consider the nuclear Zeeman Hamiltonian $\mathsf{H}_{\mathsf{Z,n}}$,

$$\mathsf{H}_{\mathsf{Z,n}} = -\gamma\hbar H_0 \mathsf{I}_z, \tag{9.4}$$

and the so-called hyperfine interaction between the spins of the proton and the electron, described by a term H_{H},

$$\begin{aligned}\mathsf{H}_{\mathsf{H}} &= a\mathsf{S} \cdot \mathsf{I} \\ &= a\,(\mathsf{S}_x\mathsf{I}_x + \mathsf{S}_y\mathsf{I}_y + \mathsf{S}_z\mathsf{I}_z).\end{aligned} \tag{9.5}$$

The isotropic hyperfine coupling constant a may be of either sign. Its values are invariably given in Gauss in the literature. They can be converted into energies by multiplication with $g\beta$. In the magnetic fields commonly used in EPR spectroscopy, the magnitude of a is typically a factor of 100 ... 1000 smaller than the electron Zeeman energy. (We note in passing that this interaction is also orientation dependent. However, as we are interested in radicals in liquid solution, the anisotropy is not observable due to rapid and random tumbling of the molecules. For the same reason, the direct dipolar interaction between the magnetic moments of the electron and the proton is averaged out completely and need not be considered here.)

As a basis for the description of this system we choose the products of the electron spin functions with the nuclear spin functions (subscripts e and n, respectively),

$$\begin{aligned}|1\rangle &= |\alpha_e\alpha_n\rangle \qquad |3\rangle = |\beta_e\alpha_n\rangle \\ |2\rangle &= |\alpha_e\beta_n\rangle \qquad |4\rangle = |\beta_e\beta_n\rangle.\end{aligned} \tag{9.6}$$

It is easily verified that $\mathsf{H}_{\mathsf{Z,e}}$ and $\mathsf{H}_{\mathsf{Z,n}}$ are diagonal in this basis,

$$\mathsf{H}_{\mathsf{Z,e}} = \frac{\hbar\omega_e}{2}\begin{pmatrix} +1 & 0 & 0 & 0 \\ 0 & +1 & 0 & 0 \\ 0 & 0 & -1 & 0 \\ 0 & 0 & 0 & -1 \end{pmatrix}$$

$$\mathsf{H}_{\mathsf{Z,n}} = \frac{\hbar\omega_n}{2}\begin{pmatrix} -1 & 0 & 0 & 0 \\ 0 & +1 & 0 & 0 \\ 0 & 0 & -1 & 0 \\ 0 & 0 & 0 & +1 \end{pmatrix}. \tag{9.7}$$

Here ω_e and ω_n are the resonance angular frequencies of electron and proton,

$$\omega_e = g\beta H_0/\hbar \qquad \omega_n = \gamma H_0. \tag{9.8}$$

Futhermore, we can decompose H_{H} into an operator $\mathsf{H}_{\mathsf{H,\,sec}}$ that is also diagonal in this basis, the so-called 'secular' part, and an operator $\mathsf{H}_{\mathsf{H,\,nonsec}}$ that only possesses off-diagonal elements (the 'nonsecular' part),

$$\begin{aligned} \mathsf{H}_{\mathsf{H}} &= \mathsf{H}_{\mathsf{H,\,sec}} + \mathsf{H}_{\mathsf{H,\,nonsec}} \\ &= a\mathsf{S_z I_z} + a(\mathsf{S_x I_x} + \mathsf{S_y I_y}) \\ &= a\mathsf{S_z I_z} + \tfrac{1}{2}a(\mathsf{S_+ I_-} + \mathsf{S_- I_+}), \end{aligned} \tag{9.9}$$

where we have used the operator relations derived in Sect. 8.3 for the last line of this equation. The matrix elements of these operators are

$$\mathsf{H}_{\mathsf{H,\,sec}} = \frac{a}{4}\begin{pmatrix} +1 & 0 & 0 & 0 \\ 0 & -1 & 0 & 0 \\ 0 & 0 & -1 & 0 \\ 0 & 0 & 0 & +1 \end{pmatrix} \qquad \mathsf{H}_{\mathsf{H,\,nonsec}} = \frac{a}{2}\begin{pmatrix} 0 & 0 & 0 & 0 \\ 0 & 0 & 1 & 0 \\ 0 & 1 & 0 & 0 \\ 0 & 0 & 0 & 0 \end{pmatrix}. \tag{9.10}$$

The complete spin Hamiltonian $\mathsf{H}_{\mathsf{Z,e}} + \mathsf{H}_{\mathsf{Z,n}} + \mathsf{H}_{\mathsf{H}}$ does not contain any off-diagonal elements for $|1\rangle$ and $|4\rangle$, so these are proper eigenfunctions. On the other hand, the states $|2\rangle$ and $|3\rangle$ are mixed to some degree by the action of the operator $\mathsf{H}_{\mathsf{H,nonsec}}$. Perturbation theory tells us that the amount of mixing is determined by the ratio of the off-diagonal matrix elements, $a/2$, to the energy difference between the two states, $\hbar(\omega_e + \omega_n)$.

We first consider the case of a strong external magnetic field, such as it is employed in EPR or NMR (nuclear magnetic resonance) spectroscopy. Since the Zeeman energies depend linearly on H_0, $\hbar(\omega_e + \omega_n)$ is much larger than $a/2$ under these conditions. Hence the degree of mixing is negligible, which means that we need not consider the nonsecular part $\mathsf{H}_{\mathsf{H,nonsec}}$ of the operator H_{H}; in high magnetic fields it is a very good approximation to retain only the secular part $\mathsf{H}_{\mathsf{H,sec}}$, so that $|2\rangle$ and $|3\rangle$ are also eigenfunctions of the Hamiltonian. We see that in this case the spin states of both the electron and the proton in all four eigenfunctions can be clearly distinguished. Varying the frequency of an

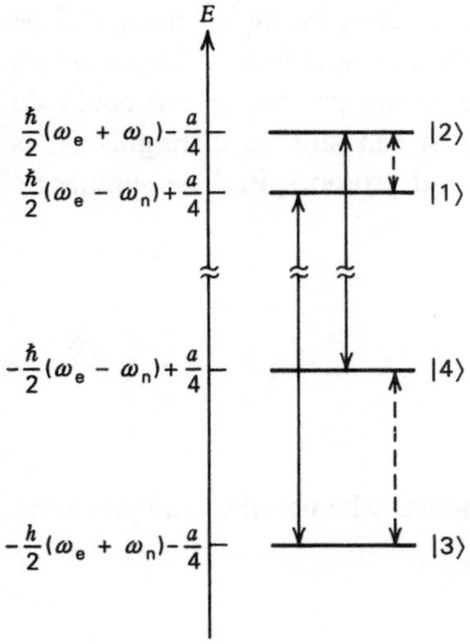

Fig. 9.1. Energy levels of a radical with one proton in a strong magnetic field for $|a| < 2\hbar\omega_n$. The eigenfunctions are given by Eq. 9.6. The EPR and NMR transitions in this system are depicted by the solid and the broken arrows, respectively

oscillating magnetic field perpendicular to H_0 one will find two EPR transitions, in which only the electron spin changes its quantum number s_z, and two NMR transitions. The two EPR lines and the two NMR lines are both separated by a frequency a/h, and the difference between the mean frequencies of the EPR and the NMR signals is $(\omega_e - \omega_n)/2\pi$. In Fig. 9.1 the energy levels of this system are displayed, together with these transitions. (Actually, this is an idealized experiment. In a real case, the NMR spectrum of such a radical will either be not observable at all, or show a single line at some intermediate frequency, for reasons beyond the scope of this chapter.)

As we are dealing with a high magnetic field, the electron Zeeman energy is much larger than a, so the order of the four states is only dependent on the relative magnitudes of the hyperfine and the nuclear Zeeman interactions. In this figure, we have assumed $|a| < 2\hbar\omega_n$. If a becomes larger than twice the nuclear Zeeman energy, the two upper levels change place, whereas the lower two have to be permuted instead, if a is negative and its absolute value larger than $2\hbar\omega_n$.

The sign of a cannot be obtained by such an EPR measurement, because we have no way of telling which of the two resonance lines belongs to which transition. If a is positive, $|3\rangle \leftrightarrow |1\rangle$, for which the spin state of the proton is $|\alpha\rangle$, has the higher frequency and $|4\rangle \leftrightarrow |2\rangle$, where the proton is in state $|\beta\rangle$, the lower. If $a < 0$, these two lines are interchanged in the spectrum.

9.2 Singlet and Triplet States

As the other extreme, we will discuss the situation of our example molecule in zero magnetic field. In this case, both ω_e and ω_n are zero, and $|2\rangle$ and $|3\rangle$ are now degenerate in energy, so the nonsecular part of the hyperfine interaction causes very strong mixing of these two states. Of course, the result must still be two different eigenfunctions, but for either of them it is now impossible to assign an individual spin state to the electron and one to the proton, as we could do in high fields. It turns out that the property which can still be distinguished is the symmetry under permutation of the electron and proton spin. The symmetric function, which we call $|t_0\rangle$,

$$
\begin{aligned}
|t_0\rangle &= \frac{1}{\sqrt{2}}(|\alpha_e\beta_n\rangle + |\beta_e\alpha_n\rangle) \\
&= \frac{1}{\sqrt{2}}(|\beta_e\alpha_n\rangle + |\alpha_e\beta_n\rangle),
\end{aligned}
\tag{9.11}
$$

is unchanged by this operation, as Eq. (9.11) shows, whereas the antisymmetric function $|s\rangle$,

$$
\begin{aligned}
|s\rangle &= \frac{1}{\sqrt{2}}(|\alpha_e\beta_n\rangle - |\beta_e\alpha_n\rangle) \\
&= -\frac{1}{\sqrt{2}}(|\beta_e\alpha_n\rangle - |\alpha_e\beta_n\rangle),
\end{aligned}
\tag{9.12}
$$

is multiplied by -1. The factors $1/\sqrt{2}$ are needed for normalization of these functions. You can convince yourselves in a homework problem that $|t_0\rangle$ and $|s\rangle$ are eigenfunctions of H_H.

It is also immediately seen that $|1\rangle$ and $|4\rangle$ are invariant with respect to exchange of electron and proton spin and, therefore, with $|t_0\rangle$ they form a group of symmetric functions. Since there are three of them altogether, they have been termed triplet functions. In contrast, we have only one antisymmetric function, which is therefore denoted as singlet. We often talk of singlet or triplet being the multiplicity of these functions. The three triplet states differ by the z-component of their total spin $(\mathsf{I} + \mathsf{S})$, which is indicated by the subscript, thus

$$
|t_{+1}\rangle = |1\rangle \qquad |t_{-1}\rangle = |4\rangle. \tag{9.13}
$$

Figure 9.2 shows the energies of the four levels in zero field. The pertaining calculations are given as a homework problem. An important point to note is that the eigenfunctions can be determined without knowledge of the magnitude of the mixing matrix element, that is the value of $a/2$. The reason for this is the degeneracy of the energies of the starting functions $|2\rangle$ and $|3\rangle$. The strength of the interaction, however, determines the eigenvalues of energy of the final states.

Now we turn to a related problem, which involves two electrons. We will discuss what happens, when we bring together two radicals, for example in the reverse of their generation reaction.

At first, we will ignore spin and concentrate on the spatial parts of the wavefunctions. In a simple picture, we assume that all the fully occupied molecular orbitals of the initially separated radicals remain unchanged in the combination process. We then have to consider only the unpaired electrons of radical 1 and

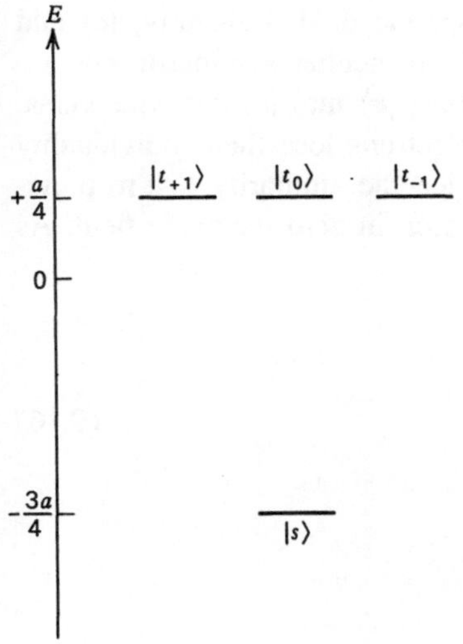

Fig. 9.2. Energy levels of a radical with one proton in zero magnetic field. The eigenfunctions are given by Eqs. 9.11–9.13

radical 2. We are tempted to think of them as residing in the orbitals ψ_1 and ψ_2 respectively, at least when the radicals are still several molecular diameters apart. Consequently, we would write the wavefunction of this system as the product of the wavefunctions of individual electrons, that is as $\psi_1(1)\psi_2(2)$, where the number in brackets denotes, which of the electrons occupies the orbital in question. However, interchanging the electrons would give us another wavefunction that also fits our model, $\psi_1(2)\psi_2(1)$, of exactly equal energy. We already know the consequences of such a degeneracy: any interaction, however slight, will mix these two states and render the two electrons indistinguishable. But there is quite an important interaction between the electrons, namely their mutual repulsion due to Coulombic forces. Hence the simple product functions are inappropriate for the description of this system, except when the separation between both radicals is so large as to make this interaction negligible. Again, the correct procedure is to form a symmetric (ψ_s) and an antisymmetric (ψ_a) linear combination of the two starting functions,

$$\begin{aligned}\psi_s &= \frac{1}{\sqrt{2}}[\psi_1(1)\psi_2(2) + \psi_1(2)\psi_2(1)] \\ \psi_a &= \frac{1}{\sqrt{2}}[\psi_1(1)\psi_2(2) - \psi_1(2)\psi_2(1)].\end{aligned} \tag{9.14}$$

These are valid eigenfunctions of the electronic Hamiltonian.

The spin part of the wavefunctions can be treated in a similar way. However, as there are two possible spin states for each electron, we have to start with four combinations,

$$\begin{aligned}|a\rangle &= |\alpha(1)\alpha(2)\rangle \qquad |c\rangle = |\beta(1)\alpha(2)\rangle \\ |b\rangle &= |\alpha(1)\beta(2)\rangle \qquad |d\rangle = |\beta(1)\beta(2)\rangle,\end{aligned} \tag{9.15}$$

which are all degenerate in the absence of a magnetic field. Evidently, $|a\rangle$ and $|d\rangle$ are symmetric functions, whereas $|b\rangle$ and $|c\rangle$ are neither symmetric nor antisymmetric, since exchanging the electrons will turn $|b\rangle$ into $|c\rangle$ and vice versa. These two states have to mix, because the two electrons lose their individuality owing to the interaction between them. You notice the similarity to the previously treated problem of one electron and one proton in zero magnetic field. As in that case, we have three triplet states,

$$\begin{aligned} |T_{+1}\rangle &= |\alpha(1)\alpha(2)\rangle \\ |T_0\rangle &= \frac{1}{\sqrt{2}}(|\alpha(1)\beta(2)\rangle + |\beta(1)\alpha(2)\rangle) \\ |T_{-1}\rangle &= |\beta(1)\beta(2)\rangle, \end{aligned} \tag{9.16}$$

and one singlet function,

$$|S\rangle = \frac{1}{\sqrt{2}}(|\alpha(1)\beta(2)\rangle - |\beta(1)\alpha(2)\rangle). \tag{9.17}$$

In a magnetic field, the triplet sublevels can be distinguished by the z-component of their total spin, which is indicated by the subscript.

Combination of the space functions of Eq. (9.14) with the spin functions of Eqs. (9.16) and (9.17) would give us eight wavefunctions altogether. However, we have to comply with the Pauli principle, which states that an acceptable wavefunction has to be overall antisymmetric. It is therefore only permissible to combine the symmetric space function ψ_s with the antisymmetric spin function $|S\rangle$, and the antisymmetric space function ψ_a with any of the three symmetric spin functions $|T_{+1,0,-1}\rangle$. For simplicity, we name the resulting four functions like their spin parts.

We thus see that some time before the two radicals are so close to each other as to be in actual contact, they already form an entity that can only exist in its singlet state, or one of its three triplet states. We denote this as a correlated radical pair.

As shown in Sect. 8.1, the squared total spin S_{tot}^2 of any system can be expressed by a single quantum number s,

$$\mathsf{S}_{tot}^2 = \hbar^2 s(s+1). \tag{9.18}$$

In our case of two electrons, we have

$$\begin{aligned} \mathsf{S}_{tot}^2 &= (\mathsf{S}_1 + \mathsf{S}_2)^2 \\ &= \mathsf{S}_1^2 + \mathsf{S}_2^2 + 2\mathsf{S}_1\mathsf{S}_2 \\ &= \mathsf{S}_1^2 + \mathsf{S}_2^2 + 2\mathsf{S}_{1z}\mathsf{S}_{2z} + \mathsf{S}_{1+}\mathsf{S}_{2-} + \mathsf{S}_{1-}\mathsf{S}_{2+}\ . \end{aligned} \tag{9.19}$$

Calculation of the expectation values of S_{tot}^2 with the spin parts of our wavefunctions is left to you as a homework problem. It is found that s equals 1 for the triplet states, whereas the singlet state is characterized by a total spin quantum number of zero, and thus does not have a magnetic moment. This is a general fact that holds for any spin system. Obviously, $|S\rangle$ cannot possess any z-component of spin as well, so the eigenvalue of its energy in a magnetic field H_0 is equal to that of $|T_0\rangle$, if we only consider the spin part of the wavefunctions. In contrast, $|T_{-1}\rangle$ is stabilized, and $|T_{+1}\rangle$ is destabilized, by an amount $(g_1 + g_2)\beta H_0$.

As the level splitting of the triplet states stems from magnetic interactions, these energy differences are minute. Even in the strong field of a superconducting magnet they are more than an order of magnitude smaller than $k_0 T$. A disproportionally larger effect is caused by the electrostatic interactions in this system, for instance those between the electrons of both radicals.

If we take the electronic Hamiltonian H_{el} as

$$\mathsf{H}_{el} = \mathsf{H}_1 + \mathsf{H}_2 + \mathsf{H}_{12}, \tag{9.20}$$

where the one-electron operators H_1 and H_2 describe the kinetic and potential energy of the electrons in the isolated radicals, and H_{12} takes into account all interactions between the two radicals, we can formally calculate the expectation values $\langle \mathsf{H}_{el} \rangle$ for the two functions ψ_s and ψ_a. With the abbreviations

$$S = \langle \psi_1(i)\psi_2(i)\rangle \tag{9.21}$$

$$E_1 = \langle \psi_1(i)|\mathsf{H}_i|\psi_1(i)\rangle \tag{9.22}$$

$$E_2 = \langle \psi_2(i)|\mathsf{H}_i|\psi_2(i)\rangle \tag{9.23}$$

$$C = \langle \psi_1(i)\psi_2(j)|\mathsf{H}_{12}|\psi_1(i)\psi_2(j)\rangle \tag{9.24}$$

$$J = \langle \psi_1(i)\psi_2(j)|\mathsf{H}_{12}|\psi_1(j)\psi_2(i)\rangle, \tag{9.25}$$

where i and j are 1 or 2, but $i \neq j$, we get

$$E_s = E_1 + E_2 + \frac{C + J}{1 + S^2} \tag{9.26}$$

$$E_a = E_1 + E_2 + \frac{C - J}{1 - S^2}. \tag{9.27}$$

As we will mostly be interested in larger separations of the two radicals than the usual bond lengths, we neglect the overlap integral S. We then find that the singlet and triplet states differ by an energy $2J$. J is called the exchange integral, since the coordinates of the electrons have been interchanged in the wavefunctions of its bra and ket. In most cases, J is negative, which causes the singlet level to be the more stable state. This is in accord with the empirical fact already mentioned, that the overwhelming majority of molecules is diamagnetic, i.e. their total electron spin is zero.

The typical dependence of the energies of singlet and triplet levels on the interradical distance r is illustrated by Fig. 9.3. For a discussion of the energy

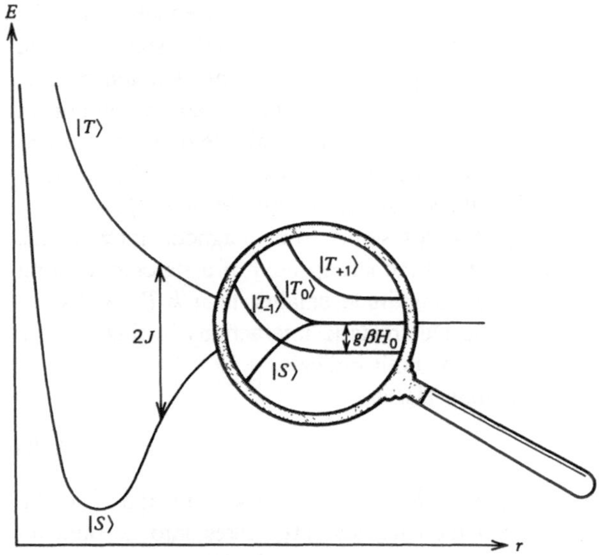

Fig. 9.3. Typical dependence of the energies of singlet and triplet states of a radical pair on the interradical distance r. The inset shows the additional splitting of the triplet levels that is caused by a magnetic field

differences between these states it is advantageous to choose the term E_1+E_2+C, which is common to all of them, as zero. When the two radicals are far apart, the electrostatic interactions between them are negligibly small, so J is zero, and singlet and triplet states are degenerate. However, in a magnetic field the Zeeman interaction splits off $|T_{+1}\rangle$ and $|T_{-1}\rangle$, leaving only the degeneracy of $|S\rangle$ and $|T_0\rangle$. As the radicals approach each other, the energy of the singlet state decreases until a minimum is reached, which corresponds to formation of a chemical bond between them. In contrast, the energy of the triplet levels rises continuously. These states are therefore called dissociative. The exchange integral J has been estimated to vary exponentially with the interradical separation r, if r is larger than the bonding distance. Typically, J decreases by two to three orders of magnitude, if r increases by one molecular diameter [1].

9.3 Intersystem Crossing by Magnetic Interactions

A conversion of states involving a change of multiplicity is called intersystem crossing. Here we will confine ourselves to the conversion of a singlet state into a triplet state and vice versa. We will discuss, how this can be effected by magnetic interactions of the unpaired electrons with an external field, or with the nuclear spins in our radical pair.

Inspection of Fig. 9.3 shows us that there is little chance of achieving intersystem crossing by these interactions as long as the radicals are close to each other, since in that case the exchange integral is much larger than the Zeeman and hyperfine energies, so the pair of radicals will be locked in a singlet or triplet state. On the other hand, the situation looks far more promising when the radicals are far apart, because then we have a degeneracy between $|S\rangle$ and $|T_0\rangle$, and we already know that degenerate states can be mixed even by very weak interactions. However, in order to be able to detect the occurrence of intersystem crossing at all, we have to start out and end up with a definite multiplicity, that means with the radicals near each other. The following scenario, which is called 'radical pair mechanism [2, 3], takes all this into account.

We begin with a stable molecule in a singlet state, into which we put an excess of energy. If we do this by heating, the molecule is excited to a higher vibrational state. The multiplicity of this state is the same as that of the ground state, since no electronic transition has taken place. If we excite the molecule by shining visible or ultraviolet light on it, it will go to a higher electronic state. Initially, this will again be a singlet state, because electric dipole transitions involving a change of multiplicity are strongly forbidden. However, organic molecules typically also possess an excited triplet state of lower energy, which can be populated by subsequent intersystem crossing from the excited singlet state. This process, which is accompanied by a decrease in energy and is therefore irreversible, can be very fast and efficient. In contrast to the radical pair mechanism, it occurs with the molecule intact. It is brought about by the interactions of the electron spins with the orbital angular momentum L. Although this type of interaction formally resembles the hyperfine interaction (its Hamiltonian is given by $\zeta \mathsf{S} \cdot \mathsf{L}$, where ζ is a constant), we will exclude this so-called spin-orbit coupling from our discussion. For intersystem crossing of a radical pair, it is unimportant.

In any case, the excited molecule reaches a metastable state of definite multiplicity after a time of the order of nanoseconds at the latest. From this state, bond cleavage can occur, leading to a radical pair. This process is also fast. Spin is conserved during it, so the multiplicity of the created correlated radical pair is the same as that of its precursor. We ignore the possibility that a singlet pair reacts back to the starting molecule immediately, because no intersystem crossing takes place at all in this instance. The two radicals will not be in contact for long, but will be separated by diffusion. Once the distance between them exceeds several molecular diameters, the exchange integral becomes negligibly small, so the degenerate energy levels can be mixed by suitable magnetic interactions. As a consequence, the spin state $|P\rangle$ of the pair is no more pure $|S\rangle$ nor, for instance, pure $|T_0\rangle$. Rather, it has to be described by a superposition

$$|P\rangle = c_s|S\rangle + c_{T_0}|T_0\rangle \tag{9.28}$$

with time dependent coefficients c_s and c_{T_0}, where $|c_s|^2 + |c_{T_0}|^2 = 1$. Since there is no preference for one state or the other as long as $J = 0$, the system oscillates between $|S\rangle$ and $|T_0\rangle$.

If the radicals lose each other for good, intersystem crossing caused by this effect cannot be detected. In liquid solution there is, however, a substantial chance that they will meet again. As their separation decreases, the exchange interaction rises sharply, so the pair can no longer exist in a superposition of two states. Instead, it is forced into one state or the other, with a probability depending on the square of the coefficient of that state in Eq. 9.28. If the pair finds itself in a singlet state upon re-encounter, it may react back to the starting compound. This is not possible for a triplet pair, given the energy scheme of Fig. 9.3. Hence one method to determine the amount of intersystem crossing that took place during the diffusive excursion of the radicals is to measure the recombination probability.

Actually, I find this spin-dependent reactivity a fascinating fact. By virtue of it, the tiny magnetic interactions, which are many orders of magnitude smaller than the energies involved in chemical reactions, can allow or forbid these reactions.

If we look closely at Fig. 9.3, we find that in addition to the degeneracy between $|S\rangle$ and $|T_0\rangle$ at large interradical separation there is also one between $|S\rangle$ and $|T_{-1}\rangle$, where these two levels cross. However, the time spent by the pair at this special distance is much too short to allow efficient intersystem crossing. In high magnetic fields, we shall therefore only have to consider mixing between $|S\rangle$ and $|T_0\rangle$. The situation is different in low fields, because then the level splitting of the triplet states is very small, so intersystem crossing involving $|T_{+1}\rangle$ and $|T_{-1}\rangle$ also becomes possible.

In a real experiment, many radical pairs are observed at the same time. The wavefunction of every pair is that of Eq. (9.28), but the coefficients c_s and c_{T_0} are different for each member of the ensemble. Hence we have an incoherent superposition of states that can best be described by a density matrix. In the next chapter, we shall therefore treat the spin dynamics of a system of two radicals by solving the equations of motion for the density matrix. Then we shall combine the results with the statistics of radical pair diffusion. However, before that we have to discuss the magnetic interactions that can effect singlet-triplet transitions.

We note that, speaking quite generally, nonvanishing matrix elements between symmetric (subscript s) and antisymmetric (subscript a) functions φ are needed for this. Such matrix elements will be $\langle\varphi_s|\mathsf{O}|\varphi_a\rangle$, and $\langle\varphi_a|\mathsf{O}|\varphi_s\rangle$, where O is the operator of the relevant interaction. The action of O is, of course, to transform the ket on which it operates into a new ket, not necessarily of the same symmetry as the old one.

We recall that the calculation of a matrix element $\langle\varphi_1|\varphi_2\rangle$ requires us first to multiply the complex conjugate of the function of its bra with the function of its ket, which just gives us some other function. We then have to sum or integrate over all the variables, depending on whether it is a discrete or continuous function. But the sum of all values of any antisymmetric function is zero, because for every term in it there is a corresponding one of the same magnitude, but opposite sign, and all these pairs cancel. The same holds for an integral, since this can be regarded as the limit of a sum.

The product of two symmetric functions, or that of two antisymmetric functions, is a symmetric function, but the product of a symmetric and an antisymmetric function is an antisymmetric function. Hence nonzero matrix elements between states of different symmetry can only result, if O changes the symmetry of the ket on which it operates. On the other hand, the condition for nonvanishing matrix elements between states of the same symmetry, which also includes all diagonal elements, is that O preserves the symmetry of its argument function.

Often, an operator in its usual form possesses no symmetry at all. However, it is always possible to decompose it into a symmetric and an antisymmetric part. This is a very simple process for an operator Op that works on two spins only. First, we construct a new operator Op′ by exchanging the spins in every term of Op, for instance turning S_{1x} into S_{2x}, or $S_{2y}S_{1z}$ into $S_{1y}S_{2z}$. The symmetric part, Op_s, and the antisymmetric part, Op_a, of the operator Op are then given by

$$Op_s = \tfrac{1}{2}(Op + Op') \qquad Op_a = \tfrac{1}{2}(Op - Op'). \tag{9.29}$$

As an example, we take the electron Zeeman interaction in our system of two radicals,

$$H_Z = (g_1 S_{1z} + g_2 S_{2z})\beta H_0. \tag{9.30}$$

In order to obtain H_Z', we have to interchange the two spin operators S_{1z} and S_{2z}, which is equivalent to an exchange of the electrons of the two radicals. The two g-factors must not be swapped in this process, because that would amount to an additional permutation of the radicals, and restore the starting configuration. We thus get our new operator

$$H_Z' = (g_1 S_{2z} + g_2 S_{1z})\beta H_0. \tag{9.31}$$

As H_Z is neither equal to H_Z' nor to $-H_Z'$, it is evident that it is neither a symmetric nor an antisymmetric operator. The prescriptions of Eq. (9.29) lead us to

$$H_{Z,s} = \frac{(g_1 + g_2)\beta H_0}{2}(S_{1z} + S_{2z}) \tag{9.32}$$

$$H_{Z,a} = \frac{(g_1 - g_2)\beta H_0}{2}(S_{1z} - S_{2z}). \tag{9.33}$$

Taking as our basis the four functions $|T_{+1}\rangle$, $|S\rangle$, $|T_0\rangle$, and $|T_{-1}\rangle$, in that order, we find that the matrix elements of the symmetric part, $H_{Z,s}$, and of the antisymmetric part, $H_{Z,a}$, of the Zeeman interaction are given by

$$H_{Z,s} = \frac{\hbar}{2}(\omega_1 + \omega_2)\begin{pmatrix} 1 & 0 & 0 & 0 \\ 0 & 0 & 0 & 0 \\ 0 & 0 & 0 & 0 \\ 0 & 0 & 0 & 1 \end{pmatrix}$$

$$H_{Z,a} = \frac{\hbar}{2}(\omega_1 - \omega_2)\begin{pmatrix} 0 & 0 & 0 & 0 \\ 0 & 0 & 1 & 0 \\ 0 & 1 & 0 & 0 \\ 0 & 0 & 0 & 0 \end{pmatrix}, \tag{9.34}$$

with

$$\omega_1 = g_1\beta H_0/\hbar \qquad \omega_2 = g_2\beta H_0/\hbar. \tag{9.35}$$

The diagonal matrix of $\mathsf{H}_{Z,s}$ describes the level splitting of the triplet states and the degeneracy of $|S\rangle$ and $|T_0\rangle$ in the absence of the exchange interaction. Intersystem crossing can be induced by the antisymmetric part of the Zeeman interaction, $\mathsf{H}_{Z,a}$, which only contains off-diagonal elements. It is seen that this can only occur between the singlet state and $|T_0\rangle$. Intersystem crossing between $|S\rangle$ and $|T_{+1}\rangle$ or between $|S\rangle$ and $|T_{-1}\rangle$ cannot be caused by this interaction, because there are no matrix elements of $\mathsf{H}_{Z,a}$ connecting these states. Since the difference of the g-factors of the two radicals determines the magnitude of the mixing matrix element, this process is often referred to as the 'Δg-mechanism' in the literature.

Intersystem crossing by this mechanism can be visualized with the aid of vector models as shown in Fig. 9.4. In these models a singlet state is depicted by two collinear vectors of opposite direction, whereas for $|T_0\rangle$ one of the vectors has been rotated by 180° around the axis of quantization. If the two radicals are well separated, so that the exchange interaction has become negligible, their spins precess around the axis of the magnetic field H_0 independently. In a coordinate system rotating with the mean Larmor frequency $(\omega_1+\omega_2)/2$, the phase difference of the two spin vectors is seen to change with angular frequency $\omega_d = \omega_1 - \omega_2$. A pair that was initially in a singlet state will therefore be in a pure triplet state after a time $\pi/|\omega_d|$, as displayed in Fig. 9.5.

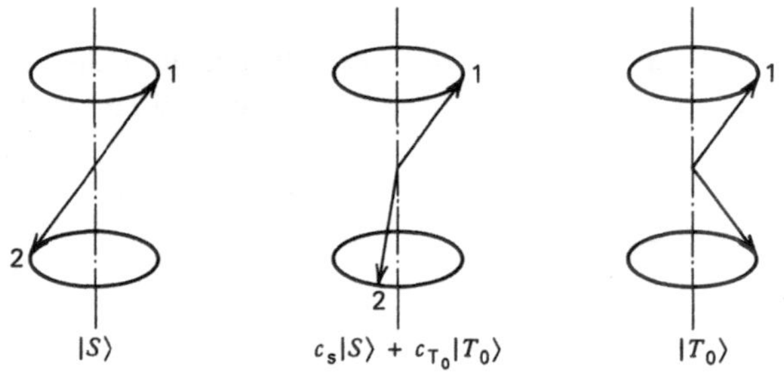

Fig. 9.4. Vector models for a radical pair in a singlet state, in a superposition state and in state $|T_0\rangle$. The vertical axis is the axis of the external magnetic field. The *numbers* indicate the electron spins of radical 1 and radical 2

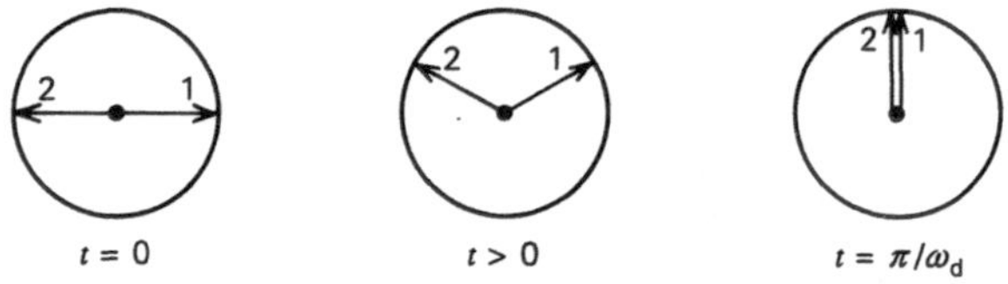

Fig. 9.5. Visualization of intersystem crossing from $|S\rangle$ to $|T_0\rangle$ by vector models. Projections of the electron spins on the xy-plane of a coordinate system rotating with the mean Larmor frequency are shown for different times

These pictures do not allow a true representation of $|S\rangle$ and $|T_0\rangle$, because four vectors would be needed for this. Nevertheless, they are valid as far as total spin and z-component of spin of these functions are concerned. The oscillations between singlet and triplet states are also correctly described.

The Hamiltonian for the hyperfine interaction of the electrons with an arbitrary number of nuclear spins in the two radicals is

$$\mathsf{H}_\mathsf{H} = \sum_i a_i \mathsf{S}_1 \mathsf{I}_\mathsf{i} + \sum_k a_\mathsf{k} \mathsf{S}_2 \mathsf{I}_\mathsf{k}, \tag{9.36}$$

where subscripts i and k are used for the nuclei in the first and second radical, respectively. The interactions between the electron spin of one radical with the nuclear spins in the other radical need not be considered, because hyperfine interactions only operate through chemical bonds, not through space.

In high magnetic fields, it is sufficient to retain only the secular part $\mathsf{H}_{\mathsf{H,\,sec}}$,

$$\mathsf{H}_{\mathsf{H,\,sec}} = \mathsf{S}_{1z} \sum_i a_i \mathsf{I}_{iz} + \mathsf{S}_{2z} \sum_k a_k \mathsf{I}_{kz}, \tag{9.37}$$

as we have seen in Sect. 9.1. This leads to a drastic simplification, because electron and nuclear spins may be separated. Hence the four basis functions $|S\rangle$ and $|T_{+1,\,0,\,-1}\rangle$ are still sufficient for the calculation of the electron spin dynamics of this system under these circumstances.

Decomposition of $\mathsf{H}_{\mathsf{H,\,sec}}$ into its symmetric and its antisymmetric part is easily possible. We get

$$\mathsf{H}_{\mathsf{H,\,s}} = \frac{1}{2}\left(\sum_i a_i \mathsf{I}_{iz} + \sum_k a_k \mathsf{I}_{kz}\right)(\mathsf{S}_{1z} + \mathsf{S}_{2z}) \tag{9.38}$$

$$\mathsf{H}_{\mathsf{H,\,a}} = \frac{1}{2}\left(\sum_i a_i \mathsf{I}_{iz} - \sum_k a_k \mathsf{I}_{kz}\right)(\mathsf{S}_{1z} - \mathsf{S}_{2z}), \tag{9.39}$$

with matrix elements

$$\mathsf{H}_{\mathsf{H,\,s}} = \frac{\hbar\omega_+}{2}\begin{pmatrix} 1 & 0 & 0 & 0 \\ 0 & 0 & 0 & 0 \\ 0 & 0 & 0 & 0 \\ 0 & 0 & 0 & 1 \end{pmatrix} \quad \mathsf{H}_{\mathsf{H,\,a}} = \frac{\hbar\omega_-}{2}\begin{pmatrix} 0 & 0 & 0 & 0 \\ 0 & 0 & 1 & 0 \\ 0 & 1 & 0 & 0 \\ 0 & 0 & 0 & 0 \end{pmatrix} \tag{9.40}$$

which are identical to those of $\mathsf{H}_{\mathsf{Z,\,s}}$ and $\mathsf{H}_{\mathsf{H,\,a}}$ except for a scaling factor. Here we have used

$$\omega_+ = \frac{1}{\hbar}\left(\sum_i a_i \mathsf{I}_{iz} + \sum_k a_k \mathsf{I}_{kz}\right) \tag{9.41}$$

$$\omega_- = \frac{1}{\hbar}\left(\sum_i a_i \mathsf{I}_{iz} - \sum_k a_k \mathsf{I}_{kz}\right) \tag{9.42}$$

as abbreviations.

It is thus seen that in high magnetic fields, the antisymmetric part of the hyperfine interaction can only mix $|S\rangle$ and $|T_0\rangle$, exactly in the same way as the antisymmetric term of the Zeeman interaction. Consequently, the vector models of Figs. 9.4 and 9.5 are also applicable to this case.

In low fields, the situation is much more complicated. The nonsecular terms

$$\mathsf{H}_{\mathrm{H,\,nonsec}} = \frac{1}{2}\left[\sum_i a_i(\mathsf{S}_{1+}\mathsf{I}_{i-} + \mathsf{S}_{1-}\mathsf{I}_{i+}) + \sum_k a_k(\mathsf{S}_{2+}\mathsf{I}_{k-} + \mathsf{S}_{2-}\mathsf{I}_{k+})\right] \tag{9.43}$$

cause simultaneous transitions of electron and nuclear spins. Hence a separation of nuclear and electronic spin spaces is not possible in this instance, and much larger basis sets are required for the computation of the spin dynamics. Even for molecules possessing only few magnetic nuclei, these calculations involve matrices of formidable dimensions.

In a homework problem, you are asked to decompose $\mathsf{H}_{\mathrm{H,\,nonsec}}$ into terms that are symmetric and antisymmetric with regard to interchange of the electrons. From this you will learn that the symmetric terms cause state conversions within the triplet manifold, whereas the antisymmetric terms effect intersystem crossing between $|S\rangle$ and $|T_{+1}\rangle$ as well as $|S\rangle$ and $|T_{-1}\rangle$. Of course, these results are to be expected from our preceding discussion of the dependence of matrix elements on operator symmetry, and because we know that s_z has to change by $+1$ or -1 in the simultaneous spin-flips of an electron and a proton that are brought about by the terms $\mathsf{S}_+\mathsf{I}_-$ and $\mathsf{S}_-\mathsf{I}_+$.

Pictorial representations of the states $|T_{+1}\rangle$ and $|T_{-1}\rangle$ are given in Fig. 9.6. With the aid of such vector models, state conversions accompanied by a change of s_z by ± 1 can also be visualized by 180°-rotations, as shown in Fig. 9.7. For intersystem crossing, e.g. from $|S\rangle$ to $|T_{+1}\rangle$, the axis of rotation is situated in the plane defined by the z-axis and the spin vectors. For conversions without a symmetry change, as between $|T_0\rangle$ and $|T_{+1}\rangle$ in the figure, the axis is perpendicular to this plane. However, quantitative results cannot be obtained from these diagrams.

In addition to the conversions induced by $\mathsf{H}_{\mathrm{H,\,nonsec}}$, we still have those caused by the secular part of the hyperfine interaction, so it is evident that in very

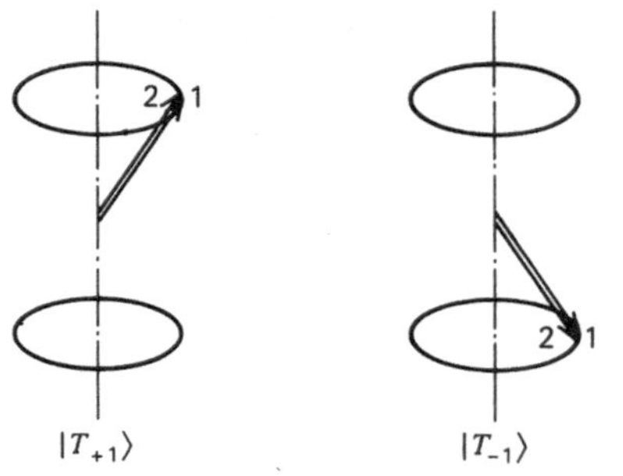

Fig. 9.6. Vector models for a radical pair in states $|T_{+1}\rangle$ and $|T_{-1}\rangle$.

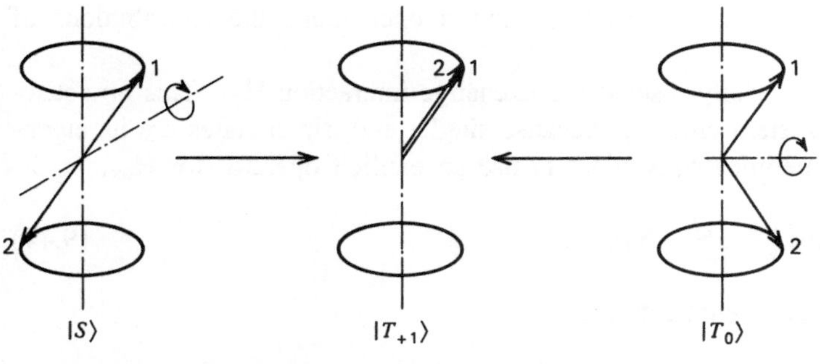

Fig. 9.7. Vector models for the visualization of intersystem crossing between $|S\rangle$ and $|T_{+1}\rangle$, as well as of state conversion between $|T_0\rangle$ and $|T_{+1}\rangle$. In both cases, only the vector 2 is rotated by 180° around the axis indicated

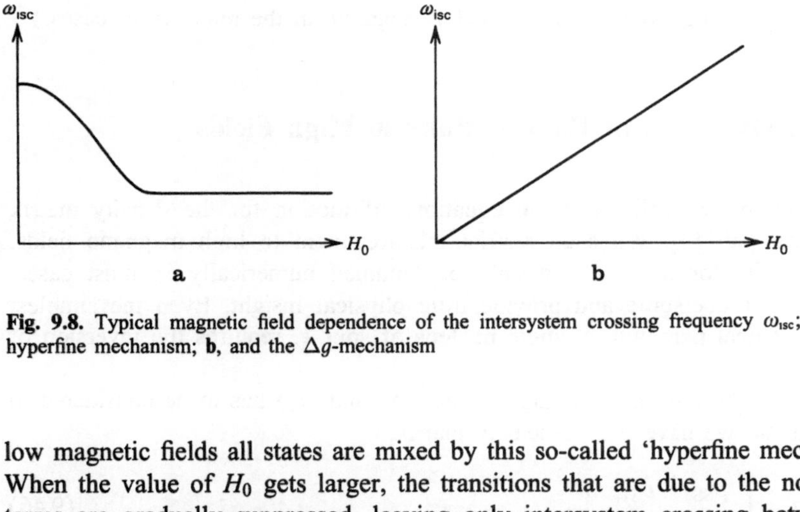

Fig. 9.8. Typical magnetic field dependence of the intersystem crossing frequency ω_{isc}; **a**, for the hyperfine mechanism; **b**, and the Δg-mechanism

low magnetic fields all states are mixed by this so-called 'hyperfine mechanism'. When the value of H_0 gets larger, the transitions that are due to the nonsecular terms are gradually suppressed, leaving only intersystem crossing between $|S\rangle$ and $|T_0\rangle$. Hence the intersystem crossing frequency ω_{isc} initially decreases with increasing strength of the static field, and then remains constant. This is illustrated in Fig. 9.8. The field dependence of the Δg-mechanism is totally different. In that case, ω_{isc} is directly proportional to H_0, as also shown in this figure.

An estimation of typical times required for intersystem crossing by these two mechanisms is put to you as a homework problem. It turns out that these are of the order of ns. As we shall see later, this is much longer than the time τ of a diffusive excursion of a radical pair. It is this fact which makes it possible to measure the dependence of ω_{isc} on the strength of the magnetic interactions after all; if ω_{isc} were much larger than $1/\tau$, equal amounts of $|S\rangle$ and $|T_0\rangle$ would invariably be found for the ensemble after the re-encounter of the pairs, because the duration of the separation is statistically distributed. On the other hand, observable effects result in spite of the very small degree of intersystem crossing occurring during one diffusive excursion, since a particular pair experiences a series of separations and re-encounters, the contributions of which are cumulative.

Finally, we will briefly discuss the exchange interaction H_{ex}. This obviously cannot cause intersystem crossing, because singlet and triplet states are its eigenfunctions. It is sometimes convenient to use an explicit operator for H_{ex},

$$\mathsf{H}_{ex} = -J\left(\tfrac{1}{2} + 2\mathsf{S}_1 \cdot \mathsf{S}_2\right). \tag{9.44}$$

In our basis, its matrix elements are

$$\mathsf{H}_{ex} = J\begin{pmatrix} -1 & 0 & 0 & 0 \\ 0 & +1 & 0 & 0 \\ 0 & 0 & -1 & 0 \\ 0 & 0 & 0 & -1 \end{pmatrix}. \tag{9.45}$$

(Remember that the exchange integral J is negative in the majority of cases.)

9.4 Spin Dynamics of Radical Pairs in High Fields

In this section, we will solve the equations of motion for the density matrix of a radical pair [4]. We shall restrict this treatment to high magnetic fields, because results for low H_0 can only be obtained numerically in most cases, or are very cumbersome and provide little physical insight. Even the simplest system, a radical pair with a single nucleus of spin $\frac{1}{2}$, requires the inversion of an 8×8 matrix.

As shown before, only mixing between $|S\rangle$ and $|T_0\rangle$ has to be considered in high fields, so we have a 2×2 density matrix,

$$\underline{\rho} = \begin{pmatrix} \rho_{SS} & \rho_{ST_0} \\ \rho_{T_0S} & \rho_{T_0T_0} \end{pmatrix}. \tag{9.46}$$

In this basis, our high-field Hamiltonian,

$$\mathsf{H} = \mathsf{H}_Z + \mathsf{H}_{H,\,sec} + \mathsf{H}_{ex}, \tag{9.47}$$

is represented by the matrix

$$\mathsf{H} = \begin{pmatrix} J & Q \\ Q & -J \end{pmatrix}, \tag{9.48}$$

where Q describes the combined influence of the Δg- and the hyperfine mechanism,

$$Q = \frac{1}{2}\left[(g_1 - g_2)\beta H_0 + \sum_i a_i \mathsf{I}_{iz} - \sum_k a_k \mathsf{I}_{kz}\right]. \tag{9.49}$$

In reality, the exchange integral J varies randomly with time as the radicals diffuse. However, we cannot solve the equations of motion of $\underline{\rho}$ in this case, so we have to keep J constant. This approximation is not as bad as it may seem at first. Owing to the exponential dependence of the exchange interaction on the separation of the radicals, J drops from a value that is much larger than Q to practically zero within a very short distance, that means during a few diffusive steps of the molecules and thus in a time of the order of ps, which is much shorter than the time required for intersystem crossing of the pair. These two limiting cases, where the spin dynamics are exclusively determined by J and Q, respectively, are correctly described by our calculations.

As we learned in Sects. 5.5 and 5.6, the time dependence of the density matrix is given by its commutator with the Hamiltonian,

$$\frac{\partial}{\partial t}\underline{\rho} = -\frac{i}{\hbar}[\mathsf{H},\underline{\rho}]. \tag{9.50}$$

We carry out the matrix multiplications of Eq. (9.50) and find

$$\frac{\partial}{\partial t}\underline{\rho} = -\frac{i}{\hbar}\begin{pmatrix} Q(\rho_{T_0S} - \rho_{ST_0}) & 2J\rho_{ST_0} + Q(\rho_{T_0T_0} - \rho_{SS}) \\ -2J\rho_{T_0S} - Q(\rho_{T_0T_0} - \rho_{SS}) & Q(\rho_{ST_0} - \rho_{T_0S}) \end{pmatrix}. \tag{9.51}$$

This is just a shorthand notation for a system of four coupled differential equations. We could write these out separately, as we did in Sect. 5.9. However, it is more convenient to cast Eq. (9.51) into the equivalent vector expression

$$\frac{\partial}{\partial t}\begin{pmatrix} \rho_{SS} \\ \rho_{ST_0} \\ \rho_{T_0S} \\ \rho_{T_0T_0} \end{pmatrix} = -\frac{i}{\hbar}\begin{pmatrix} 0 & -Q & Q & 0 \\ -Q & 2J & 0 & Q \\ Q & 0 & -2J & -Q \\ 0 & Q & -Q & 0 \end{pmatrix}\begin{pmatrix} \rho_{SS} \\ \rho_{ST_0} \\ \rho_{T_0S} \\ \rho_{T_0T_0} \end{pmatrix}. \tag{9.52}$$

The four elements of $\underline{\rho}$ are not independent. Adding the first and last row of Eq. (9.52), we find that

$$\frac{\partial}{\partial t}(\rho_{SS} + \rho_{T_0T_0}) = 0. \tag{9.53}$$

This is not a surprising result. We remember from Sect. 5.7 that the trace of the density matrix is always unity for a closed macrosystem, hence its time derivate vanishes. At this stage we are not yet concerned with chemical reactions of the radicals, so our system is indeed closed. Consequently, we have not included any kinetic terms in our master equation, Eq. (9.50).

We may use relation (9.53) to simplify Eq. (9.52). With it, we could eliminate either ρ_{SS} or $\rho_{T_0T_0}$, but we find it more useful to introduce $\rho_{SS} - \rho_{T_0T_0}$ as a new variable. This quantity amounts $+1$ for pure $|S\rangle$, that is if all radical pairs of our ensemble are in the singlet state. For pure $|T_0\rangle$ it is -1. With this change of variables, we obtain

$$\frac{\partial}{\partial t}\begin{pmatrix}\rho_{SS}-\rho_{T_0T_0}\\ \rho_{ST_0}\\ \rho_{T_0S}\end{pmatrix}=-\frac{i}{\hbar}\begin{pmatrix}0 & -2Q & 2Q\\ -Q & 2J & 0\\ Q & 0 & -2J\end{pmatrix}\begin{pmatrix}\rho_{SS}-\rho_{T_0T_0}\\ \rho_{ST_0}\\ \rho_{T_0S}\end{pmatrix}. \tag{9.54}$$

The off-diagonal elements of a density matrix are in general complex numbers, so the variables ρ_{ST_0} and ρ_{T_0S} do not correspond to any observables. Since we want to gain as much physical insight as possible by our calculations, we will change to a set of real variables in the hope of arriving at some physically significant quantity by this transformation. We utilize the fact that $\underline{\rho}$ is a Hermitian matrix, and ρ_{ST_0} and ρ_{T_0S} are thus guaranteed to be complex conjugates of each other. As in Sect. 8.8 we therefore add and subtract these off-diagonal elements, obtaining

$$\rho_{ST_0}+\rho_{T_0S}=2\mathrm{Re}(\rho_{ST_0}) \tag{9.55}$$

$$i(\rho_{T_0S}-\rho_{ST_0})=-2i\mathrm{Im}(\rho_{ST_0}). \tag{9.56}$$

We recollect (see Sect. 5.6) that the average value $\overline{\langle \mathsf{O}\rangle}$ of an observable is given by

$$\overline{\langle \mathsf{O}\rangle}=\mathrm{Tr}[\underline{\rho}\mathsf{O}]. \tag{9.57}$$

It is thus evident that the matrix of the operator O possessing an average value of $\rho_{ST_0}+\rho_{T_0S}$ has to be

$$\mathsf{O}=\begin{pmatrix}0 & 1\\ 1 & 0\end{pmatrix}. \tag{9.58}$$

Inspection of Eqs. 9.33–9.35 shows us that O is equal to $\mathsf{S}_{1z}-\mathsf{S}_{2z}$. We see that $\rho_{ST_0}+\rho_{T_0S}$ is connected with opposite z-components of the spins of both electrons.

When the radicals are created, there is no phase correlation between the individual pairs of the ensemble, so the off-diagonal elements of the density matrix are zero at that moment. This *hypothesis of random phases* has already been mentioned in chapter 5.7. Hence, $\overline{\langle \mathsf{S}_{1z}\rangle}$ and $\overline{\langle \mathsf{S}_{2z}\rangle}$ also vanish initially. The appearance of nonzero terms $\rho_{ST_0}+\rho_{T_0S}$ at a later time therefore tells us that some macroscopic z-component of the electron spins has been generated by the evolution of the density matrix.

We note that if we had two doublet states (that is two radicals without any interaction between them) in a magnetic field, we would certainly find nonvanishing values of $\overline{\langle \mathsf{S}_{1z}\rangle}$ as well as of $\overline{\langle \mathsf{S}_{2z}\rangle}$ in thermodynamic equilibrium. These are of course determined by the differences between the number of respective radicals with the electron in state $|\alpha\rangle$, and those with the electron in $|\beta\rangle$, for instance

$$\begin{aligned}\overline{\langle \mathbf{S}_{1z}(\text{doublet})\rangle} &= \rho_{\alpha\alpha}\langle\alpha|\mathbf{S}_{1z}|\alpha\rangle + \rho_{\beta\beta}\langle\beta|\mathbf{S}_{1z}|\beta\rangle \\ &= \frac{1}{2}(\rho_{\alpha\alpha} - \rho_{\beta\beta}) \\ &= \frac{1}{2Z}[\exp(-\mathscr{E}_\alpha/k_0T) - \exp(-\mathscr{E}_\beta/k_0T)],\end{aligned} \tag{9.59}$$

where $\mathscr{E}_\alpha$ and $\mathscr{E}_\beta$ are the Zeeman energies of the states $|\alpha\rangle$ and $|\beta\rangle$, and Z is the partition function. We have used Eqs. (5.46) and (5.47) in the last line of this equation. As the magnetic energies involved are much smaller than k_0T at room temperature, it is sufficient to expand the exponentials to first order. Using Eq. (9.3) we get

$$\overline{\langle \mathbf{S}_{1z}(\text{doublet})\rangle} = -\frac{g_1\beta H_0}{4k_0T}. \tag{9.60}$$

An analogous equation holds for $\overline{\langle \mathbf{S}_{2z}\rangle}$. We see that these values are negative for any doublet, in accordance with the fact already mentioned that the most stable state of an electron is $|\beta\rangle$.

On the other hand, if we have a correlated radical pair with possible states $|S\rangle$ and $|T_0\rangle$ in thermodynamic equilibrium,[1] the average values of both $\mathbf{S}_{1z}$ and $\mathbf{S}_{2z}$ are zero, as we have already seen. The creation of, for instance, nonvanishing $\overline{\langle \mathbf{S}_{1z}\rangle}$ is therefore clearly a nonequilibrium situation. Nevertheless, arbitrary values of both $\overline{\langle \mathbf{S}_{1z}\rangle}$ and $\overline{\langle \mathbf{S}_{2z}\rangle}$ are not possible. Since we do not allow any exchange of energy between our system and its surroundings, their sum has to remain zero, so any change in one term has to be compensated by a change of equal magnitude, but opposite sign in the other.

The average values of the z-component of spin are associated with population differences of energy levels, as we have seen in Eq. (9.59). A special term has been coined for the deviations of these population differences from equilibrium; they are called polarizations. Here we are dealing with electron spin polarizations. We can measure these by an EPR experiment, where they manifest themselves by anomalous signal intensities. In contrast, the frequencies of the resonance lines are unchanged, since the energies of the spin states are not perturbed. For our radical pair, we have found that a positive polarization of one radical has to be accompanied by a negative polarization of the other. Also, we know that no z-component of spin is present initially. Consequently, if we observe any EPR signals of the usual phase, i.e. absorption signals, for one radical in this system, we will necessarily find signals in emission, i.e. inverted signals, for the other. The total intensity of all these signals will be zero, in the absence of relaxation.

[1] In this treatment, thermodynamic equilibrium refers to magnetic energies only. Of course, as we have seen in Fig. 9.3, a true equilibrium state of a radical pair would not be reached before the formation of a stable diamagnetic molecule

Besides the already mentioned determination of recombination probabilities, which are directly related to the variable $\rho_{SS} - \rho_{T_0T_0}$ in Eq. (9.54), the measurement of electron spin polarization, which is given by $\rho_{ST_0} + \rho_{T_0S}$, is thus a second method to investigate magnetically induced intersystem crossing. On the other hand, the term $i(\rho_{T_0S} - \rho_{ST_0})$, which denotes some phase correlation between $|S\rangle$ and $|T_0\rangle$ in the ensemble of radical pairs, does not correspond to one of the usual observables. In a homework problem, you can try to find the operator possessing this average value.

With our new variables $\rho_{ST_0} + \rho_{T_0S}$ and $i(\rho_{T_0S} - \rho_{ST_0})$ Eq. (9.54) is turned into an equation involving only real quantities,

$$\frac{\partial}{\partial t}\begin{pmatrix} \rho_{SS} - \rho_{T_0T_0} \\ i(\rho_{T_0S} - \rho_{ST_0}) \\ \rho_{ST_0} + \rho_{T_0S} \end{pmatrix} = \frac{2}{\hbar}\begin{pmatrix} 0 & -Q & 0 \\ Q & 0 & -J \\ 0 & J & 0 \end{pmatrix}\begin{pmatrix} \rho_{SS} - \rho_{T_0T_0} \\ i(\rho_{T_0S} - \rho_{ST_0}) \\ \rho_{ST_0} + \rho_{T_0S} \end{pmatrix}. \quad (9.61)$$

Before we proceed to solve Eq. (9.61), we want to illustrate its physical meaning by the following transformation [5]. We note that the cross product of two vectors,

$$\begin{pmatrix} a \\ b \\ c \end{pmatrix} \times \begin{pmatrix} x \\ y \\ z \end{pmatrix} = \begin{pmatrix} bz - cy \\ -az + cx \\ ay - bx \end{pmatrix}, \quad (9.62)$$

can also be expressed in matrix notation, giving

$$\begin{pmatrix} a \\ b \\ c \end{pmatrix} \times \begin{pmatrix} x \\ y \\ z \end{pmatrix} = \begin{pmatrix} 0 & -c & b \\ c & 0 & -a \\ -b & a & 0 \end{pmatrix}\begin{pmatrix} x \\ y \\ z \end{pmatrix}. \quad (9.63)$$

The matrix of coefficients in this equation is identical to that of Eq. (9.61), if we choose $a = J, b = 0$, and $c = Q$. We can therefore rewrite Eq. (9.61) using a vector product,

$$\frac{\partial}{\partial t}\begin{pmatrix} \rho_{SS} - \rho_{T_0T_0} \\ i(\rho_{T_0S} - \rho_{ST_0}) \\ \rho_{ST_0} + \rho_{T_0S} \end{pmatrix} = \begin{pmatrix} 2J/\hbar \\ 0 \\ 2Q/\hbar \end{pmatrix} \times \begin{pmatrix} \rho_{SS} - \rho_{T_0T_0} \\ i(\rho_{T_0S} - \rho_{ST_0}) \\ \rho_{ST_0} + \rho_{T_0S} \end{pmatrix}. \quad (9.64)$$

Apart from the absence of relaxation terms we recognize this as a set of Bloch equations. This is a consequence of the fact that every two-level system can be described by a fictitious spin $\frac{1}{2}$. In our case the analogy to the behaviour of an ensemble of such spins in a magnetic resonance experiment is especially beautiful, since the mixing matrix element Q, which is constant, plays the role of the static field H_0, whereas the exchange integral J, which is essentially switched on or off as the radicals approach each other or diffuse apart, acts like an x-pulse in NMR. However, as the electron spin polarization corresponds to the z-magnetization, and the population difference between singlet and triplet states to one component of the transverse magnetization, the vector remaining in thermodynamic equilibrium points along the direction of the time dependent field, whereas in magnetic resonance it is orientated along the static field.

Equation (9.64) tells us that we can describe the spin dynamics of a radical pair by the precession of a vector $\mathscr{R}$,

$$\mathscr{R} = \begin{pmatrix} \rho_{SS} - \rho_{T_0T_0} \\ i(\rho_{T_0S} - \rho_{ST_0}) \\ \rho_{ST_0} + \rho_{T_0S} \end{pmatrix} \tag{9.65}$$

$$= \begin{pmatrix} \text{population difference between singlet and triplet states} \\ \text{some phase correlation not directly observable} \\ \text{opposite electron spin polarization in both radicals} \end{pmatrix},$$

around an axis $\mathscr{A}$ in the xz-plane, as shown in Fig. 9.9. The angle θ between $\mathscr{A}$ and the z-axis is given by

$$\tan(\theta) = J/Q. \tag{9.66}$$

The precession frequency depends on the resultant length A of $\mathscr{A}$,

$$A = \sqrt{Q^2 + J^2}. \tag{9.67}$$

In this model, a pure singlet state corresponds to a unit vector on the positive x-axis, and pure $|T_0\rangle$ to such a vector in the direction of $-x$. Positive and negative electron spin polarization appear on the positive and negative z-axis.

If the exchange interaction is zero, we see that there is some resemblance to our vector models of the preceding section inasmuch as in both cases intersystem crossing takes place by 180°-rotations around the axis of the mixing interaction Q. However, the model given here represents the electron spin states correctly, whereas our earlier pictures do not. On the other hand, the vectors shown in Fig. 9.9 are not linked to concepts of our physical intuition as directly as before.

We shall come back to this model later when we discuss radical pair dynamics and spin polarization effects in more detail. At the moment, we want to go

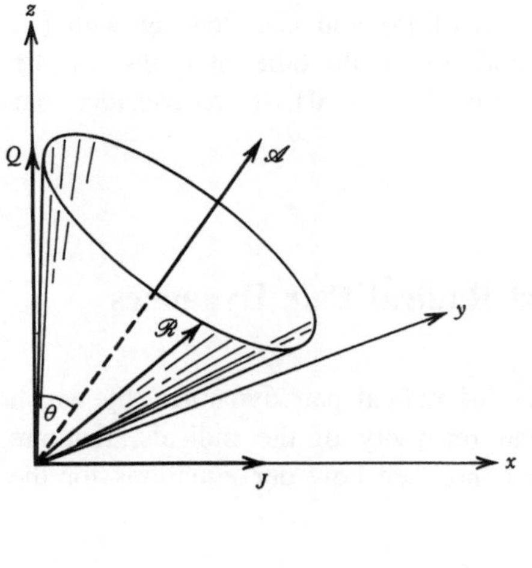

Fig. 9.9. Pictorial representation of the evolution of the density matrix of a radical pair. The components of the vector $\mathscr{R}$ (Eq. 9.65) are the elements of the density matrix. The motion of $\mathscr{R}$ is a precession around an axis $\mathscr{A}$, which is the resultant of the mixing matrix element Q given by Eq. 9.49 and the exchange integral J. In this figure, we have assumed both Q and J to be positive

on to the solution of our equation for the spin dynamics, Eq. (9.61), which can be obtained by standard methods. You can convince yourselves that the results is

$$\mathscr{R} = \begin{pmatrix} \frac{J^2}{A^2} + \frac{Q^2}{A^2}\cos(\Omega t) & -\frac{Q}{A}\sin(\Omega t) & \frac{QJ}{A^2}[1-\cos(\Omega t)] \\ \frac{Q}{A}\sin(\Omega t) & \cos(\Omega t) & -\frac{J}{A}\sin(\Omega t) \\ \frac{QJ}{A^2}[1-\cos(\Omega t)] & \frac{J}{A}\sin(\Omega t) & \frac{Q^2}{A^2} + \frac{J^2}{A^2}\cos(\Omega t) \end{pmatrix} \cdot \mathscr{R}_{t=0}, \tag{9.68}$$

where $\mathscr{R}_{t=0}$ denotes the initial value of $\mathscr{R}$, and

$$\Omega = \frac{2A}{\hbar}. \tag{9.69}$$

Although we have derived Eq. (9.68) with mixing of $|S\rangle$ and $|T_0\rangle$ in mind it is in fact quite general, because we have made no assumptions that are specific for this particular process. It describes the time evolution under a stationary perturbation P of any system of two levels $|a\rangle$ and $|b\rangle$ separated by an energy E, where the matrix element P joining these states is real. If we begin with a diagonal density matrix, as is usually the case, the population difference $\rho_{aa} - \rho_{bb}$ of the energy levels at time t is given by

$$\rho_{aa} - \rho_{bb} = \frac{1}{E^2 + 4P^2}[E^2 + 4P^2\cos(\sqrt{E^2 + 4P^2}t)] \cdot (\rho_{aa} - \rho_{bb})_{t=0}. \tag{9.70}$$

We see that $\rho_{aa} - \rho_{bb}$ oscillates harmonically around a mean value $\overline{\rho_{aa} - \rho_{bb}}$,

$$\overline{\rho_{aa} - \rho_{bb}} = \frac{E^2}{E^2 + 4P^2} \cdot (\rho_{aa} - \rho_{bb})_{t=0}. \tag{9.71}$$

Moreover, Eq. (9.70) shows that the absolute value of the population difference cannot exceed $(\rho_{aa} - \rho_{bb})_{t=0}$. Hence we have no time dependence, if initially our ensemble is composed of equal amounts of $|a\rangle$ and $|b\rangle$. Starting with the system in one pure state, the maximum population of the other state that can be attained is $4P^2/(E^2 + 4P^2)$. This follows from Eq. (9.70), if we consider that $\rho_{aa} + \rho_{bb} = 1$.

9.5 Combining Spin Dynamics and Radical Pair Dynamics

Now we will briefly describe some aspects of radical pair dynamics, by which we mean the bodily motion and chemical reactivity of the radicals, that are relevant for intersystem crossing. We shall then see how our equations for the spin dynamics, which we have derived in the preceding section, can be modified to take these points into account.

If we could look at two prospective reaction partners, e.g. a radical pair, that are close to each other in liquid phase, we would find that they are surrounded by a multitude of solvent molecules. In order to separate, the two reactants literally have to push these molecules out of their way, which requires a certain amount of activation energy. The same holds for another molecule trying to approach them. Hence a collision between the two partners of that pair is much more probable than between one of them and an outsider molecule. We denote a region of space for which this condition is valid as a 'cage'. If a radical pair reacts within the cage, we call this a 'geminate recombination', although this may not be a true recombination reaction, but also for instance a disproportionation. On the other hand, if the radicals leave the cage and react with other molecules in the bulk of the solution, we speak of 'free' radicals undergoing 'escape reactions'.

Let us now follow the typical fate of a radical pair in solution. We generate such a pair in a cage, either by cleavage of a precursor molecule, as described before, or by a chance meeting of two free radicals. In the first case, the pair will be born with a definite spin multiplicity, namely that of its precursor. Free radicals have no phase correlation of their spins, so it is equally likely that a pair formed from them will be in its singlet, or any of its triplet states. The radicals are kept together by the cage for at least 10 ps even in a solvent of very low viscosity. During this time, multiple collisions between them happen, and geminate recombination will occur with some nonzero probability, if this is allowed by the spin state of the pair. Evidence for this is, for instance, the reduced yield of photodecomposition in condensed phase as compared to the gas phase.

A more direct proof for the existence of a cage is obtained from laser flash photolysis experiments with picosecond time resolution [6]. After laser induced cleavage of I_2, recombination of some part of the iodine atoms was found to occur within 70 ps in the fairly viscous solvent hexadecane. This is much faster than the diffusion controlled reaction of free iodine radicals under the experimental conditions (which may also be observed on a longer timescale), and can only be explained by recombination within the cage.

In the cage, intersystem crossing by interactions with an external magnetic field or with the internal fields from the nuclei in the molecules is not possible, because the exchange integral is prohibitively high. Hence, if the initial spin state of the pair forbids its recombination, the radicals will eventually have to leave this so-called 'primary' cage. However, the story does not end here, because there is a definite probability that this particular pair will meet again at some later time, typically 100 ps in a non-viscous solvent, to form a 'secondary' cage. The radicals again have a chance of geminate recombination in this new cage, but the probability of this event will in general be different in comparison to the preceding cage, because in the meantime state mixing may have taken place in the region where J is negligible. Those radicals that do not react during this second encounter will separate again, and this sequence may repeat itself many times. Finally, this series of multiple re-encounters terminates either by geminate recombination, or by formation of escape products. The total lifetime of a radical pair is thus of the order of several ns.

A formal description of this scenario in quantum mechanical terms is straightforward in principle. We have to add a diffusion operator $D\Gamma$ and an operator K for the chemical reaction, the explicit expressions of which we skip here, to our density matrix master equation, Eq. (9.50), arriving at the so-called stochastic Liouville equation,

$$\frac{\partial}{\partial t}\underline{\rho} = -\frac{i}{\hbar}[\mathsf{H},\underline{\rho}] + D\Gamma\underline{\rho} + \mathsf{K}\underline{\rho}, \tag{9.72}$$

but, as you know, writing down the equations is usually not the big problem in quantum mechanics; we also have to solve them, which can only be done numerically in this case. Since this is not very satisfying intellectually, we will try to find an approximate treatment that gives us some sort of physical picture rather.

As we have seen in the last section, an exact calculation of the spin dynamics of a radical pair is possible for a constant exchange integral J, that means for constant interradical distance r. We can also separately solve the diffusion equations describing the radical pair dynamics. However, we cannot solve the coupled equations. The source of all our problems is the dependence of J on r, which makes the exchange interaction a random function of time.

We will circumvent this difficulty by assuming that during a diffusive separation and subsequent re-encounter of two radicals an abrupt transition takes place between a narrow region of very high J, which we call the exchange region, and a region extending to infinity, where J is zero. In other words, we approximate J by a step function of the distance between the radicals. As we have seen in the preceding section, this is not an unreasonable assumption, because the exchange integral falls off rapidly with r.

During a diffusive excursion, the two radicals spend most of their time outside the small exchange region. From our vector model of Fig. 9.9 we see that in this case $\mathscr{A}$ is parallel to $\underline{Q}$. Consequently, the evolution of the density matrix can be described by a rotation of $\mathscr{R}$ around the axis of $\underline{Q}$, which mixes only the variables $\rho_{SS} - \rho_{T_0T_0}$ and $i(\rho_{T_0S} - \rho_{ST_0})$. Indeed, for J equal to zero Eq. (9.68) turns into

$$\mathscr{R} = \begin{pmatrix} \cos(2Qt/\hbar) & -\sin(2Qt/\hbar) & 0 \\ \sin(2Qt/\hbar) & \cos(2Qt/\hbar) & 0 \\ 0 & 0 & 1 \end{pmatrix} \cdot \mathscr{R}_{t=0}, \tag{9.73}$$

so it is evident that electron spin polarization can neither be generated nor converted into another observable in this region, because $\rho_{ST_0} + \rho_{T_0S}$ is uncoupled from the other variables. Intersystem crossing occurs in the manner already predicted by the simple vector models of Figs. 9.4 and 9.5, and the quantitative results given there are seen to be exact.

However, we still have to take into account that the time spent outside the exchange region is not constant for all radical pairs. Instead, we have a statistical distribution of such times. Two radicals that were initially in contact, i.e. separated by a distance d at which geminate recombination may take place, have a chance of meeting at d again at some later period of time from t to $t + dt$. Let us denote the probability of this event by $f(t,d)dt$. With this, we can calculate the average density matrix $\bar{\mathcal{R}}$ for an ensemble of radical pairs, in which every pair has either completed a single diffusive excursion, or separated for good. If we neglect the dwell time of the radicals in the exchange region against the time during which they keep outside this region, we find that $\bar{\mathcal{R}}$ is given by

$$\bar{\mathcal{R}} = p \begin{pmatrix} c & -s & 0 \\ s & c & 0 \\ 0 & 0 & 1 \end{pmatrix} \cdot \mathcal{R}_{t=0}, \tag{9.74}$$

where c and s are related to the cosine and sine transforms of $f(t,d)$ at the angular frequency of intersystem crossing $2Q/\hbar$,

$$c = \frac{1}{p}\int_0^\infty \cos(2Qt/\hbar) f(t,d)dt \tag{9.75}$$

$$s = \frac{1}{p}\int_0^\infty \sin(2Qt/\hbar) f(t,d)dt, \tag{9.76}$$

and p is the total probability of at least one re-encounter,

$$p = \int_0^\infty f(t,d)dt. \tag{9.77}$$

For the sake of concreteness, we will give an explicit expression for the probability density $f(t,d)$. In general, this is not a continuous function, because diffusion occurs by discrete steps of the molecules. However, in the limit of small step sizes in comparison to the molecular diameters, we can replace this by a smooth function. If there are no significant attractive or repulsive interactions between the radicals, we have for this case [7, 8]

$$f(t,d) = \frac{p(1-p)d}{\sqrt{4\pi D}} t^{-3/2} \exp\left[-\frac{(1-p)^2 d^2}{4Dt}\right], \tag{9.78}$$

where D is the diffusion coefficient. The quantities c and s are then given by

$$c = \cos[(1-p)\delta]\exp[-(1-p)\delta] \tag{9.79}$$

$$s = \operatorname{sgn}(Q)\sin[(1-p)\delta]\exp[-(1-p)\delta], \tag{9.80}$$

with

$$\delta = \sqrt{\frac{2|Q|d^2}{D}}. \tag{9.81}$$

Intuitively, we expect the size of the diffusion steps to be comparable to the dimensions of the solvent molecules. Much larger displacements are improbable, because this would require an activation energy that is high enough to push several solvent molecules out of the way in a single step, whereas very small steps appear to be inefficient in leaving the cage. The molecular diameters of the reactants are typically larger than those of the solvent molecules, up to one order of magnitude. Theoretical calculations show that under these circumstances the re-encounter probabilities fall in the range between 0.85 and 0.95. For small mixing matrix elements Q and nonviscous solvents, i.e. high diffusion coefficients, it is therefore sufficient to expand the functions in Eqs. (9.79) and (9.80) to first order.

Inside the exchange region, J is much larger than Q, so in our vector model of Fig. 9.9 $\mathscr{A}$ is practically parallel to J. We see that in this instance $\mathscr{R}$ is rotated around the axis of J, hence the population difference between $|S\rangle$ and $|T_0\rangle$ cannot change. The fact that no intersystem crossing occurs in this region has already been discussed in Sect. 9.3. Instead, electron spin polarization and the phase correlation $i(\rho_{T_0S} - \rho_{ST_0})$ are mixed. For $J \gg Q$ Eq. (9.68) becomes

$$\mathscr{R} = \begin{pmatrix} 1 & 0 & 0 \\ 0 & \cos(2Jt/\hbar) & -\sin(2Jt/\hbar) \\ 0 & \sin(2Jt/\hbar) & \cos(2Jt/\hbar) \end{pmatrix} \cdot \mathscr{R}_{t=0}. \tag{9.82}$$

If the exchange interaction is weak, or if the radicals only stay in the exchange region for a short time τ, so that $2J\tau/\hbar \ll 1$, the matrix of coefficients in Eq. (9.82) is approximated by a unit matrix, so $\mathscr{R}$ remains essentially constant during τ. On the other hand, if $2J\tau/\hbar$ is larger than 1, more than one rotation of $\mathscr{R}$ takes place while the radicals reside in the exchange region. As the times τ are statistically distributed, and the strong distance dependence of J cannot be neglected in this region, the situation is similar to the precession of an ensemble of spins in a very inhomogeneous magnetic field. Hence, the components $i(\rho_{T_0S} - \rho_{ST_0})$ and $\rho_{ST_0} + \rho_{T_0S}$ of the average vector $\bar{\mathscr{R}}$ will be rotated through an angle that is determined by the mean exchange integral $\bar{J}$ and the average time $\bar{\tau}$ spent by the radicals inside the exchange region, but their magnitude will decay with the turning angle. As we do not know the form of the decay function, we will assume for simplicity that it can be approximated by an exponential. With this, we have an averaged density matrix for the exchange region

$$\bar{\mathscr{R}} = \begin{pmatrix} 1 & 0 & 0 \\ 0 & C & -S \\ 0 & S & C \end{pmatrix} \cdot \mathscr{R}_{t=0}, \tag{9.83}$$

with

$$C = \cos(2\bar{J}\bar{\tau}/\hbar)\exp(-\bar{\tau}/\Theta) \tag{9.84}$$

$$S = \sin(2\bar{J}\bar{\tau}/\hbar)\exp(-\bar{\tau}/\Theta), \tag{9.85}$$

where the time constant Θ is related to the spread of the precession frequencies and of τ. In the limit of very strong exchange interaction, i.e. for $2J\tau/\hbar \gg 1$, Θ becomes extremely short. We thus see that in this case all polarization and phase correlation of the electron spins is lost at each encounter of the pair.

Now we will consider chemical reactions of our radical pairs. Let us denote the probability of geminate recombination of a singlet pair in a cage by λ. Usually, λ will be approximately equal to 1. Let us further assume that for a triplet pair the probability of this reaction is zero. Evidently, the recombination influences only the population of the singlet state, that is the term ρ_{SS} of the density matrix. It will therefore be more convenient to separate the components ρ_{SS} and $\rho_{T_0T_0}$ again in the following. If at time t our system is described by a density matrix $\underline{\rho}(t)$, written as a vector,

$$\underline{\rho}(t) = \begin{pmatrix} \rho_{\mathrm{SS}} \\ \rho_{\mathrm{T_0T_0}} \\ i(\rho_{\mathrm{T_0S}} - \rho_{\mathrm{ST_0}}) \\ \rho_{\mathrm{ST_0}} + \rho_{\mathrm{T_0S}} \end{pmatrix}(t), \tag{9.86}$$

the amount F of product formed is given by

$$\begin{aligned} F &= (1 \quad 0 \quad 0 \quad 0)\,\underline{\lambda}\,\underline{\rho}(t) \\ &= (1 \quad 0 \quad 0 \quad 0)\begin{pmatrix} \lambda & 0 & 0 & 0 \\ 0 & 0 & 0 & 0 \\ 0 & 0 & 0 & 0 \\ 0 & 0 & 0 & 0 \end{pmatrix}\begin{pmatrix} \rho_{\mathrm{SS}} \\ \rho_{\mathrm{T_0T_0}} \\ i(\rho_{\mathrm{T_0S}} - \rho_{\mathrm{ST_0}}) \\ \rho_{\mathrm{ST_0}} + \rho_{\mathrm{T_0S}} \end{pmatrix}(t), \end{aligned} \tag{9.87}$$

which is just a formal matrix notation for the number $F, F = \lambda\rho_{SS}$. By the chemical reaction $\underline{\rho}(t)$ is changed to $\underline{\rho}(t')$,

$$\begin{aligned} \underline{\rho}(t') &= (\underline{\mathsf{E}} - \underline{\lambda})\underline{\rho}(t) \\ &= \begin{pmatrix} 1-\lambda & 0 & 0 & 0 \\ 0 & 1 & 0 & 0 \\ 0 & 0 & 1 & 0 \\ 0 & 0 & 0 & 1 \end{pmatrix}\begin{pmatrix} \rho_{\mathrm{SS}} \\ \rho_{\mathrm{T_0T_0}} \\ i(\rho_{\mathrm{T_0S}} - \rho_{\mathrm{ST_0}}) \\ \rho_{\mathrm{ST_0}} + \rho_{\mathrm{T_0S}} \end{pmatrix}(t), \end{aligned} \tag{9.88}$$

where $\underline{\mathsf{E}}$ is the unit matrix. (Of course, the probability that no geminate recombination takes place is $1 - \lambda$.)

A series of encounters can be treated in the following way [4]. We begin with a density matrix $\underline{\rho}_0$. Geminate recombination within the primary cage yields an amount of product F_0 that may be calculated from $\underline{\rho}_0$ with Eq. (9.87). According to Eq. (9.88), the density matrix at the beginning of the subsequent diffusive excursion is given by $(\underline{\mathsf{E}} - \underline{\lambda})\underline{\rho}_0$. During the separation, state mixing occurs, which is described by the matrix $\underline{\mathsf{M}}$, that is the matrix of coefficients in Eq. (9.74), possibly in combination with that of Eq. (9.83). Hence, at the time of the first re-encounter, we have a density matrix $\underline{\rho}_1$,

$$\underline{\rho}_1 = \underline{\mathsf{M}}(\underline{\mathsf{E}} - \underline{\lambda})\underline{\rho}_0, \tag{9.89}$$

and obtain a quantity F_1 of geminate product by multiplication of $\underline{\rho}_1$ from the left with $(1\,0\,0\,0)\underline{\lambda}$. Then, the next diffusive excursion occurs. At its end, we find that

$$\begin{aligned}\underline{\rho}_2 &= \underline{\mathsf{M}}(\underline{\mathsf{E}} - \underline{\lambda})\underline{\rho}_1 \\ &= [\underline{\mathsf{M}}(\underline{\mathsf{E}} - \underline{\lambda})]^2\underline{\rho}_0,\end{aligned} \tag{9.90}$$

and F_2 is computed as before. This sequence repeats itself over and over again.

In order to get the total amount of geminate product F_{tot}, we have to sum up the contributions from all re-encounters,

$$\begin{aligned}F_{\text{tot}} &= \sum_{k=0}^{\infty} F_k \\ &= (1\,0\,0\,0)\;\underline{\lambda}\left\{\sum_{k=0}^{\infty}[\underline{\mathsf{M}}(\underline{\mathsf{E}} - \underline{\lambda})]^k\right\}\underline{\rho}_0.\end{aligned} \tag{9.91}$$

The infinite sum of matrices in this equation can be calculated as follows. By $\underline{\mathsf{S}}_m$ we denote the mth partial sum, i.e.

$$\underline{\mathsf{S}}_m = \underline{\mathsf{E}} + \underline{\mathsf{A}} + \underline{\mathsf{A}}^2 + \cdots + \underline{\mathsf{A}}^{m-1} + \underline{\mathsf{A}}^m, \tag{9.92}$$

where in our case

$$\underline{\mathsf{A}} = \underline{\mathsf{M}}(\underline{\mathsf{E}} - \underline{\lambda}). \tag{9.93}$$

Matrix multiplication of Eq. (9.92) from the right with $\underline{\mathsf{A}}$ yields

$$\begin{aligned}\underline{\mathsf{S}}_m\underline{\mathsf{A}} &= \underline{\mathsf{A}} + \underline{\mathsf{A}}^2 + \underline{\mathsf{A}}^3 + \cdots + \underline{\mathsf{A}}^m + \underline{\mathsf{A}}^{m+1} \\ &= \underline{\mathsf{S}}_m - \underline{\mathsf{E}} + \underline{\mathsf{A}}^{m+1},\end{aligned} \tag{9.94}$$

hence

$$\underline{\mathsf{S}}_m(\underline{\mathsf{E}} - \underline{\mathsf{A}}) = \underline{\mathsf{E}} - \underline{\mathsf{A}}^{m+1}. \tag{9.95}$$

It is seen that convergence of the sum $\underline{\mathsf{S}}_m$ implies that $\underline{\mathsf{A}}^m$ has to become vanishingly small, as m tends to infinity. This condition is met in our case, because the absolute values of the elements of $\underline{\mathsf{M}}$ in Eq. (9.74) are less than one. In this case, Eq. (9.95) changes into

$$\underline{\mathsf{S}}_m(\underline{\mathsf{E}} - \underline{\mathsf{A}}) = \underline{\mathsf{E}}, \tag{9.96}$$

where $\underline{\mathsf{S}}$ is the value of the sum. Consequently, we have

$$\sum_{k=0}^{\infty}[\underline{\mathsf{M}}(\underline{\mathsf{E}} - \underline{\lambda})]^k = [\underline{\mathsf{E}} - \underline{\mathsf{M}}(\underline{\mathsf{E}} - \underline{\lambda})]^{-1}, \tag{9.97}$$

where the exponent -1 denotes matrix inversion.

In a homework problem you are asked to carry out this inversion taking the matrix of Eq. (9.74) as $\underline{\mathbf{M}}$ and assuming a triplet precursor, no initial electron spin polarization and phase correlation, as well as a singlet reactivity λ of unity. The result for this fairly typical case is [4]

$$F_{\text{tot}} = \frac{p[1 - c + p(c^2 + s^2 - c)]}{2 - p(1 + 3c) + p^2(c^2 + s^2 + c)} \,. \tag{9.98}$$

The quantities p, c and s have been defined in Eqs. (9.77), (9.79) and (9.80). In non-viscous solvents this expression can be simplified by expanding c and s and keeping only first order terms. Under these circumstances, we get

$$F_{\text{tot}} = \frac{p\delta}{2} \,, \tag{9.99}$$

where δ is given by Eq. (9.81). It is seen that F_{tot} is proportional to the square root of the intersystem crossing frequency in this instance.

Let us now cast a quick look back and take stock. We find that we are at last in a position to calculate the change of the density matrix of an ensemble of radical pairs in all circumstances relevant for us, and thus describe such a system as fully as nature permits. Our treatment has necessarily been approximate in several places. This, however, has enabled us to use concrete physical concepts rather than abstract mathematics. In the following chapters, we will go on to discuss some of the experimental effects that arise from magnetic field induced intersystem crossing in radical pairs.

9.6 Chemically Induced Nuclear Spin Polarizations

We have already learned that radical pairs in a magnetic field may develop electron spin polarization, which we can detect in EPR spectra of the intermediates. Experiments have shown that polarizations of the nuclear spins may also be generated under these circumstances. As it is usually not possible to observe radicals directly by NMR spectroscopy, these polarizations have to be measured in the diamagnetic reaction products. Both these effects only occur during chemical reactions, so they have been rightly termed 'chemically induced'. Moreover, they were named 'dynamic polarizations', because in the early investigations of these types of spin polarizations people tried to explain them by a sort of Overhauser effect (electron-nuclear cross relaxation). Later work showed this to be wrong, and led to the description of these phenomena by the radical pair mechanism. Nevertheless, since that time the tag 'dynamic' seems to have stuck irreversibly, and the two effects [9] are now universally called 'chemically induced dynamic electron polarization' (CIDEP) and 'chemically induced dynamic nuclear polarization' (CIDNP).

We will tackle CIDNP first, since we have already dealt with electron spin polarizations in several places. Besides, it is of greater practical importance than

CIDEP, simply because NMR spectroscopy is a more frequently used technique than EPR spectroscopy. In order not to let the discussion become too abstract, we will start with a real experimental example.

In Fig. 9.10 we have displayed the NMR spectrum of a chemical system recorded during illumination, i.e. during a photoreaction. By a special technique [10] the signals of unreacted molecules, that is to say the signals which we would see in a normal NMR spectrum in the dark, have been suppressed. In this example, the compound D of formula shown in the figure has been illuminated in a polar solvent in the presence of a second substance S, a so-called sensitizer, which serves to absorb the light, and initiate the chemical reaction. Apart from this function, S is of little importance to us here; it can be replaced by many other compounds without a change in the displayed part of the NMR spectrum

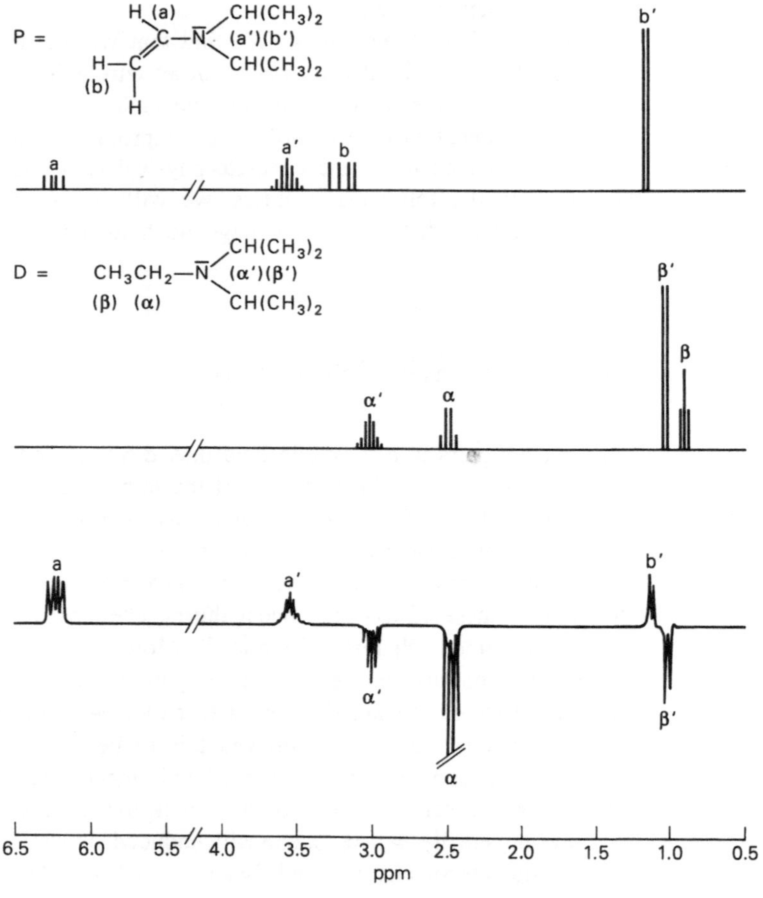

Fig. 9.10. Example of CIDNP in the sensitized photoreaction of the compound D, which yields one new reaction product P. The bottom trace shows the experimental NMR spectrum during the reaction. Stick spectra of D and P are given in the upper traces

of this reaction system. Only one new reaction product P is formed from D. The structure of this product is also shown in the figure. Schematic NMR spectra of D and P are given there as well, for your orientation.

When we look at the experimental spectrum, we immediately notice that the signal intensities are anomalous. For instance, the quartet of the protons of the starting compound C that are labelled α appears in emission, i.e. upside down. (The phase of all signals in this measurement has been adjusted so that a normal NMR spectrum would show absorption lines.) Also, the signals of these two protons are at least two orders of magnitude stronger than those of the three protons marked with β, which are undetectable in the lower spectrum. In contrast, these two signal groups have about the same signal intensity in a normal NMR spectrum, as seen in the center trace. The frequencies of all peaks are normal, as a comparison with the stick spectra shows. We have already learned that this is typical for spin polarization effects. The energy levels of the system are unchanged, but their populations deviate from those in thermal equilibrium.

Let us briefly compile some facts about the chemical reaction of this example, without going into details or presenting our evidence for them. We know for sure that there is a radical pair involved. Interestingly, in this instance this is not generated by cleavage of a molecule, but by electron transfer from D to the electronically excited sensitizer. It is also certain that the excited state of the sensitizer participating in this reaction is a triplet state. Since the spin multiplicity cannot change during electron transfer, the correlated radical ion pairs, which we have as intermediates, are born as triplet pairs. Spin polarizations are found for the starting compound D regained in the reaction. This is of necessity a cage product. With the molecules of our example geminate recombination, which in this case means back electron transfer, can only occur from the singlet state. The new product P has to be an escape product.

In both D and P, the protons that are polarized are those at the carbon atoms directly attached to the nitrogen, and the terminal protons in the branched side chains. We have labelled these protons with α, α' and β' in D, and with a, a' and b' in P. What strikes us is the fact that the groups of signals α and a, α' and a', as well as β' and b' have opposite phase in D and P. This is generally observed in all CIDNP experiments in high magnetic fields: corresponding nuclei in cage and escape products bear opposite polarizations.

In Fig. 9.11 a mechanism of our example reaction is given that is able to explain all the experimental results. This scheme is typical for every CIDNP experiment, apart from the fact that in a general case the radical pair may also

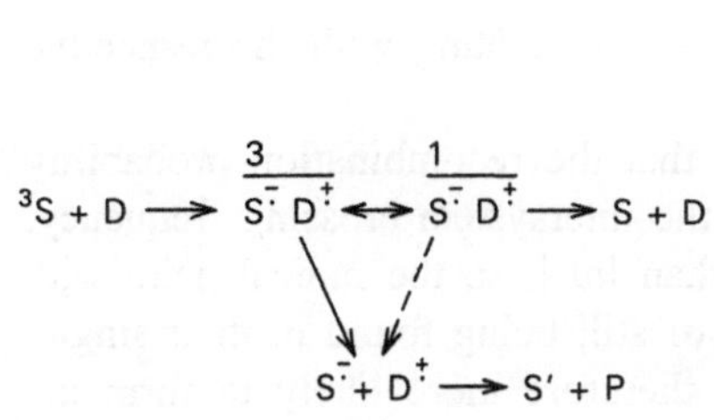

Fig. 9.11. Reaction mechanism for the photoreaction of Fig. 9.10. Correlated pairs are indicated by a *bar* over the radicals. The *double arrow* denotes intersystem crossing. S′ is the escape product formed from the sentitizer S

be generated from a singlet precursor, and by breaking a chemical bond rather than by electron transfer, which would lead to minor modifications of this figure.

We start with a radical pair in a triplet state, but obtain the geminate recombination product, which is only possible from a singlet pair. It is therefore obvious that intersystem crossing takes place in the pair. In the figure, we have symbolized this process by the double arrow. It occurs in the way described in the previous sections, by interactions of the electron spins with the external magnetic field as well as – and here the proton spins come into play – with the field due to the nuclei in the radicals. From Eqs. 9.49, 9.81 and 9.99 we see that the amount of intersystem crossing is dependent on the configurations of the nuclear spins in the radicals. As a consequence, the populations of those spin states that lead to a higher intersystem crossing probability are enhanced in the geminate recombination product, whereas they are depleted in the free radicals and their subsequent escape products. In effect, the nuclear spins are thus selectively distributed between the two kinds of reaction products.

Two experimental observations are immediately explained by this mechanism. First, corresponding protons of cage and escape products must be oppositely polarized owing to the nuclear spin sorting. Second, if we change the multiplicity of the precursor to singlet, all polarizations are inverted, because in that case more efficient intersystem crossing decreases the probability of forming geminate recombination product.

We will now discuss the dependence of the polarization phases on the magnetic parameters of the radical pair. For this purpose, we shall use the vector models of Figs. 9.4 and 9.5, because we have seen that these provide an adequate description of singlet-triplet mixing of such a pair in high magnetic fields. As a first example, we shall take the simplest system, a radical pair with one proton. Let us assume a singlet precursor for a change, a positive difference of the g-factors of the radicals, i.e. $g_1 > g_2$, and a positive hyperfine coupling constant a of the proton, which shall be contained in radical 1. The Larmor frequencies are $(g_1\beta H_0 \pm a/2)/\hbar$ for the first radical, depending on the spin state of the proton, and $g_2\beta H_0/\hbar$ for the second radical. Intersystem crossing occurs with angular frequency ω_α for the proton in state $|\alpha\rangle$ and ω_β for the proton in state $|\beta\rangle$,

$$\omega_\alpha = \frac{1}{\hbar}\left[(g_1 - g_2)\beta H_0 + \frac{a}{2}\right] \tag{9.100}$$

$$\omega_\beta = \frac{1}{\hbar}\left[(g_1 - g_2)\beta H_0 - \frac{a}{2}\right]. \tag{9.101}$$

This is depicted in Fig. 9.12 in a coordinate system rotating with the respective mean Larmor frequency.

We have learned in the previous section that the recombination probability of a pair is determined by the square root of the intersystem crossing frequency. For our choice of parameters, $|\omega_\alpha|$ is larger than $|\omega_\beta|$, so the radical pairs with the proton in state $|\beta\rangle$ have the better chance of still being found in their singlet state after their diffusive excursions, and are therefore more likely to form the

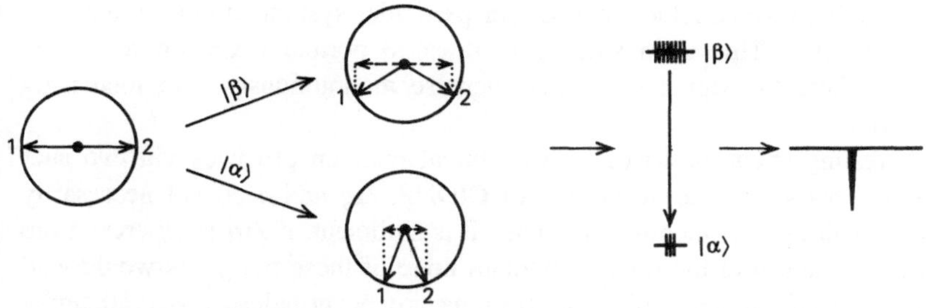

Fig. 9.12. *Left*: visualization of nuclear spin dependent mixing of $|S\rangle$ and $|T_0\rangle$ in a radical pair containing one proton in the first radical, where Δg and a are both positive. A singlet precursor has been assumed. The *broken arrows* give the projection on the singlet state, their length representing the amount of singlet character of the pair
Center: as a consequence of this nuclear spin dependent intersystem crossing and the electron spin dependent reactivity, we get an overpopulation of the nuclear spin level $|\beta\rangle$ in the cage recombination product
Right: the resulting NMR spectrum therefore shows an emission line

geminate recombination product. Consequently, the $|\beta\rangle$ level of the nuclear spins will be more populated than the $|\alpha\rangle$ level in this cage product, resulting in an emissive NMR peak.

By the way, such a polarized NMR signal is often two to three orders of magnitude stronger than the corresponding one in a normal spectrum. The reason is that for nuclear spins the population differences in thermodynamic equilibrium are minute owing to the small Boltzmann factor $\hbar\omega/k_0T$. Hence, instead of simply 'absorption' we frequently speak of 'enhanced absorption' in discussing CIDNP experiments.

In Fig. 9.12 the vector of the first electron spin is rotated counterclockwise and that of the second electron clockwise. It would make no observable difference, if we inverted the sense of rotation of both vectors simultaneously; the relevant quantity is the absolute value of the intersystem crossing frequency. It follows that by a CIDNP measurement we cannot distinguish between our given radical pair and another one in which both Δg and a are negative. However, if we change the sign of only one of the parameters Δg or a, the sign of the polarization is inverted, because in that instance $|\omega_\beta|$ is identical to $|\omega_\alpha|$ of the former case, and vice versa.

Obviously, no CIDNP can be produced in our system, if the nuclear spin does not interact with any of the electron spins, that is if a is zero. What is perhaps not immediately obvious is the fact that for $\Delta g = 0$ no nuclear spin polarization is generated either. In this case the intersystem crossing frequencies $|\omega_\alpha|$ and $|\omega_\beta|$ are identical, so radical pairs with the proton in state $|\alpha\rangle$ and pairs containing a $|\beta\rangle$ proton react to the geminate recombination product with exactly equal probability. As the populations of the nuclear spin states $|\alpha\rangle$ and $|\beta\rangle$ in the products are proportional to $\sqrt{\omega_\alpha}$ and $\sqrt{\omega_\beta}$, respectively, we see that the magnitude of the polarizations also tends towards zero, if the Δg-term, i.e. the difference in the electron Zeeman energies, becomes very large. From Eqs.

(9.100) and (9.101) we conclude that for our particular system an optimum exists, if $|\Delta g|\beta H_0 = |a|/2$. This is in striking contrast to normal magnetic resonance experiments, where the signal intensities increase monotonously with increasing strength of H_0.

An interesting situation arises, if a chemical reaction proceeds via two successive radical pairs. For the generation of CIDNP, Δg and a do not necessarily have to be nonzero simultaneously in a pair; it is sufficient, if Δg is different from zero in one pair, and a in the other. Although none of these two pairs would lead to CIDNP by itself, nuclear spin polarizations are nevertheless observed under these circumstances. The resulting signals are equal to those of a hypothetical pair with the combined magnetic parameters of the successive pairs. You can convince yourself of this in a homework problem by drawing the appropriate vector diagrams.

All the discussed qualitative predictions for CIDNP in high magnetic fields can be summed up in a simple rule [12] that gives the phase Γ_n of the polarizations ($\Gamma_n = +1$: enhanced absorption, $\Gamma_n = -1$: emission) of the nuclei in question as a function of the signs of a and Δg, of the precursor multiplicity μ, and the type of reaction product ε,

$$\Gamma_n = \text{sgn}(\Delta g) \cdot \text{sgn}(a) \cdot \mu \cdot \varepsilon \, , \tag{9.102}$$

where we take $\mu = +1$ for a triplet and $\mu = -1$ for a singlet precursor, and $\varepsilon = +1$ or $\varepsilon = -1$ for cage or escape products, respectively. We have already noted that a radical pair may be formed by an encounter of two free radicals as well. In this case we speak of a random phase precursor, which has equal probability of being in one of its four possible electron spin states. As there are three triplet states but only one singlet state, such a radical pair behaves qualitatively like a triplet born pair, so μ is also $+1$ in this instance.

The polarizations of our experimental example shown in Fig. 9.10 can be explained with this rule. In this chemical system, the g-factor of $S^{\overset{-}{\cdot}}$ is known to be larger than that of $D^{\overset{+}{\cdot}}$. When we discuss the polarizations in D and P, we therefore have to take $\Delta g < 0$. The hyperfine coupling constants of all protons in $D^{\overset{+}{\cdot}}$ are positive. With the multiplicity of the precursor being triplet, emission is predicted for the cage product, and enhanced absorption for the escape product, in accordance with the observed spectrum.

Three facts make CIDNP measurements a very interesting technique for the chemist. First and foremost, no other kind of spectroscopy except for the very similar CIDEP experiments, which we will discuss in the next section, allows one to determine the electron spin multiplicity of the precursor of a chemical reaction in such a direct way.

Second, although one only observes the reaction products – and of course one can characterize these quite well, since the detection method is high resolution NMR – one also obtains specific information about the intermediates, because the polarization intensities reflect the hyperfine coupling constants and thus the spin densities in the radicals [11]. For instance, in our example reaction of

Fig. 9.11 the initially formed radical ion of D may lose a proton from a carbon atom next to the nitrogen. If this process takes place very rapidly, i.e. faster than the cage lifetime, this proton is transferred to $S^{\overline{\cdot}}$, so that a pair of neutral radicals results, which can also give rise to CIDNP, and which eventually leads to the same product P as the radical ion pair. However, the spin density in the deprotonated radical of D is totally different from that in $D^{\dot{+}}$, so the polarization patterns in P are also totally different for these two mechanistic alternatives, as shown in Fig. 9.13. Hence, CIDNP experiments can be used to identify the intermediates in which the polarizations are created.

Third, the polarizations are generated during the lifetime of the correlated radical pair (~10 ns), but in the diamagnetic products they persist for the spin-lattice relaxation time, which is of the order of seconds for protons. CIDNP is therefore suited for the investigation of fast radical reactions, in spite of the inherently slow timescale of NMR experiments. With pulsed chemical excitation by a laser, and detection with a pulse and Fourier transform NMR spectrometer, polarizations can be sampled with submicrosecond time resolution, and kinetics of reactions in the millisecond to microsecond range can be studied.

If the reaction products contain more than one magnetic nucleus, multiplet patterns may be observed in their NMR spectra. In measurements in liquid solution, the dominant interaction between two nuclei is given by the indirect spin-spin coupling, with a Hamiltonian H_J that is very similar to that of the hyperfine interaction,

$$\mathsf{H}_J = hJ\mathsf{I}_1 \cdot \mathsf{I}_2. \qquad (9.103)$$

Convention has it that the same letter is used for the exchange integral and for the coupling constant in this relation. We have to keep aware of the danger of confusing these two quantities.

We will briefly treat the energy levels and NMR spectrum of a spin system of two protons, because we shall need these in the following discussions. As

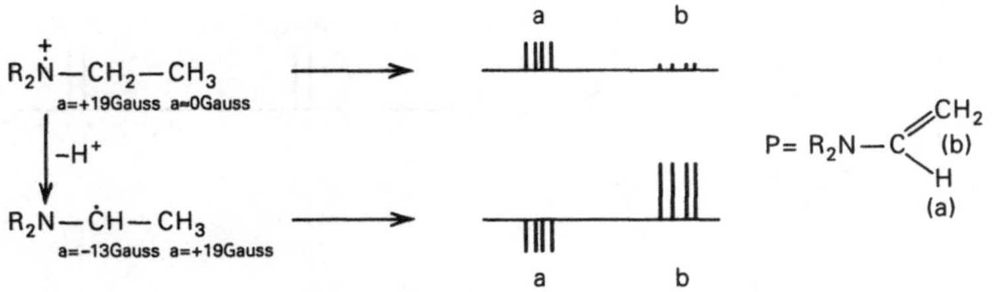

Fig. 9.13 *Left*: two possible intermediates in the sensitized photoreaction of D, a radical ion (*top*) and a neutral radical (*bottom*). The hyperfine coupling constants of the protons in both radicals are very different
Right: resulting polarizations in the NMR spectrum of the escape product P, assuming a triplet precursor and a positive difference of the *g*-factors of the radical pairs in both cases

usual, we take the case of a strong magnetic field, so that the nonsecular terms in Eq. (9.103) can be neglected. Moreover, we assume H_0 to be high enough that the difference of the Zeeman energies of the two nuclei is much larger than their coupling energy hJ. In order to symbolize this latter point, one customarily denotes the two protons by letters from far away positions in the alphabet. Thus, we have an AX-system in this instance. The energy levels of this system, as well as the allowed NMR transitions between them and the resulting spectrum are shown in Fig. 9.14. In this figure, we have chosen J to be positive and ω_A, the resonance angular frequency of nucleus A, to be larger than that of nucleus X.

It is seen that two transitions labelled A_1 and A_2 occur, in which only the spin state of nucleus A changes, and two others (X_1 and X_2) of nucleus X. With our parameters, the frequencies of the transitions with the index 1, where the spin state of the other proton is $|\beta\rangle$, are higher by J than the corresponding ones with the index 2. As NMR spectra are always displayed with frequency increasing from right to left, the peak A_1 appears to the left of A_2, and X_1 to the left of X_2. The two lines of both doublets are separated by J.

If $\omega_X > \omega_A$, the energies of the states $|\alpha\beta\rangle$ and $|\beta\alpha\rangle$ are interchanged. As a result, A- and X-transitions change places in the NMR spectrum. On the other hand, if $J < 0$, the energies of $|\alpha\alpha\rangle$ and $|\beta\beta\rangle$ are increased by $hJ/2$, and the energies of the other two states are decreased by the same amount. This corresponds to a permutation of the lines A_1 and A_2, as well as of X_1 and X_2 in the spectrum, so that now the spin state of the coupled nucleus is $|\alpha\rangle$ in the transition with the higher frequency.

In such coupled spin systems, two kinds of CIDNP effects can occur. The first is the one that we have already described, and which is present in our experimental example of Fig. 9.10. Its characteristic is that the heights of all the individual lines of a multiplet are changed by the same factor in the CIDNP spectrum as compared to the standard NMR spectrum. Thus the integrated in-

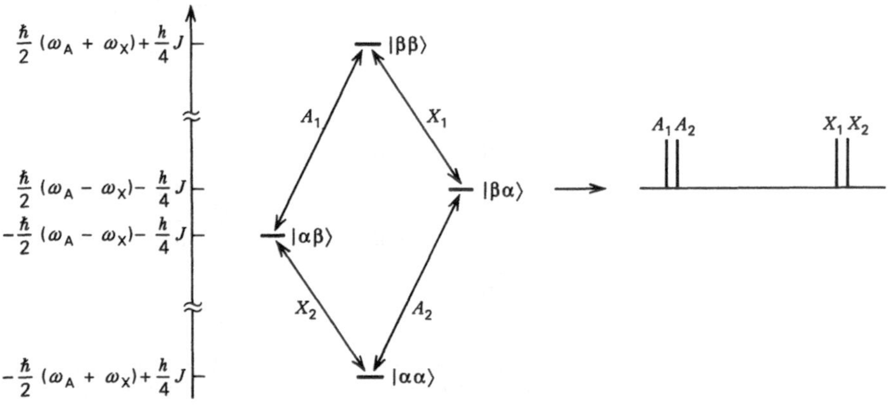

Fig. 9.14. Energy levels, eigenfunctions, allowed NMR transitions and resulting spectrum of an AX-system of nuclear spins for the case $\omega_A > \omega_X$ and $J > 0$

tensity of this multiplet has a nonzero value, which is different from the value in thermodynamic equilibrium. In this case, it is meaningful to speak of the polarization of a particular nucleus, because all its transitions are influenced to the same extent. This has been named a CIDNP 'net effect'.

In addition, there may also be a so-called 'multiplet effect', which means that both emission and enhanced absorption are found within a multiplet, so that its total integrated intensity is zero. As an example we will discuss a radical pair containing two protons in the first radical. We start with a triplet precursor and measure the geminate recombination product. We will assume that the difference of the g-factors is zero, and that the hyperfine coupling constants a_A and a_X as well as the nuclear spin-spin coupling constant J are positive. Also we take $a_A > a_X$.

Singlet-triplet mixing in the radical pairs is visualized by the vector models given in Fig. 9.15. The intersystem crossing frequencies for the four different nuclear spin configurations, which are indicated by the subscripts, are

$$\omega_{\alpha\alpha} = \frac{1}{2\hbar}(a_A + a_X) \tag{9.104}$$

$$\omega_{\alpha\beta} = \frac{1}{2\hbar}(a_A - a_X) \tag{9.105}$$

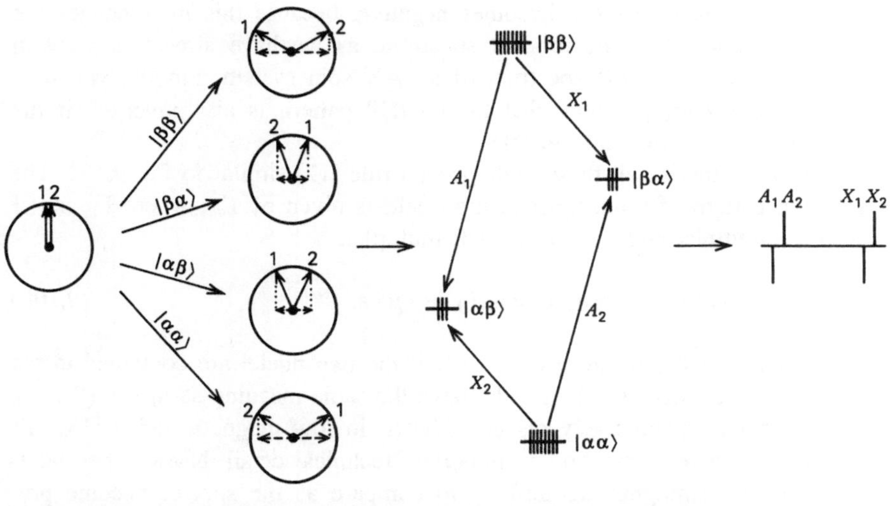

Fig. 9.15. *Left*: vector models for nuclear spin dependent intersystem crossing between $|S\rangle$ and $|T_0\rangle$ in a radical pair with $\Delta g = 0$ containing two protons in the first radical. A triplet precursor and positive hyperfine coupling constants have been assumed. The *broken arrows* give the projection of the electron spin vectors on the singlet state.
Center: as a result, the nuclear spin levels $|\alpha\alpha\rangle$ and $|\beta\beta\rangle$ are overpopulated in the cage recombination product.
Right: for positive coupling constant J, an E/A multiplet effect is observed in the NMR spectrum. The assignment of the transitions is the same as in the preceding figure

$$\omega_{\beta\alpha} = -\frac{1}{2\hbar}(a_A - a_X) \tag{9.106}$$

$$\omega_{\beta\beta} = -\frac{1}{2\hbar}(a_A + a_X). \tag{9.107}$$

It is seen that the probability of intersystem crossing is higher for the pairs in which the nuclear spin states of the protons are equal than for those where they are different. As only the absolute value of the intersystem crossing frequency is of importance, nuclear spin states $|\alpha\alpha\rangle$ and $|\beta\beta\rangle$ are overpopulated to the same degree in the geminate recombination product, whereas the populations of the states $|\alpha\beta\rangle$ and $|\beta\alpha\rangle$ are equally depleted. In consequence, both low field lines A_1 and X_1 in the NMR spectrum are polarized in emission, whereas A_2 and X_2 appear in enhanced absorption, as also shown in Fig. 9.15. Such a pattern is denoted as an E/A multiplet.

We see that the relative magnitude of the hyperfine coupling constants does not influence this result, as long as both are positive, because this can only interchange the vector diagrams for the nuclear spin states $|\alpha\beta\rangle$ and $|\beta\alpha\rangle$ in Fig. 9.15. On the other hand, inverting the sign of both a_A and a_X permutes the vector diagrams for the two other nuclear spin states in this figure, which is again of no consequence for the CIDNP signals. If, however, only one hyperfine coupling constant becomes negative, the vector diagrams for $|\alpha\alpha\rangle$ and $|\beta\alpha\rangle$ as well as those for $|\alpha\beta\rangle$ and $|\beta\beta\rangle$ are interchanged. We see at once that in this case all polarizations are inverted. We call the resulting pattern an A/E multiplet. The same thing happens, if J becomes negative, because this interchanges the lines of each doublet in the product spectrum, as we have already learned in our discussion of the NMR spectrum of an AX spin system. Finally, you may show in a homework problem, that the CIDNP pattern is also inverted, if the two nuclei reside in different radicals.

We can summarize all these findings in a rule [12] similar to Eq. 9.102. The polarization pattern of a multiplet in high field is given by Γ_m, where Γ_m is $+1$ for an E/A multiplet and -1 for an A/E multiplet,

$$\Gamma_m = \mathrm{sgn}(a_A) \cdot \mathrm{sgn}(a_X) \cdot \mathrm{sgn}(J) \cdot \sigma \cdot \mu \cdot \varepsilon. \tag{9.108}$$

In this equation, σ has to be taken as $+1$, if the two nuclei are contained in the same radical, otherwise as -1. μ and ε have the same meaning as in Eq. (9.102).

Now we will qualitatively discuss CIDNP in low magnetic fields [13]. To begin with, we have to mention an important technical detail. NMR experiments at different field strengths are difficult to compare as the spectra become progressively less resolved with decreasing H_0, and multiplet patterns totally change their appearance if the coupling constants become comparable to the frequency separations of the signal groups. In very low fields, measurements are hardly practicable; in zero field they are impossible. We avoid these problems by generating the polarizations in the field of an auxiliary magnet, and then transferring the sample into the strong magnet of a high resolution NMR spectrometer. Two constraints are imposed on the transfer time. First, in order to avoid loss of signal intensities it should be significantly shorter than the nuclear spin relaxation times. As these are of the order of several seconds for protons, this condition can be met, for instance by the use of flow systems. Second, the transfer has to proceed so slowly as to be adiabatic, which means that the populations of the nuclear eigenstates in low field are transferred without change to those high field eigenstates with which they correlate. For coupled spin systems this implies that the transfer times should not be shorter than the reciprocal of $2\pi J$, because otherwise transitions between eigenstates may be induced. As typical coupling constants J between proton are greater than one Hertz, this requirement poses no problems.

For a description of CIDNP net effects in low fields, we revert to our example of a radical containing one proton with a positive hyperfine coupling constant. We will now take a triplet precursor. Further, we will assume that intersystem crossing due to the Δg-mechanism can be neglected, i.e. we focus our attention on magnetic fields that are low enough for $\Delta g \beta H_0$ to be much smaller than a. Typically, this holds for fields below, say, 1000 Gauss. Obviously, the sign of Δg cannot have any influence on the polarization phase in this case.

As we saw in Sect. 9.3 (Eq. 9.43) simultaneous transitions of nuclear and electron spins have to be taken into account in low fields. These cause intersystem crossing between $|S\rangle$ and $T_{+1}\rangle$ as well as $|S\rangle$ and $|T_{-1}\rangle$. We therefore have to take a basis of eight spin functions for our radical pair, which we can group according to their total spin f as follows

$$
\begin{aligned}
f &= +\tfrac{3}{2}: \quad |T_{+1}\alpha\rangle \\
f &= +\tfrac{1}{2}: \quad |S\alpha\rangle \quad |T_0\alpha\rangle \ |T_{+1}\beta\rangle, \\
f &= -\tfrac{1}{2}: \quad |S\beta\rangle \quad |T_0\beta\rangle \ |T_{-1}\alpha\rangle \\
f &= -\tfrac{3}{2}: \quad |T_{-1}\beta\rangle
\end{aligned}
\tag{9.109}
$$

where $|S\rangle$ and the three functions $|T\rangle$ refer to the electron spins, and α and β to the nuclear spin. These energy levels are displayed in Fig. 9.16, where we have assumed that the distance between the two radicals is large enough to make the exchange integral negligible.

In Sect. 9.2 we saw that we could not assign individual spin states to an electron and a proton in zero magnetic field. In low fields, we might therefore expect the three product functions of electron and nuclear spins with $f = \frac{1}{2}$ and those with $f = -\frac{1}{2}$ to be inappropriate as well. However, here the situation is slightly different. As in all our previous discussions of intersystem crossing, we have to use the eigenfunctions of the radical pair that are valid, if the two radicals are close to each other. Hence, the electronic singlet and triplet states can be distinguished even in zero field owing to the exchange interaction, which is only experienced by the electrons. The remaining states with $|f| = \frac{1}{2}, T_0\alpha\rangle$ and $|T_{+1}\beta\rangle$, as well as $|T_0\beta\rangle$ and $|T_{-1}\alpha\rangle$, are no longer eigenstates only in extremely

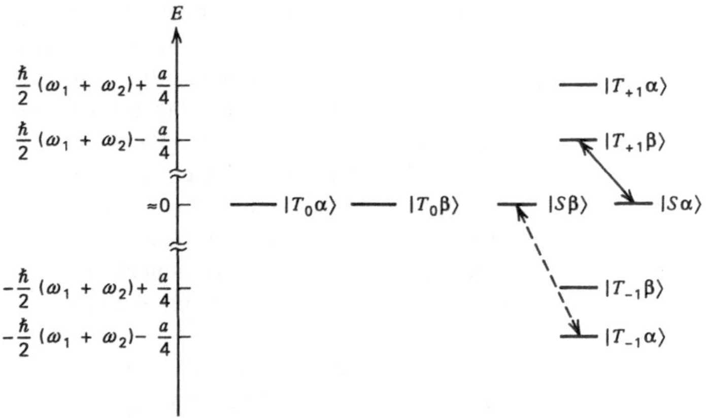

Fig. 9.16. Energy levels given by Eq. 9.109 of a radical pair containing one proton in a low magnetic field. The exchange integral has been assumed to be zero.

low fields, of several tens of Gauss, where the Zeeman splitting of the triplet levels becomes comparable to the hyperfine interaction. At the moment, we will focus our attention on magnetic fields that are higher than this.

In the absence of radio frequency fields, transitions between states of different f are not possible, so we see that the functions with $f = \pm\frac{3}{2}$ are isolated from the rest and need not be taken into account. Within the remaining two groups, the first two states are mixed by the secular part of the hyperfine interaction. This is, of course, the mechanism that is responsible for CIDNP in high magnetic fields. In our discussion of this we have found that it can only generate net nuclear spin polarizations, if Δg of the radical pair is nonzero, that is if there is a difference between the Zeeman frequencies of both radicals. However, in low magnetic fields this difference becomes extremely small, because it is proportional to H_0. Hence intersystem crossing between $|S\rangle$ and $|T_0\rangle$ can be neglected in our particular case.

The second and third function of each group with $|f| = \frac{1}{2}$ in Eq. 9.109 are mixed by the symmetric part of the nonsecular terms of H_H. As these state conversions do not involve a change of multiplicity, we will ignore them here. Thus only the remaining two transitions, $|T_{+1}\rangle \leftrightarrow |S\rangle \leftrightarrow |T_{-1}\rangle$, which are caused by the antisymmetric part of $\mathsf{H}_{\mathsf{H,nonsec}}$, are of interest for us and have been included in Fig. 9.16.

If the probabilities of intersystem crossing between $|S\alpha\rangle$ and $|T_{+1}\beta\rangle$ and between $|S\beta\rangle$ and $|T_{-1}\alpha\rangle$ were identical, radicals possessing protons with α spins and such with β spins would react to the geminate recombination product with exactly the same rate, and this product would contain as many protons with α spins as with β spins, so we would have no nuclear spin polarization at all. However, inspection of Fig. 9.16 shows that the energy differences between these two sets of states is not equal. We know that the amount of mixing will therefore also be different. The states with the larger energy separation, in our case $|T_{-1}\alpha\rangle$ and $|S\beta\rangle$, mix less than the other two, $|T_{+1}\beta\rangle$ and $|S\alpha\rangle$. Starting from a triplet pair, intersystem crossing will thus lead to a surplus of protons with α spin in the singlet pair, so enhanced absorption results for the geminate recombination product.

In contrast to the previously discussed rules for CIDNP in high H_0, the protons of the escape products are also absorptively polarized, because the population of $|T_{+1}\beta\rangle$ in the radicals is depleted more than that of $|T_{-1}\alpha\rangle$. For CIDNP net effects in low magnetic fields it is generally true that polarizations of geminate recombination products and escape products have the same phase. This fundamental difference between experiments in high and low fields is due to the fact that nuclear spins are only distributed between the two kinds of products in the former case, whereas in the latter they are flipped in the intersystem crossing process.

If the hyperfine interaction were negative, the sublevels of the upper and lower triplet state would be interchanged. It is obvious from Fig. 9.16 that this would invert all polarizations. Had we started with a singlet precursor, a surplus of protons with β spins would have been found in the escape product, accompanied by a depletion of the protons with α spins in the geminate recombination product. Hence, all polarizations would have been reversed. The dependence of the polarization phases on the sign of a and on μ in thus the same in high and in low fields.

These results lead us to a very simple rule for CIDNP net effects in low fields,

$$\gamma_n = \operatorname{sgn}(a) \cdot \mu, \qquad (9.110)$$

where $\gamma_n = +1$ signifies enhanced absorption and $\gamma_n = -1$ emission.

In order to round off our discussion of CIDNP, we will finally describe the polarizations in zero field. Generally, net CIDNP effects vanish in this case; no polarizations can be created for any number of noncoupled nuclei if there is no magnetic field. Again, we take our radical pair containing a single proton as an example. If we look at Fig. 9.16, we find that the energy differences between $|T_{-1}\alpha\rangle$ and $|S\beta\rangle$, and between $|T_{+1}\beta\rangle$ and $|S\alpha\rangle$ are exactly equal in zero field, so the probabilities of intersystem crossing between these two sets of functions do not depend on the nuclear spin states. (We have already mentioned that four of the high field eigenfunctions given in Eq. 9.109 are no longer correct eigenfunctions of our system in the absence of an external magnetic field; instead of $|T_0\alpha\rangle$ and $|T_{+1}\beta\rangle$ as well as of $|T_0\beta\rangle$ and $|T_{-1}\alpha\rangle$ we should then use linear combinations of these functions. However, as we have seen that the matrix elements connecting $|S\alpha\rangle$ and $|T_0\alpha\rangle$, and $|S\beta\rangle$ and $|T_0\beta\rangle$ can effect no nuclear spin polarization in low fields, the above argument also holds for linear combinations of the high field eigenstates.)

If we have a radical pair with more than one proton, and these protons are coupled among themselves, multiplet effects can occur also in zero field. Let us treat these with the simplest example possible, which we have already taken for the discussion of multiplet effects in high fields, a radical pair with two protons in the first radical. We assume both hyperfine coupling constants to be positive. A natural choice of basis functions is given by the products of the zero field states of the electrons with those of the protons, which we again arrange in groups of equal total spin f. We will use capital letters to denote the electron spin functions and small letters for the nuclear spin states. The resulting sixteen functions fall in four categories, which we will set out separately.

First, we have a class of nine functions made up by combining triplet states of both electron and nuclear spins,

$$\begin{aligned} f &= +2: \quad |T_{+1}t_{+1}\rangle \\ f &= +1: \quad |T_{+1}t_0\rangle \; |T_0t_{+1}\rangle \\ f &= 0: \quad |T_{+1}t_{-1}\rangle \; |T_{-1}t_{+1}\rangle \; |T_0t_0\rangle. \\ f &= -1: \quad |T_{-1}t_0\rangle \; |T_0t_{-1}\rangle \\ f &= -2: \quad |T_{-1}t_{-1}\rangle \end{aligned} \tag{9.111}$$

Of these functions, $|T_{+1}t_{+1}\rangle$ and $|T_{-1}t_{-1}\rangle$ are proper eigenstates already. All the others of equal f mix in zero field. If we would form linear combinations of appropriate symmetry, we would get one quintet, one triplet, and one singlet from these nine functions. Nevertheless, we find it more convenient to work with the high field eigenfunctions, because they are factorized into nuclear and electronic terms. Four transitions between the functions of this set can be induced by the symmetric terms of the nonsecular part of the hyperfine interaction, but as these occur within the triplet manifold, they do not concern us here.

The second class,

$$\begin{aligned} f &= +1: \quad |T_{+1}s\rangle \\ f &= 0: \quad |T_0s\rangle, \\ f &= -1: \quad |T_{-1}s\rangle \end{aligned} \tag{9.112}$$

is made up of three functions, in which the electrons are in a triplet state and the nuclei in a singlet state. Between these functions and those of Eq. (9.111), six transitions can be caused by the hyperfine interaction, but no CIDNP is generated by them, since no intersystem crossing of the electron spins takes place.

In the three functions of the third class, the electrons are found in the singlet state, and the nuclei in a triplet state,

$$\begin{aligned} f &= +1: \quad |St_{+1}\rangle \\ f &= 0: \quad |St_0\rangle, \\ f &= -1: \quad |St_{-1}\rangle \end{aligned} \tag{9.113}$$

and lastly there is one function that is antisymmetric in both electron and nuclear spins,

$$f = 0 : \quad |Ss\rangle. \tag{9.114}$$

Transitions between this group and the preceding cannot be effected by H_H, as you may prove in a homework problem.

Owing to the exchange interaction, the first two classes, which are given in Eqs. (9.111) and (9.112), are isolated from the latter two (Eqs. 9.113 and 9.114), if the radicals of the pair are near each other. Outside the exchange region intersystem crossing can occur between these two sets. However, transitions due to the hyperfine interaction are not possible between the functions of Eq. (9.112) and $|Ss\rangle$, so we are left with only three pathways for singlet-triplet mixing.

If we look at intersystem crossing between the functions of Eq. (9.111) and those of Eq. (9.113), we find six allowed transitions. (We have learned that transitions can only take place between functions of the same value of f, which excludes the states of Eq. (9.111) with $f = \pm 2$. Furthermore, mixing of $|T_0 t_0\rangle$ and $|St_0\rangle$ cannot be brought about by H_H.) The characteristic of these transitions is that the multiplicity of the nuclear spins remains unchanged in them. Hence they cause a sorting of the nuclear spin states similar to that found in high magnetic fields. Starting from a triplet radical pair, intersystem crossing by this pathway produces a population enhancement of the nuclear triplet states in the cage product, and an equally large depletion in the free radicals and thus in the escape products. The other two singlet-triplet mixing processes, i.e. between the functions of Eq. (9.111) and Eq. (9.114) as well as those of Eqs. (9.112) and (9.113), always occur simultaneously, because their dependence on the parameters of the radical pair is identical. Calculations show that they also lead to spin sorting, but in the opposite way, so in the case of a triplet precursor they result in an increase of the populations of the nuclear singlet state in the geminate recombination product and a corresponding decrease in the escape products. It is obvious, that all these polarizations are inverted, if the radical pair is generated in its electronic singlet state.

The efficiency of intersystem crossing is largest with the first process, if the two protons are contained in the same radical and the signs of their hyperfine coupling constants are the same, or likewise if they reside in different radicals and possess hyperfine coupling constants of different signs. In contrast, the second process is favoured by opposite signs of the hyperfine coupling constants, if both protons belong to the first radical, or by identical signs, if not.

In our example, the first pathway therefore dominates, and the states $|St_{+1}\rangle$, $|St_0\rangle$ and $|St_{-1}\rangle$ are overpopulated to the same degree, compared to $|Ss\rangle$. The populations of these spin states can now be projected on the nuclear spin states $|t_{+1}\rangle, |t_0\rangle, |t_{-1}\rangle$ and $|s\rangle$ of the diamagnetic recombination product. As we have used a product representation of the spin functions of the radical pair (Eqs. 9.113 and 9.114), there is, of course, a direct correspondence.

It can be shown [13] that in a general case all sublevels of one proton multiplicity are equally populated in these experiments. This is intuitively reasonable, because in the absence of anisotropy effects no preferred axis of quantization exists in zero field. Hence, the transition probabilities are only dependent on the total magnitude of the spin vector of the system of coupled protons.

A correlation diagram between the nuclear spin states of the geminate recombination product in zero field and in the high field of the observation magnet is shown in Fig. 9.17. As before, we have assumed a positive coupling constant J. With this diagram, we can immediately construct the resulting NMR spectrum, which is also depicted schematically in this figure.

As there is no population difference across two of the four transitions, the corresponding two peaks are absent. It is seen that these are the two inner lines of both signal groups. This phenomenon is observed in all CIDNP investigations of coupled nuclear spin systems in zero field and has been termed '$n-1$ multiplets'. The signal phase is denoted as before, so in this instance we have an E/A multiplet.

The dependence of the relative populations of the nuclear spin levels on the parameters of the radical pair has already been discussed. It is evident that an overpopulation of the nuclear singlet state instead of the triplet states would lead to an inversion of the NMR signals in Fig. 9.17. On the other hand, if the sign of the nuclear coupling constant J becomes negative, the correlation between the zero field states $|t_0\rangle$ and $|s\rangle$ and the high field states $|\beta\alpha\rangle$ and $|\alpha\beta\rangle$ changes, but the transitions A_1 and A_2 as well as X_1 and X_2 are also interchanged, as we have seen before. In consequence, the inner lines of the multiplets are still absent, but the CIDNP phases of the other signals are inverted.

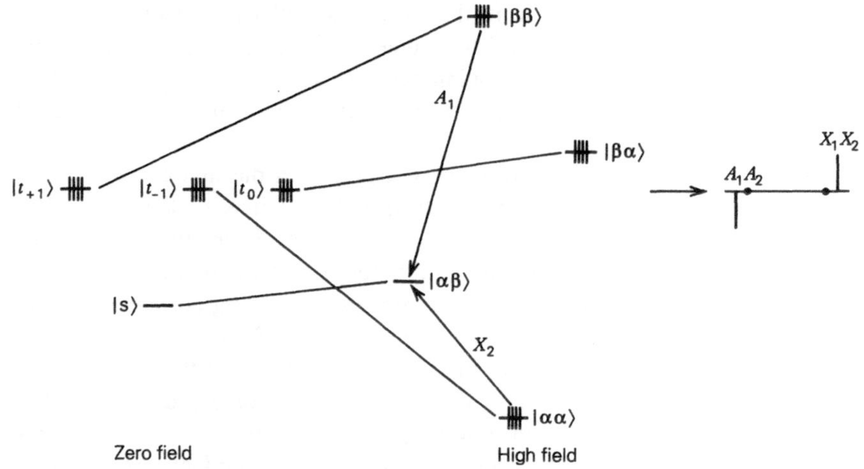

Fig. 9.17. *Left*: correlation between the zero field eigenstates and the high field eigenstates of a spin system of two nuclei. State $|\alpha\beta\rangle$ receives none of the excess population.
Right: the CIDNP spectrum of this system shows $n-1$ multiplets. A positive coupling constant J has been assumed, which leads to *E/A* multiplets. The assignment of the transitions is the same as in Fig. 9.14

All these regularities are thus described by the same rule as for the multiplet effect in high fields,

$$\gamma_m = \mathrm{sgn}(a_\mathrm{A}) \cdot \mathrm{sgn}(a_\mathrm{X}) \cdot \mathrm{sgn}(J) \cdot \sigma \cdot \mu \cdot \varepsilon. \tag{9.115}$$

where $\gamma_m = +1$ and $\gamma_m = -1$ denote $n - 1$ multiplets of phase E/A and A/E, respectively, and all other parameters have the same meaning as in Eq. (9.108).

9.7 Chemically Induced Electron Spin Polarizations

Now we turn to electron spin polarizations, which may be caused in chemical reactions by the radical pair mechanism. We already know that this phenomenon is denoted, somewhat incorrectly, as chemically induced dynamic electron spin polarization [8], CIDEP. It is directly related to the term $\rho_{ST_0} + \rho_{T_0S}$ of the density matrix of the radical pairs, as we have seen in Sect. 9.4. Hence it is in a way easier to understand than CIDNP, where the recombination probabilities of the radicals, which are proportional to the quantity ρ_{SS}, are determined as a function of the nuclear spin states by measuring the NMR signal intensities of the reaction products. In any case, these two kinds of polarizations have different origins, although they are quite similar in appearance.

We will discuss CIDEP with the vector model introduced in Sect. 9.4. As we recall, its characteristics are that we can visualize the time development of the density matrix of a radical pair in a Cartesian coordinate system by the precession of a vector $\mathcal{R}$, the components of which along the x-, y-, and z-axis are $\rho_{SS} - \rho_{T_0T_0}, i(\rho_{T_0S} - \rho_{ST_0})$ and $\rho_{ST_0} + \rho_{T_0S}$, in that order, around a vector $\mathcal{A}$ that has the exchange interaction as its x-component and the matrix element Q defined in Eq. (9.49) as its z-component. As long as we are dealing with high magnetic fields, this model is an exact representation of the spin dynamics of a radical pair.

In Fig. 9.18 we use such vector pictures to explain the generation of CIDEP. In practically all cases the exchange integral J is negative, but the models are more convenient to draw, if we take it along the positive x-axis instead. The effect of this would just be a reversal of the sense of the rotations of $\mathcal{R}$ that are brought about by this interaction. We can easily compensate this by additionally changing the sign of the z-component of $\mathcal{R}$, which is only involved in this type of rotations, so we will work with the inverted electron spin polarization $-(\rho_{ST_0} + \rho_{T_0S})$ in these figures.

Starting with a newly born radical pair, we have a diagonal density matrix at first, as required by the *hypothesis of random phases*, thus neither CIDEP nor phase correlation $i(\rho_{T_0S} - \rho_{ST_0})$ are present initially. As long as the radicals are inside the exchange region, J is much larger than Q, so the vector $\mathcal{A}$ points along the x-axis, as does $\mathcal{R}$. It is therefore obvious that in the primary cage no precession of $\mathcal{R}$ can take place at all. In particular, no CIDEP is produced at this stage.

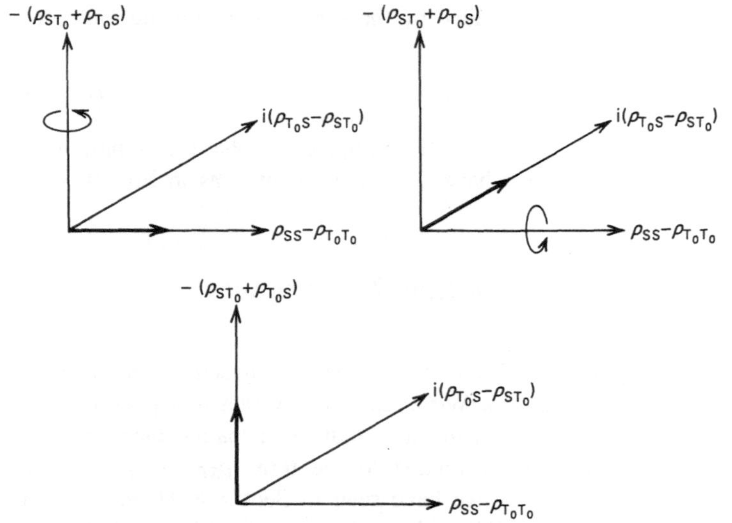

Fig. 9.18. Visualization of the two-step process leading to CIDEP by the vector model of Sect. 9.4. *Left*: we start with a singlet precursor. Once the radicals have separated, the mixing matrix element Q, which we have assumed to be positive, rotates $\mathcal{R}$ towards the $+y$-axis.
Center: upon reencounter, the phase coherence $i(\rho_{T_0S} - \rho_{ST_0})$ created in this way is rotated towards the $+z$-axis by the exchange interaction $-J$, where we have $J < 0$.
Right: the result is an electron spin polarization $-(\rho_{ST_0} + \rho_{T_0S})$

During their subsequent diffusive excursion the radicals spend most of the time outside the exchange region, where J is approximately zero. $\mathcal{A}$ is then parallel to the z-axis, so only mixing of $\rho_{SS} - \rho_{T_0T_0}$ and $i(\rho_{T_0S} - \rho_{ST_0})$ is possible, and we still do not get any electron spin polarization. However, if the radicals diffuse together again and enter the exchange region for the second time, the phase correlation created during their separation can be converted into CIDEP.

Hence the electron spin polarizations are generated in a two-step sequence, which we can write schematically as

$$\rho_{SS} - \rho_{T_0T_0} \overset{Q}{\longleftrightarrow} i(\rho_{T_0S} - \rho_{ST_0}) \overset{-J}{\longleftrightarrow} -(\rho_{ST_0} + \rho_{T_0S}), \tag{9.116}$$

and which we have depicted in Fig. 9.18.

We will now discuss the dependence of the polarization phases on the parameters of the radical pair. As CIDEP experiments in low magnetic fields are of no practical importance, we concentrate on the high field case.

First, there are CIDEP net effects. The sign of these polarizations is obviously directly given by the sign of the z-component of the vector $\mathcal{R}$ that is present after the two-step process of Eq. (9.116). The final location of this vector in our coordinate system is determined by the initial condition and the two rotations that occur sequentially. Inversion of either the direction of $\mathcal{R}_{t=0}$, which means changing the precursor multiplicity, or of the sense of one of these rotations, that is inversion of the sign of either J or Q, will therefore lead to an inversion of the polarizations.

In Fig. 9.18 we started with a singlet precursor. A positive mixing matrix element Q took us to the $+y$-axis, and the (negative) exchange interaction to the $+z$-axis, so the final z-component of the density matrix is $-(\rho_{ST_0} + \rho_{T_0S})$. By comparison with Eq. (9.58) we find this to be proportional to the average value of the operator $\mathbf{S}_{2z} - \mathbf{S}_{1z}$. As Eq. (9.59) tells us, this corresponds to an overpopulation of the electron spin state $|\beta\rangle$ in the first radical, and of $|\alpha\rangle$ in the second. Hence with these parameters the EPR signals of radical 1 appear in enhanced absorption, and those of radical 2 in emission.

If we assume that the mixing matrix element Q is dominated by the Δg term, we can thus formulate a sign rule for the phase Λ_n of a CIDEP net effect,

$$\Lambda_n = \text{sgn}(\Delta g) \cdot \text{sgn}(J) \cdot \mu \,. \tag{9.117}$$

In this equation, $\Lambda_n = +1$ denotes enhanced absorption for the first radical and emission for the second, and $\Lambda_n = -1$ the opposite. The variable μ has the same meaning as in the rules for the prediction of CIDNP phases.

For $\Delta g = 0$ there can only be CIDEP multiplet effects. In order to demonstrate them, we treat the simplest example, a radical pair with one proton in the first radical. It is evident that this system would not give rise to any nuclear spin polarizations. As we saw in the preceding section, a CIDNP net effect is ruled out, if there is no g-factor difference. CIDNP multiplet effects cannot be observed either, because at least two protons are required for their creation (compare Fig. 9.15).

We will again start with a singlet precursor, and take J to be negative. Let us assume a positive hyperfine coupling constant of the proton. Then, Q is positive for those pairs, where the protons are in state $|\alpha_n\rangle$. Figure 9.18 shows us that in the free radicals 1 the state $|\beta_e\alpha_n\rangle$ will therefore be populated more than $|\alpha_e\alpha_n\rangle$. For those pairs containing β protons, Q is negative, so the sense of the first rotation is reversed. As a result, the population of $|\alpha_e\beta_n\rangle$ is larger than that of $|\beta_e\beta_n\rangle$, by the same amount as in the previous case. Referring to the discussion of Fig. 9.1 we thus find that the doublet line with the higher frequency, which is the one to the left in the EPR spectrum, is polarized in absorption, and the other line in emission. As before, we denote this as an A/E multiplet. In a homework problem, you can show that in multiplets possessing more than two lines, the polarization phases of all lines to the left of the center frequency are equal, and opposite to those to the right of this frequency.

For the second radical we have no electron spin polarization at all. Protons of nuclear spin states $|\alpha_n\rangle$ or $|\beta_n\rangle$ in radical 1 are seen to cause an overpopulation of the electron spin states $|\alpha_e\rangle$ or $|\beta_e\rangle$ in radical 2, respectively, and these opposite contributions exactly cancel.

A triplet precursor would have led to an inversion of all population differences, hence to an E/A multiplet for the first radical. The same would also have been obtained, if the sign of the exchange integral had been changed instead. Interestingly, however, the sign of the hyperfine coupling constant is of no consequence for the polarization phases. Although a negative hyperfine coupling constant would give an overpopulation of $|\beta_e\alpha_n\rangle$ and $|\alpha_e\beta_n\rangle$ compared to $|\alpha_e\alpha_n\rangle$ and $|\beta_e\beta_n\rangle$ in our case, the order of the EPR transitions within the doublet would also be permuted, as we have seen in Sect. 9.1. These two changes compensate.

Thus we have a very simple rule for the prediction of the phase Λ_m of CIDEP multiplet effects,

$$\Lambda_m = \mathrm{sgn}(J) \cdot \mu, \tag{9.118}$$

where an A/E multiplet corresponds to $\Lambda_m = +1$, and an E/A multiplet to $\Lambda_m = -1$.

We note that CIDEP experiments have the attractive feature that they allow us to determine the sign of the exchange interaction, because the polarization phases of both net and multiplet effects depend on this parameter. The magnitude of this quantity is also of importance. If J is small, very little CIDEP is generated during the short time τ the radicals reside in the exchange region, as is obvious from our discussion of Eq. 9.82. On the other hand, there we have also found that a very large exchange integral is unfavourable, since both electron spin polarization and phase correlation are destroyed in this case . Equations (9.84) and (9.85) lead us to expect an optimum for, roughly, $(\bar{J}/h)\bar{\tau} = 0.1$.

Finally, for the sake of completeness we should like to mention that there is also another mechanism that can cause CIDEP. This so-called 'triplet mechanism', which is outside the scope of this discussion, relies on different rates of intersystem crossing in an excited molecule from its singlet state to the three sublevels of its triplet state. The resulting unequal populations, i.e. polarizations, of these triplet levels can be transferred to electron spin polarizations of free radicals by fast subsequent chemical reactions, for instance bond cleavage.

9.8 Quantum Beats

Our theoretical calculations of Sect. 9.4 have shown that in an ensemble of radical pairs created in, say, a pure singlet state, oscillations between this state and the triplet state will occur in a magnetic field. This quantum mechanical phenomenon is in many ways similar to that found in classical mechanics for a system of two identical, coupled pendulums, of which only one is excited initially. It is well known that in this case the energy will oscillate between the two pendulums with a beat frequency that is given by the difference of the frequencies of their two normal modes. In analogy to this, the oscillations between $|S\rangle$ and $|T\rangle$ have been termed 'quantum beats'. We have already gathered theoretical evidence for them, and we understand how they are related to certain measurable quantities, for instance spin polarizations. In this final section, we will briefly describe their direct observation in a time resolved experiment.

Suppose that we produce radical ion pairs in solution in a very short time, about 1 ns or less. We can do this by a short pulse of fast electrons or other ionizing radiation. Under these circumstances, the initial multiplicity of the pairs is singlet.

Frequently, the redox potentials of these ions are so high that the energy of the ion pair is larger than the electronic excitation energy of one of the parent molecules. Geminate recombination of a singlet pair will then regain this molecule in an excited singlet state and the other in its ground state. We remember that most organic compounds have a triplet state that is of lower energy than the first excited singlet state. Hence in contrast to all our previously discussed examples, a correlated ion pair in a triplet state will also be able to recombine in this case, but now one of the two molecules of the geminate product will be in an excited triplet state. We see that the total spin quantum number is conserved in this reaction ($s = 1$ before the recombination, $s = 1 + 0$ afterwards), so there is no violation of the spin selection rules.

How are we to get information about intersystem crossing, if the chemical reactivity is no longer dependent on the electron spin multiplicity? The answer to this problem is that excited singlet and triplet states lose their excess energy by different channels. A molecule in an excited singlet state may return to the singlet ground state by emitting a photon, a process that is called 'fluorescence'. This spin allowed electric dipole transition is usually rapid. Typical fluorescence lifetimes are in the range of ns. A molecule in a triplet state decays nonradiatively in liquid solution. Therefore we can get the number of recombining singlet radical pairs by measuring the amount of light emitted from our system. (There is, of course, a direct proportionality.)

This detection technique has two other advantages, which are crucial for our application. First, the sensitivity of optical spectroscopy is higher by orders of magnitude than that of magnetic resonance spectroscopy. Second, the inherent time scale of optical experiments is much faster; it is routinely possible to obtain a time resolution in the ns range.

In the described experiment, the radical pairs are born in a pure singlet state. Their initial separation is about 10 nm in this special case, so the radicals are formed far outside the exchange region. In a strong magnetic field, intersystem crossing thus takes place between $|S\rangle$ and $|T_0\rangle$ with a frequency determined by the mixing matrix element Q. Fluorescence is generated at subsequent encounters of pairs, therefore we also expect the fluorescence intensity to oscillate with this frequency. However, the pair lifetime before recombination is statistically distributed, which to some extent averages out these oscillations and gives rise to a large component of the fluorescence that falls off smoothly. The oscillations are superimposed on this decay. They can be extracted, if one divides $I(t)$, the measured fluorescence intensity, by the smoothed curve, $\tilde{I}(t)$. As typical amplitudes of the modulation amount to a few percent of $\tilde{I}(t)$, it is obvious that a high signal-to-noise ratio is essential for this procedure.

The resulting beat pattern can be very complicated, because many radical pairs are observed at the same time in these experiments. Since the mixing matrix element depends on the nuclear spin states, the individual pairs have in general different values of Q. Hence intersystem crossing cannot be described by

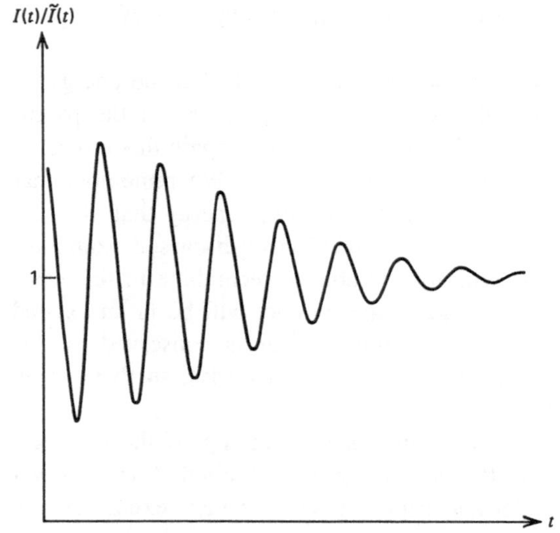

Fig. 9.19. Schematic depiction of quantum beats observed during the recombination of radical ion pairs with a large difference of their g-factors in a strong magnetic field. The ratio of the actual and the smoothed fluorescence intensities, $I(t)/\tilde{I}(t)$, is displayed as a function of the time elapsed since the generation of the pair by a pulse of fast electrons. With the parameters of Ref. [13], the maximum modulation amplitude is about 10 percent and the total time scale is 100 ns

a single frequency. If the molecules contain more than a few magnetic nuclei, the oscillations usually become undetectable. However, if the g-factors of the radicals differ greatly, the Δg-mechanism can outweigh the hyperfine mechanism, and intersystem crossing is clearly dominated by a single frequency. An experimental example with such a system has been reported in the literature [14]. The result is shown schematically in Fig. 9.19.

Well-resolved oscillations of the fluorescence intensity are found in this case. Their frequency is linearly dependent on the strength of the external magnetic field, which proves that they are due to the term $\Delta g \beta H_0$ in the mixing matrix element. In a chemical system possessing favourable hyperfine coupling constants [15], quantum beats solely due to the hyperfine mechanism have also been observed.

9.9 References

1. Adrian FJ (1972) J Chem Phys 57: 5107
2. Closs GL (1969) J Am Chem Soc 91, 4552; Closs GL, Trifunac AD (1970) J Am Chem Soc 92: 2183
3. Kaptein R, Oosterhoff LJ (1969) Chem Phys Letters 4: 195, 214
4. Pedersen JB (1977) J Chem Phys 67: 4097; and references therein
5. Monchick L, Adrian FJ (1978) J Chem Phys 68: 4376
6. Chuang TJ, Hoffmann GW, Eisenthal KB (1974) Chem Phys Letters 25: 201
7. Noyes RM (1956) J Am Chem Soc 78: 5486

8. Monchick L (1956) J Chem Phys 24: 381
9. Two comprehensive monographs about CIDNP and CIDEP are:
a) Muus LT, Atkins PW, McLauchlan KA, Pedersen JB (1977) (eds), 'Chemically induced magnetic polarization', Reidel D, Dordrecht
b) Salikhov KM, Molin YuN, Sagdeev RZ, Buchachenko AL, 'Spin polarization and magnetic effects in radical reactions', Elsevier, Amsterdam, 1984. The latter book also deals with other magnetic effects that we have not treated here
10. Goez M (1992) Chem Phys Letters 188: 451
11. Roth HD, Manion ML (1975) J Am Chem Soc 97: 6886
12. Kaptein R (1971) J Chem Soc Chem Commun 732
13. Kaptein R, den Hollander JA (1972) J Am Chem Soc 94: 6269
14. Veselov AV, Melekhov VI, Anisimov OA, Molin YuN (1987) Chem Phys Letters 136: 263
15. Anisimov OA, Bizyaev VL, Lukzen NN, Grigoryants VM, Molin YuN (1983) Chem Phys Letters 101: 131

9.10 Problems

9.1 Show that $|s\rangle$ (Eq. 9.12) and $|t_0\rangle$ (Eq. 9.11) are eigenfunctions of the operator H_H defined in Eq. 9.5

9.2 What are the eigenvalues of magnetic energy for a radical containing one proton in zero field?

9.3 Calculate the expectation values $\mathsf{S}^2_{\text{tot}}$ for the four wavefunctions of Eqs. 9.16 and 9.17

9.4 Decompose the operator $\mathsf{H}_{\mathsf{H,\,nonsec}}$ of Eq. 9.43 in the way described by Eq. 9.29 into terms that are symmetric and antisymmetric with regard to interchange of the electrons

9.5 Estimate the intersystem crossing frequency
a. for a radical pair without any magnetic nuclei in a magnetic field of 50 kGauss, a frequently used field for NMR measurements, assuming $\Delta g = 0.001$, which is a common value for organic radical pairs;
b. for a pair of radicals with equal g-factors, where one radical contains a proton with $a = 10$ Gauss, which is quite a typical hyperfine coupling constant of an organic radical

9.6 Calculate the matrix of the operator O that has an average value of $i(\rho_{T_0S} - \rho_{ST_0})$ for an ensemble of states $|S\rangle$ and $|T_0\rangle$. Express O by combinations of the operators for the electron spin

9.7 Use Eq. 9.97 to calculate the total amount F_{tot} of geminate recombination product formed from a system starting in a pure triplet state without initial electron spin polarization and phase correlation. Assume a singlet reactivity λ of unity. You should take the matrix of Eq. 9.74 as $\underline{\mathsf{M}}$. The components of interest of $\mathscr{R}$ are $\rho_{SS}, \rho_{T_0T_0}$ and $i(\rho_{T_0S} - \rho_{ST_0})$. Change $\underline{\mathsf{M}}$ accordingly. Derive an approximate expression for F_{tot} by expanding the terms c and s in the resulting formula (Eq. 9.98), keeping only first order terms

9.8 Suppose you have a radical pair X with one proton, in which Δg_X is positive and a_X is zero. After a certain time t, this pair is suddenly transformed to another radical pair Y, this time with $\Delta g_\mathrm{Y} = 0$ and $a_\mathrm{Y} > 0$. For simplicity, assume the lifetime of this pair to be t as well. Draw up vector diagrams for this case and show that the resulting polarizations are the same as those of a radical pair of lifetime t, in which Δg is equal to Δg_X and a is equal to a_Y

9.9 Construct vector diagrams for a radical pair containing two protons, one in each radical of the pair. Assume a triplet precursor, $\Delta g = 0$, positive hyperfine coupling constants and $J > 0$. Show that the resulting polarizations in the geminate recombination product are opposite to those of Fig. 9.15

9.10 Show that transitions between the functions of Eq. 9.113 and 9.114 cannot be brought about by the hyperfine interaction by calculating matrix elements of the operator H_H

9.11 Discuss CIDEP multiplet effects in a radical pair containing two protons on the basis of the vector models of Fig. 9.18 and energy level diagrams as in Fig. 9.1

10 Generalization of the Gyroscopic Model

J.D. Macomber

10.1 The Gyroscopic Model of the Interaction Process

The promises made in Chaps. 4 and 5 will be fulfilled in this chapter. In the earlier chapters assertions were made about the dynamics of the interactions between matter and radiation that lead to spectroscopic transitions. These assertions were proved for the cases of nuclear and electron magnetic resonance in Chap. 8. The theory used in the magnetic resonance case will now be generalized to cover rotational, vibrational, and electronic transitions. Relaxation processes applicable to microwave, infra-red, and optical spectroscopy will be discussed, as will the optical analogs of coherent transient effects first observed in nuclear magnetic resonance. The first step will be to establish the applicability of the gyroscopic model of the interaction process to electric dipole transitions.

A review of the gyroscopic model of a magnetic dipole transition is in order. The magnetic field, discussed in Chap. 3, is defined in terms of the torque that it exerts upon a magnetic dipole moment (**H** is the induction):

$$\mathbf{T} = \mathbf{p}_M \times \mathbf{H} . \tag{10.1}$$

In classical mechanics, there is a circular equivalent of Newton's force law: the torque is equal to the rate of change of the angular momentum,

$$\mathbf{T} = \frac{d\mathbf{S}}{dt} . \tag{10.2}$$

Equation (10.2) is the fundamental law governing the behavior of gyroscopes. Combining Eqs. (10.1) and (10.2) one finds that

$$\frac{d\mathbf{S}}{dt} = \mathbf{p}_M \times \mathbf{H} . \tag{10.3}$$

If both sides of Eq. (10.3) are multiplied by the magnetogyric ratio, γ, using the relationship between angular momentum and the magnetic moment in Eq. 8.84, it may be seen that

$$\frac{d\mathbf{p}_M}{dt} = \mathbf{p}_M \times (\gamma \mathbf{H}) . \tag{10.4}$$

The relationships expressed in Eq. (10.4) are illustrated in Fig. 10.1.

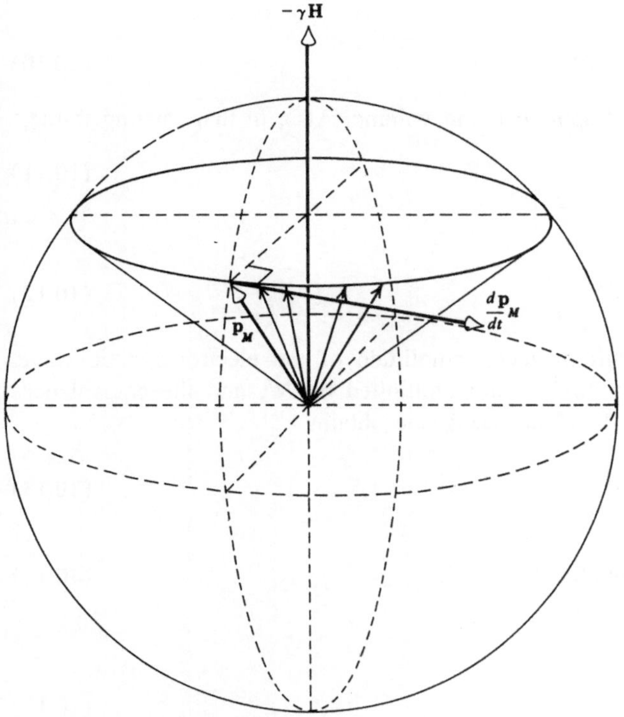

Fig. 10.1. Vector representation of the gyroscope equation for a magnetic dipole transition

From the definition of the cross product (expressed, e.g., in Eq. 8.5), Eq. (10.4) can be written component by component:

$$\frac{dp_x}{dt} = p_y \gamma H_z - p_z \gamma H_y, \tag{10.5}$$

$$\frac{dp_y}{dt} = p_z \gamma H_x - p_x \gamma H_z, \tag{10.6}$$

and

$$\frac{dp_z}{dt} = p_x \gamma H_y - p_y \gamma H_x . \tag{10.7}$$

Suppose that p_x, p_y, and p_z are replaced by their expectation values, representing the projection of the "certain component" of the magnetic moment of an individual quantum system along the three coordinate axes in the rotating frame (see Eqs. 8.163 and 8.173 to 8.175):

$$N'\langle p_x \rangle = u, \tag{10.8}$$

$$N'\langle p_y \rangle = v, \tag{10.9}$$

and

$$N'\langle p_z \rangle = M_z, \tag{10.10}$$

where N' is the number of spins per unit volume. Also, in that rotating frame,

$$H_y = 0 \tag{10.11}$$

and

$$H_x = \frac{H_1^0}{2}, \tag{10.12}$$

where H_1^0 is the maximum magnetic amplitude of the electromagnetic wave. If both sides of Eqs. 10.5 to 10.7 are multiplied by N', and the equivalences expressed in Eqs. 10.8 to 10.12 are used, one obtains

$$\frac{du}{dt} = v\gamma H_z, \tag{10.13}$$

$$\frac{dv}{dt} = M_z \frac{\gamma H_1^0}{2} - u\gamma H_z, \tag{10.14}$$

and

$$\frac{dM_z}{dt} = -v\frac{\gamma H_1^0}{2}. \tag{10.15}$$

Next, the effective value of the z component of the magnetic field in the rotating frame is defined by

$$H_z = H_0 - \frac{\omega}{\gamma}. \tag{10.16}$$

Remembering the definitions of $\omega_0, \omega_1,$ and δ from Eqs. (8.90), (8.131), and (8.157), one can use Eq. (10.16) to rewrite Eqs. (10.13) to (10.15):

$$\frac{du}{dt} = -\delta v, \tag{10.17}$$

$$\frac{dv}{dt} = \frac{\omega_1}{2} M_z + \delta u \tag{10.18}$$

and

$$\frac{dM_z}{dt} = -\frac{\omega_1}{2} v. \tag{10.19}$$

Finally, if frictional damping terms are added to each of the three equations (10.17) to (10.19), the latter will become the Bloch equations 8.177 to 8.179.

Note that Eq. (10.16) is consistent with the previous assertion that the magnetic field in the z direction vanishes at exact resonance in the rotating frame. The transformation to the new coordinate system causes H_0 to be cancelled out.

Equation (10.16) gives the formula for the effective field in the more general off-resonance case as well.

The derivation just outlined is the basis of the contention, made first in Chap. 3, that the interaction between radiation and matter is of the nature of a torque. This torque is exerted by the field component $H_1^0/2$ in the rotating frame, acting on the magnetic moment (in the usual case, $M_0\hat{z}$). The torque produces a precession of this vector about $H_1^0/2$, changing the z component of magnetization, as was shown in Fig. 8.8a. As a consequence of the fact that the magnitude of the z component is proportional to the energy of the ensemble, the rf field does work on the sample.

10.2 Electric-Dipole-Allowed Transitions

The question raised in Sect. 4.3, was "This is all very well for magnetic resonance, but how can it apply to any other form of spectroscopy?" True, there is always a dipole moment vector (usually electric) associated with any transition, which may interact with a field vector (usually electric) associated with the light wave, and these quantities may be substituted for $\mathbf{p}_M$ and $\mathbf{H}$ in Eq. (10.1). But, as was stated in Chap. 4, the transition dipole moment almost never has a z component; it is confined to the xy plane. Also, the transition dipole moment is not necessarily proportional to an angular momentum vector. If not, why should the dipole precess about the field in the rotating frame? In other words, how are the analogs to Eqs. (10.2) and (10.3) to be obtained? Finally, the oscillation frequency of the transition dipole moment in magnetic resonance, the Larmor frequency, is due to precession of the spins about a large external magnetic field in the z direction. Where is an analog to H_0 to be found in the general spectroscopic case, and, without it, how is an equation of the form of Eq. 10.16 to be obtained?

Many of these difficulties are merely apparent, not genuine, problems. The fact that the symbol H_0 represents a genuine magnetic field measurable in the laboratory (say, by a Hall-effect magnetometer) has never been used. The role of H_0 was merely to provide an energy difference between the upper and lower stationary states connected by the transition. In the magnetic resonance case the energies of both states were provided by the Zeeman Hamiltonian; in other kinds of spectroscopy, the rotational, vibrational, and electronic Hamiltonians will provide the energies, and they will serve just as well. Also, the fact that the symbol M_z represents a genuine static component of magnetization measurable in the laboratory (say, by a Göuy balance) has never been used either, $-M_z$ is merely a measure of the difference in population between any two energy levels connected by a spectroscopic transition.

The truth of the above assertions may be seen by remembering from Chap. 5 that a proper description of the dynamics during spectroscopic transitions lies in the density matrix $[\rho_{jk}]$ (or in the $[D_{jk}]$ matrices of which $[\rho_{jk}]$ is the sum).

The necessity and sufficiency of the density matrix description were established long before magnetic resonance was specifically discussed. For example, ω_0 was introduced in Eq. (2.68) as the beat frequency between the waves representing the superposed stationary states without calling it the Larmor frequency.

The treatment presented in Chap. 5 assumed that the matrix representing H_1 (the term in the Hamiltonian responsible for producing the transitions) is totally off-diagonal. This happy circumstance occurs quite naturally in magnetic resonance, where the various terms in the Hamiltonian are proportional to the Pauli spin matrices. But it was also described in Chap. 5 how any perturbation capable of producing transitions can be written in totally off-diagonal form by a suitable adjustment in the definitions of H_0 and H_1.

The off-diagonal matrix elements of the dipole moment operator in the general spectroscopic case are no longer simply $\gamma\langle S_x\rangle_{12}$ and $\gamma\langle S_x\rangle_{21}$, but are the more general $\langle p_x\rangle_{12}$ and $\langle p_x\rangle_{21}$. These latter can be calculated and will serve the same function in the equations of motion of the density matrix that the $\gamma\langle S_x\rangle$ did. The oscillations of $\langle \mathbf{p}(t)\rangle$ can be calculated from the density matrix according to the general rule, Eq. (5.19). These oscillations will occur in the xy plane as before; their ensemble average will then represent a polarization wave $\mathbf{P}$ propagating along the z axis parallel to the $\mathbf{E}$ wave that induced it.

It has just been shown why the physically observable effects of an oscillating electromagnetic field on an ensemble of two-level quantum systems will be the same regardless of the nature of the quantum transitions. One might think that at least the nice, simple picture of the interaction process in terms of a field-vector torque on a gyroscopic dipole in the rotating frame could not be used for electric dipole transitions. In fact, even that picture may be retained.

Suppose, for an electric (or even magnetic) dipole transition in the general case, the transition dipole moment, $(\mathsf{p}_x)_{12}$ is calculated using the standard methods of quantum mechanics. If this electric dipole moment had been produced by the rotation of a spherical magnetic ball (more properly, the intrinsic spin of a point magnetic monopole), $(\mathsf{p}_x)_{12}$ would be proportional to the x component of angular momentum of the ball. If the intrinsic spin of the ball were $s = \frac{1}{2}$, the angular momentum matrix element would be $\hbar/2$, just as it is for a spinning electron, proton, or neutron. The constant of proportionality should probably be called the "electrogyric ratio" by analogy with the magnetic case, and it might even be assigned the same symbol, γ. An effective γ can be calculated from

$$\gamma = \frac{2(\mathsf{p}_x)_{12}}{\hbar}, \tag{10.20}$$

even though the physical origin of $(\mathsf{p}_x)_{12}$ may have nothing directly to do with angular momentum about the x axis. A "pseudopolarization" vector can then be defined:

$$\mathbf{p} = \frac{\gamma\hbar(\rho_{11} - \rho_{22})\hat{z}}{2} + \langle\overline{p_x}\rangle\hat{x} + \langle\overline{p_y}\rangle\hat{y}. \tag{10.21}$$

If Eq. (10.21) is multiplied by N', the number of quantum systems per unit volume, it becomes

$$\mathbf{P} = \frac{\gamma\hbar(N_1 - N_2)\hat{z}}{2} + P_x\hat{x} + P_y\hat{y}. \tag{10.22}$$

The components of the pseudopolarization vector are shown in Fig. 10.2.

It should be remembered that only P_x and P_y represent genuine electromagnetic properties of the system. In contrast, the "pseudo" (false) component P_z has been introduced only to make the behavior of the quantum systems describable by means of the gyroscropic model.

To accompany Eq. (10.22), there also must be an expression for a "pseudo-electric" field vector. This quantity must be defined so as to produce the correct expression for the energy density of the system (see Eq. 8.206):

$$\begin{aligned}\mathscr{E} &= -\mathbf{P}\cdot\mathbf{E} \\ &= -(P_xE_x + P_yE_y + P_zE_z).\end{aligned} \tag{10.23}$$

The x and y components will cause no problem in finding an adequate definition for $\mathbf{E}$ for two reasons. First, they ordinarily contribute only a small part of the total energy for the ensemble. Second, these components are not "pseudo" and are therefore well defined.

The reasoning used to select an appropriate expression for E_z is as follows. One has from the definitions of $\mathscr{E}_1$ and $\mathscr{E}_2$,

$$\mathscr{E} \cong N_1\mathscr{E}_1 + N_2\mathscr{E}_2. \tag{10.24}$$

Also, since

$$\mathscr{E} \cong -P_zE_z, \tag{10.25}$$

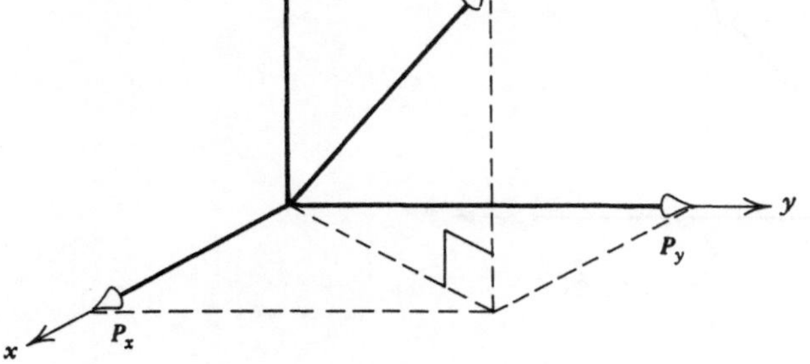

Fig. 10.2. Vector representation of the pseudopolarization vector for an electric dipole transition

a relation exists between $\mathscr{E}$ in Eq. (10.24) and P_z, given in Eq. (10.22):

$$N_1\mathscr{E}_1 + N_2\mathscr{E}_2 = \frac{\gamma\hbar(N_1 - N_2)E_z}{2}. \tag{10.26}$$

In the magnetic resonance case there was a genuine field $H_z = H_0$ instead of E_z, and, similarly, a genuine M_z instead of P_z. The magnetic energy was given by

$$\mathscr{E}_2 = -\mathscr{E}_1 = \frac{\gamma\hbar}{2}H_z. \tag{10.27}$$

The first equality in Eq. (10.27) can be satisfied in any two-level system by defining the zero of energy to lie exactly halfway between the energies of the two levels. The symbol H_z may be replaced by E_z in Eq. (10.27). The two equalities expressed therein then may be combined and solved for E_z:

$$E_z = \frac{\mathscr{E}_2 - \mathscr{E}_1}{\gamma\hbar} \tag{10.28}$$

The pseudoelectric field then becomes

$$E = \left(\frac{\mathscr{E}_2 - \mathscr{E}_1}{\gamma\hbar}\right)\hat{z} + E_x\hat{x} + E_y\hat{y}. \tag{10.29}$$

The components of the pseudoelectric field vector are shown in Fig. (10.3). Note that, when the dot product between Eqs. (10.22) and (10.29) is formed, the effective electrogyric ratio cancels out of the resulting expression.

When Eqs. (10.22) and (10.29) are used to describe the interaction between radiation and matter in the course of an electric-dipole-allowed spectroscopic transition, they give results identical with those formally derived by means of

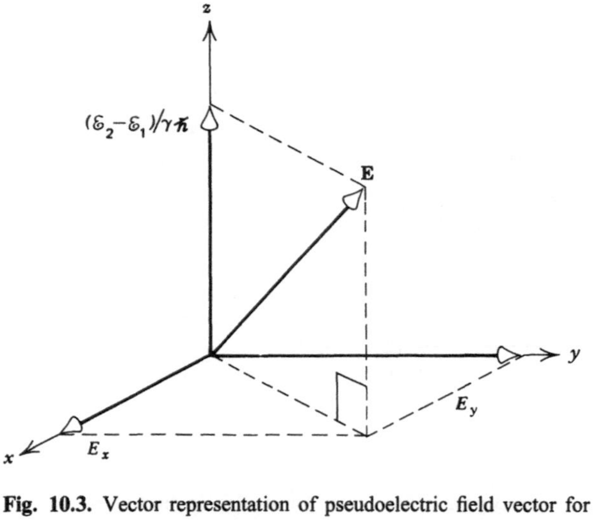

Fig. 10.3. Vector representation of pseudoelectric field vector for an electric dipole transition

the density matrix. In other words,

$$\frac{d\mathbf{P}}{dt} = \mathbf{P} \times (\gamma \mathbf{E}). \tag{10.30}$$

This equation, by analogy with Eq. 10.4 provides an accurate description of the dynamic behavior of the ensemble of quantum systems. The relationships expressed in Eq. 10.30 are shown in Fig. 10.4. For the use of the gyroscopic model to describe electric-dipole-allowed transitions see especially Ref. 3 in Chap. 5.

Why should there be a formal connection between the pseudoelectric field and pseudopolarization vectors of the type described by Eq. 10.30? The reader who suspects that this remarkable equation is not just a happy accident, but is rather a manifestation of some deeper underlying principle, is correct. The Pauli spin matrices arise naturally out of group theory as a symmetry property underlying a broad range of physical phenomena. For example, one can give an "isospin" description of the relationship between fundamental particles in which

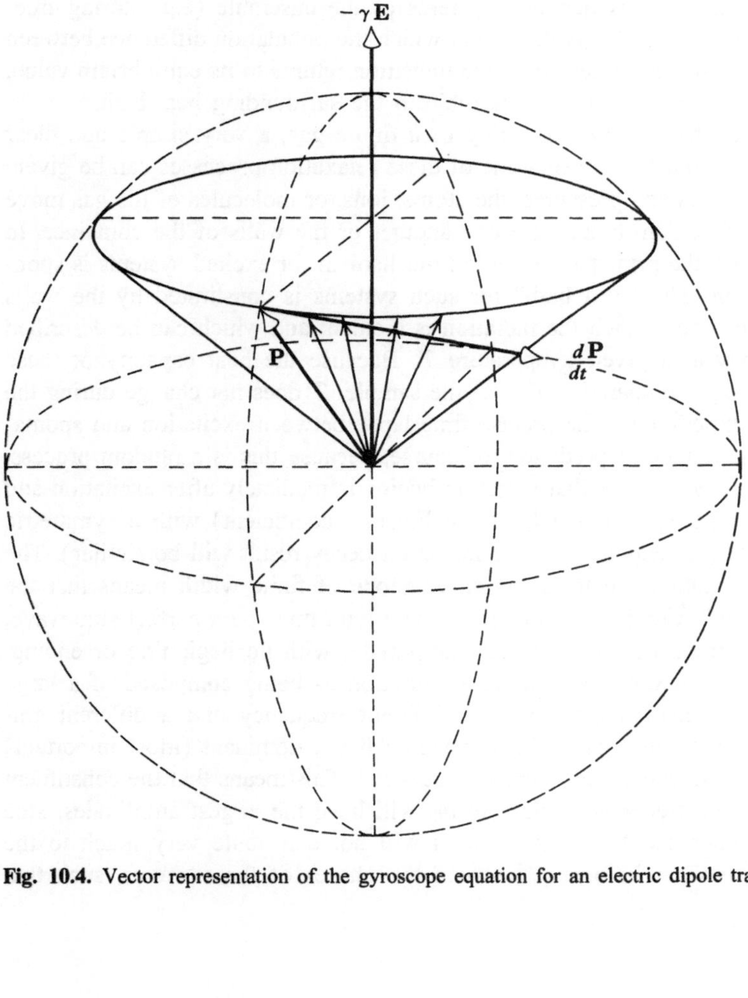

Fig. 10.4. Vector representation of the gyroscope equation for an electric dipole transition

the proton plays the part of "spin up," and the neutron, that of "spin down". Fain and Khanin [1] have shown that one can define an "energy space" analogous to "spin space" and "isospin space," in which any transition between two states of differing energy can be formally and rigorously described. Equation (10.30) is simply a special case of the energy-spin formalism, appropriate whenever the transition is brought about by means of an oscillating electromagnetic field.

10.3 Relaxation and Its Effect on Line Widths

It is to be remembered, however, that to make the description valid for times long in comparison with the relaxation times, damping terms must be added to Eq. (10.30). The same symbols as those used in magnetic resonance may be adopted for the rates of these processes, and they may be given generally the same physical interpretations. In other words, $1/T_2$ is the rate at which off-diagonal elements of the density matrix decay, through loss of phase coherence among the dipoles of the various quantum systems of the ensemble (e.g., during free-induction decay). Also, $1/T_1$ is the rate at which the population difference between the two quantum states connected by the transition returns to its equilibrium value, through flow of energy from the ensemble to the surrounding heat bath.

In the case of a sample consisting of a dilute gas, a very simple and clear description of the mechanism of some of these relaxation processes can be given. At low temperatures and pressures, the atoms, ions, or molecules of the gas move very slowly and seldom bump into one another or the walls of the container. In many such cases, the principal relaxation mechanism for excited systems is spontaneous emission. The "heat bath" for such systems is constituted by the walls of the container upon which the radiation is incident and which can be described as a black body at a given temperature T. Because the heat capacity of these walls is very large compared to that of the sample, T does not change during the course of the experiment. The precise time lapse between excitation and spontaneous emission cannot be predicted, of course, because this is a random process. However, one may imagine that radiation begins immediately after excitation and persists for a time $T_1 = 1/A_E$ (A_E is the Einstein coefficient) with a symmetric intensity profile (the exponential free-induction decay result will be similar). The fact that the emitted wave train has an envelope of finite width means that the light is not "pure". Completely monochromatic light must be a perfect sine wave, having a fixed frequency and constant amplitude, with no beginning or ending. One may think of a wave train of finite duration as being composed of a large number of sine waves, each having a different frequency and a different amplitude. In this particular case, the wave train has a dominant (most important) frequency, ω_0, also called the "carrier frequency." This means that the constituent sine waves having frequencies close to ω_0 will have the largest amplitudes; sine waves with higher and lower frequencies will not contribute very much to the sum. There exists a relationship between the phases of these waves, such that they interfere with one another constructively in the interval of time between t_0, the start of the spontaneous emission process, and $t_0 + T_1$. Because they have different frequencies, however, they must interfere destructively elsewhere along the t axis, and a wave train of finite duration results. The addition of sine waves to form a packet is shown in Fig 10.5.

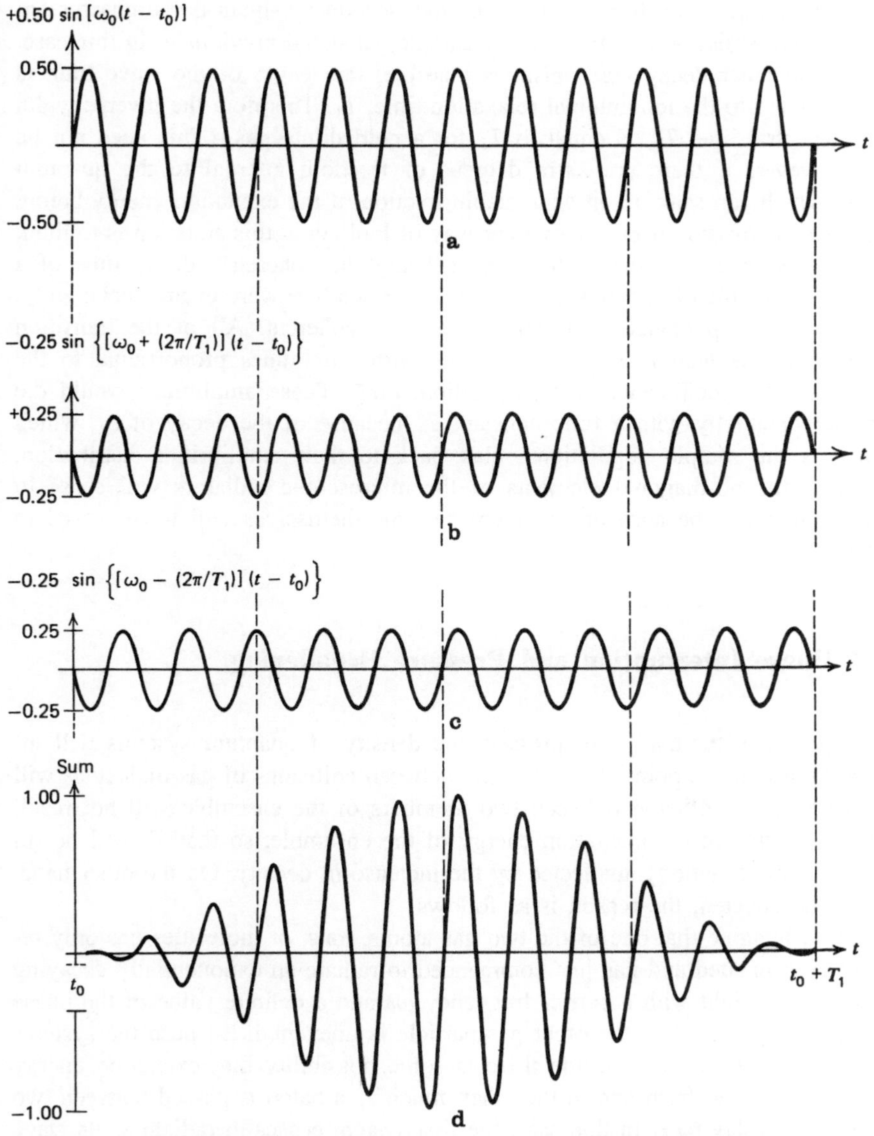

Fig. 10.5. A wave train of finite duration, expressed as a sum of three infinite wave trains of different frequencies. Note that the frequency spread (half width of spectral distribution at half height) is $2\pi/T_1$. In any physically realizable case, many more frequencies would be involved, but the spectral half width would be about the same

The mathematical technique of Fourier analysis enables one to compute the amplitudes of the sine waves constituting a finite wave train as a function of their frequencies, if the algebraic form of the wave envelope is known. The Fourier analysis of an exponentially decaying envelope produces an amplitude versus frequency curve that is Lorentzian in shape, is centered at ω_0, and has a half width (at half height) equal to the reciprocal of the decay constant. In other words, the Fourier transform of the free-induction decay signal is the unsaturated slow-passage emission spectrum of the sample, as stated previously. In this case, because no other decay mechanism is possible, the length of the wave train is simply related to the longitudinal relaxation time, T_1. Therefore the inverse width of the spectral line, T_2, is equal to T_1 for a cold dilute gas. (This need not be true, however, if there are extra degrees of freedom internal to the quantum system which can soak up an appreciable fraction of the excitation energy before spontaneous emission occurs.) Another way of looking at this situation is to think of T_2 in terms of its other definition, as being the coherence decay time of a radiating ensemble of quantum systems. If the ensembles were excited coherently, the subsequent spontaneous emission would be coherent. All of the transition dipole moments would oscillate in phase, with amplitudes proportional to the product of the coefficients of superposition, $c_1 c_2^*$. These amplitudes would die away exponentially with a time constant T_1 because of the decay of c_2, which occurs as the sample populations relax back to their equilibrium distribution. Therefore the off-diagonal elements of the microscopic radiators will cease to oscillate in phase because the quantum systems themselves will have ceased to radiate.

10.4 Phase Interruption and Pressure Broadening

If the gas is isothermally compressed, the density of quantum systems will increase until at some point the mean time between collisions of gas molecules will approach T_1. A collision between two members of the ensemble will not result in a net change in the excitation energy of the ensemble, so that T_1 will be (to the first approximation) unaffected by the increase in density. On the other hand, T_2 will be affected; the reason is as follows.

First, imagine that one of the two gas atoms, ions, or molecules has only recently been excited and has just commenced to radiate an exponentially decaying wave train of light with a carrier frequency ω_0 and a definite value of the phase constant §. Assume that the other gas particle is unexcited. Because the systems are identical, when they collide there is some possibility that excitation energy will be transferred from one to the other, much as a baton is passed between two runners in a relay race. In this case, the first system ceases to radiate as its wave train is abruptly terminated. Radiation of the second system commences at the same time and continues until all the energy has been transferred to the heat bath. The total time required for the ensemble to lose the excitation energy is still very nearly T_1, but it has been split between two wave trains of duration $T_2 < T_1$. These trains, being shorter in length, differ from a perfect sine wave in a more pronounced fashion than the one that would have been emitted if the collision had not occurred. The envelope of each of the phase-interrupted wave packets must damp down to zero along the axis more abruptly than was previously the case. A wider distribution of frequencies must be included in the mixtures that comprise both packets in order to produce more rapid destructive interference. Consequently, Fourier analysis produces a Lorentzian spectrum with a breadth greater than $1/T_1$. The transverse relaxation time is now the mean time between collisions rather than the lifetime determined solely by spontaneous emission. These facts are illustrated in Fig. 10.6.

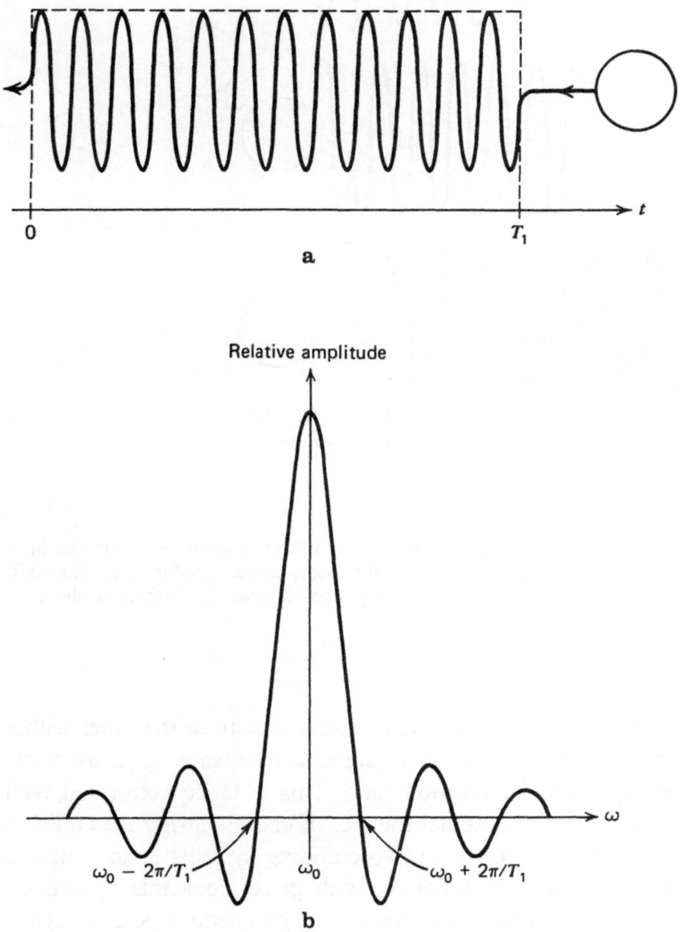

Fig. 10.6. Spin-spin relaxation. **a** An atom (size greatly exaggerated) radiating a finite wave train without interruption. **b** Fourier transform of the finite wave train in **a**. The square of this curve is customarily called the spectrum

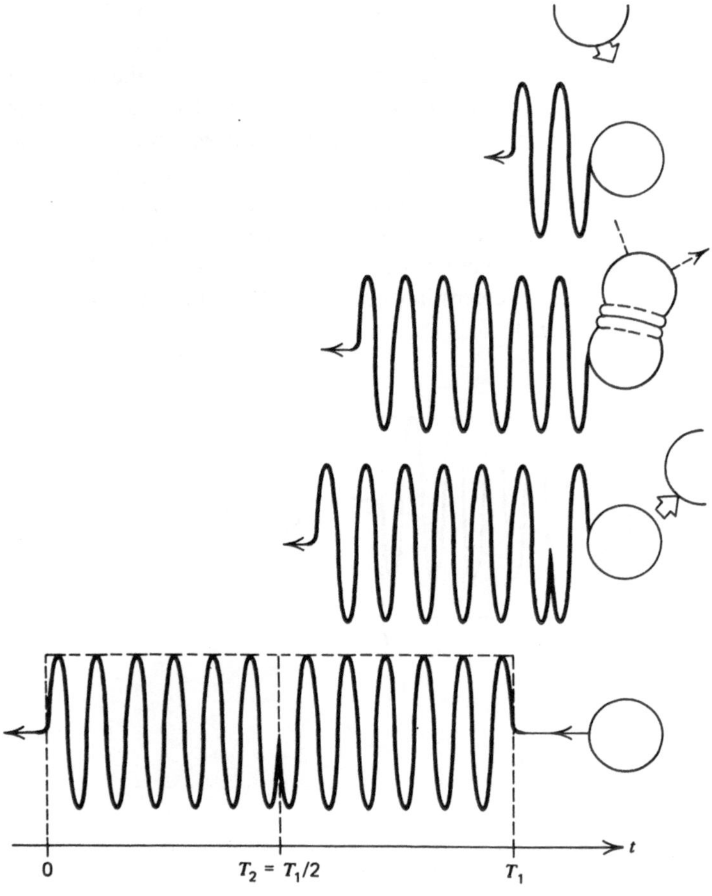

Fig. 10.6. (*Continued*) **c** An atom radiating a finite wave train of the same overall length as in **a**. At the top, another atom approaches. In the next picture, the atoms collide, producing a phase shift of 180° at $t = T_2 = T_1/2$. In the third picture, the colliding atom departs. At the bottom, the atom ceases radiating

Because energy was transferred from one quantum system to the other within the ensemble in the picture presented above, magnetic resonance spectroscopists sometimes call T_2 the "spin-spin" relaxation time. This is to be contrasted with T_1, the "spin-lattice" relaxation time. Actual transfer of energy during the collision process is not necessary, however. Imagine two colliding systems again, but this time both of them are in the act of radiating, with phase constants $\S_1$ and $\S_2$. During the act of collision, the oscillating electric or magnetic dipole moments of the two particles will exert forces on one another, which will have the effect of momentarily retarding one oscillation and accelerating the other. This, in turn, will produce a change in both $\S_1$ and $\S_2$, by an amount that depends on the details

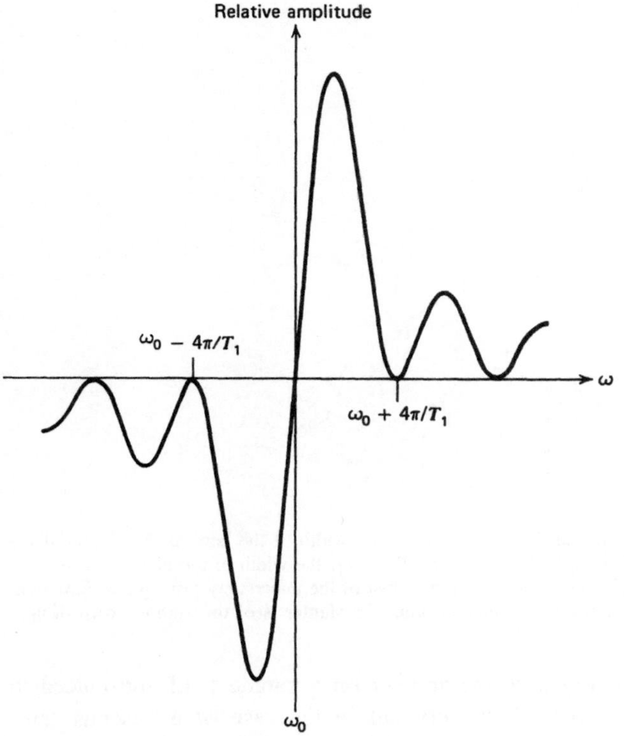

d

Fig. 10.6. (*Continued*) **d** Fourier transform of the finite wave train in **c**. Note that the first zeros of the amplitude which bound the bulk of the spectrum are twice as far away from the origin as they are in **b**

of the dynamics of the collision process. Since the various types of collisions are presumed to occur in a random and unpredictable way in a gas at mechanical equilibrium, the phases of the oscillations will be just as random during collisions of this type as if energy had actually been transferred. Both systems will resume radiating as they separate from one another after the collision, but the "damage will be done"; their wave trains will have been interrupted.

In an ensemble of identical particles, one cannot ordinarily distinguish between collisions in which energy is exchanged and collisions in which a mere phase interruption has occurred. One therefore cannot distinguish between the two corresponding T_2 mechanisms. Finally, it is also possible to view the line broadening due to phase interruption as an effect of the uncertainty principle, as illustrated in Fig. 10.7.

What will be the effect of collisions on the free-induction decay following coherent excitation? Either type of T_2 process described above will jostle the microscopic dipole moment vectors out of their proper positions in the rotating frame, so that they will not get back together properly after a 2π pulse. The

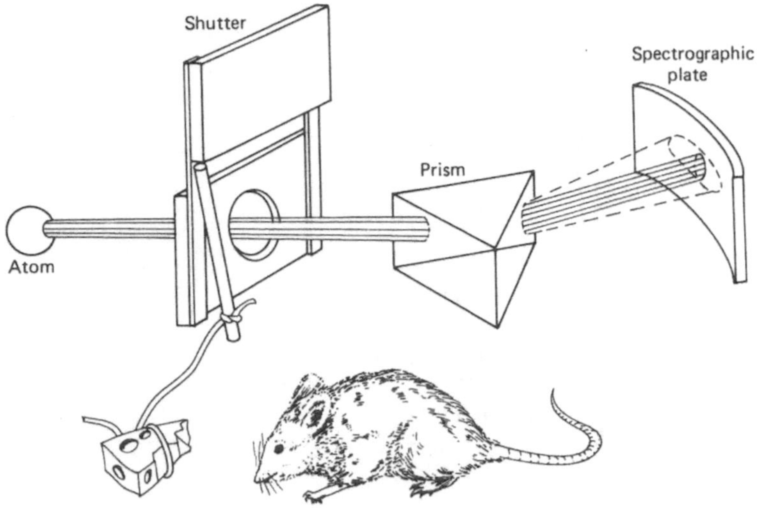

Fig. 10.7. Broadening of a spectral band with a shutter. The width of the band is $\Delta t \sim 1/T_1$ if the shutter is not used; if the shutter is open for a time $T_2, T_2 < T_1$, the width of the band is $\Delta t \sim 1/T_2$ (*dotted lines*). This may be considered to be due to the effect of the uncertainty principle, $\Delta E \Delta t \sim \hbar$, on the attempt to measure the energy of the photon, $\hbar\omega$. The shutter is of the author's own design

analogy with soldiers bumping into one another on a parade field, introduced in Chap. 5 and amplified in Chap. 8, is very apt in the case of a gaseous sample. Fluorescence spectra obtained from gases (excited by means of incoherent radiation) are also affected by T_2 processes. Increases in the widths of special lines due to molecular collisions are called "pressure broadening" in conventional spectroscopy. Since each member of the ensemble is equally likely to collide with another, each radiated wave train is shortened by the same amount on the average; pressure broadening is therefore homogeneous, and the reciprocal of the half widths of the lines are true T_2's.

10.5 Other Relaxation Processes in Gases and Solids

At any temperature above absolute zero the molecules of a gas are in motion. At mechanical equilibrium, the directions of molecular motion are distributed at random and the speeds are given by the Maxwell–Boltzmann formula. All but a few of the quantum systems will therefore possess some nonzero component of velocity along the line of sight (e.g., parallel or antiparallel to the optic axis of the spectrometer). Each such component will produce a Doppler shift in the frequency of the radiation emitted into the detector by those atoms, ions, or molecules which move appropriately. At sufficiently high temperatures, these Doppler shifts begin to exceed $1/T_2$ and therefore broaden the line. For rotational and vibrational transitions, the spectral line widths of most molecules at room temperature are largely determined by the Doppler effect. Lines associated with electronic transitions in atoms and ions in the gas phase are also frequently Doppler broadened. Molecules, on the other hand, have so many degrees of freedom that the widths of spectral lines produced by their electronic transitions are often dominated by other broadening mechanisms. Since at any other one given time each radiating species has its own particular velocity, and therefore its own particular Doppler shift, this type of broadening is called T_2^* or *inhomogeneous*. It is sometimes also called *temperature broadening*. See Fig. 10.8 for the influence of temperature on line width; the relationship between T_2 and T_2^* was given in Fig. 8.16a.

In condensed phases it is not possible to decide a priori what the dominant broadening mechanism will be, because the processes that are effective in producing relaxation depend so critically on the nature of the quantum transition and the local environments of the chromophores. The two energy levels that give rise to the R lines of ruby, for example, are due to electronic states of the chromic ion which would be degenerate in the gas phase. When small amounts of Cr_2O_3 are doped into Al_2O_3 to make a ruby crystal, the Cr^{3+} ions enter sites where the local symmetry is less than spherical. The geometrical asymmetry gives rise to an electromagnetic asymmetry, and the associated fields exert forces on the electrons of the chromic ion. The result of these forces is to produce a distortion of the stationary-state electronic eigenfunctions and a change in their energies. One such change removes the degeneracy between the states giving rise to the R lines of ruby (a process called *crystal field splitting*). There being no such things as a perfect crystal, however, each ruby sample has within it a certain amount of strain. This strain produces slight changes in the dimensions of the Al_2O_3 lattice, which are not the same in every unit cell. The amount of the distortion of the electronic wavefunction produced at any given site depends on

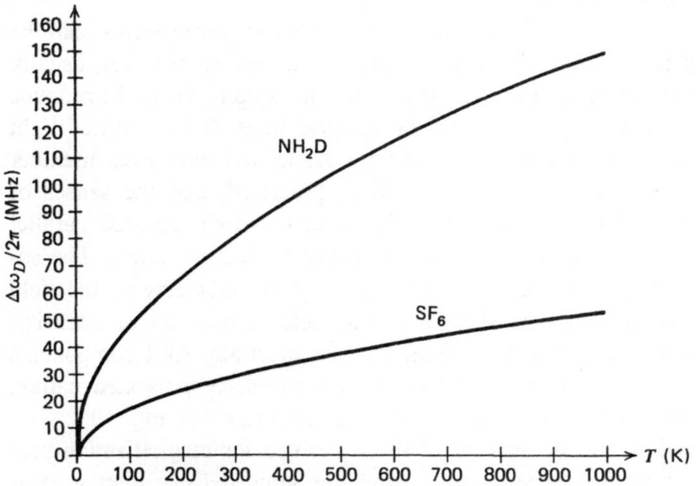

Fig. 10.8. Doppler widths as a function of temperature for NH_2D and SF_6, of their absorption bands at $\lambda = 10.6\ \mu m$ ($\omega_0/2\pi = 2.83 \times 10^{13}$ Hz)

the unit cell dimensions at that location. Therefore inhomogeneous crystal strains give rise to inhomogeneous broadening of the spectral lines associated with the transitions between these levels. Homogeneous broadening of the R lines in ruby is a complex process, involving in part a magnetic interaction between the electron spin of chromium and the nuclear spin of aluminium in adjacent lattice sites. Spontaneous emission and radiationless relaxation both contribute to T_1.

10.6 Optical Analogs to Magnetic Resonance Phenomena

One of the principal differences between magnetic resonance and other kinds of spectroscopy concerns the sign of the magnetogyric or elctrogyric ratio. The various m_s energy levels of a spinning particle are completely nondegenerate in the presence of an external magnetic field H_0, and the sign of the associated γ is a fixed property of each type of nucleus. For this reason, an ensemble of identical spinning particles always absorbs one circularly polarized component of the incident wave and, to all intents and purposes, ignores the other. By way of contrast, at least one of the energy levels associated with transitions between rotational, vibrational, and electronic states is ordinarily degenerate. For this reason, usually at least two different transitions are excited by the same incident electromagnetic wave. It is also ordinarily true that both of these transitions are associated with dipole moments having the same size but rotating in opposite directions. Therefore both circularly polarized components of the exciting wave will correspondingly be of equal amplitude, phased together in such a way that the components normal to the plane of polarization of the exciting wave always cancel vectorially. The overall susceptibility in such cases is therefore isotropic, unlike that of a spinning magnet.

The existence of the constituent circularly polarized components can be proved by removing the degeneracy of the corresponding quantum states, usually by applying an external electric or magnetic field to the sample. This will produce Stark or Zeeman splitting, respectively, of the spectral lines. If the emitted light is analyzed by means of a spectrograph, the light associated with each member of the split pair of lines is found to be circularly polarized, and the senses of polarization of the two lines are opposite. (This refers to light emitted parallel to the direction of the applied field – the normal Stark or Zeeman effect. The so-called anomalous splitting observed in light emitted perpendicularly to the field direction is more complicated.) As the perturbing field is reduced in intensity, the splitting diminishes. Finally, at zero field, the two spectrally split components coalesce in frequency, and the resulting line has no circularly polarized nature. The Zeeman effect as a function of field strength is illustrated in Fig. 10.9.

The equations below are the analogs of those derived for magnetization produced in magnetic resonance experiments, using the same density matrix treatment. They apply to a single circularly polarized component of the polarization wave induced by coherent irradiation of an ensemble of two level-systems un-

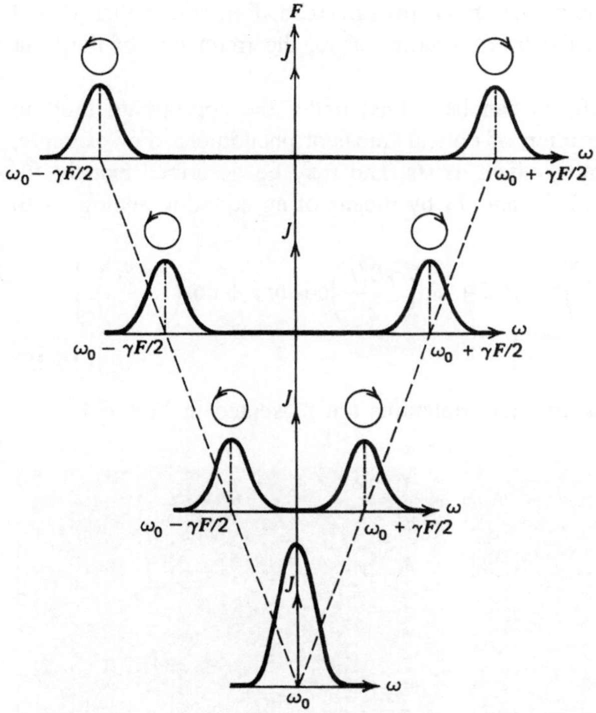

Fig. 10.9. Zeeman or Stark effect as a function of field strenth. The field F may be either magnetic or electric, γ the magneto- or electrogyric ratio, and ω_0 is the frequency of the unpolarized emission at $F = 0$ (bottom curve). For $F > 0$, the spectral line of irradiance J splits into two circularly polarized components of opposite helicity, symmetrically spaced about ω_0 (upper curves). Note that the spacing is proportional to F

dergoing electric-dipole-allowed transtions:

$$\frac{dP_{x'}}{dt} = -\delta P_{y'} - \frac{P_{x'}}{T_2}, \tag{10.31}$$

$$\frac{dP_{y'}}{dt} = -\delta P_{x'} + \frac{\gamma E_1^0}{2} P_z - \frac{P_{y'}}{T_2}, \tag{10.32}$$

$$\frac{dP_z}{dt} = -\frac{\gamma E_1^0}{2} P_{y'} - \frac{P_z - P_0}{T_1}. \tag{10.33}$$

Equations 10.31 to 10.33 are the optical analogs to the Bloch equations 8.177 to 8.179. The definition of P_0 is analogous to that of M_0 (Eq. 8.176):

$$P_0 = \frac{N'\gamma\hbar}{2} \tanh\left(\frac{\gamma\hbar E_z}{2k_0 T}\right). \tag{10.34}$$

The definition of E_z used above was given in Eq. 10.28. Remember that y' and x' are the coordinate axes in the frame rotating at ω, the frequency of the light wave.

Equations (10.31) to (10.33) can be solved under the appropriate limiting conditions to provide a description of optical transient phenomena. For example, the optical analog to the Torrey effect exists, and may be described exactly for times short in comparison with T_1 and T_2 by means of an equation analogous to Eq. 8.205:

$$\mathbf{P}(t) = P_0 \left[\sin\left(\frac{\gamma E_1^0 t}{2}\right)\sin\omega t\, \hat{x} + \sin\left(\frac{\gamma E_1^0 t}{2}\right)\cos\omega t\, \hat{y} + \cos\left(\frac{\gamma E_1^0 t}{2}\right)\hat{z}\right]. \tag{10.35}$$

Oscillograms showing optical transient nutations are presented in Figure 10.10.

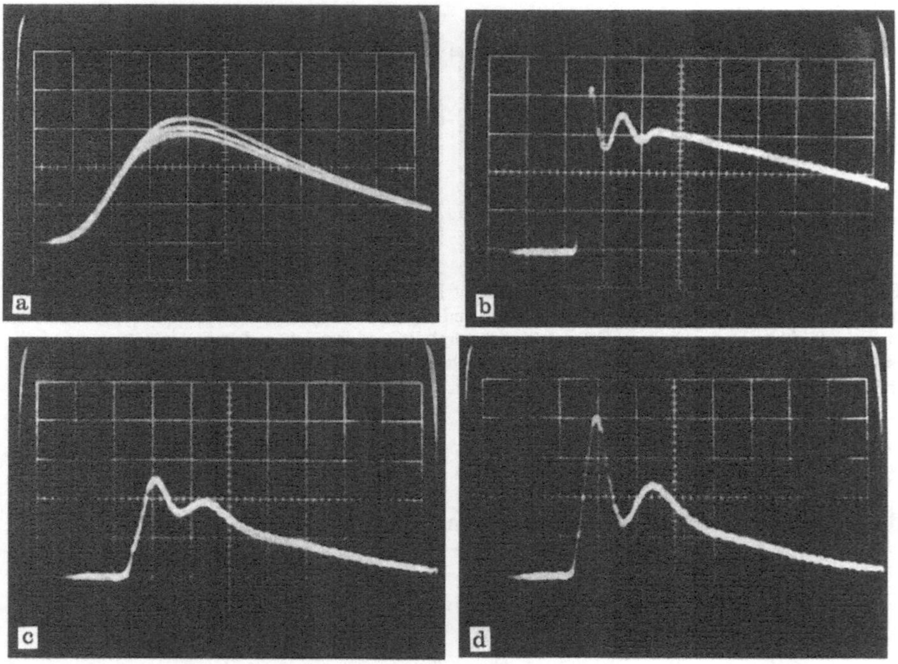

Fig. 10.10. First observation of the optical transient nutation effect, by G.B. Hocker and C.L. Tang (1969) Phys Rev 184:356, Fig 1. The sample was SF_6 gas, and the two quantum states connected by the transition have different amounts of vibrational energy in the mode conventionally designated as ν_3. The Q branch for the band (0,0,0,0,0,0) (all six modes with $v = 0$) $\rightarrow$ (0,0,1,0,0,0) lies at $\omega_0/2\pi = 2.841 \times 10^{14}$ Hz; the Q branch for the "hot band" (0,0,0,0,0,1) $\rightarrow$ (0,0,1,0,0,1) lies at $\omega_0/2\pi = 2.838 \times 10^{14}$ Hz. Since the operating frequency of the CO_2 laser is $\omega/2\pi = 2.830 \times 10^{13}$ Hz, lines from the P branches of these SF_6 bands are responsible for the interaction. Horizontal scale is 50 ns per division **a** Multiple traces of pulses without SF_6 in cell. Peak intensity could be varied from $\sim$ 35 to 6 MW m^{-2} **b** Output through cell for high-intensity pulse. $P_{SF_6} = 0.16$ torr. Vertical scale is 6.65 MW m^{-2} division **c** Output through cell for a low-intensity pulse. $P_{SF_6} = 0.12$ torr. Vertical scale is 3.33 MW m^{-2} division. **d** Same as **c** except that detector was moved slightly across the beam

The optical analog to free-induction decay exists, as in Eq. 8.221. After a $\pi/2$ pulse of duration Υ:

$$\mathbf{P}(t) = P_0\left\{\sin\omega t \exp\left(-\frac{t-\Upsilon}{T_2}\right)\hat{x} + \cos\omega t \exp\left(-\frac{t-\Upsilon}{T_2}\right)\hat{y} + \left[1 - \exp\left(-\frac{t-\Upsilon}{T_2}\right)\hat{z}\right\}. \tag{10.36}$$

Oscillograms showing microwave-induced free-induction decay are given in Fig. 10.11.

Finally, the steady-state spectrum of the coherent spontaneous emission from the sample can be expressed in an equation analogous to Eq. 8.234:

$$\mathbf{P}(t) = P_0\{(\gamma E_1^0 T_2/2)[(\sin\omega t + \delta T_2\cos\omega t)\hat{x} + (\cos\omega t - \delta T_2\sin\omega t)\hat{y}] + (1 + \delta^2 T_2^2)\hat{z}\} \div \{1 + [\gamma^2 (E_1^0)^2 T_2^2/4]T_1 T_2 + \delta^2 T_2^2\} \tag{10.37}$$

The effect of having systems with both positive and negative magnetogyric ratios present in equal numbers in the sample can be expressed by calculating the susceptibility tensor for each type of transition separately and adding the two values together [2].

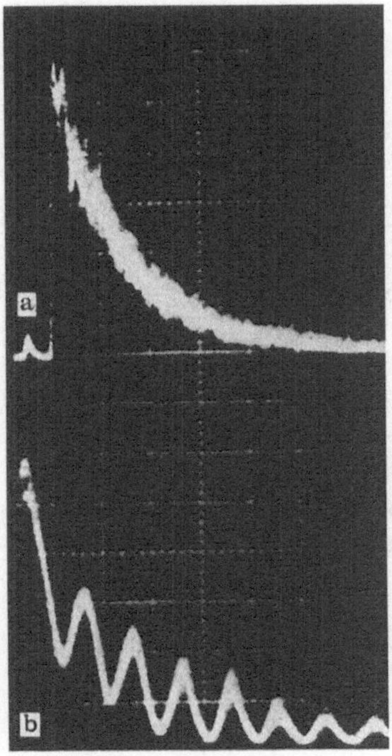

Fig. 10.11. **a** Free-induction decay in the microwave region due to a rotation transition ($J = 0$ to $J = 1$) in a gaseous sample of OCS^{32} molecules Hill et al. (1967) Phys Rev Lett 18:105, Fig. 1. The radiation field was provided by a more or less conventional microwave source producing pulses 100 ns $\rightarrow$ 1 μs in duration, 10 W peak power in the neighborhood of the resonance frequency $\omega_0/2\pi = 12.162972$ GHz. Horizontal scale = 2 μs cm^{-1}, pressure = 3×10^{-3} torr. **b** Same as **a**, but with $\sim 10^{-10}$ W of cw oscillations at $\delta = 500$ kHz added. Time scale = 0.5 μs cm^{-1}; pressure = 20×10^{-3} torr

Define **F** (which may be either **E** or **H**, depending on whether the transitions are electric- or magnetic-dipole-allowed) by the relation

$$\mathbf{F}(t) = (F_y\hat{y} + F_x\hat{x})\exp(-i\omega t). \tag{10.38}$$

The quantities F_y and F_x are complex scalars, presumed to be time independent.

The magnitude of the generalized "certain component" of the transition dipole, either electric or magnetic, is defined by

$$p \equiv \frac{\gamma\hbar}{2}. \tag{10.39}$$

The resonance denominator for the elements of the susceptibility tensor, including saturation under the combined influence of the two field components, is

$$\P \equiv 1 + \delta^2 T_2^2 + \frac{T_1 T_2 p^2(|F_y|^2 + |F_x|^2)}{\hbar^2}. \tag{10.40}$$

Another parameter, i.e. the polarization parameter o of the susceptibillity tensor, must be introduced to describe the combination of the two field components:

$$o = \frac{2T_1 T_2 p^2 |F_y||F_x|\sin(\phi_x - \phi_y)}{\P\hbar^2}. \tag{10.41}$$

In Eqs. 10.40 and 10.41, the complex amplitudes have been expressed by their magnitudes and phases according to

$$F_y = |F_y|\exp(i\phi_y) \tag{10.42}$$

and

$$F_x = |F_x|\exp(i\phi_x). \tag{10.43}$$

The magnetic susceptibility is found to be

$$\chi = \mu_0 \frac{N' p_M^2 T_2 \tanh(\hbar\omega_0/2k_0T)}{\hbar\P_M(1 - o_M^2)}\left[\begin{pmatrix}\delta T_2 & -o_M \\ o_M & \delta T_2\end{pmatrix} + i\begin{pmatrix}1 & \delta T_2 o_M \\ -\delta T_2 o_M & 1\end{pmatrix}\right]. \tag{10.44}$$

(See Eq. 8.245 for an identification of the tensor elements.)

To define the electric susceptibility, in addition to reinterpreting the symbols p and F, one must divide by the electric permittivity of the surrounding medium. The reason for including the permittivity in the numerator but the permeability in the denominator may be found from the difference in definitions between **P** and **M** (see Eqs. 3.15 and 3.21). Therefore

$$\eta = \frac{N' p_E^2 T_2 \tanh(\hbar\omega_0/2k_0T)}{\varepsilon_0\hbar\P_E(1 - o_E^2)}\left[\begin{pmatrix}\delta T_2 & -o_E \\ o_E & \delta T_2\end{pmatrix} + i\begin{pmatrix}1 & \delta T_2 o_E \\ -\delta T_2 o_E & 1\end{pmatrix}\right]. \tag{10.45}$$

For linearly polarized light, $\phi_x = \phi_y$ and $o = 0$ (see Eq. 10.41). The susceptibilities in Eqs. 10.44 and 10.45 become isotropic in this case. It is also easy to show that the intrinsic anisotropies of χ and η are unobservable if the incident light is circularly polarized, unpolarized, or incoherent, the three case that are of greatest interest [2].

10.7 Photon Echoes: Qualitative Discussion

In the case of nuclear magnetic resonance, the wavelength of the exciting radiation is ordinarily very large compared to the dimensions of the sample. For this reason, it is appropriate to ignore the phase modulation of the wave by the propagation factor $\exp(i\omega z/c)$; the sample is assumed to occupy the point (0, 0, 0). In electron spin resonance, on the other hand, the wavelength of the exciting radiation is sufficiently short ($\cong$ 1 cm) so that one may be able to observe propagation effects. These have been observed in ferromagnetic and antiferromagnetic resonance, but are hard to observe in paramagnetic samples [3]. By way of contrast, the usual conditions existing whenever rotational, vibrational, and electronic transitions are excited produce propagation effects because the associated wavelengths are ordinarily much smaller than the sample dimensions. It is of course possible to produce standing waves of very short length in a macroscopic optical cavity; a laser is an example of such a system. The result is to produce a time-independent spatial modulation of the field amplitudes (and therefore saturation parameters), periodic in $2\pi z/\lambda$, which in some cases can be ignored.

Aside from the exceptions enumerated in the preceding paragraphs, transient coherent effects in magnetic resonance ordinarily do not propogate, but their optical analogs do. This fact was utilized in a very clever way by Kurnit, Abella, and Harmann [4] (KAH), the group that first observed photon echoes.

The first step in producing the echo is to supply the sample with two pulses, the first $\pi/2$ and the second π, separated by a time interval t'. The duration of the pulses should be less than T_2^*, and t' should be greater than T_2^* but less than T_2. The pulses should be at exact resonance with the two levels in the sample that must interact with them.

Unfortunately, all of these requirements pose problems. A necessary temperature difference between the ruby crystals used as the laser material and as the sample produces a difference between their resonant frequencies, ω and ω_0. Fortunately, there are two R lines in the ruby spectrum, and KAH were able to utilize one of these lines to produce the laser light and the other to act as the absorber at the frequency $\omega = \omega_0$. Another difficulty is that the required delay between pulses, t', is so short that the laser could not be repumped to threshold in the interval. The problems of synchronization of two different lasers on that time scale are also formidable. Therefore KAH used a single pulse from a single laser and split it into two, employing a partially reflecting–partially transmitting mirror (beam splitter) located at an angle with the optic axis. A delay of one of

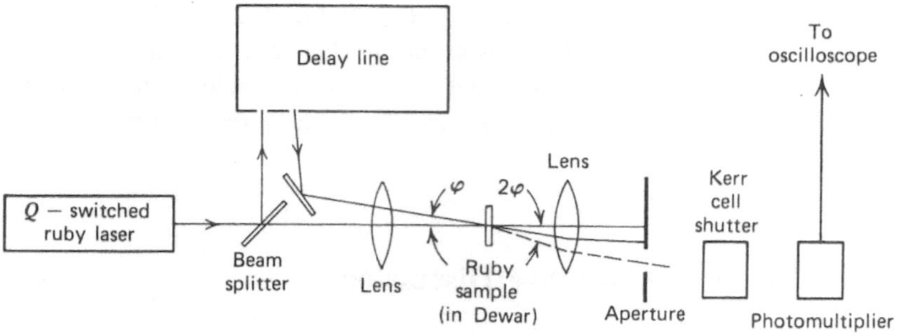

a

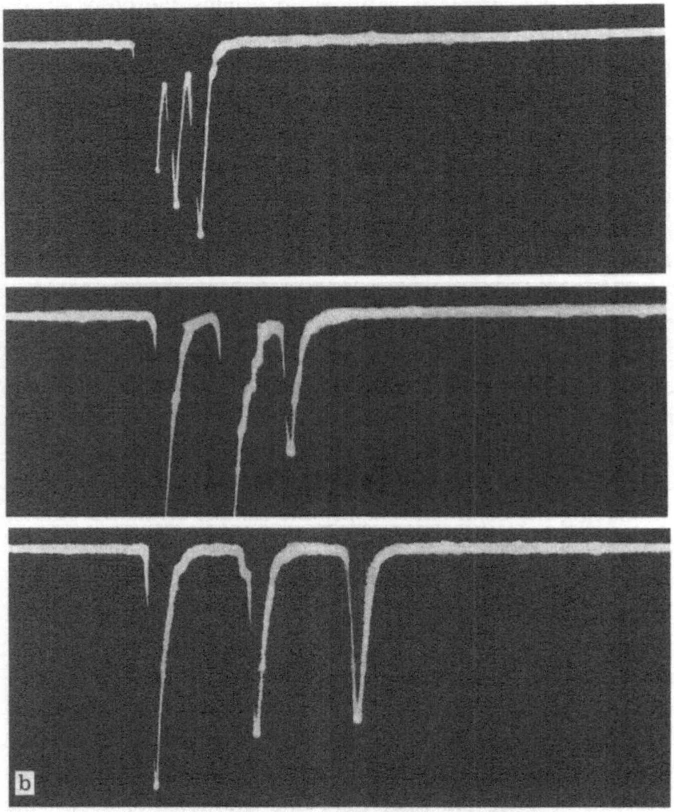

Fig. 10.12. First observation of photon echoes. **a** Experimental arrangement shown is Fig. 2 from the first paper by Kurnit et al. (1964) Phys Rev Lett 13:567. A Q-switch is a device that causes a laser to produce an intense brief pulse. **b** Oscilloscope photographs of the output from the photomultiplier shown in **a**, Ref [46]. Fig. 10. R lines of ruby at nearly exact resonance ($\delta = 0$), $E_1^0 \sim 1.2 \times 10^4$ V m^{-1}, $\omega_0/2\pi = 432$ THz (10^{12} Hz). Pulse duration $\Upsilon \sim 10$ ns (10^{-9} s), $\varphi \sim 3°$. Time increases to right at 100 ns division. The first two pulses are due to scattered light from excitation pulses. The third pulse is from the photon echo

the beams was produced by causing the light to bounce back and forth between a pair of mirrors located a fixed distance apart. The axis of this mirror pair could be adjusted so that the angle of incidence of the laser pulse on the first mirror is varied. The number of round-trip bounces of the pulse before it "walked off" the mirror pair depended on the angle of incidence, so that this part of the apparatus functioned as a variable delay line. Each 300 mm of additional beam path lengthened t' by 1 ns. Because it is very difficult to control a ruby laser, it is unlikely that either pulse corresponded exactly to $\pi/2$ or π. This is not very important, however, since the only advantage to that particular pulse sequence is that the echo amplitude is a maximum and bears a simple relationship to P_0. In general, any pair of pulses separated by t' will produce *some* echo at $t = 2t'$.

The next problem to be solved was that of saturation of the detectors used to record the arrival of the pulse. If an intense beam hits a detector at a time t', that detector will not recover immediately. If a weak pulse (e.g., an echo) arrives before recovery is complete, that pulse might not be recorded. (This problem also plagued early workers in pulsed nuclear magnetic resonance.) This difficulty was overcome by a trick based on the fact that optical coherent transients propogate. The first and second pulses were caused to be incident on the sample at slightly different directions of propagation. Under these conditions, the transmitted pulses, together with any sample luminescence produced in the direction of the beam, could be recorded on separate detectors located at some distance beyond the exit face of the sample.

The most important reason for irradiating the sample at two different angles was that under these circumstances the echo produced propagated in yet a third direction. If the first pulse was incident at an angle with respect to the optic axis $\theta = 0$ at a time $t = 0$, and the second pulse was incident at $\theta = \varphi$ and $t = t'$, the echo emerged at $\theta = 2\varphi$ and $t = 2t'$. The use of different angles of incidence not only avoided detector saturation but also eliminated the possibility that an echo signal might have been a spurious reflection occurring somewhere in the optical train, rather than a true echo.

The experimental setup used to produce photon echoes is presented in Fig. 10.12a.

10.8 Angle of Echo Propagation Using the Gyroscopic Model

Abella, Kurnit, and Hartmann gave a very elegant but highly abstract explanation of the reason for the emergence of the echo at 2φ in their first long paper [4b]. The reader is invited to compare that proof with the one given below, presented by the same authors in a subsequent work [4d]. The later explanation will be seen to be an excellent advertisement for the utility of the gyroscopic model of spectroscopic transitions.

Suppose that the electric field vector associated with the $\pi/2$ pulse is applied in such a way that $\mathbf{E}_1$ makes an angle α_1 with the x' axis in the rotating frame.

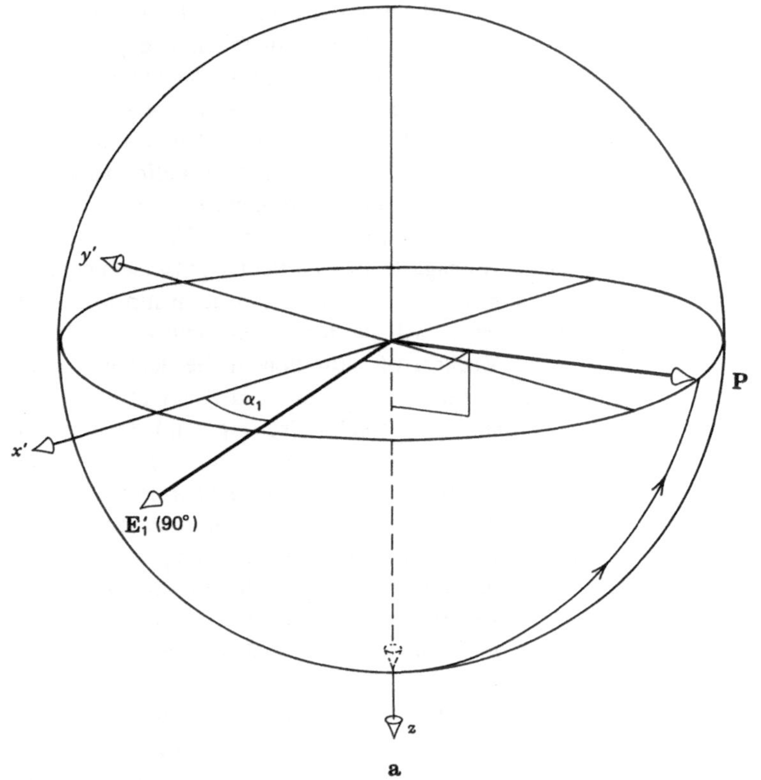

Fig. 10.13. Echo production by propagating waves, viewed in rotating coordinate system. **a** The first laser pulse, E_1', rotates the pseudopolarization vector through an angle of 90°. The angle α_1 depends in part on the direction of propagation of the pulse in the laboratory

The pseudopolarization vector will therefore be rotated into the $x'y'$ plane in such a way that **P** will make an angle $\S = \alpha_1 + \pi/2$ with the x' axis. This is shown in Fig. 10.13a. Because of dephasing processes, the microscopic dipoles $\mathbf{p}_M$ of which **P** is composed begin to spread out in the $x'y'$ plane with a characteristic rate $\sim 1/T_2^*$. Some will remain at the angle $\alpha_1 + \pi/2$ because their resonance frequencies ω_0^0 are identical with ω, the rotational frequency of the coordinate system. Other dipoles move either clockwise (f) or counterclockwise (s) away from $\alpha_1 + \pi/2$, as shown in Fig. 10.13b.

Suppose further that the electric field vector associated with the π pulse is applied along an axis that makes an angle α_2 with the x' axis in the rotating frame. This will have the effect of transferring the stationary microscopic dipoles from the axis making an angle of $\alpha_2 + (\Delta\S/2)$ with the x' axis to an axis that makes an angle of $\alpha_2 - (\Delta\S/2)$ with the x' axis, as shown in Fig. 10.13c. Since $\alpha_2 + (\Delta\S/2)$ is equal to $\alpha_1 + \pi/2$,

$$\Delta\S = 2(\alpha_1 - \alpha_2) + \pi. \tag{10.46}$$

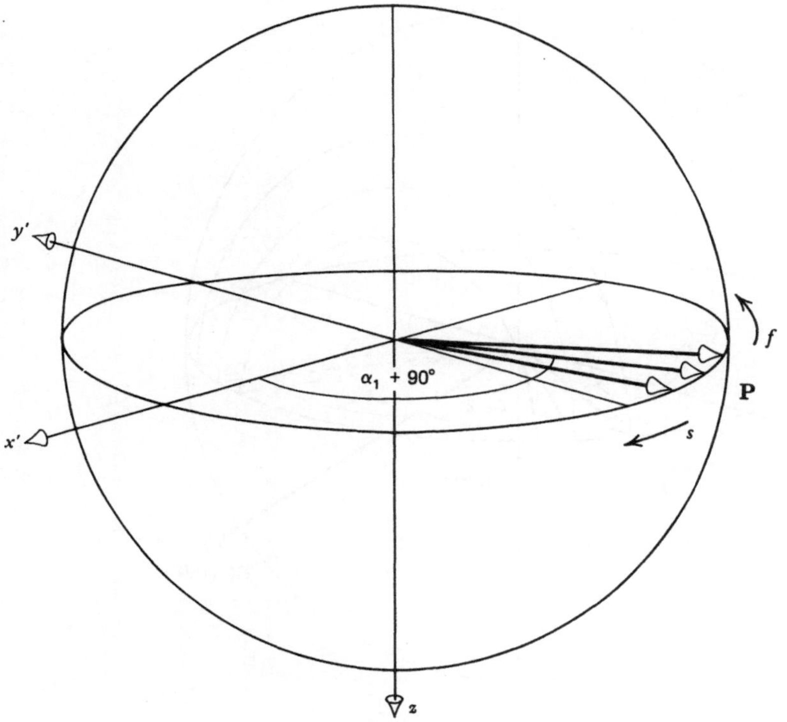

b

Fig. 10.13. (*Continued*) **b** The pseudopolarization wave induced by the first laser pulse decays because of inhomogeneous dephasing. The vector marked f precesses faster than the average in the laboratory pseudoelectric field, E_z. The vector marked s precesses slower than the average. The vector between f and s precesses at the average frequency and is therefore stationary in the rotating frame

It is easy to see from the location of the vectors labeled f and s that all of the microscopic dipoles will come back together at the new location of the stationary dipole, so that the pseudopolarization vector associated with the echo will make an angle of

$$\S_3 = 2\alpha_2 - \alpha_1 - \frac{\pi}{2} \tag{10.47}$$

with the x' axis.

What is the relationship between the angles $\alpha_1, \alpha_2,$ and $\S_3$ and the direction of propagation of the associated pulses in the laboratory frame of reference? These angles are the time-independent parts of the phases of the three oscillating fields;

$$\mathbf{E}_1\left(\frac{\pi}{2}\right) = \hat{k}_1 E_{11}^0 \cos(\omega t + \alpha_1), \tag{10.48}$$

$$\mathbf{E}_1(\pi) = \hat{k}_2 E_{12}^0 \cos(\omega t + \alpha_2), \tag{10.49}$$

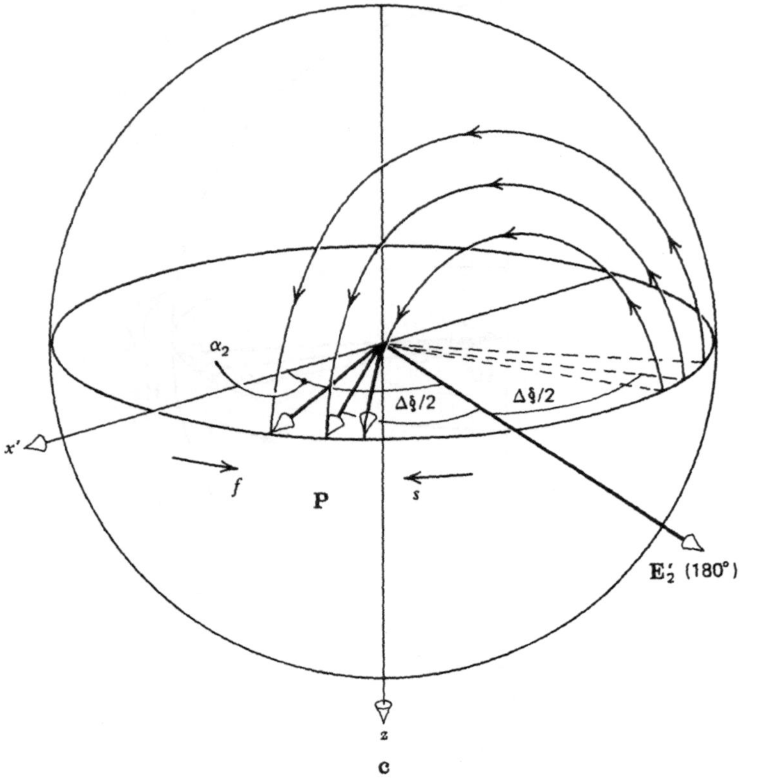

Fig. 10.13. (*Continued*) **c** The second laser pulse, E_z', rotates the pseudopolarization vector through an angle of 180°. The angle α_2 depends in part on the direction of propagation of the pulse in the laboratory. It is easy to see that the vectors will come together again. The angle $\Delta\S$ between the initial and final locations of **P** is related to the direction of propagation of the echo in the laboratory

and

$$\mathbf{P}(\text{echo}) = \hat{k}_3 P_0 \cos(\omega t + \S_3). \tag{10.50}$$

It is easily to be shown [4d] that

$$\alpha_1 = \mathbf{k}_1 \cdot \mathbf{r} + \phi_1, \tag{10.51}$$

$$\alpha_2 = \mathbf{k}_2 \cdot \mathbf{r} + \phi_2, \tag{10.52}$$

and

$$\S_3 = \mathbf{k}_3 \cdot \mathbf{r} + \phi_3, \tag{10.53}$$

Therefore, by combining Eqs. (10.51) to (10.53) with Eq. (10.47) one can find

$$\mathbf{k}_3 \cdot \mathbf{r} + \phi_3 = (\mathbf{k}_2 \cdot \mathbf{r} + \phi_2) - (\mathbf{k}_1 \cdot \mathbf{r} + \phi_1) - \frac{\pi}{2}. \tag{10.54}$$

The terms that contain no dependence on **r** can be equated independently

$$\phi_3 = 2\phi_2 - \phi_1 - \frac{\pi}{2}. \tag{10.55}$$

The **r**-dependent terms yield the relationship

$$\mathbf{k}_3 \cdot \mathbf{r} = (2\mathbf{k}_2 - \mathbf{k}_1) \cdot \mathbf{r}. \tag{10.56}$$

The information contained in Eq. (10.56) will be sufficient to calculate the angles between the propogation vectors. Let

$$\mathbf{k}_1 = k_1 \hat{z}, \tag{10.57}$$

$$\mathbf{k}_2 = k_2(\cos\phi\hat{z} + \sin\phi\hat{x}), \tag{10.58}$$

and

$$\mathbf{k}_3 = k_3(\cos\theta\hat{z} + \sin\theta\hat{x}), \tag{10.59}$$

Then, from Eq. (10.56),

$$\mathbf{k}_3 = (2k_2\cos\varphi - k_1)\hat{z} - 2k_2\sin\varphi\hat{x}. \tag{10.60}$$

Therefore, comparing Eq. (10.59) with (10.60) term by term, one obtains

$$\frac{\sin\theta}{\cos\theta} = \frac{2k_2\sin\varphi}{2k_2\cos\varphi - k_1}. \tag{10.61}$$

The sines and cosines that appear in Eq. (10.61) can be expanded in a power series:

$$\frac{\theta - (\theta^3/3!) + \cdots}{1 - (\theta^2/2!) + \cdots} = \frac{2[\varphi - (\varphi^3/3!) + \cdots]}{2[1 - (\varphi^2/2!) + \cdots] - (k_1/k_2)}. \tag{10.62}$$

If φ and θ are small angles, and if

$$k_1 = k_2, \tag{10.63}$$

Eq. (10.62) yields

$$\theta = 2\varphi + \frac{13\varphi^3}{3} + \ldots. \tag{10.64}$$

10.9 Mathematical Analysis of $\pi/2$, π Echoes

A mathematical description can be given of the free-induction decay of the signal that follows a $\pi/2$ pulse at the center of an inhomogeneously broadened line. First, consider the behavior of an assembly of quantum systems, all with absorption lines centered a distance $\Delta\omega_j^0$ from the peak resonance frequency, ω_0^0. This assembly is sometimes called an *isochromat*. The behavior of the macroscopic polarization vector due to this particular isochromat in the rotating frame will be given by the Bloch equations 10.31 to 10.33. If $t = 0$ is chosen to be the time when the $\pi/2$ pulse terminates, for all $t > 0$, $E_1^0 = 0$. Also, it must be assumed that inhomogeneous (reversible) dephasing is a faster process than homogeneous (irreversible) dephasing, if one hopes to observe an echo. Therefore all terms proportional to $1/T_2$ and $1/T_1$ may be neglected. The results are as follows:

$$\frac{dP_{x'}^j}{dt} = -\Delta\omega_j^0 P_{y'}^j, \tag{10.65}$$

$$\frac{dP_{y'}^j}{dt} = -\Delta\omega_j^0 P_{x'}^j, \tag{10.66}$$

and

$$\frac{dP_z^j}{dt} = 0. \tag{10.67}$$

The boundary conditions after a $\pi/2$ pulse are as follows:

$$P_{x'}^j(0) = 0, \tag{10.68}$$

$$P_{y'}^j(0) = P_0^j, \tag{10.69}$$

and

$$P_z^j(0) = 0. \tag{10.70}$$

Therefore the solutions to Eqs. 10.65 to 10.67 are as follows:

$$P_{x'}^j(t) = -P_0^j \sin(\Delta\omega_j^0 t), \tag{10.71}$$

$$P_{y'}^j(t) = P_0^j \cos(\Delta\omega_j^0 t), \tag{10.72}$$

and

$$P_z^j(t) = 0, \tag{10.73}$$

for all $t > 0$.

To calculate the polarization of the entire sample, one must sum Eqs. 10.71 to 10.73 over all the isochromats:

$$P_{x'} = \sum_j P_{x'}^j(t) \cong -P_0 \int_{-\infty}^{\infty} \sin(\Delta\omega^0 t) f(\Delta\omega^0) d(\Delta\omega^0), \qquad (10.74)$$

$$P_{y'} = \sum_j P_{y'}^j(t) \cong -P_0 \int_{-\infty}^{\infty} \cos(\Delta\omega^0 t) f(\Delta\omega^0) d(\Delta\omega^0), \qquad (10.75)$$

and

$$P_z = \sum_j P_z^j(t) = 0. \qquad (10.76)$$

In Eqs. 10.74 to 10.76, $f(\Delta\omega^0)$ is the normalized line shape function centered at $\Delta\omega^0 = 0$:

$$f_{\max}(u) = f(0), \qquad (10.77)$$

$$f_{\min}(u) = f(\pm\infty), \qquad (10.78)$$

and

$$\int_{-\infty}^{\infty} f(u) du = 1. \qquad (10.79)$$

Ordinarily, $f(u)$ is a Gaussian function of u. However, in the interest of mathematical simplicity, a Lorentzian will be used instead:

$$f(u) = \frac{T_2^*}{\pi[1 + u^2 (T_2^*)^2]}. \qquad (10.80)$$

The reader may verify that $f(u)$ defined in this way satisfies Eqs. 10.77 to 10.79 (The substitution $uT_2^* = \tan w$ will prove useful.) When Eq. 10.89 is substituted into Eqs. 10.74 and 10.75, it is found that $P_{x'}$ is merely the Fourier sine transform of the Lorentzian line shape function, and $P_{y'}$ is the cosine transform. Note that $f(u)$ in Eq. 10.80 is an even function of u and that $\sin ut$ is odd. For this reason, the integral on the right-hand side of Eq. 10.74 is zero:

$$P_{x'} = 0. \qquad (10.81)$$

Equation 10.81 would be true even if a Gausssian or any other symmetric line shape function had been used instead of a Lorentzian. The geometric vector model of the dephasing process also makes Eq. 10.81 obvious. The cosine transform on the right-hand side of Eq. 10.75 can be found in standard tables

$$P_{y'} = P_0 \exp(-t/T_2^*). \qquad (10.82)$$

If a π pulse is applied at time $t = t'$, all of the components of $\mathbf{P}_j$ are transformed in accordance with the following:

$$P_{x'}^j\left(t' + \frac{\Upsilon}{2}\right) = P_{x'}^j\left(t' - \frac{\Upsilon}{2}\right), \tag{10.83}$$

$$P_{y'}^j\left(t' + \frac{\Upsilon}{2}\right)\Upsilon = -P_{y'}^j\left(t' - \frac{\Upsilon}{2}\right), \tag{10.84}$$

and

$$P_z^j\left(t' + \frac{\Upsilon}{2}\right) = -P_{z'}^j\left(t' - \frac{\Upsilon}{2}\right). \tag{10.85}$$

Equations 10.83 to 10.85 may be used in the limit that the pulse width, Υ, goes to zero. Now the same Bloch equations 10.65 to 10.67 are solved with the new boundary conditions, Eqs. 10.83 to 10.85:

$$P_{x'}^j(t) = -P_0^j\{\sin(\Delta\omega_j^0 t')\cos[\Delta\omega_j^0(t - t')] - \cos(\Delta\omega_j^0 t')\sin[\Delta\omega_j^0(t - t')]\}, \tag{10.86}$$

$$P_{y'}^j(t) = -P_0^j\{\cos(\Delta\omega_j^0 t')\cos[\Delta\omega_j^0(t - t')] + \sin(\Delta\omega_j^0 t')\sin[\Delta\omega_j^0(t - t')]\}, \tag{10.87}$$

and

$$P_z^j(t) = 0. \tag{10.88}$$

Equations 10.86 to 10.88 are good for all $t > t'$. They may be used to compute the total $P_{x'}, P_{y'}$, and P_z, in a fashion analogous to that expressed by Eqs. 10.74 to 10.76:

$$P_{x'} = -P_0 \int_{-\infty}^{\infty} \sin[\Delta\omega^0(2t' - t)]f(\Delta\omega^0)d(\Delta\omega^0) = 0, \tag{10.89}$$

$$P_{y'} = -P_0 \int_{-\infty}^{\infty} \cos[\Delta\omega^0(2t' - t)]f(\Delta\omega^0)d(\Delta\omega^0), \tag{10.90}$$

and

$$P_z = 0. \tag{10.91}$$

From Eq. 10.90,

$$P_{y'} = -P_0 \exp\left(-\frac{|t - 2t'|}{T_2^*}\right). \tag{10.92}$$

This may be compared with the form of the echoes actually observed: see Figs. 10.12b and 10.14.

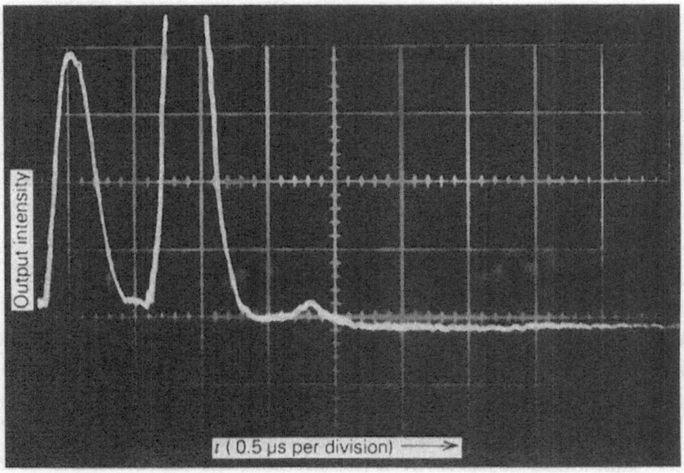

Fig. 10.14. Infrared echoes in gaseous SF_6 excited by a CO_2 laser, Patel CKN, Slusher RE (1968) Phys Rev Lett 20:1087, Fig. 1. The quantum systems and the excitation source were described previously in the caption to Fig. 7.10. Shown is a typical oscilloscope trace of output pulses from SF_6 cell at $P_{SF_6} \cong 0.015$ torr. The first two pulses are transmitted CO_2 laser pulses, and the third is the photon echo. The second laser pulse is off scale by a factor of about 4

10.10 Self-Induced Transparency: Qualitative Discussion

One very interesting example of the importance of the fact that photon echoes propagate but spin echoes ordinarily do not is provided by the self-induced transparency effect. Consider the utter futility of a 2π pulse in magnetic resonance spectroscopy. The system first absorbs and then emits the rf radiation (largely by stimulated emission, although some coherent spontaneous emission is produced and provides the signal), undergoing no net change. The same might be said of a $4\pi, 6\pi, \ldots,$ pulse as well. However, if one can produce just such a pulse for an optical (rotational, vibrational, electronic) transition, the result is no longer trivial or uninteresting.

Suppose that the spectroscopic sample is shaped into a cylinder of such a length that it is essentially opaque at the frequency ω_0; that is to say, the sample is to be much longer than the reciprocal of the Beer's law absorption coefficient. If an ordinary light wave at the frequency ω_0 is directed into one end of the sample so that it travels along the cylinder axis, its intensity will damp to practically zero long before the light rays have approached the opposite (exit) face. If, however, this ordinary light source is replaced by an intense coherent pulse of sufficient brevity so that it succeeds in producing a rotation of the pseudopolarization vector through an angle of 2π before relaxation sets in, there will be no net absorption of energy by the sample.

Imagine the sample to consist of a horizontal stack of differentially thin disks normal to the cylinder axis. The leading edge of the pulse will be strongly absorbed by the quantum system in each lamina. The trailing edge of the pulse will then produce stimulated emission and will force all of the energy back out of the matter into the radiation field. The pulse then passes into the next lamina, where the process is repeated. In this way, the burst of electromagnetic radiation chews its way through the sample, and finally exits, having suffered no net energy loss. Any time that light passes through a material without suffering a reduction in intensity thereby, the material is said to be transparent. Therefore a completely opaque sample has been rendered completely transparent! The point is that the radiation field itself created this situation, and for this reason the phenomenon is called "*self-induced transparency*." It was first predicted and observed by McCall and Hahn [6]. A mechanical analog to self-induced transparency is shown in Fig. 10.15.

There are several interesting features of a propagating 2π pulse. In the first place, it is probably not proper to call it a "light pulse." This term implies that at every instant of time, most of the pulse energy is in the form of electromagnetic radiation. By way of contrast, the 2π pulse is a mixed excitation, in the sense that the energy is shared between the quantum systems and the radiation field. The allocation of energy between these two reservoirs depends on the characteristics of the sample. In particular, the larger the Beer's law absorption coefficient, the larger is the fraction of the energy properly assigned to the quantum systems. For this reason, the propagation of a 2π pulse through matter is acccompanied by a reduction in the velocity of the disturbance. Insofar as the disturbance is purely light, it ought to propagate at the velocity $c_0 = 1/\sqrt{\varepsilon_0\mu_0}$. But if, on the other hand, the energy of the excitation spends 50% of its time residing in the (essentially motionless) quantum system of the sample, the velocity will drop

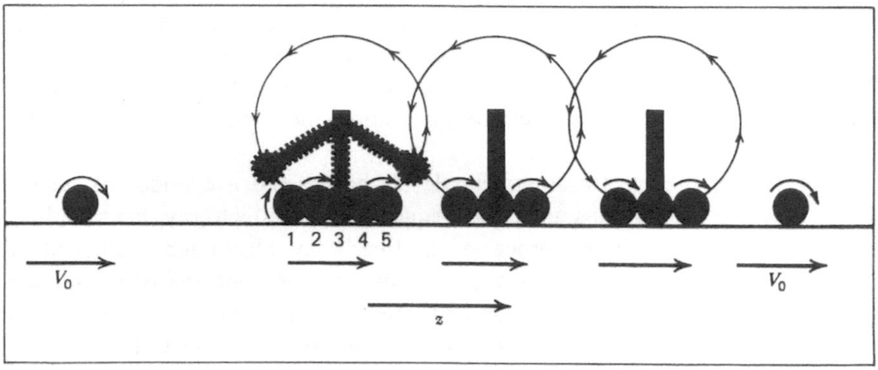

Fig. 10.15. Mechanical analog to self-induced transparency. Pendulum row (sample) is struck inelastically by rolling ball (light pulse), which gives most of its energy to first pendulum ball and slows down. Pendulum turns 360° and spanks rolling ball on backside. Second impact restores energy to rolling ball, and it moves in original direction. Process can repeat along pendulum row. Ref. [6c]

to $c/2$. In principle, there is no lower limit to the velocity. In practice, of course, one dares not permit the energy to reside in any one atom, ion, or molecule as long as T_2. If one were to permit any relaxation, the trailing edge of the pulse would not be able to recover fully the energy delivered to the sample by the leading edge, and the pulse would eventually be damped out. Reductions to velocities of the order $10^{-3}c_0$ have been observed in the laboratory.

What happens if the initial pulse does not correspond exactly to the formula $2n\pi$? If the pulse is *less* than 2π, it will eventually be totally absorbed. If it is greater than 2π (but less than 4π), it is attenuated until it corresponds exactly to a 2π pulse, which then propagates without loss. If it corresponds to more than 4π, the pulse eventually breaks up into a train of 2π pulses, and all the extra energy is absorbed. So far, it has been assumed that the pulse propagates as a plane wave – in other words, that the intensity of a beam propagating in the z direction is independent of x and y. In practice, laser beams are most intense in the neighborhood of the optic axis and die away to zero intensity rather quickly as one's observation post is moved away from the origin of the xy plane. The simplest beam profiles are those produced by a laser firing in what is called a TEM_{00} mode. The beam profile in such cases is very nearly Gausian:

$$J(x, y) = J(0,0)\exp\left[\frac{-(x^2 + y^2)}{a_G^2}\right]. \tag{10.93}$$

Not infrequently, the diameter of the sample is much greater than a_G. This means that, even though the 2π pulse condition is satisfied for any given bundle of rays (e.g., those in the neighborhood of the region $x^2 + y^2 = r^2$), it is not satisfied for any other bundles. Therefore the portion of the beam in the region $x^2 + y^2 > r^2$ is stripped off from the core, and attenuated to zero intensity. The portions in the region $x^2 + y^2 < r^2$ exceed the 2π condition and are damped until (ideally) a cylindrical beam profile is attained.

A clever variant of the self-induced transparency effect enables one to save the entire Gaussian beam. This is called "*zero π pulse propagation*". At first thought, this seems to be a silly idea; surely a pulse that produces rotation through the angle $2n\pi$ with $n = 0$ must be no pulse at all! This is not, however, the case. Suppose that one begins irradiation with a pulse having a Gaussian (or any other) intensity distribution. Suppose further that the intensity at the center corresponds to, say, a $\pi/7$ pulse. At some distance off the axis, another ray corresponds to a $2\pi/31$ pulse. At the midpoint of the pulse, then, a quantum system in the path of the central ray will have its pseudodipole rotated clockwise in the rotating frame through an angle of $\pi/14$. The off-axis ray will have produced a rotation of $\pi/62$ in the pseudodipole in another quantum system at this time. Suppose now that one suddenly shifts the phase of the $\mathbf{E}_1$ wave by 180°. In the rotating frame, this phase shift will have the effect of reversing the direction of the electric field from $(E_1^0/2)\hat{x}'$ to $(-E_1^0/2)\hat{x}'$. Since the pulse is symmetric, the second half will be exactly like the first. However, because the sign of the electrogyric ratio remains the same, the pseudodipole of the quantum system in the path of the central ray

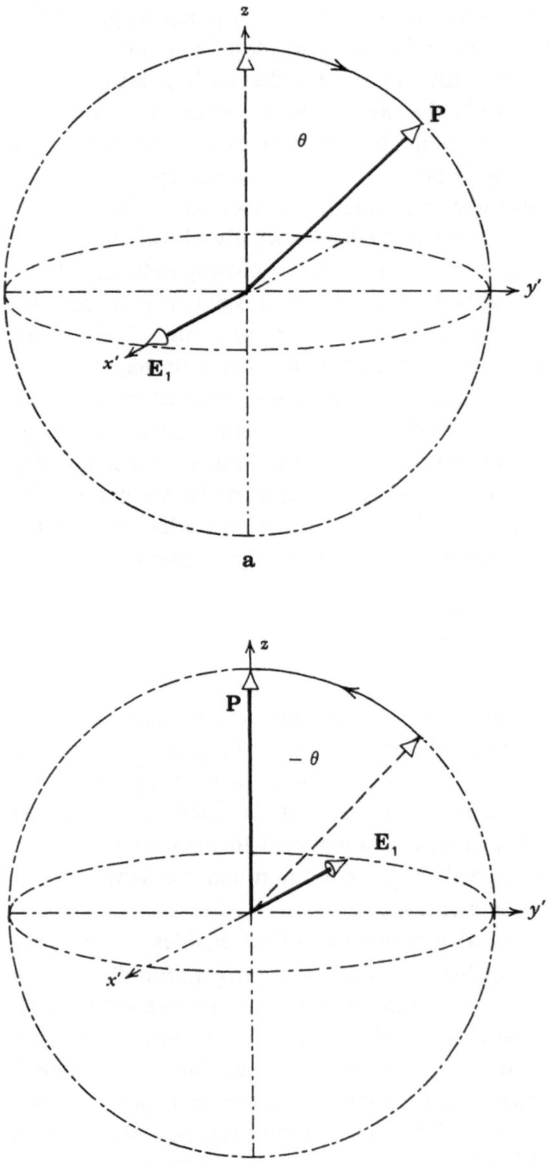

Fig. 10.16. Zero π pulse propagation in the rotating frame. **a** During the first half of the pulse, the electric field vector of the light wave, $\mathbf{E}_1$, causes the pseudopolarization vector, **P**, to rotate clockwise through the arbitrary angle θ in the $y'z$ plane. This corresponds to absorption of energy from the light wave by the sample. **b** During the second half of the pulse, the electric field vector of the light wave changes phase by 180° in the laboratory frame, which reverses its orientation in the rotating frame. This causes the pseudopolarization vector to reverse its direction of rotation in the $y'z$ plane, corresponding to stimulated emission. If the phase shift in $\mathbf{E}_1$ comes in the midpoint of a symmetric pulse, the rotation angle will be $-\theta$ and the sample will be returned to its initial state, with no net absorption of energy

will rotate through $-\pi/14$, and that of the off-axis system, through $-\pi/62$. They will therefore both end up exactly along the z axis, their equilibrium position; all of their absorbed energy will have been returned to the radiation field. See Fig. 10.16. A description of this technique can be found in the work of Grieneisen et al. [7].

Several different methods of producing the 180° phase shift have been employed. In one of them, an electro-optic phase shifter is inserted between the laser and the sample. By carefully synchronizing the electrical pulse that actuates the phase shifter with the firing of the laser, it is possible to switch on the phase shifter while the light filament is exactly halfway through. In another method, the laser pulse is split into two parts by a partially reflecting mirror. One half is directly incident on the sample. The other is bounced around with totally reflecting mirrors and tacked onto the trailing edge of the first half pulse. By properly adjusting the mirror separations, the bounced half can be made to be exactly 180° out of phase with the first half.

10.11 Other Coherent Transient Phenomena

In all experiments designed to observe optical analogs to coherent transient effects in magnetic resonance spectroscopy, one of the major experimental difficulties has been control of the laser pulse. One would like to have the laser fire in a single TEM_{00} mode (no nodes in the electromagnetic wave in any plane that includes the axis of propagation), and yet control the intensity and duration of the pulse over wide ranges. Brewer and Shoemaker [8] have devised a technique for observing transient coherent phenomena that makes this possible. They use a laser that operates continuously. Such lasers can be made to be very stable and forced to fire in practically any desired mode. Brewer and Shoemaker use a gaseous sample having a transient frequency at some $\omega_0 \neq \omega$, but also one that can be shifted in frequency by means of the Stark effect (the Zeeman effect could also be used). The sample is placed between capacitor plates, by means of which the Stark field, $\mathbf{E}_s$, may be applied. The pulse-forming network is then applied, not to the laser beam, but to the source of the Stark field. The laser is fired through the sample, which is ordinarily transparent at the frequency ω. The $\mathbf{E}_s$ pulse is then applied, causing the gaseous molecules to jump suddenly into resonance with the light beam, $\omega_0 \rightarrow \omega_0^s = \omega$. After the quantum systems have interacted with the light wave for the desired amount of time, the $\mathbf{E}_s$ pulse terminates. The Stark shift disappears, and the molecules hop back to their unperturbed frequencies, $\omega_0^s \rightarrow \omega_0 \neq \omega$. The laser beam is still on, of course, but it has ceased to interact with the sample just as it had been switched off. The Stark-shift technique is displayed schematically in Fig. 10.17; experimental results are shown in Figs. 10.18 and 10.19.

The most useful feature of this technique, however, is not that the sample is pulsed rather than the laser. Even more important is the fact that one laser can

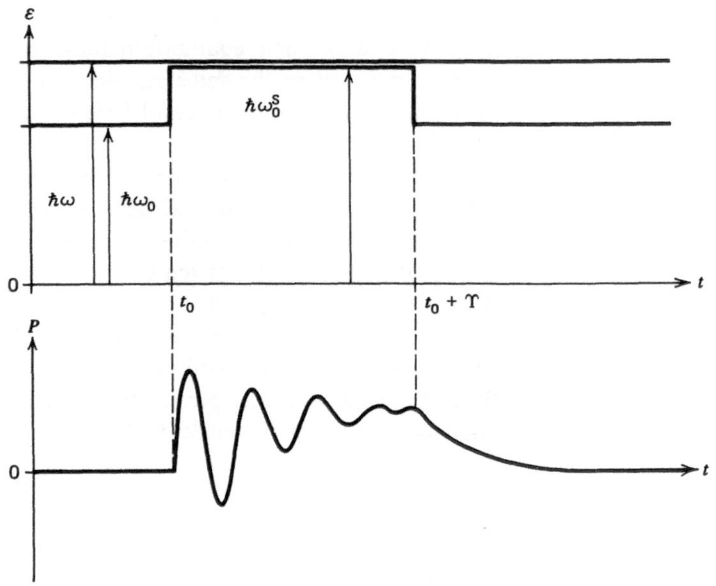

Fig. 10.17. Stark-shifted coherent transients: theory. *At the top*, the energy difference $\hbar\omega_0$ between the quantum states is boosted into resonance with the laser beam $\hbar\omega$ at time t_0 by application of a Stark pulse. The Stark field is then switched off at time $t_0+\Upsilon$. *At the bottom*, the sample polarization is zero until the pulse is applied. At that point, transient nutations begin. Relaxation processes then set in, damping these oscillations toward the asymtotic saturation value of P. (the curve shown is actually the envelope of much more rapid oscillations at the frequency ω_0^s. The drawing is not to scale.) Finally when the Stark pulse is turned off, the sample begines to undergo free-induction decay; now, P is the envelope of oscillations at the unshifted frequency ω_0

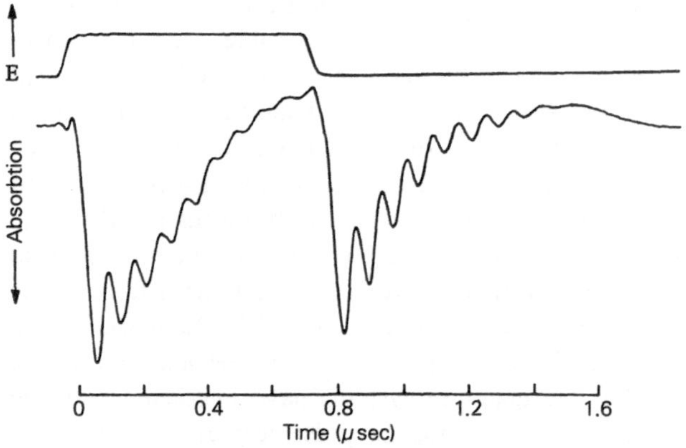

Fig. 10.18. Stark-shifted coherent transients: experiments. Brewer and Shoemaker [(1972) Phys Rev A6:2001, Fig. 4] performed essentially the same experiment as that shown in Fig. 7.17. However, they found it desirable to place one isochromat of the inhomogeneously broadened absorption band in resonance (by means of a dc Stark shift if necessary) at the outset, and then jump to the other side with an additional electric field pulse. In this case, the free-induction decay from the initial isochromat provides an exponentially diminishing base line for the transient nutations undergone by the second isochromat; when the pulse terminates, the same phenomena occur but with reversed roles for the two isochromats. The sample is gaseous NH_2D illuminated by means of a CO_2 laser; the transition is $(v_2, J, M) = (1,5,5) \rightarrow (0,4,4)$. The symbol v_2 refers to the "umbrella" vibration

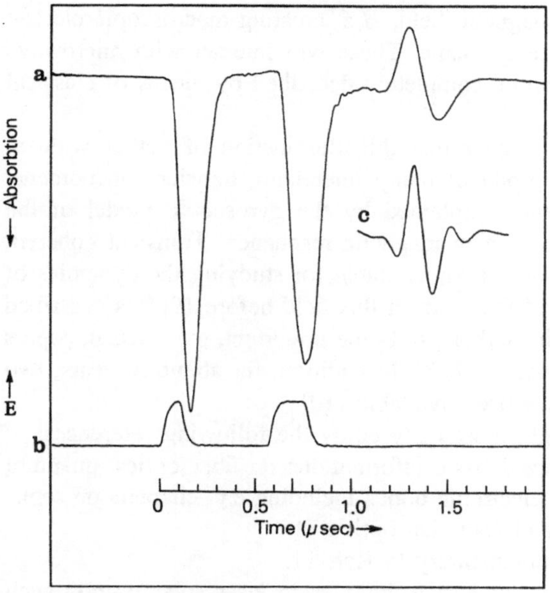

Fig. 10.19. Photon echoes observed by the Stark-shifted technique, Brewer and Shoemaker (1971) Phys Rev Lett 27:631, Fig. 2. The sample is gaseous $^{13}CH_3F$ illuminated by means of a CO_2 laser. The pulses occur at the laser frequency, $\omega_0^s = \omega$, but the echo occurs at the unshifted value of $\omega_0 \neq \omega$. The echo signal is heterodyned with the laser signal by the detector, and the interference beat notes may be seen in (*a*) ($E_S = 3.5$ kV m^{-1}). The interference beats have a higher frequency in *c*) because the Stark shift is larger ($E_S = 6.0$ kV m^{-1})

be used to observe transients in a large variety of samples. Few quantum systems are likely to be found with an ω_0 exactly coincident with the ω for any given laser, and a varying ω is usually quite difficult. On the other hand, $\omega_0^s - \omega_0$. can be varied over a wide range by simply varying the intensity of the Stark field, $\mathbf{E}_s$. One can usually find many quantum systems with ω_0 close enough to ω so that the Stark effect can bring their transitions into resonance with the radiation field.

Laser-induced transient coherent phenomena have been observed for many different kinds of transitions; those between crystal-field-split electronic states, of course, and between vibrational levels of neutral molecules have been the most popular. In addition, these effects have been produced in a large number of systems without the aid of lasers. It has been mentioned already that they were first studied in nuclear magnetic and electron paramagnetic resonance; rf and microwave oscillators are intrinsically coherent sources suitable for these studies. After photon echoes were observed with ruby, a number of people recognized that a large number of systems other than paramagnetic electron spins could be made to produce transitions simply by using ordinary microwave oscillators. For example, echoes have been observed in ferromagnetic resonance, antiferromagnetic resonance, and molecular rotational transitions using these sources. Even cyclotron resonance echoes have been produced in this fashion. In the latter, ions in a plasma circulate around a magnetic field, $H_z\hat{z}$, creating macroscopic electric dipole moments oscillating in the xy plane. These will interact with microwave pulses to produce echoes that can be completely described by means of classical electromagnetic theory.

It has been shown in this chapter that the illumination of matter with an intense coherent beam of light produces many interesting transient phenomena. These phenomena are conveniently explained by the gyroscopic model of the interaction process developed for use in magnetic resonance. Transient coherent experiments are the most dramatic means available for studying the dynamics of spectroscopic transitions. Much of the work in this field before 1968 is described in a bibilography published by the author [9]. Some important, more recent papers have been listed as references [4c, 6c, 7, 8]. In addition, for about 20 years, two fine books on these subjects have been available [10].

For further reading you might thoroughly enjoy the following references:

Ref. 11 presents well-organized basic informations on fiber optics, quantum electronics, optoelectronics, and electronic optics including e.g. chapters on semiconductor microlasers, Fourier and Gaussian optics, etc.

Ref. 12 should be used complementery to Ref. 11.

Ref. 13 describes even most modern techniques in laser spectroscopy such as tunable solid state lasers, optical parametric oscillators, and the production of ultrashort pulses.

Ref. 14 might be most helpful for studies of such solid state matter.

10.12 References

1. Fain VM, Khanin Ya I (1965) Quantum electronics, vol 1 translated by Massey HSH (MIT Press, Cambridge, Mass, 1969), pp 128–139
2. Macomber JD (1971) Appl Opt 10: 2506
3. Bloembergen N, Wang S (1954), Phys Rev 93: 72
4. a. Kurnit NA, Abella ID, Hartmann SR (1964) Phys Rev Lett 13: 567
   b. Abella ID, Kurnit NA, Hartmann SR (1966) Phys Rev 141: 391
   c. Hartmann SR, (April 1968) Sci Am 218: 32
   d. Kurnit NA, Abella ID, Hartmann SR (1966) "Photon echoes in ruby," in Physics of Quantum Electronics, Kelly PL, Lax B, Tannenwald PE, Eds (McGraw-Hill Book Company, New York), p 267
5. Anonymous (1956) Handbook of Chemistry and Physics, 38th ed, Hodgman CD Ed (Chemical Rubber Publishing Company, Cleveland, Ohio), p 275
6. a. McCall SL, Hahn EL (1967) Phys Rev Lett 18: 908
   b. Anonymous (June 1967) Sci Am 216: 57
   c. Anonymous (August 1967) Phys Today 20: 47
   d. McCall SL, Hahn EL (1969) Phys Rev 183: 457
7. Grieneisen HP, Goldhar J, Kurnit NA, Javan A, Schlossberg HR (1972) Appl Phys Lett 21: 559
8. a. Brewer RG, Shoemaker RL (1971) Phys Rev Lett 27: 631
   b. Brewer RG, Shoemaker RL (1972) Phys Rev A6: 2001
   c. Anonymous (December 1971) Phys Today 24: 17
   d. Brewer RG (1972) Science 178: 247
9. Macomber JD (1968) IEEE J Quantum Electron QE-4: 1.
10. a. Sargent M, III, Scully MO, Lamb WE, Jr (1974) Laser Physics (Addison-Wesley Publishing Company Reading, Mass,).

11. Saleh B, Teich M (1991) Fundamentals of photonics (Wiley, New York)
12. Yariv A (1989) Quantum Electronics (Wiley, New York)
13. Demtröder W (1993) Laserspektroskopie, Grundlagen und Techniken (3rd ed. Springer, Berlin)
14. Haken H (1993) Quantenfeldtheorie des Festkörpers (2nd ed. Teubner BG, Stuttgart)

10.13 Problems

10.1 For a transition described by $(\mathsf{p}_x)_{12} = q_0 a_0$, calculate the electrogyric ratio using Eq. 10.20.

10.2 Calcualte $\langle \overline{p_z} \rangle$ for a gas at a pressure of 1 torr, assuming that each molecule carries one chromophore. Let the energy level separation be hc_0/λ, with $\lambda = 1.60\,\mu$m, and let the ensemble be at thermal equilibrium with a heat bath at 300K. (Use Eq. 10.21 or 10.34/N′.)

10.3 Calculate E_z for the system described in Problem 10.2 (Use Eq. 10.28; express the answer in volts per meter.)

10.4 a. Calculate ¶ (Eq. 10.40) for a wave that is linearly polarized in the xy plane at exact resonance with the two-level chromophores. Let $T_1 = T_2 = 1.0$ ns.
b. Show that the parameter $o = 0$ in this case (Eq. 10.41).
c. Calculate the electric susceptibility η by means of Eq. 10.45. What are the units?
d. Calculate the Beer's law absorption coefficient from formulas presented in Chap. 3.

10.5 Suppose that photon echoes are produced by means of colinear $\pi/2$ and π pulses polarized in planes that make angles of 0 and ϕ, respectively, with the xy plane. Show, by a reasoning process analogous to that used in Section 10.8, that the plane of polarization of the echo will make an angle of $2\phi_0$ with the xy plane.

10.6 Perform the integration on the right-hand side of Eq. 10.90, using a Gaussian for $f(\Delta\omega^\circ)$. Compare the result with that presented in Eq. 10.92.

10.7 Find a formula for the radius of a beam having a cylindrical profile that contains the same total power as the beam described in Eq. 10.93. Assume that the peak irradiances, $J(0,0)$, are the same for the two beams.

IV. Applications of LASER and SR Techniques

11 Stimulated Scattering: Third Order Processes

M. Pfeiffer and A. Lau

11.1 Coherent Material Excitation

In this chapter we will restrict our discussion to third order processes, which will be described by Four Wave Mixing (**FWM**) mechanism. In FWM one can generally distinguish two steps: a first step in which a coherent material excitation is performed and a second step in which the excitation is interrogated, yielding a signal wave (fourth wave) in which all information about the dynamics and the energy of the excitation is inherent.

During the past two decades the investigation of dephasing processes has received increasing interest using coherent techniques. These techniques apply laser radiation that interacts via a higher order process with a material, yielding to a transfer of the coherence of the laser radiation to a material excitation that may extend over macroscopic regions.

The coherent excitation can be generated by two different methods. In the first variant, intense laser radiation interacts with the corresponding material and induces a stimulated scattering process. It is built up from random fluctuations by the interaction of the incident and the classically scattered light. This scattered light becomes laser-like after a short starting range and in the space behind this initiating range practically two laser radiations act to excite the material coherently. The laser frequencies differ just by the frequency of the excited material transition. Hereby only those transitions with the highest probability are excited.

In the second variant the process is started by two laser radiations, the frequency difference is chosen equal to the frequency of the material transition of interest. In the second method each wanted transition (even very weak ones) can be excited coherently. Therefore, this method will be considered in this chapter exclusively in investigating stimulated scattering. We have named the method the '*biharmonic pump*'.*

The biharmonic pump combined with the probe and signal waves justifies the above chosen name FWM. In the language of nonlinear optics, FWM is a third order process and it represents the lowest order nonlinear process that may occur in isotropic media.

---

* Having in mind real lasers with finite bandwidths also the notation *bichromatic pump* is applied.

The FWM-technique for investigating material excitation dynamics is also applied under steady state conditions as in the transient regime. In the steady state regime the laser pulse durations are long compared to the characteristic material relaxation. Here quasi cw lasers are applied and the analysis is carried out in the frequency domain. In this case the frequency difference of the biharmonic pump must be scanned through the resonance of the material and its response amplitude is monitored. In transient investigations, lasers with pulse durations shorter than the material relaxation-times are applied for the biharmonic pump (now the carrier frequencies of the two exciting lasers meet the material resonance) and the excitation dynamics is followed by a probing pulse with variable time delay. This time domain technique gives a direct picture of the temporal development of the material reaction process.

All these techniques using the coherence of the excitations give some principal advantages over the spontaneous incoherent spectroscopic techniques. In steady state techniques cw lasers can be applied with extremely narrow bandwidths and so very high resolution of the spectra can be obtained, which goes far beyond the linewidth of the corresponding spontaneous transition. A special advantage of the method of "Coherent Antistokes Raman Scattering" (*CARS*), is the shift of the signal wave out of the spectral range of fluorescence, so that a disturbing overlay of signals of different origin is avoided.

A principal advantage of coherent nonlinear spectroscopy derives from the appearance of interference effects between contributions from different processes of material excitations. Such interference may be used to relate parameters from the one process which is well known to unknown ones. This opens up the way for different new precise measuring techniques. As an example of this, we will discuss the information that CARS gives about the nonlinear susceptibility of electronic origin by its influence upon the observed line shape of the Raman signal. Another example is the phenomenon of beating of contributions from spectrally close neighbouring Raman signals in mixtures of similar molecules, which can be used to determine frequency differences by time resolved spectroscopy. The removal of destructive interference of two terms can be used e.g. to determine the pressure dependence of damping which one of the contributions may be subject to. The available manifold of four different laser-beam-polarizations in FWM-experiments allows one to discriminate between contributions from different scattering tensor components, or to get information about the spatial arrangement of material units in a macroscopic sample. A further example shows how the frequency resolution can be increased beyond the resolution determined by the reciprocal damping constant of the transition by prolonged interrogation.

Steady state investigations of the third order nonlinear susceptibility in condensed and gaseous materials have been carried out increasingly since 1975. Bloembergen et al. investigated different resonant processes occurring within a molecular term scheme. The manifold of terms occurring in third order perturbation theory was systematized by using double-sided Feynman type diagrams (see Sect. 11.8) by Yee and Gustafson and the group led by J.P Taran. The experimental determination of the dispersion of third order susceptibilities from CARS measurements and its relation to $X^{(1)}(\omega)$ dispersion near to molecular electronic resonances was analysed by Albrecht et al.

The transient character of stimulated scattering was first investigated for stimulated Brillouin scattering [1]. The relative long relaxation times of the sound waves in the liquid phase reveal transient behavior even in the 0.5 ns-time regime, i.e. for pulses available as far back as 1965. Much importance was given to the transient techniques concerning molecular vibrations in the condensed phase. The availability of ps-pulsed dye-lasers with tunable frequencies allowed transient investigations of the characteristic vibrational relaxation times, which lie in the range. Beginning in 1970 several theoretical papers about transient stimulated Raman scattering appeared [2–5] initiating a period of intense experimental investigation. With fs-techniques during the last decade, the range of investigations was extended to the analysis of increasingly shorter-living excitations, that are due to other physical processes. Molecular orientation relaxation, Kerr nonlinearity relaxation and effects of coherent population modulation in molecular electronic transitions have been investigated. FWM with fs-pulses has been applied to investigate phonon relaxation processes in semiconductors.

The corresponding experimental techniques are, however, complicated and expensive and are up to now available only in a limited number of laboratories. Mostly these devices have only a limited range of frequencies. Expansion to other frequency ranges by first generating a broadband continuum with help of nonlinear fs-techniques and then filtering and amplifying spectral components, can, in principle, be realized without loss of pulse quality, but needs further complicated experimental setups.

Independent of these generation difficulties, additional problems arise due to linear and nonlinear pulse broadening mechanisms, which take place in all passed mediums.

To avoid some of these problems, since 1980 new approaches have been developed which open up interesting possibilities of short time analysis and this at a lower price. These methods use broad-bandwidth incoherent (or "noisy") light beams with known statistics of time fluctuations. The applied radiation fields have pulse durations t_p much longer than their reciprocal bandwidths $\Delta\omega_p$ and the available time resolution is governed by the correlation time $\tau_c = 1/\Delta\omega_p$. It corresponds to the time durations of the noisy fluctuations. The correlation time of broad bandwidth laser pulses can be sufficiently small to provide sub-ps-time resolution. This permits one to measure short relaxation time in ps- and fs-regions using relatively simple nanosecond lasers.

This chapter will be organized to give first an introduction to the theoretical description of the excitation and probing processes in connection with FWM. Some aspects of steady state investigations will be discussed in detail mainly related to CARS. An excursion will be made to steady state application of FWM to determine longitudinal and transversal relaxation rates in molecular resonant electronic transitions. Excite and probe technique with pulsed lasers in the time domain will concentrate on molecular vibrational transitions and their relaxations for materials in the condensed state. A contribution about stimulated scattering applying incoherent lasers will demonstrate how time resolved measurements in the sub-ps range can be performed applying pulsed lasers in the ns range with a fluctuating time structure of the order of 100 fs.

11.2 Theoretical Background of Four Wave Mixing (FWM)

As pointed out in the introduction of this chapter, FWM can be described as a two-step process: first the generation of a coherent material excitation in the interference field of two traveling electromagnetic waves named a 'biharmonic pump' and secondly the probing of this excitation by a third wave, which results in a new wave, the scattered signal. The biharmonic pump is produced by two laser beams, the pump(p) - and the Stokes shifted(s) laser that can be represented as quasi-monochromatic, plane waves:

$$\mathscr{E}_{\mathrm{p}}(I,t) = A_{\mathrm{p}}(I,t)\, e^{-i(\omega_{\mathrm{p}}t - k_{\mathrm{p}}I)} + C \cdot C. \tag{11.1}$$

and

$$\mathscr{E}_{\mathrm{s}}(I,t) = A_{\mathrm{s}}(I,t)\, e^{-i(\omega_{\mathrm{s}}t - k_{\mathrm{s}}I)} + C \cdot C. \tag{11.2}$$

The A denote the amplitude vectors, slowly varying in space and time compared to the periodicity of the exponential. As material excitation we will discuss here a Raman active molecular vibrational transition characterized by the frequency ω_{vib}. The laser frequencies are chosen so that their difference is nearly ω_{vib}, admitting a small deviation δ of the order of the spectral line width of the transition γ

$$\delta = \omega_{\mathrm{vib}} - \omega_{\mathrm{p}} + \omega_{\mathrm{s}}\,. \tag{11.3}$$

The molecular excitation will be described by the off-diagonal density matrix element for the transition between levels 1 and 2. We assume a dipole forbidden vibrational transition. The Raman like biharmonic excitation proceeds according to

$$\frac{\partial \rho_{21}}{\partial t} + (i\omega_{\mathrm{vib}} + \gamma)\rho_{21} = i\kappa_1 A_{\mathrm{p}} A_{\mathrm{s}}^{*} e^{-i[(\omega_{\mathrm{p}} - \omega_{\mathrm{s}})t - (k_{\mathrm{p}} - k_{\mathrm{s}})I]} \tag{11.4}$$

κ_1 is a constant which may be derived by analyzing the Raman two-photon process via an intermediate virtual level in the molecular term scheme. Introducing the transition amplitude σ according to

$$\rho_{21} = \sigma e^{-i[(\omega_{\mathrm{p}} - \omega_{\mathrm{s}})t - (k_{\mathrm{p}} - k_{\mathrm{s}})I]} \tag{11.5}$$

the material excitation becomes

$$\frac{\partial \sigma}{\partial t} + (i\delta + \gamma)\,\sigma = i\kappa_1 A_{\mathrm{p}} A_{\mathrm{s}}^{*} n_{\mathrm{o}}\,. \tag{11.6}$$

n_{o} is the molecular ground state density.

Expression (5) can be interpreted as a material excitation in a spatial grating (see Chap. 14), given by the k-dependence of the exponential. If $(\omega_p - \omega_s) \neq 0$, it is a grating running in time.

The interaction of the probe wave with the material excitation, given by ρ_{21}, produces a dielectric polarization of 3rd order, serving as a nonlinear source term in the Maxwell equation for the signal wave. Choosing the probe as a plane wave with the parameters ω_{pr}, k_{pr}, the material polarization of the Anti-Stokes shifted component becomes proportional to

$$P_{\text{anti-Stokes}} \propto \sigma A_{pr} e^{-i[(\omega_{pr}+\omega_p-\omega_s)t-(k_{pr}+k_p-k_s)I]} . \qquad (11.7)$$

In the slowly varying amplitude approximation from the Maxwell equation for the signal wave the equation for the signal amplitude is derived

$$\left(\frac{1}{v_{gr}} \frac{\partial}{\partial t} + \frac{k_{pr}}{|k_{pr}|} \frac{\partial}{\partial I} \right) A_{sgn}(I,t) = i\kappa_2 \sigma A_{pr} e^{i\Delta k I} . \qquad (11.8)$$

The expression for the parameter κ_2 results from the linearization procedure of the Maxwellian. Its derivation is given to the reader as an exercise in Sect. 11.10 The signal wave is excited at the frequency

$$\omega_{sgn} = \omega_{pr} + \omega_p - \omega_s . \qquad (11.9)$$

The approximate equation of first order in the derivatives Eq. (11.8) holds down to pulse durations in the order of some 10 fs. We will assume here, that the group velocities v_{gr} of the different fields are equal, so that the overlap of field amplitudes along the interaction range will be maintained.

Small deviations yield a phase mismatch, given by

$$\Delta k = k_{pr} + k_p - k_s - k_{sgn} \qquad (11.10)$$

Special arrangements with properly oriented wave vector directions must be taken to suppress the influence of Δk, if a coherent interaction over an extended spatial range is to be obtained. A fulfillment of the index-matching condition ($\Delta k = 0$) leads to high signal intensities and a spatial separation of the signal beam of interest from scattered waves produced in another manner.

Examples of some typical arrangements applied in FWM experiments to maintain phase matching are given in Fig. 11.1.

The beam angles necessary to fulfill index matching in optical four-wave mixing experiments in condensed materials are about a few degrees, so that in the experiments practically nearly colinearly moving beams occur. This allows us in Eqs. (11.1–11.8) to substitute the space coordinate vector ρr by a coordinate of quasi one-dimensional propagation, taken in the z-direction.

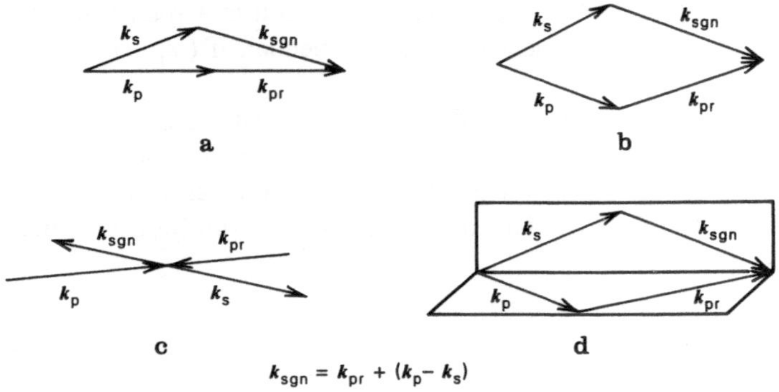

Fig. 11.1. Some typical arrangements of wave vectors used in four wave optical mixing

11.2.1 Extension of the Applicability of the Raman FWM-Equations and the Method of Their Solution

Equations analogous to Eq. (11.6) can be used to describe nonlinear effects of the third order with material reactions of different physical properties. The special form of Eq. (11.6), containing only the time derivative is bound to local processes of the material reaction, i.e. transport processes for the material excitations are ignored. Such a situation holds for many of molecular reactions leading to a non-zero polarization, e.g. rotation or libration of molecules with spatially preferred charge distribution. This results in the Kerr effect, or electronic excitation if one of the pumping frequencies is in resonance with an electric-dipole-allowed molecular transition. In Sect. 11.4 on Nonlinear Polarization Spectroscopy, we will give a short insight into an effect of the latter type. Otherwise, in this chapter we will confine ourselves to molecular vibrational excitation.

The development of the signal wave in the third order is fully described by the two equations (6) and (8). They are applicable for a large range of time-dependent amplitude functions of the lasers ranging from $A_i = \text{const}$ to time variations in the range of 10^{-14} seconds.

The differential operators in Eqs. (11.6) and (11.8) are decoupled by going over to a coordinate system moving with the group velocity of the signal wave. Introducing

$$\xi = z; \quad \eta = t - \frac{z}{v_{gr}} \tag{11.11}$$

these equations are transformed into

$$\left(\frac{\partial}{\partial \eta} + i\delta + \gamma\right)\sigma = i\kappa_1 A_p A_s^* n_{o'} \tag{11.12}$$

$$\frac{\partial}{\partial \xi} A_{sgn} = i\kappa_2 \sigma n_o A_{pr}^* e^{i\Delta\kappa\xi} . \tag{11.13}$$

The instantaneous molecular vibrational amplitude is characterized by the inhomogeneous part of the solution, Eq. (11.12). It reads

$$\sigma(\eta,\xi) = i\kappa_1 n_o \int_{-\infty}^{\eta} d\eta' e^{-(i\delta+\gamma)(\eta-\eta')} A_p(\eta',\xi) A_s^*(\eta',\xi) \quad (11.14)$$

11.2.2 FWM Under Steady State Conditions

For applying FWM for conventional spectroscopic methods, the use of narrow band lasers –, i.e. beams with field amplitudes ≈ constant over the duration of the molecular dephasing time $1/\gamma$, are desirable. The spectroscopic resolution attainable is given by the laser linewidth. Due to a lack of short-pulsed lasers in the first years of FWM-applications, these narrowband lasers conditions were applicable, even for "time-resolved" experiments.

In the steady state, the integration of Eq. (11.14) is carried out easily. One obtains

$$\sigma_{\text{st.st}}(\xi) = i\kappa_1 n_o \frac{A_p(\xi) A_s^*(\xi)}{i\delta + \gamma} \quad (11.15)$$

For the "Raman-amplification" method, one part of the pump laser radiation is used for generating and the other for probing the excitation, and the signal is measured as the intensity-alteration at the Stokes frequency. Here index matching is fulfilled automatically. Carrying out the integration over the interaction length (L) and calculating the absolute value squared, yields the amplified Stokes-Raman intensity

$$I_s(L) = I_s(0) e^{g_{\text{Ra}} I_p L} \quad (11.16)$$

Here $g_{\text{Ra}}(\delta)$ denotes the so-called Raman gain. It is given by

$$g_{\text{Ra}}(\delta) = 2\kappa_1 \kappa_2 |A_p|^2 \gamma \frac{n_o}{\delta^2 + \gamma^2} \quad (11.17)$$

11.2.3 FWM Under Stationary Conditions

Applying pulsed lasers with pulse times in the range of the molecular dephasing time, an integro-differential equation for the rise of the signal wave has to be solved. The measured signal results from the folding of the *driving* pulse amplitudes with the time behavior of the molecular reaction. It may be noted that in the case of Raman amplification it is possible to find a closed solution of the system of integro-differential equations. [1, 3]. A discussion of the excitation efficiency for a pump laser with short time phase jumps is given in Sect 11.7. It describes a mathematical procedure for the solution of this problem.

At this point we will generalize our discussion, introducing the notation of a molecular response function ϕ. This will lead to a modified Eq. (11.14):

$$\sigma(\eta) = \int_{-\infty}^{\infty} d\eta' \, \phi(\eta - \eta') e^{-i\delta(\eta-\eta')} A_p(\eta') A_s^*(\eta') \tag{11.18}$$

If dephasing dominates, ϕ in (11.18) reads:

$$\phi_{\text{Deph}}(\eta' - \eta) = \begin{cases} e^{-\gamma(\eta-\eta')} & \text{for} \quad \eta \geqslant \eta' \\ 0 & \text{otherwise} \end{cases} \tag{11.19}$$

Now we apply the considerations to gases, for which in the low pressure limit Doppler broadening may prevail. ϕ must then be substituted by

$$\phi_{\text{Doppler}}(\eta' - \eta) = \begin{cases} e^{-\left(\frac{\eta-\eta'}{\tau_d}\right)^2} & \text{for} \quad \eta \geqslant \eta' \\ 0 & \text{otherwise} \end{cases} \tag{11.20}$$

with the time constant τ_d governed by the translational temperature T_{TR} of the Maxwellian velocity distribution according to

$$\tau_d = \frac{c}{2\omega} \sqrt{\frac{2M}{kT_{TR}}} \, . \tag{11.21}$$

11.2.4 The Notation of Nonlinear Susceptibility

Equation (11.7) for the signal polarization (derived above for the case of biharmonic excitation of a molecular vibrational transition and probing it by a 3rd wave) is a special case of a process of nonlinear radiation-matter-interaction in the 3rd order. Generally, the 3rd order polarization can be expressed in a phenomenological manner by a time integral over the response function of the material under the influence of three interacting field components, as given in

$$P_i^{\text{sgn}}(I,t) = \int_{-\infty}^{t} dt_1 \int_{-\infty}^{t_1} dt_2 \int_{-\infty}^{t_2} dt_3 R^{(3)}_{ijkl}(I, t - t_1, t_1 - t_2, t_2 - t_3) \mathscr{E}_j(t_1) \mathscr{E}_k(t_2) \mathscr{E}_l(t_3) . \tag{11.22}$$

$R^{(3)}$ comprises several possible processes, which in Eq. (11.22) are assumed to be of a local character. Otherwise, the reaction at position Il would depend on radiation interactions in different space points. For each process of interest the functional form of $R^{(3)}$ may be derived by solving the density matrix-equation in 3rd order perturbation theory within an adequate molecular term scheme. Equation (11.22) is the generalization of expression (11.18) including possible resonances of the exciting and signal frequencies with molecular one-photon-transitions, respectively.

If the material is excited under steady state conditions the corresponding FWM-process can be described by plane monochromatic waves. In this case, the time integration in Eq. (11.22) can be performed. Fourier transformation of the signal polarization results in an expression of the form

$$P_i^{\mathrm{sgn}}(\omega_\mathrm{a}) = D\,\chi_{ijkl}^{(3)}(\omega_\mathrm{a};\omega_\mathrm{pr},\omega_\mathrm{p},\omega_\mathrm{s})\mathscr{E}_j(\omega_\mathrm{pr})\mathscr{E}_k(\omega_\mathrm{p})\mathscr{E}_l(\omega_\mathrm{s})\,. \tag{11.23}$$

$X^{(3)}$ is the nonlinear susceptibility. D is a prefactor bound to the definition of the susceptibility [6]. Its value depends on the degree of degeneration of the applied radiation components.

The generalization of the expression for the material reaction by a general 3rd order susceptibility is of interest in connection with investigations of vibrational spectra, as such additional terms interfere in the measurements. These terms may modify the Raman signal, and information about other molecular processes may be derived from this modification. The consideration of these terms led, e.g. to interesting results about parameters of the so-called electronic background contribution in the resonance CARS-spectra.

The general structure of the 3rd order susceptibility derived from density matrix calculation (See Sect. 11.8 and Chap. 12) has the form

$$X^{(3)} \propto \left(\frac{i}{\hbar}\right)^3 \frac{|\mu|^4}{[i\Delta_\mathrm{signal}^{(1)}+\Gamma]\,[i\Delta_\mathrm{p,s}^{(2)}+\gamma]\,[i\Delta_p^{(1)}\pm\Gamma]}\,n_\mathrm{o}\,, \tag{11.24}$$

where $\Delta^{(1)}$ denotes the frequency mismatch from a molecular one-photon-resonance, relating one to the pump- and the other to the signal-frequency, respectively. $\Delta^{(2)}$ equals the frequency mismatch of a two-photon-resonance, which may be a Raman-resonance or a resonance concerning a molecular two-photon absorption process.

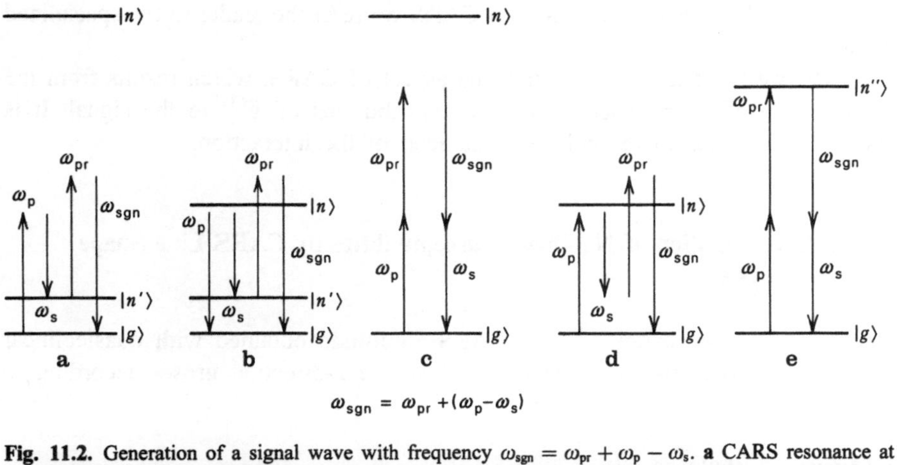

Fig. 11.2. Generation of a signal wave with frequency $\omega_\mathrm{sgn} = \omega_\mathrm{pr} + \omega_\mathrm{p} - \omega_\mathrm{s}$. **a** CARS resonance at excitation far off an electronic resonance, **b** CARS with electronic resonance, **c** non-resonant FWM **d** electronic transition under the action of two nearly resonant fields, **e** two photon absorption resonance

Some typical processes of interest in the discussion of this chapter are characterized in Fig. 11.2.

11.3 Coherent Antistokes Raman Spectroscopy

One important variant of coherent FWM is the method of Coherent Antistokes Raman Scattering (CARS). Equation (11.7) shows the time-dependence of the signal wave in this case. CARS has numerous improved experimental possibilities, compared to spontaneous Raman scattering and to other methods of non-linear Raman spectroscopy.

In comparison to incoherent spectroscopic methods, there is an improvement of the spectral resolving power that under steady state excitation conditions is given by the linewidth of the exciting and probing lasers only. Another important advantage is the Antistokes shift of the frequency of the probe-signal into a spectral region which is free from fluorescence. This allows the investigation of highly fluorescent samples.

One advantage of CARS is related to the use of a special geometry for the exciting and probing beams, which permits a spatial localization of the measured interaction range in an extended sample. Due to the spatially confined overlap region of the beams and the directional selectivity of the signal beam, it is possible to reduce the active sample volume to some $(\mu m)^3$. Figure 1d characterizes schematically the geometrical arrangement of the beams. So a local temperature analysis within flames could be performed by an analysis of rotational Raman intensity distributions [7].

The fourfold manifold of choosable beam polarizations in CARS allows the analysis of all symmetry properties of the scattering tensor, which is otherwise impossible.

For details of these advantages of CARS we refer the reader to the specialized literature [8, 9].

Here we want to treat in detail one aspect of CARS, which results from the phenomenon of interference of several contributions of $X^{(3)}$ to the signal. It is also a consequence of the coherent character of the interaction.

11.3.1 Determination of Nonlinear Susceptibilities by CARS Line Shape Analysis

Under steady state conditions the CARS amplitude obtained with quasicolinear irradiation of two monochromatic waves in the z-direction grows according to the relation

$$\frac{dA_{\mathrm{CARS}}}{dz} = i\frac{2\pi\omega_{\mathrm{CARS}}}{c}X^{(3)}A_{\mathrm{pr}}A_{\mathrm{p}}A_{\mathrm{s}}^{*}e^{i\Delta kz} \text{ with } \Delta k = k_{\mathrm{p}} - k_{\mathrm{s}} + k_{\mathrm{pr}} - k_{\mathrm{CARS}}\,. \quad (11.25)$$

The integration results in the expression for the signal intensity

$$I_{\mathrm{CARS}} \propto |A_{\mathrm{CARS}}|^2 = \left(\frac{\pi\omega}{c}\right)^2 |X^{(3)}|^2 l^2 |A_{\mathrm{pr}}||A_{\mathrm{p}}||A_{\mathrm{s}}| \left(\frac{\sin\frac{\Delta kl}{2}}{\frac{\Delta kl}{2}}\right)^2 , \qquad (11.26)$$

with l the interaction length in the sample. The sin-term expresses the signal dependence on phase-mismatch, it may be set equal to one if the phase-matching condition is fulfilled.

It is important to notice, that the CARS-signal is proportional to the squared modulus of $X^{(3)}$. This results in a characteristic difference of the observed spectral line shape of the CARS signal, compared to the two-wave techniques like spontaneous Raman scattering, Raman-gain or Raman-loss spectroscopies. In the latter methods the imaginary part of $X^{(3)}$ is responsible for the signal, yielding a Lorentzian line shape for the Raman lines, coming from $X_{\mathrm{Raman}} = \chi_{\mathrm{ra}}/(\delta - i\gamma)$.

Investigating samples in the condensed phase, usually several additional contributions to $X^{(3)}$ have considerable magnitude, which cannot be neglected. They are essentially of electronic origin and usually lumped together in X_{NR}. Mostly they result from processes with fs-relaxation times, which cause a broadband spectrally structureless feature.

The experimental findings for CARS spectra in a liquid can be described by a susceptibility of the form

$$X^{(3)} = X_{\mathrm{NR}} + \frac{\chi_{\mathrm{ra}}}{\delta - i\gamma} . \qquad (11.27)$$

At excitation far from electronic resonance both X_{NR} and χ_{ra} are positive and real valued.

The interference of both susceptibility contributions results into the characteristic CARS-line shape

$$I_{\mathrm{CARS}} = A + \frac{B\delta + C}{\delta^2 + \gamma^2} , \qquad (11.28)$$

with $A = X_{\mathrm{NR}}^2$, $B = 2X_{\mathrm{NR}}\chi_{\mathrm{ra}}$ and $C = \chi_{\mathrm{ra}}^2$. The line-shape according to Eq. (11.28) for an isolated Raman line, assuming positive real values of X_{NR} and χ_{ra}, is given in Fig. 11.3.

On a basic background line given by X_{NR}^2, there appears a dispersion-like Raman signal, the positions of the peak and the dip lie a small distance from the point of strong Raman resonance at the positions

$$\delta_{1,2} = \frac{C}{B} \pm \sqrt{\left(\frac{C}{B}\right)^2 + \gamma^2} . \qquad (11.29)$$

In solutions with a low solute concentration, the dispersion structure of the corresponding CARS-line is symmetric and the peak and dip have the distance $\delta_{\max} - \delta_{\min} = 2\gamma$ from each other. The dip is shifted to that side of the spectrum,

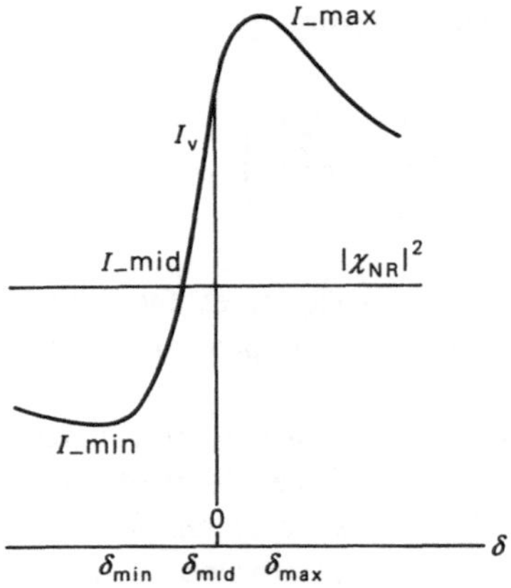

Fig. 11.3. Characteristic CARS line-shape for an isolated Raman line of a molecule in a solvent

where the probing laser is situated. If the solute contribution is stronger than that of the solvent-background, the CARS signal approximates a Lorentzian-shape. Generally, the coherent interference signal results in an alteration of the Raman line shape if the concentration ratio of solute to solvent is changed. The effect is at a maximum if both contributions are of comparable magnitudes. The alteration can be used to find out the value of one unknown contribution if the other is known. This opens up the possibility of using CARS line-shape-analysis as an effective method to find out unknown susceptibility contributions. This can be a susceptibility of an electronic nature, originating from the Raman active sample itself or from other species possibly present in a solution.

The special form of the CARS line shape, given in Fig. 3, was due to the conditions of the real values $X_{NR}, \chi_{ra} > 0$. Other forms appear, if the Raman amplitude or background-contribution become negative or complex valued, as it happens at excitation near to further molecular resonances. This fact complicates the line shape dependencies on the solute concentrations by an additional dependence on the frequency difference of the exciting lasers from a molecular resonance position. It offers, however, the possibility of obtaining much more information on molecular parameters, than is possible with other non-coherent-techniques (as e.g. spontaneous resonance Raman spectroscopy). In the past, CARS line-shape analysis was successfully applied to determine resonant susceptibility contributions as well as their dispersions in scanning the exciting- and probe-fields through a molecular one-photon resonance. Typically in these cases the solute is excited under electronic resonance conditions, whereas the solvent remains under off-resonance conditions. The CARS susceptibility then becomes

$$X_{CARS} = f_1 X_{NR}^{\text{solvent}} + f_2 \left[b' - ib'' + \frac{R - iJ}{\delta - i\gamma} \right] . \tag{11.30}$$

Here f_1 denotes the mole fractions of the two components in a binary mixture. By computer fit of the experimental line shapes for different concentrations (f_2/f_1), all the molecular parameters appearing in Eq. (11.30) can be obtained. The values of b', b'', R and J are taken approximately as constants over the spectral range of a single Raman line. Usually they change their values, when the pump- or probe-frequencies are alterated.

An example of this method is given in Fig. 11.4. It shows, for the 1654 cm^{-1} line in a Rhodamine 6G solution, how sensitive the CARS line-shape reacts under the variation of the relative concentrations and the laser frequencies.

The measurements are done for slightly different probe-laser frequencies of 608 nm and 604 nm. The exciting pump lies at 694.3 nm. For both probing wave lengths the observed line shape changes from a 'resonant'-type at low concentration into a 'non-resonant' one at higher dye concentration. This alteration goes on but in both cases over a completely different form of the intermediate symmetric line shape: an emission-form (peak) in the 608 nm case and an absorption-form (dip) in the 605 nm case. The extremely favourable constellation of parameters

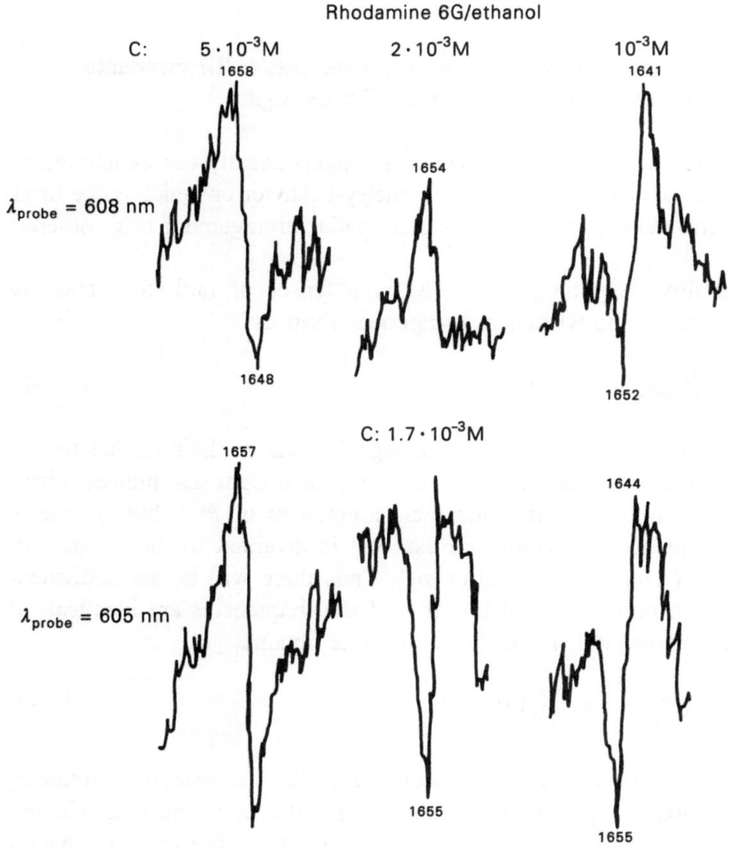

Fig. 11.4. CARS line shape alteration on concentration variation for two slightly different probe wavelenths (Rhodamine 6G in ethanol; $\lambda_p = 694.3$ nm, $\lambda_{pr} = 608$ and 605 nm, $\lambda_{abs}^{max} = 531$ nm) [10]

under the conditions of Fig. 11.4 allows a precise determination of the Rhodamine 6G susceptibility values.

This method of CARS-line-shape-analysis found several further applications. So it was applied to find out the dispersion of the modules and the phases for the nonlinear electronic susceptibilities of some organic polymer samples. The method completes standard methods in determining 3rd order susceptibilities, as 3rd harmonic generation and the Electric Field Induced Harmonic Generation (*EFISH*) [11]. Resonance conditions in these different techniques naturally are different, but the magnitudes of the obtained susceptibilities can be related to each other by an analysis of the actual resonance denominators.

The method of concentration variation in binary mixtures has been recently used to find out susceptibilities in thin layers. In this case, a double layer arrangement was applied in which the contribution from the material with known parameter values was varied by changing its layer-thickness. By this method $X^{(3)}$-values for organic molecules in layers in the range of some nm thickness could be measured with the help of the CARS technique [12].

11.3.2 Application of Special Polarization Arrangements to Discriminate Between Different Contributions to the CARS Signal

The four field polarizations which may be chosen independently can be applied to discriminate signals, which differ in their symmetry-behavior or which come from material components with a specially ordered spatial arrangement (e.g. ordered layers).

The susceptibility describing the FWM is a tensor of rank four and the material polarization can be written in component form as

$$P_i^{(3)} = X_{ijkl}^{(3)}(\omega_1, \omega_2, \omega_3) A_j^1 A_k^2 A_l^3 \tag{11.31}$$

were now 1, 2, 3 denote pump, Stokes and signal fields or their conjugates. In Eq. (11.31) the summation is carried out over all repeated Cartesian indices. There will be 27 terms in this summation for each component of $P^{(3)}$, but symmetry restrictions greatly reduce this number. First $X^{(3)}$ is invariant to the six permutations of the pairs $(j, \omega_1), (k, \omega_2)$ and (l, ω_3). Thus there will be six equivalent terms for any chosen order of the fields. If n of the frequencies are identical, $n!$ of these permutations are indistinguishable and one obtains

$$P_i^{(3)} = \frac{6}{n!} X_{ijkl}^{(3)}(\omega_1, \omega_2, \omega_3) A_j^1 A_k^2 A_l^3 \tag{11.32}$$

Further $X^{(3)}$ must be transformed according to the macroscopic symmetry properties of the medium. For a crystal, this means that $X^{(3)}$ must remain unchanged by any of the symmetry operations appropriate to the crystal. The restrictions resulting from this and the decomposition of the tensor elements into the symmetry species of the symmetry classes are given in [6, 13].

Of much interest is the case of isotropic media as liquids or gases. Here the tensor X_{ijkl} has 21 nonzero elements of which only three are independent. They are the elements with $ijkl$:

$$\begin{aligned} 1111 &= xxxx = yyyy = zzzz \\ 1122 &= xxyy = xxzz = yyxx = yyzz = zzxx = zzyy \\ 1212 &= xyxy = xzxz = yxyx = yzyz = zxzx = zyzy \\ 1221 &= xyyx = xzzx = yxxy = yzzy = zxxz = zyyz \end{aligned} \tag{11.33}$$

and the following relation holds

$$X_{111} = X_{1122} + X_{1212} + X_{1221}. \tag{11.34}$$

The combination of the permutation and isotropy restrictions greatly simplifies the summation in Eq. (11.31) and, in the general case of Four-Wave-Mixing, we obtain

$$\begin{aligned} P_i^{(3)} &= 6X_{1122}^{(3)}(\omega_1, \omega_2, \omega_3)A_i^1 A_j^2 A_j^3 + 6X_{1212}^{(3)}(\omega_1, \omega_2, \omega_3)A_j^1 A_i^2 A_j^3 \\ &+ 6X_{1221}^{(3)}(\omega_1, \omega_2, \omega_3)A_j^1 A_j^2 A_i^3 . \end{aligned} \tag{11.35}$$

For CARS with coinciding pump- and probe-waves two of the frequencies are identical. So we have to divide by 2! and also equate X_{1122} to X_{1212} because of permutation symmetry. This gives

$$P_{\mathrm{CARS}}^{(3)} = [6X_{1122}^{(3)}(\omega_\mathrm{p}, \omega_\mathrm{p}, -\omega_\mathrm{s})e_\mathrm{p}(e_\mathrm{p} * e_\mathrm{s}) + 3X_{1221}^{(3)}(\omega_\mathrm{p}, \omega_\mathrm{p}, -\omega_\mathrm{s})e_\mathrm{s}]A_\mathrm{p}^2 A_\mathrm{s}^* . \tag{11.36}$$

Considering the microscopic origin of the Raman susceptibility, different terms in Eq. (11.33) are directly related to the symmetry properties of the Raman scattering tensor of the individual Raman active molecules. The single molecule is characterized by its Raman hyperpolarizability γ_{ijkl} which is connected to the nonlinear susceptibility according to

$$X_{ijkl}^{(3)}(\omega_\mathrm{a}; \omega, \omega_1, -\omega_2) = n_\mathrm{o} L^4 \langle \gamma_{ijkl}(\omega_\mathrm{a}; \omega, \omega_1, -\omega_2) \rangle \tag{11.37}$$

Here n_o is the density of molecules forming the medium, the angled brackets denote the average of the orientations which are assumed to be isotropically distributed. $L = (n^2 + 2)/3$ is the Lorentzian factor correction for the acting internal field. γ_{ijkl} can be related to the molecular polarizability α_{kl}, parametrically depending on the normal coordinate of the vibrational motion. The normal mode describing the respective Raman transition is a harmonic oscillator. Its amplitude Q_σ is driven by the biharmonic pump field according to

$$\frac{d^2}{dt^2}Q_\sigma + 2\gamma \frac{d}{dt}Q_\sigma + \omega_{\mathrm{vib}}^2 Q_\sigma = \frac{1}{2\mathrm{M}} \frac{\partial \alpha_{kl}}{\partial Q_\sigma} A_k^1 A_1^{2*} e^{-i(\omega_1 - \omega_2)t - (k_1 - k_2)z} \tag{11.38}$$

A probe wave is scattered by this excitation yielding a polarization at the Anti-stokes frequency ω_3

$$P_{\mathrm{m}} = \frac{\partial \alpha_{\mathrm{mn}}}{\partial Q_\sigma} Q_\sigma A_{\mathrm{n}}^3 e^{-i(\omega_3 t - k_3 z)} . \tag{11.39}$$

Inserting the steady solution of Eq. (11.38) into Eq. (11.39) and performing the ensemble average results in

$$X^{(3)}_{Ra-ijkl} = \frac{n_o L^4}{48 M \omega_{\mathrm{vib}}} \left\langle \left(\frac{\partial \alpha_{ij}}{\partial Q_\sigma} \right) \left(\frac{\partial \alpha_{kl}}{\partial Q_\sigma} \right) \right\rangle \frac{1}{\delta - i\gamma} . \tag{11.40}$$

Introducing the three invariants of the molecular Raman scattering tensor as

$$\text{trace} \quad b = \frac{1}{3} Sp \left(\frac{\partial \alpha_{ik}}{\partial Q_\sigma} \right), \tag{11.41}$$

$$\text{anisotropy} \quad g^2 = \frac{3}{2} \left\{ Sp \left[\left(\frac{\partial \alpha_{ij}}{\partial Q_\sigma} \right) \left(\frac{\partial \alpha_{jk}}{\partial Q_\sigma} \right) \right] - 3b^2 \right\} \tag{11.42}$$

and the antisymmetric part of the anisotropy

$$g_{\mathrm{a}}^2 = -\frac{1}{4} Sp \left\{ \left[\left(\frac{\partial \alpha_{ij}}{\partial Q_\sigma} \right) - \left(\frac{\partial \alpha_{ji}}{\partial Q_\sigma} \right) \right] \left[\left(\frac{\partial \alpha_{jk}}{\partial Q_\sigma} \right) - \left(\frac{\partial \alpha_{kj}}{\partial Q_\sigma} \right) \right] \right\} \tag{11.43}$$

these invariants enter the Raman susceptibility according to

$$X_{ijkl} \propto \left(b^2 - \frac{2}{45} g^2 \right) \delta_{ij} \delta_{kl} + \left(\frac{1}{15} g^2 + \frac{1}{6} g_{\mathrm{a}}^2 \right) \delta_{ik} \delta_{jl} + \left(\frac{1}{15} g^2 - \frac{1}{6} g_{\mathrm{a}}^2 \right) \delta_{il} \delta_{kj} . \tag{11.44}$$

The δ_{ij} herein denote the Kronecker symbols.

For the background susceptibility we assume here an excitation far off any electronic resonance. In this case the so-called Kleinmen symmetry holds, according to which X_{NR} is invariant against permutations about all index-pairs. Then all three terms in Eq. (11.33) are identical and the relation

$$X_{\mathrm{NR}-1111} = 3 X_{\mathrm{NR}-1122} \tag{11.45}$$

holds.

11.3.3 Background Suppression by Suitably Chosen Field Polarizations

In CARS spectroscopy the detectability of a Raman signal for a solute in a solution may deteriorate due to a dominating background contribution from the solvent. Considering the different symmetry behavior of the Raman and the background susceptibilities, the latter may be suppressed by suitably chosen field polarizations and properly chosen signal analysator settings. In the case of a neglectable antisymmetric anisotropy, the symmetry of the Raman scattering tensor is characterized by the relation of the trace to anisotropy terms. For identical probe and pump frequencies, one can define a depolarization ratio ρ as in spontaneous Raman spectroscopy

$$\rho = \frac{3g^2}{45b^2 + 4g^2} . \tag{11.46}$$

As for the background contribution Kleinman's symmetry holds the material polarization at the CARS frequency can be expressed as

$$P^{(3)}(\omega_a) = [X_{NR-1111}P_{NR} + X_{Ra-1111}P_R]A_1^2A_2^* , \tag{11.47}$$

where

$$P_{NR} = 2e_1(e_1e_2^*) + (e_2^* + e_2^*(e_1e_1)); P_R = 3(1-\rho)e_1(e_1e_2^*) + 3\rho e_2^*(e_1e_1) . \tag{11.48}$$

e_i are the unit vectors of the bichromatic field components. P_{NR} is the direction that the CARS-polarization takes for $\delta \gg \gamma$. Setting the analysator perpendicular to this direction, the background contribution is suppressed. The remaining Raman contribution for this analysator setting appears as a Lorentzian, free from any background. It is, however, diminished in its maximum intensity by a factor $\sim$ 10 to 1000, depending on the depolarization of the respective Raman transition [13].

Under electronic resonance conditions, the antisymmetric contribution may be considerable. A specially designed experimental arrangement, using circularly polarized pump- and Stokes-beams, enables one to detect the g_a^2-term only. A four-frequency-CARS variant (separated pump- and probe-beams) allows a very precise determination of its value.

In the limit $\omega_p \simeq \omega_s$, let's say for $|\omega_p - \omega_s| < 50\,\mathrm{cm}^{-1}$, intramolecular vibrational excitation, as discussed in the preceding chapter, loses its significance as the source for an essential $X^{(3)}$-contribution. Other forms of material excitation become effective. Mostly they have a non-local character, i.e. they do not represent internal degrees of freedom of a molecule, but stand for larger collectives. The weaker-acting coupling strength results in lower frequency shifts. Stimulated scattering processes responsible for low frequency shifts concern the excitation of acoustical phonons (Stimulated Brillouin scattering), of molecular orientations (Stimulated Rayleigh wing scattering), concentration fluctuations (Stimulated concentration scattering) and temperature fluctuations (Stimulated thermal scattering or thermal Rayleigh scattering). Some of these forms of excitation will be discussed in the following chapter in connection with the discussion of nearly **D**egenerate **F**our **W**ave **M**ixing (**DFWM**), i.e. in the case when all four radiation frequencies in the FWM-process nearly coincide.

11.4 Nonlinear Polarization Spectroscopy

There is one type of internal molecular reaction with large spectroscopic relevance which will be treated here. It relates to the case of two energetically neighboring radiation components near to a molecular one-photon-resonance (i.e. within an absorption band). In this case the bichromatic irradiation in the 2nd order excites an oscillation of the population difference between electronic ground and first excited states, giving a term $\Delta(n_1 - n_2) \sim \cos([\omega_p - \omega_s]t)$ which is then probed in a FWM-process.

The amplitude of the population-modulation depends on the relaxation connected with the one-photon resonant transition. Here the dephasing time T_2 as well as the population relaxation time T_1 can be found by the decrease of the scattering signal intensity in enlarging the frequency difference $|\omega_p - \omega_s|$.

The acting molecular process can be characterized by the term scheme of Fig. 11.2d. The driving force exerted by pump- and Stokes-shifted beams excites a molecular transition $|g\rangle\langle-\rangle|n\rangle$. The probe of the coherently oscillating population difference will be performed by the pump-component itself. The field-vectors of the ω_p- and the ω_s-components are linearly polarized, at an angle of 45° to each other.

The signal wave coinciding in frequency with ω_s is measured with an analyzer oriented perpendicularly to the polarization of the entering ω_s-component.

Pump- and Stokes-shifted components are counterpropagating so that the signal will not be disturbed by radiation of the spectral neighbor pump-component. Figure 11.5 characterizes the applied experimental arrangement.

The measuring principle was designed by Song, Lee and Levenson [14]. It was given the name Nonlinear Polarization Spectroscopy (**NLPS**). It represents

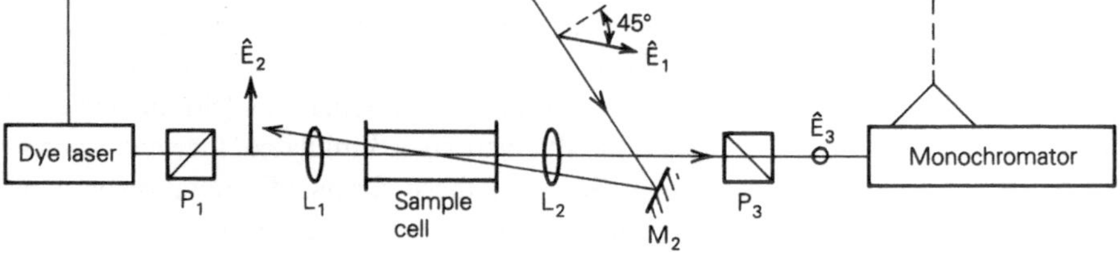

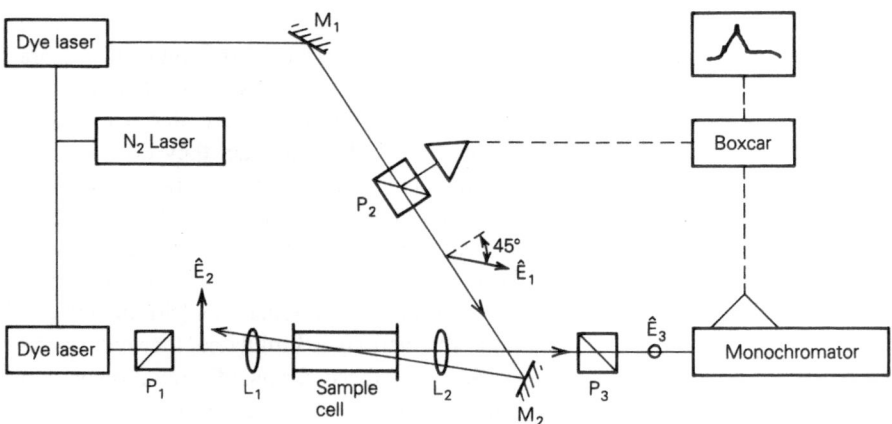

Fig. 11.5. Experimental setup for Nonlinear Polarization Spectroscopy. P, L, and M denote polarizer, lens, and mirror, respectively. The dye laser beam (ω_s) which goes through the crossed polarizers P_1 and P_3 is the signal beam

a two-frequency mixing technique for which index matching is fulfilled automatically. A signal could have been obtained also by the Antistokes-shifted wave in a CARS-like arrangement, a technique named Resonant Rayleigh type Mixing which was introduced by Yajima et al. [15].

NLPS is successfully applied to measure the longitudinal relaxation time as well as the homogeneous line width in inhomogeneously broadened molecular transitions. Values of T_1- and T_2 times in the range of 100 fs – 10 ps are available under steady state conditions. Furthermore cross relaxation times T_3, describing the population transfer between different species within an inhomogeneously broadened ensemble, can be determined by quantitatively analyzing the signal decrease in scanning the frequency difference $\omega_p - \omega_s$.

The corresponding material polarization is derived by solving the equations of motion for the elements of the density matrix for a two level absorber. The equations for the difference of the diagonal elements and for the nondiagonal elements are

$$\frac{d}{dt}(\rho_{nn} - \rho_{gg}) = -\frac{2i}{\hbar}(H_{ng}\rho_{gn} - H_{gn}\rho_{ng}) - \Gamma_{nn}[(\rho_{nn} - \rho_{gg} + \rho(g)] , \tag{11.49}$$

$$\frac{d}{dt}\rho_{ng} + (\Gamma_{ng} + i\omega_0)\rho_{ng} = -\frac{i}{\hbar}H_{ng}(\rho_{nn} - \rho_{gg}) . \tag{11.50}$$

where the interaction Hamiltonian has the nonvanishing matrix elements

$$H_{ng} = H^*_{gn} = -\mu_{ng}(A_p e^{-i\omega_p t} + A^*_p e^{i\omega_p t} + A_s e^{-i\omega_s t} + A^*_s e^{i\omega_s t}) \tag{11.51}$$

$\rho(g)$ is the equilibrium population of level $|g\rangle$, and the level $|n\rangle$ is assumed unpopulated at equilibrium.

The total material polarization is given by

$$P = n_o \int_{-\infty}^{\infty} d\omega_o g(\omega_o) \langle \mu_{gn}\rho_{ng} + \mu_{ng}\rho_{gn} \rangle , \tag{11.52}$$

where $g(\omega_o)$ is the distribution of resonant frequencies, describing the inhomogeneously broadened absorption band; n_o is the density of absorbing species. The bracket represents an orientational average over the ensemble.

The steady state solution in the 3rd order perturbation development for the ρ_{gn}-element, oscillating with ω_s, becomes

$$\rho^{(3)}_{gn}(\omega_s) = \frac{i\rho(g)|\mu_{gn}A_p|^2 \mu_{gn}A_s}{\Gamma_2 + i\Delta_2} \left\{ \frac{1}{\Gamma_1}\frac{2\Gamma_2}{\Delta_1^2 + \Gamma_2^2} + \frac{1}{i\Delta + \Gamma_1}\frac{i\Delta + 2\Gamma_2}{(i\Delta_2 + \Gamma_2)(-i\Delta_1 + \Gamma_2)} \right\} \tag{11.53}$$

where

$$\Gamma_1 = \Gamma_{nn}; \Gamma_2 = \Gamma_{ng}; \Delta_1 = \omega_o - \omega_p; \Delta_2 = \omega_o - \omega_s; \Delta = \omega_p - \omega_s . \tag{11.54}$$

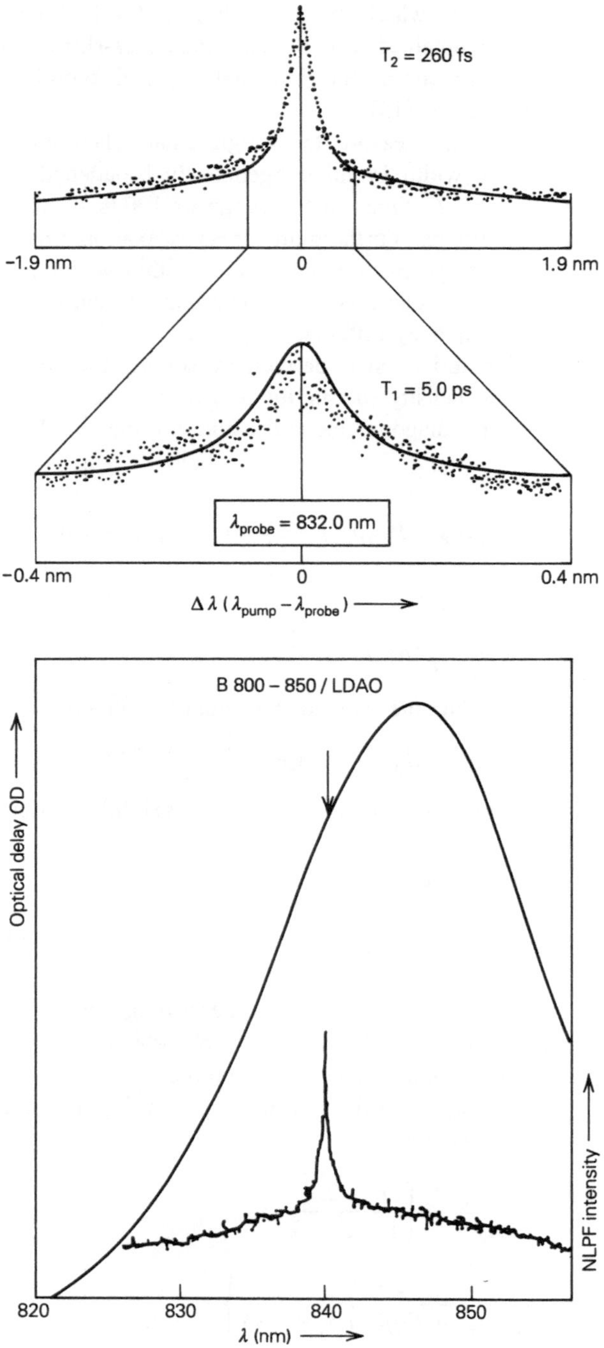

Fig. 11.6. NLPS-measurements of B 800–850 (antenna complex of purple bacteria) [16]

Assuming the inhomogeneous ω_o-distribution to be much broader than the homogeneous width (Γ_2), the integration in Eq. (11.52) can be performed in a closed manner and yields a 3rd order material polarization:

$$P_i^{(3)}(\omega_s) = X_{ijkl}^{(3)}(-\omega_s, \omega_p, -\omega_p, \omega_s) A_j(\omega_p) A_k^*(\omega_p) A_l(\omega_s) . \qquad (11.55)$$

The tensor components of $X^{(3)}$ measured in polarization spectroscopy have a common factor, describing the spectral dispersion of the signal as

$$X^{(3)}(\omega_s) \propto f(\Delta) = \frac{1}{\Gamma_2 + i\Delta} \left\{ \frac{1}{\Gamma_1} + \frac{1}{\Gamma_1 + i\Delta} \right\} \qquad (11.56)$$

Expression (11.56) contains two factors each depending on Δ. The prefactor decreases with Δ like a Lorentzian and its halfwidth gives the dephasing constant. A second factor represents the variation of the population difference. The first term in the second factor is due to 3rd order saturation by the radiation component $|A_p|^2$, the second term comes from population modulation. Both terms are equal for $\Delta = 0$.

In an interesting example the method of NLPS has been applied to investigate the homogeneous substructure of extremely inhomogeneously broadened bacteriochlorophyll samples. Figure 11.6 shows corresponding results for the B850 component of the isolated B800-850 antenna of *Rhodobacter* spheroids at room temperature.

The figure shows a special position of the exciting radiation in the molecular absorption band and the results of the NLPS-measurements with computer fitting according to Equation (11.56). The time of phase relaxation as well as that of the energy relaxation could be determined under steady state excitation conditions.

11.4.1 Pressure-Induced FWM Signals

In Eq. (11.56) the net signal results from the superposition of two different contributions which sum up constructively. Both are of equal size in the $\Delta = 0$ limit.

We will mention a related phenomenon resulting in a destructive interference at this point, as it is well suited to show the possibilities of obtaining information about molecular relaxation dynamics from coherent interference effects by FWM. This phenomenon was first demonstrated by Prior et al. [17] for a Raman like transition in Na-gas. Enhancing the pressure of the buffer gas He, a transition from spontaneous radiation damping to the collisional damping case is performed. For a three level term of 3S−3P transition of Na, with the levels a ($3S_{1/2}$), b ($3P_{1/2}$) and c ($3P_{1/2}$), the second order density matrix element under biharmonic irradiation becomes

$$\rho_{cb}(\omega_p - \omega_s) = \frac{\mu_{ca}\mu_{ab}}{\hbar^2} N_a \frac{A_p A_s^*}{(\omega_{ca} - \omega_p - i\Gamma_{ca})(\omega_{ba} - \omega_s + i\Gamma_{ab})} *$$

$$\left[1 - \frac{i(\Gamma_{ca} + \Gamma_{ab} - \Gamma_{cb}}{\omega_{cb} - (\omega_p - \omega_s) - i\Gamma_{cb}}\right]. \tag{11.57}$$

Expression (11.57) is the generalization of the term entering Eq. (11.56) if the upper level is split up.

The signal which a probe beam generates from the material excitation given by Eq. (11.57) shows a Raman like resonance with

$$\delta = \omega_{cb} - \omega_p + \omega_s \tag{11.58}$$

and the linewidth Γ_{cb}. In the limiting case of dilute Na the relation for

$$\text{spontaneous lifetime broadening} : \Gamma_{cb} = \Gamma_{ca} + \Gamma_{ab} \tag{11.59}$$

holds and no resonance occurs at the energy separation ω_{cb}. In the presence of collisions the Γ with mixed indices may be written as $\Gamma_{ca} = \Gamma_{ca}^{sp} + \gamma_{ca} p_{He}$ and the pressure dependent growth of the line intensity and line width can be used to find γ (p_{He} denotes the partial pressure of the buffer gas).

11.5 Time Resolved Technique of FWM

For the application of pulses with pulse durations equal to or shorter than the dephasing time of the excited molecular transition ($t_p \leqslant T_2 = 1/\gamma$), the steady state description of the respective molecular reaction breaks down. Now the resonant reaction on the radiation field has to be calculated with help of time integral expressions over the corresponding response function. They give the nonlinear reaction a 'transient character'. If the corresponding radiation matter interactions proceed far off any resonance, the time dependent response function undergoes high frequency oscillations with the period of the resonance mismatch ($\Delta = \omega_o - \omega_{laser}$), giving it the character of the $\delta(t)$-function and a steady state description with help of a susceptibility-expression is admissible.

If different terms with transient as well as with steady state character contribute to the nonlinear polarization expressions of the type

$$P^{nl} \propto n_o \frac{\partial \alpha}{\partial q} \langle q(t) \rangle \mathscr{E}_{pr} + X_{NR} \mathscr{E}_{pr} \mathscr{E}_p \mathscr{E}_s^* \tag{11.60}$$

result. The probe field $\mathscr{E}_{pr}$ is assumed to be a pulse of long duration. Here the equation

$$\langle q(t) \rangle = \gamma_1 \int_{-\infty}^{t} dt_1 e^{(i\omega_{vib} + \gamma)(t - t_1)} \mathscr{E}_p(t_1) \mathscr{E}_s^*(t_1), \tag{11.61}$$

describes the transient Raman scattering.

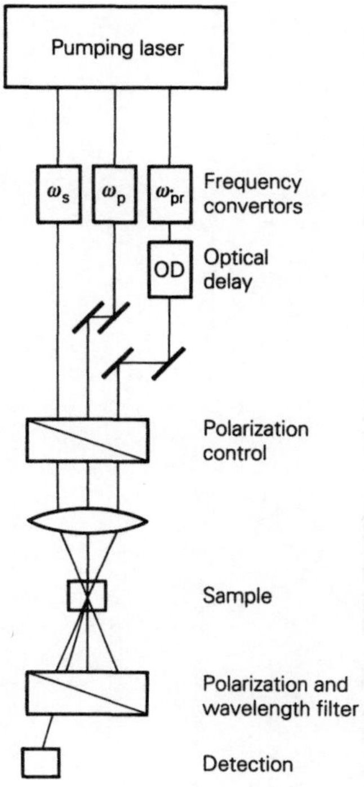

Fig. 11.7. Experimental system used for time-resolved coherent Raman scattering. Synchronized pulses at frequencies ω_p, ω_s, and ω_{pr} are generated by different frequency converters from a pumping laser source. The pulses at ω_p and ω_s excite the sample via bichromatic pumping. The coherent excitation is monitored by the delayed (optical delay OD) probing pulse at frequency ω_{pr}. The different beams cross in the sample at phase-matching angles. Polarization and wavelength filters separate the coherent signal from the exciting light

The merits of a coherent ultra-short-pulse-technique lie in the possibility to apply a definite and scannable delay time between the exciting and the probing pulses, thus following directly the dynamics of a material reaction. A scheme of an experimental setup for time resolved FWM-measurements is given Fig. 11.7.

11.5.1 Characteristic Experimental Results

According to Eq. (11.61), a coherent vibrational excitation decays with a time constant $T_2 = 1/\gamma$ (dephasing time). An example is given in Fig. 11.8.

The ν_7 (783 cm^{-1}) and ν_2 (2925 cm^{-1}) vibrations of acetone are investigated with a time resolution of approximately 80 fs. In the diagram of the ν_7 mode the influence of the non-Raman-resonant term of $X^{(3)}$ is clearly seen during the first 100 fs delay. Due to the very fast relaxation of this term its influence is negligible after a few 100 fs. The corresponding material response is often used to measure the time resolution of the arrangement.

For pulse measuring techniques with low repetition rates, the laser intensities, necessary to get detectable signals, are often higher than those applied in steady state measurements. The compression, e.g., of a 1 MW/cm^2 pulse of 10 ns duration, used for 'steady state' measurements, to a 10 ps pulse, used in time resolved technique, causes an intensity enhancement into the GW/cm^2–range.

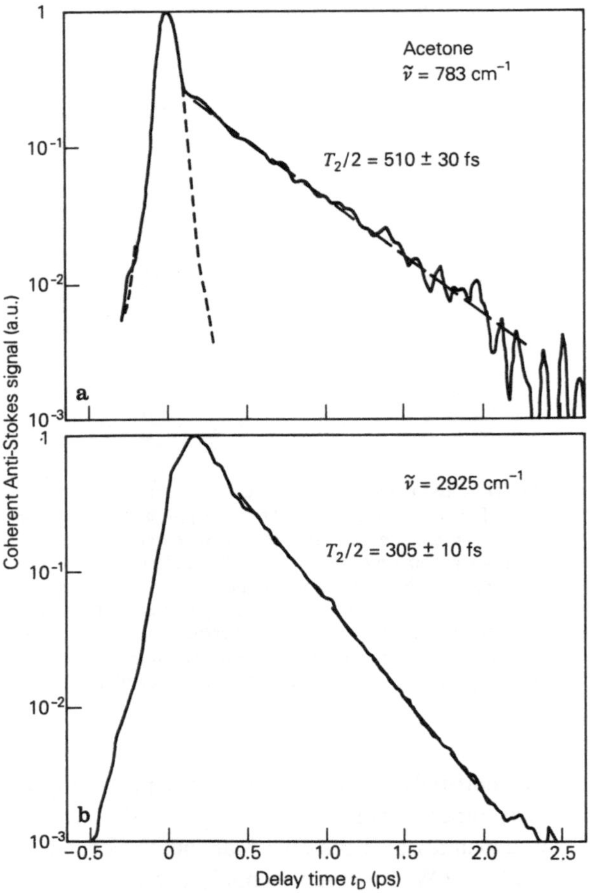

Fig. 11.8. Coherent signal from acetone measured with fs-laser pulses [18]

Thus, in short-pulse-experiments even over a short sample length a backreaction of the molecular excitation on the exciting radiation fields must be considered. This backreaction may express itself in a modification of the time- and spectral-behavior along the propagation route or in time. This behavior must be considered in the corresponding measurements.

If, e.g., the Stokes component of the bichromatic excitation for a FWM-process is produced by stimulated Raman scattering of the pump-laser in a special cuvette, the transient regime causes, for the Stokes- and pump-laser radiation, a different time behavior, as shown in Fig. 11.9.

Such time shifts must be compensated by special means before the exciting radiations enter into the sample.

Effects of unwanted shifts in the spectral distributions of the pumping radiation pair may also occur. The frequency distribution of the pump-radiation as it

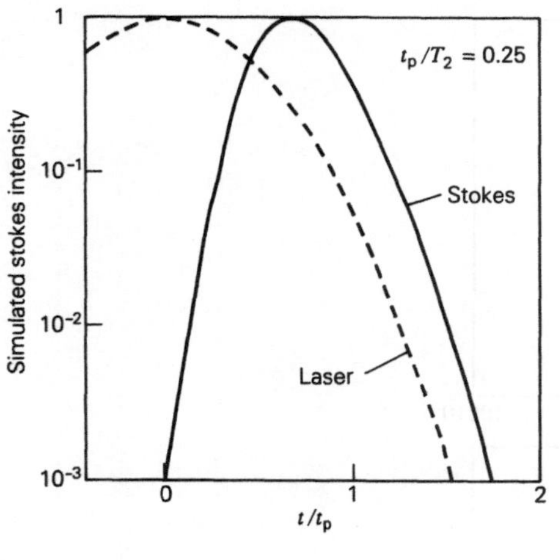

Fig. 11.9. Calculated time evolution of the Stokes-pulse, generated by stimulated Raman scattering under transient excitation conditions ($t_p/T_2 = 0.25$), applying a Gaussian pump pulse [3]

exits may deviate markedly from that of the beams entering. An example of this behavior is shown in Fig. 11.10.

This effect can be explained by the influence of the non-Raman-resonant part of $X^{(3)}$, leading to phase modulation of the Stokes shifted wave during the time evolution of the laser pulse [19].

Effects of the same kind may occur in the sample to be measured and should be avoided by lowering the applied intensities.

Another problem in time-resolved measurements may occur if there is no strict coincidence between the frequency difference of the bichromatic pump waves and the frequency of the molecular resonance. The material reaction upon the bichromatic excitation will be described by the solution of the appropriate density matrix element (Eq. (11.4)). It consists of two parts: the homogeneous solution of the differential equation, oscillating with the molecular eigenfrequency and the inhomogeneous driven part, oscillating with the difference of the carrier frequencies of the driving lasers. During irradiation, the latter part dominates but it decays quickly after the end of the excitation phase, especially if it contains a strong contributions from the non-Raman-resonant background terms. At later times the term oscillating with the molecular frequency dominates. Figure 11.11 characterizes the situation.

This result shows, that probing should be carried out after the coherent excitation is finished.

The frequency-width of fs-pulses is very broad, even for transform limited pulses (50 fs pulses of Gaussian amplitude distribution correspond to 300 cm^{-1}). If within this frequency region there are several resonances of the molecule, they are excited simultaneously. As their frequencies are different, their relative phases begin to change after some time (or space in the interaction region), which results in a beating phenomenon, modulating the T_2 decay curve. An example of corresponding terahertz beating is depicted in Fig. 11.12. From these beating curves one can obtain the difference frequencies of the excited molecular resonances.

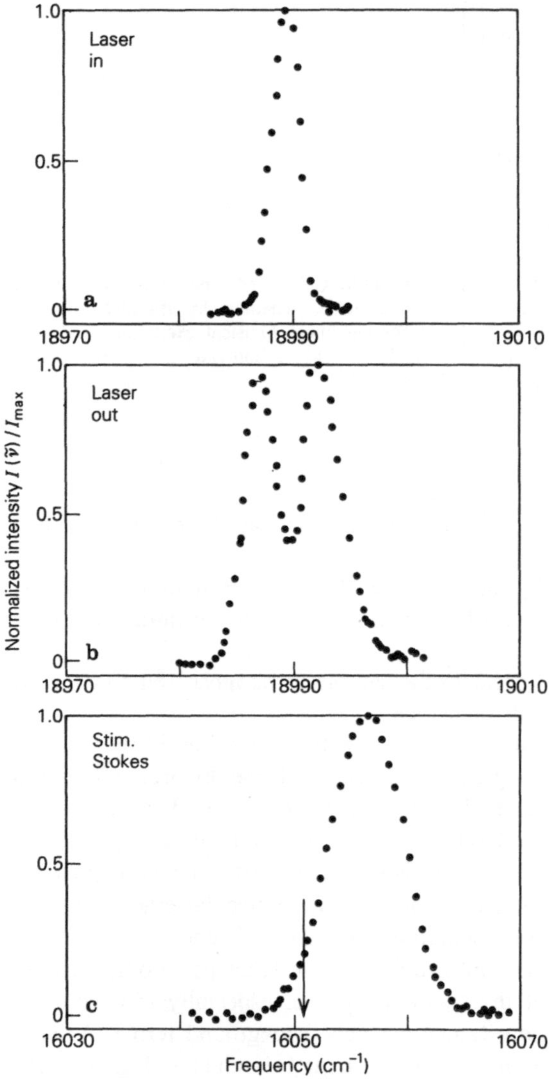

Fig. 11.10. Effect of self phase modulation on transient SRS in liquid CH_3CCl_3

The shape of the decaying excitation curve allows us to distinguish between homogeneously and inhomogeneously broadened transitions, easily. This is a further advantage of time-resolved spectroscopy in comparison to spectroscopies in the frequency domain. For the latter methods inhomogeneity can also be found. However, it needs very careful measurement of the line shapes in the wings of the spectral lines. These spectral regions are often hidden by an overlap of other lines, sometimes even preventing a corresponding determination. In the

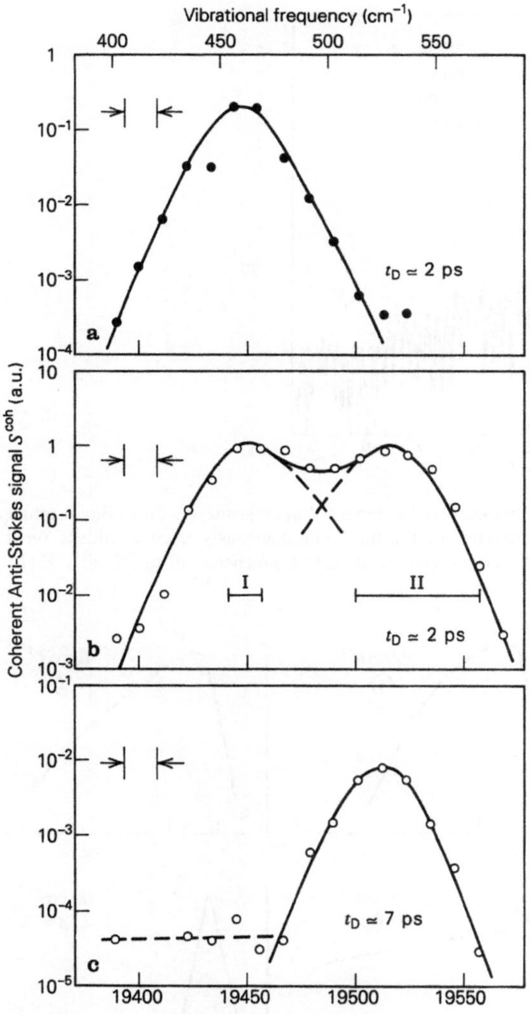

Fig. 11.11. Spectral intensity distribution of coherent Anti-Stokes probe scattering: **a** resonantly excited totally symmetric mode at 460 cm^{-1} of CCl_4 ($\Delta\omega = 0$) studied with delayed pulses: $t_D = 2$ ps. **b** ν_3-vibration of CH_3I at $\nu_0 = 525$ cm^{-1}, non-resonantly excited with $\Delta\nu = -65$ cm^{-1}; delay between excitation and probing pulses: $t_D = 2$ ps. Two spectral bands at $\omega_p - \omega_s$ and $\omega_p + \omega_0$ are observed. **c** Same as **b** for $t_D = 7$ ps; the faster decaying band at $2\omega_p - \omega_s$ has already disappeared [20]

time domain, the exponential or non exponential decay character gives this distinction clearly. In Fig. 11.13, an example of lines in the low frequency region of 1-(t)-histidine $HCl.H_2O$ crystals at 10 K is given. It is very interesting that, for the same sample, one part of the modes are homogeneously, another inhomogeneously broadened and the time constants differ by orders of magnitude. From these data, different interaction effects on the corresponding modes can be derived.

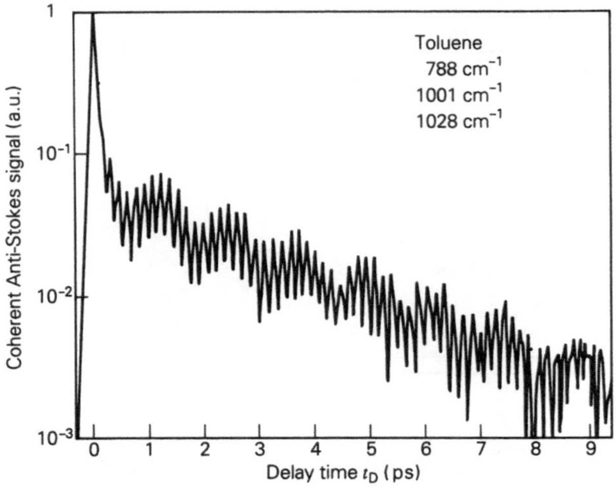

Fig. 11.12. Terahertz beats observed in time-resolved coherent Raman scattering from liquid toluene using femtosecond light pulses. The interference of the three simultaneously excited toluene modes at 788,1001, and 1028 cm^{-1} generates a rich structure with beat frequencies up to 7.2 THz [21]

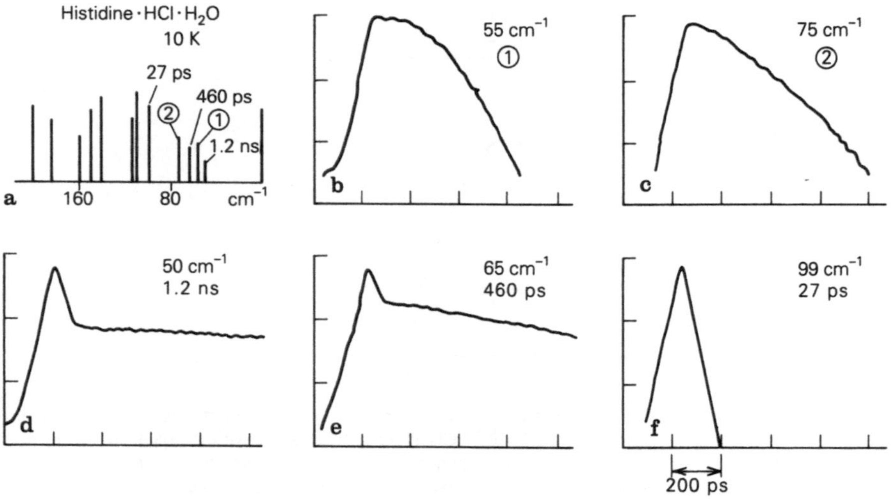

Fig. 11.13. Time resolved coherent Raman data for 1-(t)-histidine. HCl.H_2O crystals at 10 K (semilog plots). **a** positions of the low frequency modes. **b–f** time resolved data for the five low frequency modes, shown in **a**. Exponential decays are found for the totally symmetric modes at 99 cm^{-1}, 65 cm^{-1}, and 50 cm^{-1}, whereas the two modes at 55 cm^{-1} **b** and 75 cm^{-1} **c** suggest inhomogeneous line-broadening [22]

With a special time resolved excite and probe technique it is possible to determine the longitudinal relaxation time T_1 (vibrational population time). Here the pump radiations serves to produce an alteration in the population of a vibrational mode of the sample. These Raman induced populations are described as process of the fourth order, thus going beyond the description given in (Sect. 11.2, Eq. (11.6)). Vibrational population with the density n_2 of the excited molecular state builds up from the coherent Raman amplitude according to

$$\left(\frac{\partial}{\partial t} + \frac{1}{T_1}\right) n_2 = \kappa_3 A_{\mathrm{p}}^* A_{\mathrm{s}} \sigma + C \cdot C. \tag{11.62}$$

Following energy transfer to other modes destroys the coherent character of the excitation. The altered population is measured by an Antistokes shifted spontaneous Raman signal, which is detected in dependence on the time delay between the exciting and probing pulses. Those investigations give information about population transfer channels between different vibrational modes. The population of the excited state can be produced by a strong infrared pulse too. In Fig. 11.14, the spontaneous Antistokes signals of $(CH_2Cl)_2$ and CH_2ClBr are plotted versus the time delay of the probe pulse.

The energy transfer from a primarily excited 3265 cm^{-1} CH-stretching mode of acetylene to the carbon triple bond-mode and the ensuing population decay is shown in Fig. 11.15.

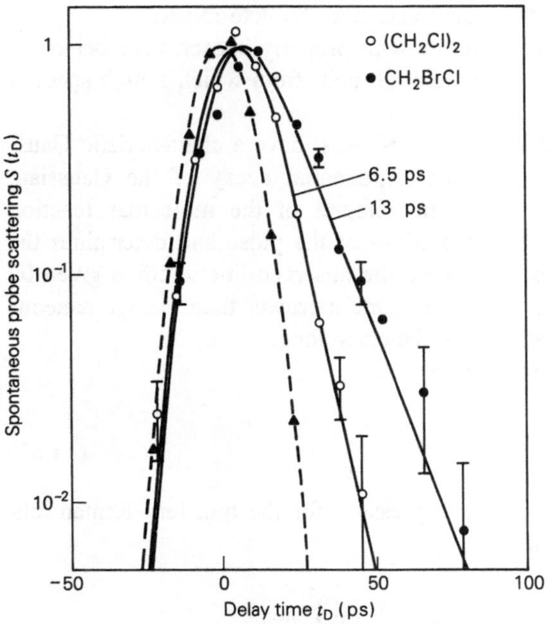

Fig. 11.14. Spontaneous Antistokes scattering signal $S(t_D)$ of the probe pulse versus delay time t_D for the symmetric CH_2-stretching modes of $(CH_2Cl)_2$ (*circles*) and CH_2BrCl (*points*); *solid*: calculated curves. *Broken line* and *triangles* represent the instrumental response of the measurement [23]

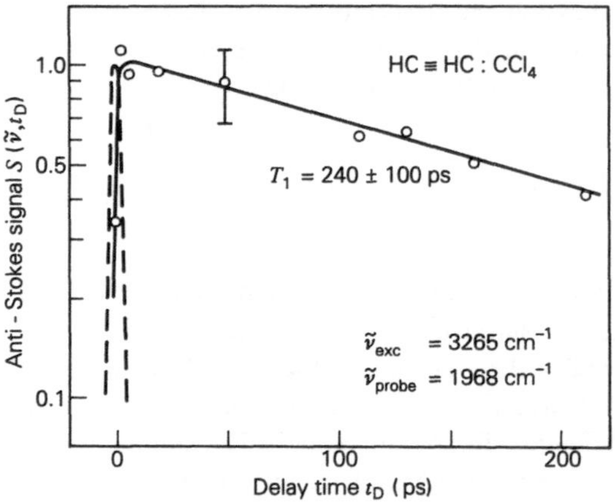

Fig. 11.15. Time dependence of the population of the C≡C-mode of acetylene in liquid CCl_4 following excitation of the ν_3CH-stretching mode at 3265 cm^{-1} After a rapid rise the population decays slowly with a decay time of 240 ps [24]

11.5.2 Increased Spectral Resolution Applying Specially Designed Pulsed FWM Techniques

In the steady state Raman experiment, the probe laser measures the spectral profile of the excited vibrational amplitude, which is given by Eq. (11.15). How good the vibrational frequency can be measured depends on the precision with which the position of the maximum of the Lorentzian may be determined.

Applying excitation and probing pulses with properly chosen time behavior, the molecule itself may act as a time integrating unit, from which a high spectral resolution of the measured signal may result.

Many laser systems, e.g. mode locked Nd-lasers, have a characteristic Gaussian shaped time behavior. The quadratic exponential decay of the Gaussian, occurring as convolution term in the time integral of the molecular reaction dominates the linear damping term at the edges of the pulse and determines the spectral cut-off for the material excitation. So the observed line width is given by the spectral width of the laser and it can be made narrower than the spontaneous line width if the laser pulse duration is sufficiently long.

For pulses of Gaussian temporal shapes

$$A(t) = A_0^{-\left(\frac{t}{t_p}\right)^2 2\ln 2} \tag{11.63}$$

the molecular reaction, as given by the expression for the transient Raman amplitude (Eq. (11.14)), takes the form

$$\sigma(t,\delta) = \kappa e^{-\left(i\delta+\frac{1}{T_2}\right)t} \times \int_{-\infty}^{t} dt' e^{\left(i\delta+\frac{1}{T_2}\right)t'} e^{-\left(\frac{t'}{t_p}\right)^2 2\ln 2}. \tag{11.64}$$

Here the exciting field amplitudes and prefactors are lumped together into κ.

Joining the real exponential contributions in the integral gives a Gaussian function with shifted maximum. The modulus squared of the material excitation becomes

$$|\sigma(T,\delta)|^2 = \kappa^2 e^{\left[-\frac{2T}{T_2}+\left(\frac{t_e}{t_p}\right)^2 4\ln 2\right]} \times \left| \int\limits_{-\infty}^{T} dt' e^{i\delta t'} e^{-\left(\frac{t'-t_e}{t_p}\right)^2 2\ln 2} \right|^2 \tag{11.65}$$

with

$$t_e = \frac{t_p^2}{4 \ln 2\, T_2} \tag{11.66}$$

The time T corresponds to the time of observation and for later times the integral in Eq. (11.64) is proportional to the Fourier transform of the Gaussian driving pulse. So the observed frequency bandwidth is determined by that of the Gaussian.

$$\Delta\omega_g = \frac{4 \ln 2}{t_p} \tag{11.67}$$

It can be made narrower than the spontaneous one, which is given by the dephasing time T_2, if $t_p > 1.4\ T_2$ holds.

Unfortunately the advantage of line narrowing is paid for by an essential decrease of the signal intensity. The narrowing to Gaussian shape is obtained for sufficiently late observation times of $T > 5t_p$. Because of the T-dependent exponential prefactor, the peak amplitude of the material reaction decreases by orders of magnitude, if essential line narrowing below the spontaneous width is to be achieved.

For practical applications the obtainable spectral resolution is determined by the signal to noise ratio of the experimental system.

A modification of this technique was developed in connection with Four-Wave-Mixing experiments by W. Zinth [25]. In the so-called **SEPI**-technique (for Short Excitation and Prolonged Interrogation) a freely decaying molecular vibration, which was previously excited by short bichromatic irradiation is interrogated by a long Gaussian pulse. According to Eq. (11.13) the signal wave $A_{sgn}(t)$ bears the Gaussian behavior of the probing wave, modified in time by the decaying material excitation process. The folding of the signal with the Gaussian occurs through time integration by the spectrometer unit. The spectrally resolved scattered signal has the form

$$I_{sgn}(\omega_{sgn}) \propto \left| \int\limits_{-\infty}^{\infty} dt e^{i\omega_s t} A_{sgn}(t) \right|^2 \tag{11.68}$$

Using a Gaussian-shaped probe pulse at carrier frequency ω_p

$$E_{pr}(t) = A_{pr} e^{-\left(i\omega_{pr}+\left(\frac{t-T_D}{t_{pr}}\right)^2 2\ln 2\right)t} \tag{11.69}$$

and introducing $\Delta\omega = \omega_{sgn} - \omega_{pr} \pm \omega_{vib}$ we obtain at a late delay time T_D a Gaussian shaped spectrum

$$I_s(T_D, \Delta\omega) \propto e^{-\frac{2T_D}{T_2} + \left(\frac{t_e}{t_{pr}}\right)^2 4\ln 2} \times \left| \int_{-T_D}^{\infty} dt e^{i\Delta\omega t - \left(\frac{t+t_e}{t_{pr}}\right)^2 2\ln 2} \right|^2 \tag{11.70}$$

centered at the frequency $\omega_p \pm \omega_o$:

11.5.3 Photon Echoes of Polyatomic Molecules in Condensed Phases

Ordinary optical line shapes of molecular systems in condensed phases are usually dominated by electronic inhomogeneous broadening, resulting from the variation of local environments of different molecules. As a result, the useful structural and dynamical information is hidden under a broad inhomogeneous envelope, which makes it impossible to extract this information using linear optical measurements.

One spectroscopic technique used to probe molecular dynamics and optical dephasing processes by eliminating inhomogeneous broadening, is the three-pulse stimulated photon echo method. The photon echo is the optical analog of the magnetic resonance spin echo (Chapt. 8.16).

We consider the echo variant in which three short laser pulses with wave vectors k_1, k_2 and k_3 are sequentially applied to the system, with time delays τ and $\tau + \tau'$. The stimulated echo pulse, which centers around $t = \tau'$ after the third pulse, is generated in the direction $k_{sgn} = k_3 + k_2 - k_1$. In the weak field limit the echo signal is given by the time integral over the modulus squared of the third order polarization, derived from the incident fields. If the excitation pulses are well-separated $P^{(3)}$ can be expressed as

$$P^{(3)}_{echo}(k_{sgn}, t) = i^3 \int_0^{\infty} dt_3 \int_0^{\infty} dt_2 \int_0^{\infty} dt_1 R(t_3, t_2, t_1)\chi(t_3 - t_1)A_3 A_2 A_1^* \\ * e^{i[(\omega_3+\omega_2-\omega_1-\omega_{ng})t_3 + (\omega_2-\omega_1)t_2 - (\omega_1-\omega_{ng})t_1]} \tag{11.71}$$

The response function here is factorized into two parts, R denoting the dynamical contribution and χ the static (inhomogeneous dephasing) contribution. The Fourier transform of the latter gives the inhomogeneous broadening.

The physical picture is simplified, if we consider an experiment with infinitely short pulses, $A_j(t) = \delta(t)$. In this case the echo amplitude is given by

$$|P^{(3)}_{echo}(k_{sgn}, t)| = |R(t, \tau, \tau')| \, |\chi(t - \tau')| . \tag{11.72}$$

The forming of the echo can be described as follows. At time $t = -(\tau + \tau')$, the initial, ground state, equilibrium density matrix ρ_g is transferred by the first impulsive pulse to a coherent optical excitation described by ρ_{eg}, which then evolves freely. At $t = -\tau$ the system interacts with the second pulse and is transferred to

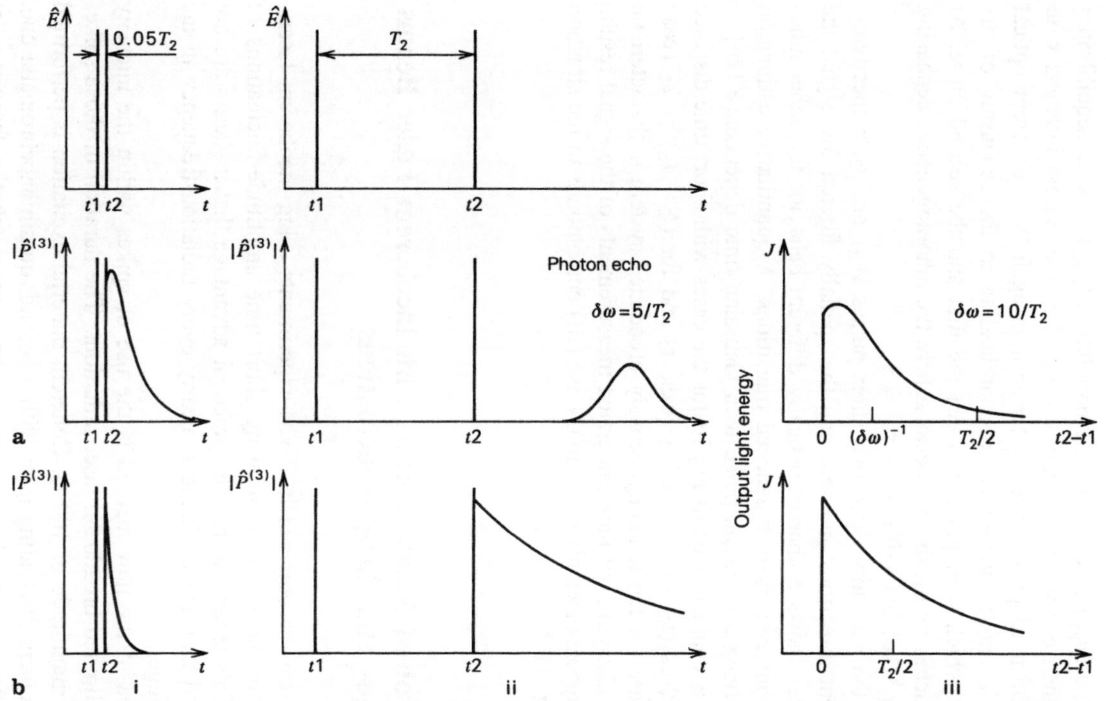

Fig. 11.16. Calculated signal behaviour in a low field (3rd-order) echo-experiment. Different behaviour of the signal for inhomogeneously **a** and homogeneously **b** broadened molecular transitions. Time resolved (i, ii) and time-integrated measurements (iii)

an electronic ground-(ρ_{gg}) or excited state population (ρ_{nn}). The nonequilibrium population states then evolve freely, until $t = 0$, when the system interacts with the third pulse. The third pulse transfers the system again to a coherent optical excitation. There is one excitation component leading to the rephasing of the dephasing process which took place between the first and the second pulse. At $t = \tau'$ the echo reaches its maximum, because here the inhomogeneous dephasing is eliminated, as χ takes the value $\chi(t - \tau') = 1$.

Carrying out the excitation with two short pulses at t_1 and t_2, respectively, entering in different directions and selecting the spatially filtered 3rd signal the time-resolved signal shows a characteristically different behavior for either inhomogeneously or homogeneously broadened transitions. A quantitative estimation can be given, applying a radiation pulse with Gaussian time dependence [26].

The situation is characterized in Fig.16 for the cases with short time distance of the exciting pulses $(t_2 - t_1) = 0.05T_2$ (case i) and for $(t_2 - t_1) = T_2$ (case ii). In the latter case for the inhomogeneously broadened system a characteristic echo-peak occurs. Carrying out time integrated measurements of the signal results into the different appearances given in (iii). Case (iii) corresponds to the situation represented in Fig. 11.8.

11.6 Time-Resolved Measurements with Incoherent Laser Beams Bearing ps- and Sub-ps-Fluctuations

fs-time resolution can be obtained in FWM-experiments using incoherent lasers, i.e. pulsed lasers in the ns-range showing short time amplitude-fluctuations of sub-ps duration. This technique has the practical advantage that ns-lasers are not very expensive and they are available at nearly every radiation-frequency in the visible- and IR-range.

The main idea of the new method is the use of spikes within the intensity of non-transform-limited broadband laser radiations. The duration of those spikes is in the order of magnitude of 100 fs. Coherent material excitation is performed combining a broadband fluctuating laser with a second, quasimonochromatic one. The frequency-differences of the laser beams have to match the frequency of the material resonance. The coherent excitation, forming a fluctuating transient grating, can be interrogated by a delayed probe pulse, which has to be a replica of the exciting broadband pulse. So there is a unique assignment between the spikes of the exciting and probing laser pulses. The time fluctuations of a broadband laser and its delayed replica are given schematically in Fig. 11.17. The probe laser generates a signal, the intensity of which decays as the coherent excitation decays, added by a signal, that is produced by accidental peak coincidences between the lasers. Because within one ns-laser pulse are thousands of fs-spikes (leading to a corresponding number of fs-measurements) T_2 can be measured with high precision.

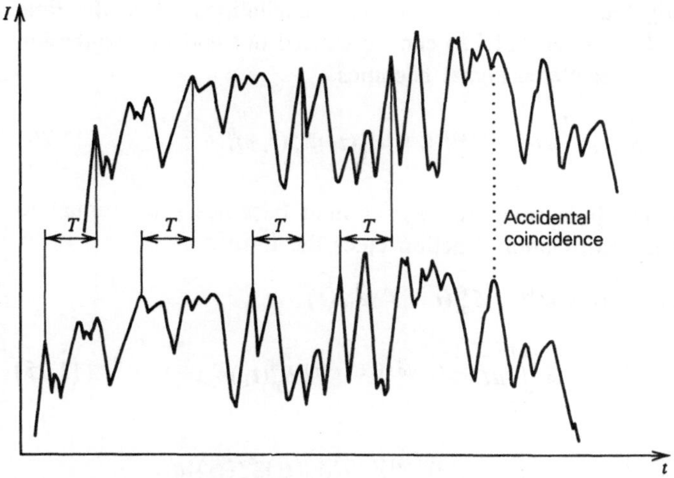

Fig. 11.17. Intensity distribution of an incoherent broadband laser radiation, split into two identical beams separated by delay time T

A closely related method, developed by van Voorst's group, uses **two** radiations for probing – the broadband and the narrowband laser. This is in close analogy to the technique known as Raman gain spectroscopy. In this method the CARS signal is measured, which is produced by the delayed radiation pair. It allows the direct measurement of a molecular vibrational frequency and its related T_2-time to be carried out on a real time scale.

The mathematics of the process treats the fluctuating laser as a stationary Gaussian process. In contrast to the methods using short-pulse-techniques, here the theory has to take into account the time-fluctuating fields. This demands that we consider the radiation interaction of excitation- and probe beams over the whole pulse duration on the basis of time integrals over the 3rd order material response function. This is necessary because for "quasistationary" processes it is impossible to distinguish between the processes of exciting and probing, as long as there is an overlap in time, lying within $\Delta t = t_{\text{autocor}} - \tau$, with τ being the delay-time between excite and probe fields.

The time domain description of the third order polarization was given in 11.2.4 (Eq. 11.22). $R^{(3)}(.\,.)$ is the material response function. For the case of the coherent Raman scattering (**CSRS**) from a vibrational transition in the electronic ground state of a molecule (cf. the term scheme of Fig. 11.2a) it is of the form

$$R^{(3)} \propto e^{-(t-t_1)\,(i\omega_{\text{ng}}+\Gamma_{\text{ng}})} e^{-(t_1-t_2)(i\omega_{\text{n}'\text{g}}+\gamma_{\text{n}'\text{g}})} e^{-(t_2-t_3)\,(i\omega_{\text{ng}}+\Gamma_{\text{ng}})} \tag{11.73}$$

The Fourier transform of Eq. (11.73) is, for monochromatic radiation fields of the Raman process, of the form of the 3rd order susceptibility, as given in Eq. (11.24).

In the following we assume, that the Raman excitation is performed with laser radiations, the frequency of which lies far away from any one photon resonance of the investigated molecules (off resonant excitation). Here the time integrations over t_1 and t_3 in Eq. (11.22) can be carried out and the expression for the polarization of the scattered signal becomes

$$P^{(3)}_{\text{sgn}}(t) \propto e^{i\omega_{\text{sgn}}t}\mathscr{E}_{\text{pr}}(t)\int_{\infty}^{t} dt_1 e^{-(\gamma+i\delta)(t-t_1)}\mathscr{E}_{\text{p}}(t_1)\mathscr{E}^*_{\text{s}}(t_1)q^0_1. \tag{11.74}$$

If the interaction length is small the signal field becomes proportional to $P^{(3)}_{\text{sgn}}(t)$ and the signal autocorrelation function takes the form

$$\begin{aligned}\langle\mathscr{E}^*_{\text{sgn}}(t+T)\mathscr{E}_{\text{sgn}}(t)\rangle \propto\ & e^{i\omega_{\text{sgn}}T}\mathscr{E}^*_{\text{pr}}(t+T)\mathscr{E}_{\text{pr}}(t)\\ & * \int_{-\infty}^{t+T} dt_1 e^{-(\gamma-i\delta)(t+T-t_1)}\mathscr{E}^*_{\text{p}}(t_1)\mathscr{E}_{\text{s}}(t_1)\\ & * \int_{-\infty}^{t} dt_2 e^{-(\gamma+i\delta)(t-t_2)}\mathscr{E}_{\text{p}}(t_2)\mathscr{E}^*_{\text{s}}(t_2)\rangle q^0_1.\end{aligned} \tag{11.75}$$

11.6.1 CSRS with Incoherent Laser Light

We will discuss here in detail the evaluation of expression (11.75) for the case of CSRS, which was the Raman process first analyzed with the incoherent technique by Kobayashi's group [27]. The material excitation in this case is performed by the combination of a 'monochromatic' and a Stokes-shifted broadband laser, a replica of the latter is applied with time delay as probe. The signal appears Stokes-shifted to the broadband component. The signal autocorrelator now can be expressed as

$$\begin{aligned}\langle\mathscr{E}^*_{\text{sgn}}(t+T)\mathscr{E}_{\text{sgn}}(t)\rangle =\ & |C|^2 e^{i\omega_{\text{sgn}}T}\int_{-\infty}^{t+T} dt_1 e^{-(\gamma+i\delta)(t+T-t_1)}\\ & * \int_{-\infty}^{t} dt_2 e^{-(\gamma-i\delta)(t-t_2)}\langle\mathscr{E}^*_{\text{s}}(t+T)\mathscr{E}_{\text{s}}(t)\mathscr{E}^*_{\text{s}}(t_2)\mathscr{E}_{\text{s}}(t_1)\rangle\end{aligned} \tag{11.76}$$

Into Eq. (11.76) enters the temporal average over the incoherent stokes field $\mathscr{E}_{\text{s}}$, expressed by the angled brackets. The broadband amplitude will be assumed to be a stationary zero-mean Gaussian process. Its autocorrelator is taken to be of Lorentzian type (a good approximation if the incoherent radiation stems from a dye laser). Under these conditions the autocorrelation of the Stokes amplitude takes the form

$$\langle A^*(t+T)A(t)\rangle = \phi(\tau) = |A_0|^2 e^{-\kappa|\tau|} = \frac{\kappa}{\pi}|A_0|^2\int_{-\infty}^{\infty} d\omega\frac{e^{i\omega\tau}}{\omega^2+\kappa^2}. \tag{11.77}$$

The Stokes field entering into Eq. (11.76) is a superposition of a broadband pulse, followed with time-delay τ by its replica:

$$\mathscr{E}_s(t) = A(t) + A(t-\tau)e^{i\omega_s\tau} \tag{11.78}$$

The signal autocorrelator will contain terms, functionally depending on τ. They result from those contributions to the signal for which the τ-shifted pulse probes the coherent material-excitation performed, by the pre-pulse. These terms give information about the material reaction dynamics in the mean time.

The signal resulting from the combination of the first and the delayed pulses derive from the following part of the autocorrelator of (11.76)

$$\langle[A^*(t+T)A^*(t_1-\tau)+A^*(t+T-\tau)A^*(t_1)][A(t)A(t_2-\tau)+A(t-\tau)A(t_2)]\rangle \tag{11.79}$$

Performing time averaging, expression (11.79) breaks into eight terms, some of which are identical. The resulting time-arguments are of the following form:

$$\begin{aligned}&2\phi(T)\phi(t_1-t_2)+2\phi(t+T-t_2)\phi(t-t_1)\\&+\phi(t-t_1+\tau)\phi(t+T-t_2+\tau)+\phi(t-t_1-\tau)\phi(t+T-t_2-\tau)\\&+\phi(T+\tau)\phi(t_1-t_2-\tau)+\phi(T-\tau)\phi(t_1-t_2+\tau).\end{aligned} \tag{11.80}$$

Each term of Eq. (11.80) if set into Eq. (11.76) gives an expression of the form

$$\sim |C|^2 e^{i\omega_{sgn}t}\int_{-\infty}^{t+T} dt_1 e^{-(\gamma+i\delta)(T+t-t_1)}\int_{-\infty}^{t} dt_2 e^{-(\gamma-i\delta)(t-t_2)}\phi_I\phi_{II} \tag{11.81}$$

and the autocorrelator entering into it may be expressed as

$$\phi_I(\theta_1)\phi_{II}(\theta_2) = \left(\frac{\kappa}{\pi}\right)^2 \int\int_{-\infty}^{\infty} d\omega_1 d\omega_2 \frac{e^{i\omega_1\theta_1}e^{i\omega_2\theta_2}}{[\omega_1^2+\kappa^2][\omega_2^2+\kappa^2]} \tag{11.82}$$

The θ_i are linear functions of t_1, t_2, T and τ, according to the special ϕ-term.

The FWM-signal is recorded by a photodetector. It may be used in the broadband regime, "white-detector", or applying a spectral discrimination by a monochromator with a slitwidth γ_{slit}.

11.6.2 White-Detector Limit

In the white-detector limit the signal of the photodetector is proportional to expression (11.76), taken at $T \Rightarrow 0$. If Eq. (11.82) is set into Eq. (11.81) the time integrations may be carried out and the I_{sgn}-expression separates into two factors, corresponding to the two integrals over ω_1 and ω_2. Each integral breaks into a number of terms exponentially depending on τ. The arguments of the exponentials are determined by the poles of the integrand. They generally have the forms

$$e^{-\kappa|\tau|} \ ; \ e^{-(\gamma|\tau| \pm i\delta\tau)} \tag{11.83}$$

The signal consists of a strong coherent peak determined by the autocorrelation function of the broadband laser, a weaker function decaying with γ, bearing the wanted information and a constant background contribution. Because of the intensity difference of the two first terms, the second one can only be observed for times longer than the correlation time of the broadband laser.
For the special case of CSRS the single contributions corresponding to the terms given in Eq. (11.80) after performing the time integration over expression (11.76) are

$$\begin{aligned} I_{\text{white detector}}(\tau) \propto\ & 2\frac{\kappa+\gamma}{\gamma[\delta^2+(\kappa+\gamma)^2]} + 2\frac{1}{\delta^2+(\kappa+\gamma)^2} + \frac{e^{-2\kappa|\tau|}}{\delta^2+(\kappa+\gamma)^2} \\ & + \left[\frac{e^{-\kappa|\tau|}}{\gamma-\kappa-i\delta} + \frac{2\kappa e^{-(\gamma+i\delta)|\tau|}}{(\delta+i\gamma)^2+\kappa^2}\right] * \left[\frac{e^{-\kappa|\tau|}}{\gamma-\kappa+i\delta} + \frac{e^{-(\gamma+i\delta)|\tau|}}{(\delta-i\gamma)^2+\kappa^2}\right] \\ & + e^{-\kappa|\tau|}\left[\frac{e^{-\kappa|\tau|}}{(\delta+i\kappa)^2+\gamma^2} + \frac{\kappa}{\gamma}\frac{e^{-(\gamma+i\delta)|\tau|}}{(\delta-i\gamma)^2+\kappa^2}\right] + CC. \end{aligned} \tag{11.84}$$

As the signal is measured without spectral decomposition the actual δ is not precisely defined and one has to take an average over $\delta \approx 0$ which may be simply performed by setting $\delta = 0$ in Eq. (11.84). So the signal intensity can be described by the expression

$$\begin{aligned} I(\tau) \propto\ & \frac{2(\kappa+2\gamma)}{\gamma(\kappa+\gamma)^2} + \frac{4[\gamma^2 e^{-2\kappa|\tau|} + \kappa^2 e^{-2\gamma|\tau|}]}{(\kappa^2-\gamma^2)^2} \\ & + \frac{2\kappa(\kappa-3\gamma)}{\gamma(\kappa+\gamma)(\kappa-\gamma)^2} e^{-(\kappa+\gamma)|\tau|}. \end{aligned} \tag{11.85}$$

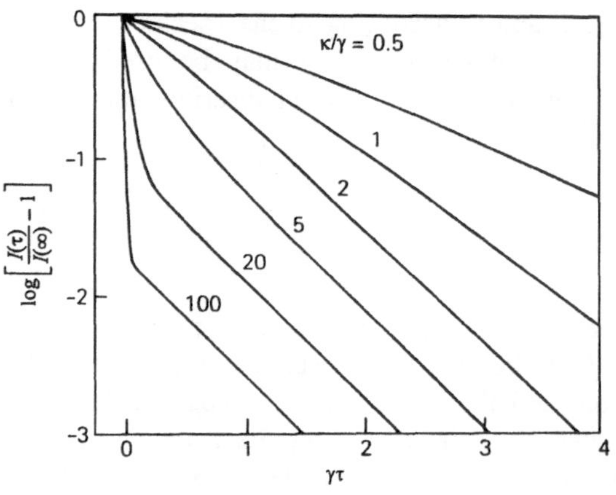

Fig. 11.18. Theoretical curves of the CSRS signal as function of the delay time in white detector limit [27]

Calculated dependencies of the CSRS-signal intensity as a function of $\gamma\tau$ for several ratios of the dephasing time to the Stokes-laser correlation time are given in Fig. 18.

11.6.3 Spectrally Resolved Detection

The group of A.C.Albrecht [28] extended the method of incoherent scattering by introducing a spectrally resolved detection through which finer spectroscopic information about the signal becomes available. Here the detector signal will be proportional to the Fourier transform of the signal field autocorrelator

$$I_\Omega \propto \frac{1}{2\pi}\int_{-\infty}^{\infty} dT e^{-(i\Omega T+\gamma_{\text{slit}}|T|)} \langle \mathscr{E}^*_{\text{sgn}}(t+T)\mathscr{E}_{\text{sgn}}(t)\rangle \tag{11.86}$$

The expression is evalued at first by carrying out the integrations over dt_1 and dt_2. Then the integration over T may be performed. The Lorentzian character of the broadband field (in the case $\gamma_{\text{slit}} \Longrightarrow 0$) transforms it into a δ-function. This δ-function relates to each spectral component of the pump-field, just one spectral component of the probe field, so that the sum of the contributions reaches the monochromator setting

$$\Omega = 2\omega_s - \omega_p + \omega_1 + \omega_2. \tag{11.87}$$

In Fig. 19, the physical meaning of this connection is shown. The spectral position in the broadband pump field and the related position in the probe-spectrum have been marked and they combine to a signal at the monochromator frequency Ω. Performing the remaining single ω-integration gives, for the signal, some τ-dependent terms oscillating with the frequency of the Raman mismatch

$$p = \Omega - \omega_{\text{sgn}}. \tag{11.88}$$

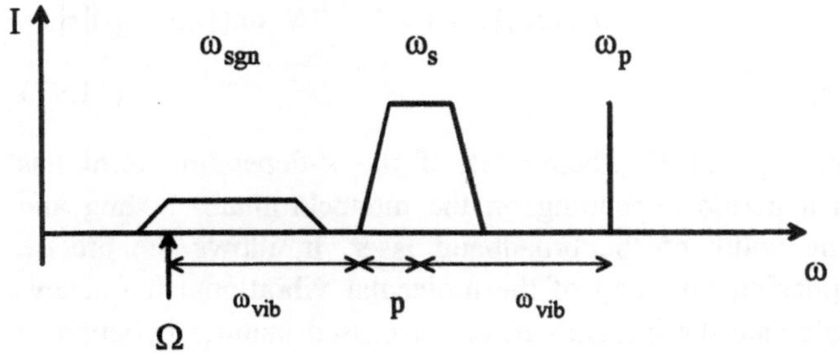

Fig. 11.19. $\boldsymbol{\Omega}$: slit frequency; ω_{sgn}: frequency of the Stokes shifted Raman line belonging to the vibrational frequency ω_v; ω_s: frequency component of the broad band pump laser; ω_p: frequency of the narrow band Stokes laser; p: detuning

The calculation procedure for the case of spectrally resolved detection shall be shown for the term number 4 in expression (11.80). The actual autocorrelator takes the form

$$\phi(\theta_1)\,\phi(\theta_2) = \left(\frac{\kappa}{\pi}\right)^2 \int\int_{-\infty}^{\infty} d\omega_1 d\omega_2 \frac{e^{i\omega_1(T+t-t_1)}e^{i\omega_2(t-t_2)}}{[\omega_1^2+\kappa^2][\omega_2^2+\kappa^2]}$$

$$\times e^{i(\omega_2-\omega_1)T}e^{i(\omega_1+\omega_2)\tau} \tag{11.89}$$

Insertion of expression (11.89) into Eq. (11.81) gives, after time integrations, the signal autocorrelator

$$\langle \mathscr{E}^*_{\text{sgn}}(t+T)\mathscr{E}_{\text{sgn}}(t)\rangle \propto e^{i\omega_{\text{sgn}}T} \int\int_{-\infty}^{\infty} d\omega_1 d\omega_2$$

$$\times \frac{e^{i(\omega_2-\omega_1)T}e^{i(\omega_1+\omega_2)\tau}}{[\omega_1^2+\kappa^2][\omega_2^2+\kappa^2][\gamma+i(\delta-\omega_1)][\gamma-i(\delta+\omega_2]} \tag{11.90}$$

Time integration of the spectrally filtered signal at monochromator setting Ω results in the limit of zero slit width into the δ-function $\delta(\omega_{\text{sgn}}-\Omega+\omega_2-\omega_1) = \delta(\omega_{\text{sgn}}-\Omega-p)$. Only combinations of $\omega_2+p=\omega_1$ are allowed. Applying this condition, the spectrally filtered signal becomes

$$I_\Omega \propto \int_{-\infty}^{\infty} d\omega_1 \frac{e^{-i(2\omega_1-p)\tau}}{[\omega_1^2+\kappa^2][(\omega_1-p)^2+\kappa^2][\gamma+i(\delta-\omega_1)][\gamma-i(\delta+\omega_1-p)]} \tag{11.91}$$

Evaluation of Eq. (11.91) may be done by Cauchy-integration around the poles at

$$\omega_1=-i\kappa;\ \ \omega_2=-i\kappa+p; \omega_3=-i\gamma+\delta;\ \ \omega_4=-i\gamma-\delta+p. \tag{11.92}$$

The terms depending exponentially on the monochromator setting ($\sim p$) take the form

$$I_\Omega(\tau) = e^{-2\kappa|\tau|}(M_1\sin(p|\tau|)+M_2\cos(p\tau))+e^{-2\gamma|\tau|}[N_1\sin((2\delta-p)|\tau|)$$

$$+N_2\cos((2\delta-p)\tau] \tag{11.93}$$

The second term in Eq. (11.93), being left if the κ-depending term has decayed, oscillates with a period depending on the monochromator setting and being independent of the width of the broadband laser. It allows the precise determination of the dephasing time and of the molecular vibrational frequency.

The possibility to calculate the integrals in the discussed manner is bound to the Lorentzian character of the incoherent radiation field. Other types of spectral distribution lead to more complicated integral expressions for the τ-dependence of the signal, but with analogous physical content.

The weights with which the different poles contribute depend on the specially chosen kind of FWM-arrangement.

11.6.4 Some Experimental Results

1. In experiments of van Voorst's group [29] with a method named Raman Fringe Decay (**RFD**) the excitation and probe are both done by irradiation of a pump- and Stokes-pair. Excitation- and probe pulses follow each other with time delay τ. The pump-field is chosen to be broadband, the Stokes component narrowband and the Anti-Stokes signal is recorded.
Confining the detection to terms oscillating $\sim (\omega_p - \omega_s)\tau$ – the so-called Raman fringes – the signal obtained with the white detector takes the form

$$I \propto (\alpha + i\beta)\, e^{-i(\omega_p - \omega_s)\tau} + C.C. \tag{11.94}$$

where

$$\begin{aligned} \alpha &= e^{-\gamma|\tau|}\,[X\cos(\delta\tau) + Y\sin(\delta|\tau|)] + Ze^{-\kappa|\tau|}, \\ \beta &= e^{-\gamma|\tau|}\,[Y\cos(\delta\tau) - X\sin(\delta|\tau|)] + Ye^{-\kappa|\tau|}, \end{aligned} \tag{11.95}$$

and

$$\begin{aligned} X &= \frac{\kappa(\kappa+3\gamma)(\kappa-\gamma)+\kappa\delta^2}{N};\; Y = \frac{4\gamma\delta}{\tau * N}; \\ Z &= \frac{(\kappa-\gamma)\,\kappa^2 - \gamma\kappa - 4\gamma^2 + \delta^2(4\gamma+\kappa)}{N}, \\ N &= \gamma[(\kappa-\gamma)^2 + \delta^2][(\kappa+\gamma)^2 + \delta^2]. \end{aligned} \tag{11.96}$$

To obtain analytical expressions, a Lorentzian distribution of the Gaussian correlator for the broadband field is assumed, with the pulse duration t_p taken to be long compared to the correlation time of the incoherent field $t_c = \kappa^{-1}$ and to the dephasing time $T_2 = 1/\gamma$ of the molecular vibration.

An experimental result obtained by RFD-technique is shown in Fig. 20. The fringe character of the signal is visible by the time-base-extension given in the inset. From the fringe oscillations the frequency of a Raman active vibration can be measured directly. A least squares fit of the formula (11.94) to the fringe amplitude gives κ= 5 ps^{-1} and γ= 0.42 ps^{-1}. According to the applied white-detector-recording, no influence of detuning off resonance on the vibrational mode is found within the experimental error.
2. Incoherent scattering in the spectrally resolved detection limit was developed by Albrecht et al. [28]. They applied this technique for the analysis of frequencies and corresponding T_2-times of single excited vibrations as well as for samples with several simultaneously excited vibrations. In the first case a single exponential damped oscillation occur, with a frequency, determined by the spectral

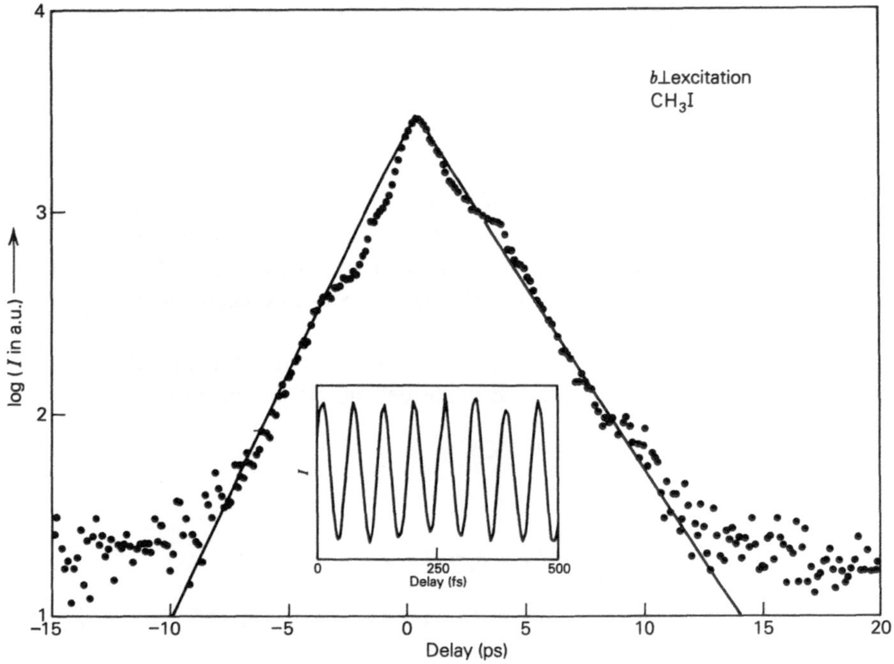

Fig. 11.20. Experimental data for RFD-measurements of methyl iodide. The *solid curve* gives the fit to formula (11.94). Data taken from [29]

position of the monochromator, the frequency of the narrowband laser and the vibrational frequency of the excited vibration. In the latter case a beating feature between all excited vibrations could be observed, caused by the sum of independent damped oscillations, corresponding to their single excitations. The theoretical description is given by Eq. (11.91), in which the integral stands for each vibration.

The situation is illustrated by measurements of Albrecht's group showing the beating structure obtained by superposition of two similar modes in a mixture of different molecules with two closely lying Raman-transitions A and B. The CSRS-signal takes the form

$$I_{\text{signal}}^{\text{AB}}(\tau) = I_{\kappa}^{\text{AB}}(\tau) + I_{\gamma}^{\text{AB}}(\tau) + I_{\text{c}}^{\text{AB}} \qquad (11.97)$$

Here I_{κ} decays with τ on the time scale of κ^{-1}, I_{γ} decays with τ on the time scale of the reciprocal material dephasing (T_2^{-1}) constants: γ_{A} and γ_{B}, and last I_{c} is the constant background signal. For the special two-component case one of the characteristic dephasing terms takes the form

$$I_{\gamma}^{\text{AB}} \propto 2\kappa^2[F(A,A) + F(B,B) + F(A,B) + F(B,A)] \qquad (11.98)$$

where under condition $\kappa \gg \gamma_A, \gamma_B$ $F(A,B)$ becomes

$$F(A,B) \approx \frac{e^{-2\gamma_B|\tau|}}{[(\Delta_1^B)^2+\kappa^2][(\Delta_2^B)^2+\kappa^2]} \tag{11.99}$$

$$* \left(\frac{(\omega_A - \omega_B)\sin(\Delta^B|\tau|) + \gamma_{AB}^{(+)}\cos(\Delta^B\tau)}{(\omega_A-\omega_B)^2 + (\gamma_{AB}^{(+)})^2} + \frac{(\Delta_1^A+\Delta_2^B)\sin(\Delta^B|\tau|) + \gamma_{AB}^{(-)}\cos(\Delta^B\tau)}{(\Delta_1^A+\Delta_2^B)^2+(\gamma_{AB}^{(-)})^2} \right)$$

In the experiment, a monochromator set at frequency Ω is applied as spectral filter. For the two molecular vibrations $Q = A$, B the expression $\Delta_1^Q = \omega_Q + \omega_s - \omega_p$ denotes the detuning on the Raman excitation process, $\Delta_2^Q = \omega_Q + \Omega - \omega_s$ the detuning on the probing process. The sum $\Delta^Q = 2\omega_Q + \Omega - \omega_p$ is the variable responsible for the sinusoidal beating. Furthermore in Eq. (11.99) appear the sum $\gamma_{AB}^{(-)} = \gamma_A + \gamma_B$ and the difference $\gamma_{AB}^{(-)} = \gamma_A - \gamma_B$ of the dephasing terms of both types of molecules. The signal intensity decreases with the time constant of only one vibration (B). A corresponding term holds for the vibration (A) too.

Figure 21 shows the results of an CSRS experiment for a 1 : 1 mixture of benzene (A) and its fully deuterated derivative (B) with spectral filtering of

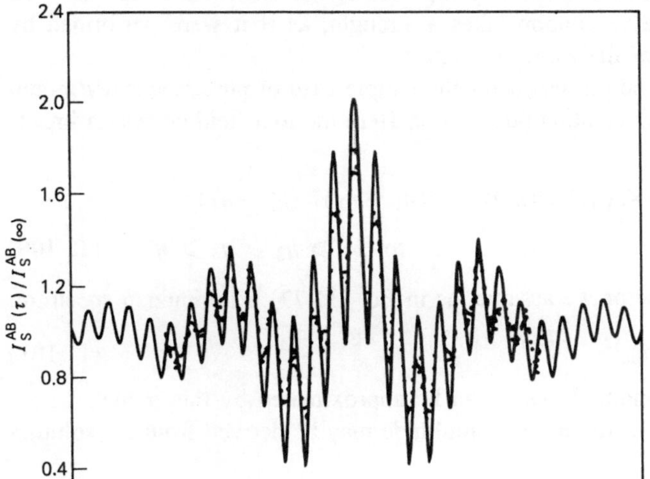

Fig. 11.21. Spectrally filtered CSRS-signal versus τ for the combination of benzene $\nu_A = 991\,\mathrm{cm}^{-1}$) - and benzene-d$_6$ $\nu_B = 946\ \mathrm{cm}^{-1}$) ring breathing modes. Monochromatic pump $\nu_p = 18776\ \mathrm{cm}^{-1}$, monochromator setting $\nu_\Omega = 16807\ \mathrm{cm}^{-1}$. The theoretical fit is with $\gamma_A^{-1} = 4.8$ ps, $\gamma_B^{-1} = 5.3$ ps, $2\pi(\Delta^A)^{-1} = 2.6$ ps and $2\pi(\Delta^B)^{-1} = 433$ fs, [28]

the signal. The intensity of the signal versus delay-time τ can be described by Eq. (11.97).

A very important application of this method was given by the Albrecht group for determining frequencies and T_2-times of a cyanine dye. In the experiment, part of the molecules were pumped to the 1st excited electronic state and several ground and excited state vibrations were simultaneously excited. By fitting the experimental curves separately for each vibration, they were able to determine the frequencies themselves and the frequency shifts between vibrations in the ground and excited states as well as their T_2 times with high accuracy [28].

11.7 Material Excitation by SRS for a Frequency Modulated Statistically Fluctuating Laser

An excitation of Raman transitions by broadband laser radiations reduces the effectivity of the excitation, if the linewidth of the material resonance (γ) is smaller than the spectral width of the laser (κ), i.e. if $\gamma < \kappa$. This is the case e.g. in applying fs-pulse lasers for the excitation of molecular vibrations. Nevertheless, the equations describing the material excitation under stimulated Raman scattering show, that there will occur a matching of the time behaviour of the exciting lasers and the material excitation. After an incipient region of lowered scattering efficiency, the nonlinear interaction of pump- and Stokes-amplitudes becomes well phased and the material excitation takes a strength, as if it were performed by monochromatic lasers of the same intensities.

This situation will be discussed for the simple case of pure frequency modulation of a statistically fluctuating pump laser. Here the four-field correlator breaks up according to

$$\langle A_p^*(\eta_1)A_p(\eta_2)A_p^*(\eta_3)A_p(\eta_4)\rangle = K(\eta_1-\eta_2)K(\eta_3-\eta_4) \quad \text{for } \eta_1 \geqslant \eta_2 \geqslant \eta_3 \geqslant \eta_4 \tag{11.100}$$

For the two-field correlator we assume, as in Eq. (11.77) a Lorentzian spectrum.

$$K(\eta-\eta') = |A_{po}|^2 e^{-\kappa|\eta-\eta'|}. \tag{11.101}$$

A phase modulated monomode laser can be approximated by this model.

The spatial growth of the Raman amplitude may be derived from the solution in the transient regime

$$\sigma(\xi,\eta) = \kappa_1 \int_{-\infty}^{\eta} d\eta_1 e^{-(i\delta+\gamma)(\eta-\eta_1)} A_p(\eta_1) A_s^*(\xi,\eta_1) \tag{11.102}$$

by forming its spatial derivative

$$\frac{\partial\sigma(\xi,\eta)}{\partial\xi} = \kappa_1 \int_{-\infty}^{\eta} d\eta_1 e^{-(i\delta+\gamma)(\eta-\eta_1)} A_p(\eta_1) \frac{\partial A_s^*(\xi,\eta_1)}{\partial\xi}. \tag{11.103}$$

Applying Eq. (11.13) we obtain

$$\frac{\partial\sigma(\xi,\eta)}{\partial\xi} = i\kappa_1\kappa_2|A_p|^2 \int_{-\infty}^{\eta} d\eta_1 e^{-(i\delta+\gamma)(\eta-\eta_1)}\sigma(\xi,\eta_1). \tag{11.104}$$

From this follows the differential equation for the σ-autocorrelator

$$\frac{\partial}{\partial\xi}K_\sigma(\xi,\eta-\eta') = \kappa_1\kappa_2|A_p|^2\left[\int_{-\infty}^{\eta} d\eta_1 e^{-(i\delta+\gamma)(\eta-\eta_1)}K_\sigma(\xi,\eta-\eta_1) + \int_{-\infty}^{\eta} d\eta_1 e^{-(i\delta+\gamma)(\eta-\eta_1)}K_\sigma(\xi,\eta'-\eta_2)\right]. \tag{11.105}$$

Assuming a monochromatic weak Stokes initial signal A_s $(0, \eta)$=A_{so} from Eq. (11.105) derives the expression for the initial value of the autocorrelator of the Raman amplitude

$$K_\sigma(0,\eta-\eta') = |\kappa_1 A_{so}A_p|^2 \int_{-\infty}^{\eta} d\eta_1 \int_{-\infty}^{\eta'} d\eta_2 e^{-(i\delta+\gamma)(\eta-\eta_1)}e^{-(-i\delta+\gamma)(\eta'-\eta_2)} \times e^{-\kappa|\eta_1-\eta_2|}$$

$$= |\kappa_1 A_{so}A_p|^2\frac{\kappa}{\pi}\int_{-\infty}^{\infty} d\omega\frac{e^{i\omega(\eta-\eta')}}{(\omega^2+\kappa^2)\ [(\delta+\omega)^2+\gamma^2]}. \tag{11.106}$$

K_σ according to Eq. (11.106) represents a stationary process. Solving the differential equation (11.105) for the Fourier transform of K_σ with the initial value from Eq. (11.106) and performing its backtransformation, yields the solution for the Raman amplitude autocorrelator

$$K_\sigma(\xi,\tau) = |\kappa_1 A_{so}A_p|^2\frac{\kappa}{\pi}\int_{-\infty}^{\infty} d\omega\frac{e^{i\omega\tau}e^{G\gamma^2/\omega^2+\gamma^2}}{(\omega^2+\gamma^2)(\omega^2+\kappa^2)}. \tag{11.107}$$

Herein δ is set equal 0 (line-center). G is the 'steady state' Raman gain $G = 2\kappa_1\kappa_2\xi|A_p|^2/\gamma$. The amplitude squared of the molecular excitation after an interaction length ξ is given by expression (11.104), taken for $\eta-\eta'=0$. It can be written as

$$\langle|\sigma(\xi,0)|^2\rangle = \frac{|\kappa_1 A_{so}A_p|^2}{\gamma^2}e^{G(\xi)} * F\left(G,\frac{\kappa}{\gamma}\right) \tag{11.108}$$

The integral F becomes

$$F\left(G,\frac{\kappa}{\gamma}\right) = \frac{\kappa}{\pi}\gamma^2\int_{-\infty}^{\infty} d\omega\frac{e^{-G\frac{\omega^2}{\omega^2+\gamma^2}}}{(\omega^2+\gamma^2)(\omega^2+\kappa^2)} \tag{11.109}$$

For monochromatic pump-laser ($\kappa \Rightarrow 0$) it follows $F(G,0) = 1$. For a small Raman gain ($G \Rightarrow 0$) it takes the value $F(0,\kappa/\eta) = \gamma/\kappa$, which means a considerable reduction of the molecular excitation amplitude. Assuming a large Raman gain, the integral may be approximated by

$$F\left(G, \frac{\kappa}{\gamma}\right) \approx \frac{1}{\pi\kappa} \int_{-\infty}^{\infty} d\omega e^{-G\frac{\omega^2}{\gamma^2}} = \frac{\gamma}{\kappa} \frac{1}{\sqrt{2\pi G}}. \tag{11.110}$$

The decrease given by Eqs. (11.109) and (11.108) can be compensated for large G-values by enlarging the interaction length l about the range of phase equalization Δl given by

$$\frac{\Delta l}{l} = \frac{\ln\left(\frac{\kappa}{\gamma}\sqrt{2\pi G}\right)}{G} \tag{11.111}$$

Taking representative values $G = 10$, $\kappa/\gamma = 10$ a value of $\Delta l \approx 0.4\ l$ is obtained.

11.8 The Description of Molecular Susceptibilities by Feynman Diagrams

The process of four wave mixing can be described by a complex third-order susceptibility $\chi^{(3)}$. There are many other nonlinear phenomena belonging to third-order processes with different structure of the corresponding $\chi^{(3)}$ expression. As was shown by several authors [30–32] there can be distinguished altogether 48 different terms in the case that 3 different waves interact with a definite molecular 4 level scheme. To get a general overview over the manifold of different contributions in each order of susceptibility it is convenient to apply graphical methods which allow to visualize each term. Such a method is the diagram technique developed by R.P. Feynman (1948–1949). It gives a graphical representation of the reaction dynamics of a molecular system which is described by the Liouville equation for the density matrix

$$\frac{d}{dt}\rho = L\rho = -\frac{i}{\hbar}[H_0 + H_{\text{int}}, \rho] - \Gamma\rho \tag{11.112}$$

In Eq. (11.112) the damping of the molecular states is included explicitly.

The perturbation is the interaction of the molecular system with the radiation field, described by the interaction Hamiltonian entering into Eq. (11.112). For many processes of nonlinear optics the radiation-matter interaction is suitably described as a dipole interaction and the interaction Hamiltonian becomes

$$H_{\text{int}} = -\mu E \tag{11.113}$$

With this radiation term as a perturbation, the polarization induced by the incident electro-magnetic waves will be expressed as a series in powers of the radiation field E. For the spectral representation of the molecular polarization(i.e.taking the

Fourier-transform of the time-dependent polarization-expression)

$$P(-\omega) = \chi^{(1)}(-\omega,\omega)E(\omega) + \chi^{(2)}(-\omega,\omega_1,\omega_2)E(\omega_1)E(\omega_2) + \chi^{(3)}(-\omega,\omega_1,\omega_2,\omega_3)E(\omega_1)E(\omega_2)E(\omega_3) + \cdots \quad (11.114)$$

each power of $E(\omega_i)$ stands either for the absorption or the emission of one photon of the frequency ω_i. Each radiation interaction act is accompanied by a transition of the molecular system from a quantum state k to another state j, coupled, according to the interaction Hamiltonian Eq. (11.113), by a non-zero transition dipole moment μ_{kj}. These radiation-interaction processes can be expressed in form of graphs, describing the entering of a photon into the molecule (a) respectively the outgoing of a photon from it (b) by graphic elements of the form given in Fig. 11.22.

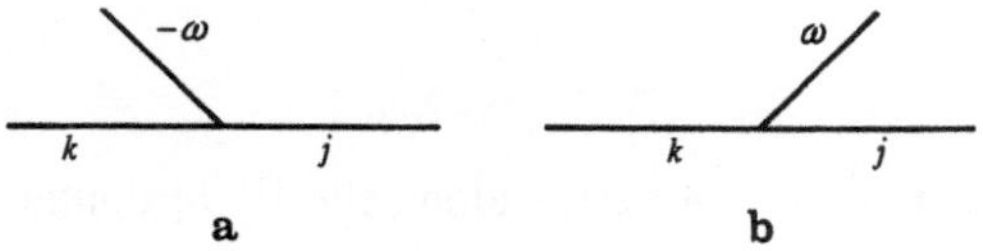

Fig. 11.22. Graphic representation of the elementary photon absorption **a** and emission **b** processes for -ket-branch(vide infra)

11.8.1 Rules for Deriving the $\chi^{(n)}$-Expressions from the Diagrams

The microscopic expression for a given diagram (e.g. Fig. 11.23) can be obtained by applying the following 6 rules to obtain the single multiplication factors [33].

1. The system starts with $|g\rangle\rho_{gg}^{(0)}\langle g|$.
2. The propagation of the ket state appears as multiplication factors on the left, and that of the bra state on the right of the susceptibility expression.
3. A vertex bringing $|a\rangle$ to $|b\rangle$ through **absorption** at ω_i on the upper (**ket**) side of the diagram is described by the matrix element

$$\left(\frac{1}{i\hbar}\right)\langle b|H_{\text{int}}(\omega_i)|a\rangle$$

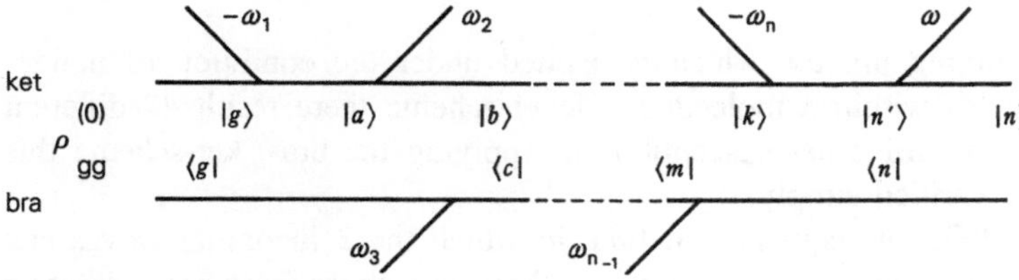

Fig. 11.23. A representative double Feynman diagram describing the term $\rho^{(n)}(\omega = -\omega_1 + \omega_2 + \cdots - \omega_n)$

with $H_{\text{int}} \propto e^{-i\omega_i t}$. If it is **emission** instead of absorption, the vertex should be described by

$$\left(\frac{1}{i\hbar}\right)\langle b|H_{\text{int}}^{+}(\omega_i)|a\rangle.$$

Because of the adjoint nature between the bra and ket sides, an absorption process on the ket side appears as an emission process on the bra side, and vice versa. Therefore, on the lower (**bra**) side of the diagram, the vertex for **emission** is described by

$$-\left(\frac{1}{i\hbar}\right)\langle a|H_{\text{int}}(\omega_i)|b\rangle,$$

that for **absorption** by

$$-\left(\frac{1}{i\hbar}\right)\langle a|H_{\text{int}}^{+}(\omega_i)|b\rangle$$

respectively.

4. Propagation from the jth vertex to the $(j+1)$th vertex along the $|l\rangle\langle k|$ double lines is described by the propagator

$$\prod_j = \frac{1}{\pm\left[i\left(\sum_{m=1}^{j}\omega_m - \omega_{lk}\right) - \Gamma_{lk}\right]}.$$

   The frequency ω_i is taken as positive if absorption of ω_i at the ith vertex occurs on the upper or emision of ω_i on the lower line; it is taken as negative if absorption of ω_i occurs on the lower or emission on the upper line.
5. The final state of the system is described by the product of the final ket and bra states, for example,
   $|n'\rangle\langle n|$ after the nth vertex in Fig. 11.23 for $\rho^{(n)}$.
6. The product of all factors describes the propagation from $|g\rangle\langle g|$ to $|n'\rangle\langle n|$ through a particular set of states in the diagram

Summation of these products over all possible sets of states yields the final result with contributions from all states.

11.8.2 The General Form of the $\chi^{(3)}$-Susceptibility Expression

For a 4 wave mixing process which is excited under the condition of nondegenerate laser fields within a molecular 4 level scheme there result 48 different contributions to the 3rd order susceptibility. Applying the bra-, ket-scheme this manifold can be verified simply:

> there are 3! different sequences in time in which the 3 incoming waves can interact with the molecule and for each of these sequences there are 8 different manners to distribute the corresponding vertices along the ket and the bra lines (cf. Fig 11.24)

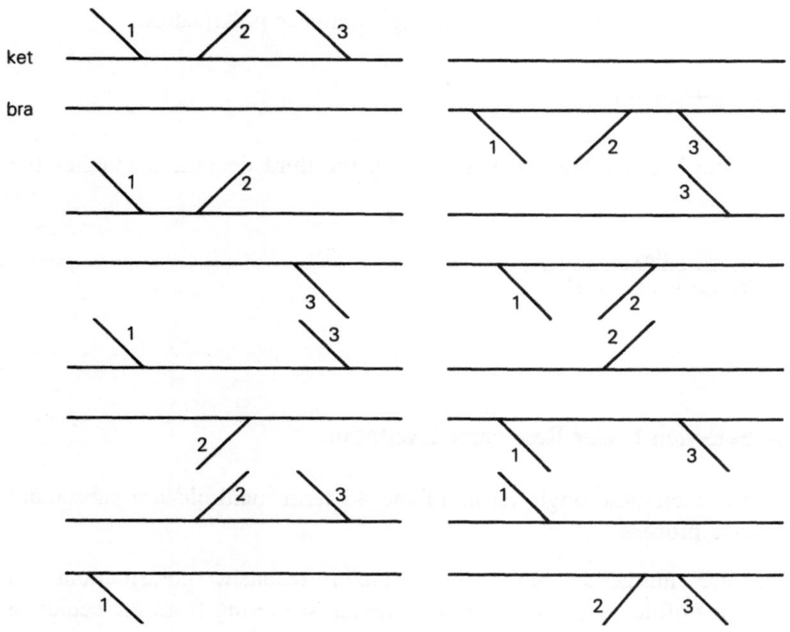

Fig. 11.24. The 8 possible different distributions of 3 incoming fields upon the ket- and bra-lines

In putting together the vertices and propagators the general structure of the terms

$$\chi^{(3)}(-\omega_4, \omega_3, -\omega_2, \omega_1)$$

can be obtained (cf. Eq. (11.24) in Sect. 11.24). All terms contain a one photon propagator describing the incidence of the first photon.
It is proportional to

$$\sim \frac{i}{\hbar} \frac{\mu_{ig}}{i(\omega_{ig} - \omega_1) + \Gamma_{ig}} \quad \text{if the photon causes a ket-transition}$$

and

$$\sim \frac{i}{\hbar} \frac{\mu_{ig}}{i(\omega_{ig} - \omega_1) - \Gamma_{ig}} \quad \text{if the photon causes a bra-transition}$$

The second interaction results into a two-photon propagator which can be written as

$$\sim \frac{i}{\hbar} \frac{1}{i\delta + \gamma}$$

with different meaning of the two photon resonance mismatch δ, depending on the special process under discussion. The third denominator is determined by the frequency resonance mismatch of the resulting nonlinear polarization

$$\omega_4 = \omega_1 + \omega_3 - \omega_c$$

Assuming the final ket and bra states as $|n'\rangle\langle n|$, the third denominator takes the form

$$\sim \frac{i}{\hbar} \frac{\mu_{nn'}}{i(\omega_{nn'} - \omega_4) + \Gamma_{nn'}}$$

11.8.3 Term Selection Under Resonance Excitation

Under resonance excitation single terms of the 48 term manifold are substantial for the resonance process.

1) So there are altogether only two Raman resonant contributions in the 48-term manifold of $\chi^{(3)}$ representing Raman scattering from molecules in the molecular ground state $|\alpha\rangle$ connected with the vibrational transition $\omega_{vib} = \omega_{\alpha\beta}$. The corresponding Feynman diagrams are displayed in Fig. 11.25. Both diagrams result in $\chi^{(3)}$-terms including the Raman resonant denominator

$$\sim \frac{1}{i(\omega_{vib} - \omega_1 + \omega_2) + \gamma}$$

2) In the case of the Song, Levenson [34] experiment all the incident radiation fields are nearly in resonance with the molecular electronic transition of interest, which may be represented as a two level system. There are only 4 terms resonantly contributing to $\chi^{(3)}$ $(-\omega_2, \omega_2, \omega_1, -\omega_1)$. Their Feynman diagrams are given in Fig. 11.25, together with the molecular term scheme with which the radiation fields resonantly interact.

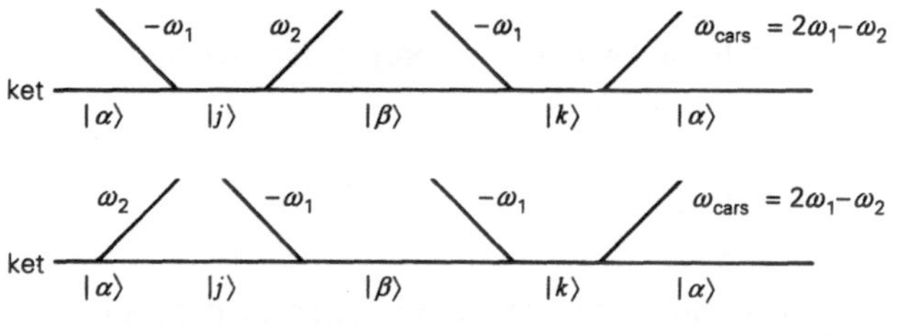

Fig. 11.25. Feynman diagrams for the ground state Raman scattering process

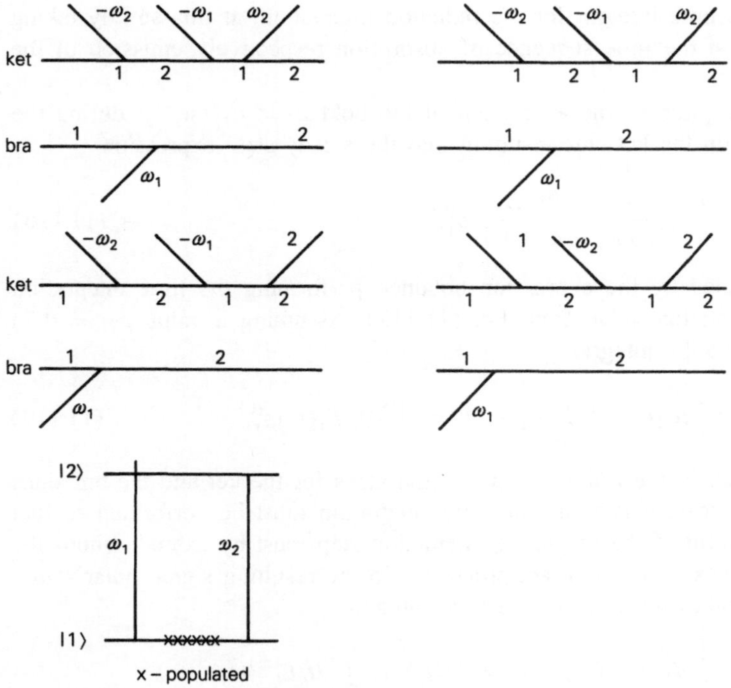

Fig. 11.26. Resonant contributions to the Song, Levenson process of Nonlinear Polarization spectroscopy

The corresponding terms can be easily written down with help of the above given rules. They are in the order of their appearance in Fig. 11.26:

$$\chi^{(3)}(-\omega_2, \omega_2, \omega_1, -\omega_1) = \frac{|\mu_{12}|^4}{\hbar^3} \frac{\rho^0{}_{11}}{[i(\omega_{21} - \omega_2) + \Gamma_{21}]}$$

$$\left\{ \frac{1}{i(\omega_1 - \omega_2) + \Gamma_2} \left[\frac{1}{-i(\omega_{21} - \omega_1) + \Gamma_{21}} + \frac{1}{i(\omega_{21} - \omega_1) + \Gamma_{21}} \right] \right.$$

$$\left. + \frac{1}{\Gamma_2} \left[\frac{1}{-i(\omega_{21} - \omega_1) + \Gamma_{21}} + \frac{1}{i(\omega_{21} - \omega_1) + \Gamma_{21}} \right] \right\} \qquad (11.115)$$

11.8.4 Determination of Molecular Polarizations by the Feynman Diagram Technique in the Case of Time Dependent Fields

The Feynman diagram method is based upon the stepwise perturbative solution of the density matrix for the molecule in the radiation field. The treatment in the case of radiation fields with time dependent amplitudes proceeds by stepwise carrying out the time integral for the radiation interaction, at this strictly taking into consideration the time sequence of absorption respectively emission of the single photons.

So the description of the absorption of the field $E_1 = E_1(t)e^{-i\omega t}$ during the first interaction on the ket side resulting into the steady state expression

$$\rho_{21}(t) = \frac{i}{\hbar}\mu_{21}\frac{e^{i(\omega_{21}-\omega_1)t}E_1}{i(\omega_{21}-\omega_1)+\Gamma}\rho_{11}^{(0)} \tag{11.116}$$

must be substituted by the expression obtained performing the time integration of the ρ_{21} expectation value from Eq. (11.112). Assuming a value $\rho_{21} = 0$ in the field-free case the integral

$$\rho_{21}(t) = \frac{i}{\hbar}\mu_{21}e^{-\Gamma_{21}t}\int_{-\infty}^{t} dt_1 e^{[i(\omega_{21}-\omega_1)+\Gamma_{21}]t_1}E_1(t_1)\rho_{11}^{(0)} \tag{11.117}$$

is easily obtained. In the following interaction steps for the ket and the bra sides a common time scale exists and the time-integration must be performed so that the upper time limit of the preceding interaction step must by extended until the time, when the next interaction act proceeds. So the resulting signal polarization $P(t) = \langle \mu\rho(t)\rangle$ takes the form of the time integral

$$P(t) \propto \int_{-\infty}^{t} dt_1 E_1^{(\pm)}(t_1)\int_{-\infty}^{t_1} dt_2 E_2^{(\pm)}(t_2)\ldots\int_{-\infty}^{t_{n-1}} dt_n E_n^{(\pm)}(t_n)$$

$$* R^{(n)}(t, t_1, t_2, \ldots, t_n) \tag{11.118}$$

with

$$t_n \leqslant \ldots t_2 \leqslant t_1 \leqslant t \tag{11.119}$$

As an example is given the expression for the Raman polarization produced by time dependent pumplaser fields, corresponding to the diagram of Fig. 11.25. It reads

$$P(t) \propto \left(\frac{i}{\hbar}\right)^3 \mu^4 e^{-i(2\omega_1-\omega_2)t}\int_{-\infty}^{t} dt_1 e^{-[i(\omega_{k\alpha}-2\omega_1+\omega_2)+\Gamma_{k\alpha}](t-t_1)}E_1(t_1)$$

$$* \int_{-\infty}^{t} dt_2 e^{-[i(\omega_{\beta\alpha}-\omega_1+\omega_2)+\gamma](t_1-t_2)}E_2^{+}(t_2)$$

$$* \int_{-\infty}^{t_2} dt_1 e^{-[i(\omega_{j\alpha}-\omega_1)+\Gamma_{j\alpha}](t_2-t_3)}E_1(t_3)\rho_{11}^{(0)} \tag{11.120}$$

For constant amplitudes of the radiation fields the integration can be performed and the steady state susceptibility results from Eq. (11.120).

Separate Feynman diagrams for the ket and bra vectors are necessary to obtain a proper treatment of the damping, so it is included as a loss term in the Liouville equation of motion for the density matrix. Compared to the corresponding one-line diagram which would describe the physical process completely in the limit of zero damping – as in this case the ket segments of the diagram fully corresponds to the equally structured bra segment (they give complex conjugated contributions of equal physical content)-, in the damped case additional terms contribute to $\chi^{(3)}$ which must compensate each other in the limit of zero damping. So some of the resonance denominators in the full 48 term expression disappear in the limit of zero damping. Such disappearance of single resonances occurs even under the condition that all the damping derives exclusively from the effect of lifetime broadening by spontaneous emission. For an example of this the reader is referred to problem number 5 in section 11.10.

11.9 References

1. Kroll N (1965) J Appl Phys 36: 34
2. Akhmanow SA, Drabovich KN, Suchorukov AP, Chirkin AS (1970) Zhournal exp i teoret Fiziki 59: 485
3. Carman RL, Shimizu F, Wang CS, Bloembergen N (1970) Phys Rev A 2, 60
4. Lauberau A, Kaiser W (1978); Rev. of Modern Physics 50,607
5. Schubert M, Wilhelmi B (1978) 'Eintührung in die nichtlineare optik qtik' Teubner Verlag, Leipzig
6. Butcher PN, (1965) Nonlinear optical phenomena, Ohio State University, Engineering Publication, Columbus
7. Attal-Trétout B, Bouchary P, Herlin N, Lefebvre M, Magre P, Péalat M, Taran JP (1992) in Springer Proceedings in physics 63: 'Coherent raman spectroscopy', Springer Verlag, p.224
8. Nibler JW, Knighten GW (1979) in Springer Topics in current physics Vol. 11, 'Raman spectroscopy of gases and liquids' ed Weber A
9. Lau A, Werncke W, Pfeiffer M (1990) Spectrochimica Acta Rev 13, 191
10. Pfeiffer M, Lau A, Werncke W (1984) Journ of Raman spectroscopy 15–1, 20
11. Kajzar F, Messier J (1987) in 'Nonlinear optical properties of organic molecules and crystals', ed Chemla DS and Zyss J, Academic Press
12. Werncke W, Pfeiffer M, Johr T, Lau A, Woggons, Schrader S, Jüpner HJ, Freyer W (1994) Nonlinear Optics B7
13. Akhmanov SA, Koroteev NI (1977) Uspechi fiz nauk 123, 405
14. Song JJ, Lee JH, Levenson MD (1978) Phys Rev A 17, 1439
15. Yajima T, Suoma H, Ishida Y (1978) Phys Rev A 17, 324 Yajima T, Morita N and Ishida Y (1984) Phys Rev A 30, 2525
16. Leupold D, Voigt B, Pfeiffer M, Bandilla M, Scheer H (1993) Photochem Photobiol 57, 24
17. Prior A, Bogdan AR, Dagenais AR, Bloembergen N (1981) Phys Rev Lett 46, 111
18. Zinth W, Leonhardt R, Holzapfel W, Kaiser W (1988) J Quant electron QE-24, 455
19. Zinth W, Kaiser W (1980) Opt Commun 32,507
20. Laubereau A, Telle HR (1984) Appl Phys B 34, 23
21. Leonhardt R, Holzapfel W, Zinth W, Kaiser W (1988) in 'Topics in applied phys Vol.60: Ultrashort laser pulses and applications', ed Kaiser W, Springer, p.258
22. Kosic TJ, Cline jr RE, Dlott DD (1984) J Chem Phys 81, 4746
23. Graener H, Lauberau A (1982) Appl Phys B 29, 213
24. Zinth W, Kolmeder C, Benna B, Irgens-Defregger A, Fischer SF, Kaiser W (1983) J Chem Phys 78, 3916
25. Zinth W, Kaiser W (1980) Opt Comm 32, 507
26. Yajima T, Taira Y (1979) Journ of the physical society of Japan 47,1620
27. Hattori T, Terasaki A, Kobayashi T (1987) Phys Rev A 35, 715 Kobayashi T, Terasaki A, . Hattori and . Kurokawa (1988) Appl Phys B47, 107
28. Dugan MA, Melinger JS, Albrecht AC (1988) Chem Phys Lett 147, 411 Dugan MA, Albrecht AC (1991) Phys Rev A 43, 3877 and 3922
29. Müller M, Wynne K, Voorst JDW (1990) JOSA B 7, 1694

30. Bloembergen N, Lotem H, Lynch, jr RT (1978) Indian J Pure appl phys 16, 151
31. Druet SAJ, Attal B, Gustafson TK, Taran JP (1978) Phys Rev A 18, 1529 Druet SAJ, Attal B, Taran JP and Bordé Ch J (1979) J Phys (Paris) 40, 890
32. Oudar JL, Shen YR (1980) Phys Rev A 22, 1141
33. Shen YR (1984) 'The principles of nonlinear optics', John Wiley & Sons, Inc
34. Song JJ, Lee JH, Levenson MD (1978) Phys Rev A17, 1439
35. Prior Y, Bogdan AR, Dagenais M and Bloembergen (1981) Phys Rev Lett 46, 111
36. Bloembergen N (1965) 'Nonlinear optics' . Benjamin A, Inc., New York, Amsterdam
37. Li W, Purucker HG, Lauberau A (1992) Optics Comm 94, 300

11.10 Problems

1. Applying the dispersion formula for the refractive index

$$n(\lambda) = n_o + \frac{C}{\lambda - \lambda_o} \tag{11.121}$$

with n_o=1.3483; c=5.55 nm; λ_o= 177 nm (ethanol) calculate the indexmatching angles for a plane CARS two color arrangement (i.e. $\omega_{pr} = \omega_p$). $\omega_{vib} = 1500\ \text{cm}^{-1}$.
Which are the angles $\langle$(p,s), $\langle$(p, as)?

2. Applying the Maxwell equation for a radiation field with nonlinear source term

$$\nabla * (\nabla * \mathscr{E}_{sgn}) + \frac{n^2}{c^2}\frac{\partial^2}{\partial t^2}\mathscr{E}_{sgn} = -\frac{4\pi}{c^2}\frac{\partial^2}{\partial t^2}P^{nl} \quad \text{with} \quad P^{nl} = X^{(3)}\mathscr{E}_p\mathscr{E}_{pr}\mathscr{E}_s \tag{11.122}$$

derive the linearized Maxwell equation for the amplitude of the signal with plane monochromatic carrier wave. Thus explicit expressions for the coupling constant κ_2 entering into Eq. (11.8) can be obtained (cf. e.g. [36]).

3. Given is an isotropic medium for which the susceptibility of the background contribution to the CARS signal shall obey Kleinman-symmetry. Applying a 3 colour variant of CARS by properly choosing the polarizations of the applied radiation fields it is possible to suppress all signal contributions except that coming from the antisymmetric part of the Raman-anisotropy. Applying Eq. (11.44) for the Raman susceptibility and assuming $X^{NR}_{1122} = X^{NR}_{1212} = X^{NR}_{1221}$ to hold, derive a suitable polarization scheme. Effective term suppression by polarization measures is 10^{-5} [37]. Which values of g_a^2/b^2 are measurable?

4. The spatial development of the autocorrelator for the Stokes-field excited by Stimulated Raman Scattering with a frequency modulated fluctuating laser be derived. The method can be analogous to the procedure with which in Sect. 11.7 the correlator of the material excitation amplitude Eq. (11.108) was calculated.

5. For the four wave mixing experiment characterized by Fig. 11.27 in which bichromatic irradiation (ω_1, ω_2) is applied to the term scheme of the lowest levels of the free Na atom under nearly resonant conditions the signal at $2\omega_1 - \omega_2$ is measured. By applying the Feynman diagram technique to determine the 4 quasi resonant contributions it is to be shown that for broadening by spontaneous emission, for which $\Gamma_{kj} = \Gamma_{kg} + \Gamma_{gi}$ holds a resulting 2 photon resonance contribution

$$\sim \frac{1}{i(\omega_{jk} - \omega_2 + \omega_1) + \Gamma_{jk}}$$

does not exist. From the expression to be derived it must be seen that a corresponding resonant contribution appears as soon as additional collisions at enhanced gas pressure cause Γ_{kj} to become unequal to $\Gamma_{kg} + \Gamma_{gi}$.

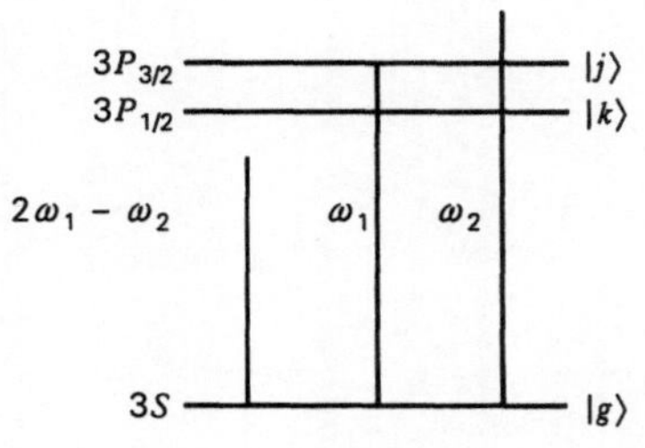

Fig. 11.27. Frequencies for the PIER4[6] [35] experiment and the relevant energy scheme of the free Na atom

12 Transient Grating Spectroscopy

F.-W. Deeg

Key to Symbols

α, α_{ijkl}	Polarizability and polarizability tensor
$\bar{\beta}_0$	Spatially averaged absorption coefficient
$\Delta\beta_0$	Grating modulation depth of absorption coefficient
C	Auger coefficient
C_p	Specific heat at constant pressure
c, c_0	Concentration of excited species
Δc	Grating modulation depth of concentration
$\chi^{(3)}$	second nonlinear term in electric susceptibility, also called third-order optical susceptibility
$\chi^{(3)}_{ijkl}$	third-order optical susceptibility tensor
d	Thickness of sample
D	Mass diffusion coefficient
D_{rot}	Rotational diffusion coefficient
D_{th}	Thermal diffusion coefficient
$\mathbf{E}_i^0$	Electric field vector in a wave propagating with wave vector $\mathbf{k}_i$ and angular frequency ω
$\mathbf{E}_{obj}$	Electric field of object wave
$\mathbf{E}^*$	Electric field complex conjugate to $\mathbf{E}$
ε_r	Electric permittivity relative to electric permittivity of vacuum
ε_R'	Real part of ε_R
ε_R''	Imaginary part of ε_R
$\bar{\varepsilon}_R$	Spatially averaged ε_R
$\Delta\varepsilon_R$	Grating modulation depth of ε_R
η_{eff}	Diffraction efficiency of optical grating
η_{eff}^0	Diffraction efficiency of optical grating for $t = 0$
η_{vis}	Shear viscosity
g, g_{klmn}	Light-material coupling constant and coupling constant tensor
Γ_{ac}	Damping constant of acoustic wave
I	Light intensity
ΔI	Grating modulation depth of light intensity

ΔI_{kl}	Grating modulation depth of light intensity tensor
$K(t)$	Correlation function for excited state population
k_B	Boltzmann constant
$k_r, k_d, k_M, k_D, k_T, k_{eff}$	Relaxation or dissipation rate
k_{heat}	Heat conductivity
$\tilde{\kappa}$	Coupling constant in coupled-wave theory
Λ	Wavelength of optical grating
λ_m	Light wavelength in material
$\Delta N_{el}, \Delta N_{vib}, \Delta N_{rot}$	Grating modulation depth of number of electronic, vibrationally and rotationally excited states
N_{exc}	Number of excited states
Δn_R	Grating modulation depth of refractive index
$\tilde{n}$	Complex refractive index
ω_{ac}	Angular frequency of acoustic wave
$\Delta\omega$	Spectral width of light pulse
p	Probability
Q, Q_{kl}	Material excitation mode
$\mathbf{q}$	Grating vetor
q	Amplitude of grating vector
R	Mean distance between molecules
$r(t)$	Correlation function for orientational motion
r_0	Value of $r(t)$ at $t = 0$
r_{eff}	Effective molecular radius in Debye-Stokes-Einstein model of translational diffusion
ρ	mass density
$\Delta\rho_{el}$	Grating modulation depth of a space charge field
T_{ni}	Nematic-isotropic phase transition temperature
T_g	Calorimetric glass transition temperature
T_c	Critical temperature in mode coupling theory
ΔT	Grating modulation depth of temperature
τ_{ac}	Acoustic wave cycle time
$\tau_r, \tau_{scat}, \tau_{ex}, \tau_{int}, \tau_{pol}$	Life or decay time
τ_p	Pulse width
τ_{12}	Delay time
$\Theta, \Theta_{in}, \Theta_{out}$	Angle of incidence for lightwave, half angle between grating-forming beams
Θ^*	Angle of grating vector with respect to sample
Θ_m	Half angle between grating-forming beams in material
Θ_n	Nutation angle
V_{eff}	Effective molecular volume in Debye-Stokes-Einstein model of rotational diffusion
v_{ac}	Velocity of acoustic wave
∇	Nabla-operator for the spatial derivative in three-dimensional space

12.1 Transient Grating and Four-Wave Interaction

In Chapter 11 a general theory of four-wave mixing (FWM) and the third order nonlinear susceptibility $\chi(3)$ was developed. In this Chapter the basics and applications for a special case of a FWM process the so-called transient grating (TG) or forced light scattering will be considered. This method has found wide-spread use in nonlinear laser spectroscopy and serves as an example for the versatility and the detailed understanding a specific technique allows in the study of light-matter interactions and material dynamics. A more thorough and exhaustive treatment of this subject can be found in the monograph by Eichler, Günter and Pohl [1].

Before describing the specific characteristics and applications the position of forced light scattering within the more general context of FWM shall be clarified in this section. An electric field $\mathbf{E} = \mathbf{E}^0 \cdot \exp[i(\mathbf{k} \cdot \mathbf{r} - \omega t)]$[1] in a material forces the atoms or molecules in the medium to oscillate at the frequency ω. These oscillating dipoles are the source of a macroscopic polarization $\mathbf{P}$ and an electric displacement $\mathbf{D}$ (see Eqs. 3.14–3.18). This process determines the well-known propagation characteristics of light in matter. If three electric fields $\mathbf{E}_i = \mathbf{E}_i^0 \cdot \exp[i(\mathbf{k}_i \cdot \mathbf{r} - \omega_i t)]$ with $i = 1,2,3$ are present at the same time they interact simultaneously with the material and through nonlinear parts in the response of the medium they induce oscillating dipoles and a macroscopic nonlinear polarization (the fourth term in the expansion 3.15)

$$\mathbf{P}^{NL}(\omega_4 = \pm\omega_1 \pm\omega_2 \pm\omega_3) = \varepsilon_0 \chi(3)(\omega_4) \cdot \mathbf{E}_1^0 \mathbf{E}_2^0 \mathbf{E}_3^0 \cdot \exp[i(\mathbf{k}_4 \cdot \mathbf{r} - \omega_4 t)] \tag{12.1}$$

at all possible sum and difference frequencies ω_4. In the direction $\mathbf{k}_4$ defined by the phase matching condition $\mathbf{k}_4 = \pm\mathbf{k}_1 \pm \mathbf{k}_2 \pm \mathbf{k}_3$ this oscillating nonlinear polarization generates an emitted coherent field with the frequency ω_4. The evolution of this field is then described by the time-dependent inhomogeneous wave equation (see Eq. 3.28)

$$\left[\nabla^2 - \frac{\varepsilon_r}{c^2}\frac{\partial^2}{\partial t^2}\right]\mathbf{E}_4 = \frac{1}{\varepsilon_0 c^2}\frac{\partial^2}{\partial t^2}\mathbf{P}^{NL} \tag{12.2}$$

with the nonlinear polarization $\mathbf{P}^{NL}$ as the source term. ε_R is the relative dielectric constant (see Eq. 12.14). All information about the interaction of the light with the specific medium and about the dynamics in this material is contained in the nonlinear susceptibility $\chi(3)$ and this parameter will be treated in the next section.

---

[1] For the convenience in some mathematical operations a notation of the electric fields with complex numbers is used: $\mathbf{E} = \mathbf{E}^0 \cdot \cos(\mathbf{k} \cdot \mathbf{r} - \omega t) = \frac{1}{2}\mathbf{E}^0 \cdot \exp[i(\mathbf{k} \cdot \mathbf{r} - \omega t)] + cc$. cc stands for the conjugate complex $\frac{1}{2}\mathbf{E}^0 \cdot \exp[-i(\mathbf{k} \cdot \mathbf{r} - \omega t)]$. The notation $\mathbf{E}^*$ is in general used for the complex conjugate wave to $\mathbf{E}$.

[2] ∇ is the Nabla-operator for the spatial derivative in three-dimensional space.

The description in Eq. 12.1 is appropriate for the steady state case, i.e. if cw laser beams are used. For the understanding of time-resolved processes (which shall be the focus of this Chapter) a description of the four-wave interaction in the time domain through the introduction of time dependent variables

$$\mathbf{P}^{\mathrm{NL}}(t) = \int_{-\infty}^{\infty} dt_3 \int_{-\infty}^{\infty} dt_2 \int_{-\infty}^{\infty} dt_1 \cdot$$

$$\varepsilon_0 \chi(3)(t - t_3, t - t_2, t - t_1) \cdot \mathbf{E}_3(t_3)\mathbf{E}_2(t_2)\mathbf{E}_1(t_1) \tag{12.3}$$

is more convenient. Eq. (12.3) can be obtained from Eq. (12.1) by a Fourier transformation.

The situation simplifies if all four electric fields considered have the same frequency ω and two of the beams are counterpropagating $\mathbf{k}_1 = -\mathbf{k}_2$. This case is in general referred to as degenerate FWM. For degenerate FWM only three components with different wavevectors remain for the nonlinear polarization [2].

$$\mathbf{P}^{NL}(\omega, \mathbf{k}_4 = \pm\mathbf{k}_1 \pm \mathbf{k}_2 \pm \mathbf{k}_3) = \mathbf{P}^{NL}(\omega, \mathbf{k}_4 = -\mathbf{k}_3) +$$

$$\mathbf{P}^{NL}(\omega, \mathbf{k}_4 = 2\mathbf{k}_1 + \mathbf{k}_3) + \mathbf{P}^{NL}(\omega, \mathbf{k}_4 = -2\mathbf{k}_1 + \mathbf{k}_3) \tag{12.4}$$

with

$$\mathbf{P}^{NL}(\omega, \mathbf{k}_4 = -\mathbf{k}_3) = \varepsilon_0 \chi(3)(\omega) \cdot \mathbf{E}_1(\mathbf{k}_1) \cdot \mathbf{E}_2(-\mathbf{k}_1) \cdot \mathbf{E}_3^*(\mathbf{k}_3) \tag{12.5}$$

$$\mathbf{P}^{NL}(\omega, \mathbf{k}_4 = 2\mathbf{k}_1 + \mathbf{k}_3) = \varepsilon_0 \chi(3)(\omega) \cdot \mathbf{E}_1(\mathbf{k}_1) \cdot \mathbf{E}_2^*(-\mathbf{k}_1) \cdot \mathbf{E}_3(\mathbf{k}_3) \tag{12.6}$$

$$\mathbf{P}^{NL}(\omega, \mathbf{k}_4 = -2\mathbf{k}_1 + \mathbf{k}_3) = \varepsilon_0 \chi(3)(\omega) \cdot \mathbf{E}_1^*(\mathbf{k}_1) \cdot \mathbf{E}_2(-\mathbf{k}_1) \cdot \mathbf{E}_3(\mathbf{k}_3) \tag{12.7}$$

The amplitude of the generated wave $\mathbf{E}_4$ can again be calculated by solving the wave equation (12.2). However, there is a much simpler physical description of the four-wave interaction, namely the grating picture (see Fig. 12.1).

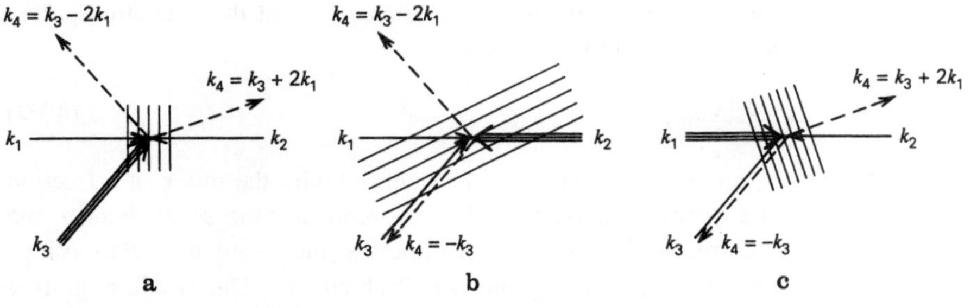

Fig. 12.1. Visualization of the four-wave mixing process as diffraction of an incident wave by the static grating formed by two other incident waves: **a** grating formed by $\mathbf{k}_1$ and $\mathbf{k}_2$ waves, **b** grating formed by $\mathbf{k}_1$ and $\mathbf{k}_3$ waves, and **c** grating formed by $\mathbf{k}_2$ and $\mathbf{k}_3$ waves (after Ref. [2])

Two of the incident waves interfere to generate an optical interference pattern and a stationary grating in the material. (There can also be contributions from moving gratings with an oscillation frequency 2ω which can in general be neglected.) The third wave is now Bragg diffracted (see Sect. 12.3 for a detailed explanation) at this grating leading to a new fourth wave $\mathbf{E}_4$. As implicated by Eqs. 12.4–12.7 and illustrated in Fig. 12.1 there are three different stationary gratings which have to be considered. $\mathbf{k}_1$ and $\mathbf{k}_2 = -\mathbf{k}_1$ interfere to generate a grating with the vector $\mathbf{q}_1 = \mathbf{k}_1 - \mathbf{k}_2 = 2\mathbf{k}_1$ and the electric field $\mathbf{E}_3$ is diffracted in the directions $\mathbf{k}_3 \pm \mathbf{q}_1 = \mathbf{k}_3 \pm 2\mathbf{k}_1$ (Fig. 12.1(a)). Similarly, there is a grating with the vector $\mathbf{q}_2 = \mathbf{k}_1 - \mathbf{k}_3$ generated by the interference of $\mathbf{E}_1$ and $\mathbf{E}_3$. The wave $\mathbf{E}_2$ is diffracted off this grating in the directions $\mathbf{k}_2 \pm \mathbf{q}_2 = -\mathbf{k}_3$ and $\mathbf{k}_3 - 2\mathbf{k}_1$ (Fig. 12.1(b)). The third grating is induced by the interference of $\mathbf{E}_2$ and $\mathbf{E}_3$ and has a vector $\mathbf{q}_3 = \mathbf{k}_3 - \mathbf{k}_2 = \mathbf{k}_3 + \mathbf{k}_1$. The Bragg diffracted electric field $\mathbf{E}_4$ in this case has the directions $\mathbf{k}_1 \pm \mathbf{q}_3$, i.e $\mathbf{k}_4 = -\mathbf{k}_3$ and $\mathbf{k}_3 + 2\mathbf{k}_1$ (Fig. 12.1(c)). The amplitude of the electric field diffracted in a certain direction is calculated by summing up the appropriate terms.

The electric field in the $-\mathbf{k}_3$-direction which always fulfills the phase matching condition has two contributions from the $\mathbf{q}_2$ and $\mathbf{q}_3$ gratings. In transient grating spectroscopy the equivalence of these two contributions is removed and a clear separation between the excitation process, i.e. the generation of an optical grating through the interference of two electric fields, and the probing process, i.e. the Bragg diffraction of a third electric field at the optical grating, is introduced. This is in general accomplished by choosing an amplitude for the exciting electric fields which is much larger than the amplitude of the probing field and in the time-resolved technique by delaying the third pulse with respect to the two exciting pulses.

Using the picture introduced above the basics of the transient grating method can be described in a simple straightforward way. Two short excitation pulses derived from the same laser with the same (center) frequency ω and wave vectors $\mathbf{k}_1$ and $\mathbf{k}_2$ are crossed in a sample to produce an optical interference pattern with a wavevector $\mathbf{q} = \mathbf{k}_1 - \mathbf{k}_2$. The spatial period Λ of the interference pattern is

$$\Lambda = 2\pi/q \tag{12.8}$$

This grating wavelength Λ depends on the wavelength λ_m of the excitation pulses and the angle $2\theta_m$ between the two pulses

$$\Lambda = \lambda_m/(2\sin\theta_m). \tag{12.9}$$

The parameters λ_m and θ_m are evaluated in a material with the index of refraction n_R. One obtains the identical equation (12.9) by replacing the parameters in the medium with the corresponding vacuum parameters and using the relationships $\lambda = n_R \cdot \lambda_m$ and $\sin\theta = n_R \cdot \sin\theta_m$ (but see Problem 2). The smallest grating period Λ_{min} is obtained for antiparallel laser beams ($\theta_m = 90°$)

$$\Lambda_{min} = \lambda_m/2 = \lambda/(2n_R), \tag{12.10}$$

an upper boundary for the grating period ist given by the diameter of the laser beams.

Using the cartesian coordinate system of Fig. 12.2 the grating vector **q** is parallel to the x-axis and the x-dependence of the electric field amplitude in this optical interference pattern is given by

$$\mathbf{E_q} = \tfrac{1}{2}(\mathbf{E}_1^0 e^{ik_x x}) + (\mathbf{E}_2^0 e^{-ik_x x}) + cc. \tag{12.11}$$

Here k_x is the x-component of the propagation vectors $\mathbf{k}_1$ and $\mathbf{k}_2$. This results in an intensity distribution of the form

$$\begin{aligned} I = 2n_R\varepsilon_0 c(\mathbf{E_q}\cdot\mathbf{E_q^*}) &= (n_R/2)\varepsilon_0[|E_1^0|^2 + |E_2^0|^2 + 2\mathbf{E}_1^0\cdot\mathbf{E}_2^0\cos(2k_x x)] \\ &= I_1 + I_2 + 2\Delta I\cos(2k_x x). \end{aligned} \tag{12.12}$$

If the two excitation beams have identical intensities $I_1 = I_2 = I_0$ and are both polarized in the y-direction the expression (12.12) simplifies to

$$I = 2I_0\,[(1+\cos(2k_x x)] = 2\,I_0\,[(1+\cos(qx)], \tag{12.13}$$

i.e., a completely modulated cosine function with minima $I = 0$ and maxima $I = 4\,I_0$.

This optical interference pattern generates (due to various light-matter interactions to be discussed in the next section) a coherent spatially periodic response

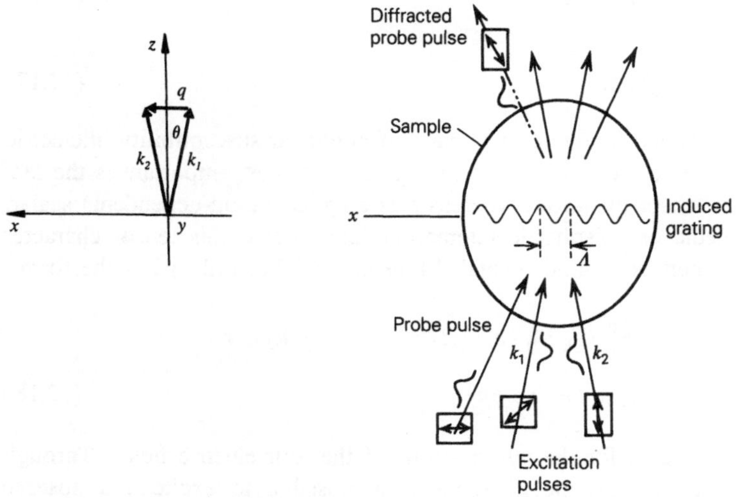

Fig. 12.2. Schematic illustration of the transient grating technique. Two crossed excitation pulses with wave vectors $\mathbf{k}_1$ and $\mathbf{k}_2$ generate a spatially periodic perturbation (wave vector **q**, perodicity Λ) of the sample in the x-direction. This induced modulation acts as a diffraction grating and scatters the probe pulse which is incident on the sample under the correct Bragg angle. As indicated by the *double arrows* the polarizations of all four pulses as well as the time delay of the probe pulse with respect to the excitation pulses can be controlled

in the material, i.e a spatially periodic modulation of the absorption coefficient β_0 and/or the index of refraction n_R. A third probe pulse entering at the correct Bragg angle is diffracted by this optical grating and the amplitude of this diffracted pulse depends on the susceptibility $\chi(3)$ of the material. This process can be illustrated in the schematic way shown in Fig. 12.2.

Whereas FWM processes are usually discussed in terms of the nonlinear susceptibility $\chi(3)$ as used above the transient grating literature in general uses the relative dielectric constant ε_R or the index of refraction and the absorption constant. Depending on the problem discussed one or the other notation is more convenient. The remaining of this Chapter will use ε_R as the central parameter to characterize the optical properties of the material and only in a few instances where necessary invoke the other variables. The relationship between these parameters which was established in Chapter 3 is repeated here:

$$\tilde{n}^2 = \varepsilon_R = 1 + \eta \tag{12.14}$$

where $\tilde{n}$ can be understood as a complex refractive index

$$\tilde{n} = n_R + i\, n_R \frac{\beta_0}{2k} \tag{12.15}$$

with n_R and β_0 being the usual refractive index and absorption coefficient, respectively. k is the magnitude of the wave vector (in the medium). For small absorption coefficients $\beta_0 \ll 2k$ the relations

$$\eta' = \varepsilon_R' - 1 = n_R^2 - 1, \tag{12.16}$$

$$\eta'' = \varepsilon_R'' = n_R^2 \beta_0 / k. \tag{12.17}$$

between the real ($'$) and imaginary ($''$) parts of nonlinear susceptibility, dielectric constant and refractive index follow from Eq. (12.14). Very important is the fact that η and ε_R are tensor quantities whereas $\tilde{n}$ is a (polarization-dependent) scalar. If anisotropic media or anisotropic interactions are studied this tensor character of the optical properties is fundamental. More general Eq. (12.1) has the form

$$P_{\mathrm{i}}^{NL} = \varepsilon_0 \chi_{\mathrm{ijkl}}^{(3)} \cdot E_{1\mathrm{j}}^0 E_{2\mathrm{k}}^0 E_{3\mathrm{l}}^0 \cdot \exp[(i(\pm\mathbf{k}_1 \pm \mathbf{k}_2 \pm \mathbf{k}_3) \cdot \mathbf{r} - (\pm\omega_1 \pm \omega_2 \pm \omega_3)t] \tag{12.18}$$

where i, j, k and l stand for the polarizations of the four electric fields. Through appropriate choice of these polarizations it is possible to excite and observe different components of the susceptibility tensor. The influence of the polarizations is illustrated in Fig. 12.3.

For parallel y-polarization of the two excitation fields there is an intensity modulation as described by Eq. (12.13) (see Fig. 12.3a), therefore this case is often referred to as an intensity grating. If one of the beams is y-polarized and

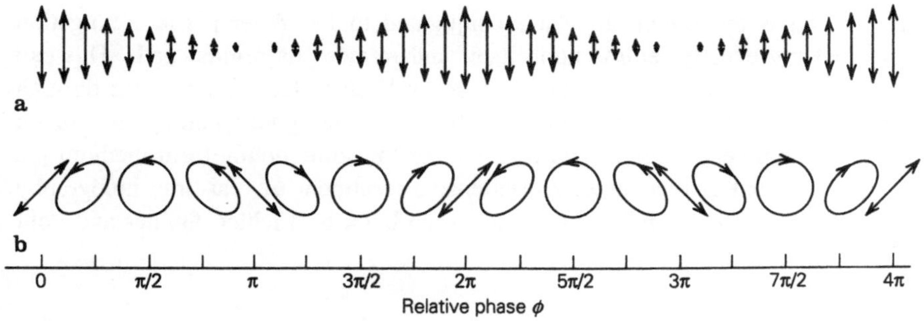

Fig. 12.3. Variation of the electric field polarization in the optical interference pattern for **a** parallel (intensity grating) and **b** perpendicular polarizations (polarization or crossed grating) of the two excitation beams

the other is x-polarized, the two electric fields are perpendicular to each other and the ΔI term in Eq. (12.12) vanishes. There is no intensity modulation due to the interference pattern. However, there is a spatial modulation of the polarization due to the optical interference as indicated in Fig. 12.3b. This alone can give rise to a modulation of the optical properties as will be seen later in the discussion of so-called polarization gratings. It is also possible to take care of these polarization effects in a formal way within Eq. 12.12 by allowing ΔI_{kl} to have tensor character [1].

In the following two sections the general mechanisms which can lead to the formation of a spatially modulated dielectric constant, i.e. the optical grating, and the probing process, i.e. the diffraction at this grating will be discussed separately before several spectroscopic applications of the technique will be treated in more detail.

Two interesting extensions of the FWM or TG approach are real-time holography and phase conjugation. In real-time holography one of the excitation beams $\mathbf{E}_{obj}$ can no longer be described by a simple plane wave. This field $\mathbf{E}_{obj}$, called object wave in holography, originates from reflective scattering at a three-dimensional object and the complex amplitude and phase dependence of $\mathbf{E}_{obj}$ contains the complete information about the form of this object. This information can be stored in the sample in the form of a complex spatial modulation of the dielectric constant and $\chi(3)$ and read out by a Bragg diffracted probe beam for times shorter than the relaxation time τ_0 associated with $\chi(3)$. The most important application is real-time interferometry which allows the characterization of spatial dynamics, e.g. in vibrating objects, with a spatial resolution of the order of the optical wavelength λ and a time resolution of τ_0. Phase conjugation describes the FWM process in which an electric field $\mathbf{E}_1$ is transformed into a field $\mathbf{E}_2$ in which all terms which are not time dependent are replaced by their complex conjugate. That is the wavefront $\mathbf{E}_1 = \mathbf{E}_1^0(\mathbf{r}) \cdot \exp[i(\mathbf{k} \cdot \mathbf{r} - \omega t)]$ incident on a sample generates a wavefront $\mathbf{E}_2 = \mathbf{E}_1^{0*}(\mathbf{r}) \cdot \exp[i(-\mathbf{k} \cdot \mathbf{r} - \omega t)]$. This can e.g. be seen in Eq. (12.5) where the nonlinear polarization $\mathbf{P}^{NL}$ and therefore the generated wavefront $\mathbf{E}_4$ is propagating along $-\mathbf{k}_3$ and has an amplitude proportional to $\mathbf{E}_3^*$, i.e $\mathbf{E}_4$ is the wavefront phase conjugated to $\mathbf{E}_3$. After phase conjugation an incident ray exactly returns upon itself and retraces its orginal path. This can be used to restore wavefronts which are severely distorted after passing through a certain optically nonuniform medium because after phase conjugation in an appropriate nonlinear medium and retraversing the same nonuniform medium the original wavefront is recovered. An extensive treatment of real-time holography and phase conjugation can e.g. be found in the book by Eichler, Günter and Pohl [1].

12.2 Grating Excitation Mechanisms

There are two basically different possibilities for the first step in the formation of a grating. In the first case the frequency ω of the exciting fields is in resonance with a transition in the material, the initial step is absorption of a photon and population of a high-frequency (in general electronic) excited state (resonant case). In the second case the frequency ω is far away from any material resonance and low-frequency excitations in the material are induced through a stimulated scattering process (non-resonant case).

In both cases the incident electric fields E change the population of specific states or modes Q in the material. The population density or excitation amplitude of these modes Q depends on their coupling to the electric field and is in the most simple linear case given by

$$Q_{\mathrm{kl}} = g_{\mathrm{klmn}} \Delta I_{\mathrm{mn}}. \tag{12.19}$$

g can be understood as a generalized optoelastic constant and absorption coefficient in the nonresonant case and resonant case respectively. In general (e.g. anisotropic crystal samples) the tensor character of g must be explicitly considered. If the active modes are linearly coupled to the dielectric susceptibility the induced change of ε_R through the excitation of the mode Q can be expressed as

$$\Delta\varepsilon_{R,ij} = \alpha_{\mathrm{ijkl}} Q_{\mathrm{kl}} \tag{12.20}$$

where in many cases α_{ijkl} can be considered as a kind of generalized polarizability associated with this material mode which is proportional to the nonlinear susceptibility $\chi^{(3)}_{\mathrm{ijkl}}$. Again in the general case this polarizability is a tensor.

From the preceding Eqs. (12.19) and (12.20) it obvious that the induced optical change in the sample depends on two parameters, namely the coupling of the excitation field to the material mode Q through an absorption or scattering process and the sensitivity of the dielectric susceptibility to this mode Q. To elucidate if and to what extent a certain material excitation can be observed in an optical grating experiment both parameters, i.e. g_{klmn} and α_{ijkl}, have to be known.

The first step in the *resonant case* is the interaction of the two time-coincident pulses with the transition dipole moment of the electronic transition setting up an ensemble of coherently oscillating electric dipoles with a periodic spatial modulation defined by the grating period Λ. The coherence decays with the transverse relaxation time T_2 on a femtosecond to nanosecond timescale. With the longitudinal relaxation time T_1 in general being much longer than T_2 this process leaves an incoherent population density grating ΔN_{el}. The relaxation of this primary excited state can give rise to a variety of new gratings as illustrated in Fig. 12.4. Radiationless relaxation can populate other metastable electronic states or vibrational and rotational degrees of freedom and thereby generate vibrational and rotational population density gratings ΔN_{vib} and ΔN_{rot}. Coupling of these states to translational degrees of freedom and thermalization induces a spatial temperature modulation ΔT in the sample. the ensuing thermal expansion induces strain, stress and density modulations in the sample and can generate acoustic waves.

Convective velocity and concentration gratings can be a consequence of the temperature modulation in fluids and fluid mixtures. Photochemical reactions can take place from the primary excited or one of the intermediates states and generate a spatial modulation Δc of new species. If the primary excited state is a weakly bound electron-hole pair, free charge carriers are produced and can establish space charge fields $\Delta\rho_{el}$ in certain materials. All these parameters couple to the dielectric susceptibility ε_R and their respective contributions are determined by the generalized polarizabilities α in Eq. (12.20). The material dynamics associated with all these parmeters can therefore be observed in a transient grating experiment. The timescale of the various processes which can be studied ranges from femtoseconds to days. Several examples will be discussed in the next sections.

So far it has been assumed that the two excitation pulses are time-coincident in the sample. In the following the case of a delay time τ_{12} of excitation pulse 2 with respect to excitation pulse 1 will be discussed. It will turn out that this situation is quite analogous to the polarization evolution in the case of the photon echo with $\pi/2$ and π pulses being not collinear (see Sects. 7.7 and 7.8). As described by the optical Bloch equations Eqs. (7.31–33) the interaction of a single pulse with an ensemble of two level-systems sets up a coherent polarization wave in the medium. If the width τ_p of this pulse is much shorter than T_2, the pseudopolarization vector has the form (see Eq. 7.35)

$$\mathbf{P}(t=0) = P_0[\sin\Theta_n \sin(\omega t - \mathbf{k}_1 \cdot \mathbf{r})\hat{x} + \sin\Theta_n \cos(\omega t - \mathbf{k}_1 \cdot \mathbf{r})\hat{y} + \cos\Theta_n \hat{z}] \tag{12.21}$$

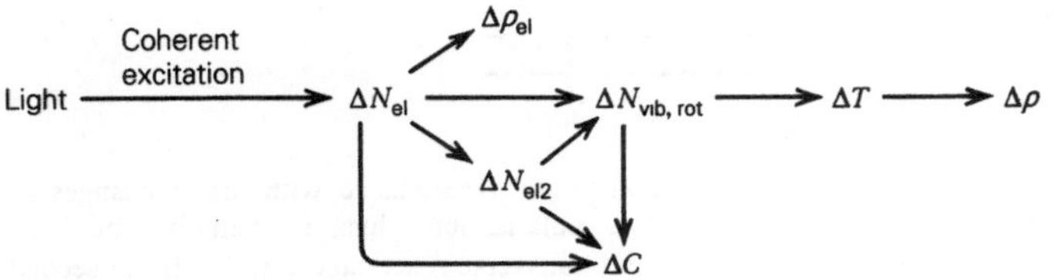

Fig. 12.4. Overview of various grating formation mechanism after population of an electronically excited state

after the pulse. The nutation angle Θ_n is given by

$$\Theta_n = \int_{-\infty}^{\infty} \frac{\gamma E_1^0}{2} dt \qquad (12.22)$$

which for a rectangular pulse of width τ_p simplifies to

$$\Theta_n = \frac{\gamma E_1^0 \tau_p}{2} \qquad (12.23)$$

The important point in Eq. 12.21 is the fact that the information about the phase of the first light pulse is stored in the polarization wave. Although the population density (the $\hat{z}$-component) is spatially homogeneous the macroscopic polarization described by P_x and P_y retains the spatially periodic character of the light pulse $\sin(\omega t - \mathbf{k}_1 \cdot \mathbf{r})$. This is illustrated in Fig. 12.5.

If the spectroscopic transition is homogeneously broadened i.e. $T_2 \ll T_2^*$, this macroscopic polarization (P_x, P_y) decays with the transverse relaxation time T_2 whereas the induced population change represented by P_z relaxes with T_1 which is assumed to be much longer than T_2 and any experimental times considered at

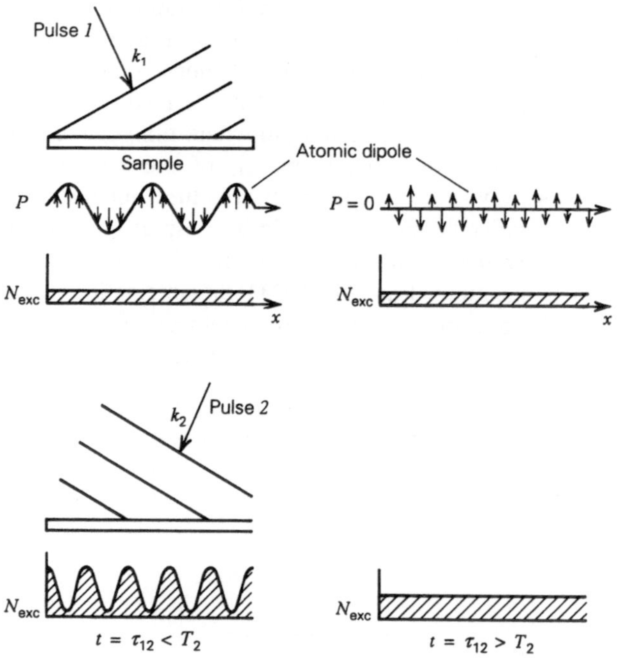

Fig. 12.5. Grating formation by two delayed excitation pulses (reproduced with minor changes of notation from [1]). The first excitation pulse induces a polarization which is spatially modulated according to $\cos(\omega t - \mathbf{k}_1 \cdot \mathbf{r})$. This pattern decays with the transverse relaxation time T_2. If the second excitation pulse hits the sample with a delay $\tau_{12} \leqslant T_2$ its wavefront $\cos(\omega t - \mathbf{k}_2 \cdot \mathbf{r})$ interacts with this polarization pattern to generate an excited state grating (left side). For $\tau_{12} \gg T_2$ this phase memory has been lost and no excited state grating can be induced

the moment. After a delay τ_{12} **P** has the form

$$P(t = \tau_{12}) = P_0[\sin\Theta_n \exp(-\tau_{12}/T_2)\sin(\omega t - \mathbf{k_1} \cdot \mathbf{r})\hat{x} \\ + \sin\Theta_n \exp(-\tau_{12}/T_2)\cos(\omega t - \mathbf{k_1} \cdot \mathbf{r})\hat{y} + \cos\Theta_n \hat{z}]. \quad (12.24)$$

A second light pulse with the wave vector $\mathbf{k_2} \neq \mathbf{k_1}$ but otherwise identical properties and coherent with the first pulse incident on the sample at the time $t = \tau_{12}$ interacts with the pseudopolarization described by Eq. (12.24). As only the population difference is of interest in this case, only the evolution of the z-component P_z during the second pulse

$$P_z(t = \tau_{12} + \tau_p) = P_z(t = \tau_{12})\cos\Theta_n \\ + \tfrac{1}{2}P_y(t = \tau_{12})\sin\Theta_n \cos(\omega t - \mathbf{k_2} \cdot \mathbf{r}) \\ + \tfrac{1}{2}P_x(t = \tau_{12})\sin\Theta_n \sin(\omega t - \mathbf{k_2} \cdot \mathbf{r}) \quad (12.25)$$

has to be calculated here. Inserting the values of **P** at $t = \tau_{12}$ (see Eq. 12.24) into Eq. 12.25 P_z turns out to be

$$P_z(t = \tau_{12} + \tau_p) = P_0[\cos^2\Theta_n + \sin^2\Theta_n \exp(-\tau_{12}/T_2) \\ \times \cos((\mathbf{k_2} - \mathbf{k_1}) \cdot \mathbf{r})]. \quad (12.26)$$

Terms proportional to cos $2\omega t$ can be neglected and $\tau_p \ll \tau_{12}$ has been assumed. As Eq. (12.26) shows, there is a spatially modulated population difference with the periodicity of the optical interference pattern as long as the delay of the two excitation pulses is comparable to T_2 even if there is no temporal overlap between the two pulses. As long as this condition is fulfilled the oscillating atomic dipoles in the sample have a fixed phase relationship with the electric field of the second pulse. If sample dipole and second electric field are in phase the number of excited states, N_{exc} is increased. If sample dipole and second electric field are out of phase, N_{exc} is decreases. For $\tau_{12} \ll T_2$ and the two fields 180 degrees out of phase the second pulse exactly cancels the effect of the first pulse ($N_{exc} = 0$). This phase relationship is described by $\cos((\mathbf{k_2} - \mathbf{k_1}) \cdot \mathbf{r})$. If $\tau_{12} \gg T_2$ (see right side in Fig. 12.5) all phase memory in the sample is lost and no spatially modulated population difference can be generated by the second pulse. The induced excited state grating decays with the excited state lifetime T_1. This means that in the TG-experiment the information about the coherent interaction between the two excitation pulses which occurs only for times $t \leqslant T_2$ is stored for a much longer time, namely T_1. The derivation above has been made for homogeneously broadened transitions. In the case of inhomogeneously broadened transitions $T_2 \gg T_2^*$ and frequencies of the electric field which are detuned from the spectroscopic resonance the situation is much more complex [3]. However, the general conclusion that a spatial grating is generated as long as the delay τ_{12} between the two excitation pulses is smaller than T_2 is still valid.

The first transient grating experiment demonstrating this influence of the phase memory of the sample has been carried out by Styrkov et al. [3] with the setup shown in Fig. 12.6. The 10 ns pulse of a Q-switched ruby laser is split into two excitation pulses which overlap in a ruby crystal which is at very low temperature to allow for a long transverse relaxation time T_2 (at $T = 2.2\,\mathrm{K}\ T_2$ is of the order of 100 ns, T_1 is 3 μs). Excitation pulse 2 can be delayed with respect to pulse 1 through an optical delay line. In this case the probe pulse is produced by reflecting excitation pulse 2 at a flat mirror behind the sample so that it satisfies the Bragg condition. The signal pulse is detected with a fast photodetector and oscilloscope. Typical results are shown in Fig. 12.7.

The upper trace shows the two excitation pulses which are delayed by ca. 50 ns. The lower trace shows two signals. The first one is the diffracted probe

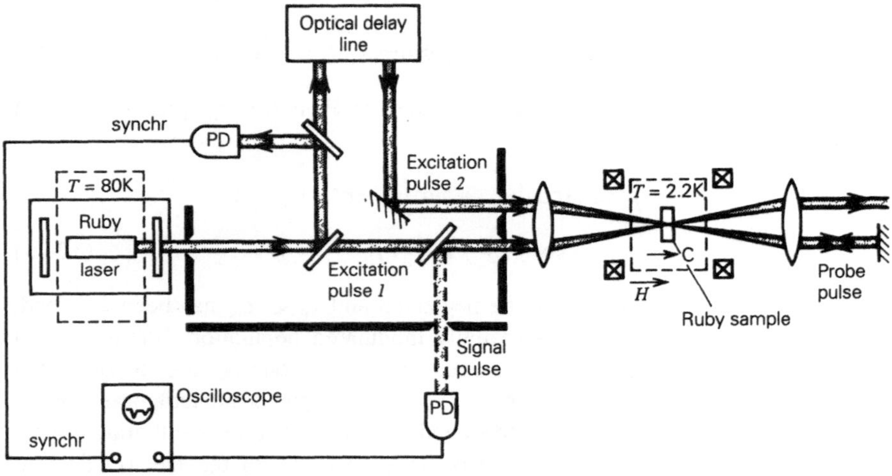

Fig. 12.6. Experimental configuration for the study of grating formation by two delayed excitation pulses in ruby (after Ref. [3])

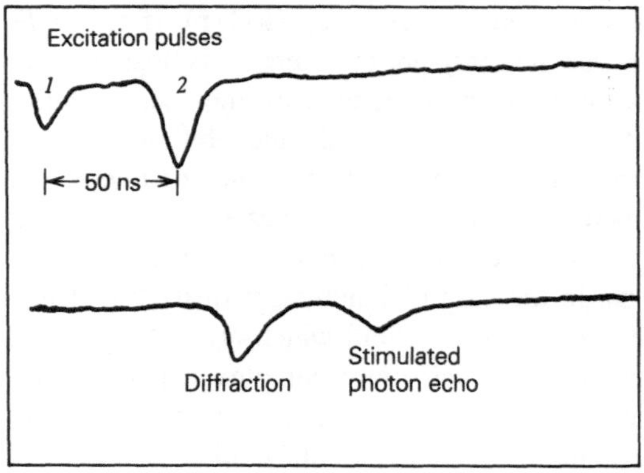

Fig. 12.7. Oscilloscope traces as obtained by a study of ruby with the setup in Fig. 6. The *upper trace* shows the two excitation pulses. The *lower trace* shows signals due to the transient grating and a stimulated photon echo (after Ref. [3])

pulse as discussed above. The second signal is due a stimulated photon echo which forms in the same direction as the transient grating signal. As demonstrated convincingly in Fig. 12.7 there is a diffracted signal although there is no temporal overlap between the two excitation pulses. This signals disappears if either one of the two excitation pulses is blocked.

The first step in the *nonresonant* case is the scattering of a photon with the frequency ω_1 out of light pulse 1 into a photon with the frequency ω_2 in light pulse 2 or vice versa and the concomitant coherent excitation of the material mode Q with the frequency $\omega_Q = \omega_1 - \omega_2$. This inelastic scattering process is not possible if the two excitation beams are purely monochromatic with the same frequency ω_0. However, in the case of excitation with pulses of a duration τ_p the uncertainty relation $\Delta\omega\tau_p \geqslant 1/2$ guarantees a finite spectral width $\omega = \omega_0 \pm \Delta\omega/2$ of the pulses and allows the excitation of modes with a frequency $\omega_Q \leqslant \Delta\omega$. That is, the spectral content of the excitation pulses determines the maximal energy of the material modes Q which can be excited through this stimulated scattering process. This process is completely analogous to the better known stimulated scattering processes in the frequency domain (see preceding Chapter) and has been labeled *impulsive* stimulated scattering (ISS) to emphasize the special characteristics of the excitation process in this time resolved technique [4].

If the energy criterium discussed above is fulfilled and an appropriate coupling (see Eq. 12.19) of the light to the mode Q is given various degrees of freedom can be excited in the sample. ISS excitation of acoustic and optic phonons, of intramolecular vibrations, of librational and reorientational motions has been experimentally demonstrated so far. As in the case of resonant excitation the evolution of this mode, i.e. propagation and relaxation, can be monitored by a time-delayed probe pulse. It should be emphasized at this point that in the impulsive limit $\tau_p \ll \omega_Q^{-1}, \tau_r$, i.e. an excitation pulse width which is much shorter than the oscillation or relaxation time of the mode Q considered, the stimulated scattering process allows a *coherent* excitation of this mode. For example, after excitation of an intramolecular vibration all molecules in the sample oscillate in phase and it is possible to observe the dephasing of this vibration in addition to its depopulation.

12.3 Diffraction at a Grating

To investigate the dynamics of one or several of the processes mentioned in the last section a probe beam has to be diffracted off the optical grating

$$\varepsilon_R(x) = \bar{\varepsilon}_R + \Delta\varepsilon_R \cos(qx) \tag{12.27}$$

associated with these processes. The characteristics of this diffraction process depend on the properties of the optical grating, especially the sample thickness

d. If d is smaller than the grating period Λ the grating is called thin, if $d \gg \Lambda$ the grating is called thick..

The diffraction at a *thin grating* with a sinusoidal modulation as given in Eq. (12.27) is a straightforward extension of the familiar treatment of Fraunhofer diffraction at an array of slits [5] (see Fig. 12.8).

A beam incident on the multiple slit pattern with an angle θ_{in} leads to the generation of a diffracted beam in the direction θ_{out} whenever the pathlength difference for light from two adjacent slits is a multiple m of the optical wavelength:

$$\Lambda(\sin\theta_{\text{out}} - \sin\theta_{\text{in}}) = m\lambda \tag{12.28}$$

That is, diffraction occurs for an *arbitrary angle* of incidence θ_{in} and various diffraction orders can be observed (two of them are shown in Fig. 12.8). The diffraction criterium (12.28) is equivalent to the relationship

$$k_{\text{out, x}} = k_{\text{in, x}} + mq \quad m = 0, \pm 1, \pm 2, \text{ etc.} \tag{12.29}$$

between the x-components of the wavevectors.

The amount of light in the various diffraction orders depends on the exact shape of the optical grating (e.g. rectangular, sawtooth, sinusoidal) and the relative contributions of phase (i.e. n_R) and amplitude (i.e. β_0) modulation. In general the diffraction efficiency is defined as

$$\eta_{\text{eff}} = \frac{I_{\text{diff}}}{I_{\text{in}}} = \frac{\mathbf{E}^2_{\text{diff}}}{\mathbf{E}^2_{\text{in}}}, \tag{12.30}$$

the ratio of diffracted light to incident light. So it can be shown that for a pure sinusoidal amplitude grating the maximum efficiency in the first order is 6.25 % whereas the same number for a sinusoidal phase grating has the value 34 %.

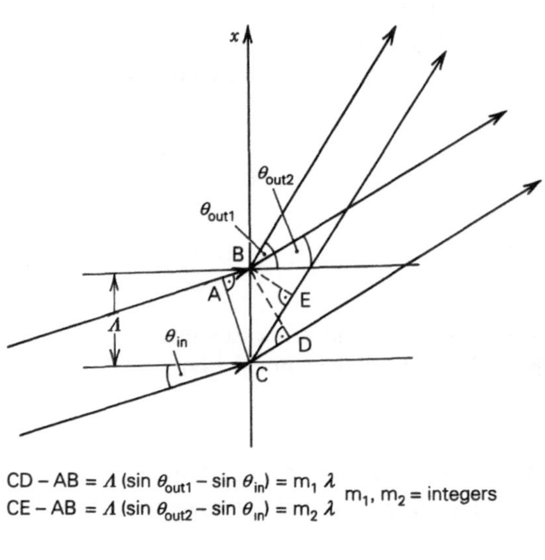

Fig. 12.8. Diffraction at an array of slits. The relations below show the conditions for constructive interference of the transmitted rays

Of more interest in the context of this Chapter is the case $d \gg \Lambda$ or *thick grating*. Under these circumstances efficient diffraction is *only* possible if the *Bragg condition*

$$\sin\theta_{\text{in}} = \frac{m\lambda}{2\Lambda} \quad m = \pm 1, \pm 2, \text{ etc.} \tag{12.31}$$

for the incident beam is fulfilled. A very thorough theory of this subject based on a so-called coupled wave analysis has been given by Kogelnik [6] considering especially the angular and wavelength sensitivities of the diffraction process. In the following Kogelnik's treatment shall be demonstrated for the simplest case of a grating with unslanted fringes being probed by a wave incident at the correct Bragg angle θ_{in} – see Fig. 12.9.

The starting point for the description of the interaction of the electric field with the optical grating $\varepsilon_R(x)$ in the sample is the so-called Helmholtz equation

$$\left[\nabla^2 + \frac{\omega^2}{c_0^2}\varepsilon_R(x)\right]\mathbf{E} = 0 \tag{12.32}$$

which follows from the time-dependent wave-equation (3.28) if a harmonic $\mathbf{E} = \mathbf{E}(\mathbf{r}) \cdot \exp(i\omega t)$ ansatz is used for the electric field. In the following it is assumed that the probing beam is y-polarised so that the electric field in the equation above can be considered a scalar quantity. Inserting Eq. (12.27) into Eq. (12.32) one obtains

$$\left[\nabla^2 + \frac{\omega^2}{c_0^2}(\bar{\varepsilon}_R + \Delta\varepsilon_R \cos qx)\right]E = 0. \tag{12.33}$$

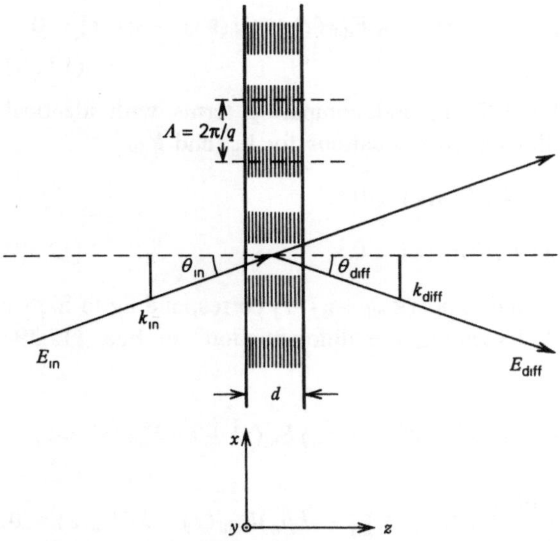

Fig. 12.9. Geometry for Bragg diffraction at a thick grating with nonslanted fringes

Using the complex notation for $\cos qx$ Eq. (12.33) transforms into

$$[\nabla^2 + (k^2 + ik\bar{\beta}_0) + k\tilde{\kappa}(\exp(iqx) + \exp(-iqx))]E = 0 \tag{12.34}$$

with

$$k^2 = \frac{\omega^2}{c_0^2}\varepsilon_R'; \quad \bar{\beta}_0 = \frac{k}{2}\varepsilon_R''; \quad \tilde{\kappa} = \frac{k}{2\varepsilon_R'}\Delta\varepsilon_R. \tag{12.35}$$

k is the magnitude of the (average) wavevector in the sample, $\bar{\beta}_0$ is the (average) absorption coefficient and $\tilde{\kappa}$ is a coupling constant which as will turn out ensures a flow (or more generally an interchange) of energy between the incident wave and the diffracted wave. For $\tilde{\kappa} = 0$ there is no diffraction.

The total electric field in the sample

$$E = E_{\text{in}}(z)\exp(i\,\mathbf{k}_{\text{in}} \cdot \mathbf{r}) + E_{\text{diff}}(z)\exp(i\,\mathbf{k}_{\text{diff}} \cdot \mathbf{r}). \tag{12.36}$$

is assumed to be a superposition of a wave E_{in} with the wavevector $\mathbf{k}_{\text{in}}$ of the incident wave and a wave E_{diff} with the wavevector $\mathbf{k}_{\text{out}}$ defined by the conservation of momentum

$$\mathbf{k}_{\text{diff}} = \mathbf{k}_{\text{in}} + \mathbf{q}. \tag{12.37}$$

$E_{\text{in}}(z)$ and $E_{\text{diff}}(z)$ describe the complex amplitudes of the waves in z-direction which vary slowly due to absorption and exchange of energy. Inserting Eq. (12.36) into Eq. (12.34) one obtains a number of terms corresponding to electric fields with different directions of the wavevector.

$$\begin{aligned}(\nabla^2 + k^2 + ik\bar{\beta}_0)E_{\text{in}}(z)\exp[i\,\mathbf{k}_{\text{in}} \cdot \mathbf{r}] + (\nabla^2 + k^2 + ik\bar{\beta}_0)E_{\text{diff}}(z)\exp[i\,\mathbf{k}_{\text{diff}} \cdot \mathbf{r}]\\ + k\tilde{\kappa}\,E_{\text{in}}(z)\exp[i(\mathbf{k}_{\text{in}} + \mathbf{q}) \cdot \mathbf{r}] + k\tilde{\kappa}\,E_{\text{in}}(z)\exp[i\,(\mathbf{k}_{\text{in}} - \mathbf{q}) \cdot \mathbf{r}]\\ + k\tilde{\kappa}\,E_{\text{diff}}(z)\exp[i\,(\mathbf{k}_{\text{diff}} + \mathbf{q}) \cdot \mathbf{r}] + k\tilde{\kappa}\,E_{\text{diff}}(z)\exp[i\,(\mathbf{k}_{\text{diff}} - \mathbf{q}) \cdot \mathbf{r}] = 0.\end{aligned} \tag{12.38}$$

Taking advantage of relationship (12.37) and comparing terms with identical exponentials one obtains the following two equations for $\mathbf{k}_{\text{in}}$ and $\mathbf{k}_{\text{diff}}$:

$$\begin{aligned}(\nabla^2 + k^2 + ik\bar{\beta}_0)\,E_{\text{in}}(z) + k\tilde{\kappa}\,E_{\text{diff}}(z) = 0,\\ (\nabla^2 + k^2 + ik\bar{\beta}_0)\,E_{\text{diff}}(z) + k\tilde{\kappa}\,E_{\text{in}}(z) = 0.\end{aligned} \tag{12.39}$$

The terms with $\exp[i(\mathbf{k}_{\text{in}} - \mathbf{q}) \cdot \mathbf{r}]$ and $\exp[\,i\,(\mathbf{k}_{\text{diff}} + \mathbf{q}) \cdot \mathbf{r}]$ corresponding to higher diffraction orders are neglected. Executing the differentiations in Eqs. (12.39) these transform into

$$\begin{aligned}\frac{d^2E_{\text{in}}(z)}{dz^2} + 2ik_{\text{in},z}\frac{dE_{\text{in}}(z)}{dz} + (k^2 - \mathbf{k}_{\text{in}}^2 + ik\bar{\beta}_0)\,E_{\text{in}}(z) + k\tilde{\kappa}\,E_{\text{diff}}(z) = 0,\\ \frac{d^2E_{\text{diff}}(z)}{dz^2} + 2ik_{\text{out},z}\frac{dE_{\text{diff}}(z)}{dz} + (k^2 - \mathbf{k}_{\text{diff}}^2 + ik\bar{\beta}_0)E_{\text{diff}}(z) + k\tilde{\kappa}E_{\text{in}}(z) = 0.\end{aligned} \tag{12.40}$$

As can be inferred from Fig. 12.9 (with $\theta = \theta_{\text{in}} = \theta_{\text{out}}$ and $k = k_{\text{in}} = k_{\text{out}}$) the components of the two wavevectors are given by

$$\mathbf{k}_{\text{in}} = k\sin\theta \cdot \hat{x} + k\cos\theta \cdot \hat{z}$$
$$\mathbf{k}_{\text{diff}} = -k\sin\theta \cdot \hat{x} + k\cos\theta \cdot \hat{z} \tag{12.41}$$

Taking into account the relations above and neglecting the second derivatives, i.e assuming a slow energy exchange between $E_{\text{in}}(z)$ and $E_{\text{diff}}(z)$ and small absorption (Eqs. 12.17), Eqs. (12.40) simplify to

$$\frac{dE_{\text{in}}(z)}{dz} = -\frac{\bar{\beta}_0}{2\cos\theta}E_{\text{in}}(z) + i\frac{\tilde{\kappa}}{2\cos\theta}E_{\text{diff}}(z)$$
$$\frac{dE_{\text{diff}}(z)}{dz} = -\frac{\bar{\beta}_0}{2\cos\theta}E_{\text{diff}}(z) + i\frac{\tilde{\kappa}}{2\cos\theta}E_{\text{in}}(z) \tag{12.42}$$

These are the famous coupled wave equations as first derived by Kogelnik [6]. Inserting the coupling constant as given in Eq. (12.35) and solving the equations for the boundary conditions $E_{\text{in}}(z=0) = E_0$ and $E_{\text{diff}}(z=0) = 0$ one obtains

$$E_{\text{in}}(z) = E_0 \exp\left(\frac{-\bar{\beta}_0 z}{2\cos\theta}\right)\cos\left(\frac{k\Delta\varepsilon_R z}{4\varepsilon_R'\cos\theta}\right),$$
$$E_{\text{diff}}(z) = i\,E_0 \exp\left(\frac{-\bar{\beta}_0 z}{2\cos\theta}\right)\sin\left(\frac{k\Delta\varepsilon_R z}{4\varepsilon_R'\cos\theta}\right). \tag{12.43}$$

$\Delta\varepsilon_R$ can be real, imaginary or complex, depending on the fact if the optical grating is a pure phase grating, a pure amplitude grating or a mixed grating respectively. For the general mixed case the diffraction efficiency can be written as the sum of a phase grating term and an amplitude grating term

$$\eta_{\text{eff}} = E^2_{\text{diff}}(d)/E_0^2 = \exp\left(\frac{-\bar{\beta}_0 d}{\cos\theta}\right)\left(\sin^2\frac{k\Delta\varepsilon_R' d}{4\varepsilon_R'\cos\theta} + \sinh^2\frac{k\Delta\varepsilon_R'' d}{4\varepsilon_R'\cos\theta}\right). \tag{12.44}$$

Inserting the relations Eqs. (12.17) and assuming that the induced change of the refractive index is small $\Delta n_R \ll \bar{n}_R$ it is possible to write down the diffraction efficiency as a function of the modulation of the refractive index Δn_R and the absorption coefficient $\Delta\beta_0$:

$$\eta_{\text{eff}}(d) = \exp\left(\frac{-\bar{\beta}_0 d}{\cos\theta}\right)\left(\sin^2\frac{\pi\Delta n_R d}{\lambda\cos\theta} + \sinh^2\frac{\Delta\beta_0 d}{4\cos\theta}\right). \tag{12.45}$$

For small values of Δn_R and $\Delta\beta_0$ the trigonometric functions in Eq. (12.45) can be replaced by their arguments. In spectroscopic applications where only the time dependence of the investigated process is of interest and the absolute value of the induced change of the dielectric constant is unimportant the signal can then be approximated by

$$\eta_{\text{eff}} \propto \Delta n_R^2 + \Delta \beta_0^2 \tag{12.46}$$

An important consequence of the discussion above is the fact that the induced optical grating can be detected with an optical wavelength which is different from the excitation wavelength. This will be important in a number of applications to be discussed later. The only prerequisite is that the correct Bragg angle for this wavelength has been chosen. Readers should be reminded that this is not the case for the detection of holographic patterns which are more complex than the sinusoidal modulation. A straightforward transient grating does not allow to distinguish between a phase and an amplitude grating. However, it is possible to perform so-called *heterodyne* TG-investigations which detect a signal which is *linear* in the phase and absorption change thereby tremendously increasing the detection limit and providing information to separate the real and imaginary parts of $\Delta \tilde{n}_R$.

12.4 Characteristics of the Transient Grating Technique

The most widespread technique in time resolved spectroscopy is the two beam pump-probe concept. In this case a single pulse induces a change of the optical properties of the sample and this change is monitored by a second beam which is, depending on the time resolution required, a CW-beam or a time delayed probe pulse. In the following the focus will be on the specific advantages of the transient grating approach with respect to the simple pump-probe technique.

In the case of the resonant pump-probe method the observable which is detected is the intensity of the probe beam. This has the consequence that there is always a signal on the detector independent of the fact if a change has been induced in the sample by the pump beam. What is monitored are *modulations* of the probe beam through the pump process in the sample, e.g. a decrease of the absorption coefficient due to excitation of molecules and therefore a reduction of the number of molecules in the ground state. So in general small changes of a large number are detected. In contrast to this the transient grating is a so-called zero-background technique. The observable is the *diffracted* probe beam. Diffraction is only possible if an optical grating with the correct grating wavelength has been generated. That is, light is incident on the detector only if certain material modes have been excited by the two pump beams. This leads in general to an increased signal-to-noise ratio and a higher sensitivity of the TG-approach.

If the wavelength used for excitation and the probe is the same the Bragg angle for the probe beam is identical with the incidence angle of the excitation beams and therefore signal and one of the excitation beams have the same direction. To ensure a spatial separation in this case and prevent any light from the excitation beam to hit the detector the beam geometry shown in Fig. 12.10, often

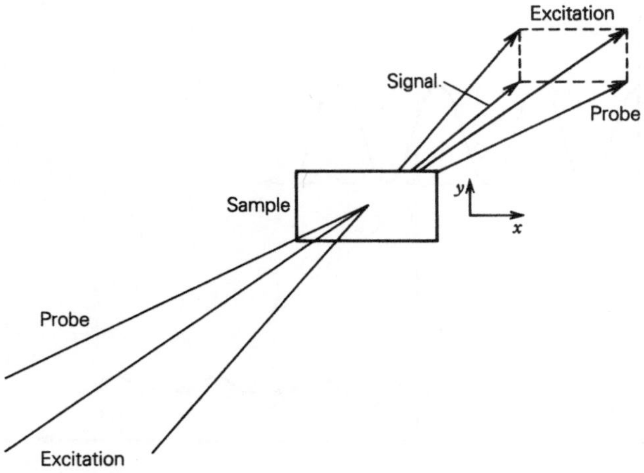

Fig. 12.10. Boxcar geometry for transient grating setup. Using different y-components for the k-vectors of the excitation beams and the probe beam allows a spatial separation of the signal from the excitation beams without violating the Bragg condition

called boxcar, is generally employed. The Bragg angle (Eq. 12.31) is defined only with respect to the grating direction (the x-direction in the notation used here). It is therefore possible to use a probe beam with a k-vector outside of the plane defined by the two excitation beams (i.e. a y-component different from the excitation beams) thereby spatially separating all four beams of the experiment without violating the Bragg condition.

Most important is the fact that the TG-method allows the investigation of a whole new class of phenomena inaccessible to the pump-probe technique. This class comprises all kinds of transport processes. This is due to the fact that the optical grating responsible for the observed signal is modified not only by local relaxation processes but as well by spatial propagation and diffusion phenomena. This is illustrated in Fig. 12.11. Fig. 12.11a shows the spatial modulation of the optical properties due to e.g. the excitation of charge carriers in a semiconductor. This modulation can now decay because of recombination of the charge carriers and relaxation back to the ground state (Fig. 12.11b) or due to a migration of the charges and a spatial equilibration of the charge density (Fig. 12.11c). The intrinsic lengthscale which is monitored in a TG-experiment is the grating wavelength Λ, i.e.. in the order of a few hundred nm to mm (see Eq. 12.9). As shown in several examples in later sections the small grating periods in general are the most interesting. As argued above the TG-signal associated with the grating in Fig. 12.11 decays due to a relaxation as well as a diffusion process. However, whereas the decay because of local processes is independent of the grating wavelength, Λ is decisive for the influence of transport processes on the grating signal. For the simple case of a material excitation which is characterized by a relaxation time τ_r and a diffusion constant D the spatial and temporal evo-

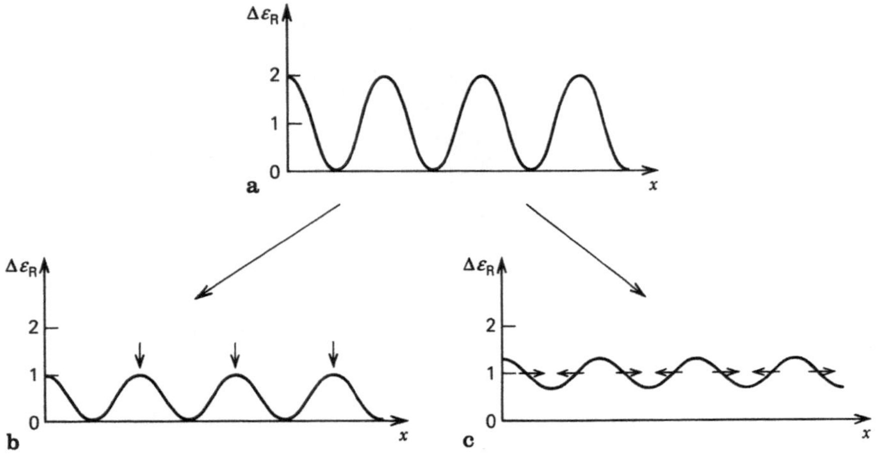

Fig. 12.11. Effect of a local relaxation **b** and a transport process **c** on the evolution of the spatially modulated optical properties

lution of the distribution $c(x, t)$ of the excited states is decribed by a modified one-dimensional diffusion equation

$$\frac{\partial c(x,t)}{\partial t} + \frac{c(x,t)}{\tau_r} = D\frac{\partial^2 c(x,t)}{\partial x^2} \tag{12.47}$$

with the initial distribution determined by the optical interference pattern (see Eq. 12.13)

$$c(x,0) = c_0 \cdot [1 + \cos(2\pi x/\Lambda)] \, . \tag{12.48}$$

For excitation pulses of negligible widths Eq. (12.47) has the solution

$$c(x,t) = c_0 \cdot [1 + \exp(-4\pi^2 Dt\,/\Lambda^2)\cos(qx)]\exp(-t/\tau_r) \, . \tag{12.49}$$

From Eq. (12.49) a grating signal of the form

$$\eta_{\mathrm{eff}}(t) = \eta^0_{\mathrm{eff}} \exp(-2Kt) \tag{12.50}$$

with

$$K = \frac{1}{\tau_r} + \frac{4\pi^2 D}{\Lambda^2} \tag{12.51}$$

can be calculated. Analysis of the grating signal for different values of Λ, i.e. different intersection angles of the excitation beams, with Eq. (12.51) allows the simultaneous evaluation of τ_r and D (for more details see Sect. 12.6).

A third difference with respect to the pump-probe technique is the inherent sensitivity of the TG-experiment to processes which only modulate the refractive index Δn_R and do not effect the absorption coefficient $\Delta\beta_0$ of the sample (see Eq. 12.45). Therefore a TG-experiment allows e.g. the straightforward observation of temperature changes and structural relaxations or density changes in a sample with high time resolution without the need for an absorbing (local) optical probe (see Sects. 12.7 and 12.8). It is also possible to follow any transient birefringence as e.g. associated with the optical Kerr effect.

The fourth major advantage of the transient grating technique or more general of every FWM technique is the fact that through the control of the polarization of *four* beams involved in the experiment it is possible to selectively excite and monitor various processes in the sample which are characterized by different symmetries of the nonlinear susceptibility $\chi(3)$. As an illustration, Fig. 12.12 shows TG-signals from a liquid pentylcyanobiphenyl (5CB) sample [7]. The only change introduced in the experimental setup is the polarization β of the monitored signal. A closer analysis of the polarization dependence shows that the fast component selected in Fig. 12.12d is due to the optical Kerr effect (OKE) and orientational dynamics whereas the slower component dominating the signal monitored in Fig.

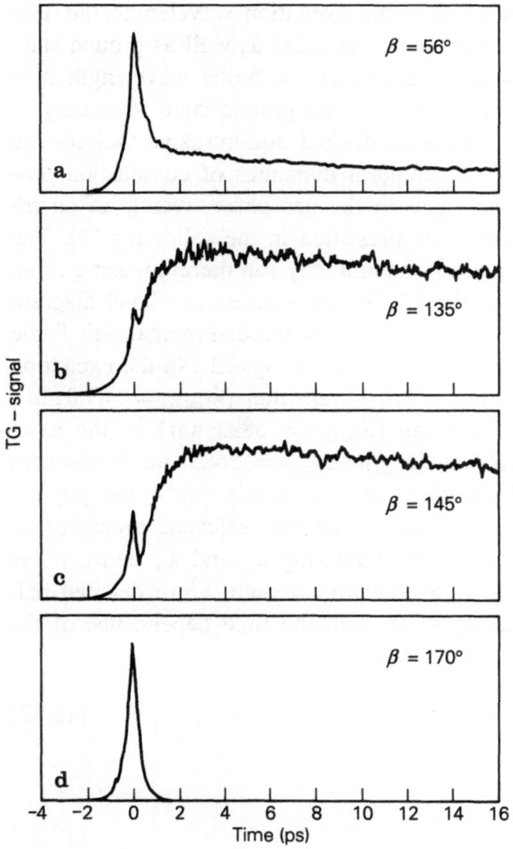

Fig. 12.12. TG-signals of a liquid pentylcyanobiphenyl sample excited and probed with $\lambda = 575$ nm. In all cases the polarization of the two excitation pulses and the probe pulse are 0°, 45°, and 90° respectively. The polarization β of the signal pulse which is monitored changes from **a** to **d** as indicated in the figure (after Ref. [7])

12.12b is due to electronically excited state dynamics. In the intermediate settings Figs. 12a and 12c both processes contribute to the signal. If the origin of the nonitored signal is ambiguous the evaluation of the specific polarization dependence of TG-signals and the associated symmetry of the nonlinear susceptibility $\chi(3)$ allows in many cases the identification of the underlying physical process.

12.5 Depopulation and Orientation Processes

In this section several illustrations for TG-investigations of local processes, i.e. dynamics which do not depend on the grating period Λ, will be given. The most simple example is the measurement of excited state lifetimes. If the excitation wavelength is tuned into a certain resonance, a spatial grating of excited states as well as (depleted) ground states is generated. It is important to keep in mind that in the general case that the excited state does not relax directly back to the ground state the temporal evolution of the excited state and the ground state grating are not identical and the TG-signal depends on the probe wavelength used. If the probe wavelength is identical to the excitation wavelength the TG-signal contains contributions from excited state relaxation as well as ground state recovery. However, through appropriate selection of the probe wavelength it is possible to monitor the dynamics of excited state and ground state separately.

Laser spectroscopy has become a well-established non-intrusive tool for the investigation of the complex spatial and temporal dynamics of combustion processes. As an example for a TG-experiment in the gas phase, results of an investigation of sodium-seeded flames will be presented in the following [8]. The properties of sodium atoms are well understood and they can therefore serve as an optical probe for the flame dynamics. Fig. 12.13 shows an energy level diagram for the Na atom. Fig. 12.14 depicts TG-signals for a Na-seeded methane/air flame with parallel y-polarization of all beams. In the case of Fig. 12.14a the excitation wavelength λ_{exc} is tuned into the $3S_{1/2} \rightarrow 3P_{1/2}$ transition ($\lambda_{NaD_1} = 589.0\,\text{nm}$), in Fig. 12.14b the $3S_{1/2} \rightarrow 3P_{3/2}$ transition ($\lambda_{NaD_2} = 589.6\,\text{nm}$) of the atom is selected. In both cases the probe wavelength $\lambda_{probe} = 568.8\,\text{nm}$ is resonant with the $3P_{3/2} \rightarrow 4D$ transition. This allows to exclusively probe the population/depopulation dynamics of the $3P_{3/2}$ state. The two relevant processes as indicated in Fig. 12.13 are the excited state scattering k_u and k_d between the 3P levels and excited state relaxation k_r to the ground state. From detailed balance arguments it can be shown that $k_u \approx 2k_d$ and the time dependence of the TG-signals in Figs. 12.14a and b should follow

$$\eta^{a}_{eff} \propto \exp(-2k_r t)[1 - \exp(-3k_d t)]^2 \tag{12.52}$$

and

$$\eta^{b}_{eff} \propto \exp(-2k_r t)[1 + 2\exp(-3k_d t)]^2 \tag{12.53}$$

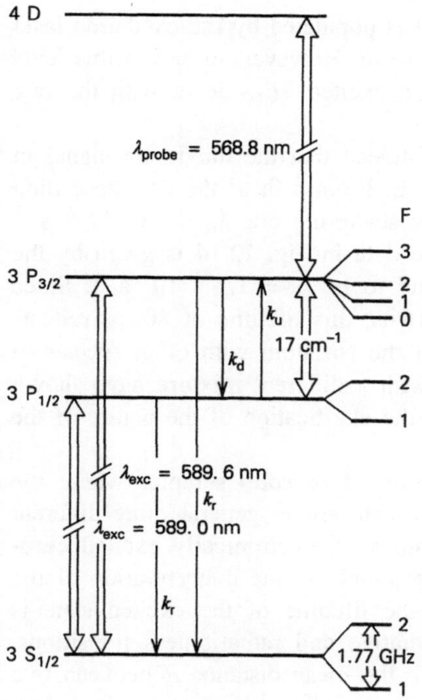

Fig. 12.13. Energy level diagram and kinetic scheme of the Na atom. The *wide arrows* represent the optical transitions accessed in the TG-experiment. The *single line arrows* represent the relevant relaxation processes

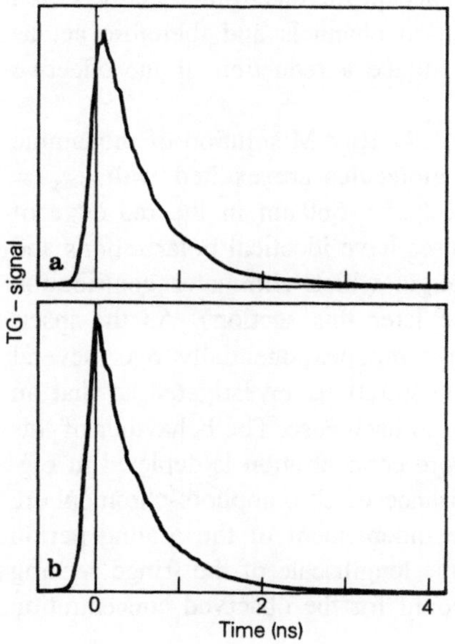

Fig. 12.14. Intensity grating signals for a Na-seeded methane/air flame. **a** Excitation from ground state to $3P_{1/2}$ and probing from $3P_{3/2}$ to $4D$. **b** Excitation from ground state to $3P_{3/2}$ and probing from $3P_{3/2}$ to $4D$. (after Ref. [8])

respectively. In case b the $3P_{3/2}$ level probed is populated by the excitation laser pulses and the rise time given by the pulse width. However, in case a this level is only populated through scattering from the excited $3P_{1/2}$ level with the rate $2k_d$ and there should be a delayed TG-signal.

A closer look at the data demonstrates indeed that the maximum signal in case a is reached at a later time than in case b. From a fit to the data according to Eqs. (12.52) and (12.53), an excited state scattering rate $k_d = 1.6 \times 10^9\ s^{-1}$ can be calculated. The long-tail decay of the data in Fig. 12.14 is given by the relaxation to the ground state which is found to be $k_r = 1.2 \times 10^9\ s^{-1}$. Since the natural Na 3P excited state lifetime is 16 ns, this lifetime of 800 ps reflects excited state quenching through collisions of the Na atom with other species in the flame. Similar investigations in flames with a different mixture have shown variations of this lifetime and are helpful in the clarification of the nature of the quenching species.

The same technique can be applied to liquid or solid samples where the processes limiting the lifetime of an excited state are in general quite different from the gas phase. For example the dynamics of electronically excited chromophores dissolved in a liquid depends very much on the concentration of the dye molecules. At very low concentration the lifetime of the excited state is essentially determined by intramolecular radiative and radiationless transitions. With increasing chromophore concentration c the mean distance R between two molecules decreases and the probability p of a transfer of the electronic energy to another chromophore through the so-called Förster dipole-dipole mechanism

$$p \propto 1/R^6 \propto c^2 \tag{12.54}$$

increases. At the same time an increasing number of dimers (i.e. aggregates of two dye molecules) is formed which can also accept the electronic energy. Often these dimers exhibit fast radiationless relaxation channels and therefore act as efficient traps for the electronic energy and induce a reduction of the effective excited state lifetime.

Figure 12.15a shows the TG-signal for a 1.7×10^{-2} M solution of rhodamine 6G (Rh6G) molecules in glycerol [9]. The molecules are excited with $\lambda_{exc} =$ 355 nm and the probe wavelength used is $\lambda_{probe} = 560$ nm in the red edge of the $S_0 \rightarrow S_1$ transition. The two excitation pulses have identical polarizations and the probe pulse is polarized at the magic angle which eliminates contribution of orientational dynamics on the signal (see later this section). As the insert in Fig. 12.15a demonstrates the signal decays monoexponentially over several lifetimes. The same is true for all Rh6G concentrations investigated so that an effective decay constant k_{eff} can be determined in each case. The behaviour of this effective decay constant versus the chromophore concentration is depicted in Fig. 12.15b offering ample evidence for the importance of chromophore-chromophore interactions. The TG-signals in all cases are independent of the grating period proving that long-range energy transport on the lengthscale of the fringe spacing is negligible (see Eq. 12.51) and cannot account for the observed concentration dependence.

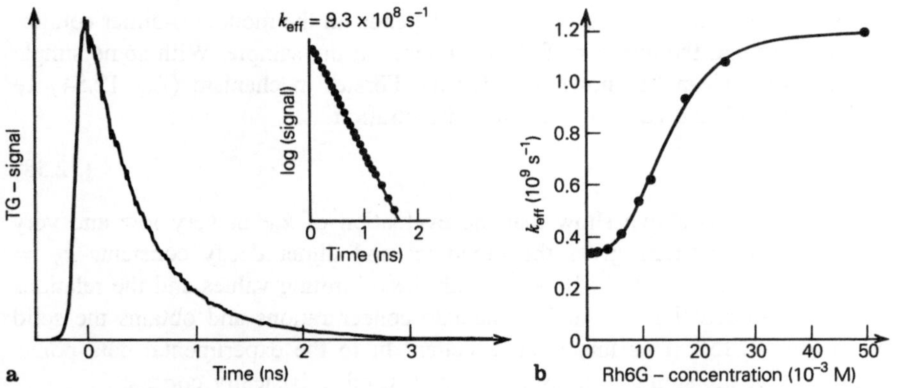

Fig. 12.15. a TG-signal for a 1.74×10^{-2} M solution of rhodamine 6G in glycerol. The molecules are excited with $\lambda_{exc} = 355$ nm and probed under the magic angle with $\lambda_{probe} = 560$ nm. The effective decay constant in this case is $k_{eff} = 9.3 \times 10^8$ s^{-1} **b** Effective decay constant vs concentration of rhodamine 6G in glycerol. The *crosses* indicate experimental data points, the *solid line* is calculated as explained in the text. (after Ref. [9])

However, a simple calculation based on the monomer/dimer model presented above can explain the results very well. In this model the rate equations governing the excited state populations are given by

$$dc_{M^*}/dt = -k_M c_{M^*} - k_T c_{M^*} \tag{12.55}$$

and

$$dc_{D^*}/dt = -k_D c_{D^*} + k_T c_{M^*} \,. \tag{12.56}$$

In these equations c_{M^*} and c_{D^*} are the concentrations of excited monomers and dimers respectively, k_M and k_D are the (intramolecular) decay constants of monomer and dimer and k_T is the trapping rate, the energy transfer rate between monomer and dimer. Now if monomer and dimer absorb at the probe wavelength with the same probability the TG-signal is proportional to the square of the sum of excited monomers and dimers (it can be shown that the following line of argumentation is also true for different absorption probabilities of monomer and dimer [9]). Solving Eqs. (12.55) and (12.56) for monomer and dimer concentration the sum of excited states has the form

$$c_{N^*}(t) \propto \exp[-(k_M+k_T)t]+[k_T/(k_M+k_T-k_D)]\cdot\{\exp(-k_D t)-\exp[-(k_M+k_T)t]\}. \tag{12.57}$$

For a better understanding of Eq. (12.57) some special cases are disscused. If trapping is negligible, i.e. k_T is small, c_{N^*} decays with the monomer rate constant k_M. If the trapping process is much faster than any intramolecular decay process, i.e. $k_T \gg k_M, k_D$, the decay is dominated by the dimer rate constant k_D. And if the dimer decay k_D is the fastest of the three processes the excited state concentration decays with the rate constant $(k_M + k_T)$.

The value of the trapping constant k_T depends on the monomer-dimer equilibrium constant and the number of chromophores in the sample. With some simple considerations it can be shown that for the Förster mechanism (Eq. 12.54) k_T should vary with the cube of the dye concentration

$$k_T \propto c^3. \tag{12.58}$$

The considerations above show that the evaluation of k_{eff} at very low and very high dye concentrations gives the monomer and dimer decay constants $k_M = 3.3 \times 10^8\,\text{s}^{-1}$ and $k_D = 1.2 \times 10^9\,\text{s}^{-1}$. With these limiting values and the relations 12.58 one can calculate k_{eff} at intermediate concentrations and obtains the solid line in Fig. 12.15b. This line is an excellent fit to the experimental data points indicating that the simple microscopic model used is basically correct.

If light pulses with a well-defined polarization are employed in the type of experiments described above then the influence of orientational motion of the excited species (we will restrict ourselves to molecules in the following) on the TG-signal must be considered explicitly. This is due to the fact that the excitation probability of a molecule is proportional to the square of the dot product $|\mathbf{p}_\mathbf{E} \cdot \mathbf{E}|^2$ of electric field amplitude $\mathbf{E}$ and transition dipole moment $\mathbf{p}_\mathbf{E}$, i.e the excited species have their transition dipole moment selectively aligned parallel to the exciting electric field. Therefore an anisotropic distribution of excited state molecules as well as depleted ground state molecules with corresponding anisotropic optical properties is created. If these molecules undergo rotational motion the anisotropic orientational distribution and the associated optical anisotropy decays. An appropriately designed polarization-selective experimental setup allows to follow the orientational dynamics of the molecules.

A detailed theory of the influence of photoselection and orientational relaxation on TG-experiments has been given by Jena and Lessing [10]. We will only discuss the results of their analysis for the simple case of a two-level system with no complicating intermediate level kinetics. The time-dependent behaviour of the anisotropic excited state population can be separated into two correlation functions $r(t)$ and $K(t)$ describing the orientational motion and the excited state population dynamics respectively. If two excitation pulses with parallel polarization are used, an intensity grating configuration (see Fig. 12.3a), a spatially modulated pattern of excited states with an anisotropic orientational distribution is generated. The efficiency of this grating depends on the polarization of the probe pulse. For a polarization parallel ($\|$) and perpendicular ($\perp$) to the excitation pulses theoretical TG-signals of the form

$$\eta_{\text{eff}}(\|) = \eta_{\text{eff}}^0 \{K(t)[1 + 2r(t)]\}^2 \tag{12.59}$$

and

$$\eta_{\text{eff}}(\perp) = \eta_{\text{eff}}^0 \{K(t)[1 - r(t)]\}^2 \tag{12.60}$$

are found. In both cases the signal contains contributions from rotational motion as well as excited state decay. By the appropriate combination of the results from the two experiments it is possible to evaluate $K(t)$ and $r(t)$ separately:

$$K(t) = 1/(3\eta^0_{\mathrm{eff}})[\eta_{\mathrm{eff}}(\|)^{1/2} + 2\eta_{\mathrm{eff}}(\perp)^{1/2}] \tag{12.61}$$

$$r(t) = 1/K(t)[\eta_{\mathrm{eff}}(\|)^{1/2} - \eta_{\mathrm{eff}}(\perp)^{1/2}] \tag{12.62}$$

It is actually possible to choose an angle between the polarization of the probe pulse and the excitation pulses, the so-called magic angle = 54.7° mentioned further above, at which the contribution of the orientational dynamics to the TG-signal disappears completely and pure excited state relaxation is observed.

Orientational motion of molecules in liquids can in general be characterized by a Brownian type rotational diffusion with small angle steps. Using a hydrodynamic approach to this problem it can be shown that the associated rotational diffusion constant D_{rot} is related to the viscocity η_{vis} of the liquid and an effective volume V_{eff} of the rotating entity through the familiar Debye-Stokes-Einstein equation:

$$D_{\mathrm{rot}} = \frac{k_B T}{6\eta_{\mathrm{vis}} V_{\mathrm{eff}}}. \tag{12.63}$$

For a molecule of complex shape rotational motion around different molecular axes and therefore several rotational diffusion constants must be considered and the correlation function $r(t)$ can be complicated. However, for a molecule with an approximately rod-like shape only rotation around axes perpendicular to this rod axis is relevant and if the transition dipole moment is parallel to this rod axis the orientational correlation function reduces to $r(t) = r_0 \exp(-t/\tau_{\mathrm{rot}})$. Describing the excited state decay also with a single lifetime τ_{ex} the TG-signals in Eqs. (12.59) and (12.60) have the simple form

$$\eta_{\mathrm{eff}}(\|) = \eta^0_{\mathrm{eff}} \exp(-2t/\tau_{\mathrm{ex}})\,[1 + 2r_0 \exp(-t/\tau_{\mathrm{rot}})]^2 \tag{12.64}$$

and

$$\eta_{\mathrm{eff}}(\perp) = \eta^0_{\mathrm{eff}} \exp(-2t/\tau_{\mathrm{ex}})\,[1 - r_0 \exp(-t/\tau_{\mathrm{rot}})]^2. \tag{12.65}$$

Figure 12.16a shows typical TG-signals from an investigation of the chromophore rhodamine B in 1-propanol [11]. In both cases the dye molecules are excited through an intensity grating configuration and then the grating is probed with parallel and perpendicular polarization respectively. At early times the signals are dominated by the decay of the optical anisotropy through orientational dynamics. After ca. 2 ns the equilibrium orientational distribution has been restored, the amplitude of the two signals is identical as expected for an isotropic distribution of excited states, and the time dependence of the signal is determined by excited state relaxation alone. Combination of the TG-signals according to Eqs. (12.61) and (12.62) and a logarithmic plot of the resulting data as shown in Fig. 12.16b demonstrate that the data can be described by a single rotational diffusion time and excited state lifetime with the values $\tau_{\mathrm{rot}} = 480\,\mathrm{ps}$

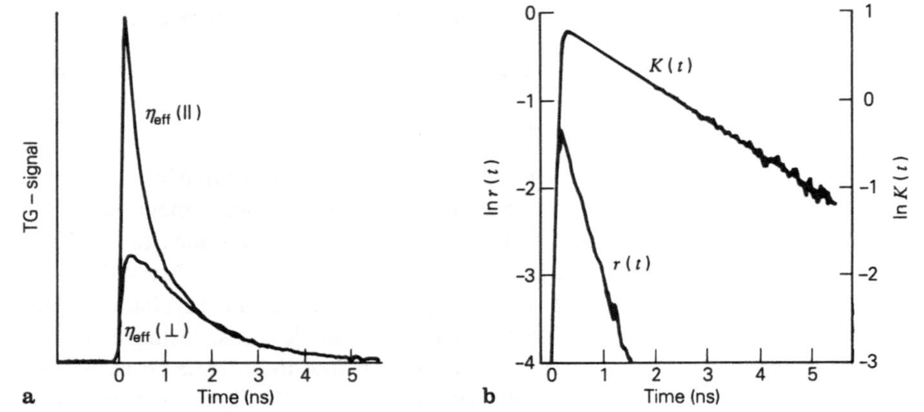

Fig. 12.16. **a** Polarization dependence of TG-signals for rhodamine B in 1-propanol. In both cases shown the sample is illuminated by two excitation pulses with parallel linear polarization and probed by a pulse with linear polarization either parallel or perpendicular to the excitation pulses. **b** Reorientational and excited state correlations functions $r(t)$ and $K(t)$ as calculated from the signals in **a**) with Eqs. (12.61) and (12.62), (after Ref. [11])

and $\tau_{ex} = 2.6\,\text{ns}$. The results for the rotational diffusion of rhodamine B in a series of different alcoholic solvent with a large range of viscosities are consistent with the Debye-Stokes-Einstein equation (12.63) and can be explained by classic hydrodynamic models [11].

12.6 Electronic Energy Transfer and Charge Carrier Dynamics

This section will focus on TG-experiments concerned with excited state transport and charge carrier diffusion. As mentioned in the last section, an early investigation [9] of the excited state dynamics of rhodamine 6G monomer and dimer molecules in liquid glycerol showed no indication of long-range excited state transport. A more recent experiment [12], however, has revealed that this long-range transport can indeed be observed if extremely small grating periods are used. The authors used disodium fluorescein as the probe molecule because there is no dimer and aggregate formation for this chromophore even at concentrations above 5×10^{-2} mol/l employed in the experiment. Therefore the trapping processes dominating the high concentration data in Ref. [9] are negligible in the fluorescein case.

The most obvious way to test for transport is an evaluation of the grating decay constant versus the grating period Λ (see Eq. 12.51). This technique was not applicable in the fluorescein /ethanol sample studied due to acoustic effects (see Sect. 12.8) obscuring the excited state dynamics. However, at the smallest grating period which can be achieved (Λ = 97.7 nm with counterpropagating beams $\theta = 90°$) these acoustic effects can be suppressed. Therefore the authors studied the TG-decay at this fixed grating period and varied the chromophore concentration. The TG-decays for two different concentrations (3×10^{-3} mol/l and 3×10^{-2} mol/l) are shown in Fig. 12.17. In the low concentration sample there is no contribution from energy transfer and these TG-data are used to determine the excited state lifetime $\tau_{ex} = 5.0$ ns of fluorescein. With this value of τ_{ex} and Eq. (12.51) the higher concentration decays allow the direct evaluation of the electronic energy diffusion constant D_e. The inset in Fig. 12.17 shows that for concentrations up to 3×10^{-2} mol/l D_e is proportional to $c^{4/3}$. This is in very good agreement with theoritical work which has shown that in a three-dimensional disordered system the Förster transition dipole-dipole mechanism implicates a diffusion constant D_e with a concentration dependence given by $c^{4/3}$ [13]. Therefore these TG-experiments are a direct confirmation of long-range electronic energy transfer in solution. The Förster transfer distance parameter, i.e. the critical distance still allowing efficient energy transfer, for fluorescein is 50 Å. The deviation at high concentration evident in the inset in Fig. 12.17 is attributed to the finite size of the molecules which is not included in the theoretical model.

Very similar TG-experiments on spatial electronic energy transport have been performed in a number of solid state materials. One example are single crystals of $Nd_xLa_{1-x}P_5O_{14}$. In these crystals an excited state of the Nd^{3+} ions is populated and the dynamics of these excited states are monitored. At room temperature the grating decay in these systems is monoexponential. Plotting the decay constant measured for various fringe spacings versus $1/\Lambda^2$ as shown in Fig. 12.18 reveals a linear behaviour as expected from Eq. (12.51). This proves that there is indeed diffusive electronic energy transport on the μm-lengthscale of the fringe spacing.

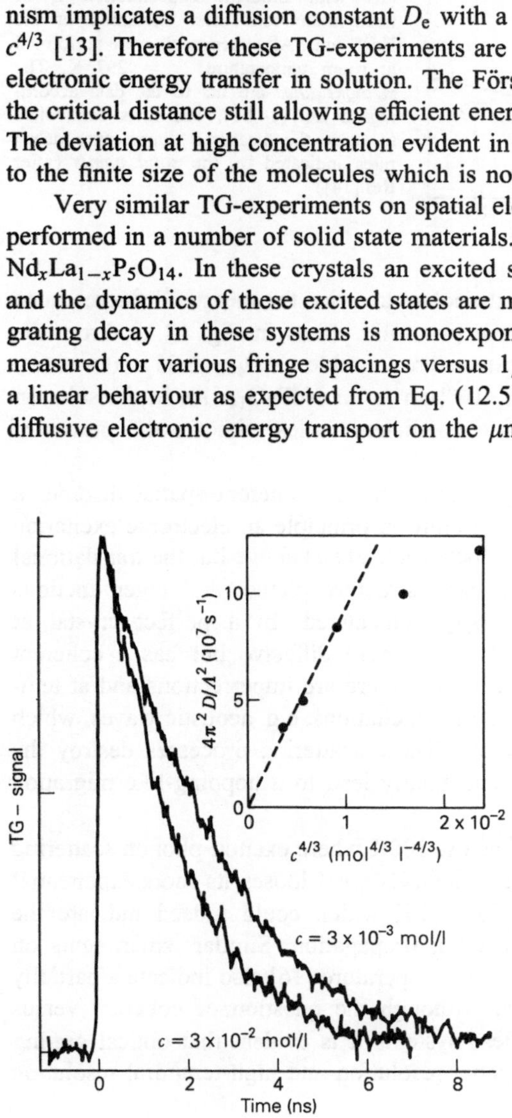

Fig. 12.17. TG-decays for two different fluorescein concentraions. The grating period in these studies has the value $\Lambda = 97.7$ nm. The *inset* shows the concentration dependence of the energy transport diffusion constant determined this way which scales with $c^{4/3}$ at low concentration consistent with theory (after Ref. [12])

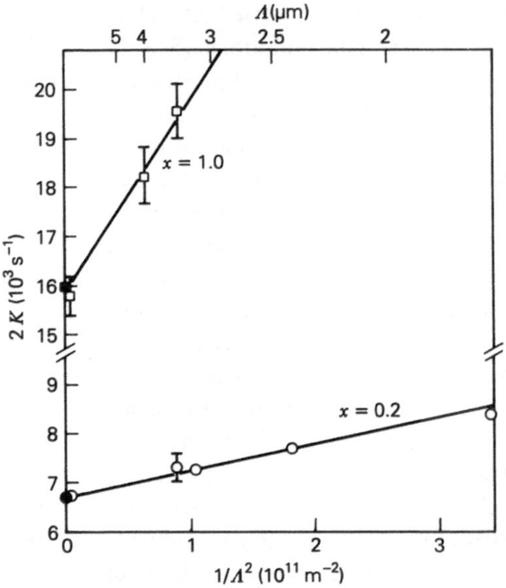

Fig. 12.18. Grating decay constant versus grating period for two $Nd_xLa_{1-x}P_5O_{14}$ crystals with different concentration x of neodymium ions after excitation of the Nd^{3+} ions. All experiments were performed at room temperature $T = 297$ K. The *straight lines* are fits to the data according to Eq. (14.51). The lines extrapolate to twice the measured fluorescence decay rates indicated by the *solid points* (after Ref.[14])

For $\Lambda \to \infty$, i.e. $1/\Lambda^2 \to 0$ the fits extrapolate to twice the measured fluorescence decay rates which are indicated by the two solid points in Fig. 12.18. From the slopes in Fig. 12.18 the diffusion constants are found to be $D_e = 6.85 \times 10^{-7}\,cm^2/s$ for $Nd_{0.2}La_{0.8}P_5O_{14}$ and $D_e = 5.09 \times 10^{-6}\,cm^2/s$ for NdP_5O_{14}. As in the solution example discussed above the diffusion coefficient varies with the 4/3 power of the concentration.

However, in contrast to the liquid solution with inherent spatial disorder a crystal exhibits spatial symmetries. Therefore in principle an electronic excitation can no longer be treated as a spatially localized state as above but the translational symmetry of the system leads to a band structure picture and eigenfunctions (so-called excitons) which are spatially delocalized. In a perfect crystal at $T = 0$ K electronic energy transport is no more diffusive but has a coherent wave-like nature. Howeever, in a real crystal there are imperfections and at temperatures above 0K there are other material excitations, e.g. acoustic waves, which act as scattering centers for the exciton. These scattering processes destroy the coherent propagation of the exciton and finally lead to a hopping-like migration as discussed above.

Cooling a NdP_5O_{14} sample well below 100K where exciton-phonon scattering processes should become less frequent the TG-signal looses its monoexponential character and exhibits clear oscillations [15] which could indeed indicate the transition to a coherent wave-like exciton propagation. Similar experiments on anthracene molecular crystals at very low temperature [16] also indicate a partially coherent electronic energy transport. Although the question of coherent versus diffusive transport is still open in these systems it is evident that optical grating techniques with their well-defined spatial resolution and high temporal resolution have made important contributions.

Through the orientation of the two excitation beams with respect to the sample the direction of the grating vector can be selected and therefore anisotropies of propagation processes can be easily characterized. This is very evident in an investigation of excitonic transport in a GaAs/AlGaAs multiple quantum well structure [17]. A quantum well is very thin layer of a semiconductor with a smaller band gap (e.g. GaAs) embedded between two semiconductor layers with a larger band gap (e.g. AlGaAs). Within the layer the material characteristics are similar to the bulk, however, perpendicular to the layer the electronic excitation is confined because of the different band gap energies. Through molecular beam epitaxy it is possible to prepare heterostructures with many alternating semiconductor layers on top of each other and create so-called multiple quantum wells with a topology as indicated in Fig. 12.19. Through appropriate orientation of the two excitation pulses spatial gradients of exciton /charge carrier concentrations can be created parallel to the semiconductor layer (Fig. 12.19a), perpendicular to the quantum wells (Fig. 12.19b), or at an intermediate angle (Fig. 12.19c). Besides the pure lifetime contribution the TG-decay is due to carrier transport within the layer in case (a), due to interwell hopping in case (b) and has contributions from both processes in case (c).
The data presented in Fig. 12.20 show (in semilogarithmic plots) TG-decays for a sample consisting of 120 periods of 65-Å thick GaAs quantum wells with 212-Å thick $Al_{0.4}Ga_{0.6}As$ barriers. The data in Fig. 12.20a are for fringe spacings Λ between 3 and 7 μm and the optical grating parallel to the layers ($\theta = 90°$). These data are consistent with a diffusive carrier transport and allow the evaluation of the carrier lifetime and intrawell transport which for this specific sample turn out to be $\tau_r = 4\,\text{ns}$ and $D = 13.8\,\text{cm}^2/\text{s}$. This diffusion coefficient is in excellent agreement with values for pure bulk GaAs demonstrating the fact that within the layer electronic charge transport is comparable to the bulk behaviour.

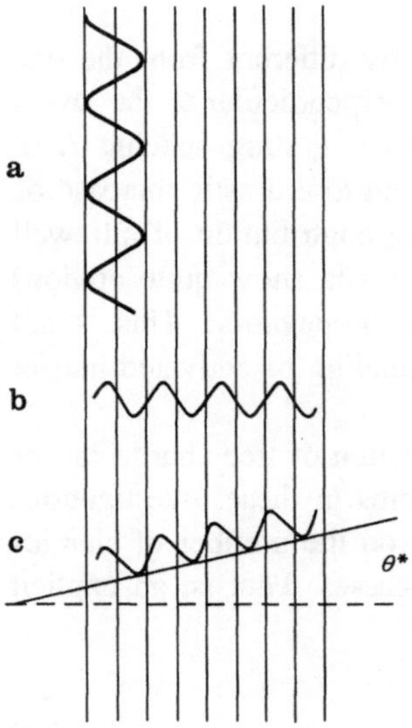

Fig. 12.19. Multiple quantum well structure with different transient grating configurations indicated by the optical interference pattern. For case **a** intrawell transport within the GaAs layer is observed, for case **b** interwell propogation is monitored. For the intermediate configuration **c** the transport has contributions from intrawell and interwell propagation (after Ref. [17])

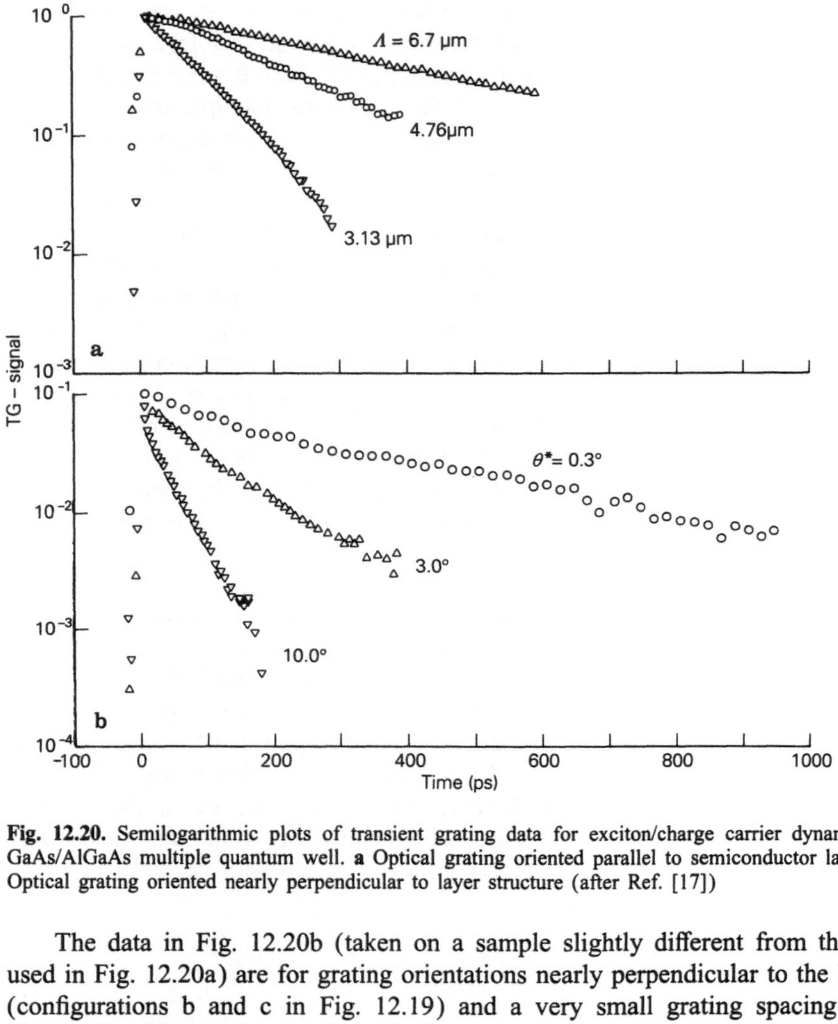

Fig. 12.20. Semilogarithmic plots of transient grating data for exciton/charge carrier dynamics in GaAs/AlGaAs multiple quantum well. **a** Optical grating oriented parallel to semiconductor layers. **b** Optical grating oriented nearly perpendicular to layer structure (after Ref. [17])

The data in Fig. 12.20b (taken on a sample slightly different from the one used in Fig. 12.20a) are for grating orientations nearly perpendicular to the layers (configurations b and c in Fig. 12.19) and a very small grating spacing $\Lambda \simeq 120\,\mathrm{nm}$. Rotation of the sample and a larger angle θ^* lead to a drastic increase of the grating decay constant. This is due to the increasing contribution of intrawell transport (see Problem 6) with increasing θ^*. So these data show quite obvious that interwell transport is much slower than intrawell propagation. This is not surprising as interwell diffusion is only possible by tunneling or activated barrier crossing.

TG-spectroscopy has also been applied to the evaluation of free charge carrier transport in pure silicon [18]. One of the major problems in these investigations is the fact that the carrier lifetime τ_r depends strongly on the number of charges in the sample due to carrier-carrier recombination processes. That is, an explicit concentration dependence

$$\frac{c(t)}{\tau_r} = Ac(t) + Bc^2(t) + Cc^3(t) + \ldots \tag{12.66}$$

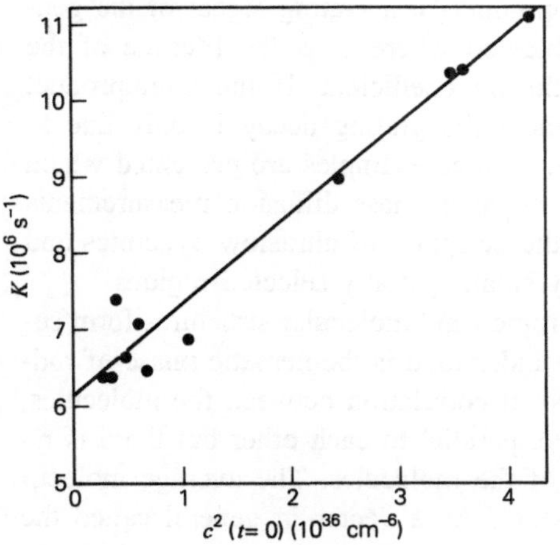

Fig. 12.21. Grating decay rate K versus the square of the initial carrier density c(t = 0) in a (111) oriented silicon sample. The carrier density was varied by controlling the energy of the two 1.064 μm excitation pulses. The grating period is 66 μm. The strainght line is a fit to the data according to Eq. (14.67). (after Ref. [18])

of the carrier lifetime must be taken into account. Investigations in silicon show that the dominating process is an Auger process where the energy set free in the recombination of electron and hole is transfered to a third charge carrier. As three carriers take part in this process the associated rate is proportional to c^3. Solving the modified diffusion Eq. (12.47) with this assumption $c(t)/\tau_r = C \cdot c^3(t)$ an exponential grating decay is found with a decay constant

$$K = \frac{4\pi^2 D}{\Lambda^2} + 3Cc^2(t = 0) \tag{12.67}$$

propotional to the square of the initial carrier density $c(t = 0)$. This initial carrier concentration can be varied in a controlled manner by changing the energy of the two excitation pulses. The concentration dependence of the grating decay time is shown in Fig. 12.21. A fit to the data according to Eq. (12.67) allows the determination of the carrier diffusion coefficient $D = 7 \pm 3\,\text{cm}^2/\text{s}$ and the Auger coefficient $C = (4 \pm 3) \times 10^{-31}\,\text{cm}^6/\text{s}$.

12.7 Mass and Heat Diffusion

The main idea in the application of optical grating techniques to mass transport phenomena is the introduction of an optical tracer whose translational motion can be optically followed. This problem is in general solved by the incorporation of molecules A which can undergo a photochemical reaction $A \xrightarrow{h\nu} B$ into the sample of interest. If the wavelength of the two excitation beams is resonant with an electronic transition of species A, a spatially modulated pattern of photoproducts B as well as depleted species A mimicking the optical interference pattern is generated. Depending on the spectra of A and B and the probe wavelength used the dynamics of A and/or B can be followed. Assuming that the probe beam is sensitive only to the photoproduct B, a grating signal of the general form Eqs. (12.50) and (12.51) is obtained where τ_r is the lifetime of the species B and D is the translational diffusion coefficient. If the photoproduct is stable on the timescale of the experiment the grating decay is only due to mass diffusion of species B. In the following three examples are presented which illustrate the main virtues of the TG-technique in mass diffusion measurements i.e. the evaluation of anisotropic motion, the detection of ultraslow dynamics and the observation of particle motion in very small spatially selected regions.

Many organic compounds with anisotropic rigid molecular structures form so-called liquid-crystalline mesophases. Best understood is the nematic phase of rod-like molecules in which there is orientational correlation between the molecules, i.e. the molecular axes are, on the average, parallel to each other but there is no correlation between the centers of mass of the molecules. The average molecular orientation of the molecules is represented by a vector in general called the director. This ordered molecular structure induces anisotropic macroscopic properties of the liquid and should also effect translational motion of particles in the mesophase.

Figures 12.22 and 12.23 show results from a TG-investigation of tracer diffusion in nematic alkylcyanobiphenyl samples [19,20]. The tracer used is the chromophore methyl red which undergoes a *trans-cis* isomerisation after excitation with $\lambda = 514$ nm from an Ar ion laser. The *cis* photoproduct has a lifetime of a few seconds. The TG-signal for pentylcyanobiphenyl (5CB) at $T = 34.9\,°C$

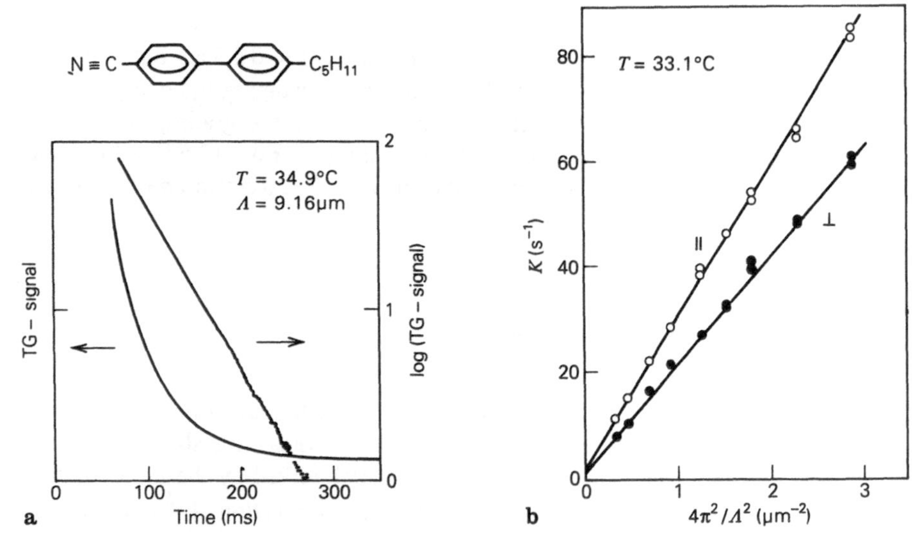

Fig. 12.22. Mass diffusion of methyl red in pentylcyanobiphenyl. **a** TG-signal after photolabeling the methyl red molecules. The decay is monoexponential as expected from the one-dimensional diffusion equation. **b** Grating period and orientational dependence of the grating decay rate K. From the slope of fits the mass diffusion constants $D_{\parallel}$ and $D_{\perp}$ can be calculated. (after Ref. [19])

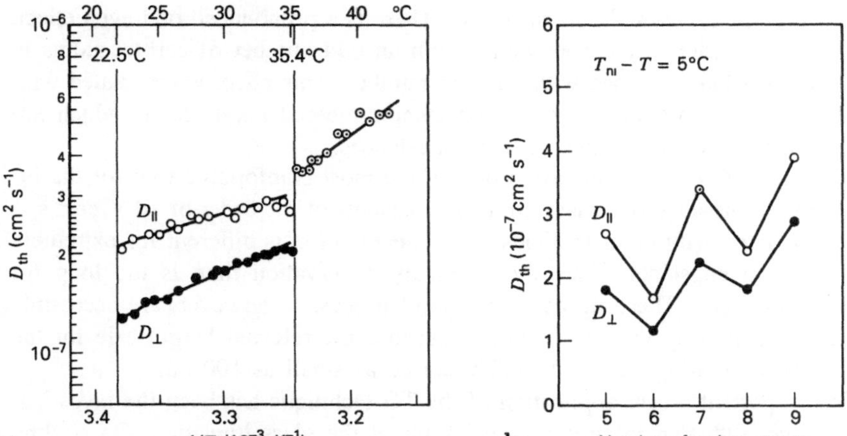

Fig. 12.23. Temperature and structure dependence of mass diffusion constant. **a** Temperature dependence for diffusion of methyl red in pentylcyanobiphenyl (5CB) in the isotropic phase and parallel and perpendicular to the director in the nematic phase. **b** Odd-even effect for probe molecule diffusion in the nematic phase of alkylcyanobiphenyls with different alkyl chain length. In each case the diffusion constants are measured 5°C below T_{ni}. (after Refs. [19 and 20])

and a grating period $\Lambda = 9.16\ \mu$m shown in Fig. 12.22a demonstrates the mono-exponential character of the decay as expected from Eq. (12.50). Note that the timescale of the decay is now in the millisecond range i.e. much slower than the electronic transport processes described in the last section. Evaluating decay constants K for different grating periods and orientation of the grating vector parallel and perpendicular to the director of the sample and plotting K versus $1/\Lambda^2$ the data in Fig. 12.22b are found. From the slope of the straight line fits (see Eq. 12.51) the diffusion constants parallel and perpendicular to the director are found to be $D_{\|} = 2.84 \times 10^{-7}$ cm^2 s^{-1} and $D_{\perp} = 2.04 \times 10^{-7}$ cm^2 s^{-1}. The intersection with the ordinate gives the cis form lifetime. The temperature dependence of the methyl red mass diffusion constants in 5CB (Fig. 12.23a) shows that the anisotropy is characteristic of the nematic phase and disappears at $T_{ni} = 35.4$ °C where the transition from the nematic to the isotropic phase takes place. The slopes of the Arrhenius plots demonstrate furtheron that the activation energies parallel and perpendicular to the director are different from each other and different from the activation energy in the isotropic phase. It should be mentioned that TG-investigations of probe molecule diffusion in the higher ordered smectic phases where the molecules are confined in parallel planes exhibit a crossover of $D_{\|}$ and $D_{\perp}$, i.e. mass diffusion perpendicular to the director is faster than parallel to it. Another proof of the intimate relationship between liquid structure and particle diffusion can be seen in Fig. 12.23b where mass diffusion constants of methyl red in the nematic phase of alkylcyanobiphenyls with different alkyl chain lengths are depicted. For the sake of comparison (the phase transition temperatures T_{ni} also depend on the alkyl chain length) the diffusion constants plotted are measured 5 °C below T_{ni} in each case. There is a pronounced odd-even effect, i.e diffusion is faster in the compounds with an odd number of carbon atoms in the alkyl chain than in those with an even number. This effect is associated with the different conformations of odd- and even-numbered alkyl chains which are also found in structural studies of these substances.

Although TG-experiments are probably the most comfortable tool for the investigation discussed above mass diffusion constants of the order of 10^{-7} cm^2 s^{-1} can be evaluated with other methods, too. The situation is different for extremely small diffusion constants where the necessary observation time is too long for traditional methods. These ultraslow transport processes, however, are accessible in an investigation by the TG-technique because the relevant lengthscale for the diffusion is the grating period Λ which can be as small as 100 nm.

One important recent application of the TG-technique has been the investigation of tracer diffusion in supercooled liquids at the glass transition. On cooling many liquids do not easily form a crystalline phase but can be supercooled considerably below the crystallization temperature and finally freeze in an amorphous glassy state at the (calorimetric) glass transition temperature T_g. In this supercooled region a dramatic increase of the viscosity over 12–14 orders of magnitude is observed in a small temperature interval. This viscosity change is accompanied by a drastic slowing down of molecular dynamics in the sample. Although this phenomenological behaviour is well-established the nature of the glass transition remains a challenging problem in condensed matter physics.

Figure 12.24a shows the temperature dependence for the diffusion of a tetrahydothiophene-indigo derivative (TTI) in the glass former tri-α-naphthylbenzene (TNB). In these experiments TTI is labeled through a photochemical *cis-trans* isomerization with a typical lifetime in the 10^3–10^4 s range. The figure illustrates the dramatic change in molecular mobility (8 orders of magnitude in a temperature interval of 80 degrees) and the extremely small diffusion constants

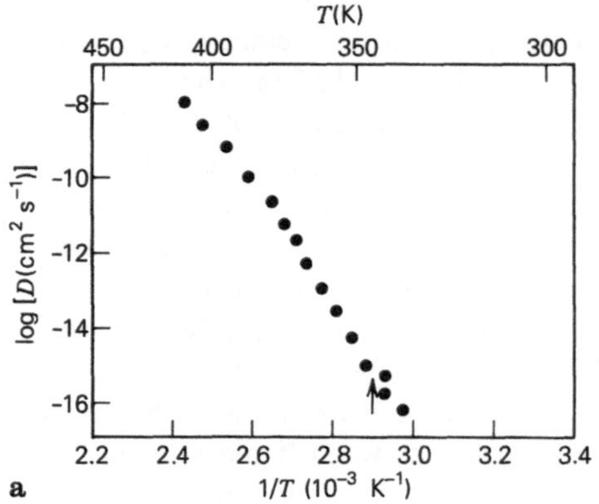

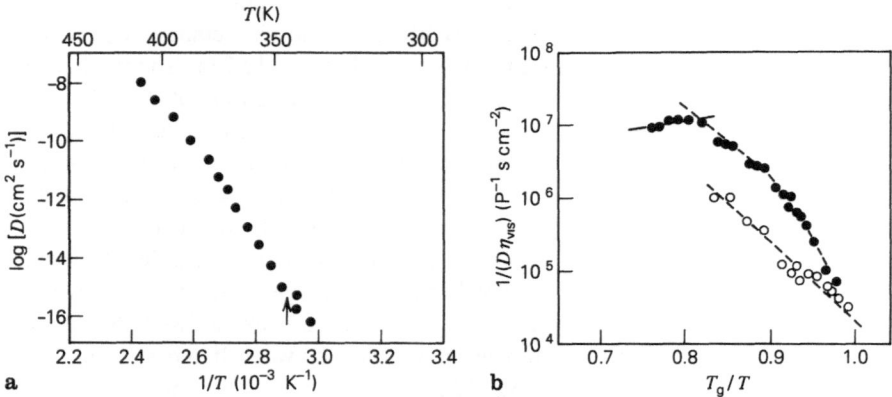

Fig. 12.24. Mass diffusion coefficient D of TTI in TNB versus 1/T. The *arrow* indicates T_g (after Ref. [21]). **b** $1/D\eta_{vis}$ as a function of the reduced temperature for diffusion of TTI in TNB and orthoterphenyl. (after Ref. [22])

accessible through the TG-method (see also problem 7). In regular liquids classical hydrodynamics can model translational diffusion satisfactorily and a Debye-Stokes-Einstein behavior quite analogous to rotational motion (see Eq. 12.63) is found, i.e.

$$D = \frac{k_B T}{\eta_{\text{vis}}\, r_{\text{eff}}} \quad \text{or} \quad \frac{1}{D\,\eta_{\text{vis}}} = \frac{r_{\text{eff}}}{k} \frac{1}{T} \tag{12.68}$$

with r_{eff} being an effective radius of the tracer molecule used. To test the Debye-Stokes-Einstein behavior in the supercooled region, the data of Fig. 12.24a are presented in a different fashion in Fig. 12.24b. Plotting $1/D\,\eta_{\text{vis}}$ versus $1/T$ should give a straight line with a positive slope according to the Debye-Stokes-Einstein relationship 12.68. Quite obviously this is not the case for TTI in TNB (○): diffusion is faster than expected for a given viscosity and this decoupling of translational motion from the viscosity increases with decreasing temperature. The same effect is found for tracer diffusion in orthoterphenyl (●) another well-characterized glass former. Here data have been gathered over a larger temperature range and indicate a change of the translational diffusion mechanism at a well-defined temperature above the calorimetric glass transition T_g. This picture is consistent with a series of other observations indicating a new critical temperature $T_c \simeq 0.8\,T_g$ [22] consistent with a dynamical phase transition as predicted from an application of mode-coupling theory to dense liquids. It should be noted that recent experiments demonstrate that rotational dynamics also decouples from the viscosity in the supercooled region although to a lesser extent than translational motion.

In the section about depopulation and orientation processes at TG-study of sodium atom dynamics in a seeded flame had been presented. With the intensity grating configuration employed in those experiments excited state relaxation constants of the Na atom due to collisional quenching as well as scattering rates between excited state levels had been evaluated (see Figs. 12.13 and 12.14). As illustrated in the following in this case a polarization grating configuration allows the investigation of particle diffusion in the gas phase. Figure 12.25a shows the TG-signal in a Na seeded methane/air flame for perpendicularly polarized excitation pulses (see Fig. 12.3) under otherwise similar experimental conditions as the intensity grating signals shown in Fig. 12.12. The contrast is striking: instead of a monoexponential decay with a $\tau_{\text{int}} = 400\,\text{ps}$ decay time a TG-signal dominated by large oscillations and an exponential envelope with a decay time of $\tau_{\text{pol}} = 2.8\,\text{ns}$ is found. Furthermore the polarization grating signal depends on the fringe spacing. A plot of the grating decay constant 2K versus the grating period Λ as implied by Eqs. (12.50) and (12.51) gives a linear dependence as expected for a diffusion process and from the slope of the plot a diffusion constant $D = 4.3\,cm^2\,\text{s}^{-1}$ for the Na atoms in the flame is found.

The theoretical basis for the shape of the polarization grating signal is quite intricate and cannot be presented here. However, the results can be understood qualitatively in the following way. In the intensity grating the resultant electric field has a single polarization and a sinusoidally modulated amplitude and the

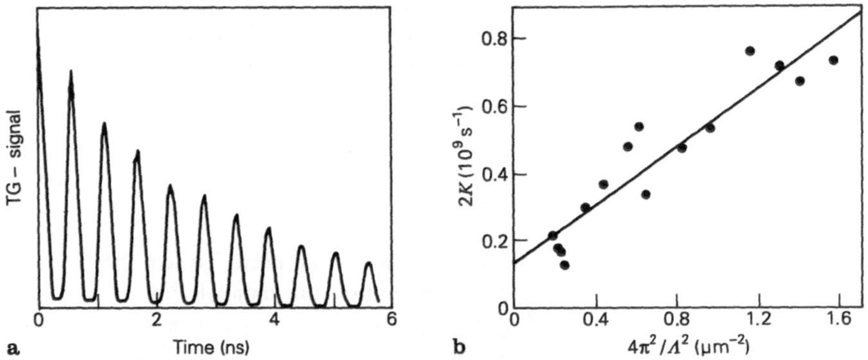

Fig. 12.25. Polarization grating signal for a Na-seeded methane/air flame (As $\lambda_{probe} = 589.0$ nm $\neq \lambda_{exc} = 589.6$ nm the probe pulse does not interact coherently with any excited state population.) **b** Grating period dependence of the polarization grating decay constant 2 K. The slope gives the Na diffusion constant D and the intercept the scattering rate between ground state magnetic sublevels. (after Ref. [8])

decay can be understood in the straightforward manner presented in Sect. 12.5. The polarization grating can be viewed as the sum of four independent intensity gratings each created by excitation beams with identical polarization (either circularly polarized or linearly polarized) but the four gratings being spatially out of phase with each other [8]. As the selection rules for the transitions between the Na magnetic sublevels depend on the polarization of the exciting light the population transferred from each sublevel has a different spatial dependence (in contrast to the case of the intensity grating) and the resulting spatial modulation of the excited state population is different for intensity and polarization grating. It can be shown that under the assumption that collisional quenching is equally probable to all magnetic sublevels this process destroys any spatially modulated sublevel population for the case of the intensity grating. In the case of the polarization grating, however, there persist spatially modulated magnetic sublevel populations with different spatial phase which decay only through scattering between these sublevels. Therefore the intensity grating configuration measures excited state collisional quenching whereas the polarization grating configuration is sensitive to the decay due to scattering between the (ground state) magnetic sublevels. The TG-experiment described here has allowed the first measurement of this sublevel scattering and found a value $\tau_{scat} = 3$ ns. This lifetime is considerably longer than the collisional quenching time and therefore the polarization grating approach is also able to measure translational diffusion constants (see Fig. 12.25b). The beating observed in the polarization configuration is caused by the constructive interference of two intensity gratings which are spatially out-of-phase and the beat frequency is given by the ground state hyperfine splitting of 1.77 GHz.

Besides giving another demonstration of the possibility to selectively monitor certain physical processes by appropriate choice of the light polarization the problem discussed above also illustrates another advantage of the TG-method: the small laser spot size allows the investigation of particle diffusion in specific selected regions within the flame on the order of tens of μm.

Heat conduction is described by the differential equation

$$\frac{\partial T}{\partial t} = D_{\text{th}} \frac{\partial^2 T}{\partial x^2} \tag{12.69}$$

where the thermal diffusivity is given by

$$D_{\text{th}} = \frac{k_{\text{heat}}}{\rho C_{\text{p}}} \tag{12.70}$$

with the mass density ρ, the specific heat C_{p} and the heat conductivity k_{heat}. If an initial spatially modulated temperature distribution of the form

$$T(x,0) = T_0 + \Delta T \cdot [1 + \cos(2\pi x/\Lambda)]. \tag{12.71}$$

has been created Eq. (12.69) has a solution analogous to the one-dimensional diffusion Eq. (12.47), and the optical grating efficiency is given by

$$\eta_{\text{eff}}(t) = \eta^0_{\text{eff}} \cdot \exp(-8\pi^2 D_{\text{th}}\, t/\Lambda^2). \tag{12.72}$$

To generate an initial temperature distribution as given in Eq. 12.71 in a TG-experiment the sample must absorb at the excitation wavelength followed by very rapid radiationless relaxation and local thermal equilibration. This local heating changes the refractive index either directly or through the volume thermal expansion setting up a phase grating which can be followed by the probe beam.

In one of the earliest examples of TG-spectroscopy in general, Eichler et al. [23] measured thermal diffusion in several liquids as well as ruby. The temperature grating in the liquids was generated through addition of dye molecules which absorbed at the excitation wavelength $\lambda_{\text{exc}} = 532$ nm. A cw Argon ion laser was used as a probe and the oscilloscope trace for the decay of the thermal grating in glycerol is shown in Fig. 12.26a. Varying the grating period and plotting the thermal grating decay time τ_{th} according to Eq. (12.72)(see Fig. 12.26b) the thermal diffusivity $D_{\text{th}} = 9.46 \times 10^{-4}\,\text{cm}^2\text{s}^{-1}$ and the heat conductivity $k_{\text{heat}} = 2.3 \times 10^{-3}$ W cm^{-1} K^{-1} are found, respectively, in very good agreement with the established values for glycerol.

A more exciting investigation using again the potential of the TG-technique to measure anisotropic transport processes, was the investigation of thermal diffusion in liquid crystals. Figure 12.27 shows as a representative example the thermal diffusivities parallel and perpendicular to the director in *p*-cyanobenzylidene-*p*-octyloxyaniline (CBOOA) [24]. The ratio $D^{\parallel}_{\text{th}}/D^{\perp}_{\text{th}}$ is larger than one throughout the nematic phase and decreases continuously as the temperature approaches the nematic-isotropic phase transition temperature T_{ni}. In the isotropic phase thermal diffusion is isotropic with a value close to $D^{\perp}_{\text{th}}$ in the nematic phase. Heat conduction is uninfluenced by the additional positional ordering in the smectic

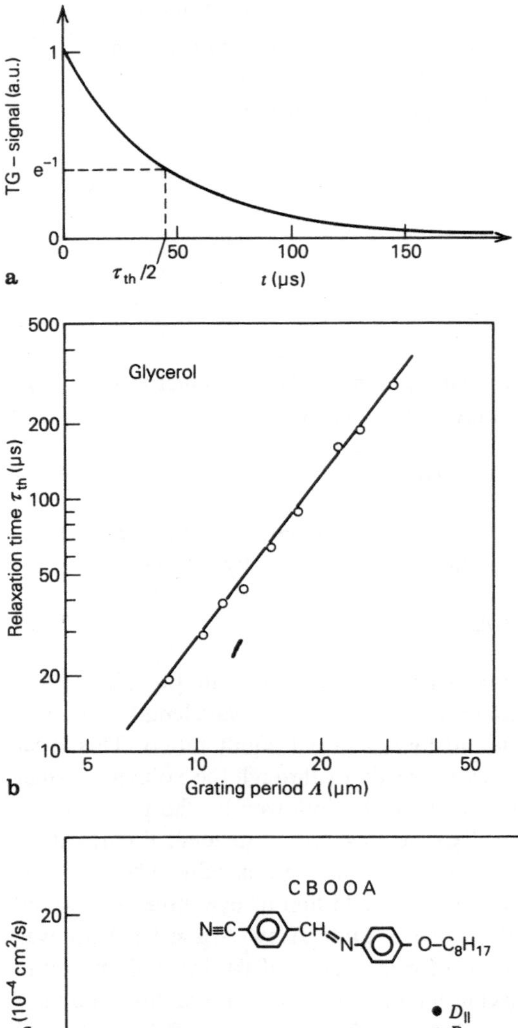

Fig. 12.26. a Time-dependent TG-signal after excitation of a temperature grating in glycerol ($\lambda_{exc} = 532$ nm, $\theta = 1.7°$). **b** Temperature grating relaxation time τ_{th} versus grating period Λ. From the slope the thermal diffusivity D_{th} = can be calculated (after Ref. [23])

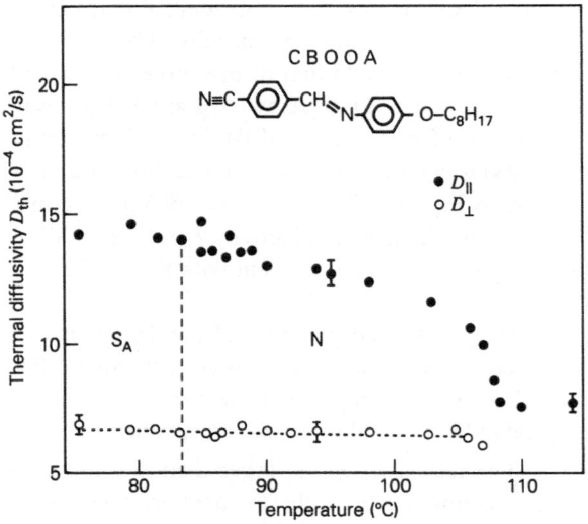

Fig. 12.27. Temperature dependence of the thermal diffusivity parallel (•) and perpendicular (∗) to the director in the nematic and smectic A phase of *p*-cyanobenzylidene-*p*-octyloxyaniline. (after Ref. [24])

A phase as demonstrated by the low temperature data points in Fig. 12.27. The authors have shown that the thermal diffusivity anisotropy is directly proportional to the orientational order parameter S in the liquid crystalline phase. The data presented above can be understood in the framework of a kinetic model where the molecular dimensions limit the mean free path of the photons responsible for the heat conduction and the anisotropic shape of the molecules therefore induces a thermal anisotropy [24].

12.8 Propagation of Ultrasonic Waves and Structural Relaxation

The optical interference pattern created in the sample by two short excitation pulses can also launch counter-propagating ultrasonic waves whose wavelength and orientation match the interference pattern geometry [25]. Two different excitation mechanisms for these ultrasonic waves are possible. In all samples there is a direct coupling of the light to the acoustic waves through the optoelastic constants, the so-called electrostrictive effect. This process corresponds to stimulated Brillouin scattering and the resulting density change in the sample and the associated refractive index modulation have the form

$$\Delta n_R = A \cdot \exp(-\Gamma_{ac}\, t) \cdot \sin(\omega_{ac}\, t). \tag{12.73}$$

Γ_{ac} is the acoustic attenuation constant and ω_{ac} is the acoustic frequency which depends on the grating period Λ and the speed v_{ac} of the ultrasonic wave:

$$\omega_{ac} = \frac{2\pi v_{ac}}{\Lambda}. \tag{12.74}$$

The electrostrictive effect is rather weak and therefore a thermal mechanism dominates the ultrasonic wave generation if the incident light is absorbed by the sample. As described in the last section fast radiationless relaxation after excited state formation induces a temperature grating in the sample. The temperature increase is followed by a volume expansion (in all materials with a positive thermal expansion coefficient) setting up a standing acoustic wave. Besides this transient acoustic wave, there is a second density grating which decays through thermal diffusion as described in the last section and the time dependence of the total density changes has the form

$$\Delta n_R = B \cdot \exp(-\Gamma_{ac} t) \cdot (1 - \cos(\omega_{ac} t)). \tag{12.75}$$

in contrast with the electrostrictive mechanism Eq. 12.73. As the grating efficiency goes as the square of the refractive index the observed acoustic frequency is half the one for the electrostrictive mechanism. In Eq. (12.75) the slow diffusive decay of the thermal grating has not been included. This kind of ultrasonic wave generation has originally been coined laser induced phonon spectroscopy (LIPS) [25] but the terms impulsive stimulated Brillouin scattering (ISBS) and impulsive stimulated thermal scattering (ISTS) can also be found in the literature. Figure 12.28 shows longitudinal acoustic waves in an ethanol solution with (from top to bottom) increasing concentration of the chromophore malachite green [26]. Pure ethanol is transparent at the excitation wavelength $\lambda_{exe} = 532$ nm and an electrostrictive mechanism is repsonsible for the generated ultrasonic wave (Fig. 12.28a). Malachite green absorbs at $\lambda_{exe} = 532$ nm and with increasing dye concentration the thermal mechanism gets more important (Fig. 12.28b) and finally completely dominates the acoustic wave excitation process (Fig. 12.28c).

The temperature and density changes necessary to produce a detectable response are extremely small. As an illustration, Fig. 12.29 shows an absorption spectrum of the v = 6 C–H stretch vibration of benzene in a 1 mm sample cell obtained by the method described here. The experimental data points were gathered by varying the excitation wavelength and dividing the amplitude B of

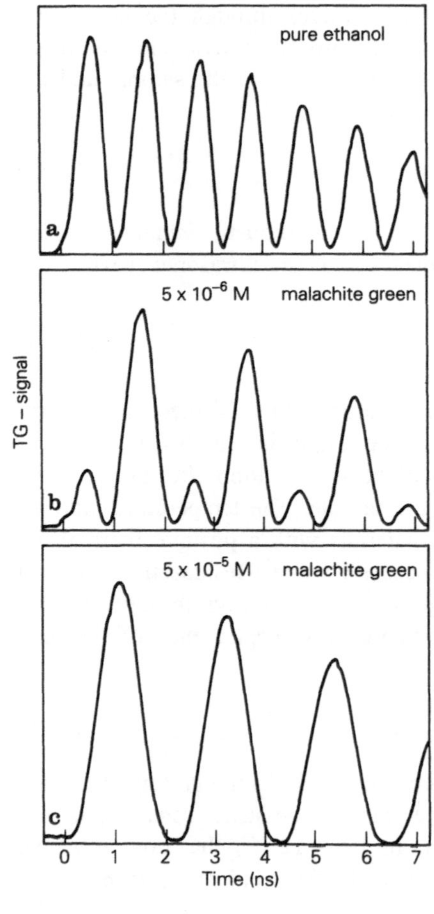

Fig. 12.28. TG-signal of ultrasonic waves in **a** pure ethanol and **b, c** solutions of malachite green in ethanol. The excitation wavelength $\lambda_{exc} = 532$ nm, the probe beam wavelength is nonresonant with any transitions $\lambda_p = 532$ nm. The grating period $\Lambda = 2.47\ \mu$m. In pure ethanol (*top*) the acoustic waves are generated via an electrostrictive mechanis,. With increasing malachite green concentration the thermal mechanism dominates. (after Ref. [26])

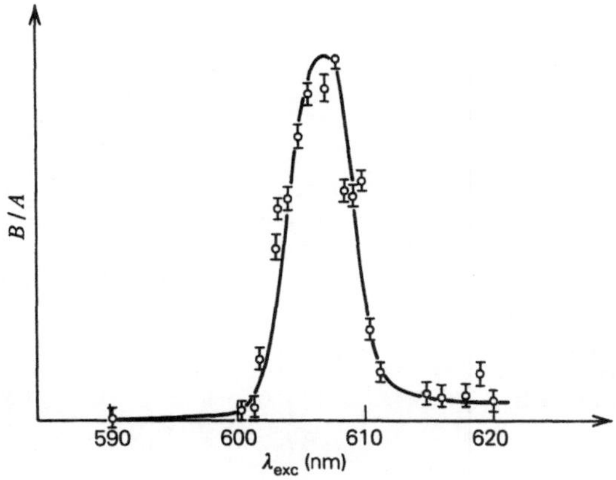

Fig. 12.29. Spectrum of absorption into the fifth vibrational overtone of the C–H stretch vibration of benzene as obtained by TG acoustic wave spectroscopy (see text). The sample cell used has a thickness of 1 mm. (after Ref. [26])

the thermally generated wave (Eq. 12.75) which is proportional to the absorption coefficient of the sample by the amplitude A of the electrostrictively launched wave (Eq. 12.73).

The LIPS technique allows the generation of acoustic waves in the 30 MHz – 30 GHz range, i.e. it covers a range which is accessible neither to classical acoustic methods nor to frequency-resolved Brillouin scattering. It is especially useful for strongly attenuating samples (as in the diffusion processes discussed in the last section the relevant lengthscale over which propagation is observed is the grating period Λ) and anisotropic materials. Figure 12.30 shows the propagation of ultrasonic waves along three different orientations in a α-perylene crystal. The two top signals arise from purely longitudinal waves which propagate along symmetry axes of the crystal. The propagation in Fig. 12.30c is between the symmetry axes, and a quasi-transverse acoustic wave is generated in addition to a quasi-longitudinal acoustic wave. Due to the different frequencies of the two phonons a beating is observed in the TG-signal.

Equation (12.75) is a correct description of the thermally generated ultrasonic wave only for the case that the induced temperature and density changes in the sample are infinitely fast or to be more specific if these changes take place on a timescale much faster than an acoustic cycle $\tau_{ac} = 2\pi/\omega_{ac}$. If e.g. there is a delayed temperature rise because of slow radiationless relaxation or the structural relaxation necessary for the thermal volume expansion is comparable or slower than τ_{ac} the TG-signal is modified in a specific and significant way. This fact can be exploited to carry out ultrafast optoacoustic calorimetry and detect structural relaxations in the ns and μs range.

A nice example [27, 28] for the use of TG-spectroscopy in calorimetry is the photoisomerization of tetraphenylethylene (TPE) – see Fig. 12.31. It is well-known that after excitation into the singlet state S_1 the molecule relaxes within a few picoseconds (rate k_1) to a twisted transition state E_p^*. From this twisted state the molecule relaxes to the electronic ground state with a smaller rate k_2. Whereas the first step k_1 induces an instantaneous temperature jump in the sense

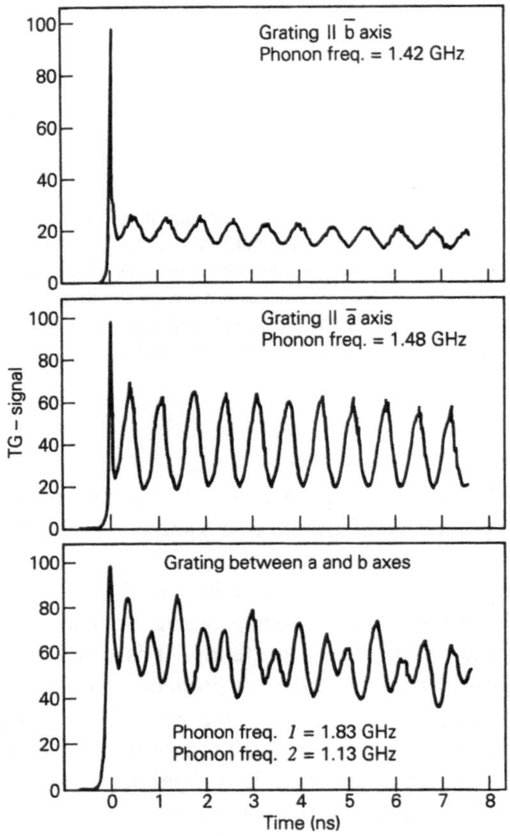

Fig. 12.30. LIPS signals in α-perylene. The optical grating is oriented **a** parallel to the crystallographic $\bar{b}$ axis of the crystal, **b** parallel to the crystallographic $\bar{a}$ axis of the crystal and **c** between $\bar{a}$ and $\bar{b}$ axes in the crystallographic $\bar{a}\bar{b}$ plane. The narrow spikes at $t = 0$ are of electronic origin and have no bearing on the acoustic data. (after Ref. [25])

discussed above and generates ultrasonic waves of the form Eq. 12.75 the heat associated with the radiationless relaxation of the excited state E_{p}^{*} is released slowly with the rate k_2 and therfore the concomitant density changes are delayed. LIPS signals after excitation of TPE in different organic solvents are displayed in Fig. 12.32. Characteristic for the delayed heat release is the strong rise of the baseline (see e.g. in comparison with Fig. 12.28c). With suitable rate equations it is possible to model the influence of the delayed density change on the grating efficiency and to simulate the time dependence of the TG-signals. With these simulations it is possible to evaluate the rate constants of the relaxation processes (in this case k_2) and the released heat (and therefore the energy of the twisted excited state E_{p}^{*}). The data in Fig. 12.32 demonstrate nearly perfect agreement between experimental data and theoretical fits. The different signal forms are due to a strong solvent dependence of the radiationless relaxation rate constant k_2: it increases from $0.6 \times 10^9\,\mathrm{s}^{-1}$ in pentane to $3.7 \times 10^9\,\mathrm{s}^{-1}$ in diethylether and $11 \times 10^9\,\mathrm{s}^{-1}$ in tetrahydrofuran. However, the energy of the intermediate twisted state E_{p}^{*} is nearly independent of the solvent polarity and can be calculated to be in the range 273–279 kJ mol^{-1}.

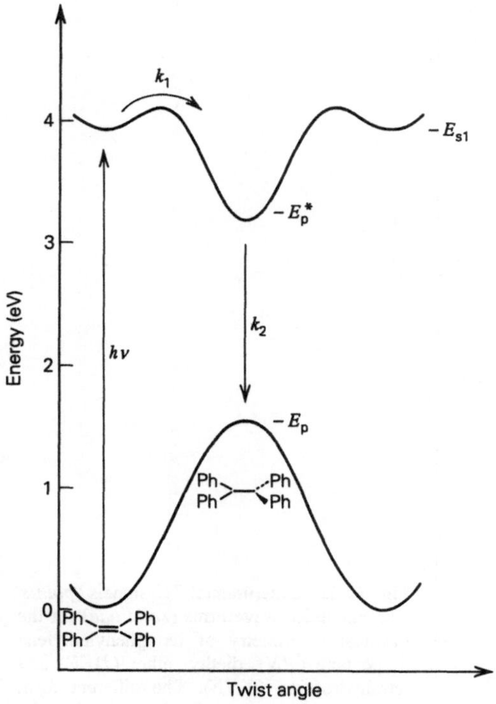

Fig. 12.31. Schematic potential energy diagram and relevant kinetic rates in the photoisomerization of tetraphenylethylene. (after Ref. [27])

TG-spectroscopy can also be a sophisticated tool to investigate structural relaxation dynamics and energy dissipation in complex systems like e.g. proteins which contain rigid regions as well as fluid-like environments[29]. Hemoglobin and myoglobin which are responsible for oxygen transport and oxygen storage in the human blood and muscle belong to the group of heme proteins. A question of basic interest is how the dissociation of the ligand (O_2 or CO) from the central heme porphyrin acceptor which can also be optically triggered is coupled to the large amplitude cooperative structural $R \to T$ transition of the whole protein. There is a multitude of detailed investigations which use the absorption of this central porphyrin chromophore as an optical probe to follow the local dynamics around the acceptor. However, there are very few studies of the induced global changes in the protein. This is only feasible with a techique like the transient grating which can detect density changes with high time resolution. A series of investigations [29] in several heme proteins has focused on energy relaxation pathways in this class of substances under different excitation conditions. Especially interesting is a comparison of TG-signals from carboxymyoglobin (MbCO) and deoxymyoglobin shown in Fig. 12.33. Both proteins are excited with $\lambda_{exe} = 532\,\text{nm}$ into a higher vibrational level of the excited electronic state and the probe pulse at $\lambda_{probe} = 1064\,\text{nm}$ being nonresonant with any electronic transitions monitors pure refractive index changes. In deoxymyoglobin where no ligand is bound to the heme porphyrin there is fast radiationless relaxation generating a

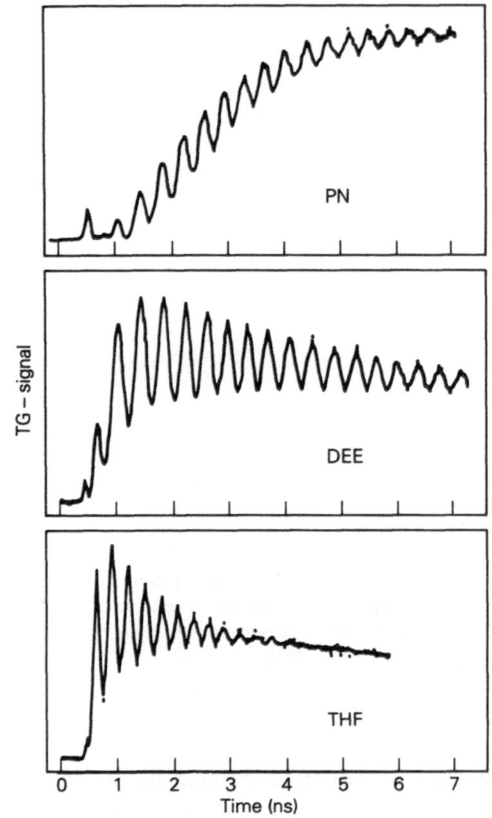

Fig. 12.32. Experimental TG-signals (*points*) and calculated waveforms (*solid line*) for the ultrafast calorimetry of tetraphenylethylene in pentane (*PN*), diethyl ether (*DEE*) and tetrahydrofuran (*THF*). The different form of the signals is essentally determined by the "slow" rate k_2 which increases from bottom to top. (after Ref. [28])

thermally induced acoustic wave with a rise time limited by the speed of sound as found e.g. in the malachite green/ethanol solution (see Fig. 12.28). The situation is very different for the MbCO sample where an instantaneous rise (within the experimental time resolution of 30 ps) of the TG-signal is found incompatible with the acoustic wave generation mechanisms discussed earlier (see Eqs. 12.75 and 12.73). It is well-known that after optical excitation the CO ligand in MbCO dissociates from the heme porphyrin on a subpicosecond timescale. However, this process cannot induce an optical grating at the probe wavelength $\lambda_{probe} = 1064\,\text{nm}$. The phase grating observed in MbCO must be due to a global structural change and concomitant material displacement in the protein triggered by the ligand dissociation. So these signals prove that the global cooperative structural changes of the protein occur within less than 30 ps on the same time scale as the local dynamics at the heme center.

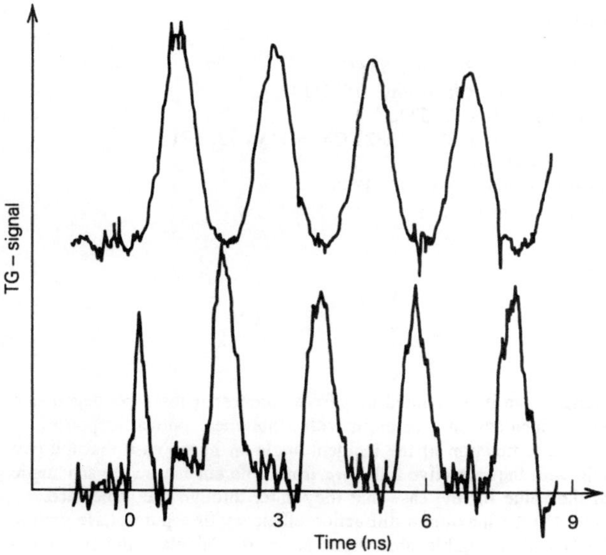

Fig. 12.33. TG-signals of deoxymyoglobin (*top*) and carboxymyoglobin (*bottom*). The deoxymyoglobin signal is due to fast vibrational relaxation in the protein and a thermally induced acoustic wave. In carboxymyoglobin a global (nonthermal) density change of the protein is triggered within less than 30 ps (the experimental time resolution) after the CO dissociation leading to an instantaneous rise in the signal. (after Ref. [29])

12.9 References

1. Eichler HJ, Günter G, Pohl DW (1986) Laser-induced dynamic gratings. Springer, Berlin, Heidelberg New York
2. Shen YR (1984) The principles of nonlinear optics. John Wiley New York
3. Shtyrkov EI, Nevelskaya NL, Lobkov VS, Yarmukhametov NG (1980) Phys Status Solidi (B) 98: 473
4. Yan Y, Nelson K (1987) J Chem Phys 87: 6240.
5. Hecht E, Zajac A (1979) Optics (Addison-Wesley, Reading)
6. Kogelnik H (1969) Bell Sys Tech J 48: 2909
7. Deeg FW, Fayer MD (1989) J Chem Phys 91: 2269
8. Fourkas JT, Brewer TR, Kim H, Fayer MD (1991) J Chem Phys 95: 5775
9. Lutz DR, Nelson KA, Gochanour CR, Fayer MD (1981) Chem Phys 58: 325
10. von Jena A, Lessing HE (1979) Opt Quant Elect 11: 419
11. Moog RS, Ediger MD, Boxer SG, Fayer MD (1982) J Phys Chem 86: 4694
12. Gomez-Jahn L, Kasinski J, Miller RJD (1986) Chem Phys Lett 125: 500
13. Förster T (1948) Ann Phys (Leipzig) 2: 55
14. Lawson CM, Powell RC, Zwicker WK (1981) Phys Rev Lett 46: 1020
15. Tyminski JK, Powell RC, Zwicker WK (1984) Phys Rev B 29: 6074
16. Rose TS, Righini R, Fayer MD (1984) Chem Phys Lett 106: 13
17. Miller A, Manning RJ, Milsom PK, Hutchings DC, Crust DW, Woodbridge K (1989) J Opt Soc Am B 6: 567
18. Eichler HJ, Massmann F, Biselli E, Richter K, Glotz M, Konetzke L, Yang X (1987) Phys Rev B 36: 3247
19. Hara M, Ichikawa S, Takezoe H, Fukuda A (1984) Jpn J Appl Phys 23: 1420
20. Hara M, Takezoe H, Fukuda A (1986) Jpn J Appl Phys 25: 1756

21. Ehlich D, Sillescu H (1990) Macromolecules 23: 1600
22. Rössler E (1990) Phys Rev Lett 65: 1595
23. Eichler H, Salje G, Stahl H (1973) J Appl Phys 44: 5383
24. Urbach W, Hervet H, Rondelez F (1983) J Chem Phys 78 5113
25. Nelson KA, Fayer MD (1980) J Chem Phys 72 5202
26. Miller RJD, Casalegno R, Nelson KA, Fayer MD (1982) Chem Phys 72: 371
27. Zimmt MB (1989) Chem Phys Lett 160: 564
28. Morais J, Ma L, Zimmt MB (1991) J Phys Chem 95: 3885
29. Miller RJD (1991) Annu Rev Phys Chem 42: 581

12.10 Problems

1. How many different output frequencies can be generated in a FWM process if three nondegenerate input frequencies $\omega_1 \neq \omega_2 \neq \omega_3$ are incident on a medium with a nonlinear optical response?
2. Calculate the grating wavelength Λ as a function of the incident angle α_0 in the case that the two interfering beams enter the sample with the refractive index n_R from different faces (see schematic figure). Compare the result with the value for the case that they enter through the same face.
3. As can be easily seen from Eq. (12.45) the maximum diffraction efficiency of a pure phase grating is 100% for $\Delta n_R = (\lambda \cos\theta)/(2d)$ and negligible absorption $\beta_0 = 0$. Calculate the maximum diffraction efficiency of a pure absorption grating. Do not neglect the influence of $\Delta\beta_0$ on $\bar{\beta}_0$.
4. Calculate the grating diffraction efficiencies η_{eff} (assume in each case sample thickness d = 1 mm, $\lambda = 600\,\text{nm}$ and $\cos\theta = 1$) for the following excitation modes: (i) the "direct" thermal grating due to a temperature modulation $\Delta T_0 = 0.01\,\text{K}$ ($\partial n_R/\partial T = 9 \times 10^{-5}\,\text{K}^{-1}$), (ii) the "indirect" thermal grating due to a temperature modulation $\Delta T_0 = 0.01\,\text{K}$ and the concomitant density change ($\partial n_R/\partial\rho = 0.5\,\text{dm}^3\,\text{kg}^{-1}$ and $\partial\rho/\partial T = 1.2 \times 10^{-4}\,\text{kg}\,\text{dm}^{-3}\,\text{K}^{-1}$), (iii) due to excitation of 0.1% of molecules with an absorption cross section $\kappa = 10^5\,\text{dm}^3\,\text{mol}^{-1}\,\text{cm}^{-1}$ in a sample with a concentration $N' = 10^{-5}\,\text{mol}\,\text{dm}^{-3}$.
5. With the technique discussed in Sect. 12.5 it is only possible to determine rotational diffusion times τ_{rot} which are smaller or comparable to the singlet excited state lifetime τ_{ex} (see Eqs. 12.64 and 12.65) which is in general in the nanosecond range. Can you think of a modified technique which allows the evaluation of orientational dynamics which is much slower?
6. Consider the example of charge carrier transport in the multiple quantum well discussed in Sect. 12.6. Assume that the motion of the carriers parallel and perpendicular to the wells is independent of each other and that interwell diffusion is given by an angular-independent lifetime $\tau_{\text{inter}} = 1\,\text{ns}$. The intrawell diffusion constant $D_{\text{intra}} = 13.8\,\text{cm}^2/\text{s}$ and the intrinsic lifetime of the excited carriers $\tau_r = 4\,\text{ns}$. Calculate the TG-decay constant for an optical grating with an orientation (see Fig. 12.19) $\theta = 0°$, $1°, 5°, 45°$ and $90°$.

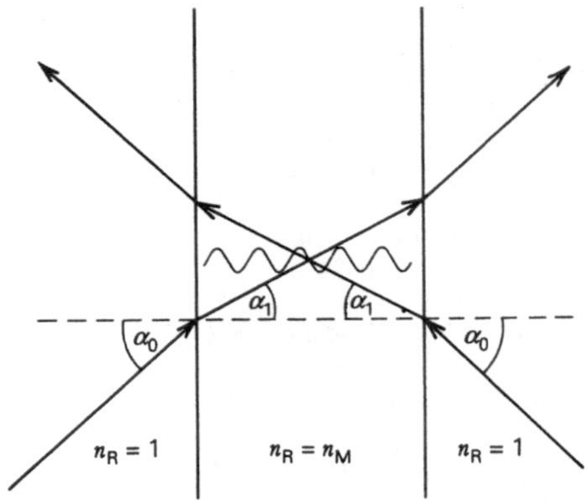

Fig. 12.34 Problem 2: Two beams entering from different faces interfere in the sample

7. Assume that the optically labeled tracer in a mass diffusion experiment has a lifetime of 10^4 s and an effective radius $r_{eff} = 20$ nm. What is the value of the maximum viscosity η_{vis} at which translational diffusion constants can be measured in a TG-experiment if a grating period $\Lambda =$ 500 nm is used?
8. Use the following physical constants: (i) dry air ($\rho = 1.2\,\mathrm{g\,dm^{-3}}$, $C_p = 29.2\,\mathrm{J\,mol^{-1}\,K^{-1}}$, $k_{heat} = 2.6 \times 10^{-6}\,\mathrm{W\,m^{-1}\,K^{-1}}$, (ii) liquid water $\rho = 1.0\,\mathrm{kg\,dm^{-3}}$, $C_p = 75.3\,\mathrm{J\,mol^{-1}}$, $k_{heat} = 6 \times 10^{-5}\,\mathrm{W\,m^{-1}\,K^{-1}}$ and (iii) solid copper ($\rho = 9.3\,\mathrm{kg\,dm^{-3}}$, $C_p = 244.4\,\mathrm{J\,mol^{-1}\,K^{-1}}$, $k_{heat} = 4 \times 10^{-2}\,\mathrm{W\,m^{-1}\,K^{-1}}$ and calculate typical (i.e. assume a grating period $\Lambda = 2\,\mu$m) decay times τ_{th} for a thermal grating in air, water and copper.
9. What is the maximum ultrasonic frequency which can be generated by the transient grating method if the speed of sound in the sample $v_{ac} = 3000\,\mathrm{ms^{-1}}$, the refractive index $n_R = 1.5$ and the laser excitation wavelength $\lambda_{exc} = 350$ nm?

13 Synchrotron Radiation and Free Electron Lasers

M. Watanabe and G. Isoyama

13.1 Introduction

The last chapter of a textbook normally is not the easiest one but, hopefully, you will find this one easy. Synchrotron radiation (SR) is a strongly recommendable light source but, unfortunately, you cannot find a description of it in most standard textbooks. In the following chapter, therefore, experimental and technical designs will also be presented; if you are going to use SR as your light source you should be able to understand what the machine operators are doing with that light source.

It is well known that an electric oscillator emits electromagnetic waves. For instance, radio waves are emitted from an antenna in which an alternating current flows with a high frequency. A charged particle also emits electromagnetic radiation when it is accelerated. Radiation emitted by a relativistic electron moving along a circular orbit under a constant magnetic field is called synchrotron radiation since it was first observed on an electron synchrotron in 1946. Synchrotron radiation had been a nuisance for high energy synchrotrons, because it takes energy away from the accelerating electron beam. However, it is a bright light with the continuous spectrum over the broad range from the far infrared to X-rays, and linearly polarized in the plane of the electron motion. Synchrotron radiation was first recognized as a superior light source in the vacuum ultraviolet and soft X-ray region (hereafter called VUV), where good light sources had not been available before. By use of synchrotron radiation from electron synchrotrons for high energy physics, spectroscopic studies were made on gaseous molecules and solids in the 1960s. Some years later, utilization of synchrotron radiation began in the X-ray region.

The intensity of synchrotron radiation from a synchrotron is not temporally stable since the electron energy varies in an acceleration cycle and the number of electrons fluctuates in acceleration cycles. Since the 1960s, storage rings have been constructed for colliding experiments of electrons and positrons in high energy physics, in which the electron beam is stored for a long time (typically several hours). Since synchrotron radiation from storage rings is more stable than that from synchrotrons, they have been also used as synchrotron radiation sources since the early 1970s, and great progress was made in studies with angle integrated and angle-resolved photoelectron spectroscopy, photofragmentation experiments, EXAFS (extended X-ray absorption fine structure), and so on. By use of the pulsed structure of synchrotron radiation, measurements of decay time of fluorescence were initiated in this period. These storage rings are called the *first* generation light source. They were not optimized for synchrotron radiation research, and could only be used parasitically by spectroscopists.

In the early 1980s, therefore, many storage rings dedicated to synchrotron radiation research were constructed. In Table 1, storage rings are shown for synchrotron radiation which are in operation or under construction. In order to obtain higher brightness of radiation than that in the first generation machines, they were designed so that the transverse size and the angular divergence of the electron beam were small. These storage rings are called the *second* generation light source. At the same time, optical components were also developed and optical systems were optimized for synchrotron radiation. A variety of monochromators were newly designed and fabricated. A number of experiments were conducted on clean and adsorbed surfaces, and new types of experiments appeared such as spin-resolved photoelectron spectroscopy and circular dichroism with circularly polarized radiation emitted out of the orbit plane. Utilization of synchrotron radiation was also demonstrated in the far infrared region.

Recently, insertion devices such as wigglers and undulators have been developed and installed at straight sections of storage rings. They produce more intense radiation than synchrotron radiation in special spectral regions. Since insertion devices are not essential components for the operation of a storage ring, they are designed to meet requirements from experiments, such as intensity, photon energy, and polarization characteristics. A superconducting wiggler with the high magnetic field produces synchrotron radiation with several times higher energy in a certain spectral region than that emitted in bending sections, and an undulator produces quasi-monochromatic radiation with extremely high brightness. In order to make best use of undulator radiation, emittance of the electron beam has to be low, which is approximately given by the product of the transverse size and the angular divergence of the electron beam.

Low-emittance storage rings equipped with a number of undulators are called the *third* generation light source. By using of intense undulator radiation, one can carry out high resolution experiments, experiments on small or dilute specimens, multiple coincidence experiments, experiments with external fields, and other novel experiments. In the early half of the 1990s, several third generation light sources have been born. To utilize undulator radiation, mirrors with ultrasmooth surfaces, multilayer mirrors, robust optical elements against heat load, gratings with high efficiency are being developed. Monochromators with gratings of a large radius have been constructed to obtain high resolution.

A *free electron laser* (FEL) is a device that converts kinetic energy of the electron beam from accelerators into high power coherent radiation. Characteristics of this radiation are the same as those of conventional lasers. The spectral linewidth is very narrow and the source size and angular divergence of radiation are determined by the diffraction limit. However, there is, in principle, no limitation in wavelength and the wavelength can be tuned continuously. In 1977,

Table 1. Storage rings for synchrotron radiation

Location	Institute	Ring	Energy (GeV)	εc (keV)	Generation	Note
DENMARK						
Aarhus	ISA	ASTRID	0.56	0.3	1	partly dedicated
FRANCE						
Orsay	LURE	DCI	1.8	3.6	1	
		Super ACO	0.8	0.67	2	
Grenoble	ESRF	ESRF	6	20.6	3	
GERMANY						
Berlin	BESSY	BESSY	0.8	0.64	2	
		BESSY II	1.7	2.6	3	construction
Bonn	Bonn Univ.	ELSA	3.5		1	partly dedicated
Hamburg	HASYLAB	DORIS III	3.5∼5.5	7.7∼29.9	1	partly dedicated
ITALY						
Frascati	LNF	ADONE	1.5	1.5	1	partly dedicated
Trieste	Syn.Trieste	ELETTRA	2.0	3.2	3	
SWEDEN						
Lund	Univ. of Lund	MAX	0.55	0.31	2	
		MAX II	1.5		3	construction
UK						
Daresbury	SRC	SRC	2.0	3.2	2	
RUSSIA						
Novosibirsk	INP	VEPP-3	2.2	2.95	1	partly dedicated
	INP	VEPP-4	5∼7		1	partly dedicated
Kharkov	KPI	N-100	0.1		2	
Moscow	Kurchatov Inst.	SIBERIA I	0.45	0.2	2	
		SIBERIA II	2.5	7.1/1.8	2	commissioning
Zelenograd	F V Lukin Inst.	TNK	1.2∼1.6		2	construction
BRAZIL						
Campinas	LNLS	LNLS-1	1.15		3	
USA						
Gaithersburg	NIST	SURF II	0.28	0.07	1	
Ithaca	CHESS	CESR	5.5	11.0	1	partly dedicated
Stanford	SSRL	SPEAR	3∼3.5	4.8∼7.6	1	
Stoughton	SRC	ALADDIN	0.8∼1.0	0.55∼1.1	2	
Upton	BNL	NSLS/VUV	0.75	0.49	2	
		NSLS/X-Ray	2.5	5	2	
Baton Rouge	Lousiana S.Univ.	CAMD	1.2	1.3	2	
Berkeley	LBL	ALS	1.5	1.5	3	
Argonne	ANL	APS	7	19.4	3	construction
CHINA						
Beijing	BSRL	BEPC	1.5∼2.8	0.7∼4.8	1	partly dedicated
Hefei	USTC	HESYRL	0.8	0.52	2	
TAIWAN						
Hsinchu	SRRC	SRRC	1.3	1.4	3	
INDIA						
Indore	CAT	INDUS-I	0.45	0.2	2	construction
JAPAN						
Tokyo	Univ.Tokyo	SOR-RING	0.38	0.11	2	
Tsukuba	ETL	TERAS	0.6	0.56	2	
	KEK	PF	2.5	4.0	2	
	KEK	AR	6.5	26.4	1	parasitic
Okazaki	IMS	UVSOR	0.75	0.43	2	
Himeji	JAERI-RIKEN	SPring-8	8	28.3	3	construction
KOREA						
Pohang	PAL	PLS	2	2.8	3	

the first FEL was constructed at Stanford for the infrared region. Since then, activities to study and construct FELs have expanded worldwide. At present, the shortest wavelength of FELs is ultraviolet. The practical use of FELs has just begun in the infrared and far infrared region. Extensive efforts are being made to shorten the wavelength.

This chapter is devoted to the introduction of radiation based on electron accelerators and techniques for its utilization with emphasis on the VUV region (0.2–200 nm). In this region, the absorption coefficient of samples is large due to excitations of not only valence electrons but also inner core electrons. The excitation relaxes with the emission of photons (fluorescence), photoelectrons and Auger electrons. Furthermore, photofragmentation follows ionization, and photodesorption occurs as a result of violent excitation by VUV radiation. Therefore VUV experiments are powerful tools for investigating the electronic structure, the relaxation process of excited states and photochemistry of gases, solids and surfaces. By analyzing the VUV spectra, one can get also more structural information. In addition, the techniques for far infrared (0.05–1.0 mm) spectroscopy and two-color experiments using synchrotron radiation and conventional lasers are briefly introduced.

13.2 Synchrotron Radiation

When an electron circulates at a constant speed much lower than that of light, it emits radiation at its revolution frequency, $\omega_0 = v/\rho$, where v is the speed of the electron and ρ is the radius of the circular motion. The angular distribution of the emitted power is given as [1]

$$\frac{\mathrm{d}P}{\mathrm{d}\Omega} = \frac{e^2}{4\pi c^3}\left|\frac{\mathrm{d}\mathbf{v}}{\mathrm{d}t}\right|^2 \sin^2\Xi, \tag{13.1}$$

where e is the electron charge, c the speed of light, $\mathbf{v}$ the velocity of the electron, and Ξ the angle between the direction of observation and that of acceleration. Fig.13.1(a) shows a schematic illustration of the motion and the angular distribution of power emitted by the non-relativistic electron. As the speed of the electron approaches that of light, the angular distribution of radiation is tipped in the forward direction due to the relativistic effect, and the radiation power is concentrated in a narrow cone with the half angular width of approximately $1/\gamma$, as shown in Fig. 13.1(b), where

$$\gamma = (1 - \beta^2)^{-1/2} = (E + mc^2)/mc^2 \approx E/mc^2, \tag{13.2}$$

for a relativistic electron with E and $mc^2 = 0.511$ MeV being the kinetic energy and the rest mass of an electron, respectively, and $\beta = v/c$. Since the circulating electron changes its direction of motion continuously and the radiated power is confined in the angular region of $1/\gamma$ (2×10^{-4} rad for an electron of 1 GeV

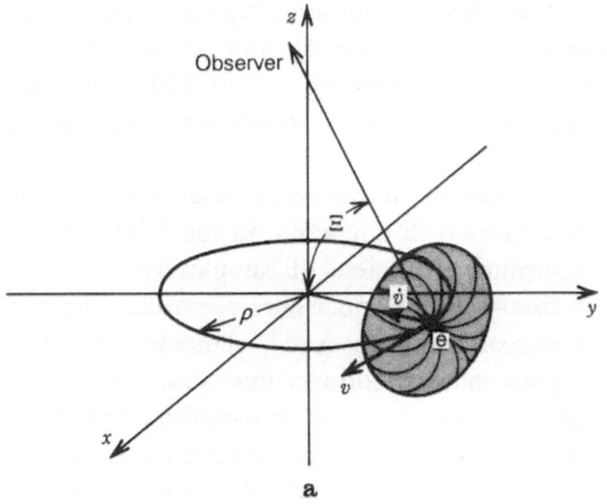

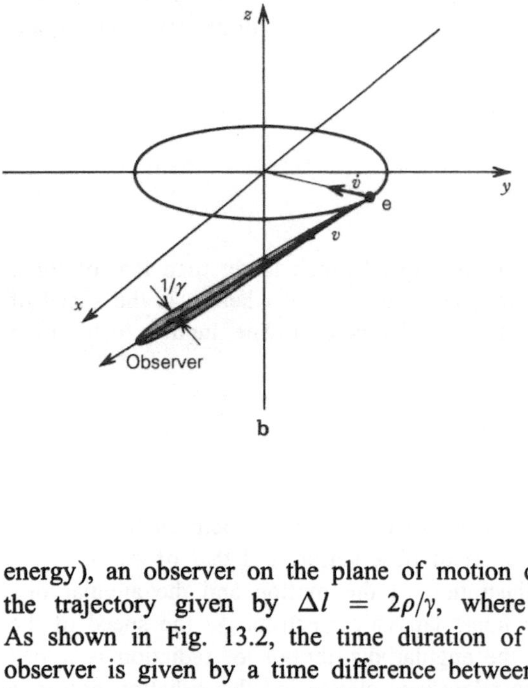

Fig. 13.1. Angular distributions of power emitted from a non-relativistic electron **a** and a relativistic electron **b** moving on a circular orbit

energy), an observer on the plane of motion can see only a very short arc of the trajectory given by $\Delta l = 2\rho/\gamma$, where ρ is the radius of curvature. As shown in Fig. 13.2, the time duration of the light pulse Δt seen by the observer is given by a time difference between the transit time of the electron in the arc, $t_e = 2\rho/v\gamma$, and that of light in the chord, $t_l = (2\rho/c)\sin(1/\gamma)$, as $\Delta t = t_e - t_l \approx 4\rho/3c\gamma^3$, which is an extremely short period of time when using the relativistic electron. From the general argument of the Fourier transformation, the short pulse with a time duration Δt contains frequency components up to the order of $\Delta\omega \sim 1/\Delta t = 3c\gamma^3/4\rho \sim c\gamma^3/\rho$. Note that the frequency of radiation becomes γ^3 times higher than the revolution frequency of the electron, $\omega_0 = c/\rho$, due to the relativistic effect. Thus, synchrotron radiation has a broad spectrum up to very high frequencies.

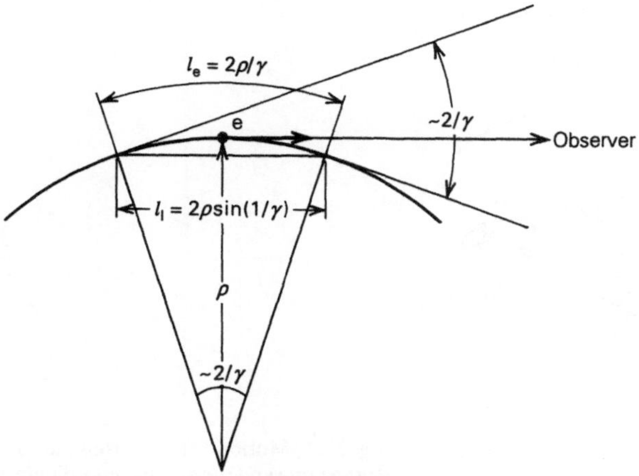

Fig. 13.2. Geometry of synchrotron radiation emitted from a relativistic electron moving along a circular path

The critical frequency is defined as [2, 3]

$$\omega_c = \frac{3}{2}\frac{c\gamma^3}{\rho}, \tag{13.3}$$

and the critical photon energy is given in practical units as

$$\hbar\omega_c(\text{keV}) \approx 2.22 \times \frac{E^3(\text{GeV})}{\rho(\text{m})}, \tag{13.4}$$

where $\hbar$ is the Planck constant.

The general formula for calculating the radiation spectrum from a moving charged particle is given by [1]

$$\frac{\mathrm{d}^2 W}{\mathrm{d}\omega \mathrm{d}\Omega} = \frac{e^2}{4\pi^2 c}\left|\int_{-\infty}^{\infty} \frac{\mathbf{n} \times [(\mathbf{n} - \boldsymbol{\beta}) \times \dot{\boldsymbol{\beta}}]}{(1 - \boldsymbol{\beta} \cdot \mathbf{n})^2} \exp\left[i\omega(t - \mathbf{n} \cdot \mathbf{r}(t)/c)\right] \mathrm{d}t\right|^2, \tag{13.5}$$

where $\mathrm{d}^2 W/\mathrm{d}\omega\mathrm{d}\Omega$ is the energy radiated per second per unit angular frequency of radiation per unit solid angle, $\mathbf{n}$ the unit vector pointing the direction of observation, $\dot{\boldsymbol{\beta}} = \mathrm{d}\boldsymbol{\beta}/\mathrm{d}t$, and $\mathbf{r}$ the position of the particle.

The motion of electrons in a storage ring is shown schematically in Fig. 13.3. Many electrons circulate in a storage ring and each electron oscillates randomly around the central orbit with a small amplitude, so that the electron beam has a finite beam size, as given in Fig. 13.3 by σ_x and σ_y in the horizontal and the vertical directions, respectively. Since the beam size is much smaller than the distance from a source point to an observer, we take the angular coordinate for synchrotron radiation as shown in Fig. 13.3. In most storage rings for synchrotron radiation, the central orbit in a storage ring is composed of a combination of

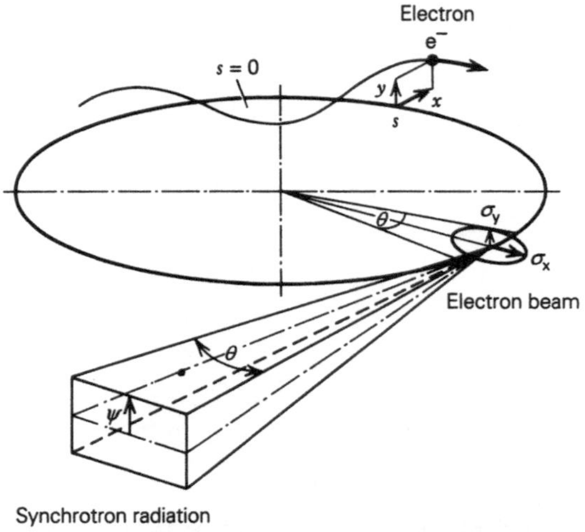

Fig. 13.3. Motion of electrons in a storage ring and the angular coordinate defined for synchrotron radiation

circular arcs and straight lines. Even in those cases, however, the above arguments as well as the following ones are valid.

The characteristics of synchrotron radiation from bending magnets in an electron or positron storage ring are as follows [4].

- It has a continuous spectrum over the wide range of wavelengths from the far infrared to X-rays.
- The intensity is high.
- The angular divergence of radiation is small in the vertical direction; typically 0.1–1 mrad. In the horizontal plane, radiation is emitted tangentially to the electron motion. Since the electron in a bending magnet changes its direction of motion continuously, radiation sweeps the horizontal plane.
- The source size of radiation is small. The shape and the size of the light source are given by those of the transverse cross section of the electron beam. The shape is elliptic, and the size is typically 0.5–1 mm high and a few mm wide.
- Synchrotron radiation is linearly polarized when it is observed on the plane of the electron motion, while it is in general elliptically polarized above and below the plane. When an angle of observation in the vertical direction is large, it is circularly polarized though the intensity is lower.
- Since the spectrum can be calculated theoretically, it can be used as a standard light source like the blackbody radiation.
- It has a pulsed time structure with a duration of $0.1 \sim 1$ ns. Although it is usually used as temporally continuous light because of a high repetition frequency, it can be used as a pulsed light source. When a long interval between pulses is required, the storage ring is operated in the single bunch operation mode, which will be described later, with an interval of 50 ns $\sim$ 5 ms between the pulses.

– The electron beam necessarily circulates in the ultra-high vacuum, so that it is a clean light source.

There are several units defining the spectral intensity, and one has to choose an appropriate one for one's purpose. Here, as units of photons per second per beam current of 1 mA at a certain wavelength or photon energy, flux, brightness and brilliance are defined as follows;

flux = phs/s/mrad/0.1%bw/mA
brightness = phs/s/mrad^2/0.1%bw/mA
brilliance = phs/s/mrad^2/mm^2/0.1%bw/mA,

where phs and bw denote the number of photons and the fractional bandwidth, respectively. The 0.1% bandwidth means utilization of light with the fractional resolution of $\Delta\lambda/\lambda = \Delta\omega/\omega = 10^{-3}$. The flux is the number of photons per horizontal angular acceptance $\Delta\theta$ of 1 mrad integrated over the vertical angle ψ. The brightness is the number of photons per both 1 mrad of $\Delta\theta$ and $\Delta\psi$, and the brilliance is the number further divided by a cross section area of the electron beam, which is the number density of photons in the four-dimensional phase space.

Let $\mathscr{I}(\hbar\omega)$ be the number of photons per second emitted at the photon energy $\hbar\omega$, α the fine structure constant, $\Delta\omega/\omega$ the fractional bandwidth, and I the stored beam current, the number of photons per unit θ (1 rad) is expressed as [2, 3]

$$\frac{\mathrm{d}\mathscr{I}(\hbar\omega)}{\mathrm{d}\theta} = \frac{\sqrt{3}}{2\pi}\alpha\gamma\frac{\Delta\omega}{\omega}\frac{I}{e}\frac{\omega}{\omega_c}\int_{\omega/\omega_c}^{\infty} K_{5/3}(x)\mathrm{d}x, \tag{13.6}$$

where $K_{5/3}$ is the modified Bessel function of the second kind. This equation is derived from Eq. (13.5). The flux may be written in the practical units as

$$\frac{\mathrm{d}\mathscr{I}(\hbar\omega)}{\mathrm{d}\theta} \approx 2.457 \times 10^{10} \times E(\mathrm{GeV})G_1(\omega/\omega_c), \ \mathrm{phs/s/mrad/0.1\%bw/mA} \tag{13.7}$$

where G_1 stands for (ω/ω_c) times the following integral in Eq. (13.6). The function G_1 is shown in Fig. 13.4. From the lower frequency side, G_1 increases gradually, reaches the maximum value at the frequency slightly lower than ω_c and decreases rapidly in the higher frequency side. Figure 13.5 shows fluxes of synchrotron radiation from bending magnets in some synchrotron radiation sources. In these storage rings, bending magnets are conventional ones and the magnetic field is around 1 T. The magnetic field in bending magnets is given by

$$B(\mathrm{T}) \approx 3.34 \times \frac{E(\mathrm{GeV})}{\rho(\mathrm{m})}. \tag{13.8}$$

Therefore, the critical photon energy given by Eq. (13.4) is proportional to the magnetic field, provided that the electron energy is constant. Since a superconducting wiggler generates the magnetic field of 4 $\sim$ 6 T, it is possible to obtain

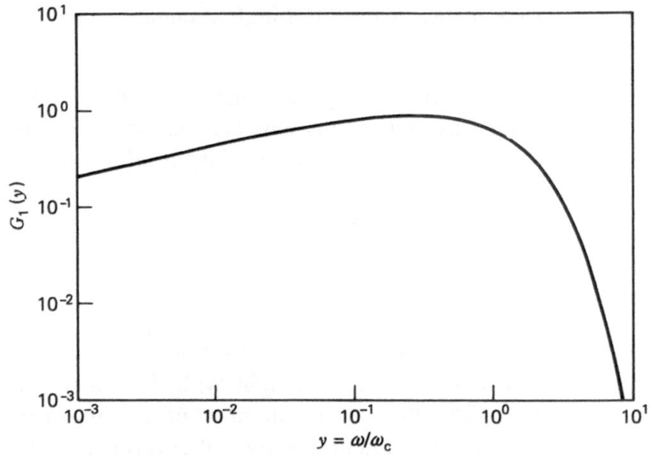

Fig. 13.4. Universal function G_1 for the flux of synchrotron radiation. The scale on the abscissa is the ratio of the frequency of light to the critical frequency

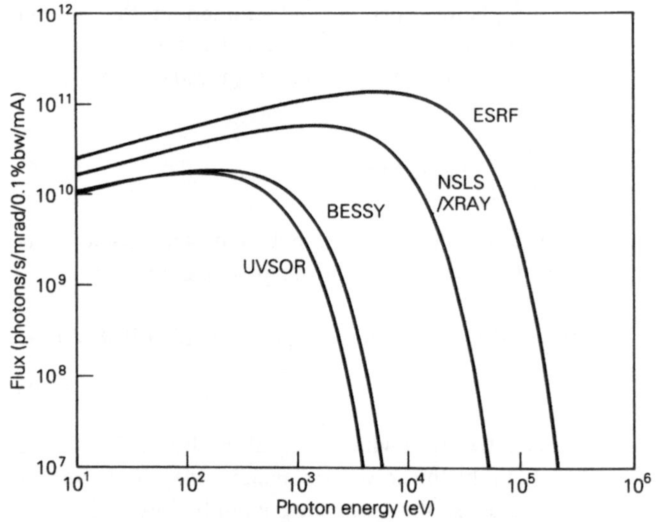

Fig. 13.5. Fluxes of synchrotron radiation from bending magnets of some well-known storage rings

synchrotron radiation of several times higher energy than that from bending magnets for the same electron energy if it is inserted in a storage ring. Hence it is called a wavelength shifter. The total number of photons incident on the first optical element is the flux times an acceptance angle in the horizontal direction $\Delta\theta$. The acceptance angle is dependent on a beamline and other experimental conditions, and it is in the region from a few mrad to tens of mrad. Furthermore, we obtain the power spectrum from the flux multiplied by $\hbar\omega$.

The angular divergence of synchrotron radiation in the vertical direction is dependent on wavelength, such that it is narrower at shorter wavelengths and wider at longer wavelengths. The normalized angular distributions for some typical frequencies in units of the critical frequency are shown in Fig. 13.6, where contribution from the angular divergence of the electron beam is neglected.

The intensity of light that has the electric field parallel to the orbit $I_{||}$ becomes maximum at $\psi = 0$, while that with the electric field perpendicular to it $I_{\perp}$ is zero at $\psi = 0$. The polarization characteristics of synchrotron radiation are evident in this figure. When the electron beam circulates clockwise (seen from above) and an observer faces the source point, the light has right elliptical polarization above the orbit plane and it has left elliptical polarization below the plane. In the vicinity of $I_{||} \cong I_{\perp}$, we may regard the light as circularly polarized. The total intensity is $I_{||} + I_{\perp}$, and the flux at each frequency is given by integrating the sum with respect to ψ. Taking into account the vertical angular divergence of the electron beam σ'_y, we obtain the vertical angular divergence of radiation as $\Sigma_{\psi} = \sqrt{\sigma_r'^2 + \sigma_y'^2}$, where σ'_r is the standard deviation of the intrinsic angular divergence of radiation in the vertical direction from a single electron. In the wavelength region longer than the critical wavelength, σ'_r is usually much larger than σ'_y.

Synchrotron radiation is transversely coherent in the wavelength region where a product of the beam size σ_y and the angular divergence of radiation Σ_{ψ} is smaller than the wavelength.

Energy radiated by an electron in one revolution is given as

$$U(\text{keV}) \approx 88.5 \times \frac{E^4(\text{GeV})}{\rho(\text{m})}\,. \tag{13.9}$$

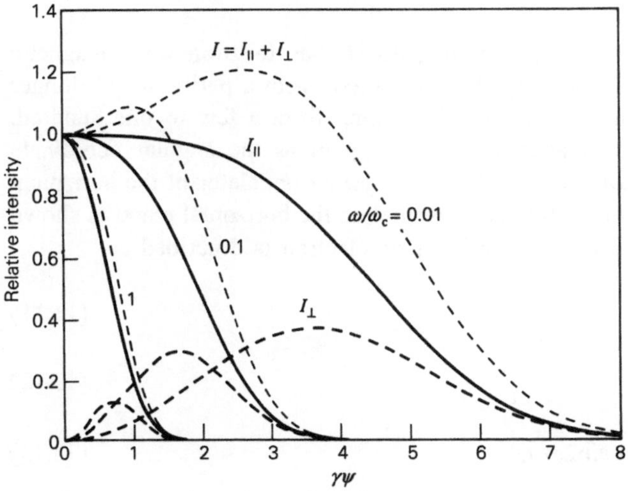

Fig. 13.6. Normalized angular distributions of synchrotron radiation in the vertical direction for some typical angular frequencies. The symbols $I_{||}$ and $I_{\perp}$ denote intensities with the polarization parallel to and perpendicular to the electron orbit, respectively

The total power radiated in the circumference of a storage ring (2π rad) is obtained by

$$P(\text{kW}) = U(\text{keV}) \times I(\text{A}), \tag{13.10}$$

where I is the beam current. The total power of light irradiating a first optical element is the power given by Eq. (13.10) times a horizontal acceptance angle of the element $\Delta\theta/2\pi$. The angular divergence (the standard deviation) of the radiated power in the vertical direction is approximately $1/\gamma$.

In the preceding paragraphs, we assumed implicitly that the intensity spectrum of synchrotron radiation from the electron beam is given by that from a single electron multiplied by the number of electrons in the beam (incoherent synchrotron radiation). This assumption is not valid when spatial dimensions of the electron beam are comparable to or less than the wavelength of radiation. The intensity at the wavelength is enhanced further due to the coherent sum of radiation emitted by individual electrons so that it is proportional to the square of the number of electrons. Since the number of electrons in a bunch ranges from 10^6 to 10^{11}, the intensity is very high compared with that of incoherent synchrotron radiation. In an electron storage ring, the longitudinal bunch length is in the region of 3 to 30 cm, and coherent effect has not been clearly observed. In an electron linac, however, the bunch length is as short as 1 mm, and coherent synchrotron radiation has been observed in the far infrared (submillimeter) and millimeter regions when the electron beam is deflected in a bending magnet [5].

13.3 Undulator Radiation

An undulator consists of an array of magnets. It has the transverse magnetic field which varies sinusoidally along the central axis with a period length longer than a few cm and the number of periods ranging from a few to one hundred. A schematic drawing of a planar undulator as well as the angular coordinate defined for radiation is shown in Fig. 13.7. In a planar undulator of the horizontal oscillation type, the electron motion is sinusoidal on the horizontal plane as shown in Fig. 13.8 (a). The trajectory of a relativistic electron is described as

$$x \approx \frac{\lambda_u K}{2\pi\gamma} \sin \omega_0 t \tag{13.11}$$

$$y = 0 \tag{13.12}$$

$$z \approx c\beta^* t - \frac{\lambda_u K^2}{16\pi\gamma^2} \sin 2\omega_0 t, \tag{13.13}$$

where λ_u is the period length of the undulator, K the deflection parameter, defined in Eq. (13.16), and $c\beta^*$ the mean longitudinal speed of the electron given

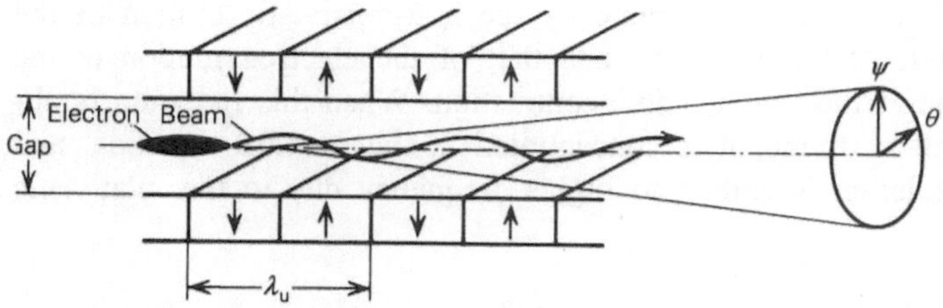

Fig. 13.7. Electron beam moving in an undulator and the angular coordinate defined for undulator radiation

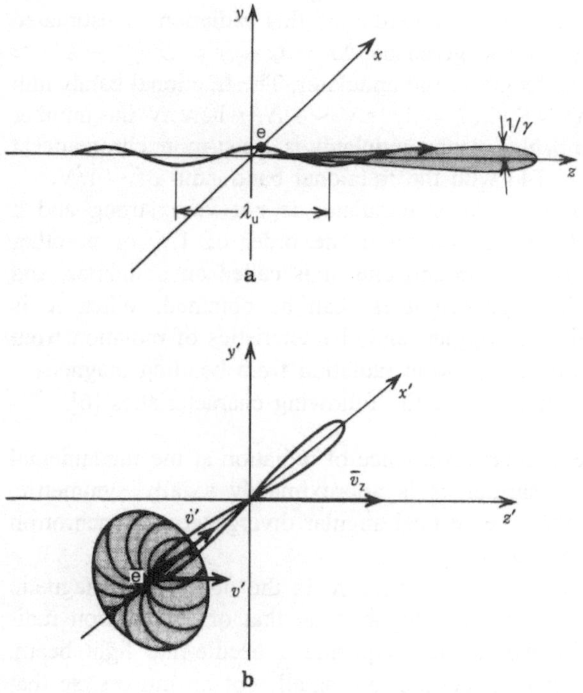

Fig. 13.8. Motions of an electron moving in an undulator and the angular distributions of emitted radiation observed in the laboratory frame **a** and in the moving frame **b**

by $\beta^* \approx 1-(1+K^2/2)/2\gamma^2$. The average speed becomes slower due to the transverse motion. The angular frequency of the transverse motion observed in the laboratory frame is given by $\omega_0 \approx 2\pi c\beta^*/\lambda_u$. When the motion is observed in the moving frame at the mean speed, it is the figure-eight motion, as shown in Fig. 13.8 (b). The amplitude of the longitudinal oscillation is small when $K \leqslant 1$. The frequency of the transverse motion ω_M in the moving frame is given by $\omega_M = \omega_0/\sqrt{1-\beta^{*2}} \approx \omega_0\gamma/\sqrt{1+K^2/2}$. In the moving frame, it emits radiation at ω_M with the angular distribution given by Eq. (13.1). Thus the frequency of radiation emitted in the moving frame by the non-relativistic motion of the electron is $\gamma/\sqrt{1+K^2/2}$ times higher than that of the electron motion in the laboratory frame due to the Lorentz time-contraction. When this radiation is observed in the laboratory frame, it is concentrated in the forward direction, and the frequency of radiation is shifted to higher frequency due to the relativistic Doppler effect.

$$\omega_L = \frac{1+\beta^*}{\sqrt{1-\beta^{*2}}}\omega_M = \frac{\omega_0}{1-\beta^*} \approx \frac{2\gamma^2}{1+K^2/2}\omega_0 \tag{13.14}$$

The frequency of radiaton is $2\gamma^2/(1+K^2/2)$ times higher than that of the motion in the laboratory frame. The spectral bandwidth of this radiation is estimated from a time duration of a light pulse given as $\Delta t = t_e - t_l = L/c\beta^* - L/c \approx L(1+K^2/2)/2c\gamma^2$, where L is the length of the undulator. The fractional bandwidth is given by $\Delta\omega/\omega_L \sim 1/\omega_L\Delta t = \lambda_u/2\pi L = 1/2\pi N \sim 1/N$, where N the number of periods. Thus, radiation from the planar undulator is quasi-monochromatic at the frequency given by Eq. (13.14) with the fractional bandwidth of $\sim 1/N$.

When the magnetic field in a planar undulator is not very strong and a maximum deflection angle of an electron is of the order of $1/\gamma$, or in other words, the deflection parameter K is around one, it is called an undulator, and quasi-monochromatic light with high brightness can be obtained. When K is much larger than one, it is called a wiggler, and characteristics of radiation from a wiggler are similar to those of synchrotron radiation from bending magnets.

Radiation from a planar undulator has the following characteristics [6].

- Directionality is high. The angular divergence of radiation at the fundamental peak as well as its higher harmonics is approximately axially symmetric, and several times smaller than the vertical angular divergence of synchrotron radiation from a bending magnet.
- The brightness is proportional to N^2, where N is the number of magnetic periods, so that it is $10^2 \sim 10^4$ times higher than that of synchrotron radiation. It displays power in experiments requiring a needle-like light beam. Furthermore, the radiation can be focused to a small spot by mirrors, so that it is advantageous for small domain experiments and high resolution experiments since a narrow slit is used in a monochromator.
- It has a series of quasi-monochromatic light, i.e., the fundamental peak and its higher harmonics, and the wavelengths are variable. When undulator radiation is used with an appropriate filter but without a monochromator, intense light may be obtained though its bandwidth is not small.

Radiation from a planar undulator has linear polarization, the direction of which is on the plane of oscillation of the electron trajectory, and the circularly or elliptically polarized light is emitted from a helical undulator where the trajectory of an electron is helical. The brightness of radiation from a wiggler is proportional to N.

The wavelength λ of the emitted radiation from a planar undulator is given by [2]

$$\lambda = \frac{\lambda_u}{2n\gamma^2}(1 + K^2/2 + \gamma^2\Theta^2), \qquad n = 1, 2, 3, \cdots \tag{13.15}$$

where K is the deflection parameter defined as

$$K = \frac{eB_0\lambda_u}{2\pi mc^2}[\text{cgs Gau}\beta] = \frac{eB_0\lambda_u}{2\pi mc}[\text{MKSA}] \approx 93.4\lambda_u(\text{m})B_0(\text{T}), \tag{13.16}$$

with B_0 being the maximum value of the sinusoidal magnetic field, and $\Theta^2{=}\theta^2 + \psi^2$. If the electron beam has no angular divergence and light is observed in the forward direction at $\Theta = 0$, then only the odd harmonics appear with n=1, 3, 5, and so on. In a helical undulator, the term $K^2/2$ in the right hand side of Eq. (13.15) should be replaced by K^2. When radiation from a helical undulator is observed on the axis for the electron beam having no angular divergence, only the fundamental peak appears. To vary the wavelength, either the deflection parameter or the electron energy is varied. Most commonly, the magnetic field B_0 is varied by changing the magnet gap of an undulator made of permanent magnets. The maximum deflection angle of an electron in the undulator is given by K/γ. Due to the interference effect, the fractional bandwidth at the n-th harmonic becomes

$$\Delta\lambda/\lambda \approx \frac{1}{nN}. \tag{13.17}$$

For a planar undulator, when the electron beam has no angular divergence, the intensity per unit solid angle at the n-th higher harmonic observed at Θ=0 is given by [2]

$$\begin{aligned} \left.\frac{d\mathcal{I}(\hbar\omega)}{d\theta d\psi}\right|_{\Theta=0} &= \alpha N^2\gamma^2\frac{\Delta\omega}{\omega}\frac{I}{e}F_n(K) \quad && n = 1, 3, 5, \cdots \\ &= 0. && n = 2, 4, 6, \cdots \end{aligned} \tag{13.18}$$

The function F_n is derived from Eq. (13.5) as

$$\begin{aligned} F_n(K) = \frac{n^2K^2}{(1+K^2/2)^2}\left\{J_{(n-1)/2}\left[\frac{nK^2}{4(1+K^2/2)^2}\right] - \right. \\ \left. J_{(n+1)/2}\left[\frac{nK^2}{4(1+K^2/2)^2}\right]\right\}^2, \end{aligned} \tag{13.19}$$

where J is the Bessel function. The brightness may be written in the practical units as

$$\begin{aligned} \left.\frac{d\mathcal{I}_n(\hbar\omega)}{d\theta d\psi}\right|_{\Theta=0} &\approx 1.744 \times 10^{11} \times N^2 E(\text{GeV})F_n(K). \\ &\text{phs/s/mrad}^2/0.1\%\text{bw/mA} \end{aligned} \tag{13.20}$$

The function $F_n(K)$ is plotted in Fig. 13.9. When the electron beam has a finite angular divergence, the bandwidth of a peak becomes larger, the brightness is decreased, and the even higher harmonics appear. Figure 13.10 shows the calculated brightness of radiation from an undulator at the Stanford Synchrotron Radiation Laboratory, in which a finite divergence of the electron beam is taken into account.

In the undulator regime where K is not very large, the angular distribution of the radiation is approximately axially symmetric with respect to the central

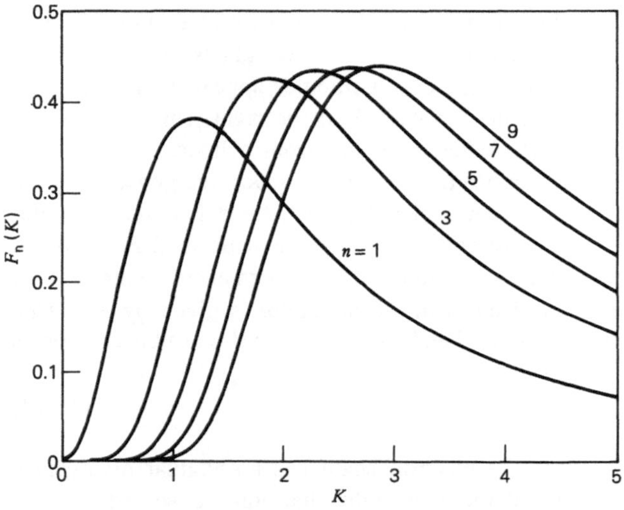

Fig. 13.9. Universal function $F_n(K)$ for the brightness of undulator radiation, where K is the deflection parameter

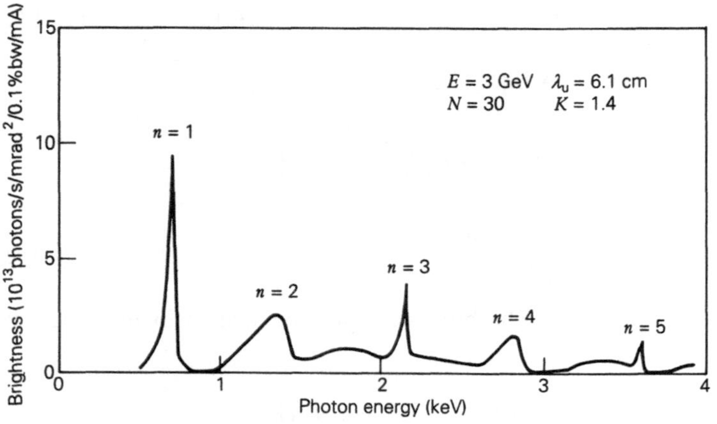

Fig. 13.10. Calculated brightness of radiation from an undulator at Stanford Synchrotron Radiation Laboratory [Winick H, Brown G, Halbach K, Harris J (1981) Physics Today: 50]

axis, and its standard deviation is given by [2]

$$\sigma_r' \approx \frac{1}{\gamma}\sqrt{\frac{(1+K^2/2)}{2nN}}. \tag{13.21}$$

When $K \sim 1$, Eq. (13.21) may be further approximated as $\sigma_r' \approx 1/(\gamma\sqrt{nN})$. Since the Gaussian distribution is assumed, one can calculate the number of photons available from the brightness multiplied by the beam current and $2\pi\sigma_r'^2$. When the beam divergence is taken into account, angular divergence of the radiation in the θ and the ψ directions are $\Sigma_\theta = \sqrt{\sigma_r'^2 + \sigma_x'^2}$ and $\Sigma_\psi = \sqrt{\sigma_r'^2 + \sigma_y'^2}$, respectively. If the beam divergence is comparable to or larger than that of radiation, the brightness is reduced. Since the angular divergence of undulator radiation is approximately $1/\sqrt{nN}$ times smaller than that of synchrotron radiation from bending magnets, the brightness of undulator radiation is more sensitive to the angular divergence of the electron beam than that of synchrotron radiation is.

The total power radiated by a planar undulator including all the higher harmonics is given by [2]

$$P(\text{kW}) \approx 0.633 \times E^2(\text{GeV})B_0^2(\text{T})L(\text{m})I(\text{A}), \tag{13.22}$$

where L denotes the total length of the undulator, $L = N\lambda_u$. If the contribution from the electron beam divergence is negligibly small, the angular divergence of the radiation power (the standard deviation) is approximately $0.5/\gamma$ in the vertical direction and $0.7K/\gamma$ in the horizontal direction. In general, the radiation power from an undulator ($K \sim 1$) is relatively low, whereas that from a wiggler is higher, so that care has to be taken to use radiation from a wiggler.

Radiation from an undulator emitted by a single electron is transversely coherent, and transverse emittance of radiation at a wavelength λ is equal to emittance ε_r given by the diffraction limit at the wavelength as

$$\varepsilon_r = \frac{\lambda}{4\pi}. \tag{13.23}$$

The effective source size at the center of the undulator and the angular divergence is given by [2, 7]

$$\sigma_r = \frac{\sqrt{\lambda L}}{4\pi}, \quad \sigma_r' = \sqrt{\frac{\lambda}{L}}, \tag{13.24}$$

where λ and L are the wavelength of light and the total length of the undulator. The angular divergence of the radiation is equivalent to that given in Eq. (13.21). The effective source size originates in the length of the light source in an undulator and the angular divergence of radiation. The product of σ_r and σ_r' is equal to the emittance given by Eq. (13.23), which ensures that the radiation is transversely coherent. When the size and the angular divergence of the electron beam at the source point are much less than those of radiation, which is realized in a very low-emittance storage ring, radiation from an undulator is transversely coherent.

Longitudinal coherence of radiation from an undulator is expressed in terms of the coherence length defined as [2, 7]

$$l_c = \frac{\lambda^2}{\Delta\lambda}, \tag{13.25}$$

where $\Delta\lambda$ is the spectral width of radiation. Since the fractional spectral width of radiation at the n-th harmonic from an undulator is given by Eq. (13.17), the coherence length is given by

$$l_c = nN\lambda, \tag{13.26}$$

where N is the number of periods of the undulator. It is equal to the length of a wave packet since nN is the number of wavelengths emitted by an electron in the undulator. By utilizing the longitudinal coherence, interference experiments such as *holography* can be made on a specimen, the thickness of which is of the order of coherent length.

13.4 Free Electron Laser

Contrary to conventional lasers, free electron lasers (FEL) do not use a gas or a solid as gain medium. It consists of (i) the electron beam from an accelerator, (ii) an undulator, and (iii) an optical resonator. Since the electromagnetic wave propagating in free space has only the transverse electric field, the electron beam cannot interact with the wave propagating along the beam. However, when the electron beam goes through an undulator, it is wiggled by the magnetic field in the undulator, so that it has a transverse velocity. Therefore, it can interact with the electromagnetic wave, or light. Figure 13.11 shows a schematic drawing of a free electron laser. In the first stage, energy of electrons is modulated in a wavelength of radiation. Electrons in the accelerating phase of light gain kinetic energy, and those in the decelerating phase lose energy. In the second stage, the energy modulation is converted to a spatial density modulation of electrons due to an energy dispersion effect in the undulator. In the final stage, the electron beam whose longitudinal density is modulated in the wavelength of light emits stimulated radiation.

Potential advantages of free electron lasers are as follows [8–10].

- No limitation on the wavelength. Since a wavelength of the laser depends only on energy of the electron beam and parameters of an undulator such as the period length and the magnetic field, one can choose any wavelength from microwave to ultraviolet at present, and possibly to X-rays in the future.
- Tunability. The wavelength can be scanned continuously by changing either the electron energy or the magnetic field of an undulator.

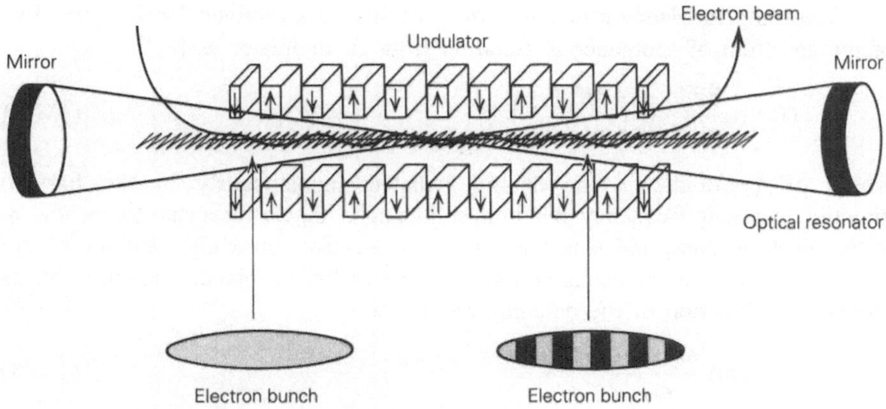

Fig. 13.11. Schematic layout of a free electron laser with an optical resonator

- High efficiency. Since it transforms the kinetic energy of the electron beam directly to energy of coherent radiation, conversion efficiency is high. The maximum conversion efficiency is a few percent.
- High power. The beam power is given by the product of kinetic energy in electron volts and the beam current in amperes. The energy of the electron beam is in the region from 1 MeV to 1 GeV, and the instantaneous peak current is in the region from 1 A to some kA. The output power is calculated from the beam power and the conversion efficiency.

For a planar undulator, the wavelength of a free electron laser is given by the same formula for a peak wavelength of radiation from the undulator, Eq. (13.15), emitted in the forward direction at $\Theta = 0$. Undulator radiation is sometimes called spontaneous emission or radiation in analogy with a conventional laser. Since the origin of the even harmonics is the longitudinal oscillation in the moving frame at the mean longitudinal speed of the electron, only odd harmonics are relevant to the free electron laser. In the field of free electron lasers, the resonant wavelength is sometimes defined by the formula used for a helical undulator, irrespective of types of undulators; the term $K^2/2$ in Eq. (13.15) is replaced by K^2. Then the definition of the deflection parameter for a planar undulator is different, and it is $1/\sqrt{2}$ times smaller than that used in the field of synchrotron radiation.

There are two regimes in free electron lasers, referred to as the Raman and Compton regimes [8, 9]. In the former, the beam current is high, so that gain is high. The collective motion of electrons is involved in the amplification process. In the latter, the beam current is relatively low, and gain is low. Physical processes involved in these regimes are quite different. Since a very high beam current is only realized in accelerators of energies lower than a few MeV presently, the wavelength of free electron lasers in the Raman regime is longer than far infrared. Therefore, we will confine ourselves to free electron lasers in the Compton regime, in which wavelengths of FELs are shorter and electron energies are higher than tens of MeV.

According to Madey's theorem, the gain per pass is given by the derivative of the spectrum of spontaneous radiation from an undulator as [8, 9]

$$G = -\frac{8\pi nc}{m\omega_L^2}\frac{\mathrm{d}}{\mathrm{d}\gamma}\left(\frac{\mathrm{d}W_L(\gamma)}{\mathrm{d}\omega\,\mathrm{d}\Omega}\right), \tag{13.27}$$

where $\mathrm{d}W_L(\gamma)/\mathrm{d}\omega\mathrm{d}\Omega$ is the energy radiated spontaneously in the forward direction per unit frequency per unit solid angle, ω_L is the laser frequency, m is the electron mass, and n is the number of electrons in a unit volume. Figure 13.12(a) shows a spontaneous emission spectrum from a planar undulator that is plotted as a function of the detuning parameter ν

$$\nu = 2\pi N\frac{\omega_0 - \omega}{\omega_0}, \tag{13.28}$$

where N is the number of magnetic periods in the undulator, ω_0 the peak angular frequency of undulator radiation, and ω the angular frequency of light, and Fig. 13.12(b) shows the gain calculated by Eq. (13.27). On the left hand side of the peak in the spontaneous emission spectrum, the gain is negative. Therefore, a part of coherent radiation is absorbed by the electron beam. On the right hand side, the gain is positive, and coherent radiation is amplified. If one knows the shape of a spontaneous emission spectrum, one can calculate where the maximum gain is located according to Madey's theorem.

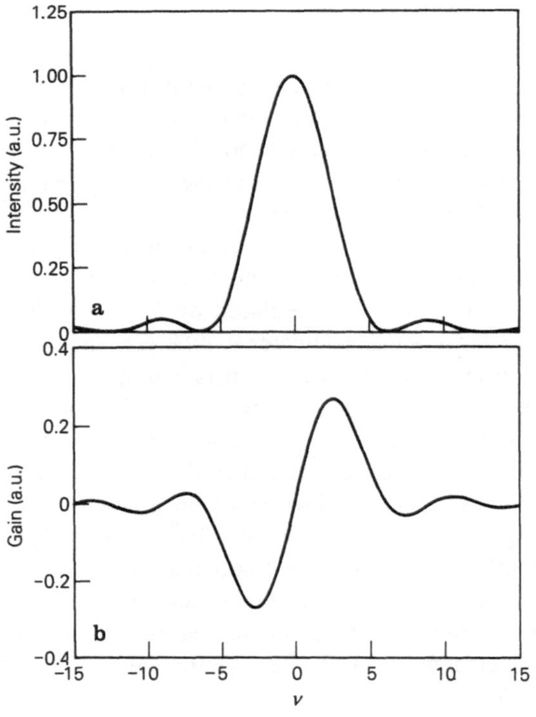

Fig. 13.12. Calculated spectrum of spontaneous radiation from a conventional undulator **a** and the gain curve calculated from it by Madey's theorem **b**

The peak gain per pass from a planar undulator is given in practical units as [9]

$$G_{\text{peak}} \approx 8.4 \times 10^{-19} \left[\frac{L(\text{m})}{E(\text{GeV})}\right]^3 \left[B_0(\text{kG})\right]^2 n(\text{cm}^{-3})\lambda_u(\text{cm})F, \qquad (13.29)$$

where L is the length of an undulator, E the electron energy, B_0 the peak magnetic field of the undulator, n the number of electrons, and F the filling factor of the electron beam to the light beam. According to Eq. (13.15), electron energy has to be higher to obtain the short wavelength. Then gain decreases according to Eq. (13.29). In order to compensate the reduced gain, a long undulator or the electron beam with a high peak current is necessary.

When the energy spread of the electron beam used for an FEL is small, the gain can be increased with a novel undulator, called an optical klystron. An optical klystron consists of two undulator sections and a dispersive section between them, which is a wiggler with one period and a large deflection parameter. Bunching of electrons in a wavelength of light is enhanced due to the strong energy dispersion in the dispersive section, so that gain becomes several times higher than that with a conventional undulator of the same total length. The radiation spectrum from the optical klystron shows many sharp peaks, the envelope of which is given by a shape of spontaneous radiation from one of the undulators.

In a gain measurement, coherent light from a laser is injected through an undulator. One can measure intensity variation of light with and without the electron beam. When gain is very low, which is the case for wavelength in the visible or shorter region, one has to use a more sensitive method since spontaneous radiation synchronizing with the electron beam is not negligible. The electron beam is temporally grouped in so-called bunches, by an RF accelerating field of an accelerator. Gain appears only when an electron bunch exists, so that gain is temporally modulated at the frequency of appearance of electron bunches, f_e. In order to distinguish the gain signal from the signal arising from spontaneous radiation, the intensity of the laser is modulated at a frequency, f_l. Then the gain signal appears at mixed frequencies of the two frequencies, $f_e \pm f_l$. After subtracting the component of the frequency f_e from the gain signal, the intensity of the signal as well as the phase to the reference signal at f_l, that is, amplification or absorption, can be measured with a phase sensitive detector such as a lock-in amplifier. The detuning parameter ν in Eq. (13.28) is scanned by varying either the magnetic field of an undulator, electron energy, or wavelength of the laser, so that one obtains a gain curve similar to that shown in Fig. 13.12(b).

When gain per pass is not very high, an optical resonator with mirrors is employed in order to obtain lasing. An optical resonator is sometimes referred to as an optical cavity. As shown in Fig. 13.11, it consists of a pair of concave mirrors with high reflectivity. Since the electron beam is bunched for higher energy accelerators, the cavity length is chosen so that the round-trip time of light in the cavity is equal to the time between successive appearances of electron bunches or to a multiple of that, which ensures that the light beam is amplified by the electron beam repeatedly. When gain per pass exceeds a reflection loss of light in a round trip, a seed of coherent light that is accidentally produced in the cavity is amplified, and lasing begins. The power of the coherent light continues to increase until it reaches a stationary value, which is called saturation. A part of the intra-cavity power is taken out with an output coupler, which may be one of the mirrors with small transmittance, a hole in one of the mirrors, or a Brewster mirror (beam splitter) inserted in the cavity for power extraction. The wavelength of light can be continuously varied by changing either the magnetic field of the undulator or energy of the electron beam. The range of tunability is, however, limited by the range of high reflectivity of the mirrors.

In order to obtain the highest gain possible, parameters of the optical cavity are chosen so that the overlap factor F in Eq. (13.29) is as close as possible to one. When the transverse size of the electron beam is small, a concentric mirror configuration is employed to reduce the size of the light beam, where a sum of the mirror radii is slightly larger than the cavity length. There are two kinds of mirrors commonly used; multilayer mirrors made of dielectric materials, and metal mirrors such as gold or aluminum. Multilayer mirrors have high reflectivity in the wavelength region down to 200 nm, but their bandwidth is relatively narrrow. Metal mirrors have high reflectivity in the infrared and the far infrared regions, but their reflectivity is relatively low in the wavelength region shorter than the visible. Mirrors may be damaged by the laser power stored in the cavity and by higher harmonics of spontaneous radiation from an undulator. Metal mirrors generally have higher durability than multilayer mirrors.

The time structure of output light from an FEL reflects that of the electron beam from an accelerator employed. In an RF linac, which is commonly used in FELs in the wavelength region longer than visible, the electron beam consists of micro pulses and macro pulses. A macro pulse contains many micro pulses. A time duration of a macro pulse ranges from several μs to tens of ms with a repetition rate of tens of Hz. The duration of micro pulses is typically a few ps with an interval ranging from 0.3 to 100 ns. The spectral linewidth of an FEL depends also on the type of the accelerator. In general, the linewidth is narrower when the temporal length of micro pulses is longer. In an FEL based on an RF linac, the linewidth is given by the Fourier transformation of the temporal profile of micro pulses as [11]

$$\frac{\sigma_\lambda}{\lambda} = \frac{\lambda}{4\pi c \sigma_t} \tag{13.30}$$

where σ_t is the standard deviation of the bunch length.

When lasing begins, the electron beam is heated in the sense that the energy spread of the electron beam becomes larger due to interaction with light. If gain is high enough, the electron beam is decelerated in an undulator due to the interaction and the average energy of the electron beam becomes considerably lower in the latter part of the undulator, so that gain becomes negative there. In both cases, the output power is saturated. The maximum power that can be extracted from the electron beam in an FEL with an conventional undulator is given as [9]

$$P_l \leqslant \frac{P_e}{2N}, \tag{13.31}$$

where P_l and P_e denote the laser power and the beam power given by the electron energy times the beam current, respectively, and N the number of periods of the undulator.

This power limitation can be overcome with one of the novel undulators. In a varied parameter undulator, either the period length or the peak magnetic field is varied along the undulator in such a way that the resonance condition given by Eq. (13.15) is satisfied for a given wavelength and energy variation of the electron beam through the undulator. Therefore, the gain does not become negative due to variation of the mean energy of the electron beam along the undulator. In a storage ring FEL, the average output power of a free electron laser is further limited by multiple use of the electron beam as [9]

$$P_l \leqslant \frac{P_{SR}}{2N}, \tag{13.32}$$

where P_{SR} is the power emitted in a storage ring by synchrotron radiation (Eq. (13.10)), which is referred to as Renieri's limit.

There is another operation regime of an FEL called self-amplified spontaneous emission [12]. As the length of an undulator used for an FEL increases, bunching of electrons in a wavelength of light is enhanced due to interaction with light for a long period of time. Note that in a conventional FEL with a relatively short undulator, coherent light interacts with electrons repeatedly by use of an optical resonator while the new electron beam comes every time that has no density modulation in the wavelength. If the undulator is long enough, gain becomes proportional not to the beam current but to the square of it. In a single pass operation, the spontaneous radiation emitted in the first part of the undulator bunches electrons in its wavelength, the bunched electrons then amplify the spontaneous emission as the electron beam moves along the undulator, and an appreciable power can be extracted from it. It is similar to coherent synchrotron radiation in the far infrared and millimeter regions as mentioned in Sect. 13.2. This operation mode opens up the possibility of short-wavelength FELs *without* mirrors.

There are many free electron laser facilities worldwide, based on electrostatic accelerators, RF linacs, and storage rings [10]. The FEL facilities and laboratories are listed in Table 2 except for those working in the microwave region. They are running, under construction, or in proposal. Some of them are user facilities, and the others are for basic research and development of FELs. Most of the user facilities are based on RF linacs for the wavelength regions of infrared and far infrared. The recent prosperity of FELs on linear accelerators has been mainly brought about by improvement of electron beam qualities such as emittance and the energy spread made with new types of electron guns. Since the electron beam is used only once, the output power is high in linac-based FELs. Average powers

Table 2. Free electron laser facilities and laboratories

Location	Project	Source	Energy (MeV)	λ (μm)	Note
France					
Orsay	Super ACO	storage ring	600~800	0.35~0.63	parasitic
	CLIO	RF linac	30~70	2~20	user facility
Bruyère	ELSA	RF linac	20	15~25	
Germany					
Dortmund	DELTA	storage ring	500~1000	UV~Vis	construction
Darmstadt	KFI	super c linac	35~50	2.5~6.0	construction
Italy					
Frascati	LISA	super c linac	25~50	2~5	construction
Frascati	ENEA	microtron	20	32	
Netherlands					
Rijnhuizen	FELIX	RF linac	15~45	6.5~100	user facility
Twente	TEU-FEL	RF linac	6	180	construction
		microtron	25	10	
Russia					
Novosibirsk	VEPP-3	storage ring	350	0.24~0.65	parasitic
	INP	microtron	35	IFR	construction
USA					
Durham	Duke/MarkIII	RF linac	35~45	1.5~8.0	user facility
	Duke/XUV-FEL	storage ring	1000	0.01~0.5	construction
Newport News	CEBAF	super c linac	45	4.5~25	construction
		(recirculation)	400	0.15~0.26	construction
Brookhaven	BNL/ATF	RF linac	50	0.5	construction
	BNL/UV/-FEL	super c linac	80~250	0.075~vis	proposal
Gaithersburg	NIST-NRL	cw microtron	17~185	0.2~10	construction/ user facility
Berkeley	CDRL-FEL	super c linac	56	3~50	proposal
Los Alamos	APEX	RF linac	46	0.37, 3~45	
	AFEL	RF linac	20	0.4~3.7	
Standord	SCA/FEL	RF linac	35~45	0.5~5.0	user facility
Tennessee	Vanderbilt	RF linac	35~45	1.0~8.0	user facility
Los Angles	UCLA-KIAE	RF linac	20	10.6	construction
Santa Barbara	UCSB-MM	Van de Graaf	6	338~2500	user facility
	-FIR			63~300	
	-MID IR			30~90	
CHINA					
Beijing	IHEP	RF linac	10~30	10.6	
JAPAN					
Tsukuba	PF	storage ring	700	0.17	preparasion/ parastic
Tsukuba	TERAS	storage ring	231	0.6	finished
	NIJI-IV	storage ring	500	0.35 ~ 0.6	
Okazaki	UVSOR	storage ring	500	0.3 ~ 0.5	parasitic
Osaka	FELI	RF linac	20~80/170	UV~IR	ind.application
Osaka	ISIR	RF linac	38	30 ~ 40	
Tokai	UT-FEL	RF linac	15	40~50	
Tokai	JAERI/FEL	super c linac	13.2	IR	construction

of these FELs are from 1 to 100 W, and peak powers are from 1 to 50 MW. In the visible and ultraviolet regions, most of the FELs are operated on storage rings in the year 1992, the shortest wavelength was 240 nm in the ultraviolet achieved with a storage-ring-based FEL [13]. The storage rings used for the FEL experiments are not solely for FELs, and most of them are synchrotron radiation sources, in which the length of a straight section for installation of an undulator is limited. The average output power of an FEL on a storage ring is limited at present below 1 W due to successive use of the electron beam (Renieri's limit).

At present, extensive efforts and studies are being made to shorten the wavelength to the VUV region. In linac-based FELs, the energy of the electron beam has to be raised to make the wavelength shorter. In storage-ring-based FELs, specialized storage rings are being constructed, in which straight sections are longer than 10 m. The shortest wavelength is expected to be tens of nano-meters. In the VUV region, the main difficulty is a lack of mirrors with high reflectivity at the normal incidence as will be described in Sect. 13.6. Therefore single pass operation of an FEL by self-amplified spontaneous emission is planned. There are proposals to operate FELs at wavelengths down to the order of 1 nm with a very large storage ring and a very large linac presently used for high energy physics.

13.5 Accelerators for Light Sources and Undulators

Synchrotron radiation sources and free electron lasers make use of high energy electron beams from accelerators. There are a variety of electron accelerators that can accelerate electrons to energies higher than 10 MeV, such as a linear accelerator (linac or RF linac), a microtron, a synchrotron and a storage ring [14]. A primary source of electrons is an electron gun. In conventional electron guns, the cathode coated with oxide materials, whose work function is low, is heated by a filament to $\sim$1000 °C, and thermal electrons are extracted and accelerated by applying the pulsed high voltage of $\sim$100 kV. These electrons are still non-relativistic. Since a synchrotron and a storage ring can accelerate relativistic electrons only, a linac or a microtron is used as an injector for them.

A linac accelerates electrons by the electric field of microwaves stored in an accelerating tube made of copper, the frequency of which is most commonly 3 GHz (S band) or 1 GHz (L band). A schematic drawing of a linac is shown in Fig. 13.13. The diameter of the accelerating tube is of the order of the wavelength of the microwaves. The electron beam from a linac is temporally bunched in the frequency of the microwave (micro pulses). The microwave power of several MW to some tens of MW is supplied by a klystron. Since it is difficult to produce a microwave of such a power level continuously, a conventional RF linac is operated intermittently. The time duration of pulses (macro pulses) is typically a few μs and the repetition rate is in the region from a few to 100 Hz. For FELs, however, conventional RF linacs are operated with longer pulse durations. On the other hand, a superconducting linac can be operated continuously since the

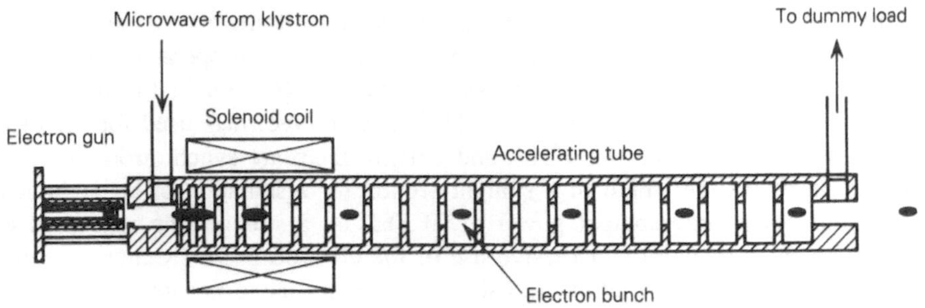

Fig. 13.13. Schematic layout of a linac

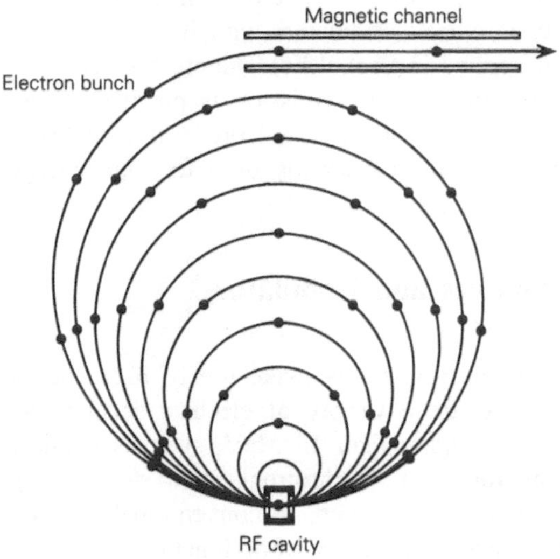

Fig. 13.14. Schematic layout of a microtron

microwave power necessary is much lower. The energy gain in a linac is typically 10 MeV/m. A number of accelerating tubes are connected serially to obtain the desired energy. Instead of increasing the length of a linac, recirculation of the electron beam through the linac is planned at some FEL facilities, by which the energy of the electron beam is increased by the number of passages through the linac.

A schematic drawing of a microtron is shown in Fig 13.14. Electrons in the constant magnetic field are accelerated with an RF cavity. An RF frequency is chosen as $f_{RF} = eB/2\pi m$, where e is the electron charge, B the magnetic field, and m the rest mass of an electron, and the energy gain of the electron per turn is equal to or a multiple of the rest mass ($mc^2 = 0.511$ MeV). The number of bunches increases turn by turn. In high energy microtrons, a magnet is divided into some parts, quadrupole magnets are installed in straight sections, and a linac is used instead of a single RF cavity.

An electron synchrotron is a circular accelerator, in which electrons circulate many times on the same orbit and they are accelerated gradually with the RF field in a cavity. The principle of operation of a synchrotron is same as that of a storage ring, which will be described in the following paragraphs. Since the speed of a relativistic electron is almost equal to that of light, the revolution period of an electron is independent of its energy. If the energy of an electron deviates from a central value by a small amount, it starts energy oscillation around the central value. Therefore, the motion of an electron with a small energy deviation is stable – this is called the phase stability. Due to this action, electrons are automatically accelerated when the magnetic field is gradually increased.

A storage ring is now used as a light source of synchrotron radiation in which high energy electrons or positrons are stored for a long period of time [3]. In order to supply electrons (or positrons) to the storage ring, a linear accelerator, a microtron, or a synchrotron is used as an injector. A storage ring is composed of the magnet system, RF cavities, and the vacuum system. The magnet system confines the electron beam in a storage ring. Bending magnets make a circular orbit, and quadrupole magnets focus the beam. The transverse size and the angular divergence of the electron beam are determined by a configuration of these magnets, which is referred to as lattice. As an example for a syndrotron radiation light source, the UVSOR storage ring at the Institute for Molecular Science is shown in Fig. 13.15. This lattice has four-fold symmetry, and each unit has the reflection symmetry at the center of a long straight section. A wiggler and two undulators are installed in the long straight sections. As in most of the storage rings, bending magnets are normal conducting ones, in which the maximum value of the magnetic field is limited below 2 T due to saturation of iron. In a few storage rings, however, superconducting bending magnets with a higher magnetic field are employed in order to make the storage ring more compact.

Due to the focusing forces of quadrupole magnets, electrons in a storage ring oscillate transversely around the central orbit, which is called the betatron oscillation, while they move at a speed close to that of light. The number of oscillations in one revolution is called the betatron number or tune; it is denoted by Q_x and Q_y in the horizontal and the vertical directions, respectively. A set of the horizontal and the vertical betatron numbers denoted by (Q_x, Q_y) is called an operating point since it is represented by a point on the tune diagram (the Q_x-Q_y plane). If the betatron number is an integer or a half-integer, the orbit diverges due to errors in the magnetic field. It is expressed by a line on the tune diagram, called a resonance line. The operating point is, therefore, chosen such that it is away from the neighborhood of resonance lines. In the coordinate shown in Fig. 13.3, the horizontal displacement x_β of an electron from the central orbit is a function of the longitudinal position s, and it is expressed as [15]

$$x_\beta(s) = \sqrt{\varepsilon_x \beta_x(s)} \cos(\phi_x(s) + \phi_{x0}) , \tag{13.33}$$

where ε_x is the emittance. It is approximately equal to the product of the oscillation amplitude and the angular divergence of the electron beam. Emittance is a

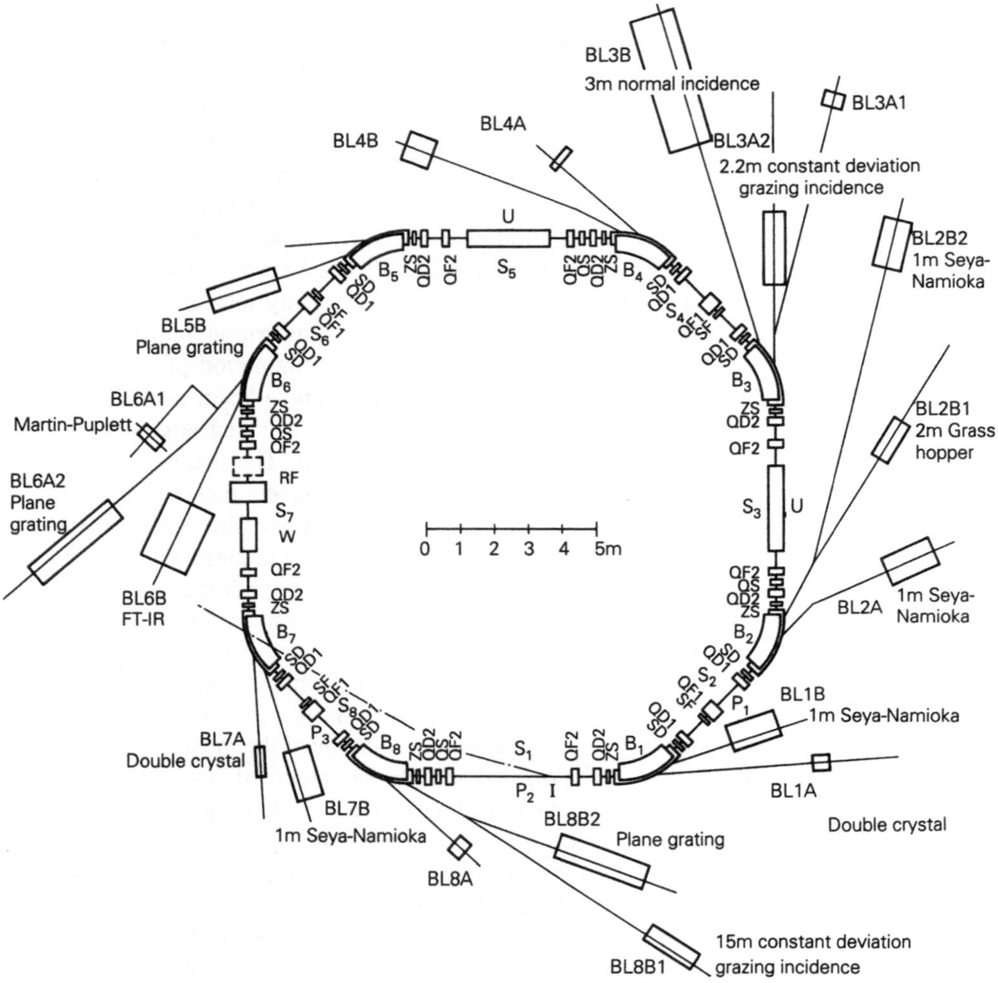

Fig. 13.15. Ground plan of the UVSOR storage ring and the beamlines at the Institute for Molecular Science. *B*: bending magnet, *S*: straight section, *QF*: focusing quadrupole, *QD*: defocusing quadrupole, *SF*: focusing sextupole, *SD*: defocusing sextupole, *QS*: skew quadrupole, *ZS*: vertical steering magnet, *RF*: RF cavity, *U*: undulator and *W*: wiggler. In each bending section two beamlines are connected. The type of a monochromator is given for each beamline. The beamline without a monochromator is a port for irradiation of white synchroton radiation or a free port. Undulator radiation from S_3 is utilized at *BL3A*1 and *BL3A*2. The undulator at S_5 is used for FEL experiments

constant of motion around the circumference of the storage ring, and its units are π m·rad. The function $\beta_x(s)$ is the betatron function that is dependent on s, and its units are meters. The function $\phi_x(s)$ is a phase advance of the betatron oscillation from $s' = 0$ to $s' = s$, and given by

$$\phi_x(s) = \int_0^s \frac{ds'}{\beta_x(s')} \, . \tag{13.34}$$

Each electron oscillates transversely with a different amplitude of $\sqrt{\varepsilon_x \beta_x(s)}$ at the position s and with a different initial phase ϕ_{x0}. This is the main reason for the size and the divergence of the electron beam. Similarly, the displacement $y(s)$, emittance ε_y, and the betatron function $\beta_y(s)$ are defined in the vertical direction. A central orbit of an electron having an energy deviation ΔE from the mean energy E is shifted horizontally as

$$x_E(s) = \eta_x(s)\frac{\Delta E}{E}, \tag{13.35}$$

and the electron executes the betatron oscillation around this central orbit. The function $\eta_x(s)$ is the dispersion function, and produced by the momentum dispersion in bending magnets. Since vertical bending magnets are not usually used in a storage ring, the vertical dispersion function $\eta_y(s)$ is zero throughout the circumference of the storage ring. Figure 13.16 shows the betatron functions and the dispersion function in a unit of the UVSOR storage ring. Due to the symmetry property of the lattice, these functions repeat themselves four times. This lattice is referred to as the double bend achromat, and employed in many storage rings for synchrotron radiation sources to make emittance lower. The dispersion function $\eta_x(s)$ is zero in long straight sections. Emittance in the horizontal direction is determined by a balance between excitation and damping of the horizontal betatron oscillation due to emission of synchrotron radiation, and given as

$$\varepsilon_{x0} = a\frac{E^2}{N_b^3}, \tag{13.36}$$

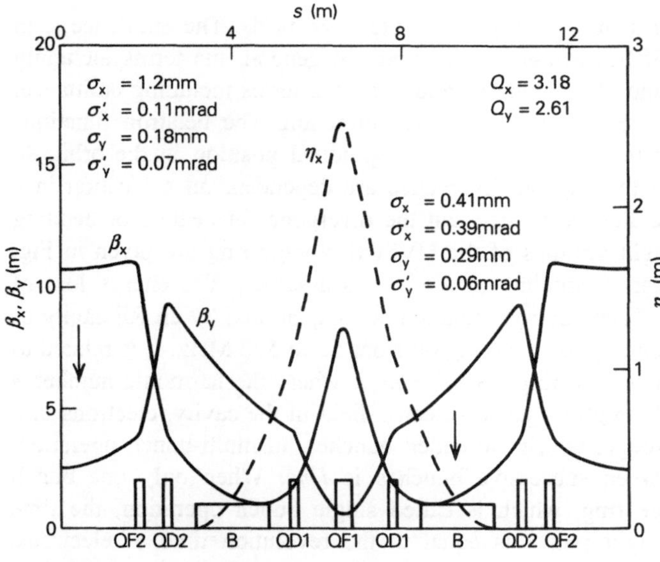

Fig. 13.16. Betatron functions and the dispersion function in a unit of the magnet lattice of the UVSOR storage ring. The calculated beam size and the angular divergence at centers of the bending magnets and the straight sections are also given, assuming the emittance coupling of $\kappa = 0.1$

where N_b is the number of bending magnets. The constant a is dependent on a type of lattice and the horizontal betatron number. Since emittance determines the size and the angular divergence of the electron beam, it is made as small as possible within the reasonable range in order to make the brilliance of radiation higher. Especially, when an undulator is used, it is required that the beam divergence is small. In the third generation light sources, the horizontal emittance ε_{x0} is in the order of $10^{-9}\pi$ m·rad [16]. Generally, the radiation excitation in the vertical direction is small, so that the vertical emittance is determined by a coupling between the horizontal and the vertical motions. The coupling constant of emittance κ, defined as $\kappa = \varepsilon_y/\varepsilon_x$, where $\varepsilon_{x0} = \varepsilon_x + \varepsilon_y$, is in the range from 0.01 to 0.1. The electron beam has an energy spread that is also determined by balance between radiation excitation and damping. The relative energy spread σ_E/E is in the order of 10^{-4} to 10^{-3}.

The transverse shape and the angular divergence of the electron beam in a storage ring have Gaussian distributions. The standard deviations of the beam size σ_i and the angular divergence σ_i' in the horizontal or vertical direction are given by [15]

$$\sigma_i(s) = \sqrt{\varepsilon_i\beta_i(s) + \left[\frac{\sigma_E}{E}\eta_i(s)\right]^2} \tag{13.37}$$

$$\sigma_i'(s) = \sqrt{\varepsilon_i\frac{1+\alpha_i(s)^2}{\beta_i(s)} + \left[\frac{\sigma_E}{E}\eta_i'(s)\right]^2}, \tag{13.38}$$

where i stands for x or y, $\alpha_i = -(\mathrm{d}\beta_i/\mathrm{d}s)/2$, and $\eta' = \mathrm{d}\eta/\mathrm{d}s$. The emittance ε_i in Eqs. (13.37) and (13.38) is in units of π m·rad. In general, the terms including σ_E/E in Eqs. (13.37) and (13.38) are smaller than the terms including emittance. Especially, η_y and η_y' are zero in the vertical direction. The betatron functions and the dispersion function depend on the longitudinal position in the orbit, so that the beam size and the angular divergence are dependent on a location in a storage ring. The calculated beam size and the divergence at centers of bending magnets and long straight sections of the UVSOR storage ring are given in Fig. 13.16, where the emittance coupling of $\kappa = 0.1$ is assumed. The energy loss of electrons by emission of synchrotron radiation is compensated by an RF cavity or cavities. The RF frequency f is in the region from 50 to 500 MHz. It is related to the circumference of a storage ring l as $f = hc/l$, where the harmonic number h is an integer. Due to the high frequency electric field in the cavity, electrons in a storage ring form h groups, which are called bunches, in multi-bunch operation. The time interval between successive bunches is l/hc. When only one bunch is stored in the storage ring, which is called single bunch operation, the time interval between two light pulses is equal to the revolution time of electrons, that is, l/c. This time interval is in the range from 50 ns to 5 ms, depending on the circumference of the storage ring. When synchrotron radiation is used as pulsed light and a long interval between light pulses is necessary, this mode of operation is adopted. The standard deviation of the bunch length σ_t is determined by the energy spread σ_E/E, the RF frequency, and the RF voltage. A time duration of light pulses is appproximately $2\sigma_t$, and it is in the reigon from 0.1 to 1 ns. As an example, the time structure of the electron beam of the UVSOR storage ring is shown in Fig. 13.17. This time structure allows e.g. the measurement of fluorescence decay times when single bunch pulses are used for excitation as will be described in Sect. 13.8.

The maximum stored beam current in an electron storage ring is in the region of 100 to 500 mA. However, there are a few storage rings that store a beam current of approximately 1 A. Electrons once injected into a storage ring are gradually lost mainly due to scattering by nuclei of residual gases in the vacuum chamber in which the electrons circulate, called Coulomb scattering, and to the two-body scattering of electrons in a bunch, called the Touschek effect. In order to increase the lifetimes, the vacuum chamber is evacuated to an ultra-high vacuum of less than 1×10^{-7} Pa and sufficient RF voltage is applied. As a result, the lifetime of the electron beam is a few hours in low energy storage rings for vacuum ultraviolet and soft X-rays, and it amounts some tens of hours in high energy storage rings for X-rays.

Electrons circulating in a storage ring oscillate around a closed orbit that more or less deviates from the central orbit. Origins of the closed orbit distortion are errors in magnetic field strength and alignment of magnets. Among them, most dominant factors are relative deviations of the magnetic field in bending magnets, which is typically $\Delta B/B < 10^{-3}$, and alignment errors of quadrupole magnets, which are approximately 0.1 mm. Due to these errors, the deviation of the closed orbit ranges from a few mm to 10 mm. Since synchrotron radiation is emitted tangentially to the electron orbit, a source point and an emission angle of light are different from those for the central orbit if there is closed orbit distortion. Therefore, the closed orbit deviation is measured by position monitors installed along the circumference of the storage ring, and corrected within

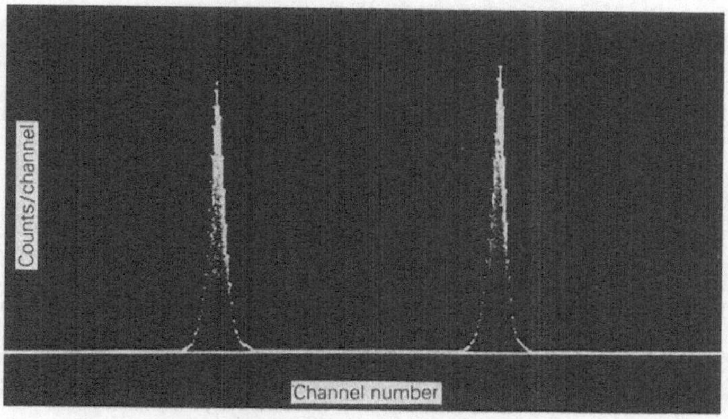

Fig. 13.17. Time structure of the electron beam of the UVSOR storage ring. Sixteen electron bunches circulate in the storage ring in multi-bunch operation, and two of them are shown. The longitudinal length of the electron bunch is 430 ps (FWHM), and the interval between bunches is 11.1 ns

approximately 1 mm with correction magnets. Strictly speaking, the closed orbit distortion thus corrected, however, does not stay constant temporally, for example, due to current fluctuations of power supplies for magnets. Since the distance from the light source to the experimental position is large in high energy storage rings, angular fluctuation of the closed orbit is serious. Stabilization of the closed orbit fluctuations by feedback methods has been developed in order to keep the position of light at the observing point as stationary as possible.

Undulators most commonly used are made of strong permanent magnets such as an alloy of neodymium, iron and boron (NdFeB) and samarium cobalt (SmCo) [17]. In a planar undulator of the horizontal oscillation type, a configuration of magnetic materials is designed such that the vertical component of the magnetic field varies sinusoidally along the electron orbit as $B_y(z) = B_0 \sin(2\pi z/\lambda_u)$, where B_0 is the peak magnetic field, and that the peak magnetic field is as high as possible. Schematic illustrations of the cross sections of two types of planar undulators are shown in Fig. 13.18. One is the Halbach type, which is sometimes referred to as the pure type since it is made of permanent magnets only, and the other is the hybrid type, which consists of permanent magnets and materials with high magnetic permeability such as iron and vanadium permendur. The peak magnetic field of a Halbach-type undulator is given by

$$B_0 \approx 1.80 \times B_r \left[1 - \exp\left(-\frac{2\pi h}{\lambda_u}\right)\right] \exp\left(-\frac{\pi g}{\lambda_u}\right), \tag{13.39}$$

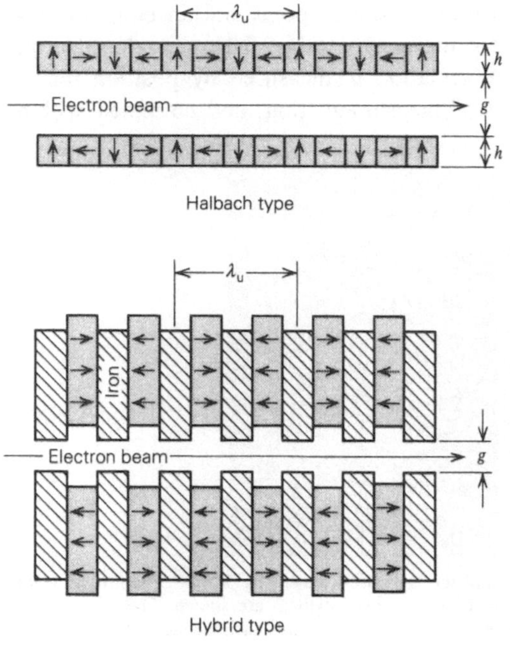

Fig. 13.18. Cross sections of two types of undulators made of permanent magnets

where B_r is the remanent field of permanent magnets, h is the height of the magnet blocks, and g is the full gap of the magnet. The typical values of B_r are 0.9 T for SmCo and 1.2 T for NdFeB. The magnetic field in the undulator is varied by changing the magnet gap mechanically. In a hybrid-type undulator, the magnetic flux produced by permanent magnets is condensed in iron, so that the higher magnetic field can be obtained for the smaller magnet gap.

There are many types of undulators and wigglers for producing circularly or elliptically polarized radiation [18]. Since circularly polarized radiation is emitted by an electron moving on a helical orbit, the magnetic field on the electron orbit should be given by $B_x = B_0\cos(2\pi z/\lambda_u)$ and $B_y = B_0\sin(2\pi z/\lambda_u)$. The magnetic field of this kind is achieved with a superconducting helical undulator in which double helical coils are wound along the central axis and current flows in the opposite direction in each coil, or in a cross-retarded field undulator in which two Halbach-type undulators are combined such that the magnetic field made by one undulator is perpendicular to the other and the phase of the field is shifted by 90°. In addition to these, other types of insertion devices have been proposed or constructed, in which the orbit of an electron is not helical or not perfectly helical [18].

13.6 VUV Optical Components

When the incident light with intensity I_0 passes through the medium with thickness l, the intensity I of the transmitted light decreases by $I = I_0 \exp(-\mu l)$, where μ is the absorption coefficient ($\mu = \sigma\rho$, where σ is the absorption cross section and ρ the particle density or σ is the mass absorption cross section and ρ the density). In the VUV region, every material has a large absorption coefficient. As mentioned below, it causes many difficulties in performing experiments. Therefore special techniques have been developed to overcome the difficulties [19]. The air has a large absorption coefficient as shown in Fig. 13.19, so that it is opaque even under one atmosphere. For instance, its absorption coefficient around 100 nm is about 10^4 cm^{-1}, so that the optical path through which the intensity decreases by $1/e$ (absorption length) is 1 μm. This is the reason why the optical path should be evacuated. Other gases are also opaque. Even rare gases become opaque below the wavelength corresponding to the ionization potential (He: 50.4 nm, Ne: 57.5 nm, Ar: 78.7 nm, Kr: 88.6 nm and Xe: 102.2 nm).

Solids have, of course, about 10^3 times larger absorption coefficients than gases. However, there are several transparent materials in the longer wavelength VUV region. Their cutoff wavelengths are as follows; LiF: 104 nm, MgF_2: 112 nm, CaF_2: 122 nm, SrF_2: 128 nm, BaF_2: 134 nm, sapphire: 141 nm and synthetic fused quartz: 160 nm (the thickness of these materials is about 1 mm). Above the cutoff wavelength, one can use prisms, lenses and polarizers made of the transparent materials. Below 104 nm, if one needs transparent materials, the thickness of the materials should be thin, that is, from several tens nm to a few

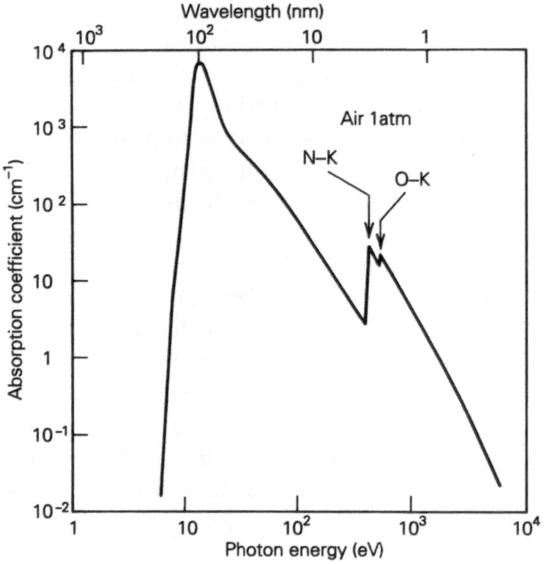

Fig. 13.19. Absorption spectrum of air (1 atm), obtained using existing data [Okabe H (1978) Photochemistry of small molecules. John Wiley & Sons, New York p 168, 177; Wainfan N, Walker WC Weissler GL (1955) Phys. Rev. 99:542; Koch E-E, Sonntag BF (1979) In: Kunz C (ed) Synchrotron radiation techniques and applications. Springer, Berlin Heidelberg New York, p 269; Henke BL, Lee P, Tanaka TJ, Shimabukuro RL, Fujikawa BK (1982) At. Data Nucl. Data Tables 27:1] assuming that the air is 4:1 mixture of N_2 and O_2. Tops of absorption peaks are connected in the line spectrum regions

μm, depending on the materials and wavelength. Figure 13.20 shows transmission spectra of typical films, which can be made self-standing. These thin films can be used as windows, evaporation substrates for transmission experiments and band pass filters. Organic films such as collodion, polypropylene and polyester films with a thickness from a few to several tens nm can be used as the window in the wavelength region longer than 10 nm and those with a thickness of a few μm can be used below 1 nm. However, below 104 nm, bulk transparent materials for lenses and prisms are not available as mentioned above, so that one needs to adopt reflection optics, that is, to use mirrors, and reflection gratings and crystals for dispersion. In one particular case, one can use a transmission grating or a zone plate made of thin Au films.

In Fig. 13.21, the reflection spectra of Au for unpolarized light in the 0.8–160 nm region are given as a function of the incident angle α. As shown in this figure, the reflectivity decreases as the wavelength becomes short and also as the incident angle becomes small. Roughly speaking, reflectivities of heavy metals are high in the long wavelength range. However, for near normal incidence, the reflectivity of Au is smaller than 20%, but that of Al overcoated with MgF_2 is about 80% above 120 nm, and that of SiC 40–50% between 60 and 200 nm. The normal incidence reflectivity generally becomes less than a few percents below 30 nm. Therefore, below 30 nm, one should use grazing incidence optics.

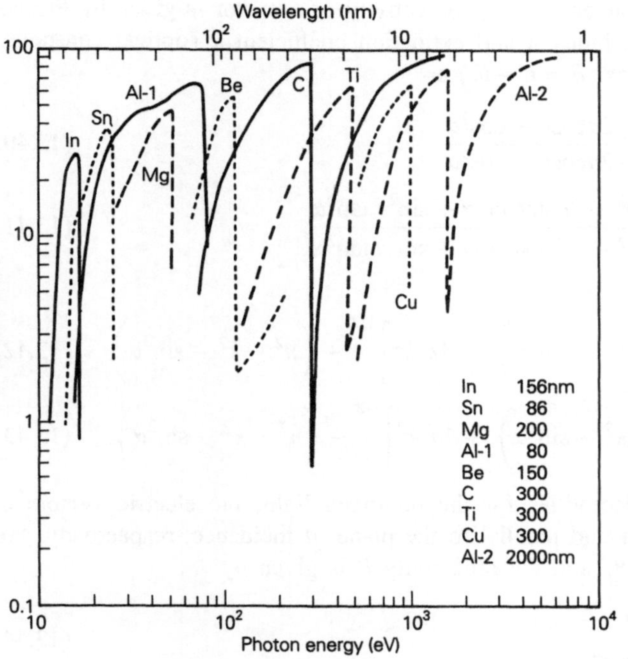

Fig. 13.20. Transmission spectra of thin films of In and Al (80 nm) [Hunter WR, Angel DW, Tousey R (1965) Appl. Opt. 4:891], Sn [Codling K, Madden RP, Hunter WR Angel DW (1966) J. Opt. Soc. Am. 56:189], Mg [Hagemann H-J, Gudat W, Kunz C (1975) J. Opt. Soc. Am. 65:742], Be [Tomboulian DH, Bedo DE (1955) Rev. Sci. Instr. 26:747], and C, Ti, Cu and Al (2000 nm) [Henke BL, Lee P, Tanaka TJ, Shimabukuro RL Fujikawa BK (1982) At. Data and Nucl. Data Tables 27:1]. (The Mg film sandwiched with two 5 nm thick Al layers is stable)

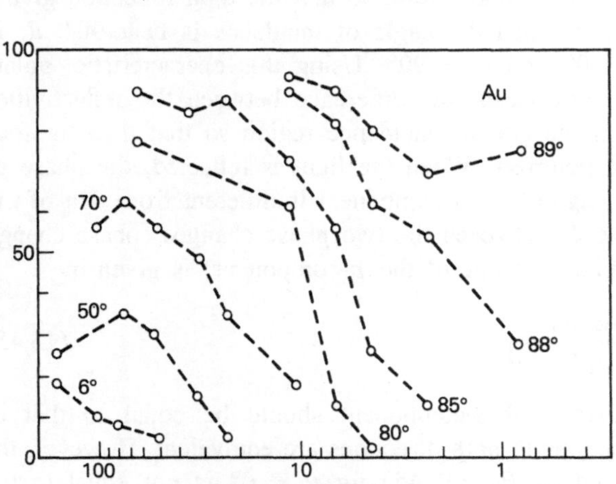

Fig. 13.21. Reflection spectra of Au for unpolarized light as a function of incident angle α compiled from existing data [Hendrick RW (1957) J. Opt. Soc. Am. 47:165; Lukirskii AP, Savinov EP, Ershov OA Shepelev Yu F (1964) Opt. Spectr. 16:168; Malina RF, Cash W (1978) Appl. Opt. 17:3309; Canfield LR, Hass G, Hunter WR (1964) J. de Phys. 25:124]

The incident angle dependence of the reflectivity of a mirror is given by Fresnel equation with refractive index n and extinction coefficient k (optical constants) (complex refractive index; $\tilde{n} = n - ik$) as

$$R_s = \frac{a^2 + b^2 - 2a\cos\alpha + \cos^2\alpha}{a^2 + b^2 + 2a\cos\alpha + \cos^2\alpha} \tag{13.40}$$

$$R_p = R_s \frac{a^2 + b^2 - 2a\sin\alpha\tan\alpha + \sin^2\alpha\tan^2\alpha}{a^2 + b^2 + 2a\sin\alpha\tan\alpha + \sin^2\alpha\tan^2\alpha}, \tag{13.41}$$

where

$$2a^2 = \left[\left(n^2 - k^2 - \sin^2\alpha\right)^2 + 4n^2k^2\right]^{1/2} + \left(n^2 - k^2 - \sin^2\alpha\right) \tag{13.42}$$

$$2b^2 = \left[\left(n^2 - k^2 - \sin^2\alpha\right)^2 + 4n^2k^2\right]^{1/2} - \left(n^2 - k^2 - \sin^2\alpha\right). \tag{13.43}$$

R_s and R_p are the reflectivities for the polarized light, the electric vectors of which are perpendicular and parallel to the plane of incidence, respectively. For $\alpha = 0°$, R_s is equal to R_p and the reflectivity R is given by

$$R = \frac{(n-1)^2 + k^2}{(n+1)^2 + k^2}. \tag{13.44}$$

The reflectivities of Au at 17.1 and 67.1 nm are plotted against the incident angle in Fig. 13.22. The reflectivity for $\alpha \approx 0°$ is smaller than that for $\alpha \approx 90°$, which is almost 1. The normal incidence reflectivity becomes small below 30 nm because n approaches 1 and k becomes very small. In the shorter wavelength region, n is slightly smaller than 1 and $k \approx 0$, so that the total reflection giving reflectivity of almost 1 occurs when the angle of incidence is near 90°. R_s is larger than R_p except $\alpha = 0°$ and $\alpha = 90°$. Using this characteristic, polarizers have been made. In many cases, the difference between the reflectivities R_s and R_p is not so large in the grazing incidence region so that three or four mirrors are used to make a polarizer. When the light is reflected, the phase of light changes. The phase change of the s-component is different from that of the p-component. The difference Δ betweeen the two phase changes (phase change of the s-component minus phase change of the p-component) is given by

$$\tan\Delta = \frac{-2a\sin\alpha\tan\alpha}{a^2 + b^2 - \sin^2\alpha\tan^2\alpha}. \tag{13.45}$$

For $\alpha = 0°$, the phase change of s-component should be equal to that of p-component, because reflections in both directions are equivalent. However, the phase difference Δ calculated by Eq. (13.45) for $\alpha = 0°$ is not equal to 0°, but 180°. This is because the direction of the light differs by 180° before and after the reflection. When the linearly polarized light is reflected by a mirror, the surface of which is neither parallel nor perpendicular to the direction of the polarization, the phases of two components of reflected light differ. When Δ is 90°, one can get elliptically (in some cases circularly) polarized light from the linearly

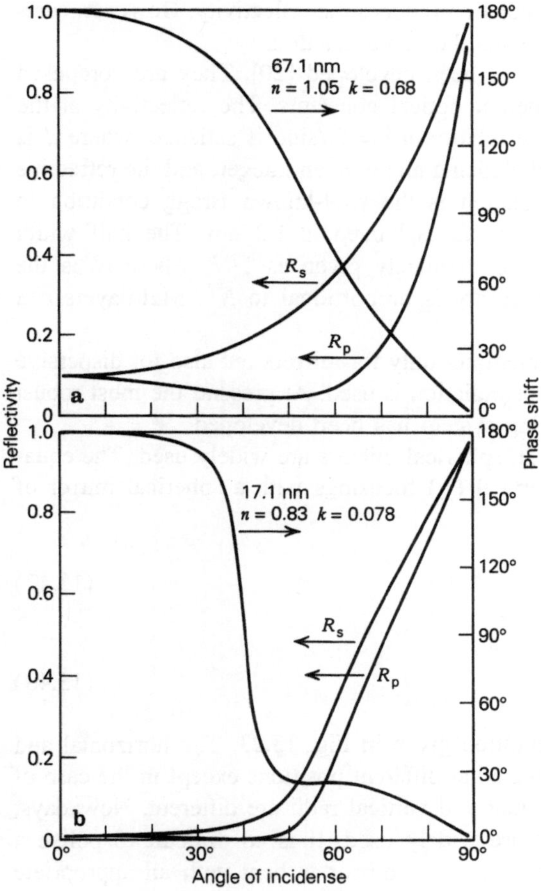

Fig. 13.22. The incident angle dependence of reflectivity of Au at 67.1 **a** and 17.1 nm **b**. *Rs* and *Rp* are the reflectivities for the polarized light, the electric vectors of which are perpendicular and parallel to the plane of incidence, respectively. The phase shift is defined as phase change of R_s minus that of *Rp*

polarized light. In contrast, using a reflector with $\Delta = 90°$ and linear polarizer one can anlayze the elliptically (or circularly) polarized light.

The surface roughness suppresses the reflectivity and increases the scattered light. The root-mean-square surface roughness should be less than 1 nm in the grazing incidence region. Recently, such an ultrasmooth surface has been achieved and the surface roughness has been measured by the contact or the optical method. From the reflectivity measurement, the surface roughness can be also estimated using the equation

$$R = R_0 \exp\left\{ -\left(\frac{4\pi\sigma\sin\theta}{\lambda}\right)^2 \right\}, \tag{13.46}$$

where σ is the root-mean-square surface roughness and θ the glancing angle $(90° - \alpha)$. By the way, the carbon contamination due to the cracked hydrocarbon generated with sychrotron radiation reduces the reflectivity. However, careful oxygen-discharge cleaning removes the contamination.

Recently, multilayer mirrors have been developed [20]. They are composed of alternating two layers with different optical constants. The reflectivity at the wavelength λ is enhanced when the relation $m\lambda = 2d\sin\theta$ is satisfied, where d is the period length of a multilayer, θ glancing angle, m an integer, and the refractive index n is regarded as 1. This relation is the well-known Bragg condition in X-ray reflection. The minimum d so far achieved is 1.2 nm. The half width $(\Delta\lambda/\lambda)$ of the reflection peak is approximately given as $1/N$, where N is the number of periods. The peak reflectivity is proportional to N^2. Multilayers can be used as efficient polarizers.

The heat load is a serious problem not only for mirrors but also for dispersive elements, especially when undulator radiation is used. At present, the most robust material is SiC. An efficient cooling system has been developed.

To gather or collimate the light, spherical mirrors are widely used. The equations of the horizontal (x) and vertical (y) focusings with a spherical mirror of the radius R are given by

$$\frac{1}{r} + \frac{1}{r'} = \frac{2}{R\cos\alpha} \tag{13.47}$$

$$\frac{1}{r} + \frac{1}{r'} = \frac{2\cos\alpha}{R}, \tag{13.48}$$

respectively, in the similar configuration given in Fig. 13.23. The horizontal and the vertical focusing points are situated at different positions except in the case of $\alpha = 0°$. In toroidal mirrors, horizontal and vertical radii are different. Nowadays, large mirrors with a large radius are widely used. It is so difficult to polish a mirror with such a large radius that a plane mirror is bent with an appropriate bender to obtain a cylindrical or elliptical mirror.

Diffraction gratings are used as dispersive elements usually in the wavelength region above 1.5 nm. When the light coming from $A(x, y, z)$ is reflected by a

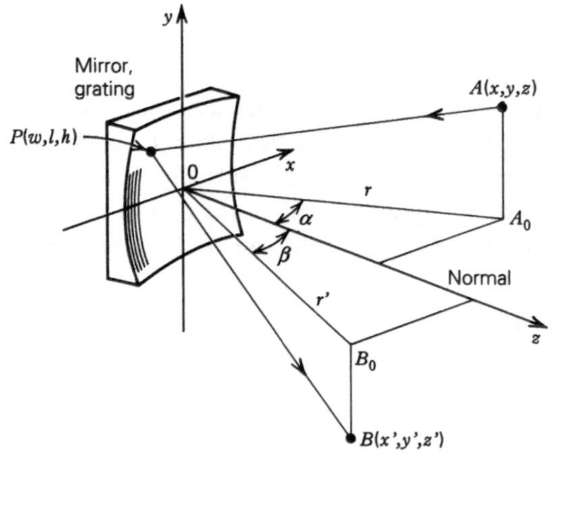

Fig. 13.23. Geometry of optical path of incident and reflected or diffracted rays

mirror at P (w, l, h) and reaches the position B (x', y', z') as given in Fig. 13.23, the path from A to B is determined by the Fermat's principle such that the path function given by $AP + BP$ is minimized. In case of a diffraction grating, the term $nm\lambda$ should be added to the path function in order to satisfy the diffraction condition that a difference between path lengths of rays diffracted by different grooves should be a multiple of wavelength λ, where n is the number of grooves between the center of the grating O and the point P, and m is an integer. Therefore the path function is given by

$$F = AP + BP + nm\lambda. \tag{13.49}$$

When the groove space d is constant, the number of grooves is given by $n = w/d$, where w is a separation between P and O in the x direction. In case of a mirror, $n = 0$. When the groove space is not constant (varied space), n in the third term of Eq. (13.49) is expressed by a series as

$$n = \frac{w}{d_0} + b_2 w^2 + b_3 w^3 + \cdots , \tag{13.50}$$

where d_0 is the nominal groove spacing at the grating center ($w = 0$) [21]. AP and BP are given by

$$AP^2 = (x - w)^2 + (y - l)^2 + (z - h)^2 \tag{13.51}$$

$$BP^2 = (x' - w)^2 + (y' - l)^2 + (z' - h)^2. \tag{13.52}$$

As seen in Fig. 13.23, $x = r\cos\alpha$, $y = r\sin\alpha$, $x' = r'\cos\beta$ and $y' = r'\sin\beta$, where $r = OA_0$ and $r' = OB_0$, and α and β are the angles of incidence and diffraction, respectively. α is always chosen to be positive. When the diffracted light comes on the same side of the incident light, the sign of β is chosen positive, but on the opposite side it is negative. In general, h is a function of w and l, $h(w, l)$, expressing the shape of the surface. For a plane surface, $h = 0$. For a spherical surface, $h = R - (R^2 - w^2 - l^2)^{1/2}$, where R is the radius of the sphere. The path function can be expanded in series of w and l as follows, when $y = y' = 0$.

$$F = r + r' + F_{10}w + F_{20}w^2 + F_{02}l^2 + F_{30}w^3 + F_{12}wl^2 + \cdots \tag{13.53}$$

From Fermat's principle

$$\frac{\partial F}{\partial w} = 0, \qquad \frac{\partial F}{\partial l} = 0. \tag{13.54}$$

From $F_{10} = 0$, the diffraction condition is obtained. From $F_{20} = 0$ and $F_{02} = 0$, the focusing conditions for horizontal and vertical directions are determined, respectively. Higher order terms give the amount of aberration.

The diffraction condition of a plane grating with the grooves equally spaced for the parallel incident light is derived as

$$d(\sin\alpha + \sin\beta) = m\lambda, \qquad m = 0, \pm 1, \pm 2 \ldots \tag{13.55}$$

When $m = 0$, that is, $\beta = -\alpha$, the diffracted light is called the 0th order light. In this case, the grating acts as a mirror without dispersion. When m is a positive integer, it is called the positive order and α is larger than $-\beta$. In the negative order, α is smaller than $-\beta$. When λ satisfies Eq. (13.55) for given α and β, and $m = 1$, wavelengths $\lambda/2, \lambda/3, \ldots$ also satisfy Eq. (13.55) for $m = 2, 3, \ldots$. The light with these wavelengths is called the higher order light. It mixes with the first order light. The amount of the higher order light is dependent on the quality of a grating. When the amount of the higher order light is not negligible, one should use a filter described above to suppress it. On the other hand, the m-th order light for given λ gives m times higher dispersion than the first order light, as seen in Eq. (13.56) given below. In this case, one should filter out light other than the nominated m-th order light.

The angular dispersion is given as follows.

$$\frac{\mathrm{d}\beta}{\mathrm{d}\lambda} = \frac{m}{d\cos\beta} \tag{13.56}$$

The smaller the groove spacing, the larger the angular dispersion. The plate factor is given by the equation

$$\frac{\mathrm{d}\lambda}{\mathrm{d}l} = \frac{1}{r'(\mathrm{d}\beta/\mathrm{d}\lambda)} = \frac{\cos\beta}{r'm(1/d)} \times 10^3\ \mathrm{nm/mm}, \tag{13.57}$$

where r' is the distance from a focusing mirror to the image formed by it and l the distance on the focal plane. The units of r' and $1/d$ are m and 1/mm, respectively. The plate factor gives the wavelength resolution on the focusing plane (originally defined for a photographic plate in a spectrograph). The resolving power is given by $\mathcal{R} = \lambda/\Delta\lambda = mN$, where the N is the number of grooves irradiated by the incident light. According to this equation the resolving power increases as the number of irradiated grooves increases. However, this equation gives the theoretical maximum value and is only valid for the ideal case when the incident light is completely parallel.

A spherical grating can focus the dispersed light by itself, so that the collimator and focusing mirrors can be omitted to minimize the reflection loss. In the case of an equally spaced spherical grating, the dispersion condition is also given by Eq. (13.55) and focusing conditions given below are derived from Fermat's principle [22]. The horizontal focusing condition is given by

$$\frac{\cos^2\alpha}{r} + \frac{\cos^2\beta}{r'} = \frac{\cos\alpha + \cos\beta}{R}, \tag{13.58}$$

where R is the radius of the grating. One of the solutions is

$$r = R\cos\alpha, \qquad r' = R\cos\beta. \tag{13.59}$$

This solution gives the well-known Rowland circle. There are other solutions (off-Rowland), but the Rowland solution gives the highest resolution, because the F_{30} term (coma) in Eq. (13.53) becomes zero [22]. The vertical focusing point is given by

$$\frac{1}{r} + \frac{1}{r'} = \frac{\cos\alpha + \cos\beta}{R}. \tag{13.60}$$

One of the solutions is

$$r = \frac{R}{\cos\alpha}, \qquad r' = \frac{R}{\cos\beta}. \tag{13.61}$$

As seen from Eqs. (13.59) and (13.61), the horizontal and vertical focusing points do not coincide except when $\alpha = \beta = 0°$. This causes astigmatism. That is, even though the light source is a point on the Rowland circle, the image of the diffracted light becomes a vertical line at the exit slit. To avoid astigmatism, toroidal gratings are used in some cases, the horizontal and vertical radii of which are different so that the vertical focusing point is also on the Rowland circle.

The resolving power of the plane grating increases with the number of irradiated grooves as mentioned before. However, in case of a spherical grating, the resolving power is limited due to the aberration. There is an optimum width W_{opt} of the irradiated area independent of the groove density [22]. When the irradiated area is smaller than W_{opt}, the resolving power increases in proportion to the number of irradiated grooves. However, when the width becomes larger than W_{opt}, the resolving power decreases. When the angle of incidence becomes large, the W_{opt} becomes small. It also becomes small when the wavelength becomes short. In the practical case, however, the band width of the dispersed light is not limited by the number of irradiated grooves (resolving power), but by the width of the slit given as

$$\Delta\lambda = \frac{w}{mR/d} \times 10^3 \text{ nm}, \tag{13.62}$$

where w is slit width given in mm, R in m and $1/d$ in 1/mm. This equation is derived from Eq. (13.57), in which dl is replaced by w and r' by $R\cos\beta$.

The cross-sectional shape of the groove of the grating is important for the intensity of the diffracted light. Many gratings are blazed gratings and their shape is shown schematically in Fig. 13.24(a). The normal of the grating is given by ON and the normal of the facet by OM. $\angle\text{NOM} = \theta_b$ is called the blaze angle. When $\alpha - \theta_b = \theta_b - \beta$, the intensity of the dispersed light is enhanced. The wavelength satisfying this condition is called the blaze wavelength λ_b, which is given by $m\lambda_b = 2d\sin\theta_b\cos k$ where $2k = \alpha - \beta$. However, the wavelength given in catalogs is the blaze wavelength at the Litrow mount ($k = 0; \alpha = \beta = 0$) and $m = 1$, that is, $\lambda_{b'} = 2d\sin\theta_b$. Therefore, the blaze wavelength is given by $\lambda_b = \lambda_{b'}\cos k$ in general. Another kind of grating used is laminar. In this case, the shape of the groove is rectangular as given in Fig. 13.24 (b). The wavelength of the enhanced diffracted light depends on the ratio of c/d and h [23]. In many cases, the coating material is Au. Recently, the laminar gratings coated with multilayers have been studied. The dispersion efficiencies of blazed and laminar gratings achieved so far are more than 10% above a few nm, but less than 10% below it.

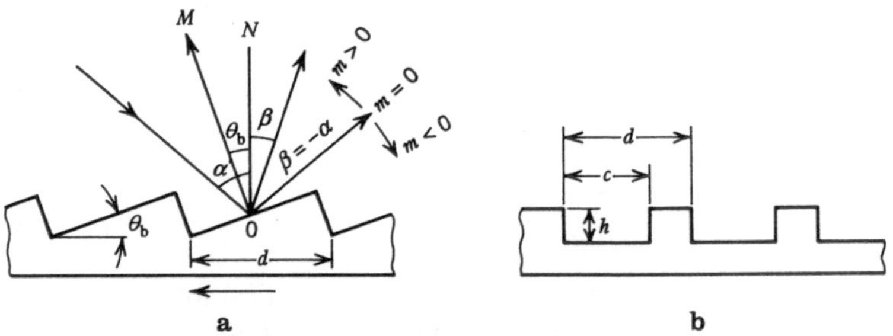

Fig. 13.24. Cross-sectional view of the reflection gratings, **a** blazed and **b** laminar

Below 1.5 nm, crystals are usually used as dispersive elements. The principle of the dispersion is Bragg reflection, which is given by $m\lambda = 2d\sin\theta$, where θ is the glancing angle ($\theta = 90° - \alpha$). Refractive indices of materials in the shorter wavelength region are almost 1 (slightly smaller than 1). The usable crystals are as follows. Beryl ($(10\bar{1}0)$: $2d = 1.595$ nm), YB_{66}((400): $2d = 1.176$ nm), quartz($(10\bar{1}0)$: $2d$=0.8512 nm), InSb((111): $2d = 0.7481$ nm), Ge((111): $2d = 0.6532$ nm), Si((111): $2d = 0.6271$ nm). KAP(potassium hydrogen phthalate (100): $2d$=2.6632 nm) has a large $2d$ value, but it is very weak against synchrotron radiation. Na β-alumina ((0002): $2d$=2.25 nm) crystal can possibly be used as a dispersive element. Multilayers can be used as dispersive elements when one does not need high resolution.

The detection of VUV light using sodium salicylate (fluorescent material) is a very common technique in the longer wavelength region. The treatment must be carried out very carefully. Especially during the baking, the temperature of the sodium salicylate should not be raised. The electron multipliers such as usual multi-dynode multipliers and channeltrons are used as the nudetype photomultipliers. Multi-channel plates (MCP) are also used. The channeltron and MCP have such a high sensitivity that they cannot be used for the intense light. Photodiodes are used as the standard detector for absolute measurement of photon number, after calibration by an ion chamber. They are aluminium oxide diodes developed by NBS and semiconductor diodes without coating material. A special charge-coupled device (CCD) can be used in the wavelength region below 10 nm for a two-dimensional detector.

13.7 VUV Monochromators

Synchrotron radiation (or undulator radiation) is extracted through a front end of a beamline composed of valves and shutters, which is attached tangentially to the vacuum chamber of the bending section of a storage ring. Then it is gathered by premirrors and introduced into a monochromator. Behind the monochromator, the dispersed light is refocused by a postmirror onto the sample in the experimental equipment. The whole instrument is called a beamline. Since synchrotron radiation is linearly polarized in the horizontal plane, the plane of incidence to optical elements is usually taken vertically to avoid the loss of the intensity and deterioration of the polarization character. That is, the entrance and exit slits, gratings and mirrors are set on the vertical plane, which is different from the illustration in Fig. 13.23. This setup, furthermore, has another advantage that large optical elements are not required to gather synchrotron radiation in comparison with those used in the horizontal setup, because the divergence of synchrotron radiation usually used is smaller in the vertical direction than in the horizontal direction.

In the early stages of synchrotron radiation research, conventional VUV monochromators [19] were used, but since then additional monochromators optimized to synchrotron radiation have been developed [24–26]. Monochromators using a spherical grating are classified into two types. One is a monochromator with a Rowland circle mounting and the other is that with an off-Rowland circle mounting. Monochromators with a Rowland circle mounting have the advantage of small aberration as mentioned before so that one can get high resolution using them. When an exit slit is fixed, a grating and an entrance slit should move so as to satisfy the Rowland circle condition in the wavelength scanning. However, this mechanism is somewhat complicated. Therefore, there are a variety of monochromators with off-Rowland circle mounting.

In the normal incidence region, monochromators with off-Rowland circle mounting are widely used. They are schematically given in Fig. 13.25(a)–(d). Figure 13.25(a) shows one of them in which the entrance and exit slits are fixed and the grating is constrained to move along the bisector of the angle subtended by both slits at the center of the grating. The Rowland circle condition is only satisfied at the 0th order. The original scanning mechanism is as follows. Two bars (length; $R/2$) are pivoted at O_o, and one of the bars is pivoted at P and the other is connected to the grating vertically. When O_o moves to O, the grating rotates and wavelength is scanned. However in the recent monochromators, the motion of the grating is not achieved by the use of such bars but by using a shorter bar and a cam. Figure 13.25(b) shows an off-plane Eagle monochromator. The entrance and exit slits are placed symmetrically above and below the Rowland circle, that is, they are apart from the Rowland plane by an amount of $\pm y_0$. The original scanning mechanism is as follows. The grating and slits are linked by two bars. The grating moves on a straight line toward the midpoint of the two slits. To get high resolution, y_0 is taken to be much smaller than R. There are several monochromators of this type with a large radius, for example, $R = 6$ m. Since α and β are small in the monochromators given in Fig. 13.25(a)–(b), the astigmatism is small. Figure 13.25(c) shows a Seya-Namioka monochromator, in which the entrance and exit slits are fixed and the grating only rotates. The angle $2k = \alpha - \beta$ is chosen as $70°30'$ and $r = r' = R\cos 35°15' = 0.8166R$. When these r and r' are substituted into Eq. (13.58), the left side term minus the right side term is quite small for any angle of incidence though it is not exactly equal

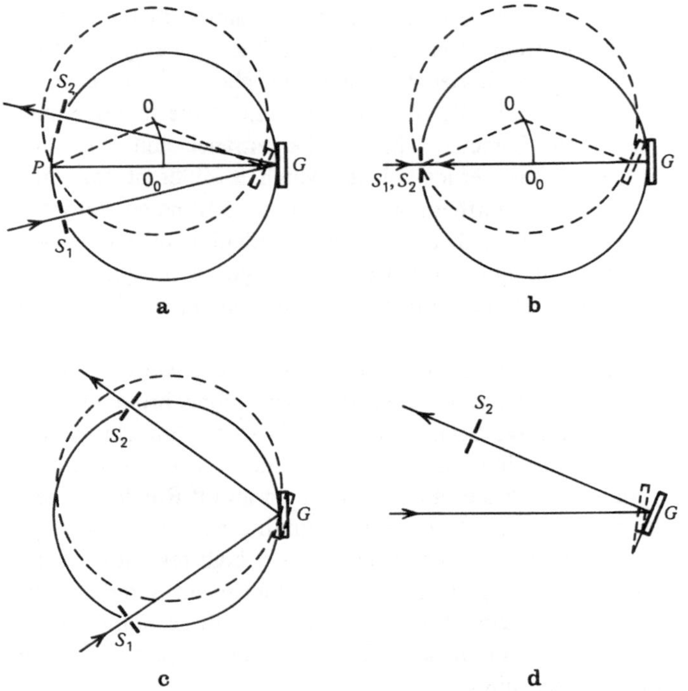

Fig. 13.25. Schematic optics of normal incidence monochromators. **a** McPherson, Robin [Robin S (1953) J. de Phys. 14:551], **b** off-plane Eagle [Ginter ML (1986) Nucl. Instr. Meth. Phys. Res. A246:474], and **c** Seya-Namioka [Namioka T (1959) J. Opt. Soc. Am. 49:951], and **d** modified Wadsworth [Skibowski M, Steinmann W (1967) J. Opt. Soc. Am. 57:112]

to zero. This approximate solution satisfies the Rowland circle condition only for the 0th order light. In a Seya-Namioka monochromator, α and β are not small so that the astigmatism is not small. Furthermore, the image at the exit slit is not linear, but curved. Even though this monochromator is not appropriate for high resolution experiment, it is used very widely because the mechanism is very simple and the fabrication is easy. Figure 13.25(d) shows a monochromator called a modified Wadsworth monochromator. There is no entrance slit and synchrotron radiation is regarded as parallel light. The grating rotates around the axis which is away from the center of the surface of the grating to give the approximate linear motion to it.

In the grazing incidence region, there are several monochromators which satisfy the Rowland circle condition exactly. The typical mount is of the Vodar type. In many cases, the exit slit is fixed. During scanning, the grating moves along the straight line connecting the grating and the exit slit, and the entrance slit moves along the straight line connecting the entrance and exit slits. The entrance slit and the grating are connected by a bar, so that the distance between the entrance slit and the grating is constant, $R\cos\alpha$. Therefore, the incident angle α is always constant and the diffracted angle β changes. Since the position of the entrance slit moves and the source point of synchrotron radiation cannot move with the entrance slit, a special premirror system is required to introduce synchrotron radiation.

A grasshopper monochromator is one of the Vodar type monochromators developed for synchrotron radiation. In Fig. 13.26(a), a schematic drawing of the grasshopper monochromator is shown. The entrance slit is a so-called Codling slit and consists of a single blade and a plane mirror. The plane mirror rotates so that the incident light always irradiates the grating. The entrance slit and a premirror are situated in a box and the box moves along the axis parallel to the synchrotron radiation. The premirror focuses the synchrotron radiation onto the entrance slit.

In the Vodar monochromator, only the exit slit is fixed, so that the scanning mechanism is complicated. Therefore, somewhat simple monochromators called constant deviation monochromators (CDM) have been developed. The angle subtended by the entrance and exit slits at the grating is constant, that is, $2k = \alpha - \beta =$ constant. (The Seya-Namioka monochromator above mentioned is one of the CDMs.) In the grazing incidence region, $2k$ is larger than 150°. They are classified into two types. One satisfies the Rowland circle condition and the other does not satisfy the condition except at special wavelengths. In Fig. 13.26(b), a schematic drawing of a CDM with a Rowland mounting is shown. The grating

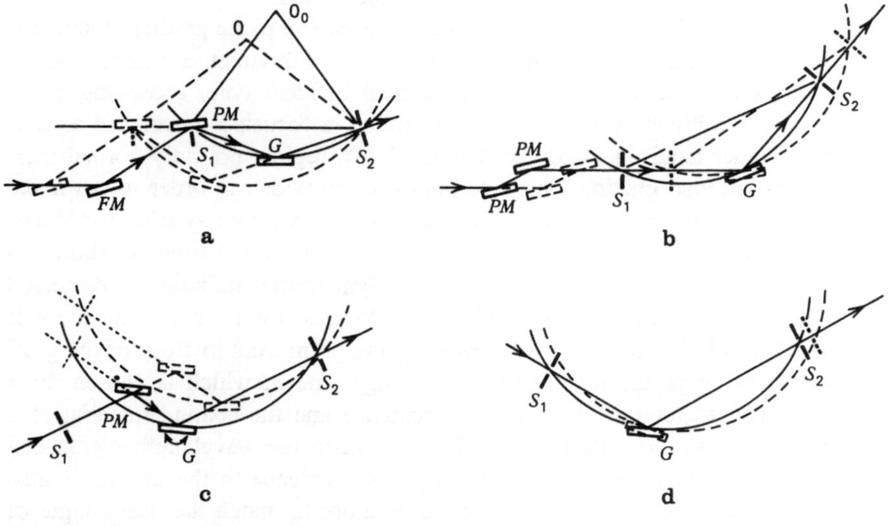

Fig. 13.26. Schematic optics of grazing incidence monochromators with a concave grating. *G*: grating, S_1: entrance slit, S_2: exit slit, *PM*: plane mirror and *FM*: focusing mirror. **a** grasshopper [Brown FC, Bachrach RZ, Lien N (1978) Nucl. Instr. Meth. 152:73]; **b** modified Rowland [Sugawara H, Sagawa T (174) In: Koch E-E, Haensel R, Kunz C (eds) Vacuum Ultraviolet Radiation Physics. Pergamon Vieweg, Braunschweig, p 790]; **c** constant length [Ishiguro E, Suzui M, Yamazaki J, Nakamura E, Sakai K, Matsudo O, Mizutani N, Fukui K, Watanabe M (1989) Rev. Sci. Instr. 60:2105]; amd **d** Dragon [Chen CT, Sette F (1989) Rev. Sci. Instr. 60:1616], TGM [Depautex C, Thiry P, Pinchaux R, Petroff Y, Lepere D, Passereau G, Flamand J (1978) Nucl. Instr. Meth. 152:101], and slitless CDM (SGM) [Padmore HA (1989) Rev. Sci. Instr. 60:1608]

only rotates and the entrance and exit slits connected by a bar move to satisfy the Rowland circle condition. Generally, the Rowland circle mounting has high resolution, but it is also possible to achieve high resolution using an off-Rowland mounting. One of the typical examples of off-Rowland mounting is given in Fig. 13.26(c), in which the entrance and exit slits are fixed so that the distance $r + r'$ is constant. A plane mirror and a grating move together along the axis of the incident light reflected by a premirror. The grating is rotated by a bar moving along a special cam. The other is given in Fig. 13.26(d), in which the entrance slit is fixed and the grating only rotates and the exit slit moves to ensure good focusing. The position of the exit slit is controlled by a computer. The monochromator of this type with the highest resolution in the short wavelength (4.5–1.5 nm) is called a Dragon, which has a grating with a large radius (57.3 m). Around 3 nm, $\lambda/\Delta\lambda$ is around 10^4. There is a similar mounting in which no entrance slit is provided and the electron beam is regarded as the light source. The above-mentioned CDMs using spherical gratings are often called SGM.

To suppress the astigmatism in grazing incidence, toroidal grating monochromators (TGM) are also used widely in many synchrotron radiation facilities. The TGM is also one of the CDMs. The positions of the entrance and exit slits and the grating are fixed. The grating only rotates. The optics for the TGM are similar to that of CDM as shown in Fig. 13.26(d), except that the exit slit is fixed. The grating with varied space is used, so that the deviation of the focusing point from the ideal point given by Eq. (13.59) is minimized.

Another type of the monochromator widely used is a plane grating monochromator (PGM). In the region from near infrared to ultraviolet, a Czerny-Turner monochromator is available. It is of the normal incident type, consisting of an entrance slit, a collimating mirror, a plane grating, a focusing mirror and an exit slit. On the other hand, in PGMs used in the VUV region, not only normal incidence optics but also grazing incidence optics are provided in order to cover the wide wavelength region. A typical one is FLIPPER, in which synchrotron radiation is regarded as a parallel light and the entrance slit is not used, as shown in Fig. 13.27(a). The plane grating only rotates. Synchrotron radiation is dispersed by the plane grating and the dispersed light is focused by a mirror onto the exit slit. The FLIPPER is equipped with several plane premirrors in front of the grating in order to cover the wide wavelength range, one of which is chosen for a certain wavelength region. The angle of incidence and the coating material of a premirror are chosen to reflect efficiently the light in the wavelength region and to suppress the higher order light. The angle of incidence to the grating is also chosen with the same consideration and furthermore to match the blaze angle of the grating. When synchrotron radiation cannot be regarded as parallel light, that is, the vertical acceptance angle is large, there are two types of PGM available. One is a PGM having a collimator mirror as shown in Fig. 13.27(b). In this case, several focusing mirrors are provided. The other is a PGM having a ellipsoidal mirror as a focusing mirror, called SX-700. The characteristic of the ellipsoidal mirror that light emitted at one focal point of ellipsoid is focused on the other focal point is utilized. As shown in Fig. 13.27(c), a plane mirror moves along

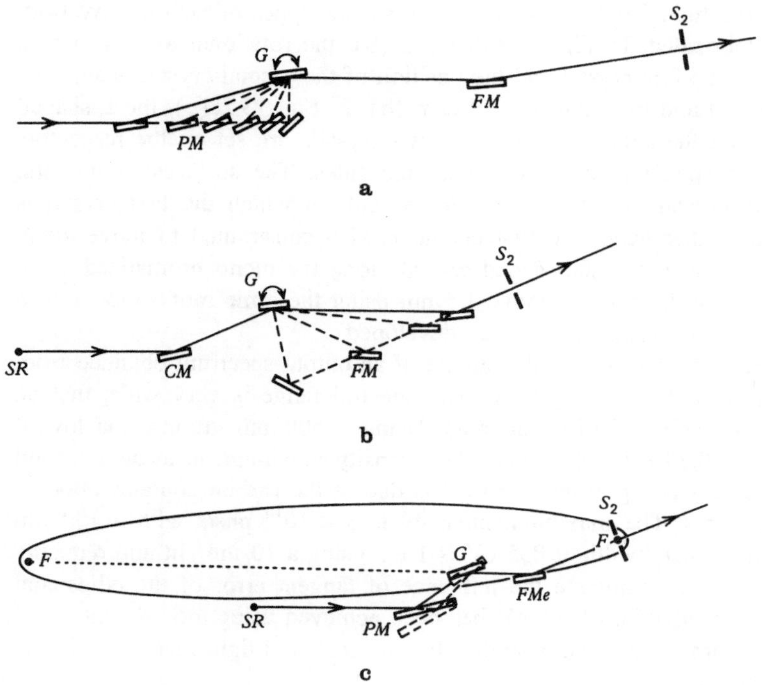

Fig. 13.27. Schematic optics of plane grating monochromators (PGM). *SR*: source point of synchrotron radiation, *G*: grating, S_2: exit slit, *PM*: plane mirror, *FM*: focusing mirror (*FMe*: ellipsoidal), *CM*: collimator mirror and *F*: focal point of ellipsoid. **a** FLIPPER [Eberhardt W, Kalkoffen G, Kunz C (1978) Nucl. Instr. Meth. 152:81]; **b** PGM with a collimator [Seki K, Nakagawa H, Fukui K, Ishiguro E, Kato R, Mori T, Sakai K, Watanabe M, (1986) Nucl. Instr. Meth. Phys. Res. A246:264]; and **c** SX-700 [Petersen H (1982) Opt. Commun. 40:402]

the axis of synchrotron radiation instead of changing several mirrors in FLIPPER. The angle of incidence always satisfies the condition that the virtual image of the electron beam is at one of the focal points of the ellipsoid.

In the above-mentioned PGMs, the electron beam acts as the entrance slit so that the resolution is limited by the beam size. Therefore, the low emittance ring has the great advantage of achieving high resolution. If one needs higher resolution, one should provide an entrance slit similar to the Czerny-Turner monochromator. Recently, PGMs with a varied space grating possessing a focusing force by itself have been developed [21, 27].

Below 1.5 nm, since up to now it is not easy to obtain high resolution using grating monochromators, crystal monochromators with resolutions of 1000–2000 are widely used. In many synchrotron radiation facilities, double crystal monochromators (DXM) are widely used. Two crystals are set parallel to each other. When one rotates two crystals, a fixed distance apart, around the center of the surface of the first crystal, the height of the monochromatized light from the second crystal changes. Therefore, to make the direction and height of the monochromatized light fixed (constant offset), several types of DXMs have been designed and fabricated. In Fig. 13.28(a) and (b), the rotational axis is located at the surface of the first crystal and the position of the second crystal is adjusted by a computer (a) and by a mechanical cam (b). In Fig. 13.28(c), the L-shaped bar rotates around the axis at the corner. Two crystals are set on the respective sides of the L-shaped bar and slide along the sides. The surfaces of the first and the second crystals are set parallel to the side at which the first crystal is situated. On the other hand, the first crystal is also constrained to move along the incident beam axis and the second crystal along the monochromatized beam axis. To obtain light below and above 1.5 nm using the same monochromator, a grating-crystal monochromator has been developed.

In Fig. 13.29 there is a typical example of an output spectrum obtained from the SX-700 using a 1200/mm grating. The spectral range is very wide, that is, from 17.5 eV to 2200 eV (70–0.56 nm). Using a 600/mm grating, the lowest photon energy is 8.75 eV (14.2 nm). The intensity maximum is located around 700 eV (1.8 nm). The dip around 280 eV is due ot the carbon contamination of the optical elements. The maximum intensity is 3×10^{10} phs/s with a 100 μm slit. The minimum bandwidth at 835 eV is 1 eV using a 10 μm slit and reducing the mirror aperture to eliminate the influence of tangent error of the ellipsoidal mirror. (Even a bandwidth of 0.5 eV has been achieved using a 5 μm slit.) The Si K-edge spectrum (inset) demonstrates the low scattered light level even in the high energy region.

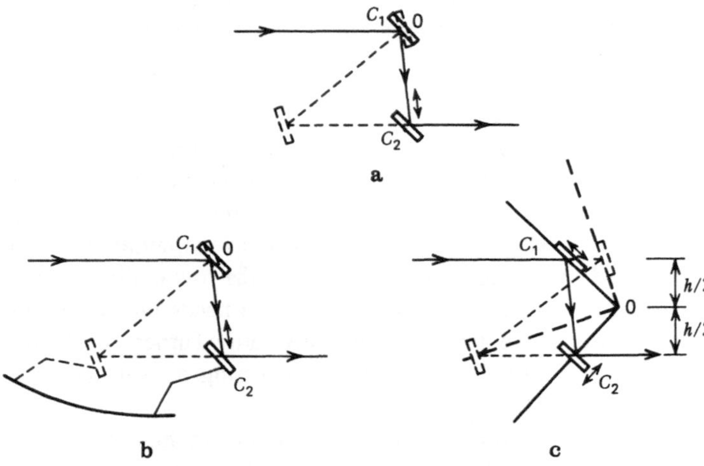

Fig. 13.28. Schematic optics of double crystal monochromators of constant offset. The rotational axis is located on the surface of the first crystal and the position of the second crystal is adjusted by a computer **a**, or by a mechanical cam **b** [Lemonnier M, Collet O, Depautex C, Esteva J-M, Raoux D (1978) Nucl. Instr. Meth. 152:109; Matsushita T, Ishikawa T, Oyanagi H (1986) Nucl. Instr. Meth. Phys. Res. A246:377]. In **c**, the L-shaped bar rotates around the corner and both the first and the second crystals slides on the sides of the L-shaped bar being constrained simultaneously to slide on axes of the coming and outgoing light [Murata T, Matsukawa T, Mori M, Obashi M, Naoe S, Terauchi H, Nishihata Y, Matsudo O, Yamazaki J (1986) J. de Phys. C8, 47:C8-135]

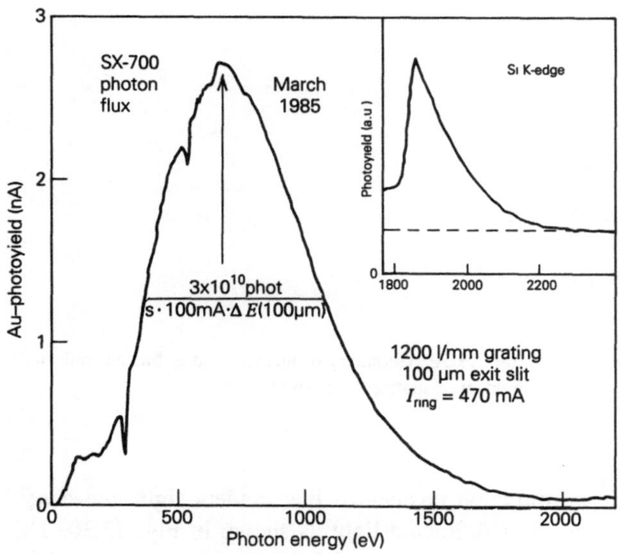

Fig. 13.29. Output signal of a PGM SX-700 at BESSY. Ordinate is photoelectric current (total yield) from a freshly evaporated Au surface [Petersen H (1986) Nucl. Instr. Meth. Phys. Res. A246:260]

The calibration of the wavelength is very important. For the calibration, the sharp peaks of rare gases in core absorption are often used. The wavelengths of the first peaks are 19.04 nm, 13.59 nm, 5.073 nm and 1.430 nm for Xe-N_5, Kr-M_5, Ar-L_3 and Ne-K core absorptions, respectively. The absorption peaks of light metals around core absorption edge are also useful. They are the absorption peaks of Al-$L_{3,2}$ (17.05, 16.95 nm), Mg-$L_{3,2}$ (25.04, 24.90 nm) and Al-K (0.7951 nm). When the output light is detected by an electron multiplier with a Cu-BeO first dynode, the sharp peaks in the spectrum originating from photoelectrons from BeO can be used for wavelength calibration. The first peaks are located at 10.44 nm and 2.259 nm for Be-K and O-K absorptions, respectively. In rare cases, a discharge lamp is installed in front of the entrance slit to calibrate the wavelength using emission lines.

To design the beamline optics, ray tracing is also very important. After the deciding on specifications of the beamline for the wavelength region, resolution, intensity, beam size etc., one should perform ray tracing to optimize the position, size, curvature and other parameters of optical components such as mirrors, gratings and slits. The relation between the directions of the incident light and reflected or diffracted light at the point $P\ (w, l, h)$ on the mirror or grating is derived from Eqs. (13.49) and (13.54) as

$$\cos\xi' = -\cos\xi - (\cos\zeta + \cos\zeta')\frac{\partial h}{\partial w} + m\lambda\frac{\partial n}{\partial w} \tag{13.63}$$

$$\cos\eta' = -\cos\eta - (\cos\zeta + \cos\zeta')\frac{\partial h}{\partial l} + m\lambda\frac{\partial n}{\partial l} \tag{13.64}$$

$$\cos^2\xi' + \cos^2\eta' + \cos^2\zeta = 1, \tag{13.65}$$

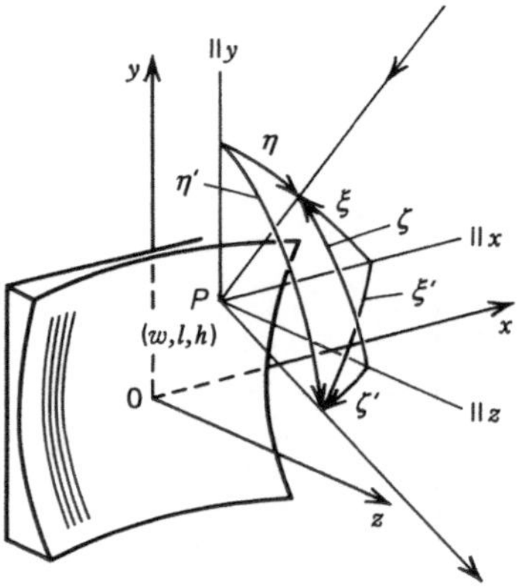

Fig. 13.30. Geometry of incident and diffracted (reflected) light by a grating (a mirror)

where $\cos\xi$, $\cos\eta$ and $\cos\zeta$ are direction cosines of the incident light and $\cos\xi'$, $\cos\eta'$ and $\cos\zeta'$, those of reflected or diffracted light as shown in Fig. 13.30. The cross section of the electron beam gives the size of the light source. It is divided into many portions and each portion is represented by the one point. The light emitted from the one point is simulated with many rays, the divergent angles of which are distributed in an acceptance angle or the natural divergence of the light discussed is Sects. 13.2 and 13.3. (One can simulate it more exactly when taking into account the longitudinal extent of the light source.) When the simulating rays reach a mirror or a grating, the reflected or diffracted rays go out according to the Eqs. (13.63–13.65). Along the optical path, one can get the spot diagram simulating the beam size. Accordingly one can design the optimum configuration of mirrors to get best focus on the entrance slit, estimate the resolution from the beam size at the exit slit and get the beam size at the sample position.

13.8 Techniques for VUV Measurements

In the absorption measurement of gases, a gas cell with a LiF window is usually used above 104 nm. Below 104 nm, such a bulk material is not available. One way to overcome this problem is to use thin films as windows. The other way is to use a differential pumping system, in which apertures with small conductance are distributed in the optical path and each section between two apertures is evacuated. To get smaller conductance, a capillary array is often used as the aperture for differential pumping. For experiments on vapors of materials, which are solids at room temperature, heat ovens of special design are used. The vapor is mixed with a buffer gas such as He. On the other hand, a supersonic free jet (molecular beam) of gas is used for experiments on cooled gases, in which high pressure gas is ejected into vacuum from a nozzle with a small aperture as shown in Fig. 13.31. The gas is ejected as pulses of 0.1–50 ms duration with a

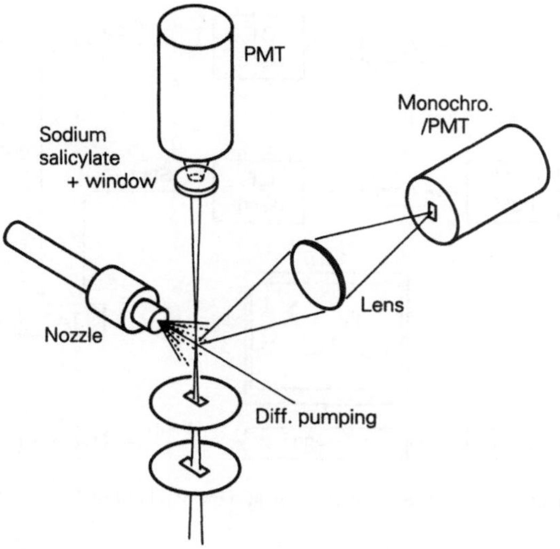

Fig. 13.31. Schematic diagram of supersonic free jet (molecular beam) apparatus

repetition frequency of a few tens Hz. In this case, the temperature of the gas becomes so low that the ground state with small vibrational quantum number can be realized. Furthermore, van der Waals molecules and clusters are obtained.

By VUV absorption valence electrons as well as core electrons are excited. The excited states are valence excited states, Rydberg states and continuum states. High resolution spectra with $\Delta\lambda/\lambda = 10^{-4}$ have been obtained above 30 nm for simple molecules, revealing the vibrational structure and Rydberg series with large n values. Recently, high resolution has been achieved in the wavelength region down to 2 nm, and K-absorptions of C, N and O have been studied.

In order to observe visible-ultraviolet fluorescence, a conventional visible-ultraviolet monochromator is used behind a quartz window, through which the fluorescence passes. However, to observe the fluorescence in the VUV region, one needs to attach a VUV monochromator directly to the sample chamber without a window. It is possible to obtain the fluorescence spectrum in a large wavelength region simultaneously. Many years ago, the spectra were taken using photographic plates. Recently, the photographic plates have been replaced by one-dimensional detectors. In the visible-ultraviolet region, photodiode arrays are available. In the VUV region, multi-channel plates (MCP) can be used. The flat focus of dispersed light can be realized using a varied space grating, in order to measure a wide spectral region with the one-dimensional detector. Measurements of fluorescence decay time are possible using the pulse structure of synchrotron radiation [28]. Figure 13.32 shows a block diagram of the time-resolved fluorescence measurement system, using single photon counting and TAC (time-to-amplitude converter).

Fluorescence experiments on gases also give information on the electronic structure and furthermore on the relaxation process of excited states including the photofragmentation of polyatomic molecules. In this case, one can assign the

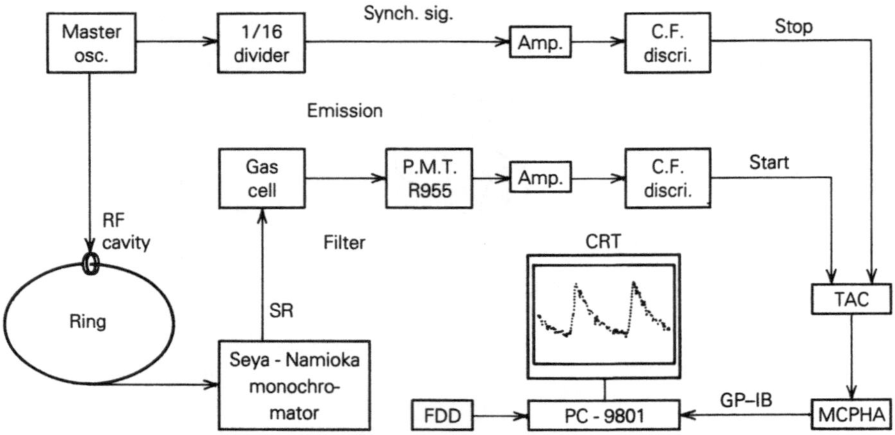

Fig. 13.32. Block diagram of a time-resolved fluorescence measurement system at UVSOR

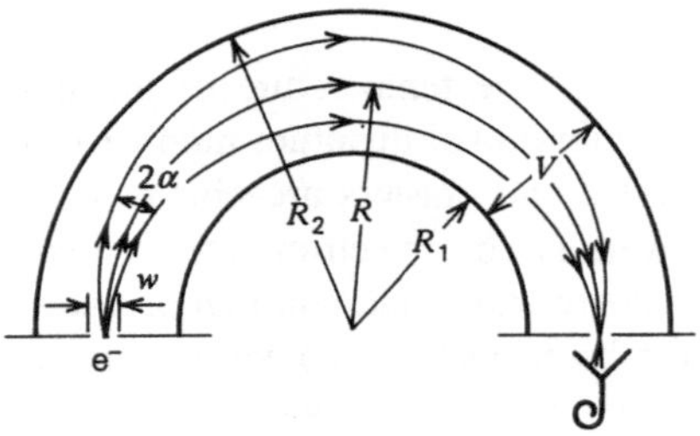

Fig. 13.33. Schematic diagram of hemisphere electron energy analyzer

photofragment of specified states by observing the fluorescence spectra. Fluorescence experiments on solids also give information on the relaxation process of excited states.

There are several types of electron energy analyzers. They are of retarding electrode type, parallel plate type, hemispherical type and cylindrical type. The hemispherical type analyzer is schematically shown in Fig. 13.33. When the applied voltage is V, electrons with the kinetic energy E_{kin} given by the following equation can pass through the analyzer,

$$E_{\mathrm{kin}} = \frac{eV}{\dfrac{R_2}{R_1} - \dfrac{R_1}{R_2}}, \tag{13.66}$$

where R_1 and R_2 are the radii of the inner and outer spheres, respectively. The energy resolution is given by

$$\frac{\Delta E}{E} = \frac{w}{2R} + \frac{\alpha_m^2}{4}, \tag{13.67}$$

where $R = (R_1 + R_2)/2$ and w is the width of the slits of the analyzer. The maximum divergence angle, α_m is determined by the practical configuration. Electrons with different energy are detected at the different applied voltage. In many cases, the photon energy ($h\nu$) is fixed and the applied voltage is swept to analyze the kinetic energy of photoelectrons. In consequence, one can get the energy distribution curve (EDC) of photoelectrons with respect to kinetic energy (E_{kin}). It can be converted easily to EDC with respect to the binding energy (E_b) from the vacuum level by the equation $E_b = h\nu - E_{\mathrm{kin}}$. Photoelectrons are emitted in the large solid angle. If a large number of them are introduced to the analyzer, one can get angle-integrated photoelectron spectra. If photoelectrons in a small solid angle are introduced, one can get angle-resolved photoelectron spectra. Recently, an analyzer which can resolve the direction of electron spin has been developed.

In the gas phase, electrons are excited from occupied states to the ionized state by photons. Therefore the EDC with respect to E_{kin} reflects the energy level of the ionized states accompanied with vibrational structure in many cases. Ionization efficiency is measured by an ionization chamber.

In many polyatomic molecules, fragmentation follows photoionization. Fragments can be detected by a mass analyzer. There are several types of mass analyzer such as quadrupole and time-of-flight mass analyzers. The right-hand part of Fig. 13.34 shows the schematic diagram of a time-of-flight mass analyzer. The mass M of the singly charged fragment can be determined according to the relation $Mv^2/2 = eV$. v is determined from $vt = l$, where l is the length of the flight tube, and V is the DC voltage applied to the acceleration section between D_2 and D_3. One of the modes to determine the mass is the pulsed mode, in which pulsed voltage is applied between D_1 and D_2 at t_1 to push ions to the acceleration section, and the ions reach the detector (MCP_1) at t_2. The flight time t is given by $t = t_2 - t_1$ so that one can assign species of the ions. Another mode is the photoelectron-photoion coincidence (PEPICO) mode, in which DC voltage is applied between D_1 and D_2 and photoelectrons from ionized molecules are pushed to the left part and detected by MCP_2. The flight time is measured by a TAC system in which the electron and the ion signals are used as start and stop signals, respectively. In photoion-photoion coincidence (PIPICO) experiments, signals from two kinds of ions are detected by MCP_1. One can clarify the relationship among the photofragments. Furthermore one can carry out experiments on ion-molecule reactions using similar apparatus to that shown in Fig. 13.34. Ions are generated by monochromatic light in the ionization

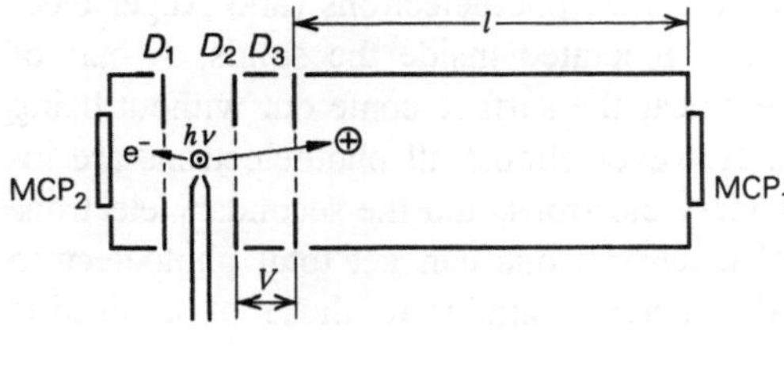

Fig. 13.34. Schematic diagram of time-of-flight mass analyzer which can be used for coincidence experiment. MCP: multichannel plate

region, and neutral molecules are introduced there. New species generated by the reaction of the ions and the molecules are detected by the mass analyzer. Mass analyzers are also used to detect molecules and ions ejected from solid surfaces by irradiation of VUV light (photodesorption).

Absorption coefficients of solids are so large that one should prepare thin films with a thickness of the order of 0.1 μm for absorption measurements on solids. These thin films are prepared by vacuum deposition onto LiF substrates or collodion films. Thin films of organic materials soluble in volatile solvents can be made from the solvent; a flat plate is soaked in the solution and dried, and then the film is peeled off.

However, it is not always possible to get thin films. If reflection spectra of near normal incidence on single crystals are available, one can get the absorption coefficient through Kramers-Kronig analysis. For normal incidence, the complex amplitude of reflected light is given by

$$r_p = r_s = \sqrt{\frac{(n-1)^2+k^2}{(n+1)^2+k^2}}e^{i\theta} \ . \tag{13.68}$$

n and k are given by

$$n = \frac{1-R}{1+R-2\sqrt{R}\cos\theta} \tag{13.69}$$

$$k = \frac{-2\sqrt{R}\sin\theta}{1+R-2\sqrt{R}\cos\theta} \ , \tag{13.70}$$

where θ is given by the Kramers-Kronig relation as

$$\theta(\omega_0) = \frac{1}{\pi}\int_0^\infty \frac{\mathrm{d}\ln\sqrt{R(\omega)}}{\mathrm{d}\omega}\ln\left|\frac{\omega+\omega_0}{\omega-\omega_0}\right| d\omega \ . \tag{13.71}$$

The absorption coefficient is given by $\mu = 2\omega k/c$. The reflectivity for normal incidence becomes less than 10^{-3} around 100 eV. However, even around 100 eV one can get the reflection spectrum using intense synchrotron radiation. In this case, k is given by $k \approx 2\sqrt{R}$ assuming that n is almost 1. If linearly polarized light is available and R_p and R_s can be measured, n and k can be obtained directly through the Fresnel equation (Eqs. 13.40–13.43).

Furthermore, even though thin films and crystals are not available, one can get the absorption spectrum (not absorption coefficient) by total photoelectron yield measurement. When photons excite solids, photoelectrons (and Auger electrons in the case of core excitation) are generated inside the solids. A part of photoelectrons from a few atomic layers near the surface come out without being scattered by other electrons in solids. However almost all photoelectrons are inelastically scattered and generate secondary electrons, and the secondary electrons come out. By detecting both kinds of electrons, one can get total photoelectron yield spectra. They give the spectral structures similar to those of absorption spectra (but do not give the correct value of the relative intensity of structures separated by a large photon energy.)

From absorption spectra due to the transition from the valence band, the joint density of states of the valence and conduction bands is obtained and from the core absorption, the density of state of the conduction band. Around the absorption edge, excition peaks are sometimes observed, which reflect the existence of the strong electron-hole interaction. The core absorptions (EXAFS) give the structural information. Furthermore, from core absorption measurements using the linear polarization, one can assign the orientation of adsorbed molecules on the surface.

Figure 13.35 shows a schematic diagram of a photoelectron spectrometer for solids. The electron energy analyzer in the main chamber can be rotated around the sample. The sample position can be adjusted horizontally and vertically by a manipulator. In many cases, the temperature of the sample can be changed from a low temperature (liquid N_2 or liquid He) to a high temperature (more than several hundred degrees). In the main chamber, a LEED-Auger spectrometer is often provided for sample characterization. The sample preparation chamber is equipped with evaporators, thickness monitors, a cleavage device, a diamond filer and view ports. Electron and ion beam cleanings are possible. The vacuum should be better than 10^{-8} Pa to avoid surface contamination.

By analyzing the energies of photoelectrons from solids which are not scattered, one can get EDC with respect to the binding energy. From the EDC in the angle-integrated mode (using angle-integrated analyzer or measuring on polycrystals) of photoelectrons from the valence band, one can get its density of state. When the crystal axis is well defined, the angle-resolved photoelectron spectroscopy gives more detailed information, such as the wave number dependence

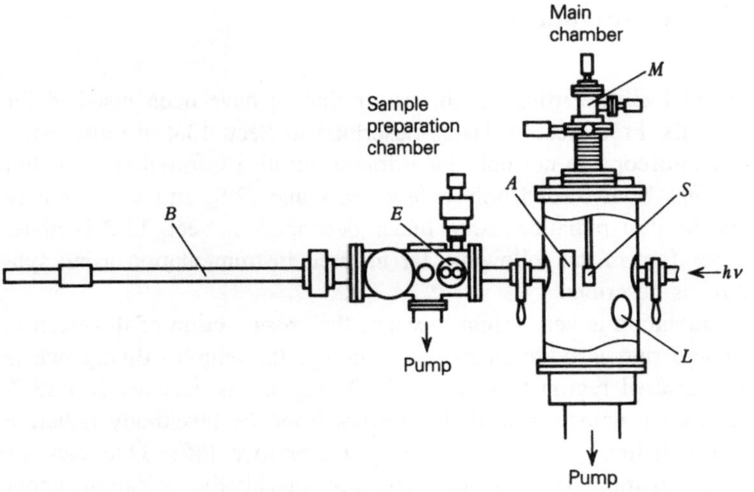

Fig. 13.35. Side view of photoelectron spectrometer for solids. *S*: sample, *A*: electron energy analyzer, *L*: Leed-Auger spectrometer, *M*: manipulator, *E*: evaporator and *B*: sample transfer bar

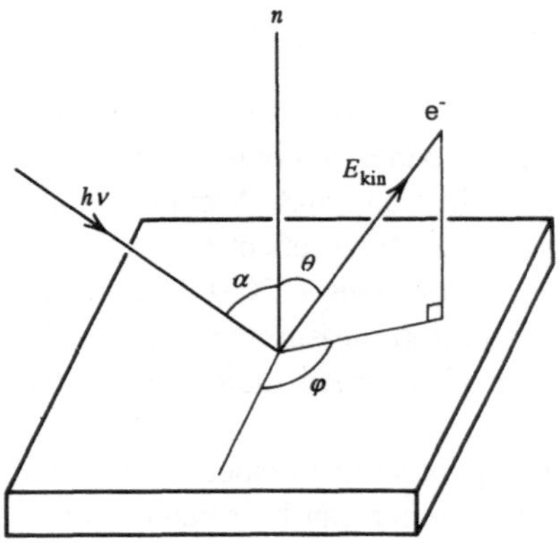

Fig. 13.36. Directions of incident light and photoelectron in polar coordinate

of energy band. For instance, in the case $\theta = 0$ in Fig. 13.36, energies of photoelectrons ejected vertically from the surface are analyzed. The EDCs are taken for various photon energies. The wave number k is given by $E_{\text{kin}} = \hbar^2 k^2/2m - V_0$ where $\hbar$ is the Planck constant, m the electron mass and V_0 the electron affinity. The binding energy E_b of a peak in EDC is given by $E_b = h\nu - E_{\text{kin}}$. Then for each $h\nu$, one can get one point in the E-k plane from one peak in EDC. Thus one can obtain the dispersion curve (E-k curve) of the valence band and compare with the band calculation. By photoelectron spectroscopy, one can furthermore clarify electronic states and get structural information for the clean and adsorbed surfaces.

13.9 Far Infrared Instrumentation

Several far infrared facilities using synchrotron radiation have been installed for practical experiments. Free electron lasers described in Sect. 13.4 are also powerful tools for spectroscopy in not only far infrared but also infrared regions, but the practical use has been started only a few years ago [29], and utilization of the coherent synchrotron radiation using linacs described in Sect. 13.2 is under investigation. Therefore in the following, far infrared instrumentation using synchrotron radiation is described.

Synchrotron radiation is very bright, because the cross section of the electron beam in the storage ring is quite small, even though the angular divergence is larger in the far infrared region than in the VUV region, as seen in Sect 13.2. Figure 13.37 shows the comparison of the powers from the blackbody radiation and synchrotron radiation at SRS, Daresbury Laboratory [30]. One can see that synchrotron radiation is stronger than the blackbody radiation above 200 μm (see theoretical curve d). Synchrotron radiation has a great advantage for far infrared spectroscopy on small samples because of its brightness. Furthermore

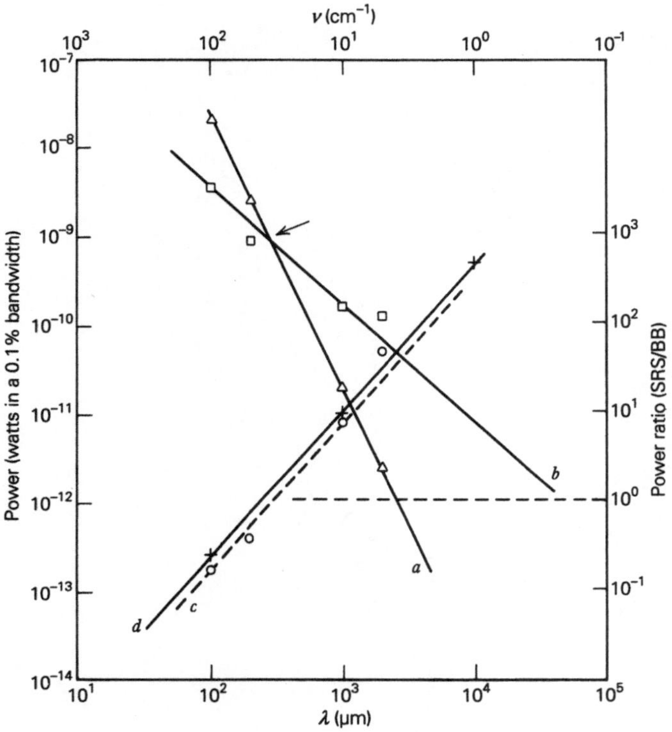

Fig. 13.37. Comparison of SRS and blackbody output powers and power ratios. *a*: blackbody curve at 1500 K with a 3 mm stop ($A\Omega = 0.045$ cm^2 sr)(△), *b*: measured SRS power output at 350 mA current and ±20 mrad vertical collection (□), *c*: power ratio measured at Daresbury in May 1984 (O), *d*: theoretical power ratio for a total vertical-angle collection and 350 mA(+) [Yarwood J, Shuttleworth T, Hasted JB, Nanba T (1984) Nature 312:742]

it has a superior polarization character and a time structure usable as pulsed light with a duration of sub ns.

Figure 13.38 shows the far infrared beamline at UVSOR equipped with a Martin-Puplett (Michelson type) interferometer [31]. Since the angular divergence of synchrotron radiation in the far infrared is large, the light collection mirrors are designed so that synchrotron radiation of 80 mrad (Hor.)×60 mrad (Vert.) can be gathered. Between the collection and the collimator mirrors, there is a focusing point. The converging synchrotron radiation passes through a wedge-shaped window made of Si or quartz before the focusing point. The window separates the ultrahigh vacuum of the mirror chamber from the rotary-pump vacuum of the collimator chamber and the interferometer. The focused beam diverges again and is made parallel again by the collimator mirror and introuced into the Martin-Puplett interferometer. The light of a 125 W Hg lamp can be also introduced into the interferometer by turning a wire-grid. The interferometer is composed of a beam splitter made of a wire-grid, and a fixed and a movable roof-top mirror. The light is chopped with a frequency between 10 and 200 Hz. Between the

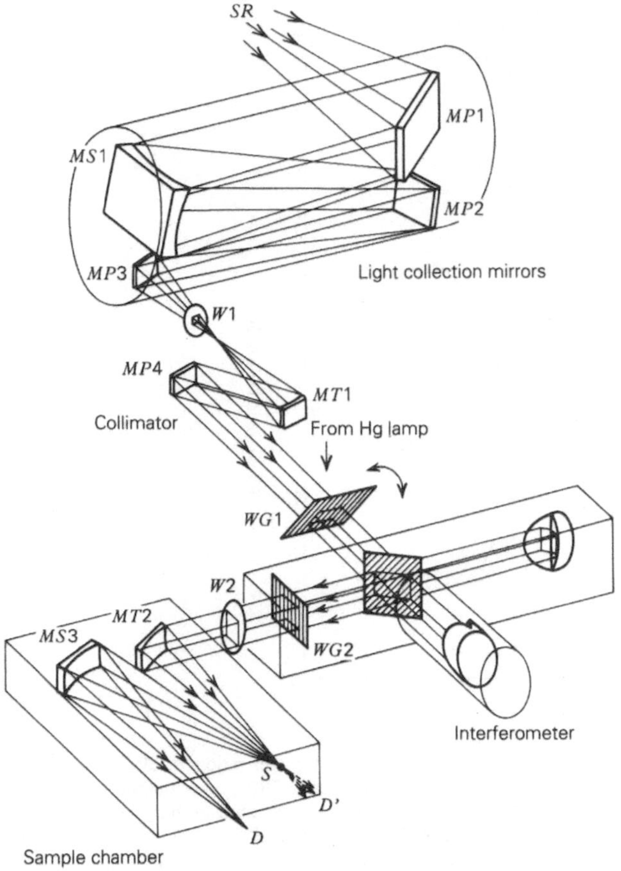

Fig. 13.38. Schematic drawing of the far infrared spectroscopic system at UVSOR. *MP*, *MS* and *MT* indicate plane, spherical and toroidal mirrors. W_1 and W_2 are silicon and polyethylene windows. *WG*s are wire-grid polarizers or reflectors whose wire spacing and thickness are 12.5 and 5 μm, respectively. Either the synchrotron radiation or light from a mercury lamp is selected by turning WG_1 through 90° [Nanba T, Urashima Y, Ikezawa M, Watanabe M, Nakamura E, Fukui K, Inokuchi H (1986) 7:1769]

interferometer and the sample chamber, there is another window to separate the ultrahigh vacuum of the sample chamber from the low vacuum of the interferometer. The parallel light from the interferometer is focused by a spherical mirror on the sample. When one makes absorption measurements, a detector is set at D' behind the sample. When one makes reflection measurements, the reflected light is gathered by another spherical mirror and detected by a detector located at D.

The interferometers of Michelson type need a beam splitter of half-mirror or wire-grid. However, a new type of interferometer has been developed recently which does not have such a half-mirror or a wire-grid, but two mirrors as shown Fig. 13.39. It is a wavefront dividing interferometer utilizing the transverse co-

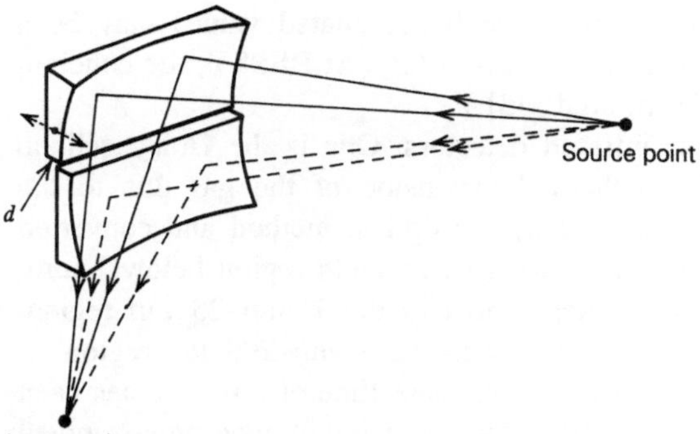

Fig. 13.39. Schematic of a wavefront dividing interferometer for a transversely coherent synchrotron radiation which reflects from the two-component mirror [Williams G (1992) Rev. Sci. Instr. 63:1535]

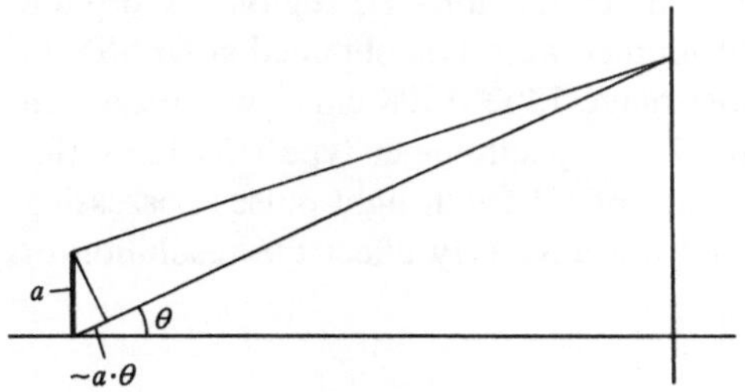

Fig. 13.40. Path difference of the lights emitted from two transverse ends of the beam

herence of synchrotron radiation [32]. When plane waves with wavelength λ pass through a slit with width a as shown in Fig. 13.40, the wave diverges due to the diffraction effect by $\theta \approx \lambda/a$. Accordingly the divergence of synchrotron radiation σ_r' gives its maximum transverse extent $s \approx \lambda/\sigma_r'$ at the source point. If s is larger than the beam size σ of the electron beam, $\sigma\sigma_r' \ll \lambda$, the light is transversely coherent. The transverse coherence does not impose any phase relations between rays emitted from different portions in the source, and guarantees small path difference between two extreme rays in comparison with λ. In the far infrared region, synchrotron radiation is transversely coherent. Therefore, one can divide the wavefront as shown in Fig. 13.39. (Using a conventional light source, wavefront dividing cannot be achieved without a large intensity loss.) All rays are recombined at an image point. As the upper mirror is scanned, a path difference, $2d$, is created between rays reflected from it and reference rays reflected from the lower mirror. Thus the instrument works like an interferometer without a beam splitter for amplitude dividing. There is a possibility that the above-mentioned technique can be extended to the VUV region using undulator radiation.

Figure 13.37 also compares output powers from the interferometer using synchrotron radiation from SRS and a Hg lamp (blackbody), which are denoted by the squares and the triangles, respectively. The power of synchrotron radiation is stronger than that of the Hg lamp, as the calculation shows. The deviation of square point at $\lambda = 2$ mm from the curve b (calculated value) may be a symptom of the coherent effect discussed in Sect. 13.2. (At BESSY, the coherent effect was not observed in middle infrared [33].)

There are several types of far infrared detectors. One is the Golay cell, in which rare gases are filled and the thermal expansion of the gas due to the absorption of far infrared light is detected by the optical method and converted to the electric signal. It can be used in the long wavelength region below 1 mm. Another detector is a Ge bolometer which works in the 1 mm–25 μm region. The InSb hot electron detector has a sensitivity in the 5 mm–200 μm region.

Recently, a fast infrared detector with the response time of a few ns has been developed using high T_c superconducting films [34], which will open time-resolved spectroscopy in the far infrared.

At present, many experiments are being carried out to study the lattice vibration of ionic crystals, molecular motion of liquid molecules, Drude absorption due to the free electron in High T_c materials and organic conductors, shallow impurity level in semiconductors and so on. In the infrared region, absorption spectra using SR instead of common light sources were first obtained at BESSY in 1984; the band of NO_2 in the wave number range 1303–1308 cm^{-1} was measured with high resolution using a commercial FT–IR spectrometer type BRUKER IFS 113v, and it turned out that the time structure of SR (with light pulses possessing a duration of about 150 ps at BESSY) did not adversely affect the resolution of the Fourier transform technique [33].

13.10 Two-Color Experiments

Figure 13.6 showed that the angular divergence of the beam increases with wavelength. This effect allows, of course, optical beam splitting for two-color experiments, i.e. no additional light sources are necessary in principle. A first experimental setup was presented at BESSY in 1983 (Fig. 13.41) [35]. Combining synchrotron radiation and laser, more than 10 experiments of several kinds have been made so far. The first two-color experiment was performed using a N_2 laser and synchrotron radiation, in which photoelectrons from excited states in solid Krypton were measured [36]. The first light (monochromatized synchrotron radiation) excites electrons to the excitonic states and the second light (the laser, $h\nu = 3.7\,\mathrm{eV}$) ionizes the excitons. Photoelectrons from the ionized excitons were detected, as the photon energy of the first light was scanned. In this experiment, the $n = 2$ excitonic state was clearly observed. This is not clearly observed by the conventional absorption measurement.

Another study is transient spectroscopy which is called the pump-probe experiment or timing experiment. Figure 13.42 shows the principle of this experiment [37]. The pump is the laser and the probe is synchrotron radiation. In this experiment, the pulse frequency of the laser is chosen to be equal to that of synchrotron radiation. When one scans the delay time (phase of laser pulses

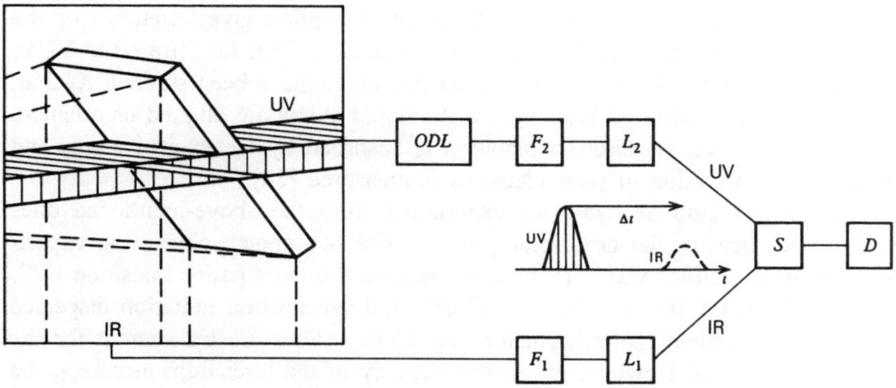

Fig. 13.41. Wavelength separation and beam splitting by means of a 45° gold plated plane mirror possessing a horizontal slit through which most of the less divergent short wavelength (e.g. UV) beam light passes, whereas the long wavelength (e.g. IR) radiation is mainly reflected. Both the IR and UV beam then possess the same time structure, e.g. pulses of about 150 ps width spaced by 2 ns at BESSY. The pulses can be delayed by an optical delay line *ODL* so that the sample *S* can be irradiated with well-defined time differences Δt. F_i and L_i denote the filters and lenses used, respectively, and *D* the detector, e.g. a fluorimeter for the measurement of the fluorescence decay and rise times of the sample [Lippert E, Prass B, Rettig W, Stehlik D (1983) BESSY Ann. Reports, p 50]

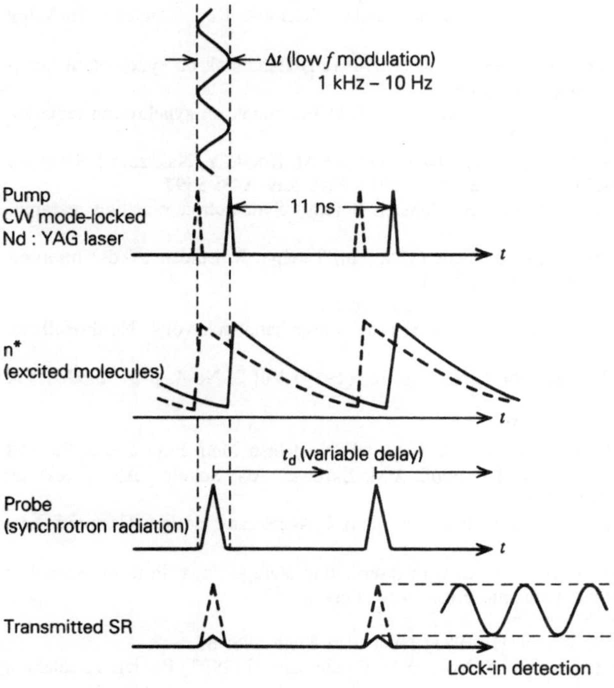

Fig. 13.42. Principle of transient absorption measurements [Mitani T, Okamoto H, Takagi Y, Watanabe M, Fukui K, Koshihara S, Ito C (1989) Rev. Sci. Instr. 60:1569]

to synchrotron radiation pulses), the transient absorption gives lifetimes of the excited states. The wavelength of the probe is tunable from far infrared to VUV.

In the gas phase, laser-excited aligned free atoms have been studied. At first, atoms absorb the polarized laser light and excited atoms are aligned or oriented. Then, the polarized synchrotron radiation is absorbed by the excited atoms and the angular distribution of photoelectrons is measured [38].

A different kind of two-color experiment from the above-mentioned ones is the experiment of the non-linear process. The two-photon absorption experiment on alkali halides was performed to observe the even parity transition [39]. For both the Nd:YAG laser light (1.17 eV) and synchrotron radiation dispersed by a normal incident monochromator, the alkali halides are transparent for the one-photon process. However, when the intensity of the laser light increases, the two-photon absorption is observed at the sum of photon energies of laser and monochromatized synchroton radiation. It is found that the two-photon absorption in alkali halides begins just above the 1s exciton absorption in one-photon process.

13.11 References

1 Jackson JD (1975) Classical electrodynamics. John Wiley & Sons, New York
2 Kim K-J (1986) In: Vaughan D (ed) X-ray data booklet. PUB-490 Rev, Lawrence Berkeley Laboratory, p 4–1
3 Krinsky S, Perlman ML, Watson RE (1983) In: Koch E-E (ed) Handbook on synchrotron radiation, vol 1A. North-Holland, Amsterdam, p 65
4 Koch E-E, Eastman DE, Farge Y (1983) In: Koch E-E (ed) Handbook on synchrotron radiation vol. 1A. North-Holland, Amsterdam, p 1
5 Ishi K, Shibata Y, Takahashi T, Mishiro H, Ohsaka T, Ikezawa M, Kondo Y, Nakazato T, Urasawa S, Niimura N, Kato R, Shibasaki Y, Oyamada M (1991) Phys Rev A43: 5597
6 Spencer JE, Winick H (1980) In: Winick H, Doniach S (eds) Synchrotron radiation research. Plenum, New York, p 663
7 Attwood DT, Kim K-J, Halbach K, Howells MR (1986) In: Tatchyn R, Lindau I (eds) Insertion devices for synchrotron sources: SPIE Proc 582: 10
8 Brau CA (1990) Free-electron lasers, Academic Press, Boston
9 Dattoli G, Renieri A (1985) In: Stitch ML, Bass M (eds) Laser handbook vol.4. North-Holland, Amsterdam, p 1
10 Ortega JM (1989) Synch Rad News Vol 2, No 3, p 18; (1989) Vol 2, No 4, p 21; (1990) Vol 3, No 1, p 26
11 Kim K-J (1991) Phys Rev Lett 66: 2746
12 Bonifacio R, Salvo Souza LD, Pierini P, Piovella N (1990) Nucl Instr Meth Phys Res A296: 358
13 Kulipanov GN, Litvinenko VN, Pinaev IV, Popik VM, Skrinsky AN, Sokolov AS, Vinokurov NA (1990) Nucl Instr Meth Phys Res A296: 1
14 Scharf W (1986) Particle accelerators and their uses: Part 1 Accelerator design. Harwood Academic, Chur London
15 Sands M (1971) In: Touschek B (ed) Physics with intersecting storage rings: Proc Int School of Phys "Enrico Fermi" Course 46. Academic Press, New York p 257
16 Jackson A (1990) Synch Rad News Vol 3, No 3, p 13
17 Brown G, Halbach K, Harris J, Winick H (1983) Nucl Instr Meth 208: 65
18 Kitamura H (1992) Synch Rad News Vol 5, No 1, p 14 Wiedemann H (1993) Particle Accelerator Physics
19 Samson JAR (1967) Techniques of vacuum ultraviolet spectroscopy. John Wiley & Sons, New York

20 Dehz P (1987) In: Koch E-E, Schmahl G (eds) Soft X-ray optics and technology: SPIE Proc 733: 308
21 Itou M, Harada T, Kita T (1989) Appl Opt 28: 146
22 Namioka T (1959) J Opt Soc Am 49: 446
23 Neviere M, Flamand J, Lerner JM (1982) Nucl Instr Meth 195: 183
24 Gudat W, Kunz C (1979) In: Kunz C (ed) Synchrotron radiation, techniques and applications. Springer, Berlin Heidelberg New York, p 55
25 Johnson RL (1983) In: Koch E-E (ed) Handbook on synchrotron radiation vol 1A. North-Holland, Amsterdam, p 173
26 West JB, Padmore HA (1987) In: Marr GV (ed) Handbook on synchrotron radiation vol 2. North-Holland, Amsterdam, p 21
27 Hettrick MC (1990) Appl Opt 29: 4531
28 Rettig W (1993) In: Wolfbeis OS (ed) Fluorscence spectroscopy. Springer, Berlin Heidelberg New York
29 Dlott DD (1992) Nucl Instr Meth Phys Res A318: 26
30 Yarwood J, Shuttleworth T, Hasted JB, Nanba T (1984) Nature 312: 742
31 Nanba T, Urashima Y, Ikezawa M, Watanabe M, Nakamura E, Fukui K, Inokuchi H (1986) Int J Infrared and Millimeter Waves 7: 1769
32 Williams GP (1992) Rev Sci Instr 63: 1535
33 Schweizer E, Nagel J, Braun W, Lippert E, Bradshaw AM (1985) Nucl Instr Meth Phys Res A239: 630
34 Carr GL, Quijada M, Tanner DB, Hirschmugl CJ, Williams GP, Etemad S, Dutta B, DeRosa F, Inam A, Venkatesan T (1990) Appl Phys Lett 57: 2725
35 Lippert E, Prass B, Rettig W, Stehlik D (1983) BESSY Ann Reports p 50
36 Saile V (1980) Appl Opt 19: 4115
37 Mitani T, Okamoto H, Takagi Y, Watanabe M, Fukui K. Koshihara S, Ito C (1989) Rev Sci Instr 60: 1569
38 Meyer M, Pahler M, Prescher Th, Raven Ev, Richter M, Sonntag B, Baier S, Fiedler W, Müller BR, Schulze M, Zimmermann P (1990) Phys Scripta T 31: 28
39 Pizzoferrato R, Casalboni M, Francini R, Grassano UM, Antonangeli F, Piacentini M, Zema N, Bassani F (1986) Europhys Lett 2: 571

13.12 Problems

13.1 Calculate the electron energies necessary to obtain synchrotron radiation with the critical photon energy of 300 eV from a bending magnet with the magnetic field of 1 T, and to obtain the same photon energy at the fundamental peak of radiation emitted in the forward direction at $\Theta = 0$ from a planar undulator. Assume that $\lambda_u = 4$ cm and $K = 1.5$.

13.2 The motion of an electron is periodic in an undulator with the period length of λ_u and the number of periods of N, and the trajectory is in general given by $\mathbf{r}(t) = (0,\ 0,\ c\beta^* t) + \mathbf{r}'(t)$, where $\mathbf{r}'(t) = \mathbf{r}'(t+T)$. The period of the motion is given by $T = \lambda_u / c\beta^*$. Show that the intensity of radiation is given by

$$\frac{d^2\mathscr{I}}{d\omega d\Omega} = \frac{e^2}{4\pi^2\hbar c\omega}\left[\frac{\sin[N\omega(1-\beta^*\cos\theta)T/2]}{\sin[\omega(1-\beta^*\cos\theta)T/2]}\right]^2$$

$$\times\left|\int_0^T \frac{\mathbf{n}\times[(\mathbf{n}-\boldsymbol{\beta})\times\dot{\boldsymbol{\beta}}]}{(1-\boldsymbol{\beta}\cdot\mathbf{n})^2}\exp[i\omega(t-\mathbf{n}\cdot\mathbf{r}(t)/c)]dt\right|^2$$

where $\mathbf{n} = (\sin\theta\cos\phi, \sin\theta\sin\phi, \cos\theta)$, and θ and ϕ are the polar angle and the azimuthal angle of the spherical coordinate, respectively. For a planar undulator, β^* is given by $\beta^* \approx 1-(1+K^2/2)/2\gamma^2$ Derive Eq. (13.15) from the above equation. Show that the intensity of radiation at a peak wavelength is proportional to N^2.

13.3 The trajectory of an electron in a planar undulator of the horizontal oscillation type is given by $x = \sin(2\pi z/\lambda_u)$ and $y = 0$, and the maximum deflection angle of the electron is K/γ . Derive Eqs. (13.11–13.13).

13.4 When the light travelling in medium 1 is reflected at the boundary between medium 1 and 2, the complex amplitudes of the light wave r_s and r_p for perpendicular and parallel components are expressed by general Fresnel equation as

$$r_s = \frac{\tilde{n}_1 \cos\alpha_1 - \tilde{n}_2 \cos\alpha_2}{\tilde{n}_2 \cos\alpha_1 + \tilde{n}_2 \cos\alpha_2}$$

$$r_p = \frac{\tilde{n}_1 \cos\alpha_2 - \tilde{n}_2 \cos\alpha_1}{\tilde{n}_1 \cos\alpha_2 + \tilde{n}_2 \cos\alpha_1}$$

where $\tilde{n}_1$ and $\tilde{n}_2$ are complex refractive indices of medium 1 and 2 respectively, and α_1 and α_2 are angles of incidence and refraction, respectively. $\tilde{n}_1 \sin\alpha_1 = \tilde{n}_2 \sin\alpha_2$ (Snell's law). By use of the above equations, derive the Eqs. (13.40–13.43), when medium 1 is vacuum and the refractive index and extinction coefficient of medium 2 is n and k ($\tilde{n}_2 = n - ik$).

13.5 From Fermat's principle (Eqs. 13.49, 13.54), derive the diffraction condition $d(\sin\alpha + \sin\beta) = m\lambda$, for a plane grating with grooves equally spaced in the case when the incident light is parallel.

13.6 Derive Eqs. (13.63–13.65) using Fermat's principle.

13.7 Derive Eq. (13.71) from the Kramers-Kronig relation (causality relation) given in complex function analysis.

13.8 Show that the total photoelectron yield spectra reflect the absorption spectra.

13.9 Check the transverse coherence of synchrotron radiation in the far infrared region in the case of UVSOR.

Subject Index

Springer-Verlag and the Environment

We at Springer-Verlag firmly believe that an international science publisher has a special obligation to the environment, and our corporate policies consistently reflect this conviction.

We also expect our business partners – paper mills, printers, packaging manufacturers, etc. – to commit themselves to using environmentally friendly materials and production processes.

The paper in this book is made from low- or no-chlorine pulp and is acid free, in conformance with international standards for paper permanency.